UTILISATION OF ELECTRICAL POWER

UTILISATION OF ELECTRICAL POWER

(Including Electrical Drives and Electric Traction)
(With a new chapter on "Universities Questions" (Latest)–with "Solutions")

SI UNITS

By

Er. R.K. RAJPUT

M.E. (Heat Power Engg.) Hons.–Gold Medallist; Grad. (Mech. Engg. & Elect. Engg.); M.I.E. (India); M.S.E.I.; M.I.S.T.E.; C.E. (India)

Recipient of : *"Best Teacher (Academic) Award"*
"Distinguished Author Award"
"Jawahar Lal Nehru Memorial Gold Medal" for an outstanding research paper (Institution of Engineers–India)

Principal *(Formerly):*
- ***Thapar Polytechnic College***
- ***Punjab College of Information Technology***

PATIALA (PUNJAB)

LAXMI PUBLICATIONS (P) LTD

(An ISO 9001:2015 Company)

BENGALURU • CHENNAI • GUWAHATI • HYDERABAD • JALANDHAR
KOCHI • KOLKATA • LUCKNOW • MUMBAI • RANCHI
NEW DELHI

UTILISATION OF ELECTRICAL POWER

Printed and bound in India
Typeset at : Goswami Associates, Delhi.
First Edition : 2006, *Second Edition* : 2013, *Reprint* : 2014, 2016, 2017, 2018, *Third Edition* : 2023
ISBN : 978-81-318-0829-0

Branches

Bengaluru	080-26 75 69 30
Chennai	044-24 34 47 26
Guwahati	0361-254 36 69
Hyderabad	040-27 55 53 83
Jalandhar	0181-222 12 72
Kochi	0484-405 13 03
Kolkata	033-40 04 77 79
Lucknow	0522-430 36 13
Ranchi	0651-224 24 64

PUBLISHED IN INDIA BY

Laxmi Publications (P) Ltd.
(An ISO 9001:2015 Company)
113, GOLDEN HOUSE, GURUDWARA ROAD, DARYAGANJ,
NEW DELHI - 110002, INDIA
Telephone : 91-11-4353 2500, 4353 2501
www.laxmipublications.com
info@laxmipublications.com

C—00231/022/06
Printed at: Ajit Printing Press, Delhi.

CONTENTS

Chapters			Pages

Chapters *Pages*

Chapters *Pages*

PREFACE TO THE THIRD EDITION

I am pleased to present the **"Third Edition"** of this book on **"Utilisation of Electrical Power"**. The warm reception which the previous edition and reprints have enjoyed all over India, has been a matter of much satisfaction to me.

The current edition of the book is thoroughly revised and updated.

Any suggestions for the improvement of the book will be warmly received and incorporated in the forthcoming edition.

— **Er. R.K. Rajput**
(Author)

PREFACE TO THE FIRST EDITION

This book on **"Utilisation of Electrical Power"** *including* ***Electrical Drives and Electric Traction*** has been written for the students preparing for B.E./B.Tech; A.M.I.E., Diploma and Competitive examinations.

It consists of *nine chapters* namely: 1. *Illumination*; 2. *Electric heating and welding*; 3. *Electrolytic processes*; 4. *Refrigeration and air-conditioning*; 5. *Electrical drives*; 6. *Electric traction*; 7. *Train movement and energy consumption*; 8. *Traction drives*; 9. *Economics of power generation, electric power supply and utilisation*.

The presentation of subject matter is very systematic and language of text is lucid, direct and easy to understand. Each chapter of the book is saturated with much needed text supported by neat and self-explanatory diagrams to make the subject self-speaking to a great extent. A large number of solved examples, properly graded, have been added in various chapters to enable the students to attempt different types of questions in examination without any difficulty. At the end of each chapter *Highlights, Objective Type Questions, Theoretical Questions* and *Unsolved Examples* have been added to make the book a complete and comprehensive unit in all respects.

Any suggestions for the improvement of this book will be thankfully acknowledged and incorporated in the next edition.

Er. R.K. Rajput
(Author)

1

Illumination

1.1. Introduction. 1.2. Definitions. 1.3. Laws of illumination or luminance. 1.4. Polar curves. 1.5. Photometry—Photometer heads—Photocells—Distribution photometry—Measurement of M.S.C.P. by integrating sphere—Measurement of brightness or luminance—Measurement of illumination. 1.6. Artificial sources of light. 1.7. Incandescent lamps. 1.8. Arc lamps. 1.9. Discharge lamps—Sodium vapour lamp—High pressure mercury vapour lamp—Mercury iodide lamps—Neon lamps— ... using and reflecting surfaces— ... Design of lighting scheme— ... for designing the lighting scheme ... tion. 1.11. Street lighting. 1.12. ... tive Type Questions—Theoretical ...

1.1. INTRODUCTION

- Light is a form of ... sensed by the human eye and the light will ... formerly a very small part ... the different colours, w...
- Light can be described as ... through space. Visible light ... of in the form of transverse ... or electromagnetic ... The speed of all ele... waves is 3×10^8 m/s in ... The wavelength and ... different for different w... velocity with which these travel is related to the ... and frequency, by the relation ...

The complete range of waves ... with their frequency and wavelength ... illustrated in Fig. 1.1.

Fig. 1.2 shows the light emitted ... gether with its colour for the wavel... within the visible spectrum.

Since the wavelengths of light are very short, smaller unit of length called Ang... strom Unit (AU) named after a Swe... scientist is used.

1

Illumination

1.1. **Introduction.** 1.2. **Definitions.** 1.3. **Laws of illumination or luminance.** 1.4. **Polar curves.** 1.5. **Photometry**—Photometer heads—Photocells—Distribution photometry—Measurement of M.S.C.P. by integrating sphere—Measurement of brightness or luminance—Measurement of illumination. 1.6. **Artificial sources of light.** 1.7. **Incandescent lamps.** 1.8. **Arc lamps.** 1.9. **Discharge lamps**—Sodium vapour lamp—High pressure mercury vapour lamp—Mercury iodide lamps—Neon lamp—Fluorescent tube (lamp). 1.10. **Lighting schemes**—Diffusing and reflecting surfaces—Requirements of good lighting—Types of lighting schemes—Design of lighting scheme—Characteristics of a good lighting scheme—Factors to be considered for designing the lighting scheme—Methods of lighting calculations—Calculation of illumination. 1.11. **Street lighting.** 1.12. **Factor lighting.** 1.13. **Flood lighting**—Highlights—Objective Type Questions—Theoretical Questions—Unsolved Examples.

1.1. INTRODUCTION

- **Light** is *a form of electromagnetic energy radiated from a body which is capable of being perceived by the human eye.* The sensation of light results from a flow of energy into the eye and the light will appear to vary if the rate of this flow of energy varies. Light radiations form only a very small part of the complete range of electromagnetic radiations. Light can be of different colours, which depend on the wavelength of the radiation causing it.
- Light can be described as a *vibratory motion*, which is transmitted in the form of waves through space. Visible light travels in the form of transverse waves of electromagnetic oscillations. The speed of all electromagnetic waves is 3×10^8 m/s in free space. The wavelength and frequency are different for different waves. The velocity with which these waves travel is related to the wavelength and frequency, by the relation : $v = \lambda f$

The complete range of waves along with their frequency and wavelength is illustrated in Fig. 1.1.

Fig. 1.2 shows the light emitted together with its colour for the wavelengths within the visible spectrum.

Since the wavelengths of light are very short, smaller unit of length called Angstrom Unit (AU) named after a Sweedish scientist is used.

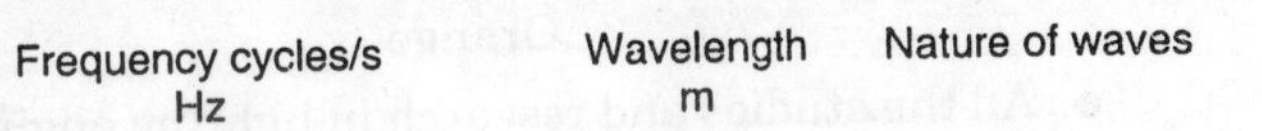

Fig. 1.1. Spectrum of electromagnetic waves.

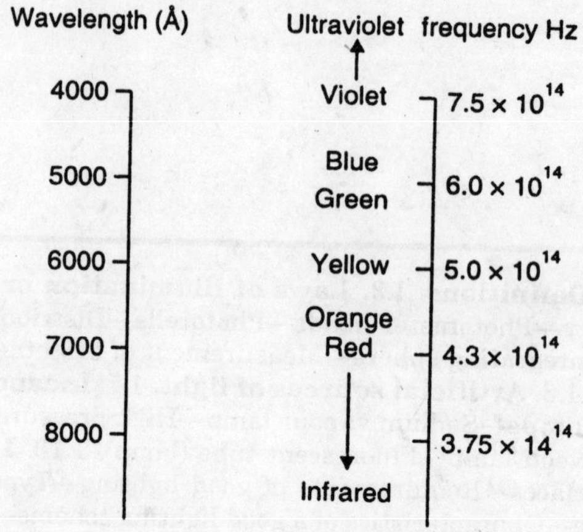

Fig. 1.2. Wavelength and colour of light.

Angstrom Unit (Å) = 10^{-10} m

Another smaller unit is *micron* (1 micron = 10^{-6} m).

A list of colours with their wavelengths is given below :

Colour	*Wavelengths*
Violet	4100 Å
Blue	4700 Å
Green	5500 Å
Yellow	5800 Å
Red	6000 Å
Orange	6100 Å

- **All the studies and research in lighting engineering try to achieve a good lighting scheme to make the occupants feel happy in the case of interior design and factory lighting and make the pedestarians and motorists and other road users comfortable in the case of highway lighting. The following are the *essentials of any good lighting system :***
  - **(*i*) Adequate illumination of suitable colour on the working surfaces.**
  - **(*ii*) Good maintenance**
  - **(*iii*) Avoidance of hard shadows**
  - **(*iv*) Avoidance of glare.**
- **The *aim of artificial lighting* is to supplement the daylight or to replace it in modern offices, homes, industries, workplaces etc. *Good illumination ensures increased production, effectivity of work and reduced accidents.***

1.2. DEFINITIONS

1. Solid angle

- **A *plane angle* is subtended at a point and is enclosed by two straight lines lying in the same plane. [Fig. 1.3 (*a*)]. Its magnitude is given by,**

$$\theta = \frac{\text{Arc}}{\text{Radius}} \text{ 'radians'}$$

The largest angle subtended at a point is 2π radians.

A *radian* is the angle subtended by an arc of a circle whose length equals the radius of the circle.

- ***Solid angle*** is the *angle generated by the surface passing through the point in space and the periphery of the area [Fig. 1.3(b)]*. It is denoted by ω, expressed in '*steradians*' and is given by the *ratio of the area of the surface to the square of the distance between the area and the point,*

i.e., $$\omega = \frac{\text{Area}}{(\text{Radius})^2} = \frac{A}{r^2}$$

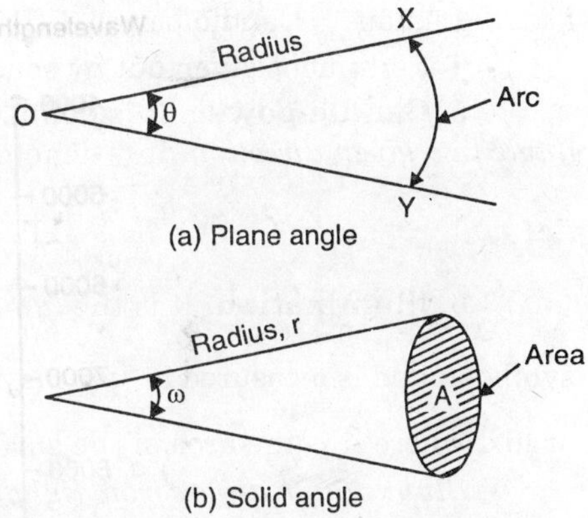

Fig. 1.3. Plane angle and solid angle.

The largest solid angle subtended at a point is due to a sphere at its centre, and is equal to $\frac{4\pi r^2 \text{ (Area of sphere)}}{r^2}$ = 4π steradians.

Relationship between ω and θ is obtained as follows :

Consider a curved surface of a spherical segment *ABC* of height *h* and radius *r* (Fig. 1.4).

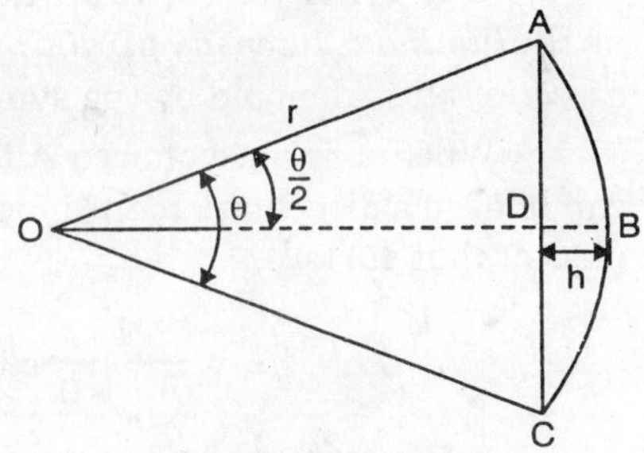

Fig. 1.4. Relation between solid and plane angles.

Surface area of segment $ABC = 2\pi rh$

Here, $h(BD) = OB - OD = r - r\cos\frac{\theta}{2} = r\left(1 - \cos\frac{\theta}{2}\right)$

$$\therefore \quad \text{Surface area of segment } ABC = 2\pi r^2\left(1 - \cos\frac{\theta}{2}\right)$$

$$\text{Solid angle, } \omega = \frac{\text{Surface area}}{(\text{Radius})^2} = \frac{2\pi r^2\left(1 - \cos\frac{\theta}{2}\right)}{r^2}$$

$$= 2\pi\left(1 - \cos\frac{\theta}{2}\right) \qquad ...(1.1)$$

1. **Light.** *The radiant energy from a hot body which produces the visual sensation upon the human eye is called* ***light.***

It is denoted by the symbol *Q*, expressed in lumen-hours (analogous to watt-hours)

2. **Luminous flux.** *The total quantity of light energy emitted per second from a luminous body is called* ***luminous flux.***

It is represented by the symbol *F* and measured in *lumens.* The concept of luminous flux assists us to specify the output and efficiency of a given light source.

3. **Luminous intensity.** Luminous intensity in a given direction is the *luminous flux emitted by the source per unit solid angle.*

It is denoted by the symbol *I* and is measured in '*candela*' (cd) or *lumens/stearadian i.e.,* $I = \frac{F}{\omega}$ lumens/steradian or candela, where ω in the solid angle).

4. **Lumen.** It is the unit of luminous flux and is defined as the *amount of luminous flux given out in a space represented by one unit solid angle by a source having an intensity of one candle power in all directions.*

i.e., Lumens = Candle power (C.P.) × Solid angle (ω)

Total lumens given out by source of one candela is 4π lumens.

5. **Candle power.** It is defined as the *number of lumens emitted by a source in a unit solid angle in a given direction.* It is denoted by symbol C.P.

i.e.,
$$\text{C.P.} = \frac{\text{Lumens}}{\omega}$$

6. **Illumination.** It is the *luminous flux received by a surface per unit area.* It is denoted by symbol E and is measured in *'lumens per square metre'* or *lux or 'metre-candle'* (*i.e.,* $E = \frac{F}{A}$ lumens/m^2 or lux, where A is the area of the surface).

Illumination differs from light very much, though generally these terms are used more or less synonymously. Strictly speaking *light is the cause and illumination is the result of the light* on the surfaces on which it falls. Thus illumination makes surfaces more or less bright with a certain colour and it is this brightness and colour which the eye sees and interprets as something useful or pleasant or otherwise.

7. **Brightness (or luminance).** Brightness of a surface is defined as the *luminous intensity per unit projected area of the surface in the given direction.* It is denoted by the symbol L.

When a surface of area A has an effective luminous intensity of I candelas in a direction θ to the normal (Fig. 1.5), then luminance or brightness of that surface,

$$L = \frac{I}{A\cos\theta} \text{ candela/square metre (Cd/m}^2\text{)}$$

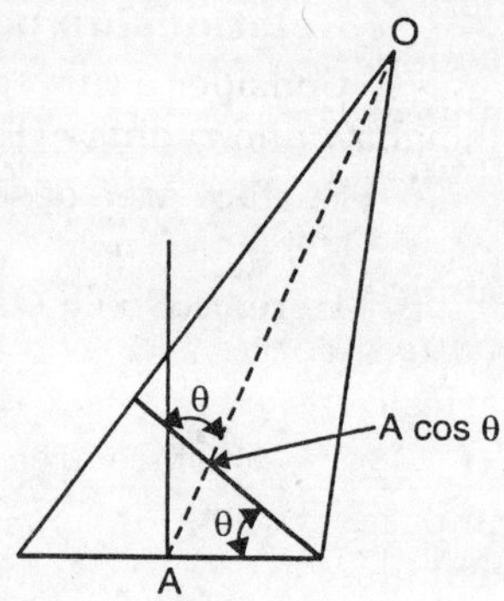

Fig. 1.5

A "*uniform diffuse source*" is one in which the *intensity per unit projected area is the same from all directions of view.*

• ***Relation between I, L and E :***

Consider a uniform diffuse spherical source with radius r metres and luminous intensity I candela.

Then,
$$L = \frac{I}{\pi r^2}$$

and,
$$E = \frac{I}{4\pi r^2} \times 4\pi = \frac{I}{r^2}$$

$$\therefore \quad E = \pi L \qquad \text{...(1.2)}$$

8. **Mean horizontal candle power (M.H.C.P.).** It is defined as the *mean of candle power in all directions in the horizontal plane containing the source of light.*

9. **Mean spherical candle power (M.S.C.P.).** It is defined as the *mean of candle powers in all directions and in all planes from the source of light.*

10. **Mean hemi-spherical candle power.** It is defined as the *mean of all candle powers in all directions above or below the horizontal plane passing through the source of light.*

11. **Reduction factor.** Reduction factor of a source of light is the *ratio of its mean spherical candle power to its mean horizontal candle power, i.e.,*

$$\text{Reduction factor} = \frac{\text{M.S.C.P.}}{\text{M.S.H.P.}}.$$

12. **Lamp efficiency.** It is defined as the *ratio of the luminous flux to the power input.* It is expressed in *lumens per watt.*

13. **Specific consumption.** It is defined as the *ratio of power input to the average candle power.* It is expressed in *watts per candle.*

14. **Space-height ratio.** It is defined as the *ratio of horizontal distance between adjacent lamps and height of their mountings.*

15. **Utilisation factor (UF).** The *ratio of total lumens reaching the working plane to total lumens given out by the lamp* is called *utilisation factor* (*or co-efficient of utilisation*).

16. **Maintenance factor (MF).** It is the *ratio of illumination under normal working conditions to the illumination when the things are perfectly clean.*

17. **Depreciation factor.** This is merely the reverse of the maintenance factor and is defined as the *ratio of initial metre-candles to the ultimate maintained metre-candles on the working plane.* It value is *more than unity.*

18. **Waste light factor.** Whenever a surface is illuminated by a number of sources of light, there is always a certain amount of waste of light on account of overlapping and falling of light outside at the edges of the surface. The effect is taken into account by multiplying the theoretical value of lumens required by 1.2 for rectangular areas and 1.5 for irregular areas and objects such as statues, monuments etc.

19. **Absorption factor.** In the places where atmosphere is full of smoke fumes, such as in foundries, there is a possibility of absorption of light. *The ratio of total lumens available after absorption to the total lumens emitted by the source of light is called the* **absorption factor.** Its values varies from unity for clean atmosphere to 0.5 for foundries.

20. **Beam factors.** The ratio of lumens in the beam of a projector to the lumens given out by lamps is called the *beam factor.* This factor takes into account the absorption of light by reflector and front glass of the projector lamp. Its value varies from 0.3 to 0.6.

21. **Reflection factor.** When a ray of light impinges on a surface it is reflected from the surface at an angle of incidence, as shown in Fig. 1.6. A certain portion of incident light is absorbed by the surface. The ratio of reflected light to the incident light is called the '*reflection factor*'. It is *always less than unity.*

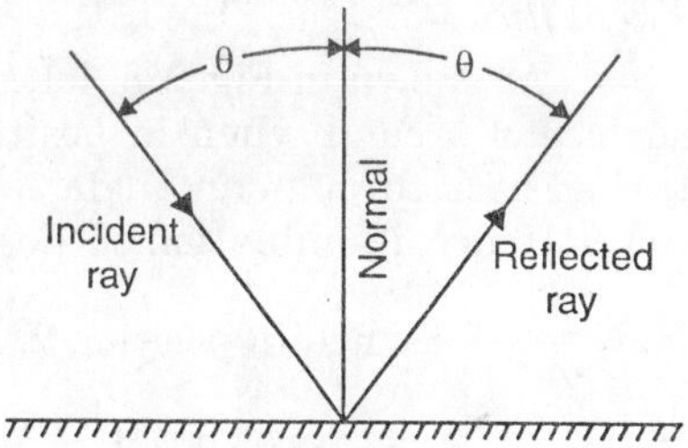

Fig. 1.6

22. **Glare.** The opening of the pupil in the human eye is controlled by the iris. If a bright object comes into the view of the eye, large amount of light produces an intense image on the retina and the iris automatically protects the eye by contracting the pupil, thus reducing the intensity of the image. When the eye is towards another object which is less bright as compared to the bright object already in the field of view, the iris will contract reducing the amount of light received on the retina from every object in the field of view and making it difficult to see the object desired. At the same time, the portion of the retina which received image of the bright object remains fatigued. This phenomenon is called ***"Glare"***, and is familiar in connection with motor-car head lights.

In other wards, **"glare"** *may be defined as the brightness within the field of vision of such a character as to cause annoyance, discomfort, interference with vision or eye-fatigue.*

1.3. LAWS OF ILLUMINATION OR LUMINANCE

The illumination (E) of a surface depends upon the following *factors* (The source is assumed to be a point source or otherwise sufficiently away from the surface to be regarded as such).

1. E is *directly proportional to the luminous intensity* (I) of the source. In other words, $E \propto I$.

2. **Inverse Square Law.** The *illumination of a surface is inversely proportional to the square of the distance of the surface from the source.*

In other words, $E \propto \frac{1}{r^2}$.

Proof: Consider surface areas A_1 and A_2 at distances r_1 and r_2 respectively from the point source S of luminous intensity I and *normal* to the rays as shown in Fig. 1.7. Let the solid angle subtended be ω.

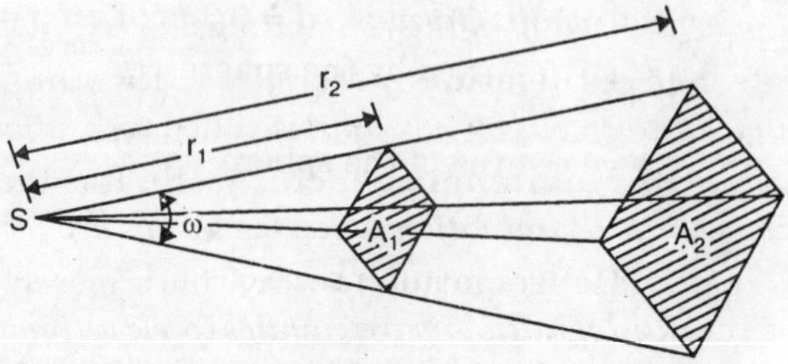

Fig. 1.7. Inverse square law.

Total luminous flux radiated

$= I\omega$ lumens

Illumination of the surface of area A_1

$$E_1 = \frac{I\omega}{A_1} = \frac{I\omega}{\omega r_1^2} = \frac{I}{r_1^2} \text{ lumens per unit area}$$

Similarly, illumination on the surface of area A_2,

$$E_2 = \frac{I\omega}{A_2} = \frac{I\omega}{\omega r_2^2} = \frac{I}{r_2^2} \text{ lumens per unit area.}$$

Hence the *illumination of a surface is inversely proportional to the square of the distance between the surface and the light source provided that the distance between the surface and the source is sufficiently large so that source can be regarded as a point source.*

3. **Lambert's Cosine Law.** According to this law, *E is directly proportional to the cosine of the angle made by the normal to the illuminated surface with the direction of the incident flux.*

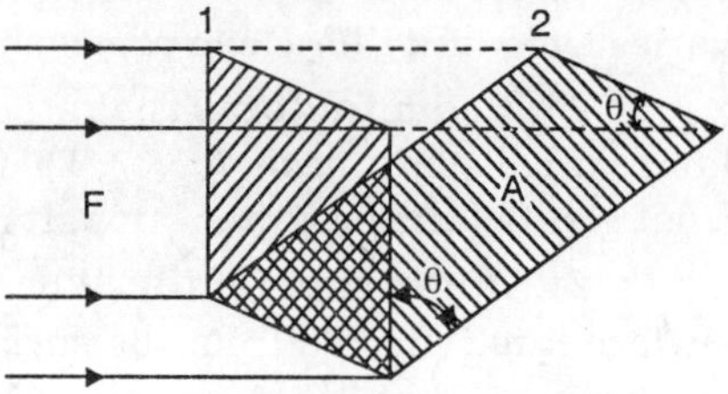

F = Luminous flux
A = Area of the surface

Fig. 1.8. Lambert's cosine law.

As shown in Fig. 1.8, let F be the flux incident on the surface of area A when in position 1. When this surface is turned back through an angle θ, then flux incident on it is $F \cos\theta$. Hence, illumination of the surface when in position 1 is

$E_1 = \frac{F}{A}$. But when in position 2, $E_2 = \frac{F\cos\theta}{A}$

$\therefore \quad E_2 = E_1 \cos\theta$

Combining all these factors together, we get

$$E = \frac{I\cos\theta}{r^2}. \text{ The unit is lumens per unit area.}$$

Example 1.1. *A 250 V lamp has a total flux of 1500 lumens and takes a current of 0.4 A. Calculate :*

(*i*) *Lumens per watt.* (*ii*) *M.S.C.P. per watt.*

Solution. Given : $V = 250$ volts ; $F = 1500$ lumens ; $I = 0.4$ A

Mean spherical candle power of lamp,

$$\text{M.S.C.P.} = \frac{F}{4\pi} = \frac{1500}{4\pi} = 119.4$$

(*i*) **Lumens per watt** $= \frac{\text{Output of lamp in lumens}}{\text{Wattage of lamp in watts}} = \frac{1500}{250 \times 0.4} =$ **15. (Ans.)**

(*ii*) **M.S.C.P. per watt** $= \dfrac{\text{M.S.C.P. of lamp}}{\text{Wattage of lamp}} = \dfrac{119.4}{250 \times 0.4}$ = **1.194. (Ans.)**

Example 1.2. *A 0.3 metre diameter diffusing sphere of opal glass having 15% absorption, encloses an incandescent lamp with a luminous flux of 4500 lumens. Calculate average luminance of the sphere.*

Solution. *Given :* $d = 0.3$ m ; Lamp = 4500 lumens

Flux 1 emitted by the sphere, $F_{sphere} = (1 - 0.15) \times 4500 = 3825$ lumens

Surface area of the sphere $= 4\pi r^2 = 4\pi \times \left(\dfrac{0.3}{2}\right)^2 = 0.2827 \text{ m}^2$

∴ Average luminance of sphere $= \dfrac{F_{sphere}}{\text{Surface area of sphere}}$

$= \dfrac{3825}{0.2827}$ = **13530 lumens/m². (Ans.)**

Example 1.3. *A filament lamp of 500 W is suspended at a height of 4.5 metres above the working plane and gives uniform illumination over an area of 6 m diameter. Assuming an efficiency of the reflector as 70% and efficiency of lamp as 0.8 watt per candle power, determine the illumination on the working plane.*

Solution. Wattage of the filament lamp = 500 W

Height of the lamp above the working plane = 4.5 m

Diameter of the uniformly illuminated area = 6 m

Efficiency of reflector = 70%

Efficiency of lamp = 0.8 W per candle power

Illumination on the working plane :

Candle power of the lamp $= \dfrac{500}{0.8} = 625$ C.P.

Luminous output of lamp $= 4\pi \times 625 = 2500\,\pi$ lumens

Flux emitted by the reflector = Efficiency of reflector × Total luminous output of the lamp

$= 0.7 \times 2500\,\pi = 1750\,\pi$ lumens

Area of working plane $= \dfrac{\pi}{4} \times (6)^2 = 9\pi \text{ m}^2.$

∴ Illumination on the working plane $= \dfrac{1750\,\pi}{9\,\pi}$ = **194.44 Lumens/m². (Ans.)**

Example 1.4. *The candle power of a lamp is 120. A plane surface is placed at a distance of 2.5 metres from this lamp. Calculate the illumination on the surface when it* (*i*) *normal,* (*ii*) *inclined to 45°, and* (*iii*) *Parallel to rays.*

Solution. (*i*) $E = \dfrac{\text{C.P.}}{d^2} = \dfrac{120}{(2.5)^2}$ = **19.2 lux. (Ans.)**

(*ii*) $E = \dfrac{\text{C.P.}}{d^2} \times \cos 45° = \dfrac{120}{(2.5)^2} \times \cos 45°$ = **13.58 lux. (Ans.)**

(*iii*) $E = 0$, since the rays of light are parallel to the surface, they cannot illuminate it. **(Ans.)**

Example 1.5. *Derive the relation to find the illumination at any point on the plane surface due to light source suspended at height h from the plane surface.*

Solution. Refer to Fig. 1.9. Consider a point P on the plane surface AB where illumination due to light source S of candle power C.P. at a height h from the surface AB is to be determined.

Let d be the distance between source S and point P.

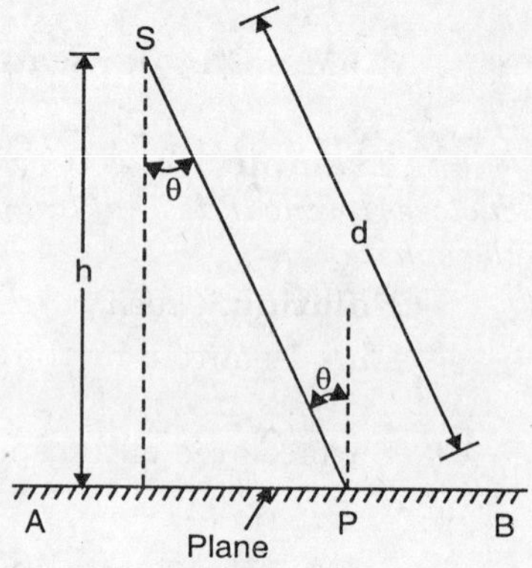

Fig. 1.9

Then, $\cos\theta = \dfrac{h}{d}$ or $d = \dfrac{h}{\cos\theta}$

Illumination at point P, by laws of illumination

$$= \frac{\text{C.P.}}{d^2}\cos\theta = \frac{\text{C.P.}}{(h/\cos\theta)^2}\cos\theta = \frac{\text{C.P.}}{h^2}\cos^3\theta,$$

and illumination at any point O, *vertically below the source of light*

$$= \frac{\text{C.P.}}{h^2}.$$

Hence *illumination at any point on a plane is* $\cos^3\theta$*-times of illumination at a point just vertically below the light source, where* θ *is the angle between the normal to the surface at the point and rays of light.*

Example 1.6. *A 500 W lamp having M.S.C.P. of 800 is suspended 3 m above the working plane :*

(*i*) *Illumination directly below the lamp at the working plane.*

(*ii*) *Lamp efficiency.*

(*iii*) *Illumination at a point 2.4 m away on the horizontal plane from vertically below the lamp.*

Solution. Wattage of the lamp = 500 W

M.S.C.P. of the lamp, $I = 800$

Height of the lamp, $h = 3$ m

(*i*) **Illumination directly below the lamp at the working plane :**

Illumination directly below the lamp,

$$E_A = \frac{I}{h^2} = \frac{800}{3^2} = \mathbf{88.9\ lux.} \quad \textbf{(Ans.)}$$

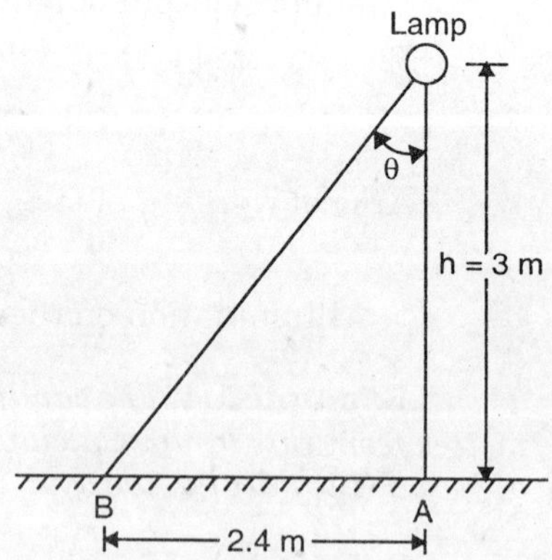

Fig. 1.10

(*ii*) **Lamp efficiency :**

$$\text{Lamp efficiency} = \frac{\text{Luminous flux}}{\text{Power input}} = \frac{4\pi \times \text{M.S.C.P.}}{500} = \frac{4\pi \times 800}{500}$$

$$= \mathbf{20.1\ lumens/watt.} \quad \textbf{(Ans.)}$$

(*iii*) **Illumination at a point 2.4 m away :**

Illumination at a point 2.4 m away on the horizontal plane from vertically below the lamp,

$$E_B = \frac{I}{h^2}\cos^3\theta$$

Here, $$\cos\theta = \frac{3}{\sqrt{3^2 + 2.4^2}} = 0.7808$$

∴ $$E_B = \frac{800}{3^2} \times (0.7808)^3 = \mathbf{42.3\ lux.}\quad \textbf{(Ans.)}$$

Example 1.7. *A lamp with reflector is mounted 10 m above the centre of a circular area of 20 m diameter. If the combination of the lamp and reflector gives a uniform C.P. of 800 over the circular area, determine the maximum and minimum illumination produced on the area.*

(Panjab University)

Solution. Candle power of the lamp, C.P. = 800

Height of the lamp, h = 10 m

Diameter of the circular area = 20 m

The *maximum illumination* will occur *directly below the lamp i.e.,* at point A and

$$= \frac{\text{C.P.}}{h^2} = \frac{800}{10^2} = \mathbf{8\ lux.}\quad \textbf{(Ans.)}$$

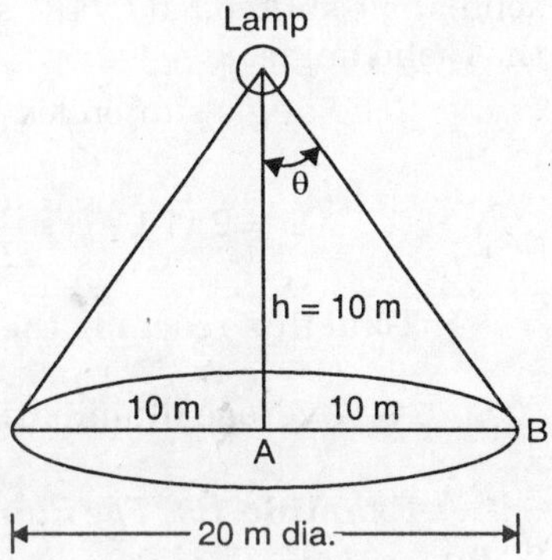

Fig. 1.11

The minimum illumination will occur at the periphery of the circular area *i.e.,* at point B and

$$= \frac{\text{C.P.}}{h^2}\cos^3\theta$$

$$= \frac{800}{10^2} \times \left(\frac{10}{\sqrt{10^2 + 10^2}}\right)^3 = \mathbf{2.83\ lux.}\quad \textbf{(Ans.)}$$

Example 1.8. *A lamp having a uniform C.P. of 300 in all directions is provided with a reflector which directs 60 per cent of the total light uniformly on to a circular area of 12 m diameter. The lamp is 5 m above the area. Calculate :*

(*i*) *The illumination at the centre and edge of the surface with and without reflector.*

(*ii*) *The average illumination over the area without the reflector.*

Solution. Candle power of the lamp C.P. = 300

Height of the lamp, h = 5 m

Efficiency of the reflector = 60%

(*i*) **The illumination at the centre without reflector.**

The illumination at the centre

$$= \frac{\text{C.P.}}{h^2} = \frac{300}{5^2} = \mathbf{12\ lux.}\quad \textbf{(Ans.)}$$

The illumination at the edge of the surface with and without the reflector :

The illumination at the edge of the surface *without reflector.*

$$= \frac{\text{C.P.}}{h^2}\cos^3\theta = \frac{300}{5^2} \times \left(\frac{5}{\sqrt{5^2 + 6^2}}\right)^3$$

= **3.15 lux. (Ans.)**

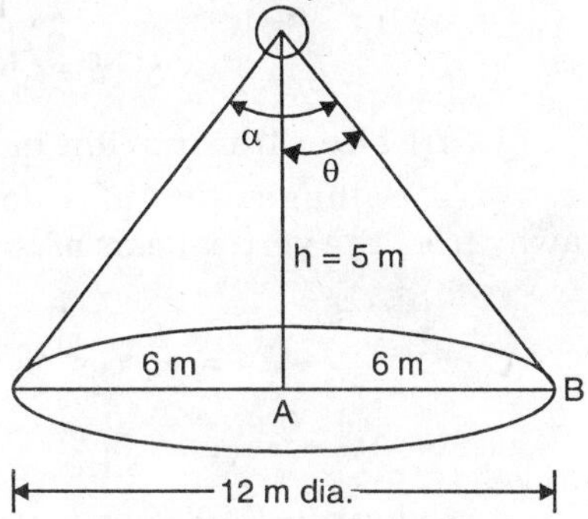

Fig. 1.12

With reflector *the illumination at the edge and at the centre will be the same since the reflector directs the light uniformly on the surface.*

Total lumens given out, $F = 4\pi \times \text{C.P.} = 4\pi \times 300 = 1200\pi$

Total lumens reaching the surface $= 0.6 \times F = 0.6 \times 1200\pi = 720\pi$

Total surface area $= \pi r^2 = \pi \times 6^2 = 36\pi \text{m}^2$

$\therefore$ Average illumination with reflector $= \dfrac{720\,\pi}{36\,\pi} =$ **20 lux. (Ans.)**

(*ii*) **Average illumination over the area without the reflector :**

Average illumination of the surface without refletor will be found by first determining the solid angle subtended by the surface at the lamp and then finding out the luminous flux emitted in that solid angle.

Solid angle subtended by the area at the lamp,

$$\omega = 2\pi\left(1 - \cos\frac{\alpha}{2}\right) = 2\pi(1 - \cos\theta) = 2\pi\left(1 - \frac{5}{\sqrt{5^2 + 6^2}}\right) = 1.28\,\pi \text{ steradians}$$

Total flux reaching the surface $= I\omega = 300 \times 1.28\pi = 384\pi$ lumens

$\therefore$ Average illumination $= \dfrac{384\,\pi}{\pi \times 6^2} =$ **10.67 lux. (Ans.)**

Example 1.9. *The illumination at a point on a working plane directly below the lamp is to be 80 lumens / m². The lamp gives 180 C.P. uniformly below the horizontal plane. Determine :*

(*i*) *The height at which the lamp is suspended.*

(*ii*) *The illumination at a point on the working plane 1.5 m away from the vertical axis of the lamp.*

Solution. Luminous intensity of the lamp, $I = 180$ C.P.

Illumination directly below the lamp, $E = 80$ lumens/m².

Refer to Fig. 1.13.

(*i*) **The height at which the lamp is suspended, h :**

We know that,

$$E_A = \frac{I}{h^2}$$

$$\therefore \quad 80 = \frac{180}{h^2}$$

or

$$h = \sqrt{\frac{180}{80}} = \textbf{1.5 m.} \quad \textbf{(Ans.)}$$

(*ii*) **The illumination at a point 1.5 m away :**

The illumination at a point on the working plane 1.5 m away from the vertical axis of the lamp,

$$E_B = \frac{I}{h^2}\cos^3\theta = \frac{180}{(1.5)^2} \times \left[\frac{1.5}{\sqrt{1.5^2 + 1.5^2}}\right]^3$$

= **2.96 lux. (Ans.)**

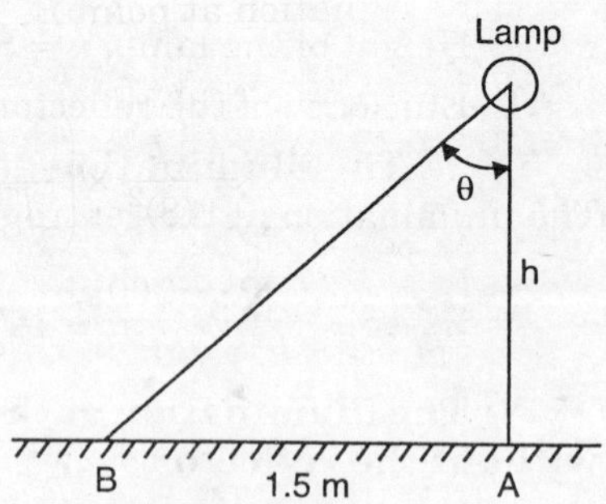

Fig. 1.13

Example 1.10. *Two lamps L_1 and L_2 are hung at a height of 9 m from the floor level. The distance between the lamps is 1 m. Lamp L_1 is of 500 C.P. If the illumination on the floor vertically below this lamp is 20 lux, find the candle power of the lamp L_2.* **(A.M.I.E)**

Solution. Candle power of lamp L_1 = 500 C.P.

Distance between the lamps = 1 m

Illumination vertically below lamp L_1 = 20 lux

Refer to Fig. 1.14.

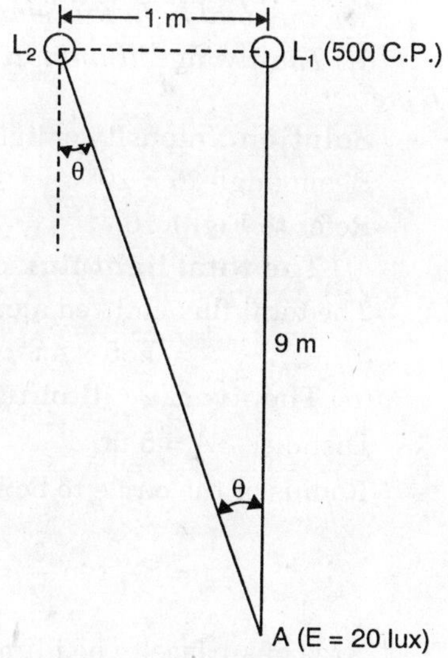

Fig. 1.14

Candle power of the lamp L_2, x :

Illumination at point A on the floor level vertically below lamp L_1

= Illumination due to lamp L_1 + Illumination due to lamp L_2

i.e.,
$$20 = \frac{500}{9^2} + \frac{x}{9^2}\cos^3\theta = \frac{500}{81} + \frac{x}{81} \times \left(\frac{9}{\sqrt{9^2+1^2}}\right)^3$$

$$= \frac{500}{81} + \frac{0.982\,x}{81} = \frac{500 + 0.982\,x}{81}$$

or,
$$x = \left(\frac{20 \times 81 - 500}{0.982}\right) = \textbf{1140.5 C.P.} \quad \textbf{(Ans.)}$$

Example 1.11. *A small light source with intensity uniform in all directions is mounted at a height of 8 metres above a horizontal surface. Two points A and B both lie on the surface with point A directly beneath the source. How far is B from A if the illumination at B is only $\frac{1}{12}$ as great as at A ?*

(Roorkee University)

Solution. Refer Fig. 1.15.

Let I = Intensity of the source, candela,

and, x = Distance between points A and B.

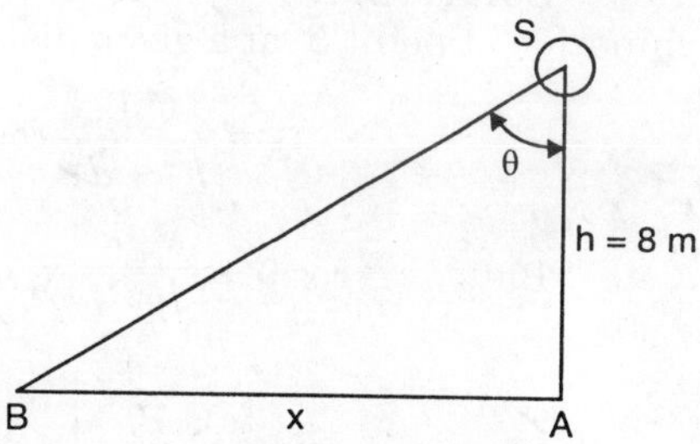

Fig. 1.15

Then, Illumination at point A, $E_A = \frac{I}{(8)^2} = \frac{I}{64}$

Illumination at point B, $E_B = \frac{I}{(h)^2} \times \cos^3\theta$

$$= \frac{I}{(8)^2} \times \left(\frac{8}{\sqrt{8^2+x^2}}\right)^3 = \frac{8I}{(8^2+x^2)^{3/2}}$$

But,
$$E_B = \frac{1}{12} E_A \qquad \text{...(Given)}$$

$\therefore$
$$\frac{8I}{(64+x^2)^{3/2}} = \frac{1}{12} \times \frac{I}{64}$$

or,
$$(64 + x^2)^{3/2} = 6144$$

or,
$$64 + x^2 = (6144)^{2/3} = (6144)^{0.667} = 336.4$$

$\therefore$
$$x = (336.4 - 64)^{1/2} = \textbf{16.5 m.} \quad \textbf{(Ans.)}$$

Example 1.12. *A light source having an intensity of 200 C.P. in all directions is fitted with a reflector so that it directs 85% of its light along a beam having a divergence 20°. Calculate :*

(i) The total light flux emitted along the beam.

(ii) The average illumination produced on a surface normal to the beam direction at a distance of 5 metres. **(Gorakhpur University)**

Solution. Intensity of light source, $I = 200$ C.P.

Beam angle, $\theta = 20°$

Refer to Fig. 1.16.

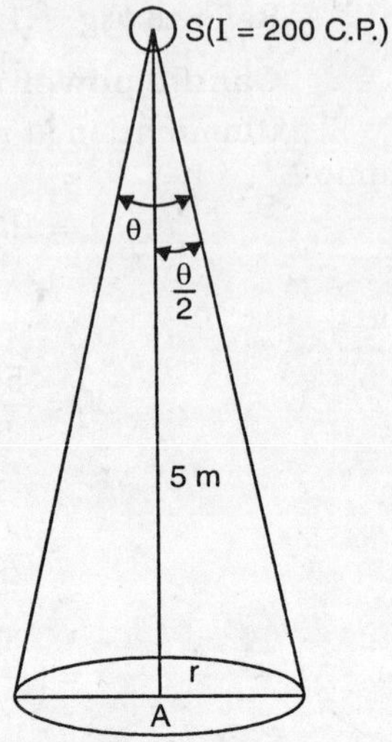

Fig. 1.16

(i) **The total light flux emitted :**

The total flux emitted along the beam

$= 0.85 \times 4\pi \times 200 =$ **2136.3 lumens. (Ans.)**

(ii) **The average illumination :**

Distance $SA = 5$ m

Radius of the circle to be illuminated,

$$r = SA \tan \frac{\theta}{2} = 5 \tan\left(\frac{20°}{2}\right) = 0.882 \text{ m}$$

Area of surface to be illuminated

$$= \pi r^2 = \pi \times (0.882)^2 = 2.444 \text{ m}^2$$

$$\therefore \quad \text{Average illumination} = \frac{2136.3}{2.444} = \mathbf{874.1\ lux.} \quad \textbf{(Ans.)}$$

Example 1.13. *Find the height at which a light source having uniform spherical distribution should be placed over a floor in order that the intensity of horizontal illumination at a given distance from its vertical line may be greatest.*

Solution. Refer to Fig. 1.17. The intensity of horizontal illumination at point B, at a given distance 'd' from the vertical line SA,

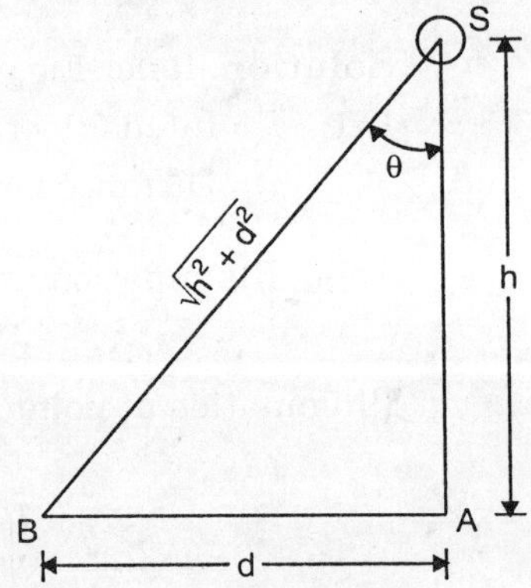

Fig. 1.17

$$I_B = \frac{\text{C.P.}}{h^2 + d^2} \cos\theta = \frac{\text{C.P.}}{h^2} \cos^3\theta$$

But, $$\cos\theta = \frac{h}{\sqrt{h^2 + d^2}}$$

$$\therefore \quad I_B = \frac{\text{C.P.}}{h^2} \times \left(\frac{h}{\sqrt{h^2 + d^2}}\right)^3 = \text{C.P.} \times \frac{h}{(h^2 + d^2)^{3/2}}$$

Differentiating I_B w.r.t. h, we have

$$\frac{dI_B}{dh} = \text{C.P.}\ [(h^2 + d^2)^{-3/2} \times 1 + h(-3/2)(h^2 + d^2)^{-5/2} \times 2h]$$

$$= \text{C.P.}\ [(h^2 + d^2)^{-3/2} - 3h^2(h^2 + d^2)^{-5/2}]$$

For maximum I_B, $\frac{dI_B}{dh} = 0$

$$\therefore \quad (h^2 + d^2)^{-3/2} = 3h^2(h^2 + d^2)^{-5/2}$$

or $$(h^2 + d^2) = 3h^2$$

or $$2h^2 = d^2 \quad \text{or} \quad h = \sqrt{\frac{d^2}{2}} = \frac{d}{\sqrt{2}} = \mathbf{0.707\ d.} \quad \textbf{(Ans.)}$$

Example 1.14. *Two lamp posts are 14 m apart and are fitted with 200 C.P. lamp each at a height of 5 m above the ground. Calculate :*

(*i*) *Illumination mid-way between them.*

(*ii*) *Illumination under each lamp.* **(Panjab University)**

Solution. Candle power of each lamp = 200 C.P.

Height of each lamp from the ground = 5 m

Distance between the two lamps = 14 m

(*i*) **Illumination midway between the lamps :**

Illumination midway between the lamps,

$$E_C = \text{Illumination due to lamp } L_1 + \text{Illumination due to lamp } L_2$$

$$= \frac{200}{5^2} \times \cos^3 \theta + \frac{200}{5^2} \times \cos^3 \theta$$

$$= 2 \times \frac{200}{5^2} \times \left(\frac{5}{8.6}\right)^3 = \mathbf{3.144\ lux.} \quad \textbf{(Ans.)}$$

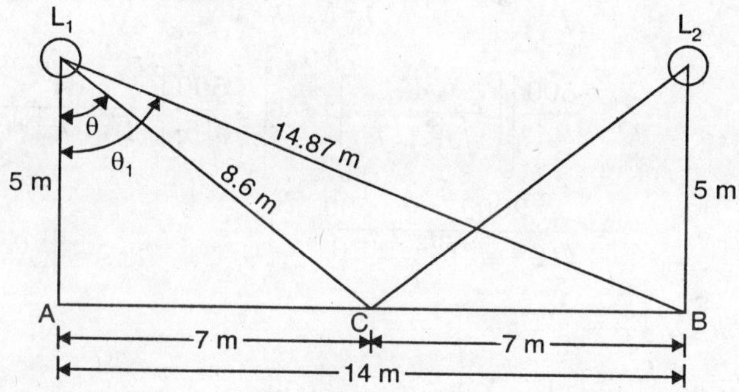

Fig. 1.18

(*ii*) **Illumination under each lamp :**

Illumination under either of the lamps, say under lamp L_2,

$$E_B = \text{Illumination due to lamp } L_1 + \text{Illumination due to lamp } L_2$$

$$= \frac{200}{5^2} \cos^3 \theta_1 + \frac{200}{5^2} = \frac{200}{5^2}(\cos^3 \theta_1 + 1)$$

$$= 8\left[\left(\frac{5}{14.87}\right)^3 + 1\right] = \mathbf{8.3\ lux.} \quad \textbf{(Ans.)}$$

Example 1.15. *Two similar lamps having uniform intensity of 500 C.P. in all directions below the horizontal are mounted at a height of 4 metres. What must be the maximum spacing between the lamps so that the illumination on the ground mid-way between the lamps shall be at least one half the illumination directly under the lamps.* **(Gorakhpur University)**

Solution. Candle power of each lamp = 500 C.P.

Height of each lamp from the ground, h = 4 m

Let the maximum spacing between the lamps be d metres.

Illumination mid-way between the lamps,

$$E_C = 2 \times \text{Illumination due to either lamp}$$

$$= 2 \times \frac{500}{h^2} \times \cos^3 \theta = 2 \times \frac{500}{4^2} \times \left(\frac{4}{\sqrt{4^2 + (d/2)^2}}\right)^3 = \frac{4000}{\left(16 + \frac{d^2}{4}\right)^{3/2}}$$

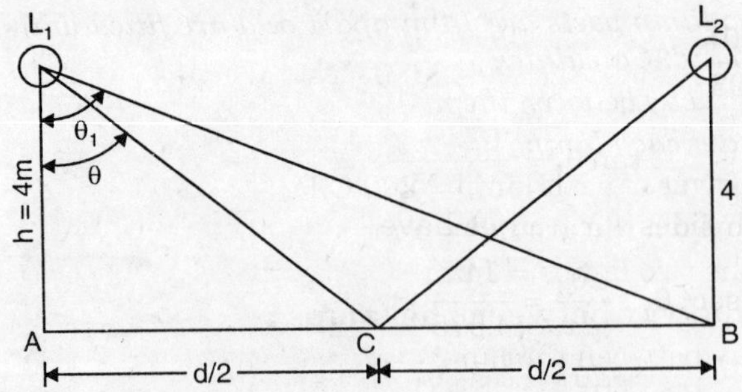

Fig. 1.19

Illumination under either of the lamps, say L_2,

$$E_C = \text{Illumination due to lamp } L_1 + \text{Illumination due to lamp } L_2$$

$$= \frac{500}{h^2}\cos^3\theta_1 + \frac{500}{h^2} = \frac{500}{h^2}(\cos^3\theta_1 + 1)$$

$$= \frac{500}{4^2}\left[\left(\frac{4}{\sqrt{4^2+d^2}}\right)^3 + 1\right] = \frac{500}{16}\left[\frac{64}{(16+d^2)^{3/2}} + 1\right]$$

$$= \frac{2000}{(16+d^2)^{3/2}} + \frac{500}{16}$$

Now, $$E_C = \frac{1}{2}E_B$$ ... *Given*

$$\therefore \quad \frac{4000}{\left(16+\frac{d^2}{4}\right)^{3/2}} = \frac{1}{2}\left[\frac{2000}{(16+d^2)^{3/2}} + \frac{500}{16}\right]$$

or $$\frac{4000}{\left(16+\frac{d^2}{4}\right)^{3/2}} = \frac{2000}{(16+d^2)^{3/2}} + 15.625$$

or **d = 9.56 m. (Ans.)**

Example 1.16. *The lamps spaced 9.15 m apart and suspended at a height of 4.575 m are lighting a corridor. If each lamp gives 200 C.P. in all directions below the horizontal, find the maximum and minimum value of the illumination on the floor along the centre line.*

Solution. Refer to Fig. 1.20. Let L_1, L_2, L_3, L_4, L_5 ...... etc. be the lamps.

The illumination at any point A due to a lamp is

$$= \frac{100}{(4.575)^2}\cos^3\theta = \frac{100}{20.93\cos^3\theta}$$

The total illumination due to all the lamps is

$$= \frac{100}{20.93}(\cos^3\theta_1 + \cos^3\theta_2 + \cos^3\theta_3 + ...)$$

In order to find out the maximum and minimum illumination, we have

$$\frac{d}{dx}\left[\frac{200}{20.93}(\cos^3\theta_1 + \cos^3\theta_2 + \cos^3\theta_3 + ...)\right] = 0$$

or, $$-\frac{200}{20.93}\times 3\left(\cos^2\theta_1 \sin\theta_1 \frac{d\theta_1}{dx}+......\right)=0 \quad ...(i)$$

Also, $$\tan\theta_1=\frac{18.3+x}{4.575}$$

Differentiating both sides w.r.t. x, we have

$$\sec^2\theta_1 \cdot \frac{d\theta_1}{dx}=\frac{1}{4.575}$$

$$\therefore \quad \frac{d\theta_1}{dx_1}=\frac{\cos^2\theta_1}{4.575}$$

Similarly, $\frac{d\theta_2}{dx_2}=\frac{\cos^2\theta_2}{4.575}$; $\frac{d\theta_3}{dx_3}=\frac{\cos^3\theta_3}{4.575}$; $\frac{d\theta_4}{dx_3}=\frac{\cos^4\theta_3}{4.575}$ and so on.

Substituting the values of $\frac{d\theta_1}{dx_1}, \frac{d\theta_2}{dx_2}, \frac{d\theta_3}{dx_3}$ etc. in eqn. (1), we get

$$-\frac{200}{20.93}\times\frac{3}{4.575}(\cos^4\theta_1 \sin\theta_1+\cos^4\theta_2 \sin\theta_2+\cos^4\theta_3 \sin\theta_3$$
$$-\cos^4\theta_2 \sin\theta_4-\cos^4\theta_5 \sin\theta_5 ......)=0$$

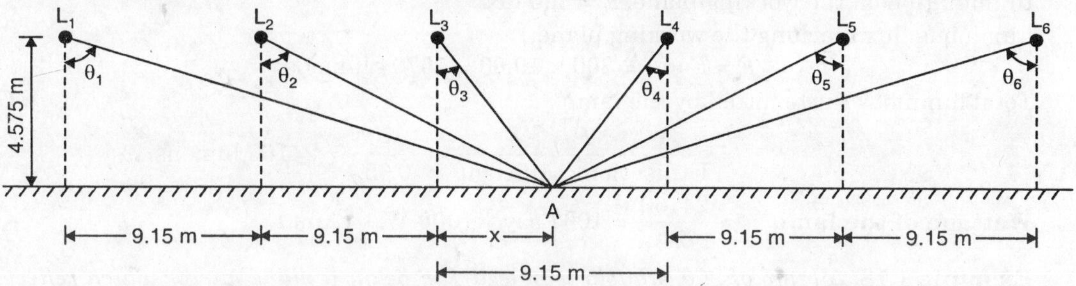

Fig. 1.20

When A is *mid-way* between L_3 and L_4, $\theta_3=\theta_4$, $\theta_2=\theta_5$ and so on, when the terms cancel out in pairs and the illumination is either a maximum or a minimum ; it is a *minimum.* When P is directly under L_3, $\theta_3=0$, $\theta_2=\theta_4$, $\theta_1=\theta_5$ etc. which gives θ_3 term as zero and rest cancel out in pairs, the illumination is a *maximum under* L_3.

The illumination under L_3 is found by putting $x = 0$ and constructing the following table :

θ	***cos θ***	***cos³ θ***
$\theta_3=0$	$\cos\theta_3=1$	$\cos^3\theta_3=1$
$\theta_4=\tan^{-1}\frac{9.15}{4.575}=63.43°$	$\cos\theta_4=0.4473$	$\cos^3\theta_4=0.0895$
$\theta_5=\tan^{-1}\frac{18.3}{4.575}=75.96°$	$\cos\theta_5=0.2426$	$\cos^3\theta_5=0.0143$
$\theta_6=\tan^{-1}\frac{27.45}{4.575}=80.54°$	$\cos\theta_6=0.1644$	$\cos^3\theta_6=0.00444$

The *maximum illumination,* $$\mathbf{E_{max}}=\frac{200}{20.93}(1+2\times 0.0895+2\times 0.0143+2\times 0.00444)$$

$$= \mathbf{11.62\ lux.} \quad \textbf{(Ans.)}$$

The minimum illumination occurs at a mid-way between L_3 and L_4 and we get :

$$\theta_3 = \theta_4 = \tan^{-1} 1 ; \ \theta_2 = \theta_5 = \tan^{-1} 3 ; \theta_1 = \theta_6 = \tan^{-1} 5 \text{ etc.}$$

We construct the following table :

θ	$\cos\theta$	$\cos^3\theta$
$\theta_4 = \tan^{-1} 1 = 45°$	$\cos\theta_4 = 0.7071$	$\cos^3\theta_4 = 0.3535$
$\theta_5 = \tan^{-1} 3 = 71.56°$	$\cos\theta_5 = 0.3163$	$\cos^3\theta_5 = 0.03164$
$\theta_6 = \tan^{-1} 5 = 78.69°$	$\cos\theta_6 = 0.1961$	$\cos^3\theta_6 = 0.00754$

The *minimum illumination*, $E_{min} = \frac{200}{20.93} \times 2(0.3535 + 0.03164 + 0.00754) =$ **7.5 lux. (Ans.)**

Example 1.17. *A small area of 5 metres in diameter is to be illuminated by a lamp suspended at a height of 4.5 m over the centre of the area. A lamp having an efficiency of 24 lumens per watt is fitted with a reflector which directs the light output only over the surface to be illuminated. If utilization coefficient is 0.65 and illumination 800 lux, determine the wattage of the lamp.*

Solution. Area of the working plane, $A = \frac{\pi}{4} \times 5^2 = 19.63 \text{ m}^2$

Efficiency of the lamp = 24 lumens/watt

Utilisation coefficient = 0.6

Illumination on the working plane, E = 800 lux

Luminous flux reaching the working plane,

$$F = E \times A = 800 \times 19.63 = 15704 \text{ lumens}$$

Total luminous flux emitted by the lamp

$$= \frac{F}{\text{Utilisation coefficient}} = \frac{15704}{0.65} = 24160 \text{ lumens}$$

Wattage of the lamp $= \frac{24160}{24} = 1007$ say **1000 W. (Ans.)**

Example 1.18. *A lamp of 200 candela is placed 1 m below a plane mirror which reflects 85 percent of light falling on it. The lamp is hung 5 m above ground. Find the illumination at a point on the ground 4 m away from the point vertically below the lamp.*

Solution. Fig. 1.21 shows the lamp and mirror arrangement.

The lamp L produces an image L' as far behind the mirror as it is in front.

Height of the image from the ground = 6 + 1 = 7 m

L' acts as the secondary source of light and its candle power = 0.85 × 200 = 170 candela.

Illumination at point B (4 m away from the point vertically below the lamp),

= Illumination due to L + Illumination due to L'

i.e., $$E_B = \frac{200}{(5)^2} \cos^3\theta + \frac{170}{(7)^2} \cos^3\theta'$$

$$= \frac{200}{(5)^2} \times \left(\frac{5}{\sqrt{5^2 + 4^2}}\right)^3 + \frac{170}{(7)^2}\left(\frac{7}{\sqrt{7^2 + 4^2}}\right)^3$$

$= (8 \times 0.476) + (3.47 \times 0.654) =$ **6.08 lux. (Ans.)**

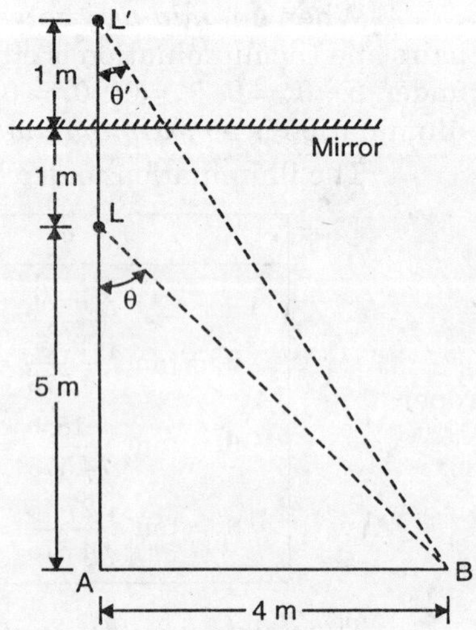

Fig. 1.21

Example 1.19. *A light is placed 4.5 m above ground and its candle power is 200 cos θ in any downward direction making an angle θ with the vertical. If A and B are two points on the ground, A being vertically under the light and the distance AB being 4.5 m, calculate :*

(*i*) *The illumination of the ground at A and also at B.*

(*ii*) *The total radiations sent down by the lamp.*

Solution. (*i*) **The illumination of the ground at A and B :**

C.P. along $LA = 100 \times \cos 0° = 100$ $\quad\therefore\quad$ $\mathbf{E_A} = \dfrac{200}{4.5^2} = \mathbf{9.87\ lux.}$ **(Ans.)**

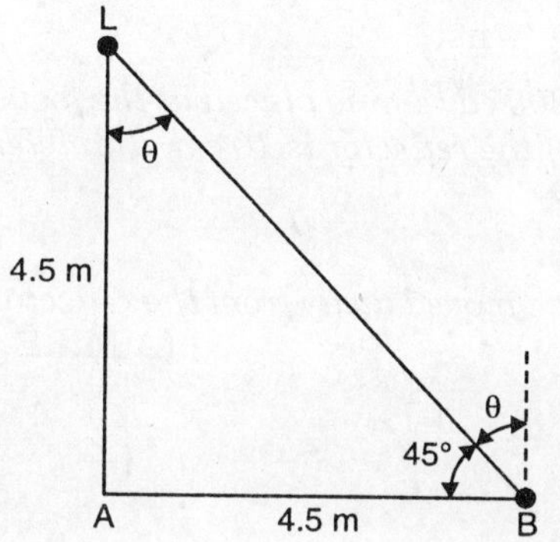

Fig. 1.22

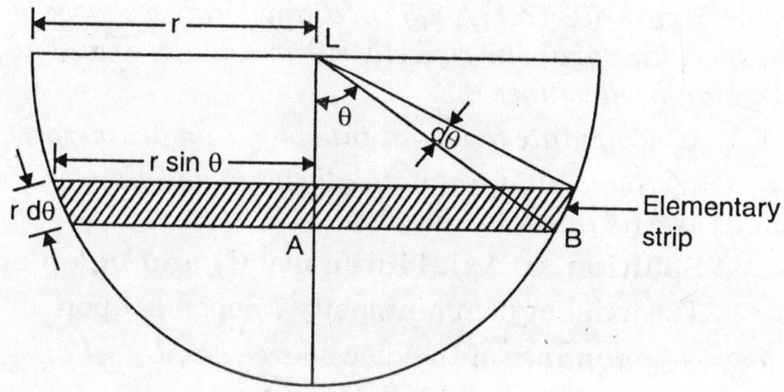

Fig. 1.23

C.P. along $LB = 200 \times \cos 45° = 141.4$ $\quad\therefore\quad$ $\mathbf{E_B} = \dfrac{141.4}{4.5^2} \times \left(\dfrac{4.5}{\sqrt{4.5^2 + 4.5^2}}\right)^3 = \mathbf{2.47\ lux.}$ **(Ans.)**

(*ii*) **The total radiation sent down by the lamp :**

Consider an imaginary hemisphere of radius *r* metre whose centre lies on the given lamp (Fig. 1.23).

C.P. along $LB = 100 \cos \theta$ $\quad\therefore\quad$ $E_B = \dfrac{200 \cos \theta}{r^2}$

The area of the elementary strip at an angular distance θ from the vertical and width of $PQ = r.d\theta$ is

$$= (2\pi r \sin \theta) \times r.d\theta = 2\pi r^2 \sin \theta.d\theta$$

Flux incident on the shaded area

$$= \frac{200 \cos \theta}{r^2} \times 2\pi r^2 \sin \theta.d\theta = 200\pi \times 2 \sin \theta \cos \theta\ .d\theta = 200\pi \sin 2\theta d\theta$$

Total flux over the hemisphere can be obtained by integrating the above expression between proper limits.

$$\therefore \quad \text{Total flux} \quad = \int_0^{\pi/2} 200\pi \sin 2\theta\, d\theta = 200\pi \left| -\frac{\cos 2\theta}{2} \right|_0^{\pi/2}$$

$= 200\pi =$ **628.3 lumens.** **(Ans.)**

Example 1.20. *A perfectly diffusing surface has a luminous intensity of 20 candelas at an angle of 60° to the normal. If the area of the surface is 50 cm², determine the brightness and total flux radiated.*

Solution. *Given* : I = 20 candelas ; $\theta = 60°$; A = 50 cm² = 50×10^{-4} m²

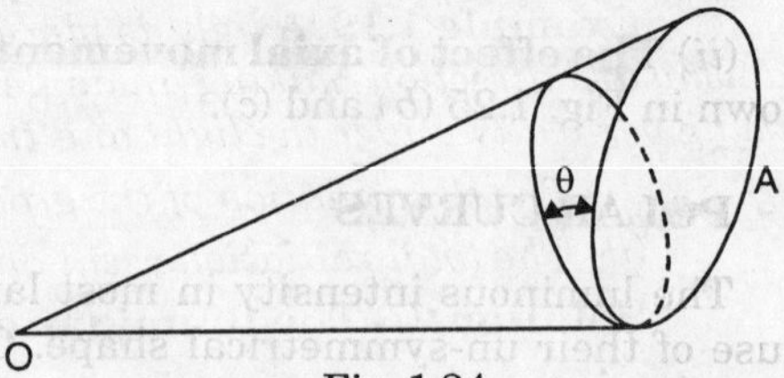

Fig. 1.24

Brightness or luminance,

$$L = \frac{I}{A \cos \theta} \text{ cd/m}^2$$

$$= \frac{20}{50 \times 10^{-4} \times \cos 60°} = \mathbf{8000\ cd/m^2}$$

$$E = \pi \times 8000 = \mathbf{25132.7\ lumens/m^2}. \quad \textbf{(Ans.)} \qquad (\because E = \pi L)$$

Total flux radiated = $25132.7 \times 50 \times 10^{-4}$ = **125.66 lumens. (Ans.)**

Example 1.21. *A 2.5 cm diameter disc source of luminance 1200 cd / cm² is placed at the focus of a specular parabolic reflector normal to the axis. The focal length of the reflector is 10 cm, diameter 40 cm and reflectance 0.8.*

(i) Calculate the axial intensity and beam-spread.

(ii) Show diagrammatically what will happen if the source were moved away from the reflector along the axis in either direction. **(A.M.I.E.)**

Solution. (*i*) **Axial intensity (I) and beam-spread (θ) :**

The axial or beam intensity I depends upon

— *luminance* of the disc source *i.e.*, L

— *aperture* of the reflector *i.e.*, A

— *reflectivity* of the reflector *i.e.*, ρ

$\therefore \quad I = \rho A L$ candela

Now, $\quad L = 1200 \text{ cd/cm}^2 = 1.2 \times 10^7 \text{ cd/m}^2$

$$A = \frac{\pi}{4} d^2 = \frac{\pi}{4} \times (0.4)^2 = 0.1257 \text{ m}^2$$

$\therefore \quad \mathbf{I = 0.8 \times 0.1257 \times 1.2 \times 10^7 = 1.207 \times 10^6\ cd.}$ **(Ans.)**

To a first approximation, the beam-spread for disc source is determined by reflector focal length and size of the disc source. If θ is the total beam-spread when the source is at the focus of the reflector [Fig. 1.25 (*a*)] then,

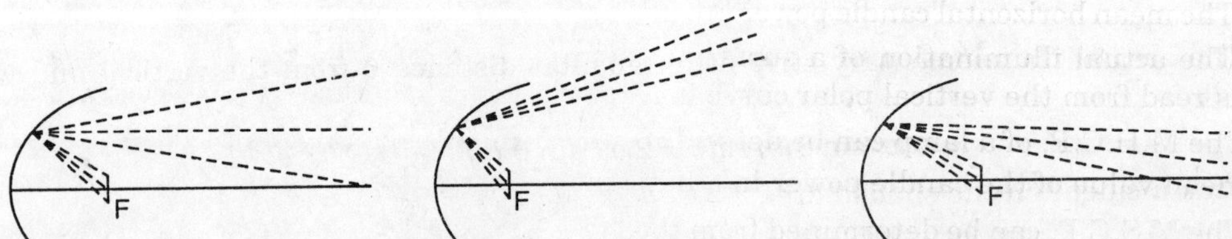

Fig. 1.25

$$\theta = 2 \tan^{-1}(r/f)$$

Here, $\quad$ Radius $r = \frac{2.5}{2} = 1.25$ cm ; Focal length, f = 10 cm ...*Given*

$$\therefore \quad \theta = 2 \tan^{-1}\left(\frac{1.25}{10}\right) = \mathbf{14.25°}. \quad \textbf{(Ans.)}$$

(*ii*) **The effect of axial movement of the source.** The effect of axial movement of the source is shown in Fig. 1.25 (*b*) and (*c*).

1.4. POLAR CURVES

The luminous intensity in most lamps or sources of light is not the same in all directions, because of their un-symmetrical shape. Often it is necessary to know the distribution of light in various directions to ascertain how the candle power of a light source varies in different directions. The luminous intensity in all the directions can be represented by ***polar curves.***

— If the luminous intensity *i.e.,* candle power is measured in a *horizontal plane* about a vertical axis and a curve is plotted between the candle power and the angular position, a "*horizontal polar curve*" or diagram is obtained.

— If the candle power is measured at angular position in a vertical plane, a polar curve in the vertical plane, called "*vertical polar curve*", is obtained.

Fig. 1.26 shows typical polar curves for an ordinary lamp.

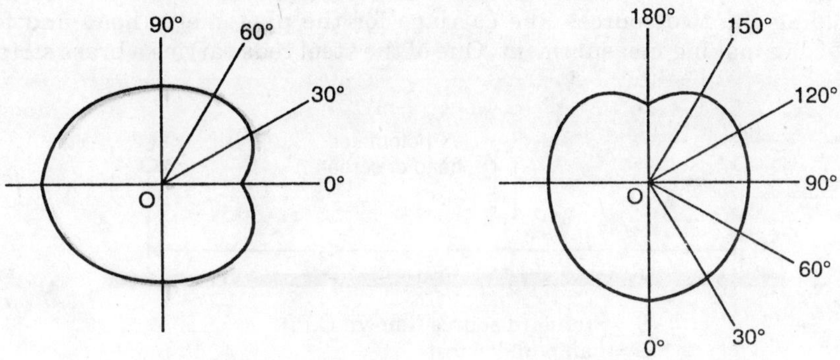

(*a*) For horizontal plane. (*b*) For vertical plane.

Fig. 1.26. Polar curves.

The *polar curves are used to determine the following :*

(*i*) The mean horizontal candle power (M.H.C.P.) and mean spherical candle power (M.S.C.P.).

(*ii*) The actual illumination of a surface by employing the candle power in that particular direction as read from the vertical polar curve in illumination calculations.

- The M.H.C.P. of a lamp can be determined from the horizontal polar curve by taking the mean value of the candle power in a horizontal direction.
- The M.S.C.P. can be determined from the vertical polar curve by *Roussean's construction.*

Roussean's construction. Suppose the vertical polar curve is in the form of two lobes symmetrical about *YOY′* axis. The Roussean's construction for this polar curve is illustrated in Fig. 1.27.

- Draw a circle with any convenient radius with centre *O* as the centre.
- Draw *PQ* parallel to *YOY′* and equal to the vertical diameter of the circle.
- Draw any line *OEA* meeting the polar curve in *E* and the circle in *A*. Let the projection be *S*.

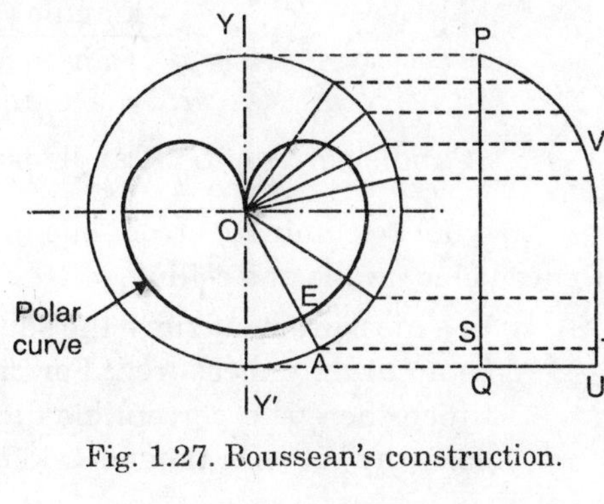

Fig. 1.27. Roussean's construction.

- At S erect an ordinate $ST = CE$.
- By similar construction draw other ordinates. The curve PSQUTVP obtained by joining these ordinates is known as *Roussean's curve*. The *mean ordinate of this curve gives the mean spherical candle power* (M.S.C.P.).

$$\text{The mean ordinate of the curve} = \frac{\text{Area PSQUTVP}}{\text{Length of } PQ}$$

The area under the curve can either be determined on a graph paper or by using Simpson's rule.

1.5. PHOTOMETRY

We shall now discuss the comparison and measurement of candle powers. The candle power of a source in any direction is measured by comparison with a standard or substandard source employing photometer bench and some form of photometer.

The "*photometer bench*" consists of two steel rods (2 to 3 metres long) which carry the stands or saddles for holding the two sources, the carriage for the photometer head and for any other apparatus employed in making measurement. One of the steel rods carries a brass strip with graduated scale in mm.

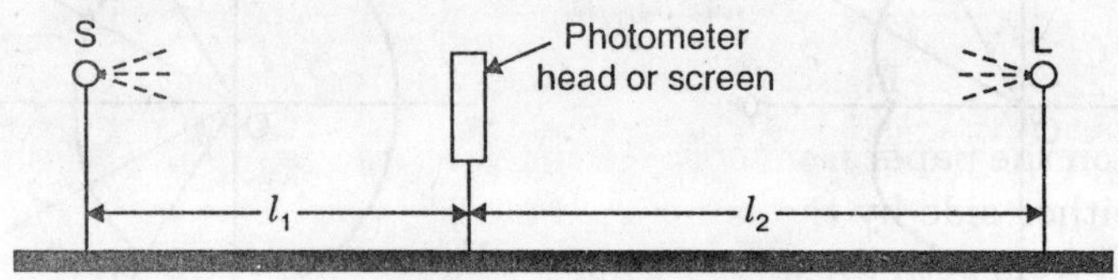

S = Standard source (known C.P.)
L = Lamp under test.

Fig. 1.28. Photometer bench for measurement of candle power.

The bench should be rigid so that source being compared may be free from vibrations and the carriage holding the photometer head should be capable of moving smoothly and with very little effort. The photometer head acts as a *screen* for comparison of illumination of the standard source and the source under test.

The photometer head or screen is moved in between the two fixed sources until the illumination on both the sides of the screen is same. If the distances of the standard source S and lamp (source) under test L from photometer head are l_1 and l_2 respectively, then applying inverse square law, we get

$$\frac{\text{Candle power of } L}{\text{Candle power of } S} = \frac{l_2^2}{l_1^2}$$

$$\therefore \quad \text{Candle power of } L = \text{Candle power of } S \times \frac{l_2^2}{l_1^2} \quad \text{...(1.3)}$$

In order to eliminate errors due to reflected light the experiment is performed in dark room with dead black walls and ceiling.

- In a measurement since the square of distances are involved, therefore, distances should be accurately measured. For obtaining distance exactly, two points are determined at which there is a perceptible difference in illumination from the two sides and the point half-way between them is taken as the position of equal illumination.

1.5.1. Photometer Heads

Although quite a few types of photometer heads are in use, yet most common in use are *Bunsen* and *Lummer Brodgun type.* These heads give quite accurate results when the lamps of the same or very similar colour are to be compared. However, when two lamps with different colours are to be compared, a *"Flicker photometer"* gives better results.

Bunsen head. It is also called the *"Grease Spot Photometer"*. It consists of a piece of thin opaque paper at the centre of which is a transluscent "spot" which is made by treating the paper with oil or wax.

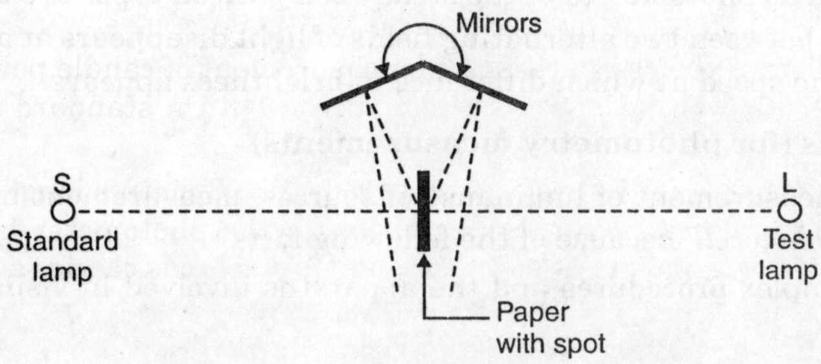

Fig. 1.29. Grease spot Photometer head.

- Light falls on the paper from both the lamps. The opaque part of the paper will be illuminated on either side by the lamp on that side only. But the transparent "spot" will be illuminated from both the lamps. The photometer head is adjusted at such positions that the *transparent spot is not perceptible.* The paper is viewed from one side first and then from the other. We get two distances l_1 and l_1' of the photometer head from the test lamp (L) and l_2 and l_2' the distances of the photometer head from the standard lamp (S). The C.P. (candle power) of the test lamp is calculated from the following relation :

$$\frac{\text{C.P. of test lamp } (L)}{\text{C.P. of the standard lamp } (S)} = \frac{l_1 l_1'}{l_2 l_2'}.$$

- In another method, one side of the paper is illuminated from a fixed lamp. One the other side of the photometer first the test lamp (L) is used and then the standard lamp (S) is used to make the spot to vanish. If l_1 and l_2 are these distances, then we have,

$$\frac{\text{C.P. of test lamp } (L)}{\text{C.P. of standard lamp } (S)} = \frac{l_1^2}{l_2^2}.$$

The *use of two mirrors with the photometer head* is *perhaps the most accurate method* (Fig. 1.29). The two sides of the spot can be viewed simultaneously and the position of the head for equal "contrast" in illumination between the opaque part and the transparent spot of the paper on the two sides is located. The C.P. of the test lamp (L) can then be calculated by the expression given earlier.

Lumner-Brodgun photometer head. There are two types of Lummer-Brodhum photometer heads :

(*i*) Equality-of-brightness type.

(*ii*) Contrast type.

Contrast type is more accurate and, therefore *more used in photometric measurements.*

Flicker photometers :

- These photometers are used *when the two sources giving light of different colours are to be compared.*
- They operate on the fact that if two illuminated surfaces are presented to the eye with rapid alternations, the *flicker disappears when the surfaces are of equal brightness.* The speed of alternation should be kept as low as possible at which the disappearance of the flicker can be obtained for the small variation in brightness.
- These photometers are not affected by the difference in colour of two lights to be compared as much as photometers of the steady comparison types are affected since the colour difference between two alternating fields of light disappears at a lower speed of alternation than the speed at which difference in brightness appears.

1.5.2. Photocells (for photometry measurements)

Except for the measurement of luminance of sources, measurement by visual comparisons have been replaced by *'photocell'* because of the following facts :

(*i*) The more complex procedures and the apparatus involved in visual measurements are avoided.

(*ii*) These cells give *more accurate and faster measurements.*

(*iii*) The measurements are consistent.

The two types of photocells used in photometry measurements are :

1. Photovoltaic cells
2. Photoemissive cells.

1. ***Photovoltaic cell.*** Fig. 1.30 shows the construction of a photovoltaic cell (also known as the barrier layer or rectified cell). It consists of a base plate made of either steel or aluminium and carries a layer of metallic selenium which is *light sensitive*. An electrically conducting layer of cadmium oxide is applied by sputtering over the selenium layer. The layer is *sufficiently thin* to allow light to reach the selenium and is electrically continuous as it acts as a negative pole. The negative contact is formed of a strip of wood's metal sprayed on to the edge of the top surface. The base plate forms the positive contact. The front surface of the cell is protected by a transparent varnish.

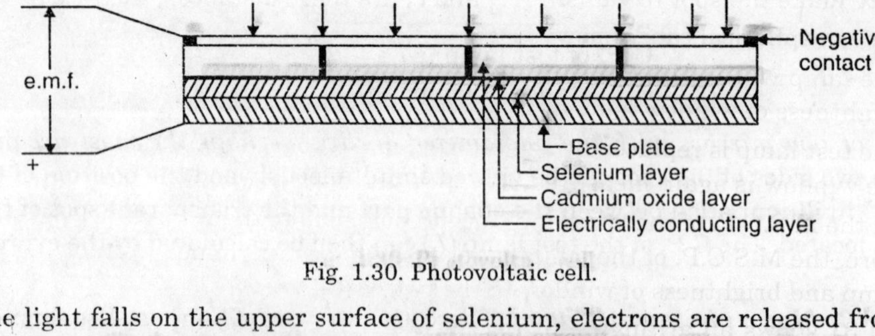

Fig. 1.30. Photovoltaic cell.

As the light falls on the upper surface of selenium, electrons are released from the surface which maintain a flow of current through the external circuit connected between the +ve and –ve contacts.

- *In photometry the current output of a photocell should be proportional to the illumination which is achieved by keeping the external resistance at a low value.* For achieving greater accuracy the illumination should not be allowed to exceed 275 lumens/m^2.

- The smaller the size of the cell (of course compatible with obtaining sufficient current to be properly measured) the better a linear relation is obtained between current and illumination. This is because, for such a cell, the resistance of the electrically conducting film is at a minimum. Also, the current being small, the voltage drop due to the circuit resistance will be kept low.

2. ***Photoemissive cell.*** In order to get greater precision in terms of linearity and stability it is better to use photoemissive cell rather than photovoltaic cell. However, the circuit involved is more complex and requires the use of some sort of valve amplifiers.

1.5.3. Distribution Photometry

- ***Distribution photometry*** *assists in finding out the variations of luminous intensity around a lighting fitting so that polar curves and other diagrams for the fitting can be obtained and also the performance of the fitting is studied.*

Following points are important for proper distribution photometry :

(*i*) The *temperature of the test area should be kept stable* (since the output flux of some sources like fluorescent tubes depends upon the temperature of the surroundings)

(*ii*) *Only that light which is be measured should be allowed to reach the photocell.*

(*iii*) *Optical path should be made sufficiently long* so that the inverse square law may be applied with a reasonable degree of accuracy.

- In a very simple ***distribution photometer***, the photocell fitting moves around the fitting on a semicircular track and its position is controlled by means of a flexible cable. A complete series of curves at different angles of azimuth can be obtained by rotating the fitting about a vertical axis. The mounting of the fitting on the apparatus is facilitated by mounting a gear system on the gentry which turns the fitting so that it can be brought down to easy access.

1.5.4. Measurement of M.S.C.P. by Integrating Sphere

The ***integrating sphere*** consists of a hollow sphere whose diameter (one or more than one metre) is large compared to the lamp to be tested, having a smooth inner surface with a uniform coating of white paint. If the lamp is hung inside the sphere, the light is so diffused that an uniform illumination is produced over the whole surface. A small window of transluscent glass provided at one side of the sphere is illuminated by reflected light from the inner surface of the sphere. A small *screen* is inserted in between the lamp and window to prevent the light from the lamp reaching the window *directly*.

The mean spherical candle power (M.S.C.P.) is measured as follows :

- The lamp whose M.S.C.P. is to be determined, is placed at the centre of the sphere and brightness of window is measured with the help of some form of illuminometer.
- The test lamp is replaced by a standard lamp whose M.S.C.P. is known and brightness of the window is again measured.

Since the M.S.C.P. of the two lamps are proportional to the respective brightness of the window, therefore, the M.S.C.P. of the lamp under test can be determined by knowing the M.S.C.P. of standard lamp and brightness of window in the two cases.

Fig. 1.31 shows an arrangement in which a Lumner-Brodgun photometer head is used in which a mirror is placed as indicated. The light from the glass window is reflected by this mirror as shown. Another lamp is used for comparison and for obtaining balance by shifting the position of the comparison lamp from the photometer head. We get two distances of comparison lamp from the photometer head : *First distance* when the glass window is illuminated by the test lamp and the *second distance* when the glass window is illuminated by the standard or sub-standard lamp. The squares of the corresponding distances of the comparison lamp from the photometer head are *inversely proportional* to the M.S.C.Ps. of the lamps.

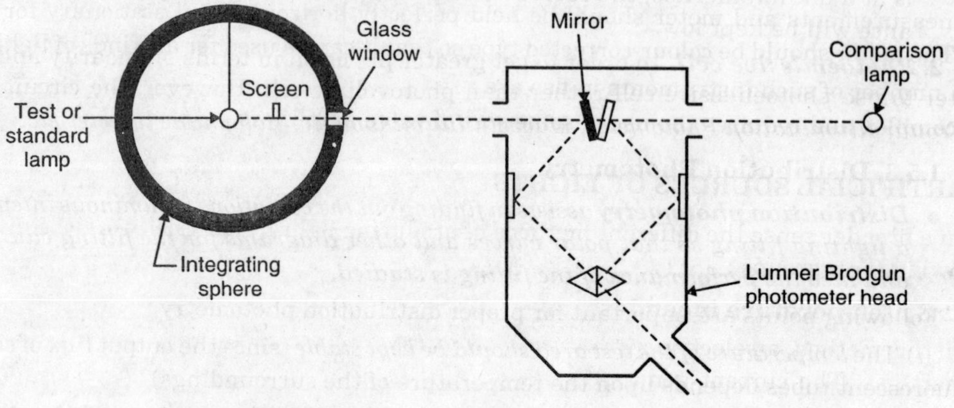

Fig. 1.31. Measurement of M.S.C.P. using the integrating sphere.

1.5.5. Measurement of Brightness or Luminance

The average brightness or luminance of a source in a given direction is given by the *ratio of its intensity to the projected area of the source in that direction.* For achieving greater sensitivity it is necessary to use some kind of optical system to collect more light flux.

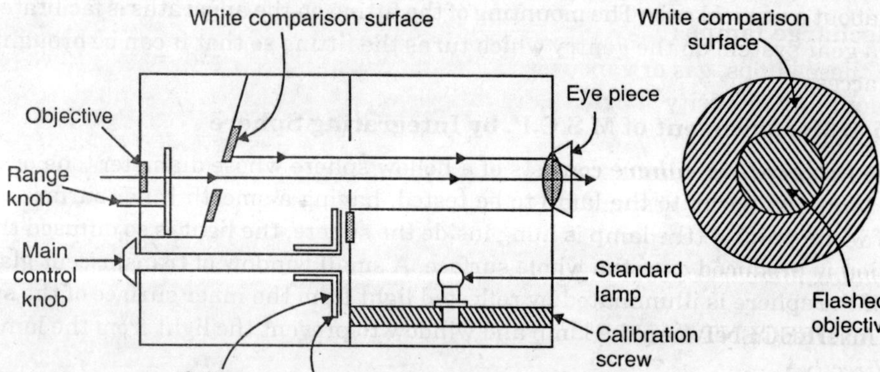

Fig. 1.32. Holophane lumeter.

Fig. 1.32 shows the schematic arrangement of the ***holophane lumeter*** wherein the light from the source falls on the objective lens which is viewed through the eye piece. The source is, therefore, seen as a uniformly bright disc having the luminance of the source (multiplied by the percentage loss in the optical system). This disc is surrounded by a ring of light reflected by the white comparison surface. The luminance of this ring is matched against the object by adjusting the range and main control knobs which are graduated.

A standard lamp placed in the whitened compartment, illuminates the white screen. This lamp is moved with the help of a calibration screw only when the instrument requires recalibration. Light reaches the white comparison surface after passing through the opal window and amount of this light is adjusted by adjusting the sector disc by means of main control knob.

1.5.6. Measurement of Illumination

The simplest and best method of measuring illumination is by a portable type illumination photometer calibrated to read directly in lux. No shadow should be allowed to fall on the meter during measurements and meter should be held perfectly horizontal and stationary for accurate results. The meter should be colour-corrected type so that it can be used for all kinds of light sources.

A number of such instruments in use are :

Trottar Illumination Photometer, Macbeth Illuminometer, Holophane lumeter etc.

1.6. ARTIFICIAL SOURCES OF LIGHTS

In a broader sense the different methods of producing light by electricity may be divided into the following *three* groups :

1. **By temperature incandescence :**

In this method, an electric current is passed through a filament of thin wire placed in vacuum or an inert gas. The current generates enough heat to raise the temperature of the filament to luminosity.

Examples. Incandescent tungsten filament lamps ; since their output depends on the temperature of their filaments, they are known as '*temperature radiators*'.

2. **By establishing an arc between two carbon electrodes :**

In this case the source of light is the incandescent electrode.

Examples. Carbon arc lamp ; flame arc lamp, magnetic arc lamp.

3. **Discharge lamps :**

- In these lamps, gas or vapour is made luminous by electric discharge through them. The colour and intensity of light *i.e.,* candle-power emitted depends on the nature of the gas or vapour only.
- These lamps are luminiscent-light lamps and *do not depend on temperature for higher efficiencies.* In this respect, they differ radically from incandescent lamps whose efficiency is dependent on temperature.

Examples. Mercury vapour lamp, sodium vapour lamp, neon-gas lamp and fluorescent lamp.

1.7. INCANDESCENT LAMPS

The incandescent or filament type lamp consists of a glass globe completely evacuated and a fine wire known as *filament* within it. The glass globe is evacuated to prevent the oxidation and convection currents of the filament and also to prevent the temperature being lowered by radiation.

Materials commonly used materials for incandescent lamps. The materials used for making filaments of incandescent lamps are *carbon, tantalum and tungsten.*

1. **Carbon :**

(*i*) Resistivity, ρ = 1000 to 7000 μ ohm cm

(*ii*) Temperature coefficient, $\alpha = -0.0002$ to -0.0008

(*iii*) Melting point = 3500°C

(*iv*) Density = 1.7 to 3.5

- To prevent the blackening of the bulb, the working temperature is 1800°C.
- The commercial efficiency of filament lamp is about 4.5 lumens per watt app.

2. **Tantalum :**

(*i*) ρ = 12.4 μ ohm cm.

(*ii*) α = 0.0036.

(*iii*) Melting point (M.P.) = 2996°C.

(*iv*) Density = 16.6.

- The efficiency is low such as 2 lumens per watt. So it is not used much now-a-days.

3. **Tungsten :**

(*i*) ρ = 5.6 μ ohms cm.

(*ii*) α = 0.0045

(*iii*) M.P. = 3400°C

(*iv*) Density = 19.3.

- The *efficiency*, when worked at 2000°C in an evacuated bulb is 18 lumens per watt. This metal is *most widely used for the purpose.*
- To prepare filament, pure tungsten powder is pressed in steel mould for small bars. Mechanical strength of bars is improved by heating electrically nearly to the melting point. These bars are then hammered at red heat and drawn into filaments. To improve efficiency, the bulb is filled with an inert gas argon with a small percentage of nitrogen. To reduce the convection currents, produced by the gas molecules in the bulb, the filament is wound into a close spiral and suspended horizontally in the form of a circular arc.
- The efficiency of the gas filled "coiled coil" tungsten filament is about 30 lumens per watt. This is due to high working temperature of 2500°C.

The *ideal material* for the filament of the incandescent lamps in one which has the following properties :

1. High melting point.
2. Low vapour pressure.
3. High resistivity.
4. Low temperature coefficient.
5. Ductility.
6. Sufficient mechanical strength to withstand vibrations during use.

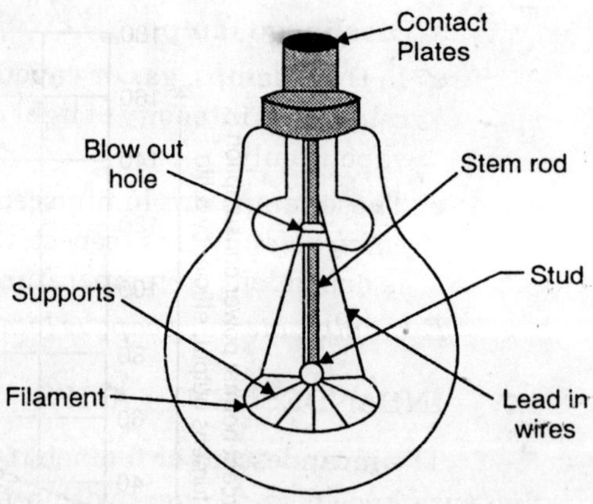

Fig. 1.33

Fig. 1.33 shows the construction of modern coiled-coil gas-filled filament lamp. The lamp has a 'wreath' filament *i.e.*, a coiled filament arranged in the form of a wreath on radial supports.

Aging effects. With the passage of time the light output of an incandescent lamp decreases due to the following *two reasons* :

(*i*) Evaporation of the filament tends to cause the bulb to blacken.

(*ii*) Evaporation makes the filament slowly decrease in diameter, which means that the resistance of the filament increases. Therefore, an old filament draw less current and operates at a lower temperature, which reduces its light output. Consequently the efficiency of the lamp (lumens output/watt input) also *decreases* with the passage of time.

Fig. 1.34 and 1.35 shows a coiled filament and coiled-coil filament respectively.

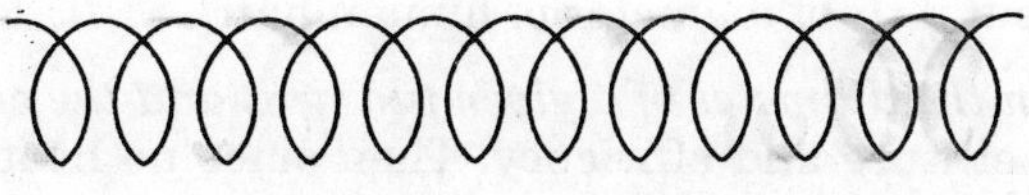

Fig. 1.34. Coiled filament.

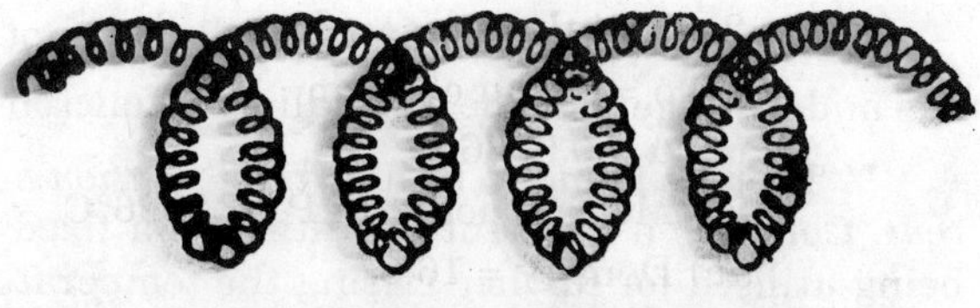

Fig. 1.35. Coiled-coil filament.

The total depreciation of the light output is roughly 15 percent over the useful life range.

Effects of voltage variation. When an incandescent lamp is subjected to voltages different from normal voltage, its operating characteristics are affected.

The efficiency of a lamp (lumens/watt) *increases with the increase in voltage* owing to increase in temperature and is *proportional to the square of the voltage.*

The various relationships are :

- *Lumens output* $\propto (V)^{3.55}$
- *Power consumption* $\propto (V)^{1.55}$
- *Luminous efficiency* $\propto (V)^2$
- *Life* $\propto (V)^{-13}$ *for vacuum lamps ;*
  $\propto (V)^{-14}$ *for gas-filled lamps.*

Fig. 1.36 shows the variation in power consumption, lumens output, efficiency and life of incandescent lamps with the variation in voltages.

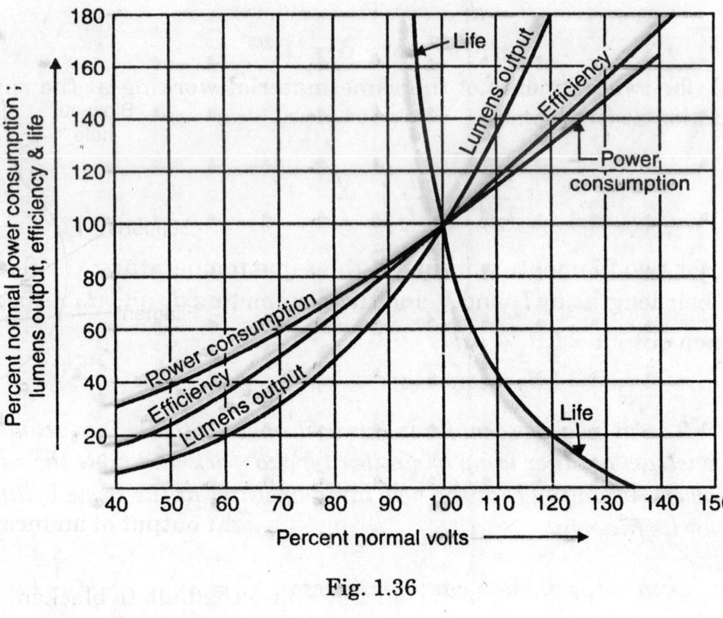

Fig. 1.36

Advantages of incandescent lamps :

1. Direct operation on standard distribution voltage.
2. Operating power factor unity.
3. Good radiation characteristics in the luminous range.
4. No effect of surrounding air temperature.
5. Availability in various shapes and shades.

Filament dimensions. The diameter of a tungsten lamp filament, depending upon the voltage and wattage, may be as small as 10 microns (about $\frac{1}{6}$th of diameter of human hair).

There is found to be a *definite relation between the diameter of a given filament and the current.* Consider a filament operating at a fixed temperature and efficiency. Then since no heat is being utilised for further raising the temperature, all the heat produced in a given time is mostly lost by radiation, if vacuum is good. In other words,

Heat produced per second = Heat lost per second by radiation.

$$\text{Now, power intake} = I^2R = I^2 \times \frac{\rho l}{a} = \frac{I^2 \rho l}{\frac{\pi}{4} d^2} = I^2 \left(\frac{4\rho l}{d^2}\right)$$

where, I = Filament current in amperes,

a = Filament cross-section

ρ = Resistivity of filament material at the working temperature,

l = Filament length, and

d = Filament diameter.

Heat radiated per second from the surface is proportional to the area of the surface and emissivity of the material.

$\therefore$ Heat lost/second $\propto$ surface area × emissivity e

or,
$$I^2 \left(\frac{4\rho l}{d^2}\right) \propto l \times \pi d \times e$$

or,
$$I^2 \propto d^3 \quad \text{...(i)}$$

$$\therefore \quad I \propto (d)^{3/2} \quad \text{or} \quad \mathbf{d \propto (I)^{2/3}}$$

In general, for two filaments of the same material working at the same temperature and efficiency, the relation at (*i*) becomes

$$\left(\frac{I_1}{I_2}\right)^2 = \left(\frac{d_1}{d_2}\right)^3 \quad \text{...(1.4)}$$

Moreover, for two filaments working at the same temperature, the flux per unit area is the same. Denoting their lengths by l_1 and l_2 and their diameters d_1 and d_2 respectively, we have

$$\text{Lumen output} \propto l_1 d_1 \propto l_2 d_2$$

or,
$$l_1 d_1 = l_2 d_2 = \text{constant} \quad \text{...(1.5)}$$

Example 1.22. *An incandescent lamp has a filament of 0.0045 cm diameter and 90 cm length. It is required to construct another lamp of similar type to work at double the supply voltage and give half the candle power. Assuming that the new lamp operates at the same brilliancy, determine suitable dimensions for its filament.* **(Utkal University)**

Solution. *Given :* $d_1 = 0.0045$ cm ; $l_1 = 90$ cm ; $V_2 = 2V_1$; $\frac{I_2}{I_1} = \frac{1}{2}$ (where I_1 and I_2 are the luminous intensities of the two lamps).

Dimensions of the filament of the second lamp ; $\mathbf{d_2, l_2}$:

Since $\quad I_1 \propto l_1 d_1 \quad$ and $\quad I_2 \propto l_2 d_2$

$$\therefore \quad \frac{I_2}{I_1} = \frac{1}{2} = \frac{l_2 d_2}{l_1 d_1} \quad \text{or} \quad l_2 d_2 = \frac{1}{2} l_1 d_1$$

Assuming that the power intertakes of the two lamps are proportional to their outputs, we have

$$I_1 \propto V_1 i_1, \quad \text{and} \quad I_2 \propto V_2 i_2$$

$$\therefore \quad \frac{V_2 i_2}{V_1 i_1} = \frac{I_2}{I_1}$$

$$\therefore \quad \frac{i_2}{i_1} = \left(\frac{V_1}{V_2}\right)\left(\frac{I_2}{I_1}\right) = \frac{1}{2} \times \frac{1}{2} = \frac{1}{4}$$

Now, $$i_1 \propto (d_1)^{3/2}, \quad \text{and} \quad i_2 \propto (d_2)^{3/2}$$

$$\therefore \quad \left(\frac{d_2}{d_1}\right)^{3/2} = \left(\frac{i_2}{i_1}\right) = \frac{1}{4}$$

or, $$\frac{d_2}{d_1} = \left(\frac{1}{4}\right)^{2/3} = 0.3968$$

$$\therefore \quad \mathbf{l_2} = \frac{1}{2} l_1 \frac{d_1}{d_2} = \frac{1}{2} \times 90 \times \frac{1}{0.3968} = \mathbf{113.4\ cm.} \quad \textbf{(Ans.)}$$

and, $$\mathbf{d_2} = 0.3968 \times 0.0045 = \mathbf{0.0017856\ cm.} \quad \textbf{(Ans.)}$$

Example 1.23. *A 60 candle power, 250 V metal filament lamp has a measured candle power of 71.5 candle at 260 V and 50 candle at 240 V. Calculate :*

(i) The constant for the lamp in the expression $C = aV^b$ where C = candle power and V = voltage.

(ii) The change of candle power per volt at 250 V.

(iii) The percentage variation of candle power due to a voltage variation of 4 percent from the normal value. **(A.M.I.E.)**

Solution. (*i*) **The constant for the lamp :**

$$C = aV^b \quad \text{...Given}$$

$$\therefore \quad 71.5 = a \times (260)^b \quad \text{...(i)}$$

$$50 = a \times (240)^b \quad \text{...(ii)}$$

Dividing (*i*) by (*ii*), we get

$$\frac{71.5}{50} = \left(\frac{260}{240}\right)^b$$

or, $$1.43 = (1.083)^b$$

or, $$\ln(1.43) = b \ln(1.083)$$

or, $$b = \frac{\ln(1.43)}{\ln(1.083)} = 4.5$$

Substituting this value of *b* in (*i*), we have

$$71.5 = a \times (260)^{4.5}$$

$$\therefore \quad a = 9.7 \times 10^{-10}$$

Hence, $$\mathbf{C = 9.7 \times 10^{-10}\ (V)^{4.5}\ candela.} \quad \textbf{(Ans.)}$$

(*ii*) **The change of candle power per volt at 250 V**

Now, $$C = 9.7 \times 10^{-10}\ (V)^{4.5}$$

Differentiating the above expression and putting $V = 250$ V, we get

$$\frac{dC}{dV} = 9.7 \times 10^{-10} \times 4.5(250)^{3.5} = \mathbf{1.078 \text{ candela per volt.}} \quad \textbf{(Ans.)}$$

(*iii*) **The percentage variation of candle power :**

When voltage *increases* by 4%, $\frac{C_2}{C_1} = (1.04)^{4.5}$

% age change in candle power $= \frac{C_2 - C_1}{C_1} \times 100$

$= (1.04^{4.5} - 1) \times 100 = \mathbf{19.3\%.}$ **(Ans.)**

When the voltage falls by 4%, $\frac{C_2}{C_1} = (0.96)^{4.5}$

$\therefore$ % age change in candle power $= [(0.96)^{4.5} - 1] \times 100 = \mathbf{16.78\%.}$ **(Ans.)**

Clear and inside-frosted gas-filled lamps

- Clear glass-filled lamps facilitate light control and are necessary for use in lighting units where *accurate distribution is required such as in flood-lights for buildings, projectors and motor-car headlights.* However they produce *hard shadows* and *glare from filaments.*
- Inside-frosted gas lamps have luminous output nearly *2% less than* clear glass lamps of the same rating, but they *produce softer shadows and practically eliminate glare from filaments.* Such lamps are ideal for use in industrial open fittings located in the line of sight at low mounting heights and in diffuse fittings of opal glass type in order to avoid the presence of filament striations on the surface of glassware etc.

Inside-silica coated lamp has high diffusion of light output, due to the fine coating of silica on the inside of its bulb. Such lamps are *less glaring, soften shadows* and *minimize the brightness of reflections from* shiny surfaces.

Halogen lamp. As the life and efficiency of an incandescent lamp falls off with use-partly due to slow evaporation of the filament and partly due to black deposit formed on the inside of the bulb, the addition of a small amount of halogen vapour to the filling gas restores part of the evaporated tungsten vapour back to the filament by means of a chemical reaction *i.e.*, there is a sort of '*regenerative cycle*'.

Following are the *advantages* of halogen lamps :

(*i*) Long life–2000 hours.

(*ii*) High operating temperature with increased luminous efficiency varying from 22 to 33 lumens/watt.

(*iii*) No blackening of lamp, hence no depreciation of lumens output.

- These lamps, being manufactured in sizes upto 5 kW, are suitable for outdoor illumination of buildings, playing fields, large gardens fountains, car parks etc.

1.8. ARC LAMPS

These lamps are used in *searchlights, projection lamps and other special purpose lamps like those in flash cameras.*

In an arc lamp electric current is made to flow through two electrodes in contact with each other which are drawn apart. The result is an arc being struck. The arc maintains the current, and is very efficient source of light.

The various forms of arc lamps are :

1. Carbon arc lamp
2. Flame arc lamp
3. Magnetic arc lamp.

- The carbon rods used with A.C. supply are of the same size as that used with D.C. supply. The positive rod is of larger size than the negative rod.
- The craters in the arc at the positive and negative rods are of the same size (with A.C. supply) while with D.C. supply, the positive crater is bigger and gives 85 per cent light at a temperature of 3500°C, while the negative crater is of smaller size. The efficiency of the lamp is 9 lumens/watt.
- The positive electrode gets consumed earlier than the negative electrode, if the size of the former is same as the latter. Hence the positive electrode is of twice the diameter than that of the negative.
- A resistance is used to stabilise the arc.
- The voltage drop across the arc is about 60 V (Fig. 1.37) and supply voltage is upto 100 V.

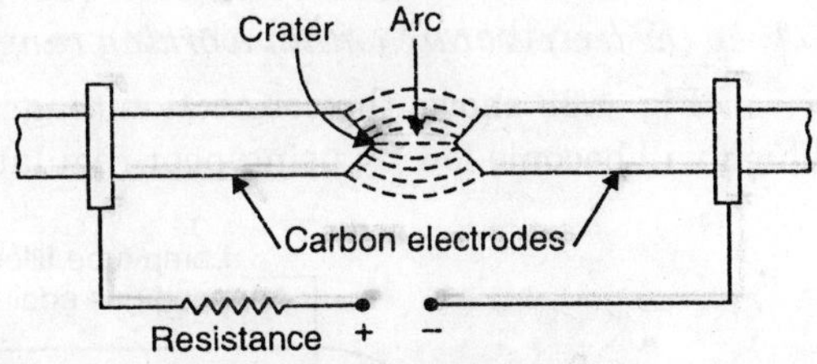

Fig. 1.37. Carbon arc lamp.

1.9. DISCHARGE LAMPS

- In all discharge lamps, *an electric current is passed through a gas or vapour which renders it luminous.* In this process of producing light by gaseous conduction, the most commonly used elements are *neon*, *mercury* and *sodium vapours.*
- The colours (*i.e.* wavelength) of light produced depends on the *nature of gas or vapour.*

— Neon discharge fields orange-red light ;

— Mercury-vapour light is always bluish ;

— Sodium vapour light is orange-yellow.

Types of discharge lamps. Discharge lamps are of the following *two* types :

Type-1. Those lamps in which *colour of light is the same as produced by the discharge through the gas or vapour.*

Example. Sodium vapour, mercury vapour and neon gas lamps.

Type-2. Those lamps which *use the phenomenon of fluorescence* ; these are known as *fluorescent lamps.* In these lamps, the discharge through the vapour produces ultra-violet waves which cause fluorescence in certain materials called as *phosphor.* The inside of the fluorescent lamp is coated with a phosphor which absorbs invisible ultra-violet rays and radiate visible rays.

Example. Fluorescent mercury-vapour tube.

In general, the discharge lamps are *superior* to metal filament lamps. However, they have the following **demerits :**

(*i*) High initial cost.

(*ii*) Poor power factor.

(*iii*) Starting, being somewhat difficult, requires starters/transformers in different cases.

(*iv*) Time is needed to attain full brilliancy.

(*v*) Since these lamps have negative resistance characteristic ballasts are necessary to stabilise the arc.

(*vi*) The flicker (caused due to the fluctuation of light output at twice the supply frequency) causes *stroboscopic effect.*

(*vii*) They are suitable only for a particular position.

1.9.1. Sodium Vapour Lamp

Construction and working of a sodium vapour lamp are described below.

Construction : Refer to Fig. 1.38.

This type of lamp is of *low luminosity*, so the length of this lamp is *large.* To get the required length it is made in the form of a *U-tube.* Two oxide-coated electrodes are sealed with the ends. The tube contains a little sodium and neon gas. The U-tube is enclosed in a *double-walled vacuum flask to keep the temperature within working range.*

Fig. 1.38 shows the connection diagram. *Capacitor is connected to improve the power factor* which will become low by using poor regulation transformer.

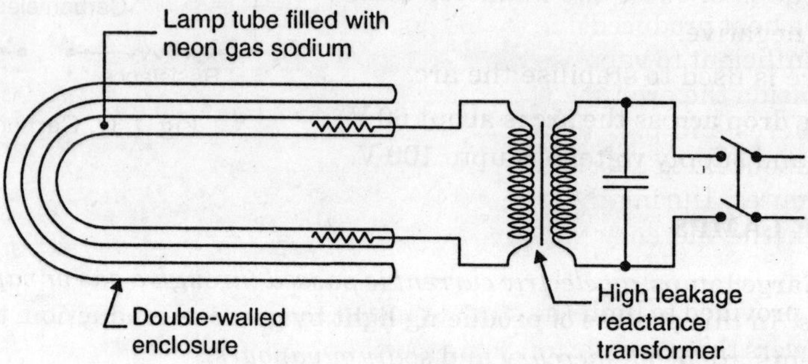

Fig. 1.38. Sodium vapour lamp.

Working :

Before the lamp starts working, the sodium is in the form of a solid, deposited on the sides of the tube walls. In the beginning when the switch is on, it operates as a low pressure neon lamp with pink colour. The lamp gets warmed, sodium is vaporised and it radiates yellow light and then, after sometime, about 10-15 minutes, the lamp starts giving full light.

In order to start the discharge lamp, a striking voltage of 380 V is required for 40 W lamp and 450 V for 100 W lamp. These voltages are obtained from a high reactance transformer or auto-transformer. At no load the voltage is very high which falls down as the lamp starts giving light, since the regulation of transformer is poor.

The lamp *fails to operate* when (*i*) the filament breaks or burns out, (*ii*) the cathode stops to emit electrons, (*iii*) the sodium particles may concentrate on one side of the tube, and (*iv*) the lamp is blackened owing to sodium vapour action on the glass, in which case the output will be reduced.

- The efficiency of a sodium vapour lamp under practical conditions is about 40-50 lumens/watt. Such lamps are manufactured in 45, 60, 85 and 140 watts ratings. The average life is about 3000 hours and is *not affected by voltage variations.* At the end of this period the light output will be reduced by 15 percent due to aging.
- This type of lamp is mainly used for highway and general outdoor lighting where colour discrimination is not required.

This lamp should be hung vertical otherwise sodium will blacken the inside of the tube.

1.9.2. High Pressure Mercury Vapour Lamp

Fig. 1.39 shows a high pressure mercury vapour lamp.

Construction :

It consists of two bulbs-an arc-tube containing the electric discharge and outer bulb which protects the arc-tube from changes in temperature. The inner tube or arc-tube is made of quartz (or hard glass) and the outer bulb of hard glass.

The arc-tube contains a small amount of mercury and argon gas. In addition to two main electrodes, an auxiliary starting electrode connected through a high resistance (about 50 kΩ) is also provided. The main electrodes consists of tungsten coils with electron-emitting coating or elements of thorium metal.

Working :

When the supply is switched on, initial discharge for the few seconds is established in the argon gas between the auxiliary starting electrode and the neighbouring main electrode and then in argon between the two main electrodes. The heat produced due to this discharge through the gas is sufficient to vaporise mercury. Consequently, pressure inside the arc-tube increases to about one to two atmospheres and p.d. across the main electrodes grows from about 20 to 150 V, the operation taking about 5 to 7 minutes. During this time, discharge is established through the mercury vapours which emit greenish-blue light.

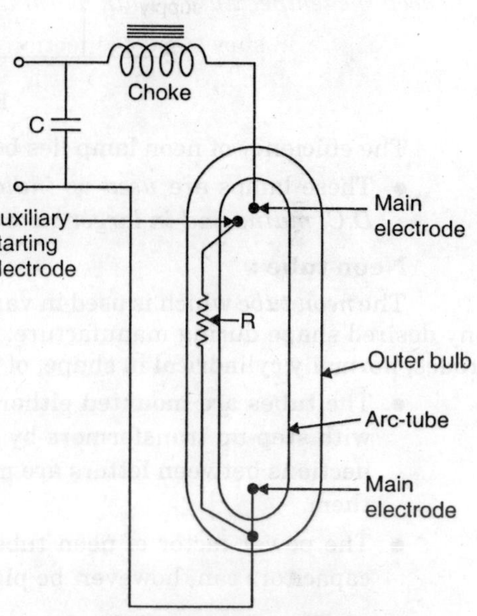

Fig. 1.39. High pressure mercury vapour discharge lamp.

The choke is provided to limit the current to a safe value. This choke lowers the power factor, so a capacitor *C* is connected across the circuit to improve the power factor.

The efficiency of this type of lamp is 30-40 lumens/watt. These lamps are manufactured in 250 W and 400 W ratings for use on 200-250 V A.C. supply mains.

These lamps are *used* for *general industrial lighting, railway yards, ports, work areas, shopping centres etc.* where greenish-blue colour light is not objectionable.

1.9.3. Mercury Iodide Lamps

- These lamps are similar in construction to high pressure mercury vapour lamps but in addition to mercury, a number of iodides are added which fill the gaps in the light spectrum, and thus, improve the colour characteristic of light. A *separate ignition device*, in addition to the choke, is required for such a lamp.
- Their *efficiency* is comparatively higher (75-90 lumens/watt)
- Such lamps are suitable for application in the fields of *flood-lighting, industrial lighting* and *public lighting*.

1.9.4. Neon Lamp

Neon lamps belong to *cold-cathode category.*

The electrodes are in the form of iron shells and are coated on the inside.

The colour of light emitted is red. If the helium gas is used in place of neon, pinkish white light is obtained. Helium and neon through coloured glass tubing produce a variety of effects.

Fig. 1.40 shows a circuit for a neon lamp. The transformer has a high leakage reactance which stabilizes the arc in the lamp. A capacitor is used for power factor improvement. High voltage is used for starting.

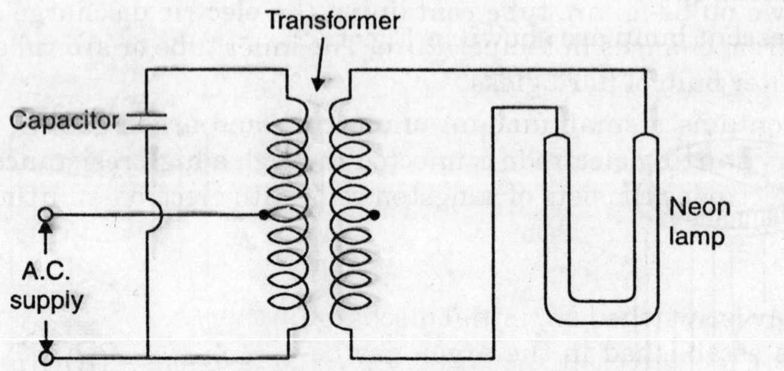

Fig. 1.40. Neon lamp.

The efficiency of neon lamp lies between 15-40 lumens/watt.

- These lamps are *used* as *indicator lamps, night lamps for determination of polarity of D.C. mains and in larger sizes on neon tubes for the purpose of "advertising"*.

Neon tube :

The *neon tube* which is used in varying lengths upto about 8 meters, may be bent into almost any desired shape during manufacture. It consists of a length of glass tubing containing two electrodes, normally cylindrical in shape, of iron, steel, or copper.

- The tubes are mounted either on a wooden frame or a metal base. These are matched with step-up transformers by connecting suitable tappings for the rated current. Connections between letters are made by nickel wires, the glass tubings being slipped over them.
- The power factor of neon tubes is quite low and is improved by using capacitors. The capacitors can, however, be placed on the low voltage side of the transformer.

1.9.5. Fluorescent Tube (lamp)

Fluorescent lighting has a great advantage over other light sources in many applications. The tubes can be obtained in a variety of lengths, with illumination in a variety of colours. It is possible to achieve quite high lighting intensities without excessive temperatures rise and, owing to the nature of light sources, the danger of glare is minimised.

The efficiency of fluorescent lamp is *about 40 lumens/watt about three times the efficiency of an equivalent tungsten filament lamp.*

Construction and working of a fluorescent lamp (tube) are described below :

Construction : Refer Fig. 1.41.

- It is a *low pressure mercury vapour lamp.* Due to low pressure, the lamp is in the form of a long tube, coated inside with phosphor.
- The tube contains a small amount of mercury and a small quantity of argon gas at a pressure of 2.5 mm of mercury.
- At each end of the tube the electrodes are of spiral form made of tungsten and coated with an electron emitting material.

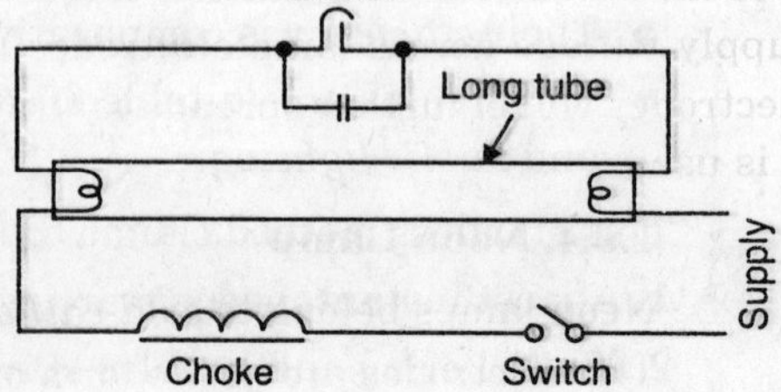

Fig. 1.41. Fluorescent tube (lamp).

- A *choke* is connected in series with the tube filament. It provides a *voltage impulse for starting* the lamp and *acts as a ballast later on when the lamp is running.*
- The filament is connected to a *starter switch* which is small cathode glow lamp with bimetal strip at the electrodes.

Parts of a fluorescent lamp are shown in Fig. 1.42.

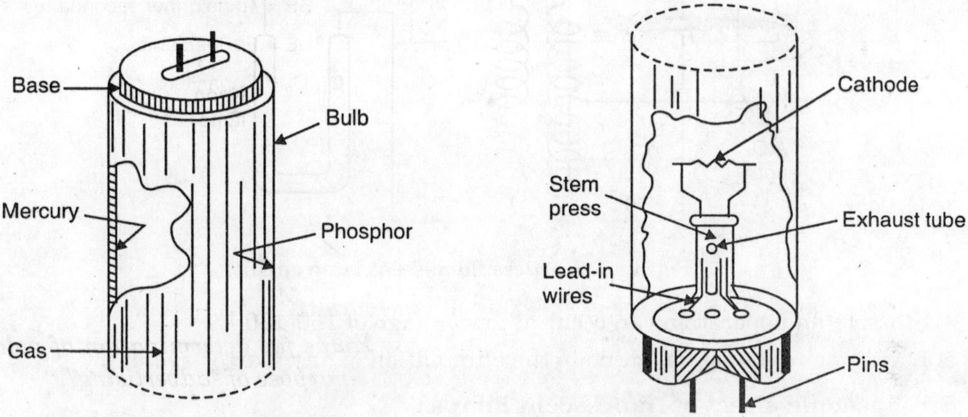

Fig. 1.42. Parts of a fluorescent lamp.

Working :

- When the starter is cold, the electrodes are open. When supply is given, full voltage acts on the starter. A glow discharge is set up in the starter which warms the electrodes and causes the bimetal strip to bend and touch the electrodes. The circuit becomes a complete series. Current flows and causes emission of free electrons from filaments. At the same time voltage at the starter falls to zero and the bimetal strip cools down. The electrodes of the starter switch then open and interrupt the current in the circuit. Its effort is to induce high voltage surge of about 1000 volts in the choke. This voltage produces the flow of electrons between the lamp electrodes and the lamp lights up immediately. Then starting contacts, are left open.
- In order to improve the power factor, usually a condenser of 4 μF capacity is connected across the supply.

Startless fluorescent lamp circuit :

Fig. 1.43 shows a startless fluorescent lamp circuit which does not require the use of a starter switch and is commercially known as '*instant start*' or '*quick-start*'. In this case, the normal starter is replaced by a filament heating transformer whose secondaries *SS* heat up the lamp electrodes *A* and *B* to incandescence in a fraction of a second. This combination of preheating and application of full supply voltage across lamp electrodes is sufficient to start ionization in the neighbourhood of the electrodes which further spreads to the whole tube. To ensure satisfactory starting an earthed strip *E* is used.

This startless method claims the following **advantages :**

1. Almost instantaneous starting.
2. No flickering and no false starts.
3. Lamp life is lengthened.

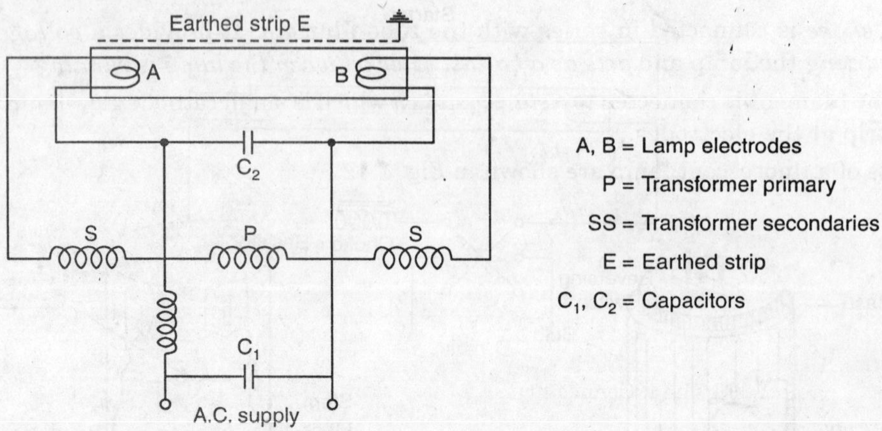

Fig. 1.43. Startless fluorescent lamp circuit.

4. Starting and operation can occur at low voltage of 160-180 V.
5. Lower maintenance cost (due to the elimination of any starter-switch).

Stroboscopic effect of fluorescent lamps :

"Stroboscopic (or flickering) effect" produced by fluorescent lamps is *due to the periodic fluctuations in the light output of a lamp caused by the cyclic variations of the current on A.C. circuits.* This phenomenon creates *multiple-image appearance* of moving objects and makes the movement appear *jerky.*

— This effect is *more pronounced at lower frequencies.*

— The frequency of such flickers is *twice the supply frequency.*

— This effect is reduced to some extent due to after-glow, as the fluorescent powder used in the tube is slightly phosphorescent.

This effect is very troublesome in the following cases : (*i*) When an operator has to move objects very quickly particularly those having polished finish. These objects would appear to move with *jerky motion* which over a long period would produce *visual fatigue* ; (*ii*) In the case of rotating machines whose frequency of rotation happens to be a multiple of flicker frequency, the machines appear to decrease in speed of rotation or be stationary. Sometimes machines may even seem to rotate in the opposite direction.

The stroboscopic effect can be *minimised* as follows :

1. By using three lamps on the separate phases of a 3-phase supply.

2. By using a 'twin lamp' circuit on a single-phase supply, one of the chokes having a capacitor in series with it and the lamp.

3. By operating the lamp from a high frequency supply (obviously, stroboscopic effect will entirely disappear on D.C. supply).

Fluorescent lamp for D.C. supply :

Fig. 1.44 shows the connections of a fluorescent tube with *D.C. mains.* For making a fluorescent tube to work on D.C. supply a *resistance is connected in addition to the usual choke.*

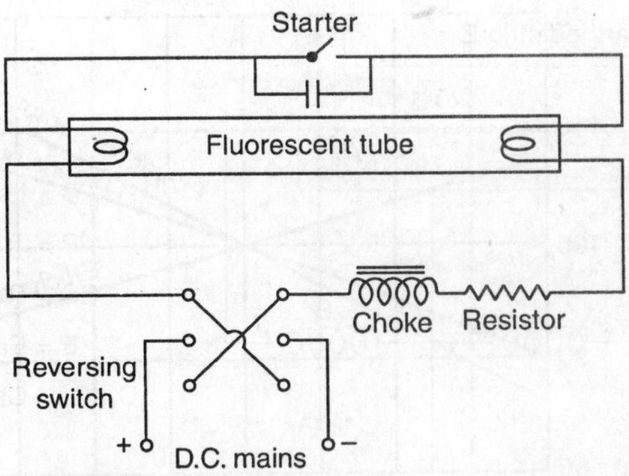

Fig. 1.44. Connections of a fluorescent tube with D.C. mains.

When the tube works for some time, the positive end of the tube gets blackened due to migration of ionized mercury vapour to the negative end. To decrease this effect, a change over switch is used. Generally a rotary switch if placed ensures the polarity reversal, every time the tube is switched on. *The use of a resistance results in increased power consumption and lesser efficiency. But there is no stroboscopic effect.*

In D.C. operation of fluorescent tube there is *no problem of power factor correction and stroboscopic effect.* Its **demerits** are :

(*i*) Low efficiency due to power loss in ballast series-resistance.

(*ii*) Increased cost of the ballast resistance and reversing switch.

(*iii*) Less life of the tube (about 80 percent of that with A.C. operation).

- Now-a-days with D.C. supplies inverters using solid state circuitry are used for use in buses, aircraft etc.

Useful fluorescent lamp life :

— The normal life of a fluorescent lamp is 7500 hours.

— The active life may vary from 5000 to 10000 hours depending upon the operating conditions.

— Light output is reduced by 15-20 per cent after 4000 hours operation and it is, therefore, advisable to replace the fluorescent lamp after 4000-5000 hours burning on economical grounds.

Performance curves :

Fig. 1.45 shows the performance curves of fluorescent lamps.

- In case of fluorescent lamps, the effect of voltage variation is *less marked* as compared to the incandescent lamp. However, their life and performance are *adversely affected both by low and high voltages* (with increased voltage there is a greater heating of electrodes and they lose emissive material by evaporation, while with reduced voltage, the current reduces causing sputtering at the electrodes shortening their life).
- The fluorescent lamps give the best performance at 20-25°C temperature. It decreases rapidly when a lamp is operated at a lower temperature or is exposed to cold with drafts. For operation at high temperature, the fittings with provision for air circulation should be employed.

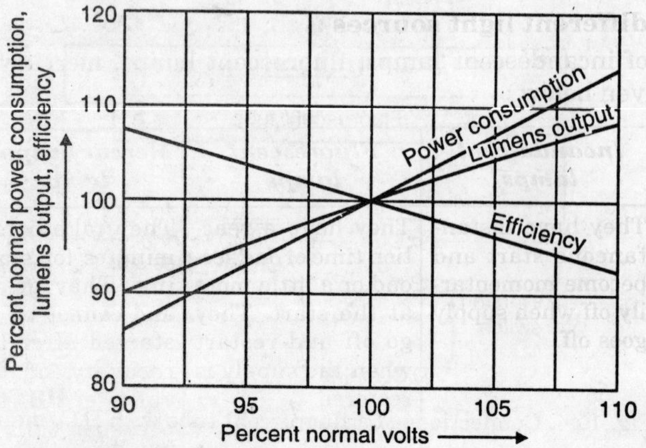

Fig. 1.45. Performance curves of fluorescent lamps.

Merits and demerits of fluorescent-lamps :

Merits :

(*i*) High luminous efficiency.

(*ii*) Long life.

(*iii*) Low running cost.

(*iv*) Low glare level.

(*v*) Less heat output.

Demerits :

(*i*) Stroboscopic effect.

(*ii*) Small wattage requiring large number of fittings.

(*iii*) Magnetic hum associated with choke causing disturbance.

Comparison between Fluorescent tube and Filament lamp :

The comparison between fluorescent tube and filament lamp is given below :

S. No.	*Aspect*	*Fluorescent tube*	*Incandescent lamp*
1.	*Luminous efficiency*	40 lm/W	10 lm/W
2.	*Cost*	Initial cost is more but more economical.	Initial cost is less.
3.	*Starting*	Starting troubles may be there.	No starting trouble.
4.	*Heating effect*	No heat is evolved.	Lot of heat is evolved.
5.	*Life*	10000 hours	1000 hours
6.	*Illumination*	Diffused light	Shadows obtained.
7.	*Variety colours*	Large variety of colours obtained.	Can produce only a few colours.
8.	*Brightness*	Less	More
9.	*Stroboscopic effect*	Yes (objectionable)	No
10.	*Maintenance cost (overall)*	Low	High

Comparison of different light sources :

The comparison of incandescent lamps, fluorescent lamps, mercury vapour lamps and sodium vapour lamps is given below :

S. No.	*Aspect*	*Incandescent lamps*	*Fluorescent lamps*	*Mercury vapour lamps*	*Sodium vapour lamps*
1.	*Starting*	They have instantaneous start and become momentarily off when supply goes off.	They have a reaction time of one second or a little more at the start. They go off and restart when the supply is restored.	They take 5 to 6 minutes for starting. They go off and cannot be restarted after the recovery of the voltage till the pressure falls to normal.	They have a starting time of 5 to 6 minutes. They go off and cannot be restarted after the recovery of the voltage till its value falls to the normal value.
2.	*Colour of light*	Very near the natural.	Varies with the phosphor coating.	They suffer from colour distortion.	Yellowish, colour distortion is produced.
3.	*Installation cost ; running cost*	Minimum ; maximum.	maximum ; minimum.	High but lesser than that of fluorescent lamps ; Much less than incandescent lamps but higher than fluorescent tubes.	Maximum, less than for filament lamps but more than for fluorescent lamps.
4.	*Average life*	1000 hours.	4000 hours.	3000 hours.	3000 hours.
5.	*Efficiency*	10 lm/W.	40 lm/W.	40 lm/W.	60 – 70 lm/W.
6.	*Stroboscopic effect*	No.	Yes.	Yes	Yes
7.	*Applications*	Automobiles, trains, emergency lights, aeroplanes, signals for railways ; domestic, industrial street lighting and flood lights etc.	• Semidirect lighting, domestic, industrial, commercial, roads and halls etc. • Their use is confined to mains voltage or complicated inverter circuits which convert 12 V D.C. into high voltage D.C.	• Suitable for open space like yards, parks and highway lighting etc. • Suitable for vertical position of working.	Very suitable for street lighting purposes ; their position of working is horizontal (*Not* suitable for local lighting).

1.10. LIGHTING SCHEMES

1.10.1. Diffusing and Reflecting Surfaces

When light falls on polished metallic surfaces or silvered surfaces, then most of it is reflected back according to the laws of reflection [Fig. 1.46 (*a*)] *i.e.,* the angle of incidence is equal to the angle of reflection. Only a small portion of the incident light is absorbed and there is always the image of the source. Such reflection is known as ***specular reflection.***

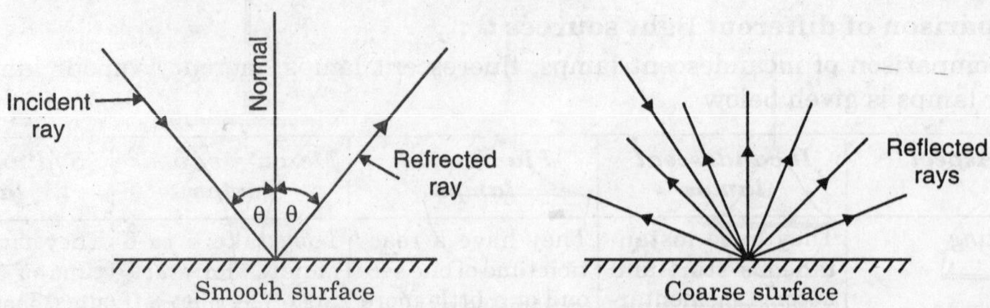

(*a*) Specular reflection (*b*) Diffuse reflection

Fig. 1.46. Types of reflections.

If the light is incident on coarse surfaces like paper, frosted glass, painted ceiling etc. then it is scattered in all directions [Fig. 1.46 (*b*)], hence no image of the source is formed. Such reflection of light is called ***diffuse reflection.*** A perfect diffuser is one that scatters light uniformly in all directions and hence appears equally bright from whatever direction it is viewed. *A white blotting paper is the nearest approach to a 'diffuser'.*

The *ratio of reflected light energy to the incident light energy is known as* **reflection factor** (also known as *reflection ratio* or coefficient of reflection of a surface).

- Perfect mirror surfaces and perfect diffusing surfaces are ideals that do not exist in nature. *The reflection from any surface is partly specular and partly diffuse*, the proportion varying widely. A surface that is almost free from mirror reflection is called a *mat* surface.

1.10.2. Requirements of Good Lighting

Good lighting is one which provides *visual comfort.* Visual comfort enhances the efficiency of a workman.

Usually good lighting is often confused with high illumination levels. The factors which are to considered are *minimum glare* and *brightness-contrast.*

- Light sources should be properly shielded by luminaries and mounted above the normal line of sight. Reflected glare is to be avoided, by mounting luminaries with respect to equipment, so that the reflected glare is directed away from the observer. Use of diffusing absorbing fixers reduce glare.

1.10.3. Types of Lighting Schemes

A good lighting scheme results in an attractive and commanding presence of objects and enhances the architectural style of the interior of a building. Due to the comfortable illumination, people would be in a position to do their work quickly, accurately and easily.

Different lighting schemes may be *classified* as follows :

1. Direct lighting
2. Semi-direct lighting
3. Semi-indirect lighting
4. Indirect lighting
5. General diffusing lighting.

Fig. 1.47 shows different types of lighting arrangements.

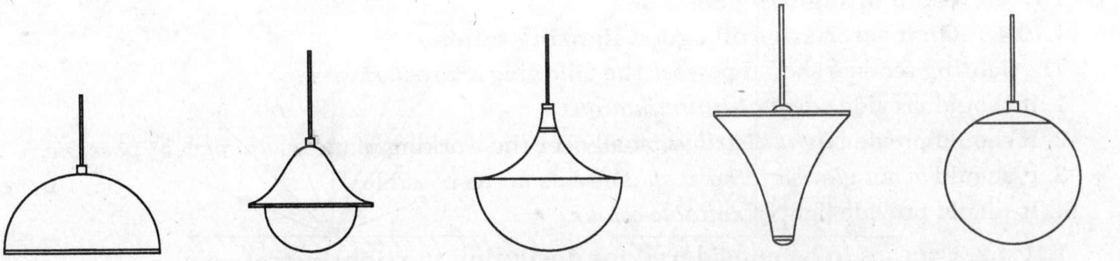

Fig. 1.47. Different types of lighting arrangements.

1. **Direct lighting.** Refer to Fig. 1.47 (*a*).

- It is most commonly used type of lighting scheme.
- In this system more than 90% of total light flux is made to fall directly on the working plane with the help of deep reflectors.
- Direct lighting, though most efficient, is *liable to cause glare and hard shadows.*
- It is mainly used for *industrial and general outdoor lighting.*

2. **Semi-direct lighting.** Refer to Fig. 1.47 (*b*).

- In this system 60 to 90% of the total light flux is made to fall downwards directly with the help of semi-direct reflectors, remaining light is to be used to illuminate the ceiling and walls.
- Such a system is best suited to rooms *with high ceilings* where a high level of uniformly-distributed illumination is desirable.
- Glare in such units is avoided by using diffusing globes which not only improve the brightness towards the eye level but improve the efficiency of the system with reference to the working plane.

3. **Semi-indirect lighting.** Refer to Fig. 1.47 (*c*).

- In this system 60 to 90% of the total light flux is made to fall downwards directly with the help of semi-direct reflectors.
- It is mainly used for *indoor light decoration purposes.*

4. **Indirect lighting.** Refer to Fig. 1.47 (*d*).

- In this system more than 90% of total light flux is thrown upwards to the ceiling for diffuse reflection by using inverted or bowl reflectors. In such a system the ceiling acts as the light source, and the glare is reduced to minimum.
- This system provides a *shadowless illumination* which is very useful for *drawing offices, composing rooms and in workshops* especially where large machines and other obstructions would cast troublesome shadows if direct lighting were used.

5. **General diffusing lighting.** Refer to Fig. 14.7 (*e*)

- In this lighting system lamps made of diffusing glass are used which give nearly equal illumination in all directions.

1.10.4. Design of Lighting Scheme

1.10.4.1. Characteristics of a good lighting scheme

The lighting scheme should possess the following *characteristics* :

1. It should provide *adequate illumination.*
2. It should provide *light distribution* all over the working plane as *uniform as possible.*
3. It should *avoid glare and hard shadows* as far as possible.
4. It should provide light of *suitable colour.*

1.10.4.2. Factors to be considered for designing the lighting scheme

While designing the lighting scheme the following *factors* should be considered :

1. Intensity of illumination.
2. Selection of luminaires.
3. Size of the room.
4. Mounting height and spacing of fittings.
5. Conditions of use.

1. **Intensity of illumination.** The intensity of illumination required for different types of work differ. The recommended levels of illumination for various occupancies is shown in Table 1.1 below :

Table 1.1. Recommended levels of illumination

S. No.	*Occupancy*	*Illumination (lux)*
1.	*Factories and Workshops*	
	(*i*) Rough work, *e.g.,* frame assembly of heavy machinery	150
	(*ii*) Medium work, *e.g.,* machined parts, engine assembly, vehicle body assembly	300
	(*iii*) Fine work, *e.g.,* radio and telephone equipment, type-writer and office machinery assembly	700
	(*iv*) Very fine work, *e.g.,* assembly of very small precision mechanisms, instruments	1500
2.	*Power Houses*	
	(*i*) Boiler house, turbine house, conveyor house, switchgear and transformer chambers	100
	(*ii*) Control rooms	300
3.	*Offices*	
	(*i*) Reception	150
	(*ii*) Conference room, general offices, typing rooms	300
	(*iii*) Drawing offices	400
4.	*Schools and Colleges*	
	(*i*) Classrooms, lecture halls, workshops, library reading tables, laboratories	300
	(*ii*) Sewing rooms, drawing halls, art rooms	500
	(*iii*) Common room, stairs	150
5.	*Hospitals*	
	(*i*) Waiting rooms, wards, casualty	150
	(*ii*) Dispensaries, laboratories, operation theatres (general)	300
	(*iii*) Operation table	Special lighting

6.	*Hotels and Restaurants*	
	(*i*) Reception, dining rooms, bedrooms, lounges, stairs	150
	(*ii*) Accounts, writing desk, dressing table	300
7.	*Shops*	
	(*i*) General areas	300
	(*ii*) Stock areas	150
	(*iii*) Shop window	Special lighting
8.	*Houses*	
	(*i*) Living room-general	150
	(*ii*) Living room-home work or sustained reading	300
	(*iii*) Kitchen, bedrooms, bathrooms etc.	150
9.	*Storage places*	
	(*i*) Loading and unloading	40
	(*ii*) General stores	100
	(*iii*) Stores of very small items	300
10.	*Sport grounds*	
	(*i*) Stadium	300
	(*ii*) Football field	200
	(*iii*) Tennis court	400
11.	*Canteens*	200

2. **Selection of luminaires :**

A luminaire (light fitting) is the apparatus which distributes, filters or transforms the light given by a lamp or lamps. It includes all the items necessary for fixing and protecting these lamps and for connecting them to the supply circuit.

- The choice of lamps for different types of occupancies differ.
  - — Tubular fluorescent lamps or tungsten filament lamps can be used when lighting is to be done in small premises.
  - — In case of large premises, the lighting can be carried out by using high intensity sources such as mercury or sodium discharge lamps.
- The linear output of the lamps can also be modified by using suitable reflectors and diffusers. Depending upon the type of illumination required (direct, indirect, diffusing etc.) the type of reflector is decided.

3. **Size of the room :**

The lumen output of the sources is not fully utilized at the work place. Part of it is lost in the fittings. Some part is directed to the walls and ceiling where part will be absorbed and part reflected. This is taken into account by a factor known as **"utilisation factor"** or **"coefficient of utilisation".**

Coefficient of utilization depends on the following *factors :*

(*i*) Lumen output of the fitting,

(*ii*) Size and shape of the room,

(*iii*) Reflection factors of walls and ceiling,

(*iv*) Height of the ceiling,

(*v*) Arrangement of the fitting etc.

Co-efficient of utilisation for different fittings is given in Table 1.2.

Table 1.2. Co-efficeint of utilization for different fittings

S. No.	*Type of fittings*	*Utilisation factor*	
		Big rooms with light coloured walls and ceilings	***Small rooms with dark walls and ceilings***
1.	*Standard direct reflectors*	0.64	0.24
2.	*Fluorescent lamp fittings*	0.64	0.24
3.	*Semi-direct fittings*	0.56	0.20
4.	*Enclosed sphere or diffusing fittings*	0.56	0.15
5.	*Indirect fittings*	0.40	0.09

4. **Mounting height and spacing of fittings.** The term "general lighting" implies that the illumination at working level should not vary substantially throughout the room. Therefore, it is apparent that the fitting for general lighting should be so placed that the illumination received from each fitting overlaps and builds up that of its neighbours. *The distance of a light source from the wall should be equal to one half the distance between two adjacent light sources. Also distance between light fittings should not exceed 1.5 times the mounting height.*

5. **Conditions of use.** Conditions of use of light fitting vary with different types of installations. Dust and dirt of the surroundings may get deposited on the light fittings and hence deteriorate the lamp efficiency. If regular periodic cleaning is adopted and assuming good atmospheric conditions the value of ***"maintenance factor"*** may be taken as 0.8. But for dusty and dirty atmosphere, the factor may be as low as 0.4. Another term ***"depreciation factor"*** is also used which is the *reciprocal of maintenance factor* (*i.e.*, Depreciation factor = 1/Maintenance factor).

1.10.5. Method of Lighting Calculations

Out of several methods employed for lighting calculations, some of them are :

1. Watt per square metre method.
2. Lumen or light flux method.
3. Point-to-point or Inverse-square law method.

1. **Watt per square method :**

- This method is *very handy for rough calculation or checking.*
- It consists in making an allowance of watts per square metre of area to be illuminated according to the illumination desired on the assumption of an average figure of overall efficiency of the system.

2. **Lumen or light flux method.** This method is applicable to those cases where the sources of light are such as to produce an approximate uniform illumination over the working plane or where an average value is required.

Total lumens received on working plane

= No. of lamps × wattage of each lamp × efficiency of each lamp (in terms of lumens/watt) × coefficient of utilisation × maintenance factor.

3. **Point-to-point or inverse-square law method :**

- *This method is applicable where the illumination at a point due to one or more sources of light is required, the candle power of the sources in the particular direction under consideration being known.*

- When a polar curve of lamp and its reflector giving candle powers of the lamp in different directions is known, the illumination at any point within the range of the lamp can be calculated from the inverse square law. If two and more than two lamps are illuminating the same working plane, the illumination due to each can be calculated and added.
- This method is not much used (because of its complicated and cumbersome applications) ; it is employed only in some special problems, such as *flood lighting, yard lighting etc.*

1.10.6. Calculation of Illumination

The following empirical formula can be used to calculate the illumination :

$$N = \frac{E \times A}{O \times UF \times MF} \quad ...(1.6)$$

where, N = Number of fittings needed,
E = Required illumination (lux),
A = Working area (square metres),
O = Luminous flux produced per lamp (lumens),
UF = Utilisation factor (or co-efficient of utilisation), and
MF = Maintenance factor.

The luminous flux of different types of lamps is given in the table 1.3 given below :

Table 1.3. Luminous flux of various types of lamps

Discription of the lamp		*Lumen efficiency per watt.*	*Lumen output at 230 volts*
1. *Fluorescent lamp* :	80 watts—5 ft. warm white	58	4640
	40 watts—4 ft. warm white	60	2400
	20 watts—2 ft. warm white	46	920
2. *Incandescent lamp* :	40 watts	10	400
	60 watts	12	720
	100 watts	13.80	1380
	150 watts	14	2100
	200 watts	14.75	2950
	300 watts	16	4800
	500 watts	16.9	8450
	1000 watts	19	19000
3. *Mercury discharge lamps* :	80 watts	31	2480
	125 watts	31	3875
	250 watts	35	8750
	400 watts	39	15600
4. *Sodium discharge lamps* :	45 watts	50	2250
	60 watts	57	3420
	85 watts	65	5525
	140 watts	70	9800

Example 1.24. *A small assembly shop 16 m long, 10 m wide, and 3 m upto trusses is to be illuminated to a level of 200 lux. The utilization and maintenance factors are 0.74 and 0.8 respectively. Calculate the number of lamps required to illuminate the whole area if the lumen output of the lamp selected is 3000 lumens.*

Solution. Working area, $A = 16 \text{ m} \times 10 \text{ m} = 160 \text{ m}^2$

Required illumination, $E = 200$ lux

Lumens output of one lamp, $O = 3000$ lumens

Utilization factor, $UF = 0.74$

Maintenance factor, $MF = 0.8$.

Number of lamps required, N :

$$N = \frac{E \times A}{O \times UF \times MF} \quad \text{...[Eqn. (1.6)]}$$

$$= \frac{200 \times 160}{3000 \times 0.74 \times 0.8} = \mathbf{18.} \quad \textbf{(Ans.)}$$

Example 1.25. *An office 25 m × 12 m is illuminated by 40 W incandescent lamps of lumen output 2700 lumens. The average illumination required at the work place is 200 lux. Calculate the number of lamps required to be fitted in the office. Assume utilization and depreciation factors as 0.65 and 1.25 respectively.*

Solution. Working area, $A = 25 \text{ m} \times 12 \text{ m} = 300 \text{ m}^2$

Required illumination, $E = 200$ lux

Lumens output of the lamp, $O = 2700$ lumens

Utilization factor, $UF = 0.65$

Depreciation factor, $= 1.25$

$\therefore$ Maintenance factor, $MF = \frac{1}{1.25} = 0.8$

Number of lamps required, N :

$$N = \frac{E \times A}{O \times UF \times MF} = \frac{200 \times 300}{2700 \times 0.65 \times 0.8} = \mathbf{4.3.} \quad \textbf{(Ans.)}$$

Example 1.26. *An illumination of 25 lux is to be produced on the floor of a room 12 m × 9 m. 18 lamps are required to produce this illumination in the room, if 50% of the emitted light falls on the floor. What is the power of the lamp in candela ?*

Assume maintenance factor as unity.

Solution. *Given* : $E = 25$ lux ; $A = 12 \times 9 = 108 \text{ m}^2$; $N = 18$ lamps ; $UF = 0.5$; $MF = 1$.

We know that, $N = \frac{E \times A}{O \times UF \times MF}$ or $18 = \frac{25 \times 108}{O \times 0.5 \times 1}$

$\therefore$ Lumen output the lamp, $O = \frac{25 \times 108}{18 \times 0.5 \times 1} = 300$ lumens

Hence, *power of the lamp in candela* $= \frac{300}{4\pi} = \mathbf{24\ cd.}$ **(Ans.)**

Example 1.27. *A football pitch 120 m × 60 m is to be illuminated for night play by similar bank of equal 1000 W lamps supported on twelve towers which are distributed around the ground to provide approximately uniform illumination of the pitch. Assuming that 40 percent of the total light emitted reaches the playing pitch and that illumination of 1000 lm/m² is necessary for television purposes, calculate the number of lamps on each tower. The overall efficiency of the lamp is to be taken as 30 lm/W.*

(Bombay University)

Solution. *Given* : $A = 120 \times 60 = 7200$ m²; Rating of each lamp = 1000 W ; No. of towers = 12 ; Percent of total light reaching the working plane = 40% ; E = 1000 lm/m² ; luminous efficiency of each lamp = 30 lm/W.

Number of lamps on each tower :

Total luminous flux required to be produced $= \dfrac{E \times A}{0.4} = \dfrac{1000 \times 7200}{0.4} = 18 \times 10^6$ lm

Luminous flux contributed by each tower $= \dfrac{18 \times 10^6}{12} = 1.5 \times 10^6$ lm

Output of each 1000 W lamp $= 30 \times 1000 = 30000$ lm

$\therefore$ No. of 1000 W lamps on each tower $= \dfrac{1.5 \times 10^6}{3000} = $ **50. (Ans.)**

Example 1.28. *It is desired to illuminate a drawing hall with an average illumination of about 250 lux. The area of the hall is 30 m × 20 m. The lamps are to be fitted at 5 m height. Find out the number and size of incandescent lamps required for an efficiency of 12 lumens/watt. Utilisation factor = 0.4 and maintenance factor = 0.85.*

Solution. *Given* : E = 250 lux ; $A = 30 \times 20 = 600$ m² ; $UF = 0.4$; $MF = 0.85$

Gross lumens required $= \dfrac{E \times A}{UF \times MF} = \dfrac{250 \times 600}{0.4 \times 0.85} = 441176$

$\therefore$ Total wattage required $= \dfrac{441176}{12} = 36765$ W

Taking 8 lamps along the length giving a spacing of $\dfrac{30}{8}$ m *i.e.*, 3.75 m lengthwise and 5 lamps along width giving spacing of $\dfrac{20}{5}$ m *i.e.*, 4 m width wise.

$\therefore$ Total number of lamps $= 8 \times 5 =$ **40. (Ans.)**

Wattage of each lamp $= \dfrac{36765}{40} \approx$ **1000 W. (Ans.)**

These 40 lamps will be arranged as shown in Fig. 1.48.

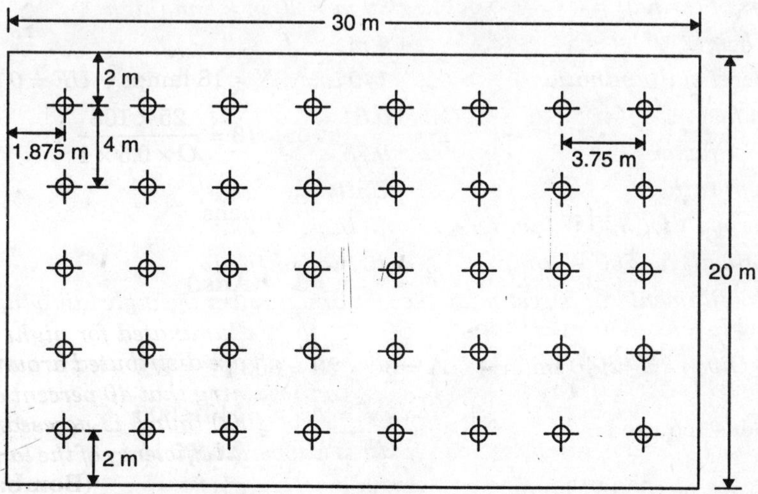

Fig. 1.48

Example 1.29. *Estimate the number and wattage of lamps which would be required to illuminate a workshop space 60 m × 15 m by means of lamps mounted 5 metres above the working plane. The average illumination required is 100 lux. Coefficient of utilisation = 0.42 ; Maintenance factor = 0.8 ; Luminous efficiency = 16 lm/W ; space-height ratio = unity.* **(Madras University)**

Solution. *Given* : $A = 60 \times 15 = 900$ m^2 ; $E = 100$ lux ; $UF = 0.42$; $MF = 0.78$;

Luminous efficiency = 16 lm/W

$$\text{Gross lumens required} = \frac{E \times A}{UF \times MF} = \frac{100 \times 900}{0.42 \times 0.78} = 274725$$

$$\text{Total wattage required} = \frac{274725}{16} = 17170 \text{ W}$$

For a space-height ratio of unity, only three lamps can be mounted along the width of the room. Similarly, 12 lamps can be arranged along the length of the room.

Total number of lamps $= 12 \times 3 = 36$

$$\text{Wattage of each lamp} = \frac{17170}{36} = 477 \text{ W}$$

We will take the nearest standard lamp of **500 W.** The arrangement of the lamps will be as shown in the Fig. 1.49.

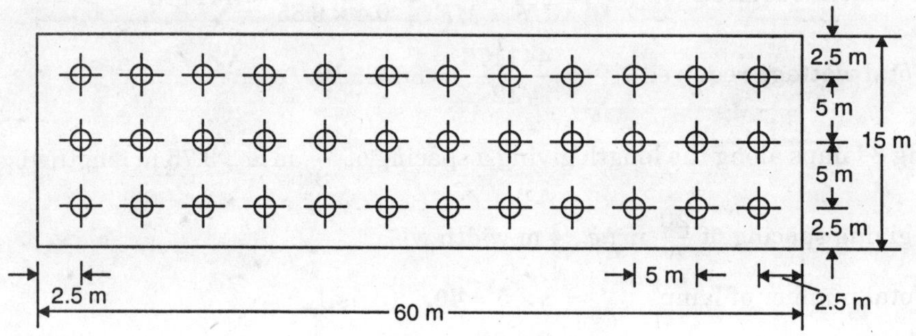

Fig. 1.49

Example 1.30. *The following data relate to the lighting scheme of an engineering college :*

Dimensions of the hall	...	*30 m × 20 m × 8 m (high)*
Mounting height	...	*5 m*
Required level of illumination	...	*140 lm / m^2*
Utilisation factor	...	*0.6*
Maintenance factor	...	*0.75*
Space-height ratio	...	*Unity*
Lumens per watt for 300 W lamp	...	*13 lumens / watt*
Lumens for watt for 500 W lamp	...	*16 lumens / watt*

Using metal filament lamps, estimate the size and number of single lamp luminaries and also draw their spacing layout.

Solution. *Given* : $E = 140$ lm/m^2 ; $A = 30 \times 20 = 600$ m^2 ; $UF = 0.6$; $MF = 0.75$

$$\text{Gross lumens required} = \frac{E \times A}{UF \times MF} = \frac{140 \times 600}{0.6 \times 0.75} = 186667 \text{ lm}$$

Lumen output per 500 W lamp $= 500 \times 16 = 8000$

$\therefore$ No. of 500 W lamps required $= \dfrac{186667}{8000} = 24$

Similarly, no. of 300 W lamps required $= \dfrac{186667}{300 \times 13} = 48$

The 300 W lamps cannot be used because their number cannot be arranged in a hall of 30 m × 20 m with a space-height ratio of unity. However 500 W lamps can be arranged in 4 rows of 6 lamps each with a spacing of 5 m both in length and width of the hall as shown in Fig. 1.50.

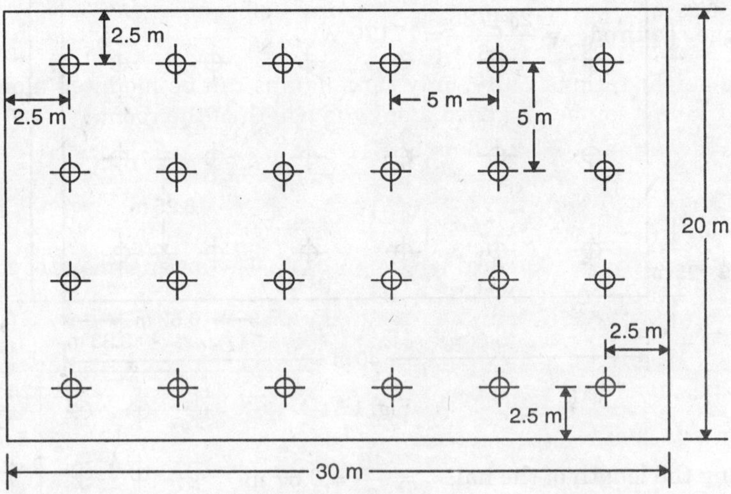

Fig. 1.50

Example 1.31. *A drawing hall 40 m × 25 m × 6 m is to be illuminated with metal filament gas-filled lamps to an average illumination of 90 lm / m² on a working plane 1 metre above the floor. Estimate suitable number, size and mounting height of lamps. Sketch the spacing layout. Assume coefficient of utilisation of 0.5, depreciation factor of 1.2 and space-height ratio of 1.2.*

Size of lamps	*200 W*	*300 W*	*500 W*	
Luminous efficiency (lm/W)	*16*	*18*	*20*	**(Bombay University)**

Solution. *Given* : $A = 40 \times 25 = 1000$ m² ; $E = 90$ lm/m² ; $UF = 0.5$, $MF = \dfrac{1}{DF} = \dfrac{1}{1.2}$;

Space-height ratio = 1.2.

Grass lumens required $= \dfrac{E \times A}{UF \times MF} = \dfrac{90 \times 1000}{0.5 \times (1/1.2)} = 216000$ lm

Lumen output of each 200 W lamp $= 200 \times 16 = 3200$ lm

Lumen output of each 300 W lamp $= 300 \times 18 = 5400$ lm

Lumen output of each 500 W lamp $= 500 \times 20 = 10000$ lm

No. of 200 W lamps required $= \dfrac{216000}{3200} = 67$

No. of 300 W lamps required $= \dfrac{216000}{5400} = 40$

No. of 500 W lamps required $= \dfrac{216000}{10000} = 22$

With a space-height ratio of 1.2, it is impossible to arrange both 200 W and 300 W lamps. Hence, the choice falls on **500 W** lamps.

Let us use **24 lamps** (instead of calculated value 22) of 500 wattage, arranging them in four rows each having six lamps as shown in Fig. 1.51.

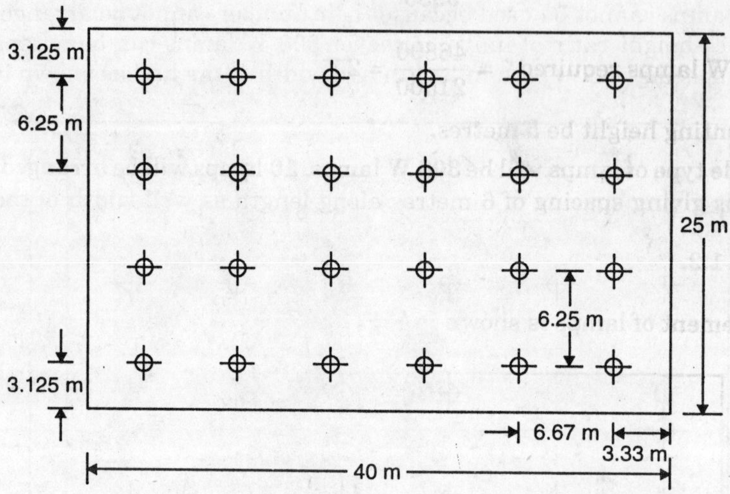

Fig. 1.51

Spacing along the length of the hall $= \frac{46}{6} = 6.67$ m

Spacing along the width of the hall $= \frac{25}{4} = 6.25$ m

Since mounting height of the lamps is 5 m above the working plane, it gives a space-height ratio of $\frac{6.67}{5} = 1.33$ along the length and $\frac{6.25}{5} = 1.25$ along the width of the wall.

Example 1.32. *A hall 30 m long and 12 m wide is to be illuminated and illumination required is 50 metre-candles. Five types of lamps having lumen outputs as given below are available :*

Watts	*100*	*200*	*300*	*500*	*1000*
Lumens	*1615*	*3650*	*4700*	*9950*	*21500*

Taking a depreciation factor of 1.3 and utilisation coefficient of 0.5, calculate the number of lamps needed in each case to produce required illumination. Out of the above five types of lamps select most suitable type and design a suitable scheme and make a sketch showing location of lamps. Assume a suitable mounting height and calculate space-height ratio of lamps. **(A.M.I.E.)**

Solution. *Given :* $A = 30 \times 12 = 360$ m^2 ; $E = 50$ metre-candles ; $UF = 0.5$; $MF = \frac{1}{D.F.} = \frac{1}{1.3}$

Gross lumens required $= \frac{A \times E}{UF \times MF} = \frac{360 \times 50}{0.5 \times (1/1.3)} = 46800$ lumens

No. of 100 W. lamps required $= \frac{46800}{1615} = 29$

No. of 200 W lamps required $= \frac{46800}{3650} = 13$

$$\text{No. of 300 W lamps required} = \frac{46800}{4700} = 10$$

$$\text{No. of 500 W lamps required} = \frac{46800}{9950} = 5$$

$$\text{No. of 1000 W lamps required} = \frac{46800}{21500} = 2$$

Let the mounting height be 5 metres.

Most suitable type of lamps will be **300 W** lamps. 10 lamps will be arranged in two rows, each row having 5 lamps giving spacing of 6 metres along length as well width of the hall with space-height ratio $\frac{6}{5}$ *i.e.*, 1.2.

The arrangement of lamps is shown in Fig. 1.52.

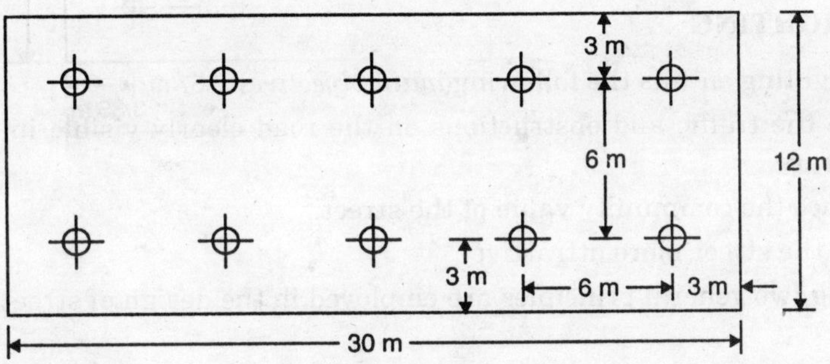

Fig. 1.52

Example 1.33. *A hall 30 m long and 15 m wide with a ceiling height of 5 metres is to be provided with a general illumination of 120 lumens / m². Taking a coefficient of utilisation of 0.5 and depreciation factor of 1.42, determine the number of fluorescent tubes required, their spacing, mounting height and total wattage. Take luminous efficiency of fluorescent tube as 40 lumens / watt for 80 W tube.* **(A.M.I.E.)**

Solution. *Given* : $A = 30 \times 15 = 450 \text{ m}^2$; $E = 120 \text{ lumens/m}^2$; $UF = 0.5$, $MF = \frac{1}{1.42}$; Luminous efficiency of fluorescent tube = 40 lumens/watt.

$$\text{Gross lumens required} = \frac{A \times E}{UF \times MF} = \frac{450 \times 120}{0.5 \times 1/1.42} = 153360 \text{ lumens}$$

$$\text{Total wattage required} = \frac{153360}{40} = 3834 \text{ W}$$

$$\text{No. of fluorescent tubes required} = \frac{3834}{80} = 48$$

Let us use 50 tubes in 5 rows, each row having 10 tubes giving spacing of 3 metres along the length as well as width of the hall. These tubes may be mounted at height of 3 metres giving space-height ratio of units. The arrangement of the fluorescent tubes is given in Fig. 1.53.

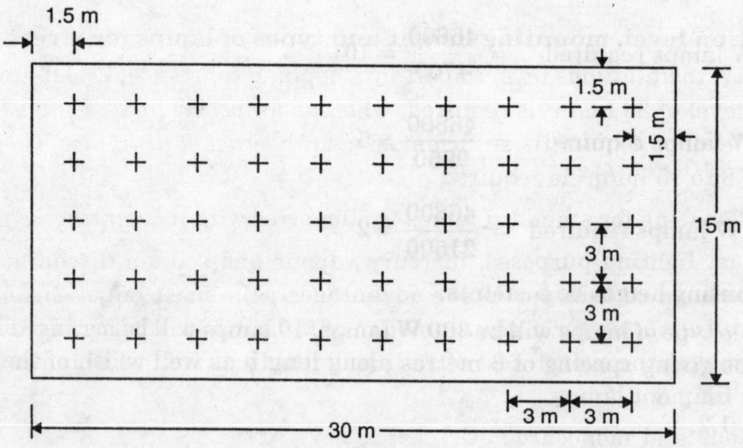

Fig. 1.53

1.11. STREET LIGHTING

The street lighting entails the following *main objectives :*

(*i*) To make the traffic and obstructions on the road clearly visible in order to promote safety and convenience.

(*ii*) To enhance the community value of the street.

(*iii*) To make the street more attractive.

The following two general principles are employed in the design of street lighting installations :

1. Diffusion principle.
2. Specular reflection principle.

1. **Diffusion principle :**

- In this case the lamps fitted with suitable reflectors are employed. The design of the reflectors is such that they may *direct the light downwards and spread as uniformly as possible over the surface of the road.* In order to avoid glare the reflectors are made to have a cut-off between 30° to 45° so that the filament is not visible except underneath it.
- The diffusing nature of the road surface causes the reflection of a certain proportion of the incident light in the direction of the observer, and therefore the road surface appears bright to the observer.
- For calculating the illumination at any point on the road surface, point-to-point or inverse-square law method is employed. Over certain proportions of the road the surface is illuminated from two lamps and the resultant illumination is the sum of the illuminations due to each lamp.

2. **Specular reflection principle :**

- Here, the reflectors are curved upwards so that the light is thrown on the road at a very large angle of incidence. In this method, the requirement of a pedestrian, who requires to see objects in his immediate neighbourhood, is also fulfilled.
- This method is *more economical*, in comparison to diffusion method of lighting. However, it has the demerit that it *produces glare* for the motorists.

Illumination level, mounting height and types of lamps for street lighting :

- In *class A installations* (*e.g.*, important shopping centres and road junctions) the illumination level of 30 lm/m^2 is required, whereas for *poorly lighted suburban streets*, illumination level of 4 lm/m^2 is sufficient. For an *average well lighted street* an illumination level of 8 to 15 lm/m^2 is required.
- Normal spacing for standard lamps is 50 metres with a *mounting height of 8 metres.*
- For street lighting purposes, mercury vapour and sodium discharge lamps have been found to have certain particular advantages ; the most important of these is the *low power consumption for a given amount of light*, which inspite of the higher cost of the lamps, makes the overall cost of an installation with discharge lamps less than that employing filament lamps.

— The colour and monochromatic nature of light produced by discharge lamps does not matter much in street light installations.

- Lamp posts should be fixed at the junction of roads.

1.12. FACTORY LIGHTING

In an industrial establishment an adequate amount of light produces the following *good effects :*

1. The productivity of labour is increased.
2. The quality of work is improved.
3. Number of work stoppages are reduced.
4. Accidents are reduced.

A factory-lighting installation, in common with indoor equipments, should provide the following :

(*i*) Adequate illumination on the working plane ;

(*ii*) Good distribution of light ;

(*iii*) Simple and easily cleaned fittings ;

(*iv*) Avoid glare (from the lamp itself as well as from any polished surface, which may be within the line of vision)

General, local and emergency lighting :

- In factories and workshops the usual scheme is to mount a number of lamps at a sufficient height so that uniform distribution of light over the working the plane is obtained. In large machine shops the height is governed by the necessity of keeping the lamps above the travelling crane.
- On some points fairly intense illumination is required. For this purpose local lighting can be provided by means of adjustable fittings attached to the machine or bench in question or mounted on portable floor standards. Such lamps should be mounted in deep reflectors to avoid the glare.
- It is very desirable to provide auxiliary lighting from the sources other than the main electric supply preferably from batteries or from small petrol driven generator set. If however, emergency light circuits are operated from main electric supply, these should be completely separated from main lighting circuit

Reflectors for industrial purposes must be simple in design and easily cleaned ; these requirements of most of the installations can be met by one of the following types of fittings :

(*i*) Industrial lighting fittings ; (*ii*) Standard reflectors ; (*iii*) Diffusing fittings ; (*iv*) Concentrating reflectors ; (*v*) Enclosed diffusing fittings ; (*vi*) Angle reflectors.

1.13. FLOOD LIGHTING

The flooding of large surfaces with light from powerful projectors is called ***flood lighting.*** It may be employed for the following *purposes* :

(*i*) To enhance the beauty of ancient monuments by night.

(*ii*) To illuminate advertisement boards and show-cases.

(*iii*) To illuminate railway yards, sports stadiums, car parks, construction sites, quarries etc.

For small buildings, rather uniform flood lighting is used. Flood lights can be placed on other buildings nearby or on suitable posts at distances of not more than about 60 metres. Light should fall nearly perpendicular to the building.

Large or tall buildings are illuminated non-uniformly. Flood lights should be so located that contours and features of the building are well defined. If any shadows are cast, they should enhance the beauty of the building or movement.

As far as possible the projectors should not be visible to the passers by. In some cases the projectors may be housed in ornamental stands.

According to the beam spread, the projectors are *classified* as follows :

(*i*) ***Narrow beam projectors***—Beam spread between 12-25°. These are used for distance beyond 70 m.

(*ii*) ***Medium angle projector***—Beam spread between 25-40°. These are employed for distance between 30-70 m.

(*iii*) ***Wide angle projectors***—Beam spread between 40-90°. These are used for distance below 30 m.

- From view point of economy, use of wide angle projector with high wattage lamp is encouraged over narrow beam projector with low wattage lamp.
- In medium and wide angle projectors, standard gas-filled tungsten filament lamps of 250, 500 or 1000 W are used.

Following terms are used in flood lighting calculations :

(*i*) *Waste light factor.* What a surface is illuminated by a number of projectors there is certain waste of light. This effect is taken into account by multiplying the theoretical value of light (in lumens) by waste light factor which 1.2 for rectangular area and 1.5 for irregular objects like statues etc.

(*ii*) *Depreciation factor.* It is defined as the ratio of illumination under ideal condition to the illumination under normal conditions. The actual amount of light to be provided by the source is greater by 50 to 100% on account of dirt and dust depositing on the reflector surface etc. :

(*iii*) *Coefficient of utilisation.* It is also called *"beam factor"* and is defined as the ratio of beam lumens to the lamp lumens. Its value lies between 0.3 and 0.5.

For any desired intensity over a definite surface the number of projectors required is obtained from the following relation :

$$N = \frac{A \times E \times \text{Waste light factor} \times \text{Depreciation factor}}{\text{Utilisation factor} \times \text{Wattage of lamp} \times \text{Luminous efficiency of lamp}}$$

where, N = No. of projectors,

A = Area of surface to be illuminated, m², and

E = Illumination level required, lm/m².

Example 1.34. *The front of a building 45 m × 20 m is illuminated by twenty 1000 W lamps arranged so that uniform illumination on the surface is obtained. Assuming a luminous efficiency of 18 lumens/watt, coefficient of utilization 0.4, waste light factor 1.2 and depreciation factor 1.3, determine the illumination on the surface.* **(Madras University)**

Solution. *Given :* $A = 45 \times 20 = 900\ m^2$; $N = 20$; Luminous efficiency = 18 lumens/watt ; Utilisation factor = 0.4, Waste light factor = 1.2, Depreciation factor = 1.3.

Illumination on the surface, E :

$$N = \frac{A \times E \times \text{Waste light factor} \times \text{Depreciation factor}}{\text{Utilisation factor} \times \text{Wattage of lamp} \times \text{Luminous efficiency of lamp}}$$

$$20 = \frac{900 \times E \times 1.2 \times 1.3}{0.4 \times 1000 \times 18}$$

$$\therefore \quad E = \frac{20 \times 0.4 \times 1000 \times 18}{900 \times 1.2 \times 1.3} = \mathbf{102.56\ lm/m^2}. \quad \textbf{(Ans.)}$$

Example 1.35. *A building frontage 50 m × 16 m is to be illuminated by flood lighting projectors situated 25 metres away. If the illumination is 100 lux, coefficient of utilisation 0.5, depreciation factor 1.5, waste light factor 1.2, estimate the number and size of projectors. Sketch the projectors recommended indicating the usual adjustments provided.* **(A.M.I.E.)**

Solution. *Given :* $A = 50 \times 16 = 800\ m^2$; $E = 100$ lux ; Coefficient of utilisation = 0.5 ; Depreciation factor = 1.5 ; Waste light factor = 1.2.

Assuming that 1000 W lamps having a luminous efficiency of 18 lumens/watt are used, we have :

$$N = \frac{A \times E \times \text{Waste light factor} \times \text{Depreciation factor}}{\text{Utilisation factor} \times \text{Wattage of lamp} \times \text{Luminous efficiency of lamp}}$$

$$= \frac{800 \times 100 \times 1.2 \times 1.5}{0.5 \times 1000 \times 18} = \mathbf{16.} \quad \textbf{(Ans.)}$$

In order to get uniform illumination, overlapping of illuminated circles is essential. As such we will choose two projector illuminated circles heightwise each of 9 m diameter. Dividing the area in equal squares we will have 8 illuminated circles lengthwise. We will therefore, need 16 projectors [Fig. 1.54 (*a*)]. Knowing the diameter of the illuminated circle and distance of the projector from the surface, we can find out the angle of spread (θ) as follows :

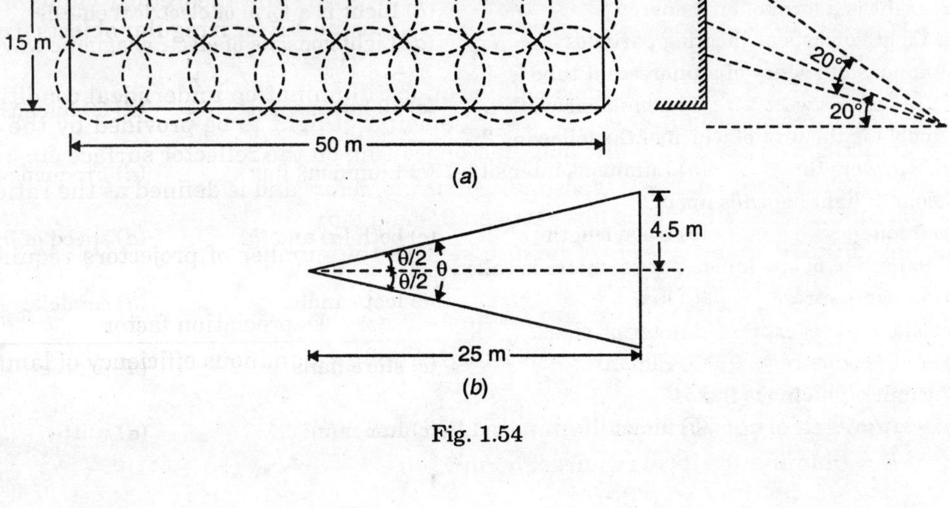

Fig. 1.54

Angle of spread, $\theta = 2 \tan^{-1}\left(\frac{4.5}{25}\right)$ = **20° [Fig. 1.54 (*b*)]. (Ans.)**

Hence 16 *projectors of 1000 W each with beam angle of 20° will be required.*

HIGHLIGHTS

1. *Light* is defined as the radiant energy from a hot body which produces the visual sensation upon the human eye.
2. *Law of inverse squares.* The illumination of a surface is inversely proportional to the square of the distance between the surface and the light source provided that the distance between the surface and the source is sufficiently large so that source can be regarded as a point source.
3. *Lambert's cosine law.* According to this law the illumination of any point on a surface is proportional to the cosine of the angle between the normal at that point and the direction of luminous flux.
4. *Types of lighting systems.* (*i*) Direct lighting ; (*ii*) Semi-direct lighting ; (*iii*) Semi-indirect lighting ; (*iv*) Indirect lighting ; (*v*) General reflection.
5. *Methods of lighting Calculations :*
   (*i*) Watt per square meter method ;
   (*ii*) Lumen or light flux method ;
   (*iii*) Point to point or Inverse-square law method.
6. The following formula can be used to calculate the illumination :

$$N = \frac{E \times A}{O \times UF \times MF}$$

where N = Number of fittings needed,
E = Required illumination (lux),
A = Working area (sq. m.),
O = Luminous flux produced per lamp (lumens),
UF = Utilization factor, and
MF = Maintenance factor.

OBJECTIVE TYPE QUESTIONS

Choose the Correct Answer :

1. Which of the following statements is *correct* ?
   (*a*) Light is a form of heat energy (*b*) Light is a form of electrical energy
   (*c*) Light consists of shooting particles (*d*) Light consists of electromagnetic waves.
2. Luminous efficiency of a fluorescent tube is
   (*a*) 10 lumens/watt (*b*) 20 lumens/watt (*c*) 40 lumens/watt (*d*) 60 lumens/watt.
3. Candela is the unit of which of the following ?
   (*a*) Wavelength (*b*) Luminous intensity (*c*) Luminous flux (*d*) Frequency.
4. Colour of light depends upon
   (*a*) frequency (*b*) wave length (*c*) both (*a*) and (*b*) (*d*) speed of light.
5. Illumination of one lumen per sq. metre is called .....
   (*a*) lumen metre (*b*) lux (*c*) foot candle (*d*) candela.
6. A solid angle is expressed in terms of ......
   (*a*) radians/metre (*b*) radians (*c*) steradians (*d*) degrees.
7. The unit of luminous flux is ......
   (*a*) watt/m² (*b*) lumen (*c*) lumen/m² (*d*) watt.

8. Filament lamps operate normally at a power factor of
(*a*) 0.5 lagging (*b*) 0.8 lagging (*c*) unity (*d*) 0.8 leading.

9. The filament of a GLS lamp is made of
(*a*) tungsten (*b*) copper (*c*) carbon (*d*) aluminium.

10. Fine diameter tungsten wires are made by
(*a*) turning (*b*) swaging (*c*) compressing (*d*) wire drawing.

11. What percentage of the input energy is radiated by filament lamps ?
(*a*) 2 to 5 percent (*b*) 10 to 15 percent (*c*) 25 to 30 percent (*d*) 40 to 50 percent.

12. Which of the following lamps is the cheapest for the same wattage ?
(*a*) Efluorescent tube (*b*) Mercury vapour lamp (*c*) GLS lamp (*d*) Sodium vapour lamp.

13. Which of the following is *not* the standard rating of GLS lamps ?
(*a*) 100 W (*b*) 75 W (*c*) 40 W (*d*) 15 W.

14. In houses the illumination is in the range of
(*a*) 2—5 lumens/watt (*b*) 10—20 lumens/watt
(*c*) 35—45 lumens/watt (*d*) 60—65 lumens/watt.

15. "The illumination is directly proportional to the cosine of the angle made by the normal to the illuminated surface with the direction of the incident flux".
Above statement is associated with
(*a*) Lambert's cosine law (*b*) Planck's law
(*c*) Bunsen's law of the illumination (*d*) Macbeth's law of illumination.

16. The colour of sodium vapour discharge lamp is
(*a*) red (*b*) pink (*c*) yellow (*d*) bluish green.

17. Carbon arc lamps are commonly used in
(*a*) photography (*b*) cinema projectors (*c*) domestic lighting (*d*) street lighting.

18. Desired illumination level of the working plane depends upon
(*a*) age group of observers
(*b*) whether the object is stationary or moving
(*c*) size of the object to seen and its distance from the observer
(*d*) whether the object is to be seen for longer duration or shorter duration of time
(*e*) all above factors.

19. On which of the following factors does the depreciation or maintenance factor depend ?
(*a*) Lamp cleaning schedule (*b*) Agening of the lamp
(*c*) Type of work carried out at the premises (*d*) All of the above factors.

20. In lighting installation using filament lamps 1% voltage drop results into
(*a*) no loss of light (*b*) 1.5 percent loss in the light output
(*c*) 3.5 percent loss in the light output (*d*) 15 percent loss in the light output.

21. For the same lumen output, the running cost of the fluorescent lamp is
(*a*) equal to that of filament lamp (*b*) less than that of filament lamp
(*c*) more than that of filament lamp (*d*) any of the above.

22. For the same power output
(*a*) high voltage rated lamps will be more sturdy (*b*) low voltage rated lamps will be more sturdy
(*c*) both low and high voltage rate lamps will be equal sturdy.

23. The cost of a fluorescent lamp is more than that of incandescent lamp because of which of the following factors ?
(*a*) More labour is required in its manufacturing (*b*) Number of components used is more
(*c*) Quantity of glass used is more (*d*) All of the above factors.

24. Filament lamp at starting will take current
(*a*) less than its full running current (*b*) equal to its full running current
(*c*) more than its full running current (*d*) none of these.

25. A reflector is provided to
(*a*) protect the lamp (*b*) provide better illumination
(*c*) avoid glare (*d*) do all of the above.

26. The purpose of coating the fluorescent tube from inside with white power is
(*a*) to improve its life (*b*) to improve the appearance
(*c*) to change the colour of light emitted to white
(*d*) to increase the light radiations due to secondary emissions.

27. ...... will need lowest level of illumination.
(*a*) Auditoriums (*b*) Railway platform (*c*) Displays (*d*) Fine engravings.

28. Due to moonlight, illumination is nearly
(*a*) 3000 lumens/m^2 (*b*) 300 lumens/m^2 (*c*) 30 lumens/m^2 (*d*) 0.3 lumen/m^2.

29. Which of the following instruments is used for the comparison of candle powers of different sources ?
(*a*) Radiometer (*b*) Bunsen meter (*c*) Photometer (*d*) Candle meter.

30. ...... photometer is used for comparing the lights of different colours ?
(*a*) Grease spot (*b*) Bunsen (*c*) Lummer brodhum (*d*) Guilds flicker.

31. In the fluorescent tube circuit the function of choke is primarily to
(*a*) reduce the flicker (*b*) minimise the starting surge
(*c*) initiate the arc and stabilize it (*d*) reduce the starting current.

32. ...... cannot sustain much voltage fluctuations.
(*a*) Sodium vapour lamp (*b*) Mercury vapour lamp
(*c*) Incandescent lamp (*d*) Fluorescent lamp.

33. The function of capacitor across the supply to the fluorescent tube is primarily to
(*a*) stabilize the arc (*b*) reduce the starting current
(*c*) improve the supply power factor (*d*) reduce the noise.

34. ...... does not have separate choke
(*a*) Sodium vapour lamp (*b*) Fluorescent lamp
(*c*) Mercury vapour lamp (*d*) All of the above.

35. In sodium vapour lamp the function of the leak transformer is
(*a*) to stabilize the arc (*b*) to reduce the supply voltage
(*c*) both (*a*) and (*b*) (*d*) none of the above.

36. Most affected parameter of a filament lamp due to voltage change is
(*a*) wattage (*b*) life (*c*) luminous efficiency (*d*) light output.

37. In electric discharge lamps for stabilizing the arc
(*a*) a reactive choke is connected in series with the supply
(*b*) a condenser is connected in series to the supply
(*c*) a condenser is connected in parallel to the supply
(*d*) a variable resistor is connected in the circuit.

38. For precision work the illumination level required is of the order of
(*a*) 500-1000 lumens/m^2 (*b*) 200-400 lumens/m^2
(*c*) 50-100 lumens/m^2 (*d*) 10-25 lumens/m^2.

39. ...... is a cold cathode lamp.
(*a*) Fluorescent lamp (*b*) Neon lamp
(*c*) Mercury vapour lamp (*d*) Sodium vapour lamp.

40. In case of ...... least illumination level is required.
(*a*) skilled bench work (*b*) drawing offices (*c*) hospital wards (*d*) fine machine work.

41. For normal reading the illumination level required is around
(*a*) 20-40 lumens/m^2 (*b*) 60-100 lumens/m^2 (*c*) 200-300 lumens/m^2 (*d*) 400-5000 lumens/m^2.

42. In electric discharge lamps light is produced by
(*a*) cathode ray emission (*b*) ionisation in a gas or vapour
(*c*) heating effect of current (*d*) magnetic effect of current.

43. A substance which change its electrical resistance when illuminated by light is called ......
(*a*) photoelectric (*b*) photovoltaic (*c*) photoconductive (*d*) none of the above.

44. In case of ...... power factor is the highest.
(*a*) GLS lamps (*b*) mercury arc lamps (*c*) tube lights (*d*) sodium vapour lamps.

45. A mercury vapour lamp gives ...... light.
(*a*) white (*b*) pink (*c*) yellow (*d*) greenish blue.

46. Sometimes the wheels of rotating machinery, under the influence of fluorescent lamps appear to be stationary. This is due to the
(*a*) low power factor (*b*) stroboscopic effect (*c*) fluctuations (*d*) luminescence effect.

47. Which of the following bulb operates on least power ?
(*a*) GLS bulb (*b*) Torch bulb (*c*) Neon bulb (*d*) Night bulb.

48. The flicker effect of fluorescent lamps is more pronounced at
(*a*) lower frequencies (*b*) higher frequencies (*c*) lower voltages (*d*) higher voltages.

49. Which of the following applications does not need ultraviolet lamps ?
(*a*) Car lighting (*b*) Medical purposes
(*c*) Blue print machines (*d*) Aircraft cockpit dashboard lighting.

50. Which gas can be filled in GLS lamps ?
(*a*) Oxygen (*b*) Carbon dioxide (*c*) Xenon (*d*) Amy inert gas.

51. The gas filled in vacuum filament lamps is
(*a*) nitrogen (*b*) argon (*c*) air (*d*) none.

52. Luminous flux is
(*a*) the light energy radiated by sun
(*b*) the part of light energy, radiated by sun, which is received on the earth
(*c*) the rate of energy radiation in the form of light waves
(*d*) none of the above.

53. The vapour discharge tube used for domestic lighting has
(*a*) no filament (*b*) one filament (*c*) two filaments (*d*) three filaments.

54. In an incandescent lamp bird cage filament is usually used in vacuum bulb so as to
(*a*) reduce the oxidation phenomenon (*b*) reduce the convection losses
(*c*) have uniform radiations (*d*) increases the life span of the filament.

55. Stroboscopic effect due to use of discharge lamps in workshops results in moving machinery appearing
(*a*) stationary (*b*) stationary running slow
(*c*) stationary running in reverse direction (*d*) all of the above.

56. Co-efficient of utilisation depends upon ......
(*a*) colour of the walls (*b*) colour of ceiling (*c*) size of the room (*d*) all of the above.

57. Glare is reduced by
(*a*) using diffusers (*b*) increasing the height of the lamp
(*c*) using reflectors to cut-off the light at certain angle
(*d*) all of the above.

58. Which of the following is present inside the fluorescent tube ?
(*a*) Argon and neon (*b*) Argon and CO_2 (*c*) Mercury vapour (*d*) Helium and oxygen.

59. When an electric bulb is broken it produces bang, this is due to

(*a*) vacuum inside the bulb
(*b*) pressure of air in the bulb
(*c*) pressure inside is equal to that outside
(*d*) none of the above.

60. Due to which of the following reasons the light of a tube appears cooler than that of a bulb ?

(*a*) Tungsten is not used in the tube
(*b*) Tube is painted with milky colour
(*c*) Tube consumer less power
(*d*) Surface area of the tube is more than that of bulb
(*e*) none of the above.

61. Sky appears blue because of

(*a*) refraction
(*b*) reflection
(*c*) radiation
(*d*) scattering of light over dust particles.

62. A poor man for his kitchen will make use of

(*a*) fluorescent tube
(*b*) incandescent lamp
(*c*) sodium vapour lamp
(*d*) high pressure mercury vapour lamp.

63. Soft shadows are produced by

(*a*) using surface source of light instead of point source of light
(*b*) increasing the number of lamps
(*c*) both (*a*) and (*b*)
(*d*) none of the above.

64. Reflectors are provided with slits at the top to

(*a*) introduce chimeny effect for cleaning
(*b*) reduce colour contrast
(*c*) reduce heating effect
(*d*) do all of the above functions.

65. In sodium vapour lamp neon gas

(*a*) acts as a shield around the filament
(*b*) assists in developing enough heat to vaporize the sodium
(*c*) change the colour of light
(*d*) prevents the vaporization of filament

66. In fluorescent tubes ballast resistance is connected in series with the choke

(*a*) to reduce stroboscopic effects
(*b*) when tube operates on D.C. supply
(*c*) when supply frequency is low
(*d*) to reduce radio interference.

67. In candescent lamps, coiled coil filaments are used for

(*a*) coloured lamps
(*b*) higher wattage lamps
(*c*) gas filled lamps
(*d*) low wattage lamps
(*e*) none of the above.

68. When a fluorescent lamp is to be operated on D.C. which of the following additional devices must be incorporated in the circuit ?

(*a*) Inductance
(*b*) Transformer
(*c*) Resistance
(*d*) Condenser.

69. The tungsten filament lamps when compared with fluorescent tubes have all the following advantages *except*

(*a*) simple installation
(*b*) longer life
(*c*) less costly
(*d*) more brightness.

70. The level of illumination on a surface *least* depends on

(*a*) ambient temperature
(*b*) candle power of the source
(*c*) distance of the source
(*d*) type of reflector used.

ANSWERS

1. (*d*)	**2.** (*d*)	**3.** (*b*)	**4.** (*c*)	**5.** (*b*)
6. (*c*)	**7.** (*b*)	**8.** (*c*)	**9.** (*a*)	**10.** (*d*)
11. (*b*)	**12.** (*c*)	**13.** (*b*)	**14.** (*d*)	**15.** (*a*)
16. (*c*)	**17.** (*b*)	**18.** (*e*)	**19.** (*d*)	**20.** (*c*)
21. (*b*)	**22.** (*b*)	**23.** (*d*)	**24.** (*c*)	**25.** (*d*)
26. (*d*)	**27.** (*b*)	**28.** (*d*)	**29.** (*c*)	**30.** (*d*)
31. (*c*)	**32.** (*c*)	**33.** (*c*)	**34.** (*a*)	**35.** (*c*)

36. (*b*)	**37.** (*a*)	**38.** (*a*)	**39.** (*b*)	**40.** (*c*)
41. (*b*)	**42.** (*b*)	**43.** (*c*)	**44.** (*a*)	**45.** (*d*)
46. (*b*)	**47.** (*b*)	**48.** (*a*)	**49.** (*a*)	**50.** (*d*)
51. (*d*)	**52.** (*c*)	**53.** (*c*)	**54.** (*c*)	**55.** (*d*)
56. (*d*)	**57.** (*d*)	**58.** (*c*)	**59.** (*a*)	**60.** (*a*)
61. (*d*)	**62.** (*b*)	**63.** (*c*)	**64.** (*d*)	**65.** (*b*)
66. (*b*)	**67.** (*c*)	**68.** (*c*)	**69.** (*b*)	**70.** (*a*).

THEORETICAL QUESTIONS

1. What is the basic nature of light ? Explain.
2. Define the following terms :
   Luminous flux, Lumen, Illumination, Lamp efficiency.
3. Explain briefly the following :
   (*i*) Space-height ratio (*ii*) Utilization factor
   (*iii*) Depreciation factor.
4. What is a solid angle ?
5. State the laws of illumination.
6. Enumerate the various types of electric lamps in common use.
7. Explain briefly the materials commonly used for incandescent lamps.
8. Explain with a neat sketch, the construction and working of a sodium vapour lamp.
9. Give the construction and working of a 'fluorescent tube'.
10. Explain briefly the various types of lighting systems.
11. Enumerate the characteristics which the lighting scheme should possess.
12. Revive expressions for illumination on a surface (*i*) when it is normal and (*ii*) when it is inclined to the axis of a beam of incident light.
13. What is a polar curve ? How is it useful to an illumination engineer ?
14. What do you understand by polar curves ? Explain Rousseau's construction for calculating m.s.c.p. of a lamp.
15. Why is tungsten selected as the filament material and on what factors does its life depend ?
16. Prove that in a filament lamp the diameter of filament is directly proportional to $I^{2/3}$, where I is the current flowing in the filament.
17. Explain the advantages of using inert gas in filament lamps and the purpose to get the filament as coiled coil.
18. Describe the metal filament lamps high lighting the effect of temperature and the choice of filament materials.
19. Explain the effects of voltage variation on the life and illumination as regards tungsten lamp and fluorescent lamp.
20. Compare the metal filament lamp with discharge lamp.
21. Compare the metal filament lamp with discharge lamp. What is the advantage of coiled coil ? Describe principle of operation, construction and working of a sodium discharge lamp.
22. Describe with a neat sketch the principle of electric discharge lamp. State the advantages and disadvantages of discharge lamps over the filament lamp and give their applications.
23. Explain with a neat diagram the principal of operation of a sodium vapour lamp. Mention its use.
24. Explain the principle and operation of a sodium vapour lamp giving its neat sketch. Mention its uses.
25. Describe with the help of a neat diagram the construction and working of a high pressure mercury vapour lamp.
26. Explain the construction and operation of a fluorescent tube and compare it with tungsten filament lamp.

27. Explain the working of a high pressure mercury vapour lamp giving its circuit diagram. Give its advantages, disadvantages and applications. What is the usual value of power factor for this lamp ?
28. Explain the working of a fluorescent tube with the help of the circuit diagram. What is the function of the choke and a starter in fluorescent tube lighting.
29. What do you mean by 'Stroboscopic effect' while using gas discharge lamps ?
30. Distinguish between "fluorescence" and "phosphorescence" properties of materials used in lighting equipments.
31. Describe with neat sketches, the construction of a tubular fluorescent lamp. Draw a diagram of connections and explain the action of tube (*i*) at starting (*ii*) when operating normally.
    Compare the spectrum of such a tube lamp with that of (*i*) tungsten filament lamp (*ii*) mercury discharge lamp, and explain why the luminous efficiency of the fluorescent lamp is so much higher than that of the tungsten filament lamp.
32. Explain the construction and working of a low pressure fluorescent lamp and state its advantages.
33. Compare the merits of alternative lighting schemes using incandescent lamps, sodium vapour lamps or fluorescent lamps for industrial lighting.
34. With the help of circuit diagrams explain the working of the following light sources :
    (*a*) High pressure mercury vapour lamp
    (*b*) Fluorescent tube
    (*c*) Carbon arc lamp.
    What are the usual values of power factors for the above lamps ? Indicate the systems of illumination where above lamps are used.
35. Explain the working of a fluorescent tube with the help of the circuit diagram giving the function of various parts. How stroboscopic effect is eliminated in fluorescent tube lighting ?
36. Describe with neat sketches, various types of electric light fittings used for illumination.
37. What do you understand by direct, indirect, and semi-indirect lighting ?
38. State and describe various types of lighting schemes.
39. What are the requirements of good lighting ? Explain in detail.
40. Describe in brief the requirements of good lighting. Enumerate the factors to be considered while designing an indoor lighting scheme.
41. How does an interior lighting design differ from external lighting design ? Explain briefly the design procedure in each.
42. Discuss various factors which have to be considered while designing any lighting scheme.
43. Explain what is meant by 'point to point method' of lighting calculations.
44. What are the requirements of good street lighting ?
45. What are the general principles that are usually employed in the design of street lighting ? Explain.
46. Explain the specular reflection principle of street lighting ?
47. Enumerate the various factors to be considered while designing street lighting.
48. What is flood lighting and where it is used ?
49. Explain how flood lighting is provided and the design considerations involved.
50. What are the aims of flood lighting and how are they achieved ?
51. Differentiate the following :
    (*i*) Illumination and luminous intensity.
    (*ii*) Lamp efficiency and specific consumption.
    (*iii*) Maintenance factor and depreciation factor.
    (*iv*) Specular reflection and diffuse reflection.
    (*v*) Direct lighting and indirect lighting.

UNSOLVED EXAMPLES

1. A 250 V lamp has a total flux of 3000 lumens and takes a current of 0.8 A. Calculate (*i*) lumens/watt, and (*ii*) M.S.C.P./watt **[Ans.** (*i*) 15 ; (*ii*) 1.2]
2. A 0.4 metre diameter diffusing sphere of opal glass, having 20% absorption, encloses an incandescent lamp with a luminous flux of 4850 lumens. Calculate the average luminance of the sphere. **[Ans.** 7720 lumens/m^2]
3. A filament lamp of 500 W is suspended at a height of 5 metres above working plane and gives uniform illumination over an area of 8 m diameter. Assuming efficiency of reflector as 60% and efficiency of lamp as 0.9 watt per candle power, determine the illumination on the working plane. **[Ans.** 83.33 lumens/m^2]
4. The candle power of a lamp is 100. A plane surface is placed at a distance of 2 metres from this lamp. Calculate the illumination on the surface when it is (*i*) normal, (*ii*) inclined to 45°, and (*iii*) parallel to the rays. **[Ans.** (*i*) 25 lux, (*ii*) 17.67 lux (*iii*) zero]
5. A 500 W lamp having M.S.C.P. of 1000 is suspended 2.7 metres above the working plane. Calculate :
   (*i*) Illumination directly below the lamp at the working plane.
   (*ii*) Lamp efficiency.
   (*iii*) Illumination at a point 2.5 metres away on the horizontal plane from vertically below the lamp.
   [Ans. (*i*) 137.17 lux ; (*ii*) 25.13 lumens/watt ; (*iii*) 54.2 lux]
6. A lamp with reflector is mounted 12 m above the centre of a circular area of 24 m diameter. If the combination of the lamp and reflector gives a uniform C.P. of 1000 over the circular area, determine the maximum and minimum illumination produced on the area. **[Ans.** 6.94 lux ; 2.46 lux]
7. A lamp having a uniform C.P. of 200 in all directions is provided with a reflector which directs 60% of the total light uniformly on to a circular area of 10 m diameter. The lamp is hung 6 m above the area. Calculate.
   (*i*) The illumination at the centre and edge of the surface with and without the reflector.
   (*ii*) The average illumination over the area without the reflector.
   [Ans. (*i*) 5.56 lux, 2.52 lux (without reflector), 19.2 lux (without reflector ; (*ii*) 3.7 lux]
8. The illumination at a point on a working plane directly below the lamp is to be 100 lumes/m^2. The lamp gives 256 C.P. uniformly below the horizontal plane. Determine the height at which lamp is suspended. Also find illumination at a point on the working table 1.2 metres away from the vertical axis of the lamp. **[Ans.** 1.6 m ; 51.2 lux]
9. A small light source with intensity uniform in all directions is mounted at a height of 10 m above a horizontal surface. Two points *A* and *B* both lie on the surface with point *A* directly beneath the source. How far is *B* from *A* if the illumination at *B* is only $\frac{1}{10}$ as great as at *A* ? **[Ans.** 19.1 m]
10. A light source having an intensity of 400 C.P. in all directions is fitted with a reflector so that it directs 80% of its light along a beam having a divergence of 15°. Determine the total light flux emitted along the beam. Also determine the average illumination produced on a surface normal to the beam direction at a distance of 8 m. **[Ans.** 4021 lumens ; 1154 lux]
11. Two lamp posts are 16 m apart and are fitted with a 100 C.P. lamp each at a height of 6 m above the ground. Calculate the illumination on the ground under each lamp and midway between the lamps. **[Ans.** 2.9 lux ; 1.2 lux]
12. A small area of 6 metres in diameter is to be illuminated by a lamp suspended at a height of 4.0 metre over the centre of the area. A lamp having an efficiency of 20 lumens per wall is fitted with a reflector which directs the light output only over the surface to be illuminated. Assuming, utilisation coefficient 0.5 and illumination level 750 lux, determine the wattage of the lamp. **[Ans.** 2120 W]
13. A lamp of 100 candela is placed 1 m below a plane mirror which reflects 90 percent of light falling on it. The lamp is hung 4 m above ground. Find the illumination at a point on the ground 3 m away from the point vertically below the lamp. **[Ans.** 5 lux]

14. A perfectly diffusing surface has luminous intensity of 10 candelas at angle of 60° to the normal. If the area of the surface is 100 cm^2, determine the brightness and total flux radiated.

[**Ans.** 6283.2 lm/m^2 ; 62.83 lm]

15. A shop measuring 15 m × 35 m is illuminated by 20 lamps of 500 W each. The luminous efficiency of each lamp is 15 lumens/watt. Allowing a depreciation factor of 0.7 and coefficient of utilisation of 0.5, determine the illumination on the working plane. [**Ans.** 100 lux]

16. It is required to provide an illumination of 100 lux in a factory hall 30 m × 12 m. Assumed that the depreciation factor is 0.8, the coefficient of utilisation 0.4, and the efficiency of proposed lamps 11 lumens/watt. Calculate the number of lamps and their disposition.

[**Ans.** 40 lamps of 200 W each in 4 rows, each row having 10 lamps]

17. A hall measuring 27.45 m × 45.75 m is to be illuminated using 200 watt filament lamps. The luminous efficiency of the 200 watt filament lamp is 14.4 lumens/watt. Inside the hall an average illumination of 108 lumens/m^2 is to be provided on the working plane. The walls and ceiling are brightly painted. Take coefficient of utilisation as 0.35 and depreciation factor as 0.9. Calculate the number of lamps required for this. [**Ans.** 150 lamps]

18. A minimum illumination of 80 lumens/m^2 is required in the factory shed of 50 m × 12 m. Calculate the number, the location and wattage of the units to be used. Assume that the depreciation factor is 0.8, coefficient of utilisation is 0.4 and the efficiency of lamp units is 14 lumens/watt.

[**Ans.** 36 lamps of 300 watts in 3 rows of 12 lamps in each row]

19. It is desired to flood-light the front of a building 42 metres wide and 16 metres high. Projectors of 30° beam spread and 1000 W lamps giving 20 lumens/watt are available. If the desired level of illumination is 75 lumens/m^2 and if the projectors are to be located at ground level 17 metres away, design and show a suitable scheme. Assume the following :

Coefficient of utilisation = 0.4 ; depreciation factor 1.3 ; Waste light factor = 1.2. [**Ans.** 10]

2

Electric Heating and Welding

2.1. **Electric heating**—Introduction—Advantages of electric heating—Modes of heat transfer—Methods of electric heating—Resistance heating—Arc heating—Arc furnaces—Induction heating—Dielectric heating—Choice of frequency—Infrared or radiant heating. 2.2. **Electric welding**—Definition of welding—Welding processes—Resistance electric welding—Electric arc welding—Submerged arc welding—Tungsten Inert—Gas (TIG) welding—Metal Inert-Gas (MIG) welding—Electro-slag and electro-gas welding—Electron-beam welding—Ultrasonic welding—Plasma arc welding—Laser beam welding—Electrodes—general aspects—Welding of various metals—Rebuilding—Hard facing—Defective welds—Under-water welding—Defects in welding—Testing of welded joints—Highlights—Objective Type Questions—Theoretical Questions—Unsolved Examples.

2.1. ELECTRIC HEATING

2.1.1. Introduction

Heating is required for :

(*i*) ***Domestic purposes :***

- Hot plates for cooking
- Room heaters
- Immersion heaters for water heating
- Electric irons
- Electric toasters
- Electric ovens for bakeries
- Pop-corn plants etc.

(*ii*) ***Industrial purposes :***

- Melting of metals
- Heat treatment processes
- Moulding of glass
- Baking of insulators
- Enamelling of copper wires
- Welding etc.

The use of electrically produced heat is always economical proposition on account of the present low cost and availability of electrical energy.

Practically all heating requirements can be met by some form of electric heating equipment.

2.1.2. Advantages of Electric Heating

There are various methods of heating a material, but electric heating is considered to be far superior for the following *reasons* :

1. The electric heating system is *free from dirt.* It is a clean system requiring *minimum cost of cleaning*.

2. The system *does not produce any flue gas.* Since no flue gases is produced in electric heating, no provision has to be made for their exit.

3. *Simple and accurate temperature control* can be made either by manual or fully automatic switches.

4. Electric heating is *economical* as electric furances are cheaper in initial cost as well in maintenance cost.

5. *Automatic protection against overcurrents or overheating* can be provided through suitable switch-gears.

6. *Special type of heating can be done very accurately* by electric heating system.

7. The *overall efficiency* of electric heating is *much higher*.

8. Electric heating system provides *better working conditions* (since this system produces no irritating noise and also the radiating losses are low).

9. Electric heating is *quite safe and responds quickly*.

10. There is *no upper limit to the temperature obtainable except the ability of the material to withstand heat*.

2.1.3. Modes of Heat Transfer

Heat transfer which is defined as the *transmission of energy from one region to another as a result of temperature gradient* takes place by the following *three modes* :

(*i*) Conduction ; (*ii*) Convection ; (*iii*) Radiation.

Heat transmission, in majority of real situations, occurs as a result of combinations of these modes of heat transfer. *Example* : The water in a boiler shell receives its heat from the fire-bed by conducted, convected and radiated heat from the fire to the shell, conducted heat through the shell and conducted and convected heat from the inner shell wall, to the water. *Heat always flows in the direction of lower temperature.*

The above three modes are similar in that a temperature differential must exist and the heat exchange is in the direction of decreasing temperature ; each method, however, has different controlling laws.

2.1.3.1. Conduction

"Conduction" is the *transfer of heat from one part of a substance to another part of the same substance, or from one substance to another* in *physical contact with it, without appreciable displacement of molecules forming the substance.*

In *solids*, the heat is conducted by the following *two mechanisms* :

(*i*) *By lattice vibration* (the faster moving molecules or atoms in the hottest part of a body transfer heat by impacts some of their energy to adjacent molecules).

(*ii*) *By transport of free electrons* (Free electrons provide an energy flux in the direction of decreasing temperature—For metals, especially good electrical conductors, the electronic mechanism is responsible for the major portion of the heat flux except at low temperature).

In case of *gases*, the mechanisam of heat conduction is simple. The kinetic energy of a molecule is a function of temperature. These molecules are in a continuous random motion exchanging energy and momentum. When a molecule from the high temperature region collides with a molecule from the low temperature region, it loses energy by collisions.

In liquids, the mechanism of heat is nearer to that of gases. However, the molecules are more closely spaced and intermolecular forces come into play.

Fourier's law of heat conduction :

Fourier's law of heat conduction is an empirical law based on observation and states as follows :

"The rate of flow of heat through a simple homogeneous solid is directly proportional to the area of the section at right angles to the direction of heat flow, and to change of temperature with respect to the length of the path of the heat flow".

Mathematically, it can be represented by the equation :

$$Q \propto A \cdot \frac{dt}{dx}$$

where, Q = Heat flow through a body per unit time (in watts), W,

A = Surface area of heat flow (*perpendicular to the direction of flow*), m^2,

dt = Temperature difference of the faces of block (homogeneous solid) of thickness 'dx' through which heat flows, °C or K, and

dx = Thickness of body in the direction of flow, m.

Thus,
$$Q = -k \cdot A \frac{dt}{dx} \quad \text{...(2.1)}$$

where, k = constant of proportionality and is known as *thermal conductivity of the body*.

The – ve sign of k [eqn. (2.1)] is to take care of the decreasing temperature alongwith the direction of increasing thickness or the direction of heat flow. The temperature gradient $\frac{dt}{dx}$ is *always negative along positive x* direction and, therefore, the value as *Q becomes* +ve.

Assumptions :

The following are the *assumptions* on which Fourier's law is based

1. Conduction of heat takes place under *steady state conditions.*
2. The heat flow is unidirectional.
3. The temperature gradient is *constant* and the temperature profile is *linear.*
4. There is no internal heat generation.
5. The bounding surfaces are isothermal in characters.
6. The material is homogeneous and isotropic (*i.e.*, the value of thermal conductivity is *constant in all directions*).

Some essential features of Fourier's law :

Following are some essential features of Fourier's law :

1. It is applicable to all matter (may be solid, liquid or gas).
2. It is based on experimental evidence and cannot be derived from first principle.
3. It is a vector expression indicating that heat flow rate is in the direction of decreasing temperature and is normal to an isotherm.
4. It helps to define thermal conductivity 'k' (transport property) of the medium through which heat is conducted.

Thermal conductivity of materials :

From eqn. (2.1), we have

$$k = \frac{Q}{A} \cdot \frac{dx}{dt}$$

The value of $k = 1$ when $Q = 1, A = 1$ and $\frac{dt}{dx} = 1$

Now
$$k = \frac{Q}{1} \cdot \frac{dx}{dt} \text{ (unit of } k : \text{W} \times \frac{1}{m^2} \times \frac{m}{K(\text{or } °\text{C})} = \text{W/mK} \text{ or } \text{W/m°C)}$$

Thus, the *thermal conductivity of a material is defined* as follows :

"The amount of energy conducted through a body of unit area, and unit thickness in unit time when the difference in temperature between the faces causing heat flow is unit temperature difference".

It follows from eqn. (2.1) that materials with high thermal conductivities are good conductors of heat, whereas materials with low thermal conductives are good thermal insulator. *Conduction of heat occurs most readily in pure metals, less so in alloys, and much less readily in non-metals*. The very low thermal conductivities of certain thermal insulators *e.g.*, cork is due to their porosity, the air trapped within the material acting as an insulator.

Thermal conductivity (a property of material) depends essentially upon the following factors :

(*i*) Material structure (*ii*) Moisture content

(*iii*) Density of the material (*iv*) Pressure and temperature (operating conditions).

Thermal conductivities (average values at normal pressure and temperature) of some common materials are as under :

Material	*Thermal conductivity (k) (W/mK)*	*Material*	*Thermal conductivity (k) (W/mK)*
1. Silver	410	8. Asbestos sheet	0.17
2. Copper	385	9. Ash	0.12
3. Aluminium	225	10. Cork, felt	0.05 – 0.10
4. Cast iron	55–65	11. Saw dust	0.07
5. Steel	20–45	12. Glass wool	0.03
6. Concrete	1.20	13. Water	0.55–0.7
7. Glass (window)	0.75	14. Freon	0.0083

2.1.3.2. Convection. *"Convection" is the transfer of heat within a fluid by mixing of one portion of the fluid with another*.

- Convection is possible only in a fluid medium and is *directly linked* with the *transport of medium itself*.
- Convection constitutes the *macroform* of the heat transfer since macroscopic particles of a fluid moving in space cause the heat exchange.
- The effectiveness of heat transfer by convection depends largely upon the mixing motion of the fluid.

This mode of heat transfer is met with in situations where energy is transferred as heat to a flowing fluid at any surface over which flow occurs. This mode is *basically conduction in a very thin fluid layer at the surface and then mixing caused by the flow*. The heat flow depends on the properties of fluid and is independent of the properties of the material of the surface. However, the shape of the surface will influence the flow and hence the heat transfer.

Free or normal convection. Free or natural convection occurs when the fluid circulates by virtue of the natural differences in densities of hot and cold fluids ; the denser portions of the fluid move downward because of the greater force of gravity, as compared with the force on the less dense.

Forced convection. When the work is done to blow or pump the fluid, it is said to be *forced convection*.

The rate equation for the convective heat transfer (regardless of particular nature) between a surface and an adjacent fluid is prescribed by *Newton's law of cooling (Refer Fig. 2.1)*

$$Q = hA\,(t_s - t_f) \quad ...(2.2)$$

where Q = Rate of conductive heat transfer,

A = Area exposed to heat transfer,

t_s = Surface temperature,

t_f = Fluid temperature, and

h = Coefficient of conductive heat transfer.

The units of h are,

$$h = \frac{Q}{A(t_s - t_f)} = \frac{\text{W}}{\text{m}^2\,°\text{C}} \quad \text{or} \quad \text{W/m}^2°\text{C} \quad \text{or} \quad \text{W/m}^2\text{K}$$

The coefficient of convective heat transfer 'h' (also known as film heat transfer coefficient) may be defined as *"the amount of heat transmitted for a unit temperature difference between the fluid and unit area of surface in unit time."*

The valve of 'h' depends on the following factors :

(*i*) Thermodynamic and transport properties (*e.g.*, viscosity, density, specific heat etc.) ;

(*ii*) Nature of fluid flow ;

(*iii*) Geometry of the surface ;

(*iv*) Prevailing thermal conditions.

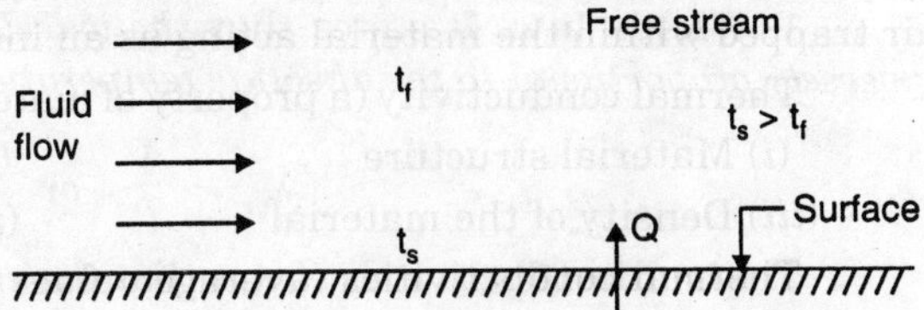

(*a*) Physical configuration

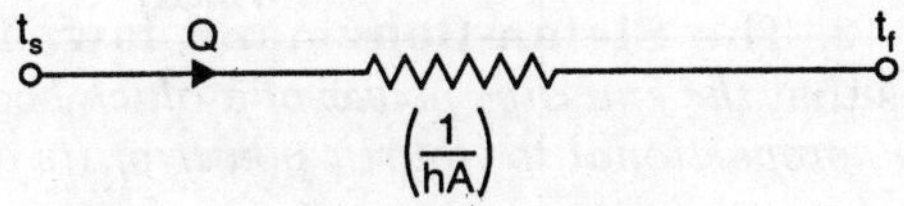

(*b*) Equivalent circuit

Fig. 2.1. Convective heat-transfer.

Since 'h' depends upon several factors, it is difficult to frame a single equation to satisfy all the variations, however, by dimensional analysis an equation for the purpose can be obtained.

The *mechanisms of convection* in which phase changes are involved lead to the important *fields of boiling and condensation.* Refer Fig. 2.1 (*b*). The quantity $\frac{1}{hA}\left[Q = \frac{t_s - t_f}{(1/hA)} \text{......Eqn. (2.2)}\right]$ is called *convection thermal resistance* [$(R_{th})_{conv.}$] to heat flow.

2.1.3.3. Radiation. *"Radiation" is the transfer of heat through space or matter by means other than conduction or convection.*

Radiation heat is thought of as *electromagnetic waves or quanta* (as convenient) an emanation of the same nature as light and radio waves. *All bodies radiate heat ; so a transfer of heat by radiation occurs because hot body emits more heat than it receives and a cold body receives more heat than it emits.* Radiant *energy* (being electromagnetic radiation) *requires no medium for propagation and will pass through vacuum.*

Note. The rapidly oscillating molecules of the hot body produce electromagnetic waves in hypothetical medium called *ether*. These waves are identical with light waves, radio waves and X-rays, differ from them only in *wavelength* and travel with an approximate velocity of 3×10^8 m/s. These waves carry energy with them and *transfer it to the relatively slow-moving molecules of the cold body* on which they happen to fall. The molecular energy of the later increases and results in a rise of its temperature. Heat travelling by radiation is known as *radiant heat.*

The properties of radiant heat in general, are similar to those of light. Some of the properties are :

(*i*) It does not require the presence of a material medium for its transmission.

(*ii*) Radiant heat can be reflected from the surfaces and obeys the ordinary laws of reflection.

(*iii*) It travels with velocity of light.

(*iv*) Like light, it shows interference, diffraction and polarisation etc.

(*v*) It follows the law of inverse square.

The wavelength of heat radiations is longer than that of light waves, hence they are invisible to the eye.

Laws of radiation :

1. **Wien's law.** It states that *the wavelength λ_m corresponding to the maximum energy is inversely proportional to the absolute temperature T of the hot body.*

i.e., $$\lambda_m \propto \frac{1}{T} \quad \text{or} \quad \lambda_m T = \text{constant} \qquad ...(2.3)$$

2. **Kirchhoff 's law.** It states that the *emissivity of the body at a particular temperature is numerically equal to its absorptivity for radiant energy from body at the same temperature.*

3. **The Stefan-Boltzmann law.** The law states that *the emissive power of a black body is directly proportional to fourth power of its absolute temperature.*

i.e., $$Q \propto T^4 \qquad ...(2.4)$$

Refer Fig. 2.2 (*a*)

$$Q = F\sigma A(T_1^4 - T_2^4) \qquad ...(2.5)$$

where F = A factor depending on geometry and surface properties

σ = Stefan-Boltzmann constant

= 5.67×10^{-8} W/m²K⁴

A = Area, m²

T_1, T_2 = temperature, degrees kelvin (K)

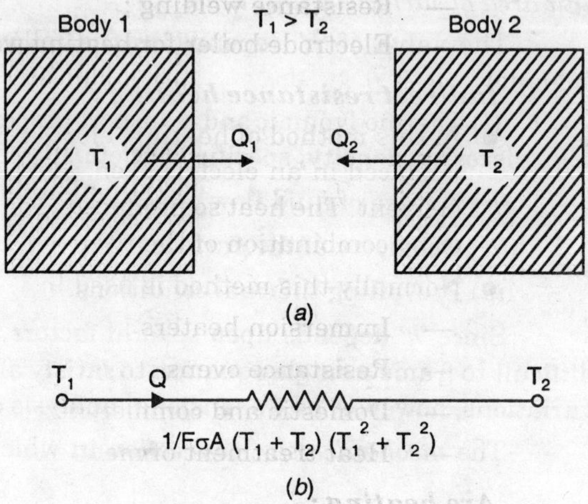

Fig. 2.2. Heat transfer by radiation.

This equation can also be written as :

$$Q = \frac{(T_1 - T_2)}{1/F\sigma A(T_1 + T_2)(T_1^2 + T_2^2)} \qquad ...(2.6)$$

where denominator is radiation thermal resistance, $(R_{th})_{rad}$. [Fig. 2.2(*b*)]

i.e., $$(R_{th})_{red} = 1/F\sigma A(T_1 + T_2)(T_1^2 + T_2^2)$$

The values of F are available for simple configurations in the form of charts and tables.

$F = 1$ ... for simple cases of black surface enclosed by other surface

F = emissivity (ε) ... for non-black surface enclosed by other surface.

2.1.4. Methods of Electric Heating

Different methods of producing heat for general industrial and domestic purposes may be *classified* as follows :

1. **Power frequency heating :**

(*i*) *Resistance heating*

(*a*) Direct resistance heating (*b*) Indirect resistance heating

(*ii*) *Arc heating*

(*a*) Direct arc heating (*b*) Indirect arc heating

2. **High frequency heating :**

(*i*) *Induction heating*

(*a*) Direct core-type induction heating (*b*) Coreless type induction heating (*ii*) *Dielectric heating*

(*iii*) *Infrared heating.*

Direct resistance heating :

- In this method electric current is passed through the body to be heated.
- This method of heating has *high efficiency* since the heat is produced in the material/charge *itself*.
- This principle of heating is *employed in* :
  - — Resistance welding ;
  - — Electrode boiler for heating water etc.

Indirect resistance heating :

- In this method of heating, electric current is passed through a resistance element which is placed in an electric oven. Heat produced is proportional to I^2R losses in the heating element. The heat so produced is delivered to the charge either by radiation or convection or by combination of the two.
- Normally this method is used in :
  - — Immersion heaters ;
  - — Resistance ovens ;
  - — Domestic and commercial cooking
  - — Heat treatment of metals etc.

Arc heating :

- The arc drawn between two electrodes develop high temperature (about 3000–3500°C) depending upon material of the electrode.
- The electric arc may be used in the following different ways :

(*i*) By *striking the arc between the charge and electrode* or *electrodes*. In this method *heat is directly conducted and taken by the charge.*

— The furnaces operating on this principle are known as *direct arc furnaces*.

(*ii*) By *striking the arc between the two electrodes*. In this method the heat is transferred to the charge by radiation.

— The furnaces operating on this principle are known as *indirect arc furnaces*.

(*iii*) By *striking an arc between an electrode and the two metallic pieces to be joined,* as in *arc welding*.

Direct induction heating :

In this method of heating the *current is induced by electro-magnetic action in the body to be heated*. The induced currents when flowing through the resistance of the body to be heated develop the heat and thus raise the temperature.

- In *induction furnace* heat is used to melt the charge.
- *Eddy current heaters* are employed for heat treatment of metals.

Indirect induction heating :

In this heating method the *eddy currents arc induced in the heating element by electro-megnetic induction*. Eddy currents set up in the heating element produce the heat *which is transferred to the body to be heated up, by radiation and convection*.

- This principle is employed in *certain ovens which are employed for heat treatment of metals*.

***Dielectric heating* :**

It is also called *high-frequency capacitive heating* and is used for heating *insulators* like wood, plastics and ceramics etc. which *cannot be heated easily and uniformly by other methods.*

- The supply frequency required for dielectric heating is between *10–50 MHz* and the applied voltage is *20 kV*.
- The *overall efficiency* of dielectric heating is *about 50%.*

***Infrared or radiant heating* :**

In this method of heating, heat energy from an *incandescent lamp is focused upon the body to be heated up in the form of electromagnetic radiations.*

- This method is employed to *dry the wet paints on an object.*

2.1.5. Resistance Heating

This method of heating is *based upon I^2R effect* and has wide applications such *as heat* treatment of metals (*e.g.,* annealing, normalising, hardening, tempering etc.), drying and baking of potteries, domestic cooking etc. In oven where wire resistances are employed for heating, temperature to the time of 1000°C can be obtained.

Following are the two methods of heating :

1. Direct resistance heating
2. Indirect resistance heating.

2.1.5.1. Direct resistance heating

In this method the *material or charge to be heated is treated as a resistance and current is passed through it.* The charge may be in the form of powder, small solid pieces or liquid. The electrodes are inserted in the charge and connected to either A.C. or D.C. supply. In case of D.C. or single-phase A.C. supply two electrodes will be required, while in case of 3-phase supply three electrodes will be used .

When the charge is *in the form of small pieces, a powder of high resistivity material is sprinkled* over the surface of the charge to *avoid short circuit. Heat is produced when current passes through it.*

- This method of heating has *high efficiency* because the heat is produced in the charge itself.
- This heating method is employed in :
  - — Salt bath furnaces ;
  - — Resistance welding ;
  - — Electrode boiler for heating water.

***Salt bath furnaces* :**

A salt bath furnace consists of two electrodes immersed in molten salt (like sodium chloride) having fusion point of 1000°C and can be heated to temperatures of 1500°C or more by passage of current between electrodes. Care should be taken to see that the electrodes are so placed with respect to the material (metal) to be heated that the current flows through the salt and not through the metal to be heated. A.C. supply is used, since *D.C. would cause electrolysis*. A step-down transformer with tapping arrangements is used to supply the bath. The *secondary voltage is of the order of 20 V* with currents varying from a few amperes to as high as 3000 amperes *depending upon the furnace and charge to be heated.* With increase in temperature of bath, its resistance decreases. Taps are, therefore, used to adjust the voltage so that a constant power input can be maintained during the heating process.

- Salt bath furnace is employed *for hardening steel tools and prevents oxidation during hardening*.

2.1.5.2. Indirect resistance heating

In this method the current is passed through a high resistance wire known as *heating element*. The heat produced due to I^2R loss in the element is transmitted to the body to be heated by one or more modes of heat transfer *viz.* conduction, convection and radiation. This method of heating is used in :

— Room heaters ;
— Bimetallic strips used in starters ;
— Immersion water heaters ;
— Various types of resistance ovens used in domestic, and commercial cooking ;
— Salt bath furnace ;
— For industrial purposes, where a large amount of charge is to be heated, the heating element is kept in a cylinder surrounded by jacket containing the charge. This arrangement provides uniform temperature. Moreover, automatic temperature control can also be provided.

2.1.5.3. Properties of a good heating element

A *good heating element* should process the following *properties* :

1. High specific resistance.
2. High melting temperature.
3. Low temperature coefficient of resistance.
4. High oxidising temperature.
5. Positive temperature coefficient of resistance.
6. High ductility and flexibility.
7. High mechanical strength of its own.

Every heating element ; with passage of time ; breaks open and becomes unserviceable. Some of the factors responsible for this failure are :

(*i*) Formation of hot spots which shine brighter during operation.
(*ii*) Oxidation.
(*iii*) Corrosion.
(*iv*) Mechanical failure.

2.1.5.4. Materials of heating elements

1. The materials commonly employed for *low and medium temperature services are* :

Alloy of ***nickel*** and ***chromium*** ...... Ni = 80%, Cr = 20%

or

Alloy of ***nickel, chromium and iron*** .... Ni = 65% ; Cr = 15%, Fe = 20%.

The addition of *iron* to the alloy *reduces the temperature at which oxidation takes place, and the cost of product is also reduced.*

— Ni-Cr alloy is suitable for temperatures upto 1150°C and for *work in severe conditions.*

— Ni-Cr-Fe alloy is recommended for use upto 850°C.

2. For operating temperatures *above* 1150°C resistors are made of *silicon carbide, molybdenum, tungsten* and *graphite*.

(*i*) "***Silicon carbide***" is the basis of a resistor material for operating in air for temperatures upto about 1500°C. The material is formed into rods of diameters and lengths for combination into circuits of the required electrical rating.

(ii) "**Molybdenum**" *resistors* are suitable for temperatures *upto 1650°C*. This metal is ductile enough at room temperature for drawing into wire for resistor windings. For the protection of these resistors, a hydrogen atmosphere is usually used. Owing to its high vapour pressure, molybdenum is not suitable for resistors of vacuum furnaces.

(iii) "**Tungsten**" *resistors* can be employed for temperatures upto 2000°C. The maximum temperature is limited by the refractory supports of the resistor. The low vapour pressure of tungsten makes it useful for resistors of vacuum furnaces.

(iv) "**Graphite**" *resistors* are suitable for any temperature that can be used. The resistors require protection against oxidation above about 600°C. Due to the chemical activity of carbon, special considerations need be given to the surrounding atmosphere.

Table 2.1 gives relevant properties of some of the commercial heating elements (alloys).

Table 2.1. Properties of some commercial heating elements

S. No.	*Particulars*	*Ni-Chromium*	*Nickel-Cr-Fe*	*Ni-Cu*	*Fe-Cr-Al*
1.	*Composition*	80% Ni 20% Cr.	60,16,24	45, 55	70, 25, 5
2.	*Commercial name*	Nichrome	—	Eureka or Constantan	Kanthal
3.	*Maxm. working temperature*	1150°C	950°C	400°C	1200°C
4.	*Specific resistance* at 20°C	109 $\mu\Omega/cm^3$	110 $\mu\Omega/cm^3$	49 $\mu\Omega/cm^3$	140 $\mu\Omega/cm^3$
5.	*Specific gravity*	8.36	8.28	8.88	7.2

2.1.5.5. Resistance furnaces or ovens

- Resistance furnaces or ovens are suitably-insulated closed chambers with a provision for ventilation. These are *used for heat treatment of metals, pottery work, commercial and domestic heating.* Normally power frequency voltage is utilized as the supply source.
- Temperatures upto 1000°C can be obtained by using heating elements made of *nickel, chromium* and *iron.* Ovens using heating elements made of *graphite* can produce temperatures upto 3000°C. Heating elements may consist of *circular wires* or *rectangular ribbons.*
- The oven generally is of firebrick construction or some other heat insulating material and is supported by the framework of metal. The heating elements are placed on the top or sides of the oven.

An enclosure for charge which is heated by radiation or convection or both is called a *heating chamber.*

— The design of the chamber, apart from mechanical consideration, is related primarily to temperature and the major mode of heat transfer to be used. The chambers are used to : (*i*) control the distribution of heat within the chamber ; (*ii*) control the cooling rate of charge, if required ; (*iii*) confine the atmosphere around the charge ; and (*iv*) store as much of the heat supplied as may be practicable and economical.

— Heating chambers may be of '*batch*' or '*continuous*' type. In the *former* the charge remains stationary during the heat application. The cycle may include cooling the charge in the chamber. In *continuous type* chambers the charge is heated as it moves through the chamber. In some cases the chamber is extended for more or less cooling of the charge before it leaves the chamber. A *continuous type* of heating chamber is recommended where flow of material is reasonably *uniform and continuous i.e.,* mass production conditions. In some cases batch heating chambers with automatic charging and discharging are essentially of continuous type.

2.1.5.6. Temperature control of resistance furnaces

Following are the three ways by which the temperature $\left(I^2Rt \text{ or } \frac{V^2}{R} t\right)$ can be controlled :

(*i*) Voltage ; (*ii*) Time ; (*iii*) Resistance.

1. **Tapped transformer.** *Voltage can be varied by using tapped transformer* for supply to the oven or by using a series resistance so that some voltage is dropped across this series resistor. The latter method is, of course, utilized for controlling temperatures of smaller capacity oven.

2. **On-off switch.** An on-off switch can be employed to control the temperature. The *time* for which the oven is *connected* to the supply and the *time* for which it *remains isolated* from supply will determine the temperature. Hence, by this simple method, the furnace temperature can be limited between two limits.

3. **Variation in circuit configuration.** The temperature can be controlled by *switching in various combinations of groups of resistances* used in the oven.

— In *single-phase* supply, various series and parallel combinations along with some resistances being in the circuit, others out of the circuits will give various temperatures.

— For 3-phase ovens, however different connection with star-delta arrangements will give different temperatures.

- If the *temperature is to be controlled automatically* some form of *thermostat* should be used in the circuit so that it operates and switches out or switches in the oven whenever the temperature goes above or below a certain predetermined value respectively (switching is carried by *contactors*).

Protective equipment. An instantaneous overload relay is provided and set to trip the circuit at 10 or 15% above normal current against damage. In addition to above, *fuses* are also provided either in the main circuit of the oven or in the hold-on coil of the energising contactor *to provide protection in case of failure of automatic control system.*

Maximum operating voltage. *Maximum operating voltage is limited by electrical insulation at high temperatures and from safety consideration to 600 V.* This value may *exceed* in some cases. Resistor winding may be in one, two or more circuits and may be connected either to single or to polyphase supply system.

Efficiency and losses :

The heat produced in the heating elements, *besides raising the charge to the required value,* is also to overcome the losses mentioned below :

(*i*) *Heat used in raising the temperature of oven or furnace* :

This loss can be calculated by knowing the mass of the refractory material, its specific heat and rise of temperature ($m \times c \times \Delta T$). This loss becomes negligible if the oven is used continuously.

(*ii*) *Heat used in raising the temperature of the containers or carriers* :

This loss is calculated exactly the same way as for oven or furnace. The container usually *has to be heated up a fresh for each charge.*

(*iii*) *Heat conducted through walls* :

This source of heat loss is most important since the heat is continuously conducted through the walls.

Heat loss by conduction through walls = $\dfrac{kA(T_1 - T_2)}{t}$ watts

where, k = Thermal conductivity of walls, W/m K,

A = Area, m^2,

T_1, T_2 = Inside and outside temperatures, K, and

t = Thickness of the walls, m.

(*iv*) *Escapement of heat due to opening of door* :

Although there is no specific formula for determination of loss occurring due to opening of door for inspection of the charge, however, this loss may be taken as 0.6 to 1.2 MJ/m^2 of the door area if the door is opened for a period of 20 to 30 seconds.

$$\textit{Efficiency of the oven} = \frac{\text{Heat required to raise the temperature of the charge to the required value}}{\text{Heat required to raise the temperature of the charge to the required value + losses}}$$

The heat required to raise the temperature of the charge to the required value,

$$Q = m \times c \times \Delta T \text{ joules}$$

where, m = Mass of charge, kg,

c = specific heat of charge, J/kg K, and

ΔT = Temperature rise, K.

The efficiency lies between 60 and 80%.

2.1.5.7. Design of heating element

The *heating elements* are normally made of *wires of circular cross-section or rectangular conducting ribbons*. Under steady-state conditions, a heating element dissipates as much heat from its surface as it receives the power from the electric supply. If P is the power input and H is the heat dissipated by radiation, then $P = H$ under steady-state conditions.

Heat radiated by a body, as per Stefan's law of radiation, is given by

$$H = 5.67\, \eta\, e \left[\left(\frac{T_1}{100}\right)^4 - \left(\frac{T_2}{100}\right)^4 \right] \text{W/m}^2 \qquad ...(2.7)$$

where, η_{rad} = Radiating efficiency,

e = Emissivity,

T_1 = Temperature of hot body, K, and

T_2 = Temperature of cold body (or cold surroundings), K.

Now, $$P = \frac{V^2}{R}, \text{ and } R = \frac{\rho l}{a} = \frac{\rho l}{\frac{\pi}{4}d^2} = \frac{4\rho l}{\pi d^2}$$

$$\therefore \quad P = \frac{V^2}{\frac{4\rho l}{\pi d^2}} = \frac{\pi d^2 V^2}{4\rho l} \qquad ...(2.8)$$

$$\text{or,} \quad \frac{l}{d^2} = \frac{\pi V^2}{4\rho P} \qquad ...(2.9)$$

If H is the heat dissipated by radiation per second per unit surface area of the wire, then,

Heat radiated per second = $(\pi d) \times l \times H$ ...(2.10)

Equations (2.8) and (2.10), we get

$$P = (\pi d) \times l \times H$$

or, $$\frac{\pi d^2 V^2}{4\rho l} = (\pi d) \times l \times H$$

or, $$\frac{d}{l^2} = \frac{4\rho H}{V^2} \quad ...(2.11)$$

From eqns. (2.9) and (2.11), we can find the values of l and d.

Ribbon type element :

If w and t are the width and thickness of the ribbon respectively, then

$$P = \frac{V^2}{R} = \frac{V^2}{\rho l/a} = \frac{V^2}{\rho l/(w \times t)} = \frac{V^2 wt}{\rho l} \quad ...(2.12)$$

or, $$\frac{l}{wt} = \frac{v^2}{\rho P} \quad ...(2.12(a))$$

Heat lost from ribbon surface

$= 2w\, lH$ (neglecting the side area $2tl$, as thickness is negligible) ...(2.13)

Equating eqns. (2.12) and (2.13), we have

$$\frac{V^2 wt}{\rho l} = 2wlH$$

or, $$\frac{t}{l^2} = \frac{2\rho H}{V^2} \quad ...(2.14)$$

The values of l and w can be evaluated by solving eqns. [2.12(*a*)] and (2.14).

Example 2.1. *A resistance oven employing nichrome wire is to be operated from 220 V single-phase supply and is to be rated at 16 kW. If the temperature of the element is to be limited to 1170°C and average temperature of the charge is 500°C find the diameter and length of the element wire. Radiating efficiency = 0.57, Emissivity = 0.9, Specific resistance of nichrome = 109 × 10⁻⁸ Ωm.*

(Panjab University)

Solution. *Given* : $V = 220$ V ; $P = 16$ kW ; $T_1 = 273 + 1170 = 1443$ K ;

$T_2 = 273 + 500 = 773$ K ; $\eta_{rad} = 0.57$; $e = 0.9$; $\rho = 109 \times 10^{-8}$ Ωm.

***l, d* :**

We know that, $$\frac{l}{d^2} = \frac{\pi V^2}{4\rho P} = \frac{\pi \times 220^2}{4 \times 109 \times 10^{-8} \times (16 \times 10^3)} = 2179660 \quad ...(i)$$

Now, $$H = 5.67\, \eta_{rad}\, e \left[\left(\frac{T_1}{100}\right)^4 - \left(\frac{T_2}{100}\right)^4\right] \text{ W/m}^2$$

$$= 5.67 \times 0.57 \times 0.9 \left[\left(\frac{1443}{100}\right)^4 - \left(\frac{773}{100}\right)^4\right] = 115729 \text{ W/m}^2$$

Now, total heat dissipated/sec. =Electrical power input

$\therefore \quad (\pi d) \times l \times 115729 = 16000 \qquad \therefore \quad dl = 0.044$

or, $\quad d^2 l^2 = 0.001936$ ...(*ii*)

Multiplying (*i*) and (*ii*), we have

$$0.001936 = 4219.8$$

∴ **l = 16.16 m. (Ans.)**

and, $$\mathbf{d} = \frac{0.044}{16.16} = 2.723 \times 10^{-3} \text{ m} = \mathbf{2.723 \text{ mm.}} \quad \textbf{(Ans.)}$$

Example 2.2. *A 27 kW, 3-phase, 400 V resistance oven is to employ nickel-chrome strip 0.25 mm thick for the three star-connected heating elements. If the temperature of the strip is to be 1000° C and that of the charge be 600°C estimate a suitable width for the strip. Assume emissivity = 0.9 and radiating efficiency to be 0.5 and resistivity of the strip material is 101.6 × 10⁻⁸ Ωm.* **(A.M.I.E.)**

Solution. *Given* : $P_{ph} = \frac{27}{3} = 9$ kW ; $V_{ph} = \frac{400}{\sqrt{3}} = 231$V ; $t = 0.25$ mm $= 0.25 \times 10^{-3}$ m

$T_1 = 273 + 1000 = 1273$ K ; $T_2 = 273 + 600 = 873$ K ; $e = 0.9$; $\eta_{rad.} = 0.5$;

$\rho = 101.6 \times 10^{-8}$ Ωm.

Width of strip, w :

If R is the resistance of the strip, then

$$R = \frac{V_{ph}^2}{P} = \frac{(231)^2}{9 \times 10^3} = 5.93\ \Omega$$

Resistance of the strip, $$R = 5.93 = \frac{\rho l}{a} = \frac{\rho l}{w.t}$$

or, $$\frac{l}{w} = \frac{5.93 \times (0.25 \times 10^{-3})}{101 \times 10^{-8}} = 1467.8 \quad ...(i)$$

Heat dissipated from surface of the strip,

$$H = 5.67 \times \eta_{rad} \times e \left[\left(\frac{T_1}{100}\right)^4 - \left(\frac{T_2}{100}\right)^4\right]$$

$$= 5.67 \times 0.5 \times 0.9 \left[\left(\frac{1273}{100}\right)^4 - \left(\frac{873}{100}\right)^4\right] = 52185 \text{ W/m}^2$$

Surface area of the strip $= 2\,wl$

∴ Total heat dissipated $= 2\,wl \times 52185$

∴ $2wl \times 52185 = 9 \times 10^3$

or, $wl = 0.0862$ ...(ii)

Inserting the value of l (= 1467.8 w) from (i) in (ii), we get

$w \times 1467.8\, w = 0.0862$

or, $w = 7.66 \times 10^{-3}$ m or **7.66 mm. (Ans.)**

Example 2.3. *A cubic water tank has surface area of 5.4 m² and is filled to 92 percent capacity five times daily. The water is heated from 15°C to 60°C. The losses per square metre of tank surface per 1°C temperature difference are 5.9 W. Calculate :*

(i) Loading in kW. *(ii) Efficiency of the tank.*

Assume specific heat of water = 4.186 kJ/kg°C and 1 kWh = 3600 kJ.

Solution. *Given* : Surface area of the tank = 5.4 m²,

% Capacity to which the tank is filled with water = 92%

Rise in temperature of water, $t_2 - t_1 = 60 - 15 = 45$°C

Loss per square metre of tank surface per 1°C temperature difference = 5.9 W (*i*) **Loading in kW :**

Let, l = Side of the tank,

Then, total surface area of the tank $= 6\,l^2$

$\therefore \quad 6l^2 = 5.4 \quad \text{or} \quad l = 0.9487 \text{ m}$

Volume of tank $= l^3 = (0.9487)^3 = 0.8538 \text{ m}^3$

Volume of water to be heated daily $= 5 \times 0.8538 \times 0.92 = 3.927 \text{ m}^3$

$\therefore$ Mass of water to be heated daily $= 3.927 \times 1000 = 3927$ kg

($\because$ Mass density of water = 1000 kg/m^3)

Heat required to raise the temperature of water

$$= m \times c \times (t_2 - t_1)$$

$$= 3927 \times 4.186 \times 45 = 739729 \text{ kJ} = \frac{739729}{3600} = 205.48 \text{ kWh}$$

Daily loss from the tank surface $= 5.9 \times 5.4 \times (60 - 15) \times 24/1000 = 34.41$ kWh

Energy supplied per day $= 205.48 + 34.41 = 239.89$ kWh

$\therefore$ Loading in kW $= \dfrac{239.89}{24} \simeq$ **10 kW. (Ans.)**

(*ii*) **Efficiency of the tank, η_{tank}:**

$$\eta_{tank} = \frac{\text{Output}}{\text{Input}} = \frac{205.48}{239.89} = 0.8566 \quad \text{or} \quad \mathbf{85.66\ \%.} \quad \textbf{(Ans.)}$$

2.1.6. Arc Heating—Arc Furnaces

On the application of high voltage across an air gap, the air in the gap gets ionized under the influence of electrostatic forces and becomes conducting medium. Current flows in the form of a continuous spark, called the *arc.* When electrodes are made of carbon/graphite, the temperature obtained is in the range of 3000°C–3500°C. The high voltage required for striking an arc can be obtained by using a *step-up transformer fed from a variable A.C. supply.*

An arc can also be obtained by using low voltage across two electrodes *initially in contact with each other.* The low voltage required for this purpose can be obtained by using a *step-down transformer.* Initially, the low voltage is applied, when the two electrodes are in contact with each other. Further, when the two electrodes are *gradually separated from each other, an arc is established between the two.*

The electric arc furnaces make use of the above principle.

The arc furnaces are usually of following *two shapes :*

1. Cylindrical shape.
2. Conical shape.

The *conical shape* entails the following *advantages* :

(*i*) Large surface area per unit bath volume.

(*ii*) Consumes less power.

(*iii*) Reduced melting time.

(*iv*) Reduced radiation losses.

— Bottom, side walls and roof of the furnace are lined with fire bricks, magnesite bricks and silica bricks. The bottom is lined with magnesite mix or ground ganister mix depending upon whether the basic acidic lining is required.

— Each furnace is provided with charging door and tap hole for introducing the charge and taking out the molten metal. These doors and tap-holes are also to be suitably lined. In order to minimize the heat losses the size of the doors/tap holes and the opening time has to be reduced.

— The furnace rests on platform which can be tilted for pouring the molten metal through the door opening in the side of the shell. Tilting mechanism consists of a motor which drives a pinion that meshes with a semicircular toothed rack at the bottom of the platform.

— The furnaces may be *door-charge type* or the *top charge type. 'Large furnace installations'* are often equipped with *'side door charging"*. *"Top charging"* is growing in favour of *'medium size furnaces.'* In this method for changing the roof of the furnace is removed, and a complete charge is placed in the chamber by a drop bucket handled by an overhead crane. This arrangement saves time and labour.

Electrodes used in arc furnaces : The following three types of electrodes are used in arc furnaces :

(*i*) *Carbon electrodes :*

These electrodes are used with small furnaces for manufacture of ferro-alloys, aluminium, calcium carbide, phosphorus etc.

- These are made of anthracite coal and coke.
- These are very cheap and cost less than one half as much for same weight as graphite electrodes.
- The large area of carbon electrodes allows for more uniform heating.

(*ii*) *Graphite electrodes :*

- These electrodes are obtained by heating the carbon electrodes to a very high temperature.
- Owing to lower resistivity of graphite (one fourth that of carbon), graphite is required half in size for the same current resulting in easy replacement.
- Graphite begins to oxidise at about 600°C whereas carbon at about 400°C. Under average conditions the consumption of graphite electrodes is about one half that of carbon electrodes.

(*iii*) *Self-baking electrodes :*

These electrodes are employed in ferro-alloys and electro-chemical furnaces and in electrolytic production of aluminium.

- These electrodes are made of special paste, whose composition depends upon the types of process for which it is used, contained in thin steel cylinder. The flow of current produces heat and the paste is baked and formed into an electrode.

2.1.6.1. Types of arc furnaces

Arc furnaces are of the following three types :

1. Direct arc furnace
2. Indirect arc furnace
3. Submerged arc furnace.

1. Direct arc furnace :

In this type of furnace *arc is formed between the two electrodes and the charge in such a way that electric current passes through the body of the charge.* Such furnaces produce *very high temperatures.*

Fig. 2.3 shows *1-phase direct arc furnace.*

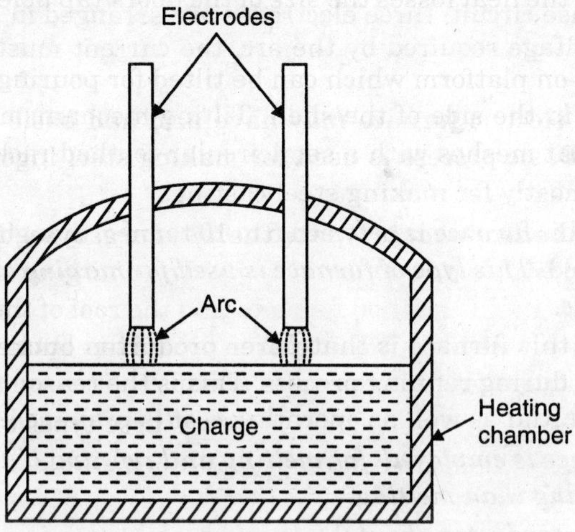

Fig. 2.3. 1-phase direct arc furnace.

Fig. 2.4 shows a *3-phase direct arc type furnace.*

- It consists of a circular steel casting lined inside with refractory material. The roof is removable and a spare is usually kept for rapid replacement. The roof is provided with three holes through which pass the electrodes. The electrodes may be of graphite or amorphous carbon ; the former material has double the conductivity and will carry $2\frac{1}{2}$ times the current ; hence, the graphite electrodes are usually about two thirds the diameter of amorphous carbon electrodes. Though graphite electrodes are costly yet the above advantages are sufficient to account for their choice for most arc furnaces.

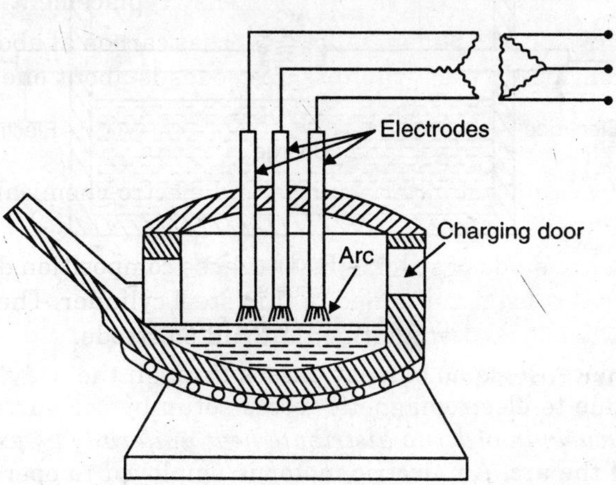

Fig. 2.4. 3-phase direct arc furnace.

To maintain a desired length of arc the electrodes are raised and lowered individually by electric motors operated by automatic regulators. The voltage between steel and electrodes may be 40–145 volts ; the longer the arc, higher the voltage required and the less the input of heat to the furnace. Electric power is supplied in bulk in the form of three-phase alternating current at 6.6 or 10 kV. A transformer set up close to the furnace reduces the voltage down to that required for the arcs and its primary windings having tappings to allow for adjustments to the arc voltage. As the power supply is a three phase circuit, three electrodes are arranged in an equilateral triangle over the metal. Owing to low voltage required by the arc, the current must be very high to obtain the desired output.

- The hearth of an electric furnace may have acid and basic lining depending upon the process adopted. Basic process is used for making steel ingots and some castings while the acid process mostly for making steel castings.
- The usual size of this furnace is between 5 to 10 tonnes, though 50 and 100 tonnes furnaces have been produced. *This type of furnace is used for making alloy steels such as stainless, high speed steel etc.*
- The *advantage* of this furnace is that purer product is obtained and composition can be exactly controlled during refining process. This is the reason that direct arc furnace even being costlier in initial as well as operating cost is preferred.
- Though *this furnace is employed for melting and refining but due to higher cost its use is restricted to refining than melting.*
- *It operates at a power factor about 0.8 lagging.*

2. **Indirect arc furnace :**

In an indirect arc furnace *arc is formed between two electrodes above the charge and heat is transmitted to the charge by radiation.*

Fig. 2.5 shows a single-phase indirect arc furnace which is cylindrical in shape. The arc is struck by short-circuiting the electrodes manually or automatically for a moment and then, *withdrawing them apart.* The heat from the arc and the hot refractory lining is transferred to the top *layer of the charge by radiation.* The heat from the hot top layer of the charge is further transferred to *other parts by conduction.*

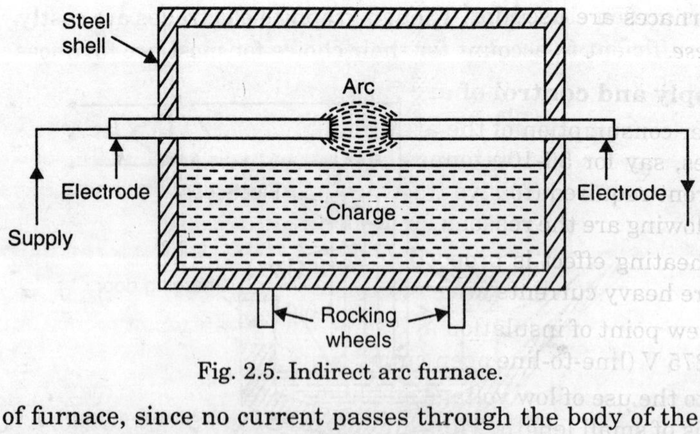

Fig. 2.5. Indirect arc furnace.

In this type of furnace, since no current passes through the body of the charge, there is *no inherent stirring action* due to electromagnetic forces setup by the current. Hence such furnaces have to be *rocked continuously in order to distribute heat uniformly* by exposing different layers of the charge to the heat of the arc. An electric motor is employed to operate suitable grinders and rollers to impart rocking motion to the furnace. *Rocking motion provides not only thorough mixing of the charge, it also increases the furnace efficiency in addition to increasing the life of the refractory lining material.*

In this furnace, since *the charge is heated by radiation only, its temperature is lower than that obtainable in a direct arc furnace.*

- Power input is regulated by adjusting the arc length by moving the electrodes.
- The power factor is about 0.85 lagging.
- The capacity of furnace varies from 0.25 tonne to 3 tonnes.
- These furnaces are *mainly employed for melting 'non-ferrous metals'*. However they can be used in iron foundaries where small quantities of iron are required frequently.

Advantages of indirect arc furnaces :

Following are the advantages of indirect arc furnaces :

(*i*) *Lower overall production cost* per tonne of molten metal.

(*ii*) Sound castings in thin and intricate designs can be produced.

(*iii*) Metal losses due to oxidation and volatilisation are quite low.

(*iv*) *Flexible in operation.*

3. Submerged arc furnace :

A submerged arc furnace is a cylindrical furnace in which *arc is formed between the carbon electrodes (from the top) and hearth electrodes.* The hearth lining is of magnetite which becomes comparatively good electrical conductor when hot. It is also mixed with coke or graphite. Sometimes a conducting hearth is used as electrode.

The number of electrodes taken from the roof depends on the type of supply. *One* for 1-phase, *two* or *four* for 2-phase and *three* for 3-phase supply, bottom conductor being connected to the neutral. The *current from the top electrode passes through the arc to the charge and returns through the electrode at the bottom of the charge.*

Power is controlled by varying distance between electrodes or by varying the voltage applied to the electrodes.

In this type of furnace better distributed heating is obtained since charge behaves as the resistance. Similarly, better mixing of charge takes place. The current under short-circuit is limited due to charge, which otherwise in indirect furnace is very high.

- The power factor is about 0.8 lagging.
- These furnaces are used for the manufacture of *ferro-alloy like ferro-chrome and ferro-manganese.*

Power supply and control of arc furnaces :

As the power consumption of the arc furnaces is *very high* (*e.g.*, about 200 kW per tonne for very large furnaces, say for 50-100 tonnes capacity) and the arc voltage lies between 50–150 V, therefore, the current required (to give the above mentioned power) is of the order of several hundred amperes. Following are the reasons for low voltage high current power supply for arc furnaces :

(*i*) As the heating effect is proportional to the square of the current, therefore, to achieve higher temperature heavy currents are essential.

(*ii*) From view point of insulation and safety considerations the maximum secondary voltage is also limited to 275 V (line-to-line open circuit voltage).

(*iii*) Owing to the use of low voltage and high current the electrodes are kept very near to the charge as the arc is of small length. Thus arc remains away from the roof and, therefore, *life of the roof refractory is increased.*

(*iv*) The use of higher voltage causes higher voltage gradient between the electrode and the charge causing nitrogen of furnace atmosphere ionised and absorbed by the charge, which produces *embrittlement*.

A typical specification for a 3-phase arc furnace transformer includes an extended primary winding with taps there in for the secondary voltage range 235-220-205-190-175-160 V, with primary winding connected in delta. This voltage range is extended by changing the connections of the primary windings from delta to star giving 58 percent voltage from each tap.

Fig. 2.6 shows the *equivalent circuit* of an arc furnace.

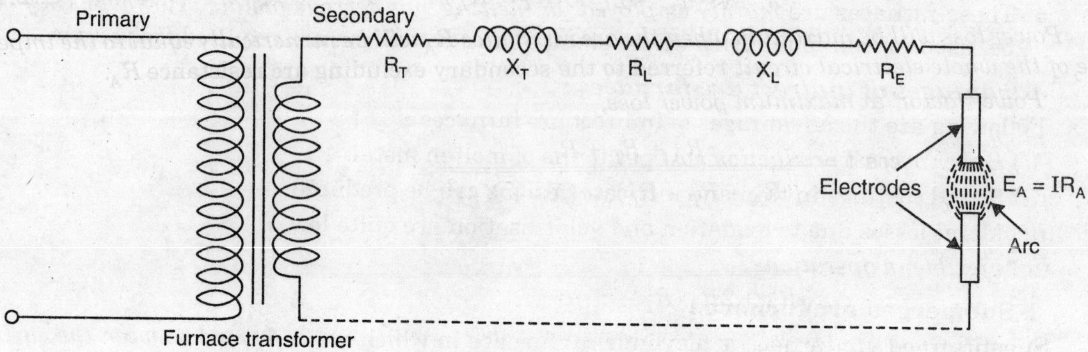

T = Transformer ; L = Load ; E = Electrode, A = Arc

Fig. 2.6. Equivalent circuit of an arc furnace.

R_T = Equivalent resistance of the furnace transformer (referred to secondary side)
X_T = Equivalent reactance of the furnace transformer (referred to secondary side)
R_L = Resistance of the load,
X_L = Reactance of the load,
R_E = Resistance of the electrodes,
R_A = Arc resistance, and
E_A $(= IR_A)$ = Voltage drop across the arc.

In order to exercise complete control of furnace temperature and to achieve best-operating conditions *both voltage and electrode controls are employed* (The power input can be varied by raising or lowering the electrodes, resulting in variation of R_A and by changing the transformer tapping, resulting in variation of voltage across the furnace).

Condition for maximum output :

$$\text{Arc current, } I = \frac{V}{Z} = \frac{V}{\sqrt{(R_T + R_L + R_A)^2 + (X_T + X_L)^2}} \quad ...(2.15)$$

$$\text{Power loss in the arc} = I^2 R_A = \frac{V^2}{(R_T + R_L + R_A)^2 + (X_T + X_L)^2} \times R_A$$

$$= \frac{V^2 . R_A}{R_A^2 + 2\,R_A\,(R_T + R_L) + (R_T + R_L)^2 + (X_T + X_L)^2}$$

$$= \frac{V^2}{R_A + 2(R_T + R_L) + \dfrac{(R_T + R_L)^2}{R_A} + \dfrac{(X_T + X_L)^2}{R_A}}$$

Power loss will be maximum when denominator is minimum

$$\therefore \quad \frac{d}{dR_A}\left[R_A + 2(R_T + R_L) + \frac{(R_T + R_L)^2 + (X_T + X_L)^2}{R_A}\right] = 0$$

$$1 + 0 - \frac{(R_T + R_L)^2 + (X_T + T_L)^2}{R_A^2} = 0$$

or, $$R_A = \sqrt{(R_T + R_L)^2 + (X_T + X_L)^2} \quad ...(2.16)$$

i.e., ***Power loss will be maximum when the arc resistance R_A will be numerically equal to the impedance of the whole electrical circuit*** **referred to the secondary excluding arc resistance R_A.**

'*Power factor*' at *maximum power loss*,

$$\cos\phi = \frac{R_A + R_T + R_L}{\sqrt{(R_A + R_T + R_L)^2 + (X_T + X_L)^2}}$$

$$= \frac{R_A + R_T + R_L}{\sqrt{R_A^2 + 2R_A(R_T + R_L) + (R_T + R_L)^2 + (X_T + X_L)^2}}$$

Substituting $(R_T + R_L)^2 + (X_T + X_L)^2 = R_A{}^2$, we get

$$\cos\phi = \frac{R_A + R_T + R_L}{\sqrt{2R_A^2 + 2R_A(R_T + R_L)}} = \frac{R_A + R_T + R_L}{\sqrt{2R_A(R_A + R_T + R_L)}}$$

or, $$\cos\phi = \sqrt{\frac{R_A + R_T + R_L}{2R_A}} = \frac{1}{\sqrt{2}}\sqrt{\frac{R_A + R_T + R_L}{R_A}} = \frac{1}{\sqrt{2}}\sqrt{1 + \frac{R_T + R_L}{R_A}} \quad ...(2.17)$$

or, $$\cos\phi = \frac{1}{\sqrt{2}} \quad \text{..... neglecting } (R_T + R_L) \text{ in comparison with } R_A.$$

However, it *not economical* to operate an arc furnace with primary side power factor below 0.8 lagging.

Fig. 2.7 shows the performance characteristics of a typical arc furnace.

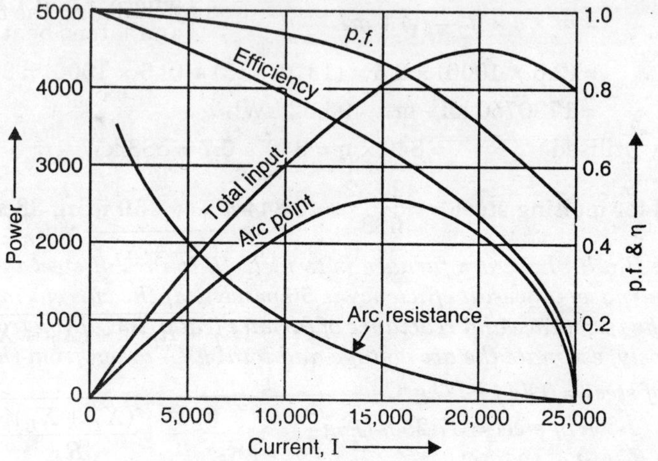

Fig. 2.7. Performance characteristics of a typical arc furnace.

- **Initially when the electrodes are short-circuited, it is as good as short-circuiting the secondary of a transformer. The total input to the furnace is almost zero (copper losses in the transformer winding).**

- When the electrodes are far apart arc is extinguished and there is no power drawn from supply.

In between the *above limits* there is a particular loading when the power input to the furnace is maximum.

Example 2.4. *The following data relate to a 4-phase electric arc furnace :*

Current drawn = *4000 A*

Arc voltage = *60 V*

Resistance of transformer referred to secondary = *0.0025* Ω

Reactance of transformer referred to secondary = *0.0050* Ω

(i) Calculate the power factor and kW drawn from the supply.

(ii) If the overall efficiency of the furnace is 70 percent, find the time required to melt 2.5 tonnes of steel if latent heat of steel = 37.2 kJ/kg, specific heat of steel = 0.5 kJ/kg K, melting point of steel = 1370° C and initial temperature of steel = 15°C.

Solution. Voltage drop due to transformer resistance = 4000 × 0.0025 = 10 V

Voltage drop due to transformer reactance = 4000 × 0.0050 = 20 V

Since arc voltage drop is resistive in nature, it is vectorially added to the transformer reactance drop.

Open circuit transformer secondary voltage/phase = $\sqrt{(60+10)^2 + 20^2}$ = 72.8 V

(*i*) **Power factor and kW drawn by supply :**

Power factor of supply $= \frac{(60+10)}{72.8} =$ **0.9615. (Ans.)**

Power drawn/phase by the secondary = 4000 × 72.8 × 0.9615 = 2799.88 W ≃ **280 kW**

Total power drawn = 3 × 280 = **840 kW (Ans.)**

(*ii*) **Energy required to melt 2.5 tonnes of steel :**

Energy required to melt 2.5 tonnes of steel

$= m \times c \times (t_2 - t_1) + mL$ (where, c and L are specific heat and latent heat of steel respectively)

= (2.5 × 1000) × 0.5 × (1370 – 15) + (2.5 × 1000) × 37.2

= 1786750 kJ or 496.32 kWh

Power actually utilised = 840 × η = 840 × 0.7 = 588 kW

Time required for melting steel $= \frac{496.32}{588} = 0.844\ h$ or **50 min, 38 sec. (Ans.)**

Example 2.5. *If a 3-phase arc furnace is to melt 10 tonnes of steel in 2 hours, estimate the average input to the furnace, if overall efficiency is 50 percent. If the current input is 9000 A with the above kW input and the resistance and reactance of furnace leads (including transformer) are 0.003* Ω *and 0.005* Ω *respectively, estimate the arc voltage and total kVA taken from the supply.*

Specific heat of steel = 0.444 kJ/kg°C ;

Latent heat of fusion of steel = 37.25 kJ/kg

Melting point of steel = 1370°C.

Assume initial temperature of steel = 20°C. **(Panjab University)**

Solution. *Given* : m = 10 tonnes = 10000 kg ; Time = 2 hours ; η = 50% ; I = 9000 A ; R = 0.003 Ω, X = 0.005 Ω ; c = 0.444 kJ/kg°C ; L = 37.25 kJ/kg ; Melting point = 1370°C ; Initial temperature of steel = 20°C.

Average input to the furnace :

Energy required to melt 10 tonnes of steel

$$= m \times c \times (t_2 - t_1) + mL$$

$$= m\,[c\,(t_2 - t_1) + L]$$

$$= 10000\,[0.444\,(1370 - 20) + 37.25] = 6366500 \text{ kJ} = \mathbf{1768.5 \text{ kWh}}$$

Average output $= \dfrac{\text{Total energy in kWh}}{\text{Time of melting in hours}} = \dfrac{1768.5}{2} = 884.25$ kN

$\therefore$ *Average input* $= \dfrac{\text{Output}}{\eta} = \dfrac{884.25}{0.5} = \mathbf{1768.5\ kW.}$ **(Ans.)**

Arc voltage :

Voltage drop due to resistance of furnace leads (including transformer)

$$= 9000 \times 0.003 = 27 \text{ V}$$

Voltage drop due to reactance of furnace leads (including transformer)

$$= 9000 \times 0.005 = 45 \text{ V}$$

Let the arc drop be V_A volts, resistive in nature.

From Fig. 2.8 :

Open circuit phase voltage of transformer secondary

$$= \sqrt{(V_A + 27)^2 + (45)^2}$$

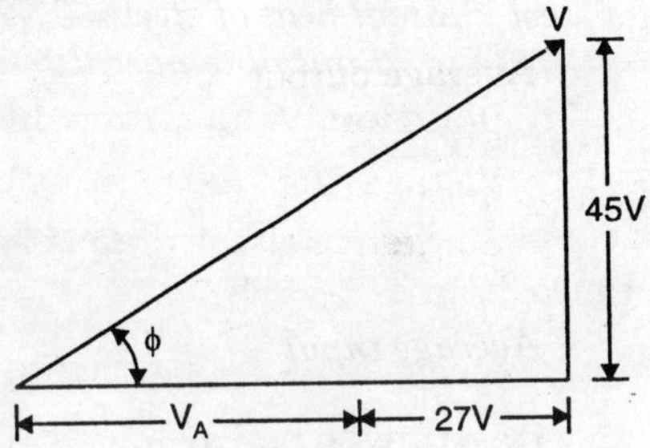

Fig. 2.8

Power factor, $\cos\phi = \dfrac{V_A + 27}{\sqrt{(V_A + 27)^2 + (45)^2}}$

Total power input = 3 × power drawn per phase

= 3 × current drawn per phase × secondary phase voltage × p.f.

$$1768.5 \times 1000 = 3 \times 9000 \times \sqrt{(V_A + 27)^2 + (45)^2} \times \frac{(V_A + 27)}{\sqrt{(V_A + 27)^2 + (45)^2}}$$

or, $$V_A + 27 = \frac{1768.5 \times 1000}{3 \times 9000} = 65.5 \text{ V}$$

$\therefore$ Arc voltage, $V_A = 65.5 - 27 = \mathbf{38.5\ V.}$ **(Ans.)**

Open circuit phase voltage (secondary)

$$= \sqrt{(V_A + 27)^2 + (45)^2} = \sqrt{(38.5 + 27)^2 + (45)^2} = 79.5 \text{ V}$$

Total kVA taken from the supply

$$= 3 \times 79.5 \times 9000 \times 10^{-3} = \mathbf{2146.5\ kVA.}\ \textbf{(Ans.)}$$

Example 2.6. *The following data relate to a 3-phase are furnace :*

Quantity of steel to be melted in one hour	*= 4.3 tonnes*
Specific heat of steel	*= 0.5 kJ/kg°C*
Latent heat of steel	*= 37.2 kJ/kg*
Melting point of steel	*= 1370°C*
Initial temperature of steel	*= 19.1°C*
Overall efficiency of steel	*= 50%*
Input current	*= 5700 A*
Resistance of transformer referred to secondary	*= 0.008 Ω*
Reactance of transformer referred to secondary	*= 0.014 Ω.*

Determine the following :

(i) Average kW input to the furnace, *(ii) Arc voltage,*

(iii) Arc resistance,

(iv) Power factor of the current drawn from the supply, and

(v) Average kVA input to the furnace. **(A.M.I.E.)**

Solution. *Given :* m = 4.3 tonnes = 4300 kg ; Time = 1 hour ; c = 0.5 kJ/kg°C ; L = 37.2 kJ/kg ; Melting point, t_2 = 1370°C ; Initial temp. of steel, t_1 = 19.1°C ; η = 50%, I = 5700 A ; R = 0.008 Ω ; X = 0.014 Ω

(i) **Average kW input to the furnace :**

Energy required to melt 4.3 tonnes of steel

$$= m \times c \times (t_2 - t_1) + mL = m\,[c(t_2 - t_1) + L]$$

$$4300\,[\,0.5 \times (1370 - 19.1) + 37.2] = 3064395 \text{ kJ} = \frac{3064395}{3600} = 851.2 \text{ kWh}$$

$$\text{Average output} = \frac{\text{Total energy required in kWh}}{\text{Time of melting in hours}}$$

$$= \frac{851.2}{1} = 851.2 \text{ kW}$$

$$\textit{Average input} = \frac{\text{Average output}}{\text{Overall efficiency}} = \frac{851.2}{0.5} = \mathbf{1702.4\ kW.} \quad \textbf{(Ans.)}$$

(ii) **Arc voltage, V_A :**

Voltage drop due to transformer resistance = 5700 × 0.008 = 45.6 V

Voltage drop due to transformer reactance = 5700 × 0.014 = 79.8 V

From Fig. 2.9 :

Open circuit secondary voltage

$$= \sqrt{(V_A + 45.6)^2 + (79.8)^2} \text{ volts}$$

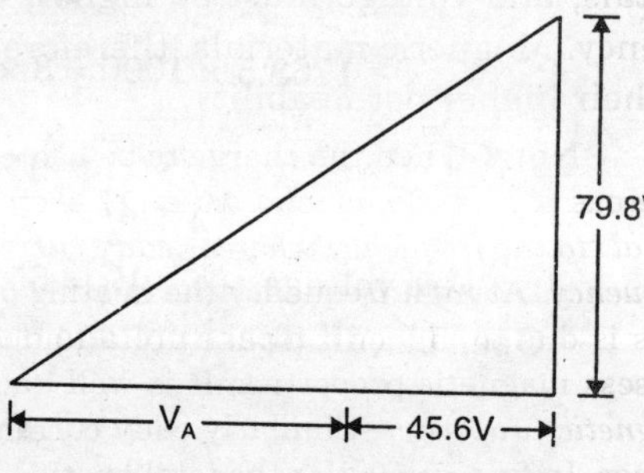

Fig. 2.9

$$\text{Power factor, } \cos\phi = \frac{V_A + 45.6}{\sqrt{(V_A + 45.6)^2 + (79.8)^2}}$$

Total power input = 3 × power drawn per phase
= 3 × current drawn per phase
× secondary voltage × p.f.

$$1702.4 \times 10^3 = 3 \times 5700 \times \sqrt{(V_A + 45.6)^2 + (79.8)^2} \times \frac{V_A + 45.6}{\sqrt{(V_A + 45.6)^2 + (79.8)^2}}$$

$$V_A + 45.6 = \frac{1702.4 \times 10^3}{3 \times 5700}$$

or, *Arc voltage,* $V_A \simeq$ **54 V. (Ans.)**

(iii) **Arc resistance, R_A :**

$$R_A = \frac{V_A}{I} = \frac{54}{5700} = \mathbf{0.00947\ \Omega.} \quad \textbf{(Ans.)}$$

(*iv*) **Power factor, cos φ :**

$$\cos \phi = \frac{V_A + 45.6}{\sqrt{(V_A + 45.6)^2 + (79.8)^2}}$$

$$= \frac{(54 + 45.6)}{\sqrt{(54 + 45.6)^2 + (79.8)^2}} = \mathbf{0.7804.} \quad \textbf{(Ans.)}$$

(*v*) **Average kVA input to the furnace :**

$$\text{kVA input to the furnace} = \frac{\text{kW input}}{\text{Power factor}}$$

$$= \frac{1702.4}{0.7804} = \mathbf{2181.4\ kVA.} \quad \textbf{(Ans.)}$$

2.1.7. Induction Heating

The process of induction heating makes use of the *currents induced by the electro-magnetic action* in the charge to be heated. Induction heating, in fact, is based on the principle of transformer working. The primary winding which is supplied from an A.C. source is magnetically coupled to the charge which acts as a *short-circuited secondary of a single turn* . When A.C. voltage is applied to the primary, it induces voltage in the secondary *i.e.,* charge. The secondary current heats up the charge in the same way as any electric does while passing through a resistance. If V in the voltage induced in the charge and R is the resistance of the charge, then heat produced = $\frac{V^2}{R}$. So to develop heat sufficient to melt the change, the resistance of the charge must be low, which is possible only with metals, and voltage must be higher, which is obtained by employing higher flux and higher frequency. Magnetic materials, therefore, can be easily treated than non-magnetic materials because of their higher permeability.

Note. When the charge to be heated is non-magnetic the heat generated is due to eddy current losses, whereas if *it is a magnetic material, there will be hysteresis losses in addition. The hysteresis loss is proportional to the frequency whereas eddy current loss is proportional to square of frequency when operating at low frequency.* At high frequency the heating due to hysteresis becomes very small as compared to eddy currents. This is due to higher temperature attained by the charge at higher frequencies when the material ceases to possess magnetic properties. It is well known that *above curie temperature the magnetic materials loss their magnetic properties.* Similarly eddy current losses do not follow f^2 law as frequency is increased higher and higher. In fact, at 10 kHz the total heating may vary directly as the frequency and even drop to one-half power at frequencies of the order of 500 kHz. It has been observed that *higher the supply frequency the lower the depth of penetration.* The depth is α $\sqrt{\frac{1}{f}}$.

Types of induction furnaces :

Basically, the two types of induction furnaces are :

1. Core type or low frequency induction furnace
   (*i*) Direct core type (*ii*) Vertical core type
   (*iii*) Indirect core type.
2. Coreless type or high frequency induction furnace.

2.1.7.1. Core type or low frequency induction furnace

It operates just like a two winding transformer.

1. **Direct core type induction furnace :**

A direct core type induction furnace is shown in Fig. 2.10. It consists of a transformer in which charge to be heated forms a single-turn short-circuited secondary and is magnetically coupled to the primary by iron core. The furnace consists of a circular hearth which contains the charge to be melted in the form of an annular ring.

When there is no molten metal in the ring, the secondary becomes open-circuited thereby cutting off the secondary current. Hence, *to start the furnace, molten metal has to be poured in the annular*.

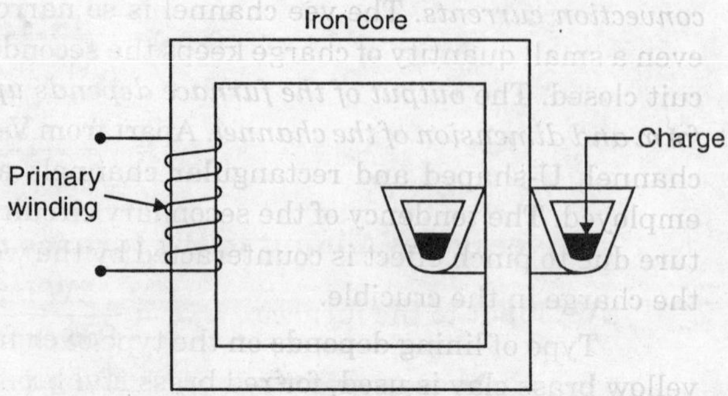

Fig. 2.10. Direct core type induction furnace.

Since the magnetic coupling between the primary and secondary is very poor, it results in high leakage and low power factor. In order to nullify the effect of increased leakage reactance, low primary frequency of the order of 10 Hz or so is used.

- The *melting is rapid* and *clean and the charge is capable of accurate control as far as temperature and alloying elements are concerned.*
- The *inherent stirring action of the melt ensures a greater uniformity of the end product.*
- However, if the current density exceeds about 500 A/cm^2, the current flowing around the melt, interacts with the alternating magnetic field and exerts *constricting forces* on the cross-section of the metal which may squeeze it to the extent that a complete interruption of the secondary circuit takes place. This is known as **Pinch effect** (formation of bubbles and voids). When the current is interrupted these forces vanish and the metal may flow as air. *The pinch effect depends upon frequency and power consumed.*

This type of furnace has the following ***drawbacks :***

(*i*) Leakage reactance is high and consequently the power factor is low on account of poor magnetic coupling.

(*ii*) Low frequencies have to employed as normal frequency causes turbulence of the charge. This requires a motor-generator set or a frequency converter.

(*iii*) The crucible for the charge is of odd shape and not convenient from the metallurgical point of view.

(*iv*) The furnace cannot function if the secondary circuit is not closed. This requires a complete ring of the charge around the core. For starting the furnace, either molten metal is poured into the crucible or sufficient molten metal is allowed to remain in the crucible from a previous operation. Also in order to close the secondary circuit, an iron ring may be placed in the crucible or the lining may be of graphite.

(*v*) It suffers from *"pinching effect."*

Such furnaces are *not suitable for intermittent services.* On account of the above drawbacks these furnaces have become *obsolete* these days.

2. **The vertical core type induction furnace :**

The furnace is also known as *"Ajax-Wyatt furnace"* and represents an improvement over the core type furnace discussed above. The furnace makes use of vertical crucible instead of a horizontal one for the charge (Fig. 2.11). The shell of the furnace is of heavy steel. The top is closed by an insulated cover which can be removed for charging.

Molten metal is *kept circulated round the vee by convection currents*. The vee channel is so narrow that even a small quantity of charge keeps the secondary circuit closed. The *output of the furnace depends upon the form and dimension of the channel*. Apart from V-shaped channel, U-shaped and rectangular channels are also employed. The tendency of the secondary circuit to rupture due to pinch effect is counteracted by the weight of the charge in the crucible.

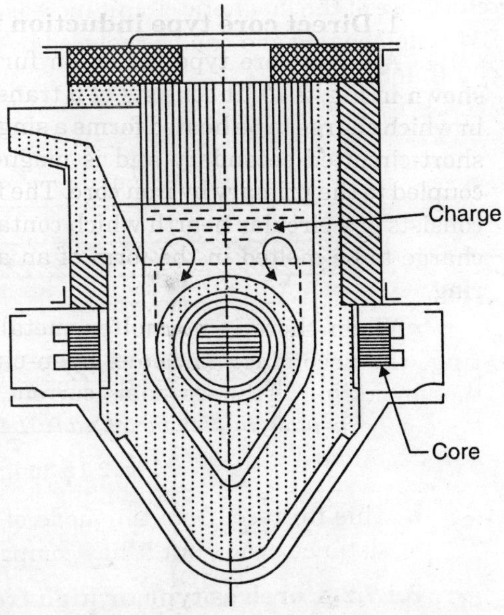

Fig. 2.11. Vertical core type furnace.

Type of lining depends on the type of charge. For yellow brass clay is used, for red brass and bronze mixture of magnesia and alumina or corundum which has a high alumina content is employed.

Tilting of the furnace for pouring out the metal is done by hydraulic means.

- This furnace is suitable for *continuous operation*.
- With normal supply frequency its efficiency is about 75%.
- Its standard size varies from 60-300 kW, all single-phase. The furnace is widely *used for melting and refining of brass and other non-ferrous metals*.

Advantages :

(*i*) Consistent performance and simple control.

(*ii*) Accurate temperature control, uniform castings, reduced metal losses and reduction of rejects.

(*iii*) Highly efficient heat, low operating costs and improved production.

(*iv*) High power factor (0.8 – 0.85) comparatively.

(*v*) Local working conditions in a cool atmosphere with no dirt, noise or fuel.

(*vi*) Absence of crucibles.

(*vii*) Absence of combustion gases resulting in elimination of the most common source of metal contamination.

3. Indirect core type induction furnace :

In this type of furnace, a *suitable element is heated* by *induction which, in turn, transfers the heat to the charge by radiation.*

Fig. 2.12 shows an indirect core type induction furnace. The secondary consists of a metal container which forms the walls of the furnace proper. The primary winding is magnetically coupled to this secondary by iron core.

When primary winding is connected to A.C. supply secondary current is induced in the metal container by transformer action which heats up the container. The metal container transfers this heat to the charge. It is advantageous in respect of temperature control without the use of external control equipment. The part *LM* of the magnetic circuit situated inside the oven chamber consists of a special alloy which loses its magnetic properties at a particular temperature but regains them when cooled back to the same temperature. As soon as the chamber attains the critical temperature, reluctance of the magnetic circuit increases manifold there by cutting off the supply of heat. The bar *LM* is detachable and can be replaced by other bars having different critical temperature (between 400°C and 1000°C).

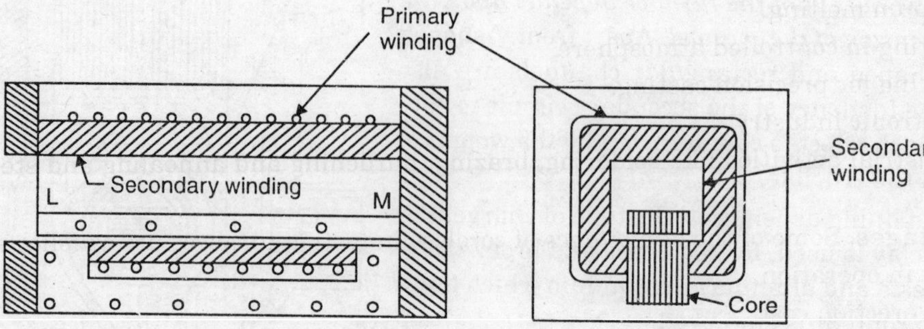

Fig. 2.12. Indirect core type induction furnace.

- This furnace, from the mode of transmission of heat, is directly in competition with resistance oven ; but it has comparatively poor p.f. (0.8 appr.).

2.1.7.2. Coreless type or high frequency induction furnace

Construction. Fig. 2.13 shows a simplified schematic design of a coreless type induction furnace. It essentially consists of three main parts :

(*i*) *Primary coil*, (*ii*) *Ceramic crucible* containing charge which forms the secondary and (*iii*) *Frame* which includes *supports* and *tilting mechanism*.

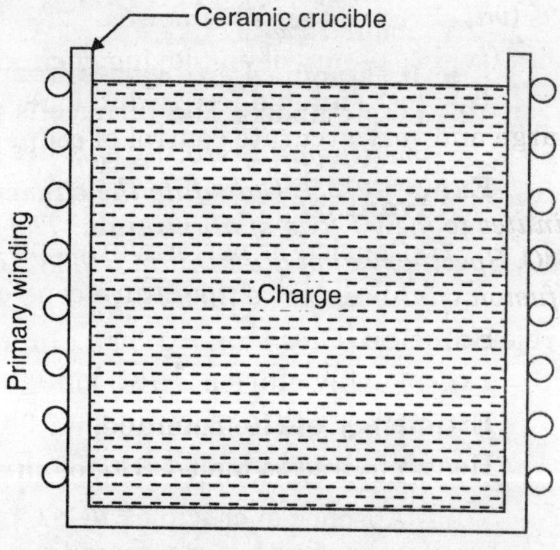

Fig. 2.13. Coreless induction furnace.

The distinctive feature of this furnace is that it contains *no heavy iron core* with the result there is *no continuous path for the magnetic flux*. The crucible and the coil are relatively light in construction and can be conveniently tilted for pouring.

Working. The charge is put into the crucible and primary winding coil is connected to high frequency A.C. supply. The flux created by primary winding sets up eddy currents in the charge which tend to flow concentrically with those in the inductor. These eddy currents heat up the charge to its melting point and also set up electro-magnetic forces producing *stirring action* which is essential for obtaining uniform quality of metal.

Since flux density is low (due to the absence of the magnetic core) high frequency supply has to be used because eddy current loss, $P_e \propto B^2 f^2$. However this high frequency increases the resistance of the primary winding due to skin effect, thereby increasing primary copper losses. Hence the primary winding is *not* made of copper wire but consists of hollow copper tubes which are cooled by water circulating through them.

As the magnetic coupling cap between the primary and secondary windings is low, the furnace *p.f.* lies between 0.1 and 0.3. *Static capacitors* are, therefore, invariably employed in parallel with such a furnace in order to *improve the p.f.* Since in this type of furnace the p.f. does not remain constant, so capacitance in the circuit during heat cycle is varied to maintain approximately unity power factor.

Applications. These furnaces find *applications* in the following fields :

(*i*) Steel production (Energy consumption is 600 to 1000 kWh per tonne of steel).

(*ii*) Melting non-ferrous metals like brass, bronze, copper and aluminium etc., along with various alloys of these elements.

(*iii*) Vacuum melting.

(*iv*) Melting in controlled atmosphere.

(*v*) Melting for precision casting.

(*vi*) Electronic industry.

(*vii*) Industrial activities like soldering, brazing, hardening and annealing and sterlizing instruments etc.

Advantages. Some of the *advantages* of coreless induction furnaces are as follows :

(*i*) Fast in operation.

(*ii*) Low erection cost.

(*iii*) Low operating cost.

(*iv*) Can be operated intermittently (as no time is lost in warming up).

(*v*) Operation is free from smoke, dirt, dust and noises.

(*vi*) Charging and pouring is simple.

(*vii*) Less melting time.

(*viii*) Precise control of power.

(*ix*) Possibility of employing vacuum heating necessary for precious metal melting.

(*x*) Most suitable for production of high grade alloy steels (due to non-contamination of the charge and very accurate control of composition).

Example 2.7. *Calculate the efficiency of a high frequency induction furnace which takes 10 minutes to melt 1.8 kg of aluminium. The input to the furnace being 4.8 kW and initial temperature 15°C. Specific heat of aluminium = 0.88 kJ/kg°C ; melting point of aluminium = 660°C ; latent heat of fusion of aluminium = 32 kJ/kg ; 1 kJ = 2.78 × 10⁻⁴ kWh.* **(Panjab University)**

Solution. *Given* : m = 1.8 kg ; Input to the furance = 4.8 kW ; t_1 = 15°C ; t_2 = 660°C,
c = 0.88 kJ/kg°C ; L = 32 kJ/kg ; 1 kJ = 2.78×10^{-4} kWh.

Efficiency of the furnace, η :

Heat required to melt 1.8 kg of aluminium

$$= m \times c \times (t_2 - t_1) + m \times L$$

$$= m\,[c \times (t_2 - t_1) + L] = 1.8\,[0.88\,(660 - 15) + 32] = 1079.28 \text{ kJ}$$

$$= 1079.28 \times 2.78 \times 10^{-4} = 0.3 \text{ kWh}$$

Energy input $= 4.8 \times \dfrac{10}{60} = 0.8$ kWh

$$\therefore \qquad \eta = \frac{\text{Output}}{\text{Input}} = \frac{0.3}{0.8} = 0.375 \quad \text{or} \quad \mathbf{37.5\%.} \quad \textbf{(Ans.)}$$

Example 2.8. *A low frequency induction furnace operating at 12 V in secondary circuit takes 480 kW at 0.5 p.f. when hearth is full. If the secondary voltage be maintained at 12 V, estimate the power factor and the power absorbed when hearth is half-full. Assume the resistance of the secondary circuit to be thereby halved and the reactance to remain the same.*

Solution. *Given* : Secondary voltage = 12 V ; Power drawn when hearth is full = 480 kW ;
p.f. = 0.5 ;

Secondary current, $I = \dfrac{480 \times 1000}{12 \times 0.5} = 80000$ A

Impedance of the secondary circuit when hearth is full,

$$Z = \frac{V}{I} = \frac{12}{80000} = 1.5 \times 10^{-4}\ \Omega$$

Resistance of the secondary circuit when hearth is full,

$$R = Z \cos \phi = 1.5 \times 10^{-4} \times 0.5 = 0.75 \times 10^{-4}\ \Omega$$

Reactance of the secondary circuit when hearth is full,

$$X = Z \sin \phi = 1.5 \times 10^{-4} \sin (\cos^{-1} 0.5) = 1.3 \times 10^{-4}\ \Omega$$

Power factor and power drawn when hearth of half full :

Resistance of the secondary circuit when hearth is *half full,*

$$R' = \frac{1}{2} R = \frac{1}{2} \times 0.75 \times 10^{-4} = 0.375 \times 10^{-4}\ \Omega$$

Reactance of the secondary circuit,

$$X' = X = 1.3 \times 10^{-4} = 1.3 \times 10^{-4}\ \Omega \qquad (\because\ X' = X\ \ ...\textit{Given})$$

Impedance of the secondary circuit,

$$Z' = \sqrt{(R')^2 + (X')^2} = \sqrt{(0.375 \times 10^{-4})^2 + (1.3 \times 10^{-4})^2} = 1.353 \times 10^{-4}\ \Omega$$

$\therefore$ *Power factor when hearth is half full*

$$= \frac{R'}{Z'} = \frac{0.375 \times 10^{-4}}{1.353 \times 10^{-4}} = \mathbf{0.277.} \quad \textbf{(Ans.)}$$

$$\text{Secondary current,} \quad I' = \frac{12}{1.353 \times 10^{-4}} = 88691.8\ \text{A}$$

$$\therefore \ \textit{Power drawn,} \quad P = (I')^2 R' = (88691.8)^2 \times 0.375 \times 10^{-4}$$

$$= 294984\ \text{W} \simeq \mathbf{295\ kW.} \quad \textbf{(Ans.)}$$

Example 2.9. *A low frequency furnace, whose secondary voltage is maintained constant at 10 V, takes 420 kW at 0.6 power factor when the hearth is full. Assuming the resistance of the secondary circuit to vary inversely as the height of the charge and reactance to remain constant, find the height upto which the hearth should be filled to obtain maximum heat.* **(Gorakhpur University)**

Solution. *Given* : Secondary voltage = 10 V, Power taken when hearth is full = 420 kW, 0.6 p.f.

Height to which hearth should be filled :

$$\text{Secondary current}\ ; I = \frac{420 \times 1000}{10 \times 0.6} = 70000\ \text{A}$$

Impedance of the secondary circuit when hearth is *full,*

$$Z = \frac{V}{I} = \frac{10}{70000} = 1.428 \times 10^{-4}\ \Omega$$

Resistance of the secondary circuit when hearth is *full,*

$$R = Z \cos \phi = 1.428 \times 10^{-4} \times 0.6 = 0.8568 \times 10^{-4}\ \Omega$$

Reactance of the secondary circuit when hearth is *full,*

$$X = Z \sin \phi = 1.428 \times 10^{-4} \times \sin (\cos^{-1} 0.6) = 1.1424 \times 10^{-4}\ \Omega$$

Let the height of the charge be x times of full hearth *i.e.,* $h = xH$

Since resistance varies inversely as the height of the charge (*Given*),

$\therefore$ Resistance of the charge with height of charge as $xH = \dfrac{R}{x} = \dfrac{0.8568 \times 10^{-4}}{x}\ \Omega$

Power drawn or heat produced will be maximum when resistance of secondary circuit will be equal to reactance of secondary circuit

i.e., $$\frac{0.8568 \times 10^{-4}}{x} = 1.1424 \times 10^{-4}$$

or, $$x = \frac{0.8568 \times 10^{-4}}{1.1424} = \mathbf{0.75.} \quad \textbf{(Ans.)}$$

(*i.e.*, Maximum heat will be obtained with the height of charge as $\frac{3}{4}$th of height of hearth).

2.1.7.3. High frequency eddy current heating

In order to heat an article by eddy currents, it is placed inside a high frequency A.C. current-carrying coil (Fig. 2.14). The alternating magnetic field produced by the coil sets up eddy currents in the article, which consequently, gets heated up. Such a coil is known as *heater coil* or *work coil* and the material to be heated is known as *charge* or *load*.

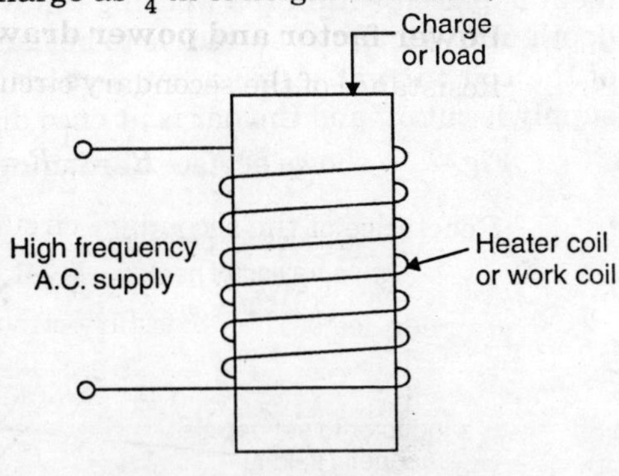

Fig. 2.14. High frequency eddy current heating.

Primarily, it is the eddy-current loss which is responsible for the production of heat although hysteresis loss also contributes to some extent in the case of magnetic materials.

As the eddy current loss $P_e \propto B^2 f^2$, this loss can be controlled by controlling flux density B and the supply frequency f. This loss is greatest on the surface of the material but decreases as we so deep inside. The depth of penetration (d) of eddy currents into the charge is given by :

$$d = \frac{1}{2\pi} \sqrt{\frac{\rho \times 10^9}{\mu_r \cdot f}} \text{ cm} \qquad ...(2.18)$$

where, ρ = Resistivity of the molten metal,

f = Supply frequency, and

μ_r = Relative permeability.

Thus, since $d \propto \frac{1}{\sqrt{f}}$, therefore, eddy current heating can be restricted to any desired depth of the material to be heated by judicious selection of frequency of the heating. The supply frequency is usually employed between 10000 and 40000 Hz.

Advantages of eddy current heating :

(*i*) Temperature control is very easy.

(*ii*) The heat can be made to penetrate into the metal surface to any desired depth.

(*iii*) This heating method is quick, clean and convenient.

(*iv*) Very less wastage of heat (as heat is produced in the body to be heated up directly).

(*v*) The equipment can be operated even by unskilled labour.

(*vi*) The surface area over which heat is produced can be accurately controlled.

(*vii*) The amount of heat produced can be accurately controlled by suitable timing devices.

(*viii*) It can easily take place in vacuum or other special atmosphere (whereas other conventional types of heating are not possible in such places).

(*ix*) The work coils are not required to fit closely around the object being treated.

Demerits :

(*i*) The generation of heat is costly.

(*ii*) Efficiency of equipment is quite low (less than 50%).

(*iii*) Initial cost of the equipment is high.

Applications of eddy current heating :

1. ***Surface hardening.*** In this case the bar whose surface is to be hardened by heat treatment is placed within the working coil which is connected to an A.C. supply of high frequency. The depth upto which the surface is to be hardened can be obtained by the proper selection of frequency of the coil current. After a few seconds, when surface has reached the proper temperature, A.C. supply is cut off and the bar is at once dipped in water.

Fig. 2.15 shows surface hardening of a bearing journal by eddy current heating.

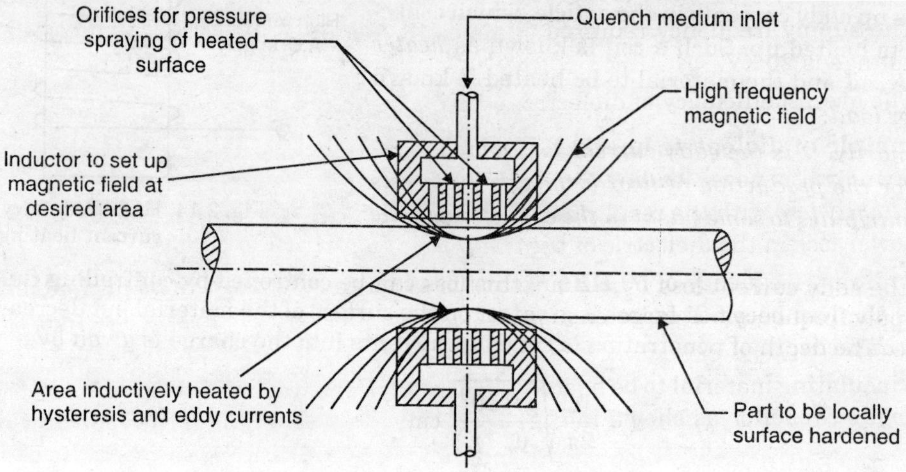

Fig. 2.15. Surface hardening by eddy current heating.

2. ***Annealing.*** In conventional method of annealing the process takes long time resulting in scaling of the metal which is undesirable. But in eddy current heating, time taken is much less so that no scale formation takes place. By this method a temperature of the order 750°C can be attained in one minute (appr.) upto a depth of 25 mm.

3. ***Soldering.*** Eddy current heating can be economically employed for soldering precisely for high temperature soldering where silver, copper and their alloys are used as solders.

Other applications of eddy current heating include the following :

(*i*) Drying of paints.

(*ii*) Welding.

(*iii*) Melting of previous metals.

(*iv*) Sterilization of surgical instruments.

(*v*) Forgings of bolt heads and rivet heads.

Example 2.10. *In the case hardening of a steel pulley, the depth of penetration required is 1.4 mm. The relative permeability is unity and the specific resistivity of steel is 5×10^{-7} Ω m*

Determine the frequency required.

Solution. *Given* : Depth of penetration, $d = 1.4$ mm $= 0.14$ cm ;

$\mu_r = 1$; $\rho = 5 \times 10^{-7}$ Ω m.

Frequency, f :

$$d = \frac{1}{2\pi}\sqrt{\frac{\rho \times 10^9}{\mu_r \cdot f}} \quad \text{...[Eqn. (2.18)]}$$

Squaring both sides and simplifying, we have

or, $$f = \frac{\rho \times 10^9}{d^2 \times 4\pi^2 \times \mu_r} = \frac{(5 \times 10^{-7} \times 100) \times 10^9}{(0.14)^2 \times 4\pi^2 \times 1} = \mathbf{64618\ Hz.} \quad \textbf{(Ans.)}$$

2.1.8. Dielectric Heating

- Dielectric heating (also sometimes called *High frequency capacitive heating*) is employed for heating insulators like wood, plastics and ceramics etc. which cannot be heated easily and uniformly by other methods.
- The supply frequency required of dielectric heating is between 10-50 MHz and applied voltage is 20 kV.
- The overall efficiency of dielectric heating is about 50 percent.

Principle of dielectric heating. When a capacitor is subjected to a sinusoidal voltage, the current drawn by it is never leading the voltage by exactly 90°. The angle between the current and voltage is *slightly less* with the result that there is a small in-phase component of the current which produces power loss in the dielectric of the capacitor.

At ordinary frequency of 50 Hz such loss may be small enough to be negligible but at high frequencies the loss becomes large enough to heat the dielectric. It *is this loss that is utilised* in *heating the dielectric*.

The insulating material to be heated is placed between two conducting plates in order to form a *parallel-plate capacitor* as shown in Fig. 2.16. The vector diagram of the capacitor is shown in Fig. 2.17.

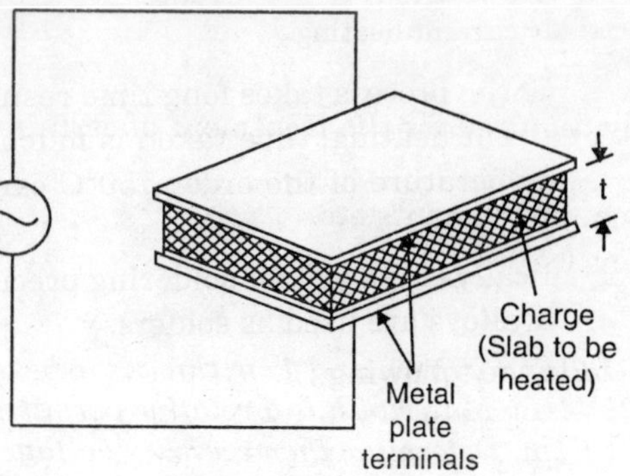

Fig. 2.16. Parallel-plate capacitor -Dielectric heating.

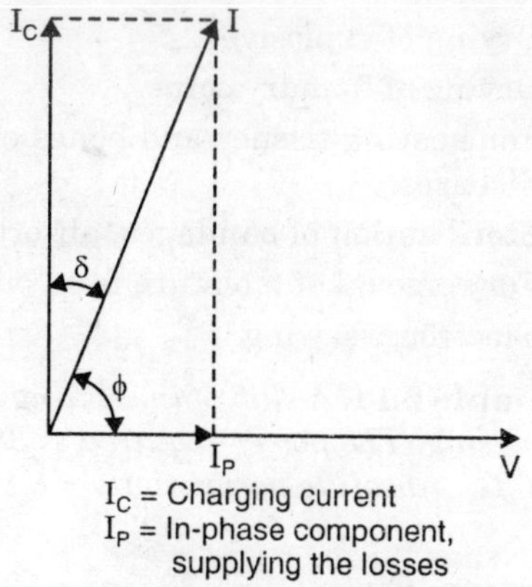

Fig. 2.17. Vector diagram of the parallel-plate capacitor

Power drawn from supply = $VI \cos \phi$

Now, $$I_C = I = \frac{V}{X_C} = \frac{V}{1/2\pi fC} = 2\pi fCV.$$

$\therefore$ $$P = V(2\pi fCV) \cos \phi = 2\pi fCV^2 \cos \phi \quad \text{...(2.19)}$$

Now, $$\phi = (90° - \delta),\ \cos \phi = \cos (90° - \delta) = \sin \delta = \tan \delta = \delta$$

where δ is very small and is expressed in radians

$$P = 2\pi fCV^2 \delta \text{ watts} \quad ...(2.20)$$

Here, $$C = \frac{\varepsilon_0 \varepsilon_r A}{t},$$

where t and A are the thickness and area of the dielectric slab respectively ; ε_r is the relative permittivity of dielectric and ε_0 is the absolute permittivity of vaccum (= 8.854×10^{-12} F/m).

This power is converted into heat. Since for a given insulation material C and δ are constant, the *dielectric loss* $\alpha V^2 f$.

That is why *high-frequency voltage is used in dielectric heating.*

Advantages of dielectric heating :

(*i*) Heating is uniform since heat is generated within the dielectric medium itself.

(*ii*) With the increase in frequency the heating becomes faster.

(*iii*) Only method for heating bad conductor of heat.

(*iv*) In this method of heating, heating is fastest.

(*v*) Heating can be stopped immediately as and when desired.

(*vi*) As no naked flame appears in the process, inflammable articles like plastics and wooden products etc. can be safely heated.

Applications of dielectric heating :

As the cost of dielectric heating is high, thus it is employed where other methods are not possible or are too slow. Some of the applications of dielectric heating are listed below :

(*i*) Drying and gluing of wood.

(*ii*) Drying of rayon cakes in textile manufacture.

(*iii*) Dehydration of foods.

(*iv*) Gluing of laminated glass.

(*v*) Rubber vulcanizing.

(*vi*) Drying of explosives.

(*vii*) Drying of foundry cores.

(*viii*) For heating tissues and bones of the body required for the treatment of certain types of pains and diseases.

(*ix*) Sterilization of bandages, absorbent cotton, instruments etc.

(*x*) For removal of moisture from oil emulsions.

(*xi*) Electronic sewing.

Example 2.11. *A slab of insulating material 130 cm² in area and 1 cm thick is to be heated by dielectric heating. The power required is 380 W at 30 MHz. Material has a relative permittivity of 5 and p.f. of 0.05. Absolute permittivity = 8.854 × 10⁻¹² F/m. Determine the necessary voltage.*

(Panjab University)

Solution. *Given* : $A = 130 \text{ cm}^2 = 130 \times 10^{-4} \text{ m}^2$; $t = 1 \text{ cm} = 0.01 \text{ m}$; $P = 380$ W ;
$f = 30$ MHz ; $\varepsilon_r = 5$; p.f. = 0.05 ; $\varepsilon_0 = 8.854 \times 10^{-12}$ F/m

Voltage, V :

Capacitance, $$C = \frac{\varepsilon_0 \varepsilon_r A}{t} = \frac{8.854 \times 10^{-12} \times 5 \times 130 \times 10^{-4}}{0.01} = 57.55 \times 10^{-12} \text{ F}$$

$$P = 2\pi fCV^2 \cos\phi \quad ...[\text{Eqn. } (2.19)]$$

$$380 = 2\pi \times 30 \times 10^6 \times 57.55 \times 10^{-12} V^2 \times 0.05$$

or, $$V^2 = \frac{380}{2\pi \times 30 \times 10^6 \times 57.55 \times 10^{-12} \times 0.05} = 700595$$

$\therefore$ $V = \mathbf{837\ V.}$ **(Ans.)**

Example 2.12. *A piece of an insulating material is to be heated by dielectric heating. The size of the piece is 12 cm × 12 cm × 3 cm. A frequency of 20 MHz is used and the power absorbed is 450 W. If the material has a relative permittivity of 5 and a power factor of 0.05, calculate the voltage necessary for heating and current that follows in the material.*

If the voltage were limited to 1700 V, what will be the frequency to get the same loss ?

Solution. *Given* : $A = 12 \times 12 = 144\ cm^2 = 144 \times 10^{-4}\ m^2$; $t = 3\ cm = 0.03\ m$;

$f = 20$ MHz ; $P = 450$ W ; $\varepsilon_r = 5$; $\cos \phi = 0.05$.

Voltage and current :

The capacitance of the parallel plate condenser that the material forms is given by,

$$C = \frac{\varepsilon_0 . \varepsilon_r\, A}{t} = \frac{8.854 \times 10^{-12} \times 5 \times 144 \times 10^{-4}}{0.03} = 21.25 \times 10^{-12}\ F$$

$$X_C = \frac{1}{2\pi f C} = \frac{1}{2\pi \times 20 \times 10^6 \times 21.25 \times 10^{-12}} = 374.5\ \Omega$$

Power, $P = 2\pi fCV^2 \cos \phi$

or $$450 = 2\pi \times 20 \times 10^6 \times 21.25 \times 10^{-12} \times V^2 \times 0.05$$

or $$V^2 = \frac{450}{2\pi \times 20 \times 10^6 \times 21.25 \times 10^{-12} \times 0.05} = 3370340$$

or *Voltage*, $\mathbf{V = 1836\ V.}$ **(Ans.)**

Current, $\mathbf{I} \simeq \mathbf{I_C} = \frac{V}{X_C} = \frac{1836}{374.5} = \mathbf{4.9\ A.}$ **(Ans.)**

Heat produced $\propto V^2 f$

$\therefore$ $$V_2^2 f_2 = V_1^2 f_1$$

or $$\mathbf{f_2} = f_1 \times \left(\frac{V_1}{V_2}\right)^2 = 20 \left(\frac{1836}{1700}\right)^2 = \mathbf{23.33\ MHz.}$$ **(Ans.)**

Example 2.13. *A plywood board 0.5 m × 0.25 m × 0.02 m is to be heated from 15°C to 135°C in 10 minutes by dielectric heating employing a frequency of 30 MHz. Determine the power required in the heating process. Assume specific heat of wood 1500 J/kg °C ; weight of wood 600 kg/m³ and efficiency of process 55 percent.* **(B.T.E. U.P.)**

Solution. *Given* : Dimensions of the plywood board = 0.5 m × 0.25 m × 0.02 m ;

$t_1 = 15°C$, $t_2 = 135°C$; Time = 10 minutes ; $f = 30$ MHz ; $c = 1500$ J/kg°C ;

$w = 600\ kg/m^3$; $\eta = 0.55$.

Power input required, P :

Volume of the plywood to be heated $= 0.5 \times 0.25 \times 0.02 = 0.0025\ m^3$

Weight of plywood $= 0.0025 \times 600 = 1.5$ kg

Heat required to raise the temperature from 15°C to 135°C

$= m \times c \times (t_2 - t_1) = 1.5 \times 1500 \times (135 - 15) = 270000$ J or $Ws = \dfrac{270000}{60 \times 60} = 75$ Wh

Since it is to be done in 10 minutes, so power required $= \dfrac{75}{10/60} = 450$ W

Power input required, $P = \dfrac{450}{\eta} = \dfrac{450}{0.55}$ = **818.2 W. (Ans.)**

Example 2.14. *A small piece of plywood having dimensions, length 5 cm, width 2 cm and thickness 1 cm is placed in between two electrodes having dimensions, length 25 cm, width 2 cm and with 2 cm distance between them. Electric current at frequency of 10 MHz is passed through electrodes for heating of wood by the process of dielectric heating. If the power consumption for heating the piece of wood is 1 kW, determine :*

(i) Voltage applied across the electrodes, and

(ii) Current through the plywood piece.

Assume that the plywood is an imperfect dielectric and has relative permittivity 5 and p.f. 0.04.

(A.M.I.E)

Solution. *Given* : Dimensions of plywood piece : length = 5 cm, width = 2 cm, thickness = 1 cm ; Dimensions of the electrodes : Length = 25 cm, width = 2 cm, distance between them = 2 cm ; f = 10 MHz ; P = 1 kW ; $\varepsilon_r = 5$; p.f. = 0.04.

(*i*) **Voltage applied across the electrodes, V :**

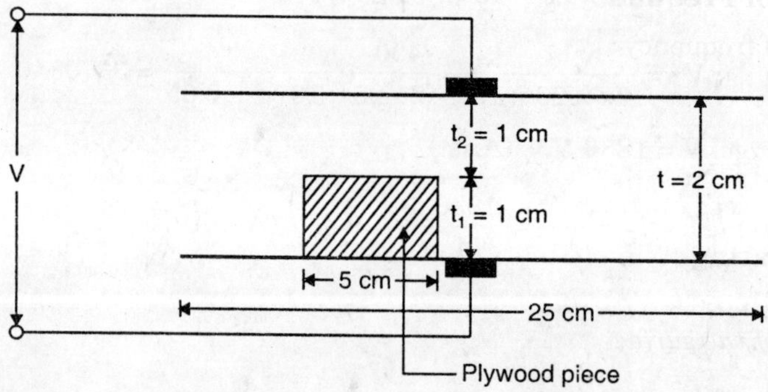

Fig. 2.18

The arrangement of heating the metarial is shown in Fig. 2.18. With this configuration the capacitance of the parallel plate condenser formed with two dielectrics, plywood and air is given by the equation,

$$C = \varepsilon_0 \left[\frac{A_1 \varepsilon_{r_2}}{t} + \frac{A_2}{\dfrac{t_1}{\varepsilon_{r_1}} + \dfrac{t_2}{\varepsilon_{r_2}}} \right]$$

where, $A_1 = (25 - 5) \times 2 = 40 \text{ cm}^2 = 0.004 \text{ m}^2$

$A_2 = 5 \times 2 = 10 \text{ cm}^2 = 0.001 \text{ m}^2$

$t = 2 \text{ cm} = 0.02 \text{ m} \; ; \; t_1 = 1 \text{ cm} = 0.01 \text{ m}$

$t_2 = 2 - 1 = 1 \text{ cm} = 0.01 \text{ m}$

ε_{r_1} = Relative permittivity of plywood = 5

ε_{r_2} = Relative permittivity of air = 1

ε_0 = Absolute permittivity = 8.854×10^{-12} F/m

∴ Capacitance of the capacitor,

$$C = 8.854 \times 10^{-12} \left[\frac{0.004 \times 1}{0.02} + \frac{0.001}{\frac{0.01}{5} + \frac{0.01}{1}} \right]$$

$$= 8.854 \times 10^{-12} (0.2 + 0.08333) = 2.509 \times 10^{-12} \text{ F}$$

Now, Power consumption, $P = 2\pi fCV^2 \cos\phi$

or $$V = \sqrt{\frac{P}{2\pi fC \cos\phi}}$$

$$= \sqrt{\frac{1 \times 10^3}{2\pi \times (10 \times 10^6) \times 2.509 \times 10^{-12} \times 0.04}} = \mathbf{12593\ V.} \quad \textbf{(Ans.)}$$

(*ii*) **Current through the plywood piece, I :**

$$I\,(\simeq I_C) = \frac{P}{V \cos\phi} = \frac{1 \times 10^3}{12593 \times 0.04} = \mathbf{1.985\ A.} \quad \textbf{(Ans.)}$$

2.1.9. Choice of Frequency

The selection of frequency is a significant factor for heating since it immensely affects work to be heated and method of its heating (whether by induction or dielectric heating).

- Furnaces running on *power frequency of 50 Hz* can be of *1 MW capacity*, those running on *medium frequencies (500 Hz to 1000 Hz)* have a *capacity of 50 kW* and those running on *high frequencies* (*1 MHz to 2 MHz*) have capacities ranging from 200 kW to 500 kW.

Induction heating. While selecting frequency for induction heating the following factors need be considered :

(*i*) *Thickness of the surface* to be heated–*Higher the frequency, thinner the surface that will get heated.*

(*ii*) The *time of continuous heating–Longer* the duration, *deeper* the penetration of heat in the work due to conduction.

(*iii*) The *temperature to be obtained–Higher the temperature, higher the capacity of generator* required.

Dielectric heating. During dielectric heating the power consumed, $P = 2\pi fCV^2 \cos\phi$ watts, where V is the supply voltage, f is the supply frequency and C is the capacitance of the condenser formed, which depends up the relative permittivity ε_r of the material.

Thus rate of heat production or power consumed $\alpha V^2 \,.\, f \,.\, \cos\phi \,.\, \varepsilon_r$. The voltage across any specimen is limited by its thickness *i.e.*, potential gradient, breakdown voltage, insulation and safety consideration.

Normally, voltages ranging from 600 V to 3000 V are employed for dielectric heating, however, sometimes voltages of the order 20 kV are also used.

Heat production rate can also be increased by applying *high frequencies* but it is also limited because of the following considerations :

(*i*) At higher frequencies it is difficult for tuning inductance resonate with the charge capacitance.

(*ii*) With the use of higher frequencies there is a possibility of formation of standing waves between the surface of two electrodes having wavelength nearly equal to or more than one quarter of the wavelength of the particular frequency used.

(*iii*) At higher frequencies it is almost impossible to get uniform voltage distribution.

(*iv*) As higher frequencies disturb near-by radio station services, special arrangement need be made to stop radiations from the high-frequency generator employed for the purpose.

(*v*) At higher frequencies it is essential to employ special matching circuit due to the fact that maximum power transfer takes place when the oscillator impedance equals the load impedance.

Table 2.2 shows the frequencies required for various heating purposes and the type of equipment used.

Table 2.2. Type of heating, frequency and source used

Type of heating	*Frequency used*	*Source used*
(1) ***Induction heating***		
(*i*) *Low temperature heating of metal, annealing*	*50 to 500 Hz*	*Rotating generator or a diode frequency converter.*
(*ii*) *Melting, deep heat penetration, 'through' heating.*	*500 HZ to 10 kHz*	*–do–*
(*iii*) *Surface heating of metals* *Hardening*	*10 to 200 kHz* *100 to 500 kHz*	*Spark gap generator.* *Spark gap generator, Vacuum tube oscillator.*
(*iv*) *Heating metal pieces, wire and metal strips*	*400 to 1000 kHz*	*Vacuum tube oscillator.*
(2) ***Dielectric Heating***	*1 to 50 MHz*	*Vacuum tube oscillator.*

• SCR inverter circuits can also be used for all these applications.

2.1.10. Infrared or Radiant Heating

In this type of heating, the elements are of tungsten filament lamps operating at about 2300°C as at *this temperature a greater proportion of infrared radiation is given* (Heating effect on the charge is greater since the temperature of the heating element is greater than in case of resistance heating). With the help of suitable reflectors (plated with rhodium) these infrared radiations are focussed on the surface to heated.

The lamps employed have ratings varying from 250 W to 1000 W operating at 115 V. *Lower voltage results in robust filaments.* With this arrangement, charge temperature obtained is between 200°C and 300°C. The heat emission intensity obtained is about 7000 W/m^2, which is much higher than that obtained with ordinary resistance furnace (1500 W/m^2).

In infrared heating, *heat absorption remains practically constant* whatever the charge temperature whereas it falls rapidly as the temperature of the charge rises in the ordinary resistance furnance.

- For getting best results, the infrared lamps should be located at a distance of 25-30 cm from the objects to be heated.

Applications :

(*i*) Drying of paints.

(*ii*) Drying of radio-cabinets and wood furniture.

(*iii*) Drying of pottery, paper, textiles etc. where moisture content is not large.

(*iv*) Low temperature heating of plastics.

(*v*) For various dehydration and other processes.

Advantages :

(*i*) Compactness of heating units.

(*ii*) Rapid heating.

(*iii*) Flexibility.

(*iv*) Safety.

2.2. ELECTRIC WELDING

2.2.1. Definition of Welding

Welding is the *process of joining two pieces of metal or non-metal at faces rendered plastic or liquid by the applicaton of heat or pressure or both. Filler material* may be used to *effect the union.*

All metals are weldable provided proper process and technique(s) are used. If either of these two aspects or both are overlooked, no good weld can be obtained.

2.2.2. Welding Processes

Welding processes are *classified* as follows :

I. **Fusion Welding :**

A. *Based on heat produced by "Electrical energy"*

1. Carbon arc welding :

(*i*) Shielded

(*ii*) Unshielded.

2. Metal arc welding :

(*i*) Shielded

(*a*) Shielded metal arc welding

(*b*) Inert gas welding

— TIG welding

— MIG welding

— MAG welding

(*c*) Submerged arc welding.

(*d*) Atomic hydrogen welding.

(*ii*) Unshielded

(*iii*) Other processes

(*a*) Plasma

(*b*) Laser.

B. *Based on heat produced by "Chemical energy"*

1. Gas welding (Oxy-acetylene)

2. Thermit welding.

II. **Non-fusion welding :**

A. *Based on heat produced by "Chemical energy"*

1. Forge welding

2. Oxy-acetylene non-fusion welding.

B. *Based on heat produced by "Mechanical energy"*

1. Friction welding
2. Ultrasonic welding.

C. *Based on heat produced by "Electrical energy"*

Resistance welding :

(*i*) Spot welding
(*ii*) Seam welding
(*iii*) Projection welding
(*iv*) Butt welding
(*a*) Flash welding
(*b*) Upset welding
(*c*) Stud welding.

Proper selection of the welding process depends on the following factors :

(*i*) Types of metals to joined.
(*ii*) Cost involved.
(*iii*) Nature of products to be fabricated.
(*iv*) Techniques of production adopted.

In welding, electricity is used to generate heat at the point of welding in order to melt the material which will subsequently fuse and form the actual weld joint. In order to produce this localised heat, the two most common methods (out of several ways) are :

1. ***Resistance welding.*** In this method *current is passed through the inherent resistance of the joint to be welded thereby generating the heat* as per the equation I^2Rt kilojoules.

2. ***Arc welding.*** In this method of producing localised heat *electricity is conducted in the form of an arc which is established between the two metallic surfaces.*

2.2.3. Resistance Electric Welding

It is the method of uniting two pieces of metal by the passage of a heavy electric current while the surfaces are pressed together. The fusing temperature is obtained by placing the surfaces to be joined in contact with one another, and passing a current of two to eight volts, at a high amperage through them. The *heat is developed around the point to which they touch, forcing them together* (by pressure mechanically applied), and at the same time *switching off the current, completes the weld.*

Some important *resistance welding processes* are :

1. Butt welding.
2. Flash welding.
3. Spot welding.
4. Seam welding
5. Projection welding.

Advantages. Some of the *advantages of resistance welding* are as under :

1. Both similar and dissimilar metals can be welded.
2. Rapid welding action.
3. Heat is localised when required.
4. Parent metal is not harmed.
5. Comparatively lesser skill is required.
6. Difficult shapes and sections can be welded.

7. Suitable for large quantity production.
8. No filler metal is required.

Disadvantages :

1. High initial cost.
2. High maintenance cost.

2.2.3.1. Butt welding

Refer to Fig. 2.19. In this type of welding which is employed to join bars and plates together end to end, one bar is held in a fixed clamp in the butt welding machine and the other bar in a movable clamp, the clamp being electrically insulated, the one from the other, and being connected to a source of current. When the two ends to be joined are brought into contact and current is switched on, the resistance at the joint causes the ends to heat up to welding temperature. Current is then switched off and the movable clamp forced up, so that a weld is made. The voltage applied across the clamps is a low one, from 2 to 6 volts, and the current is usually alternating.

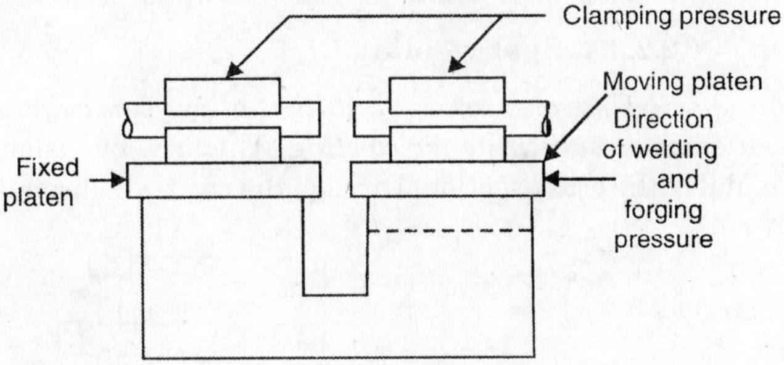

Fig. 2.19. Butt welding.

If the bars being joined are different in cross-section, the amounts they project from their clamps may have to be adjusted so as to modify the heat losses and ensure both bars being brought to the welding temperature simultaneously.

- *This process is being used for welding such things as steel rails whose cross-sectional area is as much as 6.25 cm*^{2}.

2.2.3.2. Flash welding

Refer to Fig. 2.20. In this process, the parts to be welded are clamped to the electrode fixtures, as in butt welding *but the voltage is applied before the parts are butted together.* As the *parts touch each other, an arc is established which continues as long as the parts advance at the correct speed.* This arc bursts away a portion of the material from each piece. When the welding temperature is reached, the speed of travel is increased, the power switched off and weld is upset.

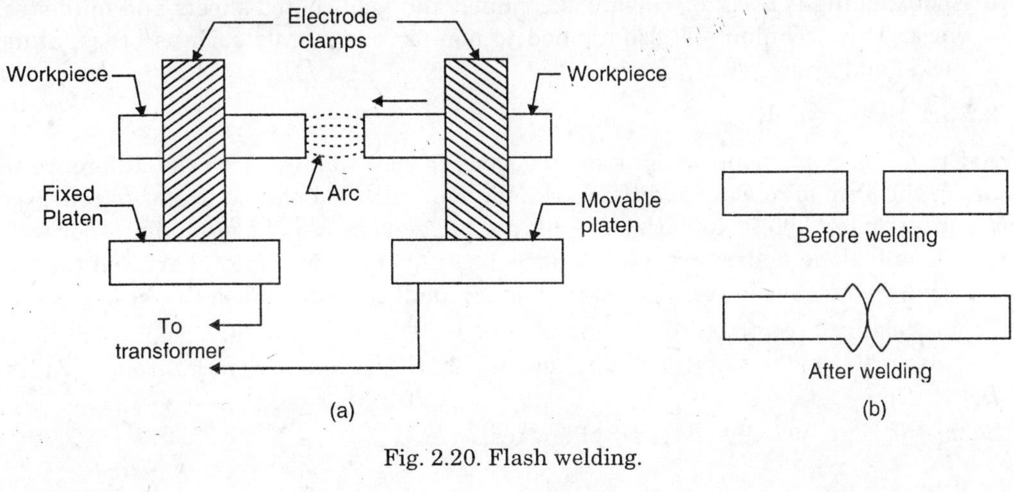

Fig. 2.20. Flash welding.

Flash butt welding claims the following *advantages* over upset method of welding ;

(*i*) Power consumed is less once the arc creates more heat with a given current.

(*ii*) The weld is made in clean virgin metal as the surfaces are burned away.

(*iii*) More quicker.

- *It is widely used in automobile construction on the body, axles, wheels, frames and other parts. It is also employed in welding motor frames, transformer tanks and many types of sheet steel containers such as at barrels and floats.*

2.2.3.3. Spot welding

Steel, brass, copper and light alloys can be joined by this method, which forms a cheap and satisfactory substitute for riveting. The area of fusion at each spot weld, in fact, is approximately equal to the cross-sectional area of the rivet which would be employed for a similar gauge of material.

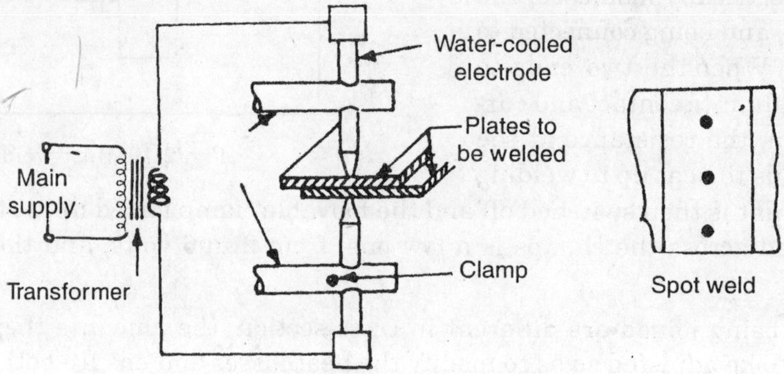

Fig. 2.21. Spot welding.

Refer Fig. 2.21. Spot welding, as the name implies, is carried out by overlapping the edges of two sheets of metal and fusing them together between copper electrode tips at suitably spaced intervals by means of a heavy electrical current. The resistance offered to the current as it passes through the metal raises the temperature of the metal between the electrodes to welding heat. The current is cut off and mechanical pressure is then applied by the electrodes to forge the weld. Finally the electrodes open.

When sheets of unequal thickness are joined, the current and pressure setting for the thinner sheets are used. Similarly four thickness may be welded, using the same settings as for two thickness.

- Spot welding is *used* for galvanized, tinned and lead-coated sheets and mild steel sheet work. This technique is also applied to non-ferrous metals such as brass, aluminium, nickel and bronze etc.

2.2.3.4. Seam welding.

Refer to Fig. 2.22. Seam welding is analogous to spot welding with the difference the electrodes are in the form of *rollers* ; and the work moves in direction perpendicular to roller axis. The current is interrupted 300 to 1500 times a minute to give a series of overlapping spot welds. The welding is usually done under water to keep the heating of the welding rollers and the work to a minimum, and thus to give lower roller maintenance and less distortion of the work.

Seam welding is confined to welding of *thin materials* ranging in thickness from 2 mm to 5 mm. It is also restricted to metals having *low hardenability rating* such as hot-rolled grades of low-alloy steels.

Seam welds are usually tested by pillow test.

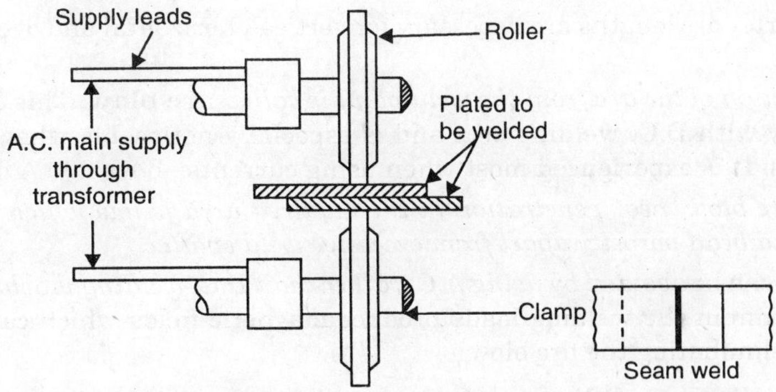

Fig. 2.22. Seam welding.

- *It is employed on many type of pressure (light or leak proof) tanks, for oil switches, transformers, refrigerators, evaporators and condensers, aircraft tanks, paint and varnish containers etc.*

2.2.3.5. Projection welding

Refer to Fig. 2.23. It is in effect, a form of *multi-spot welding* in which a number of welds are made simultaneously. The pieces to be welded are arranged between two flat electrodes which exert pressure as the current flows. The projections, and the areas with which they make contact, are raised to welding heat and are joined by the pressure exerted by the electrodes. The projections are flattened during the welding.

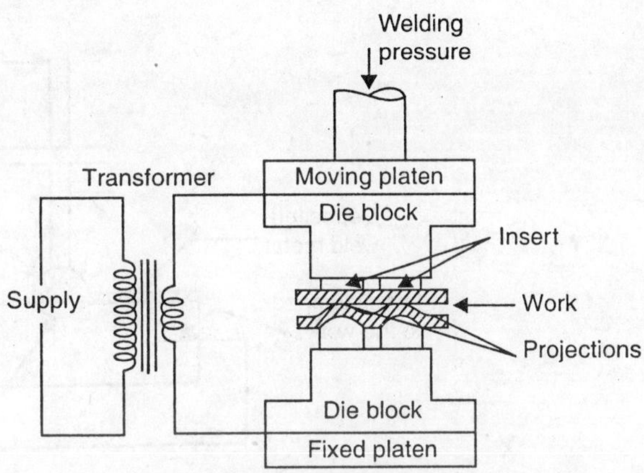

Fig. 2.23. Projection welding.

2.2.4. Electric Arc Welding

Electric arc welding *is the system in which the metal is melted by the heat of an electric arc.* It can be done with the following *methods* :

1. Metallic arc welding.
2. Carbon arc welding.
3. Atomic hydrogen welding.
4. Shielded arc welding.

- In metal arc welding , a *fairly short arc length is necessary for getting good welds. Short arc length permits the heat to be concentrated on the workpiece, is more stable because effect of magnetic blow is reduced and the vapours from the arc surround the electrode metal and the molten pool thereby preventing air from destroying the weld metal.*
- The length of arc required for welding depends on :
  - — kind of electrode used, its coating, its diameter,
  - — position of welding ;
  - — amount of current used.

Usually, shorter arc lengths are necessary for vertical, horizontal and overhead welding than for flat welding.

- *The deflection of the arc from the weld point is called* **arc blow.** This condition is encountered only with D.C. welding sets and is especially noticeable when welding with bare electrodes. It is experienced most when using currents above 200 A or below 40 A.

  Due to arc blow, heat penetration in the required area is low which leads to incomplete fusion and bead porosity apart from excessive weld spatter.

  Arc blow can be *avoided by using A.C. rather than D.C. welding machines* because reversing currents in the welding leads produce magnetic fields which cancel each other out thereby eliminating the arc blow.

2.2.4.1. Metallic arc welding

Refer to Fig. 2.24. In metallic arc welding an arc is established between work and the filler metal electrode. The intense heat of the arc forms a molten pool in the metal being welded, and at the same time melts the tip of the electrode. As the arc is maintained, molten filler metal from the electrode tip is transferred across the arc, where it fuses with the molten base metal.

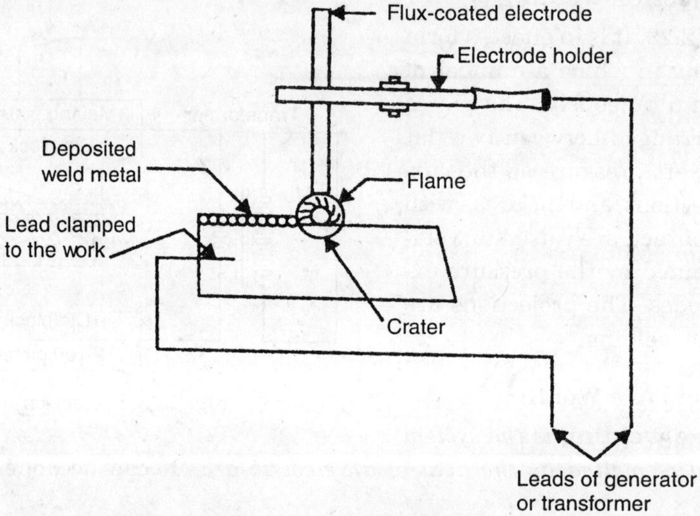

Fig. 2.24. Metallic arc welding.

Arc may be formed with direct or alternating current. Petrol or diesel driven generators are widely used for welding in open, where a normal electricity supply may not be available. D.C. may also be obtained from electricity mains through the instrumentality of a transformer and rectifier. A simple transformer is, however widely employed for A.C. arc welding. *The transformer sets are cheaper and simple having no maintenance cost as there are no moving parts.* With arc system, the covered or coated electrodes are used, whereas with D.C. system for cast iron and non-ferrous metals, base electrodes can be used. In order to strike the arc an open circuit voltage of between 60 to 70 volts is required. For maintaining the short arc 17 to 25 volts are necessary ; the current required for welding, however, varies from 10 amp. to 500 amp. depending upon the class of work to be welded.

The great *disadvantage* entailed by D.C. welding is the presence of *arc blow* (distortion of arc stream from the intended path owing to magnetic forces of a non-uniform magnetic field). With A.C. arc blow is considerably reduced and use of higher currents and large electrodes may be restored to enhance the rate of weld production.

- *The field of application of metallic arc welding includes mainly low carbon steel and the high-alloy austenitic stainless steel.* Other steels like low and medium-alloy steels can however be welded by this system but many precautions need be taken to produce ductile joints.

The *comparison* between A.C. and D.C. arc welding is given in the table below :

S. No.	Aspects	A.C. welding	D.C. welding
1.	*Power consumption*	Low	High
2.	*Arc stability*	Arc unstable	Arc stable
3.	*Cost*	Less	More
4.	*Weight*	Light	Heavy
5.	*Efficiency*	High	Low
6.	*Operation*	Noiseless	Noisy
7.	*Suitability*	Non-ferrous metals cannot be joined	Suitable for both ferrous and non-ferrous metals
8.	*Electrode used*	Only coated	Bare electrodes are also used
9.	*Welding of thin sections*	Not preferred	Preferred
10.	*Miscellaneous*	Work can act as cathode while electrode acts as anode and *vice versa*	Electrode is always negative and the work is positive

Specifications :

A.C. transformer : Step down, oil cooled = 3 phase, 50 Hz ; Current range = 50 to 400 A ; Open circuit voltage = 50 to 90 V ; Energy consumption = 4 kWh per kg of metal deposit ; Power factor = 0.4 ; Efficiency = 85%.

D.C. generator : Motor generator–3 phase, 50 Hz ; Current range = 125 to 600 A ; Open circuit voltage = 30 to 80 V ; Arc voltage = 20 to 40 V ; Energy consumption = 6 to 10 kWh/kg of deposit ; Power factor = 0.4 ; Efficiency = 60%.

Welding positions :

The following are the *four basic positions* in which manual arc welding is done :

1. Flat position
2. Horizontal position
3. Vertical position
4. Overhead position.

1. ***Flat position.*** Refer to Fig. 2.25.

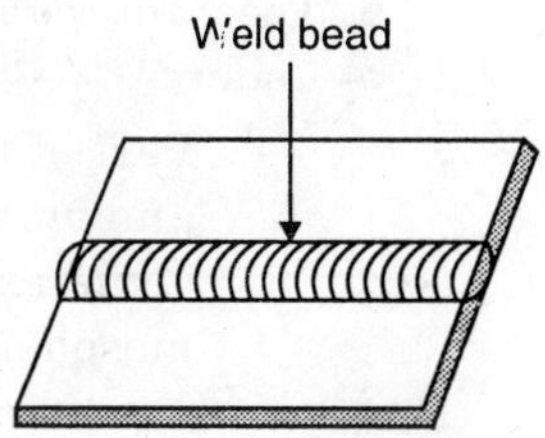

Fig. 2.25. Flat position.

- Of all the positions, flat position is the *easiest, most economical and the most used for all shielded arc welding*.
- Weld beads are exceedly smooth and free of slag spots.
- It provides the strongest weld joints.
- This position is most adaptable for *welding of both ferrous and non-ferrous metals particularly for cast iron.*

2. ***Horizontal position.*** Refer to Fig. 2.26

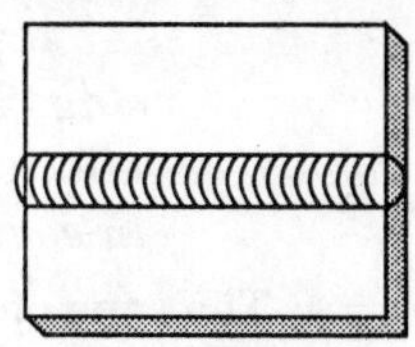
Fig. 2.26. Horizontal position.

- This position is the second most popular position.
- It also requires a short arc length because it helps in preventing the molten puddle of the metal from *sagging*.
- While welding in horizontal position the major errors that occur are : undercutting and over-lapping of weld zone.

3. ***Vertical position.*** Refer to Fig. 2.27.

- In this position, the welder can deposit the bead either in uphill or downhill direction.
- *Uphill welding* is suited for *thick metals* because it produces stronger welds.
- *Downhill welding* is preferred for *thin metals* because it is faster than uphill welding.

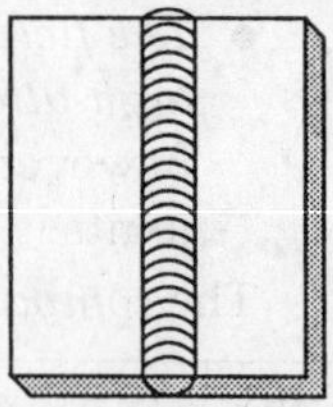

Fig. 2.27. Vertical position.

4. ***Overhead position.*** Refer to Fig. 2.28.

- In this position, the welder has to be very cautious otherwise he may get burnt by drops of falling metal.
- This position is thought to be most dangerous but not the most difficult one.

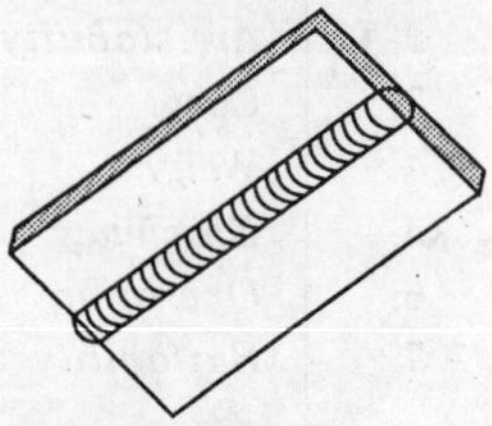

Fig. 2.28. Overhead position.

Electrodes :

An ***electrode*** *is a filler metal in the form of a wire or rod which is either bare or coated uniformly with the flux.* As per IS : 814-1970, the contact end of the electrode (Fig. 2.29) is left bare and clean to a length of 20-30 mm for inserting it into electrode holder.

Depending on the thickness of the flux coating, coated electrodes way be *classified* as follows :

1. *Lightly-dusted (or dipped)* electrodes.
2. *Semi-coated (or heavy coated)* electrodes.

The following materials are commonly used for coating :

(*i*) Titanium oxide (*ii*) Ferromanganese

(*iii*) Silica, flour (*iv*) Asbestos clay

(*v*) Calcium carbonate

(*vi*) Cellulose with sodium silicate often used to hold ingredients together.

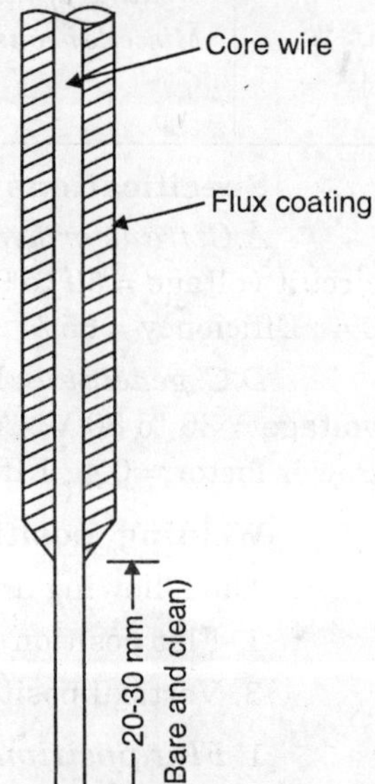

Fig. 2.29. Electrode for metallic arc welding.

- The *electrode coating helps improving the weld quality* as under :
  — Part of the coating burns in the intense heat of the arc and *provides a gaseous shield around the arc* which prevents oxygen, nitrogen and other impurities in the atmosphere from combining with the molten metal to cause a poor quality brittle and weak weld.
  — Another portion of the coating flux *melts and mixes with the impurities* in the molten metal causing them to float to the top of the weld where they cool in the form of *slag*. This slag *improves the bead quality* by protecting it from the contaminating effects of the atmosphere and causing it to cool down more uniformly. It also *helps in controlling the basic shape of the weld bead.*
- The type of electrode used depends on the following *factors* :

(*i*) The type of the metal to be welded. (*ii*) The welding position.

(*iii*) The type of electric supply (A.C. or D.C.). (*iv*) The polarity of the welding machine.

Advantages of coated electrodes :

The following are the principle *advantages* of using electrode coating :

(*i*) Arc is *stabilized* (since the coating contains compounds of sodium and potassium).

(*ii*) The impurities present on the surface being welding are *fluxed away*.

(*iii*) *Slag is formed* over the weld which protects it from atmospheric contamination, makes it cool uniformly thereby reducing the changes of brittleness and provides a smoother surface by reducing 'ripples' caused by the welding operation.

(*iv*) *Sputtering* of metal during welding is *prevented.*

(*v*) Welding operation becomes faster due to increased melting rate.

(*vi*) The coating makes it possible for the electrode to be used on A.C. supply.

- Since the efficiency of all coated or covered electrodes is *impared by the dampness*, these must always be *stored in a dry place*. In case, dampness is suspected the electrodes should be dried for a few hours in a warm cabinet.

Types of joints and types of applicable welds :

B.I.S. (Bureau of Indian Standards) has recommended the following types of joints and the welds applicable to each one of them (Fig. 2.30).

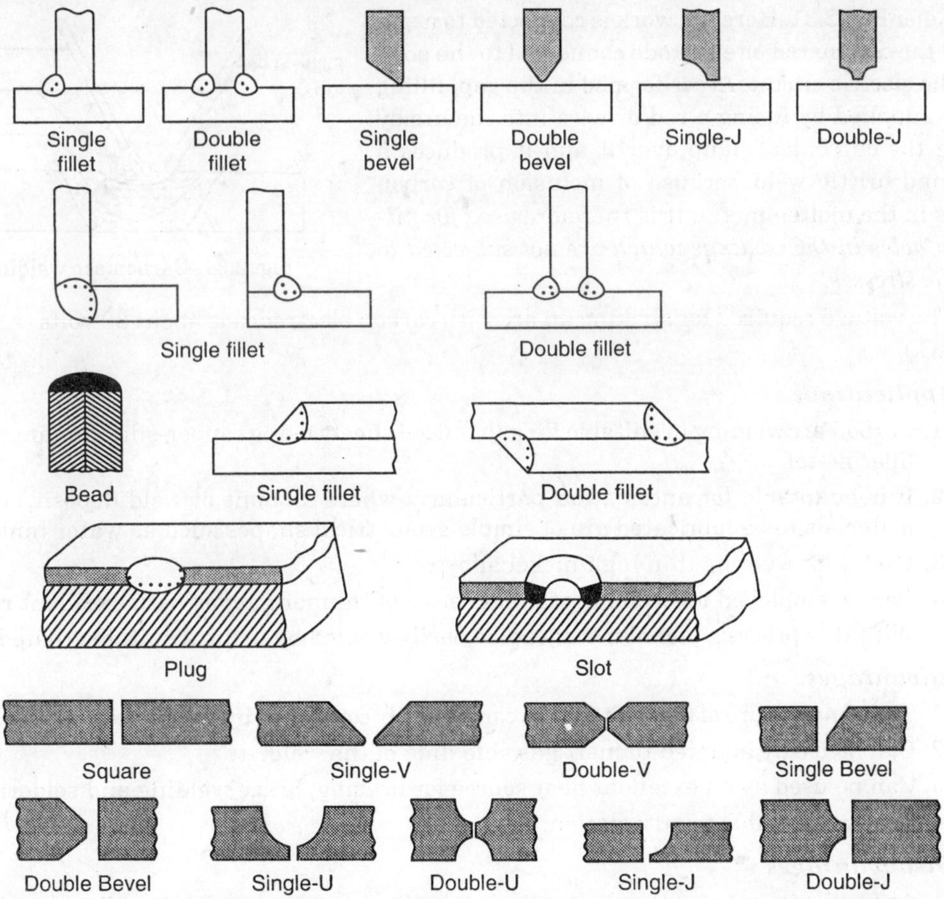

Fig. 2.30. Types of joints and types of applicable welds.

1. *Tee joint*—with six types of welds.
2. *Corner joint*—with two types of welds.
3. *Edge joint*—with one type of weld.
4. *Lap joint*—with four types of welds.
5. *Butt joint*—with nine types of welds.

Arc welding machines :

Following are the two general types of *arc welding machines* :

1. *D.C. welding machines* :

(*i*) Motor-generator set

(*ii*) A.C. transformers with rectifiers

2. *A.C. welding machines.*

2.2.4.2. Carbon arc welding

Carbon arc was the first electric welding process developed by French inventor Auguste de Meritens in 1881.

"Carbon arc welding" differs from common shield metal arc welding in that it uses non-consumable carbon-graphite electrodes instead of consumable flux-coated electrodes.

Refer Fig. 2.31. Here the work is connected to negative and the carbon rod or electrode connected to the positive of the electric circuit. Arc is formed in the gap, filling metal is supplied by fusing a rod or wire into the arc by allowing the current to jump over it and it produces a porous and brittle weld because of inclusion of carbon particles in the molten metal. It is therefore *used for filling blow holes in the castings which are not subjected to any of the stresses.*

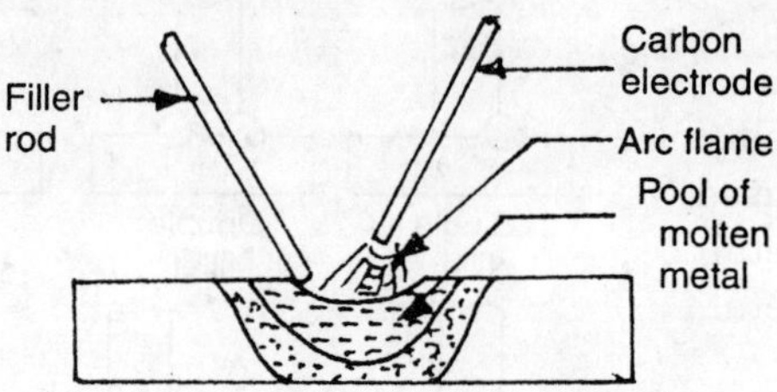

Fig. 2.31. Carbon arc welding.

The voltage required for striking an arc with carbon electrodes is about 30 volts (A.C.) and 40 volts (D.C.).

Applications :

1. Carbon arc welding in suitable for galvanised sheets using copper-silicon-manganese alloy filler metal.
2. It is adaptable for automation particularly where amount of weld deposit is large and materials to be fabricated are of simple geometrical shapes such as water tanks.
3. Useful for welding thin high-nickel alloys.
4. Can be employed for welding stainless steel of thinner gauges with excellent results.
5. With this process, monel metal can be easily welded by using a suitable coated filler rod.

Advantages :

1. The temperature of the molten pool can be easily controlled by simply varying the arc length.
2. Can be easily adapted to inert gas shielding of the weld.
3. Can be used as an excellent heat source for brazing, braze welding and soldering etc.
4. Easily adaptable to automation.

Disadvantages :

1. Approximately twice the current is required to raise the work to welding temperature as compared with metal electrode, while a carbon electrode can only be used economically on D.C. supply.

2. A separate filler rod has to be used if any filler material is required.
3. Blow holes occur due to magnetic arc blow especially when welding near edges of the workpiece.
4. Since arc serves only as a heat source it does not transfer any metal to help reinforce the weld joint.

Electrodes :

- Electrodes are made of *either carbon or graphite*, are usually 300 mm long and 2.5 mm to 12 mm in diameter.
- Graphite electrodes are harder, more brittle and last longer than carbon electrodes. They can withstand higher current densities but their arc column is harder to control. Though considered non-consumable, they do disintegrate gradually due to vapourisation and oxidisation.

2.2.4.3. Atomic hydrogen welding

Refer to Fig. 2.32. In this system heat is obtained from an alternating current arc drawn between *two tungsten electrodes in an atmosphere of hydrogen.* As the hydrogen gas passes through the arc, the hydrogen molecules are broken up into atoms and they recombine on contact with the cooler base metal generating intense heat sufficient to melt the surfaces to be welded, together with the filler rod, if used. The envelop of hydrogen gas also shields the molten metal from oxygen and nitrogen and thus prevents weld metal from deterioration.

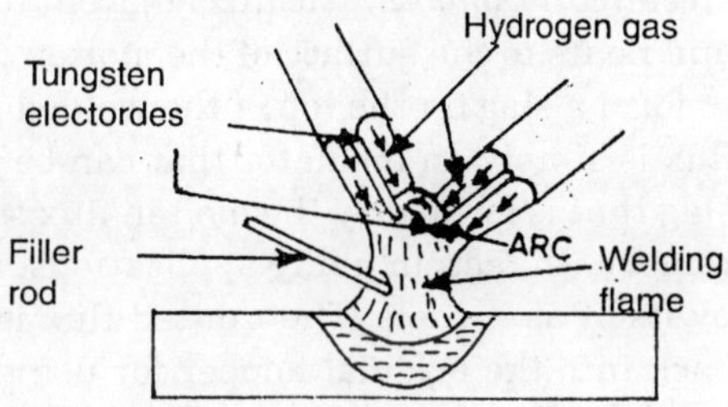

Fig. 2.32. Atomic hydrogen welding.

Applications. Atomic hydrogen welding, *being expensive is used mainly for high grade work on stainless steel and most non-ferrous metals.*

Advantages :

1. Gives strong, ductile and sound welds.
2. Quite thick sections can be welded.
3. Arc and weld zone arc shrouded by burning hydrogen which, being an active reducing agent, protects them from atmospheric contamination.
4. Can be employed for materials too thin for gas welding.

2.2.4.4. Shielded arc welding

Refer to Fig. 2.33. In this system molten weld metal is protected from the action of atmosphere by an envelope of chemically reducing or inert gas.

As molten steel has an affinity for oxygen and nitrogen, it will, if exposed to the atmosphere, enter into combination with these gases forming oxides and nitrides. Due to this injurious chemical combination metal becomes weak, brittle and corrosion resistance. Thus several methods of shielding have been developed. The simplest (Fig. 2.33) is the use of a flux coating on the electrode which in addition to producing a slag which floats on the top of the molten metal and protects it from atmosphere, has organic constituents which turn away and produce an envelope of inert gas around the arc and the weld.

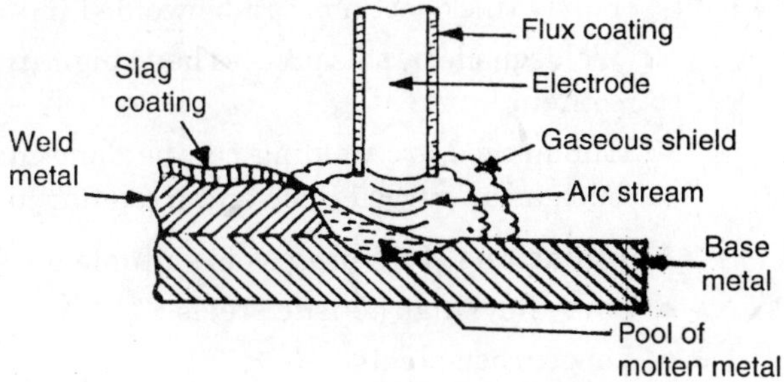

Fig. 2.33. Shielding arc welding.

- Welds made with a completely shielded arc are more superior to those deposited by an ordinary arc.

2.2.5. Submerged Arc Welding

The *submerged arc process* (which may be done manually or automatically) *creates an arc column between a base metallic electrode and the workpiece.* The arc, the end of the electrode, and the molten weld pool are *submerged in a finely divided granulated powder that contains appropriate deoxidizers, cleansers and any other fluxing elements.* The fluxing powder is fed from a hopper that is carried on the welding head. The tube from the hopper spreads the powder in continuous mount in front of the electrode along the line of the weld. This *flux mound is of sufficient depth to submerge completely the arc column so that there is no splatter or smoke,* and the weld is shielded from all effects at atmospheric gases. As a *result of this unique protection, the weld beads are exceptically smooth.* The flux adjacent to the arc column melts and floats to the surface of the molten pool ; then it solidifies to form a slag on the top of the welded metal. The rest of the flux is simply an insulator that can be reclaimed easily. The slag that is formed by the molten flux solidifies and is easy to remove. In fact, in many applications, the slag will crack off by itself as it cools. The unused flux is removed and placed back into the original hopper for use the next time. Granulated flux is a complex, metallic silicate that can be used over a wide range of metals.

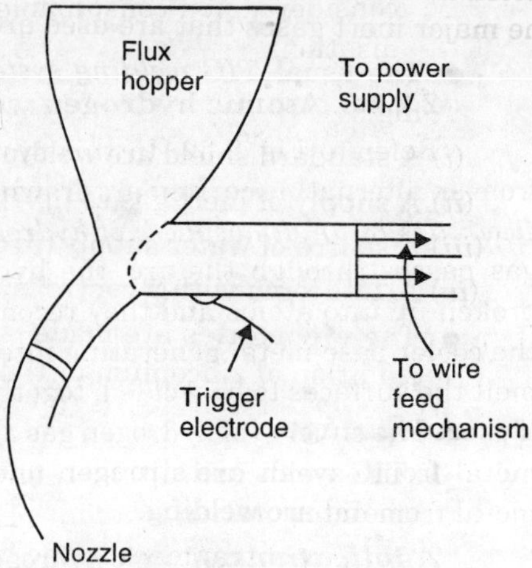

Fig. 2.34. Apparatus used in manual submerged arc welding.

The process is characterised by high welding currents. The current density in the electrode is 5 to 6 times that used in ordinary manual stick electrode arc welding, consequently the melting rate of the electrode as well as the speed of welding is much higher than in the manual stick electrode process.

Fig. 2.34 shows an apparatus used in manual submerged arc welding.

Both normal and automatic submerged arc processes are most suited for flat and slightly downhill welding positions.

Advantages :

1. Fairly thick sections can be welded in a single pass without edge preparation.
2. Welds made by this process have high strength and ductility with low hydrogen or nitrogen content.
3. Submerged arc welding can be done manually where automatic process is not possible such as on curved lines and irregular joints.

Applications. This process is suitable for *welding* :

- Low-alloy, high-tensile steels ;
- Low carbon steels ;
- Nickel, monel and other non-ferrous metal like copper ;
- High-strength steels and corrosion resistance steels etc.
- This process is also capable of welding *fairly thin gauge materials.*
- "*Industrial applicatons*" include *fabrication* of *pipes, boiler, pressure vessels, railroad tank cars, structural shapes* etc. which demand *welding in a straight lines.*

2.2.6. Tungsten Inert-Gas (TIG) Welding

Refer to Fig. 2.35. In this process the heat necessary to melt the metal is provided by a very intense electric arc which is struck between a virtually *non-consumable tungsten electrode* and *metal workpiece*. The *electrode does not melt* and become a part of the weld. On joints where filler metal is required, a welding rod is fed into the weld zone and melted with base metal in the same manner as that used with oxyacetylene welding. The weld zone is shielded from the atmosphere by an inert-gas (a gas which does not combine chemically with the metal being welded) which is ducted directly to the weld zone where it surrounds the tungsten. The major inert gases that are used are *argon* and *helium*.

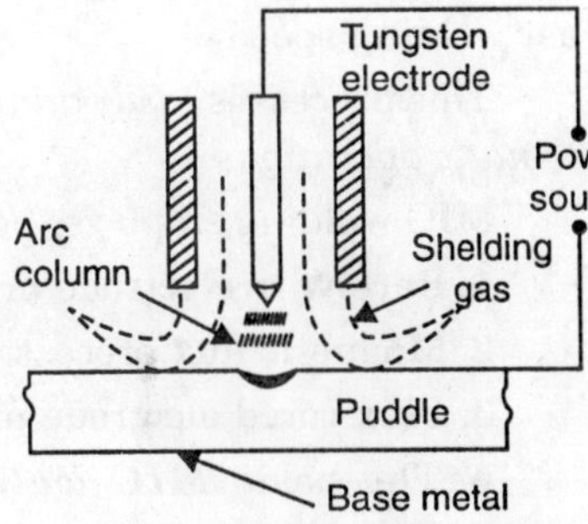

Fig. 2.35. Tungsten inert-gas (TIG) welding.

- The usual *TIG welding system* consists of the following :

(*i*) A standard shield are welding machine complete with cables etc.

(*ii*) A supply of inert gas complete with hose, regulators etc.

(*iii*) A source of water supply (in case of water-cooled torches).

(*iv*) A TIG torch with a control switch to which all the above are connected.

- The **electrodes** are made of either pure tungsten or zinconiated or thoriated tungsten. Addition of zinconium or thorium (0.001 to 2%) greatly improves electron emission.

Advantages :

1. TIG welds are stronger, more ductile and more corrosion-resistant than those of shield metal arc welding.
2. No flux entrapment in the bead, since no flux is used.
3. A wide veriety of joint design can be used, since no flux is required.
4. TIG welding provides protection to weld bead from atmospheric contamination.
5. No post-weld cleaning is necessary.
6. No weld splatter or sparks that could damage the surface of the base metal.
7. Suitable for all welding positions (*i.e.*, flat, horizontal, vertical and overhead).
8. Suitable for welding food or medical containers where entrapment of any decaying organic matter could be extremely harmful.
9. Relatively fast welding speeds.

Applications :

1. The TIG process lends itself ably to the fusion welding of *aluminium and its alloys, stainless steel, magnesium alloys, nickel base alloys, copper base alloys, carbon steel and low alloy steels.*
2. TIG welding can also be used for the combining of dissimilar metals, hard facing, and the surfacing of metals.

2.2.7. Metal Inert-Gas (MIG) Welding

Refer to Fig. 2.36. The inert-gas *consumable electrode process*, or the MIG process is a refinement of the TIG process, however, in this process, the *tungsten electrode has been replaced with a consumable electrode.* The electrode is driven through the same type of collet that holds a tungsten electrode by a set of drive wheels. The *consumable electrode in MIG process acts as the source for the arc column as well as the supply for the filler material.*

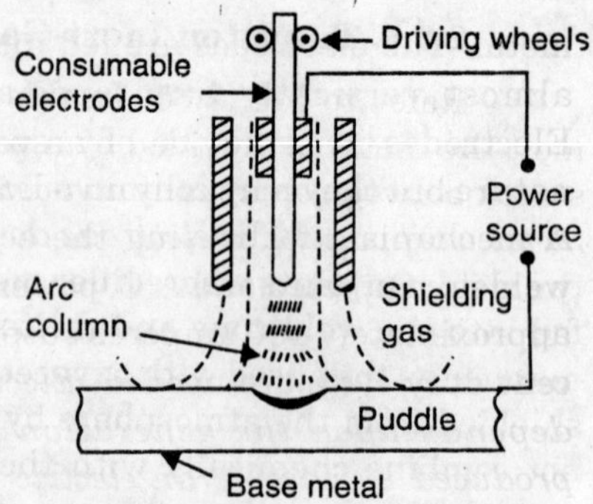

Fig. 2.36. Metal inert gas (MIG) welding.

This process *can deposit large quantities of weld metal* at a fast welding speed.

This process is *easily adaptable to semi-automatic or fully automatic operations.*

MIG welding employs the following *three basic processes.*

1. Bare-wire electrode process.
2. Magnetic flux process.
3. Flux-cored electrode process.

- The basic ***MIG welding system*** consists of the following :

(*i*) Welding power supply.

(*ii*) Inert gas supply with a regulator and flow meter.

(*iii*) Wire feed unit containing controls for wire feed, gas flow and the ON/OFF switch for MIG torch.

(*iv*) MIG torch.

(*v*) A water cooling unit, depending on amperage.

Advantages :

1. This welding process requires no flux.
2. Gives high metal deposit rates (varying from 2 to 8 kg/h).
3. Adaptable for manual and automatic operations.
4. Requires no post-welding cleaning.
5. Easy to operate (requiring comparatively much less operating skill).
6. Can be employed for a wide range of metals both ferrous and non-ferrous.
7. Provides complete protection to weld bead from atmospheric contamination.
8. Amply suited for horizontal, vertical and overhead welding positions.

Applications :

(*i*) Practically all commercially available metals can be welded by this method.

(*ii*) It can be used for deep groove welding of plates and castings, just as the submerged arc process can, but it is more advantageous on light gauge metals where high speeds are possible.

2.2.8. Electro-Slag and Electro-Gas Welding

These methods are employed to *fuse two sections of thick metal,* forming a seam in a *single pass.* Elimination of the need for making multiple passes and special joint preparations make these methods commonly used welding processes when heavy ferrous metals are to be joined. These processes have reduced costly time in fabrication of *large vessels and tanks.* There is theoretically no limit to the thickness of the weld bead.

2.2.8.1. Electro-slag welding

Refer to Fig. 2.37. This process is a vertical and uphill, two copper shoes, dams, or moulds must be placed on either side of the joint that is to be welded in order to keep the molten metal in the joint area. One or more electrodes may be used to weld a joint, depending upon the thickness of the metal. The electrodes are fed into the weld joint almost vertically from special wire guides. Electrodes need not be of a special deoxidized nature but they may contain a flux, if it is needed. A mechanism for raising the equipment as the weld is completed and A.C. power source that has approximately 100 amperes output and a 100 per cent duty cycle are needed. *Electro-slag welding depends upon the generation of heat that is produced by passing an electric current through molten* slag.

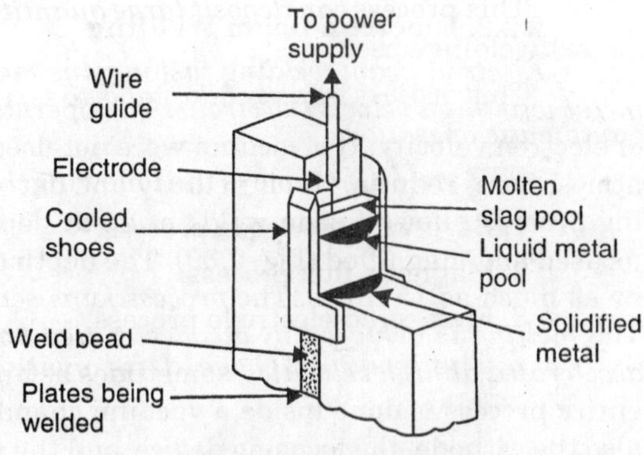

Fig. 2.37. Electro-slag welding.

Advantages **:**

1. Lower flux consumption.
2. High deposit rate (upto 20 kg of weld metal per hour).
3. Owing to uniform heating of the weld area, distribution and residual stresses are reduced to the minimal amounts.
4. No special joint preparation is required.
5. Welding is accomplished in a single pass rather than in costly multiple passes.
6. Theoretically, there is no maximum thickness of the plate it can weld.
7. There is also no theoretical upper limit to the thickness of the weld bead.
8. Requires less electrical power per kg of deposited metal than either the submerged arc welding process or the shield arc process.

2.2.8.2. Electro-gas welding

Refer to Fig. 2.38. Electro-gas welding works on the same general principle as electro-slag welding, with *the addition of some of the principles of submerged arc welding.* The major difference between electro-slag and electro-gas welding is that an inert gas, such as CO_2, is *used to shield the weld from oxidation, and there is continuous arc, such as in submerged arc welding, to heat the weld pool.* The joints and the use of flux to cleanse the weld are the same as in electro-slag process. The shoes that are used to form the weld, as in electro-slag process, are also used in the electro-gas process to control the weld zone through water cooling. However, the flux, instead of being issued to the weld zone through a hopper mechanism, is incorporated within the electrode itself in the form of cored wires.

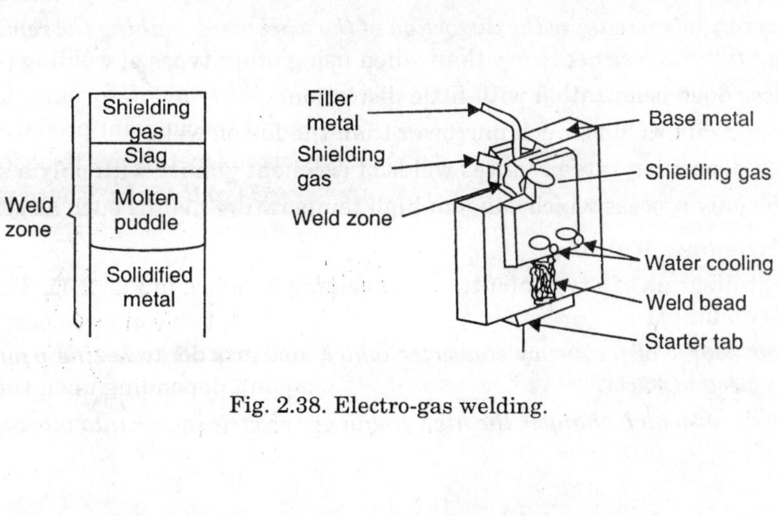

Fig. 2.38. Electro-gas welding.

2.2.9. Electron-beam Welding

Electron-beam welding fusion joins metal by bombarding a specific confined area of the base metal with high velocity electrons. The operation is performed in a vacuum to prevent the reduction of electron velocity. If a vacuum were not used, the electrons would strike the small particles in the atmosphere, reducing their velocity and decreasing their heading ability. The electron beam welding process allows fusion welds of great depth with a minimum width because the beam can be focused and magnified (Fig. 2.39). The depth of the weld bead can exceed the width of the weld bead by as much as 15 times. The *process joins separate pieces of base metal by fusing of molten metals.* The *melting is achieved by a concentrated bombardment of a dense stream of electrons, which are accelerated at high velocities,* sometimes as high as the speed of light. Under most circumstances the entire process is done inside a vacuum chamber. Most chambers house not only the workpiece but also the cathode, the focusing device and the remainder of the gun, preventing contamination of the weldment and the electron-beam gun itself (Fig. 2.40).

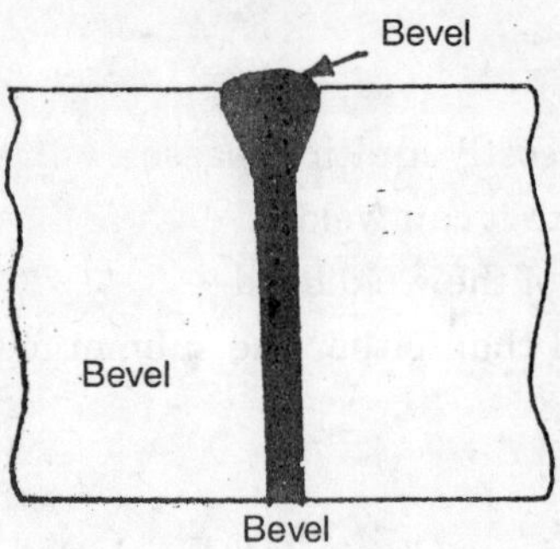

Fig. 2.39. Electro-beam welding.

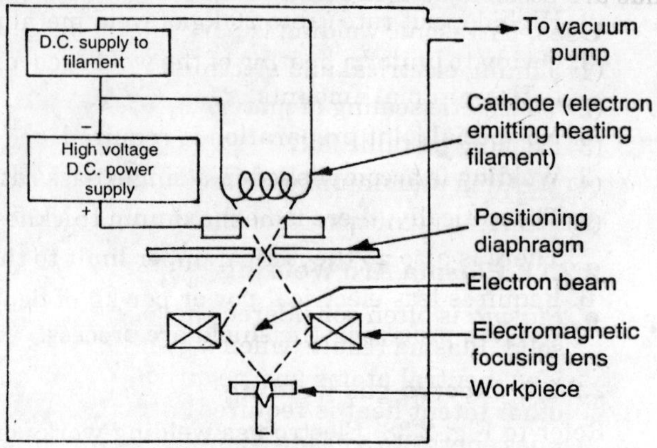

Fig. 2.40. Electron-beam gun.

Advantages :

1. The greatest advantage of electron-beam welding is that *it eliminates contamination of both the weld zone and the weld bead* because of the vacuum in which the weld is done because of the electrons doing the heating.
2. Even though initial costs are high, *operating costs are low* due to the low power usage. Many of the more costly fabrication methods could be replaced by electron beam process.
3. The narrow beam *reduces the distortion of the workpiece, making the replacement of costly jigs and fixtures less necessary* than when using other types of welding processes.
4. Produces deep penetration with little distortion.
5. Electron-beam weld is much narrower than the fusion weld.
6. Its high deposition rate produces welds of excellent quality with only a single pass.
7. It is the only process which can join high temperature metals such as columbium.

2.2.10. Ultrasonic Welding

A schematic diagram of a typical ultrasonic welding is shown in Fig. 2.41. The welding equipment consists of two units :

(*i*) *A power source of frequency converter which converts 50 cycle line power into high frequency electric power.*

(*ii*) *A transducer which changes the high frequency electric power into vibratory energy.*

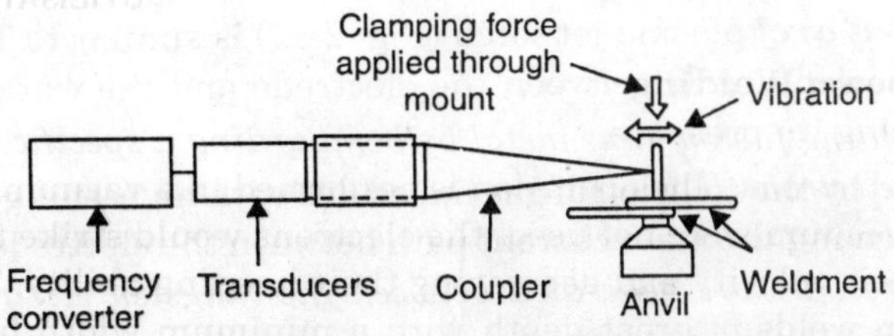

Fig. 2.41. Ultrasonic welding.

The components to be joined are simply clamped between a welding tip and supporting anvil with just enough pressure to hold them in close contact. The high frequency vibratory energy is then transmitted to the joint for the required period of time. The bonding is accomplished without applying external heat, filler rod or melting metal. Either spot-type welds or continuous-seam welds can be made on a variety of metals ranging of thickness from 0.000425 mm (aluminium foil) to 0.25 mm. Thicker sheet and plate can be welded if the machine is specifically designed for them. High strength bonds are possible both in similar and dissimilar metal combinations.

Uses. Ultrasonic welding is particularly adaptable for :

(1) Joining electrical and electronic components.

(2) Thermetic sealing of materials and devices.

(3) Splicing metallic foil.

(4) Welding aluminium wire and sheet.

(5) Fabricating nuclear fuel elements.

2.2.11. Plasma Arc Welding

- *Plasma* is often considered the *fourth state of matter.* The other three are gas, liquid and solid. Plasma results when a gas is heated to high temperature and changes into positive ions, neutral atoms and negative electrons. When a matter passes from one state to another latent heat is required to change into steam, and similarly, the plasma torch supplies energy to a gas to change it into plasma. When the *plasma changes back to a gas, the heat is released. Any high current arc is composed of plasma, which is nothing more than an ionized conducting gas.* The plasma gas is forced through the torch, surrounding the cathode. The main function of the plasma gas is shielding the body of the torch from the extreme heat of the cathode. Argon and argon mixtures are most commonly used (since they do not attack tungsten or copper cathode).
- *Plasma arc consists of an electronic arc plasma gas, and gases used to shield the jet column. The equipment necessary for plasma arc welding includes a conventional D.C. power supply with a drooping volt ampere output and with 70 open line volts.*
- The two main types of torches for welding and cutting with plasma arc are :

  (*i*) Transferred arc, and

  (*ii*) Non-transferred arc.

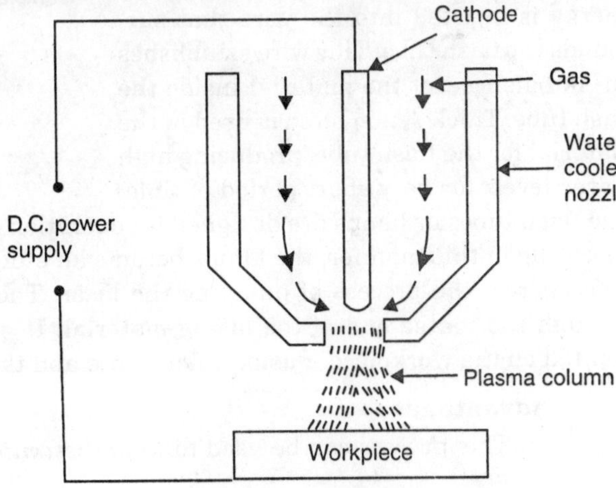

Fig. 2.42. Transferred arc plasma jet torch.

- The ***'transferred arc'*** plasma jet torch (Fig. 2.42) is similar to TIG torch, except that it has the water-cooled nozzle between the electrode and the work. This nozzle constricts the arc, increasing its pressure.

The plasma, caused by the collision of gas molecules with high-energy electron, is then swept out through the nozzle, forming the main current path between the electrode and the workpiece. *The plasma arc and transferred arc are generated between the tungsten electrode or cathode and the workpiece, or anode.*

- The ***'non-transferred arc'*** torch extends the arc from the electrode, or the cathode, to the end of the nozzle. *The nozzle acts as the anode. This type of plasma jet is completely independent of the workpiece, with the power supply contained with the equipment.*

Kay-hole method is used in actual process of welding with plasma jet. Jet column burns a small hole through the materials to be welded. As the torch progresses along the material the hole progresses also. However, it is filled by molten metal as the torch passes. Hundred per cent penetration is ensured by this method.

2.2.12. Laser Beam Welding

- The "*laser welding process*" is the *focusing of a monochromatic light into extremely concentrated beams.* It employs a carefully *focused beam of light that concentrates tremendous amount of energy on a small area to produce fusion.*

Refer to Fig. 2.43. The laser welding system comprises the following :

1. Electrical storage unit
2. Capacitor bank
3. Triggering device
4. Flash tube that is wrapped with a wire
5. Lasing material
6. Focusing lens mechanism
7. Work-table (operatable in three axes X, Y and Z).

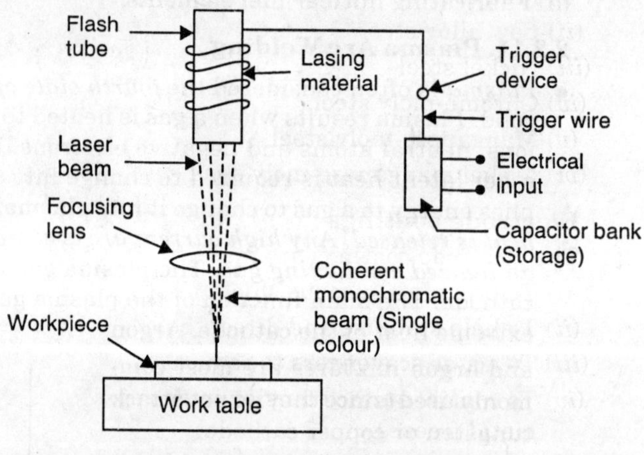

Fig. 2.43. Laser beam welding.

When capacitor bank is triggered energy is injected into the wire that surrounds the flash tube. This wire establishes an imbalance in the material inside the flash tube. Thick xenon often is used in the material for the flash tube, producing high power levels for very short period of time. The flash tubes or lamps are designed for operation at a rate of thousands of flashes per second. By operating in this manner, the lamps become an efficient device for converting *electrical energy into light energy,* the process of pumping the laser. The laser is then activated. The beam is emitted through the coated end of the lasing material. It goes through a focusing device where it is pinpointed on the workpiece. Fusion takes place and the weld is accomplished.

Advantages :

(*i*) This process can be used to weld *dissimilar metals with widely varying physical properties.*

(*ii*) *Metals with relatively high electrical resistance* and parts of *considerably different sizes and mass can be welded.*

(*iii*) Because the laser is simply a beam, *no electrode is required,* so that any part in a particular position can be welded if there is a direct line of sight from beam to the workpiece.

(*iv*) Welds can be made with a *high degree of precision* and on material that is only a few thousands of a centimetre thick.

(*v*) Laser welding holds thermal distortion and shrinkage to a minimum.

Disadvantage. Energy losses are high.

2.2.13. Electrodes—General Aspects

- Proper choice of electrodes is a vital point in getting good welds. Good preliminary precautions are of no use by improper choice of electrodes.
- The electrodes have two parts. One is *core wire* which contains the metal to be deposited. The second one is *coating. The heat causes the coating to emit gases, which cover the weld and prevent bare metal from oxidation.* Contamination with water impares the effectiveness of coating and hence the *electrodes must be kept in fairly dry places.* Some electrodes do not have any coating.
- *Hand operated electrodes* are normally "extruded wire" with coating over it. The coating may contain ingredients like SiO_2, TiO_2, FeO, MgO, Al_2O_3 and cellulose in various proportions.

Electrode materials :

Depending upon job material, the following *electrode materials* are in use :

(*i*) Mild steel
(*ii*) Low alloy steel
(*iii*) Nickel steel
(*iv*) Chrome-moly steel
(*v*) Manganese-moly steel
(*vi*) Nickel manganese-moly steel
(*vii*) Nickel-moly-vanadium steel
(*viii*) Aluminium
(*ix*) Copper-aluminium
(*x*) Lead-bronze
(*xi*) Phosphor bronze.

Electrode coatings :

The electrode coatings are made to serve the following *purposes* :

(*i*) To provide a gaseous shield to prevent atmospheric contamination.
(*ii*) To help stabilise and direct the arc for effective penetration.
(*iii*) To act as scavengers to reduce oxides.
(*iv*) To control surface tension in the pool to influence the shape of the bead formed when the metal freezes.
(*v*) To add alloying elements to the weld.
(*vi*) To insulate the electrode electrically.
(*vii*) To minimise splatter of the weld metal.
(*viii*) To form a plasma to conduct current across the arc.
(*ix*) To form a slag to carry off impurities, protect the hot metal and slow the cooling rate.

Electrodes' designation :

- The *electrodes are designated by numbers which indicate grade and by sizes of the core wire.*
- The sizes are given by absolute sizes (*e.g.,* $\frac{1''}{8}, \frac{5''}{16}, \frac{7''}{16}$ etc.) or by standard wire gauge (S.W.G) sizes. Most common sizes are 4, 6, 8, 10 and 12. The electrodes with higher SWG size number are thinner.

- Although any size electrode can be used for the job, but *for quick welding, the electrode size should be as large as permissible. Larger size electrodes permit heavier currents to be used hence, the metal deposition is faster and the job is done quicker*. It must be, however, emphasised that *choice of the electrode size does not determine the speed of the work. Good welds come out rapidly if consistent higher currents are used.*
  — Normally 12 SWG electrodes for welding can be used upto 3 mm plates or sections, 10 SWG upto 5 mm sections, 8 SWG upto 10 mm, 6 SWG upto 10/15 mm and 4 SWG for sections beyond 20 mm thickness.
- As the jobs get heated during welding, they should not be handled by naked hands. After the welding is over, the part may be put in *water for quick cooling* ; otherwise it may be allowed to *cool in open air*. After this, the scum at the *surface may be knocked out by welding hammer*.
  — If *rebuilding* is being done, it is good practice to *beat the weld so that oversize grains are broken and this increases the strength of the joint.*

Typical data on use of "Mild steel electrodes" :

(*a*) ***Recommended currents :***

Electrode size		*Current range in amps.*	
SWG	***mm.***	***Min.***	***Max.***
10	3.25	95	125
8	4.06	140	170
6	4.88	160	220
4	5.89	200	275

(*b*) ***Composition (Chemical) :***

Carbon 0.09%
Manganese 0.47%
Silicon 0.20%
Sulphur 0.016%
Phosphorus ...... 0.34%.

2.2.14. Welding of Various Metals

Some welding methods/techniques as applied to different metals/alloys are briefly given below :

Metal /Alloys and Welding methods/Techniques :

1. Low carbon steels (or mild steel)

- *Welded by* :
  — Forge welding ;
  — Resistance welding ;
  — Arc welding ;
  — Gas welding.
- No flux required.
- Welding rods are made of pure iron or mild steel
- No preheating is required even for welding thicker sections.
- In gas welding, a *neutral* flame is used in order to minimise oxidation of steel.

2. **Medium carbon steels**

(Carbon content between 0.3 and 0.5%)

- *Welded by*
  - — Arc welding ;
  - — Resistance welding ;
  - — Gas welding ;
  - — Thermit welding.
- The preheating temperature (varying from 100°C to 400°C) depends upon the carbon content in the steel.
- As this steel is harder and brittle than mild steel, therefore, it is necessary to *normalise* the components after the welding in order to relieve the residual stress present in them.
- In gas welding, a slight *carbonising flame* is used.

3. **High carbon steels**

(High %age of carbon)

- *Welded by* : Same as at (2)
- Preheating to about 400°C is essential.
- Sudden cooling should be avoided to avoid cracking along weld metal.
- Heat treatment of these steels after welding is necessary to a relieve residual stresses set up during welding.

4. **Alloy steels**

(In addition to carbon, these steels contain small amount of nickel, chromium, molybdenum, manganese, silicon, copper. These steels may be low, *medium* and *high alloy steels*).

- The welding of low alloy steels is similar to the welding of medium or high carbon steels. *Shielded metal arc* and *submerged arc welding* are oftenly used.
- Preheating and slow cooling after welding is essential to obtain crack free welds.

  Welding of "stainless steel" :
  - — Can be welded by oxy-acetylene and metal arc welding methods ; best method being *electric butt welding* followed by prompt annealing between 730° to 800°C.
  - — A suitable electrode should be selected as per manufacturer's advice.
  - — For obtaining a sound weld, cleaning of edges to be welded and removal of slag after each run is necessary.
  - — Flux may or may not be used.

5. **Cast steel**

- The *welding operation for the repair of defective casting is termed as 'welding of cast steel'.*
- Gas welding with a *neutral flame* is usually used.
- Preheating of heavy casting is essential to minimise straining of metal due to local heating at the time of welding.

6. **Cast-iron**

- *Welded by* :
  - — Metal arc welding ;
  - — Oxy-acetylene welding ;
  - — Braze welding.

- The weldability of cast iron usually decreases as the amount of free carbon in cast iron increases.
- The cast iron parts are generally preheated to a dull red heat and then welded.
- After welding, it is very important to anneal the casting.

7. **Aluminium and its alloys**

(cast and wrought forms)

- *Welded by* :
  — Welding of all types including inert-gas and atomic hydrogen welding.
  — The inert gas-tungsten arc welding is extensively used than other arc welding processes.
- In welding cast aluminium, preheating is necessary and after welding it should be cooled slowly.

8. **Copper and its alloys**

- *Welded by* :
  — Metal-arc welding
  — Carbon-arc welding.
- The direct current with straight polarity is usually employed for welding these metals.
- In gas welding of copper, a neutral flame and a filler rod of copper and silver alloy is used.

9. **Nickel and its alloys**

- *Welded by* :
  — May be easily welded by most of the major welding processes like metal-arc welding, resistance welding and oxy-acetylene welding.
- Pre-heat treatment and post-heat treatment are rarely used except for special cases.
- For welding nickel flux is not required.

2.2.15. Rebuilding

- Welding is used not only for fabrication of new parts but is used considerably for rebuilding of worn out parts, repair of *broken or cracked parts also.*
- *The worn out parts can be build up by use of proper electrodes. Where the high temperature required for welding might change the composition of the parent metal, low temperature or 'Eutectic' electrodes may be used.* After building up sufficient material the part can be re-machined to desired dimensions.
- This is done for *shafts, housings* etc.
- *Repair of cracked parts is also possible.* The cracked castings, bearing housings, broken gears etc. can be repaired this way. There is no difficulty of welding cracked mild steel, cast steel, manganese steel spares. However, cast iron and aluminium spares are difficult to weld. These require preheating and post heating so that the welding stresses do not weaken the repaired parts.

Thermit welding requires a special mould to be built up. A wax pattern is made around the gap and sand/clay mould is rammed outside it. A magnesite crucible contains charge and is mounted above the weld. After igniting crucible, the reaction takes about 30 seconds for completion. Metal is poured into the joint by removal of the tapping pin at the bottom of the crucible.

2.2.16. Hard Facing

- Welding is considerably used to reduce the wear on parts subjected to heavy abrasion, such as *jaws of stone crushers, digging teeth of shovels, surfaces of rollers* etc. If austenitic steels are used for manufacturing, then the difficulty arises when any machining is required. Further these steels may not have sufficient shock absorbing capacity. To obtain it, the usual practice is to *hard the face of mild steel with harder deposits.* These may be machinable or may be completely unmachinable.
- This hard facing may be even used for austenitic manganese steel wearing parts *used in cement or stone crushing industries. These layers should not be thick and must not be used for building up.*
- *If the hard facing is to be done on worn out parts, these may have to be built up by ordinary electrodes, before hard facing is done.* This is carried out in layers and a good pining (hammering out) should be carried out before second layer is built up. Many electrodes are available in the market, by which longitudinal joints are welded in uniform fashion. This is called seam welding and is valuable in pipe manufacturing. Water cooling is used to keep the electrodes cool.

2.2.17. Defective Welds

A weld not properly welded is a *defective weld.* A properly made weld should have the following *characteristics* :

(*i*) The weld *should not crack in the bend test.*

(*ii*) It should not contain scum or slag imbedded in the well.

(*iii*) Its appearance should be ripple like and not spongy.

(*iv*) It should not have cavities and, and the grain size should be uniform.

- *Overcurrent tries to dissolve scum in the weld while undercurrent tries to give cracks in the weld.*
- If electrode distance from the weld is varying this will cause the uneveness of the weld.

2.2.18. Under-water Welding

- Occasionally, welding has to be carried out under water and this is known as **"submerged welding".** Obviously gas welding is not possible as the heat necessary for fusion cannot be generated under water by gas welding. However, submerged welding is possible by electric arc. *Heat generated by an electric arc is much lower than in open air welding hence special electrodes are used which have low fusion points and are not affected under water.*

  — Under electric welding, the welding may be by 'arc' method or it may by 'resistance' method. In the *arc method* the arc passes between the electrode held in the holder with the welder and the job which is connected to the second terminal of the electric supply. The arc drawn between an electrode and the workpiece completes the electrical circuit. In *resistance welding*, current is passed through the two pieces held together under pressure. The *electric current causes heat in the joint due to eddy currents and loss of energy in the resistance of the circuit*. This is also known as *Thomson process.*

  — Hard facing gives 2 to 3 times the life in terms of wear as compared to plain wearing parts. This is very economical and requires less idle time for machines.

2.2.19. Defects in Welding

Some important *welding defects* are :

1. Cracked welds [Fig. 2.44 (*i*)]
2. Porous welds [Fig. 2.44 (*ii*)]

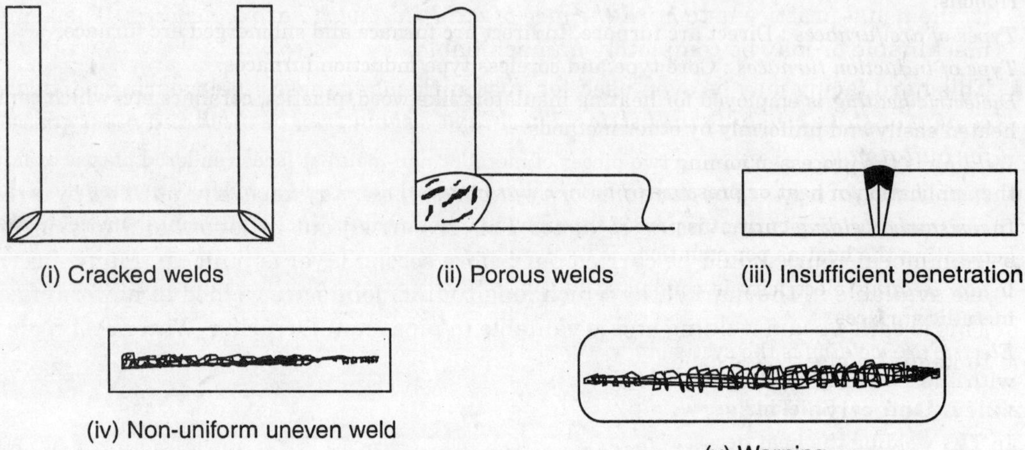

Fig. 2.44. Welding defects.

3. Insufficient penetration [Fig. 2.44 (*iii*)]
4. Non-uniform uneven weld [Fig. 2.44 (*iv*)]
5. Warping [Fig. 2.44 (*v*)]

- Cracks such as these Fig. 2.45 (*a*) in castings or in fabricated jobs can be repaired by *gauging* and *welding* as shown in Fig. 2.45 (*b*).

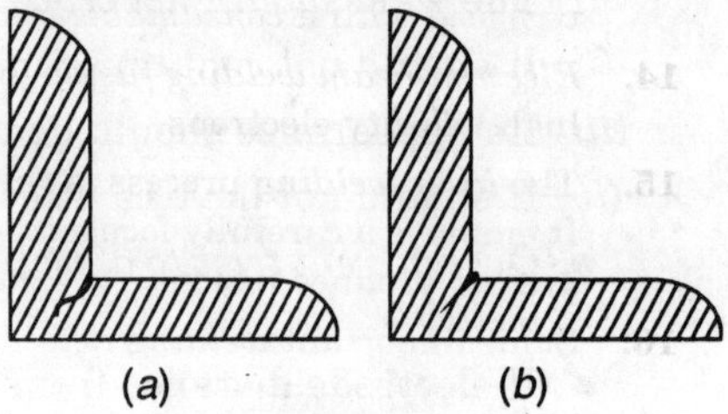

Fig. 2.45

2.2.20. Testing of Welded Joints

- It is very essential to have careful examination of the component at each stage of manufacture. Thus by inspection or quality control, the welding defects are located and preventive measures are devised to reduce or eliminate them.
- It is very difficult to ascertain whether the finished weld is up to the expected standard or not. In order to locate the defects and hidden flaws the welded portion may be inspected visually or examined with instruments.
  — The welded joints may be subjected to **destructive testing** like *tensile test, impact test, bending test, hardness* etc.
  — The **non-destructive tests** include *X-ray test, magnetic test, spark test, hydraulic test and air-pressure test for pressure vessels.*

HIGHLIGHTS

1. The use of electrically produced heat is always economical proposition on account of the present low cost and availability of electrical energy.
2. Modes of heat transfer are : *Conduction, Convection* and *Radiation*.

3. *Methods of electric heating* :
   (*i*) Power frequency heating : Resistance heating, Arc heating.
   (*ii*) High frequency heating : Induction heating, Dielectric heating, Infrared heating.
4. The *heating elements* are normally made of wires of circular cross-section or rectangular conducting ribbons.
5. *Types of arc furnaces* : Direct arc furnace, Indirect arc furnace and submerged arc furnace.
6. *Type of induction furnaces* : Core type and coreless type induction furnaces.
7. *Dielectric heating* is employed for heating insulators like wood, plastics, ceramics etc. which cannot be heated easily and uniformly by other methods.
8. *Welding* is the process of joining two pieces of metal or non-metal at faces rendered plastic or liquid by the application of heat or pressure or both. Filler material may be used to effect the union.
9. In *resistance welding* current is passed through the inherent resistance of the joint to be welded thereby generating the heat as per equation I^2Rt kilojoules.
10. In *arc welding* electricity is conducted in the form of an arc which is established between the two metallic surfaces.
11. *Electric arc welding* is the system in which metal is melted by the heat of an electric arc. It can be done with the following methods : (*i*) Metallic arc welding (*ii*) Carbon arc welding (*iii*) Atomic hydrogen welding and (*iv*) Shielded arc welding.
12. In TIG welding the heat necessary to melt the metal is provided by a very intense electric arc which is struck between a *virtually non-consumable tungsten electrode and metal wokepiece.*
13. MIG process is a refinement of TIG process, however, in this process, the tungsten electrode has been replaced with a *consumable electrode.*
14. *Electron-beam welding fusion* joins metal by bombarding a specific confined area of the base metal with high velocity electrons.
15. The *laser welding* process is the focusing of a monochromatic light into extremely concentrated beams. It employs a carefully focused beam of light that concentrates tremendous amount of energy on a small area to produce fusion.
16. Some important *welding defects* are :
   (*i*) Cracked welds, (*ii*) Porous welds,
   (*iii*) Insufficient penetration, (*iv*) Non-uniform uneven weld, and
   (*v*) Warping.

OBJECTIVE TYPE QUESTIONS

A. Choose the Correct Answer :

HEATING

1. Which of the following is an advantage of heating by electricity ?
   (*a*) Quicker operation (*b*) Higher efficiency
   (*c*) Absence of flue gases (*d*) All of the above.
2. .... has the highest value of thermal conductivity.
   (*a*) Copper (*b*) Aluminium (*c*) Brass (*d*) Steel.
3. Which of the following heating methods has maximum power factor ?
   (*a*) Arc heating (*b*) Dielectric heating (*c*) Induction heating (*d*) Resistance heating.
4. ...... method has leading power factor
   (*a*) Resistance heating (*b*) Dielectric heating (*c*) Arc heating (*d*) Induction heating.
5. ...... is used for heating non-conducting materials
   (*a*) Eddy current heating (*b*) Arc heating
   (*c*) Induction heating (*d*) Dielectric heating.

6. Which of the following methods of heating is not dependent on the frequency of supply ?
(*a*) Induction heating (*b*) Dielectric heating
(*c*) Electric resistance heating (*d*) All of the above.

7. When a body reflects entire radiation incident on it, then it is known as :
(*a*) white body (*b*) grey body (*c*) black body (*d*) transparent body.

8. For the transmission of heat from one body to another
(*a*) temperature of the two bodies must be different
(*b*) both bodies must be solids
(*c*) both bodies must be in contact
(*d*) at least one of the bodies must have some source of heating.

9. Heat transfer by condition will not occur when
(*a*) bodies are kept in vacuum (*b*) bodies are immersed in water
(*c*) bodies are exposed to thermal radiations (*d*) temperatures of the two bodies are identical.

10. A perfect black body is one that
(*a*) transmits all incident radiations (*b*) absorbs all incident radiations
(*c*) reflects all incident radiations
(*d*) absorbs, reflects and transmits all incident radiations.

11. Heat is transferred simultaneously by condition, convection and radiation
(*a*) inside boiler furnaces (*b*) during melting of ice
(*c*) through the surface of the insulated pipe carrying steam
(*d*) from refrigerator coils to freezer of a refrigerator.

12. The process of heat transfer during the re-entry of satellites and missiles, at very high speeds, into earth's atmosphere is known as
(*a*) ablation (*b*) radiation (*c*) viscous dissipation (*d*) irradiation.

13. Which of the following has the highest values of thermal conductivity ?
(*a*) Water (*b*) Steam (*c*) Solid ice (*d*) Melting ice.

14. Induction heating process is based on which of the following principles ?
(*a*) Thermal ion release principle (*b*) Nucleate heating principle
(*c*) Resistance heating principle (*d*) Electro-magnetic induction principle.

15. Which of the following insulating materials is suitable for low temperature applications ?
(*a*) Asbestos paper (*b*) Diatomaceous earth
(*c*) 80 percent magnesia (*d*) Cork.

16. A non-dimensional number generally associated with natural convection heat transfer is
(*a*) Prandtl number (*b*) Grashoff number
(*c*) Pecelet number (*d*) Nusselt number.

17. The temperature inside a furnace is usually measured by which for the following ?
(*a*) Optical pyrometer (*b*) Mercury thermometer
(*c*) Alcohol thermometer (*d*) Any of the above.

18. Which of the following will happen if the thickness of refractory wall of furnace is increased ?
(*a*) Heat loss through furnace wall will increase (*b*) Temperature inside the furnace will fall
(*c*) Temperature on the outer surface of furnace walls will drop
(*d*) Energy consumption will increase.

19. The material of the heating element for a furnace should have
(*a*) lower melting point (*b*) higher temperature co-efficient
(*c*) high specific resistance (*d*) all of the above.

20. In a resistance furnace the atmosphere is ....
(*a*) oxidising (*b*) deoxidising (*c*) reducing (*d*) neutral.

21. By which of the following methods the temperature inside a resistance furnace can be varied ?
(*a*) By disconnecting some of the heating elements
(*b*) By varying the operating voltage
(*c*) By varying the current through heating elements
(*d*) By any of the above method.

22. In induction heating .... is abnormally high.
(*a*) phase angle (*b*) frequency (*c*) current (*d*) voltage.

23. By the use of which of the following, high frequency power supply for induction furnaces can be obtained ?
(*a*) Coreless transformers (*b*) Current transformers
(*c*) Motor-generator set (*d*) Multi-phase transformer.

24. Induction furnaces are employed for which of the following ?
(*a*) Heat treatment of castings (*b*) Heating of insulators
(*c*) Melting of aluminium (*d*) None of the above.

25. In an electric room heat convector the method of heating used is
(*a*) arc heating (*b*) resistance heating (*c*) induction heating (*d*) dielectric heating.

26. In a domestic cake baking oven the temperature is controlled by
(*a*) voltage variation (*b*) thermostat
(*c*) auto-transformer (*d*) series-parallel operation.

27. In an electric press mica is used
(*a*) as an insulator (*b*) as a device for power factor improvement
(*c*) for dielectrics heating (*d*) for induction heating.

28. Induction heating takes place in which of the following ?
(*a*) Insulating materials (*b*) Conducting materials which are magnetic
(*c*) Conducting materials which are non-magnetic
(*d*) Conducting materials which may or may not be magnetite.

29. For heating element high resistivity material is chosen to
(*a*) reduce the length of heating element (*b*) increase the life of the heating element
(*c*) reduce the effect of oxidation (*d*) produce large amount of heat.

30. In resistance heating highest working temperature is obtained from heating elements made of ......
(*a*) nickel copper (*b*) nichrome (*c*) silicon carbide (*d*) silver.

31. For intermittent work which of the following furnaces is suitable ?
(*a*) Indirect arc furnace (*b*) Core less furnace
(*c*) Either of the above (*d*) None of the above.

32. Due to which of the following reasons it is desirable to have short arc length ?
(*a*) To achieve better heating (*b*) To increase the life of roof refractory
(*c*) To have better stirring action (*d*) To reduce problem of oxidation
(*e*) All of the above.

33. In the indirect resistance heating method, maximum heat-transfer takes place by
(*a*) radiation (*b*) convection (*c*) conduction (*d*) any of the above.

34. Property of low temperature co-efficient of heating element is desired due to which of the following reasons ?
(*a*) To avoid initial rush of current (*b*) To avoid change in kW rating with temperature
(*c*) Both (*a*) and (*b*) (*d*) Either (*a*) or (*b*).

35. Which of the following methods is used to control temperature in resistance furnaces ?
(*a*) Variation of resistance (*b*) Variation of voltage
(*c*) Periodical switching on and off the supply (*d*) All of the above methods.

36. It is desirable to operate the arc furnaces at power factor of

(*a*) zero (*b*) 0.707 lagging (*c*) unity (*d*) 0.707 leading.

37. Radiations from a black body are proportional to

(*a*) T^1 (*b*) T^2 (*c*) T^3 (*d*) T^4.

38. In arc furnace the function of choke is

(*a*) to stabilize the arc (*b*) to improve, power factor
(*c*) to reduce severity of the surge (*d*) none of the above.

39. Ajax Wyatt furnace is started when

(*a*) it is filled below core level (*b*) it is filled above core level
(*c*) it is fully empty (*d*) none of the above.

40. In electric press, mica is used because it is .... conductor of heat but/and .... conductor of electricity.

(*a*) bad, good (*b*) bad, bad (*c*) good, bad (*d*) good, good.

41. Resistance variation method of temperature control is done by connecting resistance elements in

(*a*) series (*b*) parallel
(*c*) series-parallel connections (*d*) star-delta connections
(*e*) all of the above ways.

42. Hysteresis loss and eddy current loss are used in

(*a*) induction heating of steel (*b*) dielectric heating
(*c*) induction heating of brass (*d*) resistance heating.

43. In heating the ferromagnetic material by induction heating, heat is produced due to

(*a*) induced current flow through the charge
(*b*) hysteresis loss taking place below curie temperature
(*c*) due to hysteresis loss as well as eddy current loss taking place in the charge
(*d*) none of the above factors.

44. Radiant heating is used for which of the following ?

(*a*) Annealing of metals (*b*) Melting of ferrous metals
(*c*) Heating of liquids in electric kettle (*d*) Drying of paints and varnishes.

45. Which of the following devices is necessarily required for automatic temperature control in a furnace ?

(*a*) Thermostat (*b*) Thermocouple
(*c*) Auto-transformer
(*d*) Heating elements of variable resistance material

46. For radiant heating around 2250°C, the heating elements are made of

(*a*) copper alloy (*b*) carbon (*c*) tungsten alloy (*d*) stainless steel alloy.

47. Which of the following is an advantage of eddy current heating ?

(*a*) The amount of heat generated can be controlled accurately
(*b*) Heat at very high rate can be generated
(*c*) The area of the surface over which heat is produced can be accurately controlled
(*d*) All of the above.

48. The electrode of a direct arc furnace is made of

(*a*) tungsten (*b*) graphite (*c*) silver (*d*) copper.

49. Direct arc furnaces have which of the following power factors ?

(*a*) Unity (*b*) Low, lagging (*c*) Low, leading (*d*) Any of the above.

50. In direct arc furnace, which of the following has high value ?

(*a*) Current (*b*) Voltage (*c*) Power factor (*d*) All of the above.

WELDING

51. During resistance welding heat produced at the joint is proportional to

(*a*) $I^2 R$ (*b*) kVA (*c*) current (*d*) voltage.

52. Grey iron is usually welded by .... welding
(*a*) gas (*b*) arc (*c*) resistance (*d*) MIG.
53. The metal surfaces, for electrical resistance welding must be ....
(*a*) lubricated (*b*) cleaned (*c*) moistened (*d*) rough.
54. In a welded joint poor fusion is due to which of the following ?
(*a*) Improper current (*b*) High welding speed
(*c*) Uncleaned metal surface (*d*) Lack of flux.
55. For arc welding, D.C. is produced by which of the following ?
(*a*) Motor-generator set (*b*) Regulator
(*c*) Transformer (*d*) None of the above.
56. .... welding process uses consumable electrodes.
(*a*) TIG (*b*) MIG (*c*) Laser (*d*) All of the above.
57. Which of the following equipment is generally used for arc welding ?
(*a*) Single phase alternator (*b*) Two phase alternator
(*c*) Three phase alternator (*d*) Transformer.
58. Which of the following is not an inert gas ?
(*a*) Argon (*b*) Carbondioxide (*c*) Helium (*d*) All of the above.
59. Electronic components are joined by which of the following methods ?
(*a*) Brazing (*b*) Soldering (*c*) Seam welding (*d*) Spot welding
(*e*) None of the above.
60. Resistance welding cannot be used for
(*a*) dielectric (*b*) ferrous materials (*c*) non-ferrous metals (*d*) any of the above.
61. Electric arc welding process produces temperature upto
(*a*) 1000°C (*b*) 1500°C (*c*) 3500°C (*d*) 5550°C.
62. Increased heat due to shorter arc is harmful on account of
(*a*) under-cutting of base material (*b*) burn through
(*c*) excessive porosity (*d*) all of the above.
63. Arc blow results in which of the following ?
(*a*) Non-uniform weld beads (*b*) Shallow weld puddle giving rise to weak weld
(*c*) Splashing out of metal from weld puddle (*d*) All of the above defects.
64. In seam welding
(*a*) the work piece is fixed and disc electrodes move
(*b*) the work piece moves but rotating electrodes are fixed
(*c*) any of the above (*d*) none of the above.
65. In arc welding major personal hazards are
(*a*) flying sparks (*b*) weld spatter
(*c*) harmful infrared and ultra-violet rays from the arc
(*d*) all of the above.
66. In spot welding composition and thickness of the base metal decides
(*a*) the amount of squeeze pressure (*b*) hold time
(*c*) the amount of weld current (*d*) all above.
67. Helium produces which of the following ?
(*a*) Deeper penetration (*b*) Faster welding speeds
(*c*) Narrower heat affected zone in base metal (*d*) All of the above.
68. Due to which of the following reasons aluminium is difficult to weld ?
(*a*) It has an oxide coating (*b*) It conducts away heat very rapidly
(*c*) Both (*a*) and (*b*) (*d*) None of the above.

69. Welding leads have
(*a*) high flexibility (*b*) high current handling capacity
(*c*) both (*a*) and (*b*) (*d*) none of the above.

70. Air craft body is
(*a*) spot welded (*b*) gas welded (*c*) seam welded (*d*) riveted.

71. For arc welding current range is usually
(*a*) 10 to 15 A (*b*) 30 to 40 A (*c*) 50 to 100 A (*d*) 100 to 350 A.

72. Spot welding is used for
(*a*) thin metal sheets (*b*) rough and irregular surfaces
(*c*) costings only (*d*) thick sections.

73. Galvanising is a process of applying a layer of
(*a*) aluminium (*b*) lead (*c*) copper (*d*) zinc.

74. A seamless pipe has
(*a*) steam welded joint (*b*) spot welded joint (*c*) arc welded joint (*d*) no joint.

75. Motor-generator set for D.C. arc welding has generator of
(*a*) series type (*b*) shunt type
(*c*) differentially compound type (*d*) level compound type.

76. Plain and butt welds may be used on materials upto thickness of nearly
(*a*) 5 mm (*b*) 10 mm (*c*) 25 mm (*d*) 50 mm.

77. In argon arc welding argon is used as a
(*a*) flux (*b*) source of heat
(*c*) agent for heat transfer (*d*) shield to protect the work from oxidation.

78. During arc welding as the thickness of the metal to be welded increases
(*a*) current should decrease, voltage should increase
(*b*) current should increase, voltage remaining the same
(*c*) current should increase, voltage should decrease
(*d*) voltage should increase, current-remaining the same.

79. In D.C. arc welding
(*a*) electrode is made positive and workpiece negative
(*b*) electrode is made negative and workpiece positive
(*c*) both electrode as well as workpiece are made positive
(*d*) both electrode as well as workpiece are made negative.

80. The purpose of coating on arc welding electrodes is to
(*d*) stabilise the arc (*b*) provide a protecting atmosphere
(*c*) provide slag to protect the molten metal (*d*) all of the above.

81. 50 percent duty cycle of a welding machine means
(*a*) machine input is 50 percent of rated input (*b*) machine efficiency is 50 percent
(*c*) machine works on 50 percent output
(*d*) machine works for 5 minutes in a duration of 10 minutes.

82. During carbon arc welding if electrode is connected to positive
(*a*) are will be dull (*b*) arc will not strike
(*c*) metal will not strike
(*d*) carbon will have tendency to go into the weld joint.

83. In which of the following methods of welding the molten metal is poured for joining the metals ?
(*a*) Thermit welding (*b*) Gas welding (*c*) TIG welding (*d*) Arc welding.

84. In atomic hydrogen welding the electrode is made of
(*a*) carbon (*b*) graphite (*c*) tungsten (*d*) mild steel.

85. The porosity of welded joint may be caused by
(*a*) low welding current (*b*) incorrect size of electrodes
(*c*) poor base metal (*d*) any of the above.

86. In electrical resistance welding the greatest resistance is offered by which of the following ?
(*a*) Metal surface (*b*) Contact point of electrode with metal top
(*c*) Contact point of electrode with metal bottom (*d*) Contact layer of metals to be welded.

87. The tips of the electrodes, for spot welding, are made of
(*a*) copper alloy (*b*) porcelam (*c*) mica (*d*) carbon.

88. During spot welding the current flows for
(*a*) fraction of a minute to several minutes (*b*) fraction of a second to several seconds
(*c*) few milliseconds (*d*) few microseconds.

89. .... welding is not a resistance welding process.
(*a*) Projection (*b*) Seam (*c*) Flush (*d*) Carbon arc.

90. The power factor of a spot welding machine is expected to be around
(*a*) Unity (*b*) 0.8 lagging (*c*) 0.3 to 0.5 lagging (*d*) 0.8 leading.

91. Which of the following methods is normally *not* preferred for welding of chromium molybdenum steels ?
(*a*) Oxyacetylene welding (*b*) Resistance welding
(*c*) Thermit welding (*d*) Submerged arc welding.

92. The welding of aluminium alloy entails which of the following problems ?
(*a*) High thermal conductivity-quick dissipating heat
(*b*) On melting becomes thin liquid-tendency to run off
(*c*) Low specific resistance requiring heavy current
(*d*) All of the above.

93. During carbon arc welding
(*a*) electrodes is connected to neutral if A.C. is used
(*b*) electrode is not connected to any voltage source when A.C. is used
(*c*) electrode is negative with respect to the work if D.C. is used
(*d*) electrode is positive with respect to the work if D.C. is used.

94. In welding weld spatter defect is generally the result of
(*a*) too high voltage during welding (*b*) too high current during welding
(*c*) low voltage during welding (*d*) low current during welding.

95. Arc blow (a welding defect) is generally encountered in
(*a*) arc welding using A.C. current (*b*) arc welding using D.C. current
(*c*) thermit welding (*d*) gas welding.

96. Steel pipes are manufactured by
(*a*) argon arc welding (*b*) thermit welding (*c*) resistance welding (*d*) arc welding.

97. A rectifier for welding has voltage/current characteristic as
(*a*) static (*b*) variable (*c*) drooping (*d*) rising.

98. Which of the following is net a welding accessory ?
(*a*) Hand screen (*b*) Cable (*c*) Electrode holder (*d*) Gloves.

99. Gray iron is usually welded by
(*a*) Gas welding (*b*) TIG welding (*c*) MIG welding (*d*) Arc welding.

100. The welding load is always
(*a*) continuous but varying (*b*) continuous and constant
(*c*) intermittent (*d*) none of the above.

101. TIG welding is
(*a*) thorium iodine gas welding (*b*) thermally induced gas welding
(*c*) temperature insulated gas welding (*d*) tungsten insert gas welding.

102. Flux used in TIG welding is
(*a*) borax (*b*) ammonium chloride (*c*) ash (*d*) none.

103. For the welding of aluminium alloys which of the following methods would you recommended ?
(*a*) Acetylene-oxygen gas welding (*b*) D.C. arc welding
(*c*) A.C. arc welding (*d*) Tungsten arc welding.

104. Which of the following automatic welding processes is likely to give maximum rate of metal deposition ?
(*a*) Continuous flux covered electrode (*b*) Multiple power submerged arc
(*c*) Gas shielded bare wire (*d*) Submerged arc (single wire).

105. A 10 SWG electrode usually operates in the current range
(*a*) 300 to 400 amperes (*b*) 95 to 135 amperes
(*c*) 50 to 65 amperes (*d*) 20 to 30 amperes.

106. Which of the following electrodes will have least diameter ?
(*a*) 20 SWG (*b*) 16 SWG (*c*) 14 SWG (*d*) 4 SWG.

107. The danger of shock is maximum
(*a*) before welding (*b*) after arcing
(*c*) during arcing (*d*) while inserting electrode into the holder.

108. The welding electric circuit is
(*a*) never earthed (*b*) always earthed
(*c*) through cables only.

109. Steel rails are welded by
(*a*) Gas welding (*b*) Thermit welding (*c*) Resistance welding (*d*) Argon arc welding.

110. In ultrasonic welding the frequency range is generally
(*a*) 2000 to 3000 Hz (*b*) 4000 to 20000 Hz
(*c*) 30000 to 40000 Hz (*d*) 50000 to 80000 Hz.

B. Fill in the Blanks/Say 'Yes' or 'No' :

111. Heating element in an incandescent lamp is of ....

112. Eddy current heating is suitable for hardening. (Yes/No)

113. Only insulating material is heated by dielectric heating. (Yes/No)

114. Indirect arc furnaces are usually made in .... sizes than direct arc furnaces.

115. Indirect arc furnaces are usually of single phase type and direct arc furnaces are usually of three phase type. (Yes/No)

116. Heat produced in dielectric heating is directly proportional to .... and square of ....

117. Rheostat wire is made up of tungsten. (Yes/No)

118. Heating elements used in household appliances are made of nichrome (Yes/No)

119. Stirring action by rocking the furnace is achieved in .... phase arc furnace and by electro-magnetic force is achieved in .... phase arc furnace.

120. Heat transfer by conduction will not occur when temperatures of the two bodies are identical. (Yes/No)

121. In case of boiler furnaces heat is transferred by all the three modes, *viz.* conduction, convection and radiation. (Yes/No)

122. A perfect black body is one that reflects all incident radiations. (Yes/No)

123. Nichrome can be used for furnace temperatures upto 1000°C (Yes/No)

124. Grey iron is usually welded by .... welding.

125. For spot welding the tips of the electrodes are made of .... alloy.

126. In .... welding the molten metal is poured for joining the metals.

127. The porosity of the welded joint may be caused by poor base metal. (Yes/No)

128. The range of open circuit voltage for arc welding is generally 40--90 V. (Yes/No)

129. Spot welding is used for thin metal sheets. (Yes/No)

130. Resistance welding cannot be used for dielectrics. (Yes/No)

ANSWERS

A. Choose the Correct Answer :

(Electric Heating and Welding)

1. (*d*)	**2.** (*a*)	**3.** (*d*)	**4.** (*b*)	**5.** (*d*)	**6.** (*c*)	**7.** (*a*)
8. (*a*)	**9.** (*d*)	**10.** (*b*)	**11.** (*a*)	**12.** (*a*)	**13.** (*c*)	**14.** (*d*)
15. (*b*)	**16.** (*b*)	**17.** (*a*)	**18.** (*c*)	**19.** (*c*)	**20.** (*a*)	**21.** (*d*)
22. (*b*)	**23.** (*c*)	**24.** (*a*)	**25.** (*b*)	**26.** (*b*)	**27.** (*a*)	**28.** (*d*)
29. (*a*)	**30.** (*c*)	**31.** (*a*)	**32.** (*e*)	**33.** (*a*)	**34.** (*c*)	**35.** (*d*)
36. (*b*)	**37.** (*d*)	**38.** (*a*)	**39.** (*b*)	**40.** (*c*)	**41.** (*e*)	**42.** (*a*)
43. (*c*)	**44.** (*d*)	**45.** (*b*)	**46.** (*c*)	**47.** (*d*)	**48.** (*b*)	**49.** (*b*)
50. (*a*)	**51.** (*a*)	**52.** (*a*)	**53.** (*b*)	**54.** (*a*)	**55.** (*a*)	**56.** (*b*)
57. (*d*)	**58.** (*b*)	**59.** (*b*)	**60.** (*a*)	**61.** (*d*)	**62.** (*d*)	**63.** (*d*)
64. (*c*)	**65.** (*d*)	**66.** (*d*)	**67.** (*d*)	**68.** (*c*)	**69.** (*c*)	**70.** (*d*)
71. (*d*)	**72.** (*a*)	**73.** (*d*)	**74.** (*d*)	**75.** (*d*)	**76.** (*c*)	**77.** (*d*)
78. (*b*)	**79.** (*b*)	**80.** (*d*)	**81.** (*d*)	**82.** (*d*)	**83.** (*a*)	**84.** (*c*)
85. (*c*)	**86.** (*d*)	**87.** (*a*)	**88.** (*b*)	**89.** (*d*)	**90.** (*c*)	**91.** (*b*)
92. (*d*)	**93.** (*c*)	**94.** (*b*)	**95.** (*b*)	**96.** (*c*)	**97.** (*c*)	**98.** (*b*)
99. (*a*)	**100.** (*c*)	**101.** (*d*)	**102.** (*d*)	**103.** (*d*)	**104.** (*b*)	**105.** (*b*)
106. (*a*)	**107.** (*d*)	**108.** (*b*)	**109.** (*b*)	**110.** (*b*)		

B. Fill in the blanks/Say 'Yes' or 'No' :

111. tungsten	**112.** Yes	**113.** Yes	**114.** smaller	**115.** Yes
116. frequency, voltage		**117.** No	**118.** Yes	**119.** single, three
120. Yes	**121.** Yes	**122.** No	**123.** Yes	**124.** gas
125. copper	**126.** thermit	**127.** Yes	**128.** Yes	**129.** Yes
130. Yes.				

THEORETICAL QUESTIONS

Electric heating

1. What are the advantages of electric heating ?
2. Give classification of various electric heating methods along wi2th brief account of their working principle.
3. Explain briefly the following modes of heat transfer :
   (*i*) conduction, (*ii*) Convection, (*iii*) Radiation.
4. List the properties of a good heating element.
5. Explain briefly the materials of heating elements.
6. Discuss the methods of temperature control of resistance ovens.
7. Explain the design procedure of the heating elements when the power and voltage of the oven is known.
8. Explain the working principle of arc furnaces and describe with the help of a sketch the construction and working of any one type of arc furnace.
9. Describe the construction and working of a 3-phase arc furnace.
10. Describe the conditions for maximum output for an electric arc furnace.
11. Mention the advantages of dielectric heating.
12. Discuss the relative merits and demerits of direct and indirect electric arc furnaces.
13. Explain different methods of induction heating. Give some applications of induction heating.

14. Explain the method of induction heating and describe coreless type of induction furnace.

15. Describe with diagram the working principle of

(*i*) vertical core type furnace and (*ii*) coreless induction furnace.

16. Explain with the help of a neat sketch the working of Ajax Wyatt furnace. What is its field of application ?

17. Briefly describe the principle of induction heating at high frequency and highlight a few applications of the eddy current heating.

18. What is dielectric heating ? Explain the factors on which the dielectric loss in a dielectric material depends.

19. What are the factors which decide the frequency and voltage of the dielectric heating ? Derive an expression for the heat produced in a dielectric material.

20. Discuss the following applications in dielectric heating :

(*i*) Heating of raw plastics (*ii*) Gluig of wood ; (*iii*) Food processing.

Electric Welding

21. Define the term 'welding'.

22. Enumerate the various welding processes.

23. What is the fundamental difference between electric arc welding and resistance welding ?

24. Explain with a neat sketch how the spot welding is carried out by a spot welding machine.

25. Discuss the principle of arc welding and the difference between carbon and metallic arc welding and their relative merits. Compare the A.C. and D.C. systems of metallic arc welding.

26. Describe with neat sketches the various methods of electric resistance welding. Give its merits and demerits with respect to arc welding.

27. What is resistance welding ? What are its limitations ?

28. With necessary figures, explain the processes of carbon arc welding and metallic arc welding.

29. Discuss in detail the principle of operation of (*i*) Ultrasonic welding and (*ii*) Laser welding.

30. Compare A.C. and D.C. welding.

31. Describe briefly the following types of welding.

(*i*) TIG welding, (*ii*) MIG welding and (*iii*) Electron-beam welding.

32. Explain briefly the following :

(*i*) Ultrasonic welding (*ii*) Atomic hydrogen welding

(*iii*) Electro-slag welding.

33. Write short notes on the following :

(*i*) Rebuilding (*ii*) Hard facing

(*iii*) Defects in welding (*iv*) Under-water welding.

UNSOLVED EXAMPLES

1. A 20 kW, 230 V, single-phase resistance oven employs nickel-chrome wire for its heating elements. If the wire temperature is not to exceed 1200°C and the temperature of the charge is to be 600°C, calculate the diameter and length of the wire. Radiating efficiency = 0.6, emissivity = 0.9 and resistivity of nickel-chrome = 101.6×10^{-8} Ωm. **[Ans.** 2.9 mm ; 17.21 mm]

2. A 30 kW, 3-phase, 400 V resistance oven is to employ nickel-chrome strip 0.254 mm thick for the three star-connected heating elements. If the wire temperature is to be 1100°C and that of charge to be 700°C, estimate a suitable width for the strip. Assume emissivity = 0.9 and radiating efficiency to be 0.5 and resistivity of the strip material is 101.6×10^{-8} Ωm. **[Ans.** 7.4 mm]

3. A 50 kW, 3-phase, 440 V, resistance oven is to provide nickel-chrome strip 0.3 mm thick, for three star-connected heating elements. If the temperature of the wire is to be 1500°C and that of the charge is to be 1000°C, calculate a suitable width of the strip. Take, emissivity = 0.91, radiating efficiency = 0.6 and resistivity of the strip material = 101.6×10^{-8} Ωm. **[Ans.** 5.67 mm]

4. A cubic water tank has surface area of 6 m^2 and is filled to 90% capacity six times daily. The water is heated from 20°C to 65°C. The losses per square metre of tank surface per 1°C temperature difference are 6.3 W, Find :

(*i*) Loading in kW.

(*ii*) Efficiency of the tank.

Assume specific heat of water = 4.2 kJ/kg°C and 1 kWh = 3600 kJ. **[Ans.** (*i*) 3.5 kW ; (*ii*) 87.4%]

5. A 4-phase electric arc furnace has the following data :

Current drawn = 5000 A ; Arc voltage = 50 V ; Resistance of transformer referred to secondary = 0.002 Ω ; Reactance of transformer referred to secondary = 0.004 Ω.

(*i*) Calculate the power factor and kW drawn from the supply.

(*ii*) If the overall efficiency of the furnace is 65 percent, find the time required to melt 2 tonnes of steel if latent heat of steel = 37.2 kJ/kg, specific heat of steel = 0.5 kJ/kg K, melting point of steel = 1370°C and initial temperature of steel = 20°C. **[Ans.** (*i*) 0.9487, 900 kW ; (*ii*) 40 min 46 sec]

6. A low frequency induction furnace operating at 10 V in the secondary circuit takes 500 kW at 0.5 p.f. when the hearth is full . If the secondary voltage be maintained at 10 V, estimate the power factor and the power absorbed when the hearth is half full. Assume the resistance of the secondary circuit to be thereby halved and reactance to remain the same. **[Ans.** 0.277, 307.8 kW]

7. In the case hardening of a steel pulley, the depth of penetration is 1.5 mm. The relative permeability is unity and resistivity of steel is 5×10^{-7} Ωm. Determine the frequency required. **[Ans.** 56290 Hz]

8. A slab of insulating material 150 cm^2 in area and 1 cm thick is to be heated by dielectric heating. The power required is 400 W at 30 MHz. Material has relative permittivity of 5 and p.f. of 0.05. Absolute permittivity = 8.854×10^{-12} F/m. Determine the necessary voltage. **[Ans.** 800 V

9. A piece of an insulating material is to be heated by dielectric heating. The size is 10 cm × 10 cm × 3 cm. A frequency of 20 MHz is used and power absorbed is 400 W. Calculate the voltage necessary for heating and the current that flows in the material. The material has a relative permittivity of 5 and a power factor of 0.05.

If the voltage were limited to 1800 V, what will be the frequency to get the same loss ?

[Ans. 2065 V ; 3.82 A ; 26.32 MHz]

10. A piece of plastic material of size 4 cm × 2 cm × 1 cm is heated by being placed between two electrodes, each having an area of 20 cm × 2 cm and the distance of separation being 1.6 cm. The frequency of voltage impressed across electrodes is 20 MHz . If the power consumed is 80 W, find :

(*i*) The voltage applied across the electrodes.

(*ii*) The current drawn through the material. **[Ans.** (*i*) 3770 V, 4.2 A]

3

Electrolytic Processes

3.1. INTRODUCTION

The processes based on the fact that electrical energy can produce chemical changes are called **Electrolytic processes.** These processes are widely used for :

(*i*) *Extraction* of pure metals from their ores (*e.g.*, copper, zinc, aluminium, magnesium etc.) ;

(*ii*) *Refining* of metals (*e.g.*, gold, silver, copper, nickel etc.) ;

(*iii*) *Manufacturing* of various chemicals (*e.g.*, caustic soda, potassium permanganate, chlorine etc.) ;

(*iv*) *Electro-deposition* of metals including electro-plating, electro-typing, electro-forming ;

(*v*) *Building* up of worn parts in metallurgical, chemical and other industries.

All the processes mentioned above, though they appear differently in apparent detail, are based on the *principle of electrolysis.*

3.2. ELECTROLYSIS—THE BASIC PRINCIPLE

When a compound formed by *electrovalent bond is dissolved in water which has high dielectric constant results in the *weakening* of the electrostatic force of attraction between the ionized atoms.

This results in the charged ions to lead an independent existence. Consider the case of a copper sulphate ($CuSO_4$) dissolved in water. It dissociates into positively-charged copper ions (Cu^{++}) and negatively-charged sulphate ions (SO_4^{--}) moving freely in the solution. If two electrodes are placed in the electrolyte (*i.e.*, $CuSO_4$ solution) and one of them is made positive and the other negative, the positively-charged ions travel towards the cathode and the negatively-charged ions travel towards the anode. Each of the positively-charged copper ions (cations) reaching the cathode will take two electrons from it and become a metallic atom of copper, and similarly each of the negatively-charged sulphate ions (anions) reaching the anode will give up two electrons to it and cease to be anion. Thus the copper is deposited at the cathode as metal. The sulphate ions collect at the anode and react with water giving out oxygen :

$$2H_2O + 2SO_4 = 2H_2SO_4 + O_2$$

*Electrovalent bond is responsible for the formation of inorganic compound and is due to transfer of electron in the outer orbit of one atom to the outer orbit of another atom.

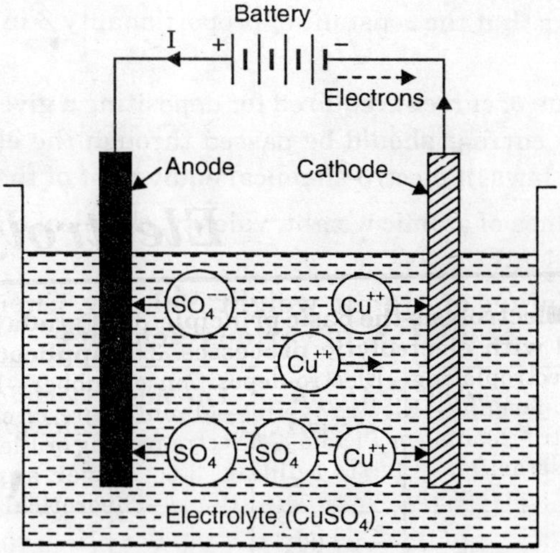

Fig. 3.1. Electrolysis.

Oxygen is liberated as gas at the anode and H_2SO_4 is formed. If the cathode is made of Cu, the sulphuric acid attacks it forming copper sulphate and liberating hydrogen :

$$H_2SO_4 + Cu = CuSO_4 + H_2$$

Thus the *copper of the anode goes into solution and copper from copper sulphate is deposited on the cathode.*

During the process there is *no accumulation of charge at any point in the circuit and the mass of copper deposited at the cathode is exactly equal to that removed from the anode.*

The whole process described above is called **electrolysis.**

3.3. FARADAY'S LAWS OF ELECTROLYSIS

Michael Faraday (an English scientist) formulated the laws governing the electrolytic processes, which are stated below :

Faraday's First Law. It is stated as follows :

"The mass of a substance liberated from an electrolyte in a given time is proportional to the quantity of electricity passing through the electrolyte."

That is $$m \propto Q \propto It \qquad ...(3.1)$$

$\therefore$ $$m = ZIt$$

where, Z = A constant called the *electrochemical equivalent*,

I = The steady current in amperes, and

t = Time (second) for which current I flows through the electrolyte.

If I = 1A, t = 1s and Z = m

Thus *electrochemical equivalent*, Z, of a substance is defined as the amount of the substance deposited on passing a steady electric current of 1A for one second through its solution. The S.I. unit of Z is kilogram per coulomb (kg/C).

Faraday's Second Law. This law states as under :

*"When the same quantity of electricity is passed through several electrolytes, the masses of the substances deposited are proportional to their respective *chemical equivalents or equivalent weights.*

---

* Chemical equivalent = $\dfrac{\text{Atomic weight}}{\text{Valency}}$

From this law it follows that the constant of proportionality Z in eqn. (3.1) is proportional to the chemical equivalent.

- The theoretical value of current required for depositing a given quantity of metal and the time for which this current should be passed through the electrolyte can be calculated from the Faraday's laws, if electro-chemical equivalent of the metal is known.

Table 3.1 gives the values of atomic weight, valency, chemical equivalent and electro-chemical equivalent of various elements.

Table 3.1. Values of Atomic weight, Valency, Chemical equivalent and E.C.E. of various elements

S. No.	*Elements*	*Symbol*	*Atomic weight*	*Valency*	*Chemical equivalent*	*Electro-chemical-equivalent kg/coulomb*
1.	*Aluminium*	Al	27.00	3	9.00	9.4×10^{-8}
2.	*Cadmium*	Cd	112.4	2	56.2	58.2×10^{-8}
3.	*Calcium*	Ca	40.08	2	20.04	20.75×10^{-8}
4.	*Chlorine*	Cl	35.45	1	35.37	36.74×10^{-8}
5.	*Chromium*	Cr	52.01	3, 6	17.3, 8.6	17.9×10^{-8}, 9.1×10^{-8}
6.	*Copper (Cupric)*	Cu	63.54	2	31.8	32.9×10^{-8}
7.	*Gold*	Au	197	3	65.4	67.68×10^{-8}
8.	*Hydrogen*	H	1.008	1	1.0	1.045×10^{-8}
9.	*Ferrous*	Fe	55.9	2	27.95	29.02×10^{-8}
10.	*Ferric*	Fe	55.5	3	18.63	19.35×10^{-8}
11.	*Lead*	Pb	207.21	2	103.2	107.16×10^{-8}
12.	*Magnesium*	Mg	24.32	2	11.97	12.43×10^{-8}
13.	*Nickel*	Ni	58.7	2	29.3	30.43×10^{-8}
14.	*Nitrogen*	N	14	3	4.67	4.85×10^{-8}
15.	*Oxygen*	O	16	2	7.98	8.296×10^{-8}
16.	*Potassium*	K	39.1	1	39.04	40.54×10^{-8}
17.	*Silver*	Ag	107.88	1	107.67	111.81×10^{-8}
18.	*Sodium*	Na	22.99	1	22.99	23.87×10^{-8}
19.	*Tin*	Sn	118.7	2	59.3	61.4×10^{-8}
20.	*Zinc*	Zn	65.38	2	32.45	33.88×10^{-8}

3.4. TERMS CONNECTED WITH ELECTROLYTIC PROCESSES

Following terms are used in electrolytic processes :

1. Current efficiency
2. Voltage
3. Energy efficiency.

1. **Current efficiency.** Due to impurities which cause secondary reactions, the quantity of the substance(s) liberated is *slightly less* than that calculated from Faraday's laws. This is taken into account by employing *a factor, called the* ***"current efficiency"***.

The ***current efficiency*** *is defined as the ratio of the actual quantity of substance liberated or deposited to the theoretical quantity, as calculated from Faraday's laws.*

i.e., $$\text{Current efficiency} = \frac{\text{Actual quantity of substance liberated or deposited}}{\text{Theoretical quantity of substance liberated or deposited}}.$$

- The value of current efficiency lies between *90 to 98 percent.*
  — In certain cases this efficiency is very low. For example in chromium plating it 12 to 15% (appr.).

2. **Voltage.** The voltage that is essentially required to pass the current through an electrolyte depends upon the potential drops at the electrodes and in the electrolyte. It is, therefore, desirable that these drops are made as small as possible. This can be achieved, in many cases, by adding *special conducting agents* to the electrolyte to make it (electrolyte) a good conductor. For example dilute sulphuric acid is added to copper sulphate bath in copper plating.

- The normal voltage required to pass current through most electrolytes is 1 to 2 V.

3. **Energy efficiency.** Owing to secondary reactions, the voltage actually required for the deposition or liberation of metal is *higher* than the theoretical value which *increases the actual energy required.*

Energy efficiency *is defined as the ratio of theoretical energy required to the actual energy required for depositing a given quantity of* metal.

i.e., $$\text{Energy efficiency} = \frac{\text{Theoretical energy required}}{\text{Actual energy required}}.$$

Example 3.1. *It is required to repair a worn out circular shaft 15 cm in diameter and 32 cm long by coating it with a layer of 1.6 mm of nickel. Determine the theoretical value of quantity of electricity required and time taken if the current density used is 210 A/m². Electro-chemical equivalent of nickel is 30.4 × 10⁻⁸ kg/C of electricity and density of nickel is 8.9 × 10³ kg/m³.*

Solution. *Given :* $d = 15$ cm $= 0.15$ m ; $l = 32$ cm $= 0.32$ m ; Thickness of coating = 1.6 mm = 0.0016 m ; Current density, $\delta = 210$ A/m² ; $Z = 30.4 \times 10^{-8}$ kg/C ; $\rho = 8.9 \times 10^3$ kg/m³.

Quantity of electricity required, Q :

Surface area of the shaft to be repaired,

$$A_s = \pi d \times l = \pi \times 0.15 \times 0.32 = 0.1508 \text{ m}^2$$

Mass of nickel to be deposited,

$$m = \text{Surface area} \times \text{thickness of coating} \times \text{density of nickel}$$
$$= 0.1508 \times 0.0016 \times 8.9 \times 10^3 = 2.147 \text{ kg}$$

Theoretical value of quantity of electricity required,

$$Q = \frac{M}{Z} = \frac{2.147}{30.4 \times 10^{-8}} \, A\text{-}s = \frac{2.147}{30.4 \times 10^{-8} \times 3600} \, A\text{-}h$$

$$= \mathbf{1961.8 \text{ A-h.}} \quad \textbf{(Ans.)}$$

Time taken, t :

$$\text{Current density,} \quad \delta = \frac{\text{Current } (I)}{\text{Surface area } (A_s)}$$

or, $$210 = \frac{I}{0.1508}$$

∴ $$I = 31.67 \text{ A}$$

Also, $$Q = I \times t$$

∴ $$t = \frac{Q}{I} = \frac{1961.8}{31.67} = \mathbf{61.94 \text{ hours.}} \quad \textbf{(Ans.)}$$

Example 3.2. *In a copper sulphate voltameter, the copper cathode is increased in weight by 0.055 kg in 2 hours, when the current was maintained constant. Calculate the value of this current. Given : Atomic weight of copper = 63.5 ; Atomic weight of hydrogen = 1 ; Atomic weight of silver = 108 ; Electro-chemical equivalent of silver = 111.8 × 10⁻⁸ kg/C.* **(A.M.I.E.)**

Solution. Increase in the weight of copper cathode, m = 0.055 kg (in 2 hours)

Atomic weight of copper = 63.5

Atomic weight of hydrogen = 1

Atomic weight of silver = 108

Electro-chemical equivalent (E.C.E.) of silver = 111.8×10^{-8} kg/C

Value of current, I :

Chemical equivalent of silver $= \dfrac{\text{Atomic weight}}{\text{Valency}} = \dfrac{108}{1} = 108$

Chemical equivalent of copper $= \dfrac{63.5}{2} = 31.75$

E.C.E. of copper, $Z = \text{E.C.E. of silver} \times \dfrac{\text{Chemical equivalent of copper}}{\text{Chemical equivalent of silver}}$

$= 111.8 \times 10^{-8} \times \dfrac{31.75}{108} = 32.867 \times 10^{-8}$ kg/C

Value of the current, $I = \dfrac{m}{Z \times t} = \dfrac{0.055}{32.867 \times 10^{-8} \times (2 \times 3600)} =$ **23.24 A. (Ans.)**

Example 3.3. *Estimate the ampere-hours required to deposit a coating of silver 0.04 mm thick on a sphere of 4.5 cm radius. Electro-chemical equivalent of silver = 111.8 × 10⁻⁸ kg/C and relative density of silver = 10.5.*

Solution. *Given* : Thickness of silver coating = 0.04 mm = 0.04×10^{-3} m

Radius of sphere, r = 4.5 cm = 0.045 m

Electro-chemical equivalent of silver, $Z = 111.8 \times 10^{-8}$ kg/C

Relative density of silver = 10.5

Ampere-hours required :

Surface area of the sphere, $A_s = 4\pi r^2 = 4\pi \times (0.045)^2 = 0.0254 \text{ m}^2$

Mass of silver to deposited, $m = A_s \times$ thickness of coating × density of silver

$= 0.0254 \times (0.04 \times 10^{-3}) \times (10.5 \times {}^*1000) = 0.01067$ kg

$$\left[\begin{array}{l} \because \text{Relative density} = \dfrac{\text{Density of metal}}{\text{Density of water}} \\ {}^*\text{Density of water} = 1000 \text{ kg/m}^3 \end{array}\right]$$

E.C.E. of silver, $Z = 111.8 \times 10^{-8}$ kg/C

$= 111.8 \times 10^{-8} \times 3600 = 0.004$ kg/A-h

∴ Ampere-hours required $= \dfrac{m}{Z} = \dfrac{0.01067}{0.004} =$ **2.66. (Ans.)**

Example 3.4. *Calculate the thickness of copper deposited on a plate area of 2.2 cm² during electrolysis if a current of 1 A is passed for 90 minutes. E.C.E. of copper = 32.95 × 10⁻⁸ kg/C and density of copper = 8900 kg/m³.*

Solution. Plate area, $A_s = 2.2 \text{ cm}^2 = 2.2 \times 10^{-4} \text{ m}^2$

Current strength = 1 A

Duration of passage of 1 A current, t = 90 minutes = 90 × 60 = 5400 s

E.C.E. of copper, $Z = 32.95 \times 10^{-8}$ kg/C

Density of copper $= 8900$ kg/m^3

Thickness of copper deposited, t :

Mass of copper deposited, $m = ZIt$

$= 32.95 \times 10^{-8} \times 1 \times 5400 = 0.001779$ kg

Volume of copper deposited, $V = \frac{\text{Mass}}{\text{Density}} = \frac{0.001779}{8900} = 0.1999 \times 10^{-6}$ m^3

Thickness of copper deposited, $t = \frac{V}{A_s} = \frac{0.1999 \times 10^{-6}}{2.2 \times 10^{-4}} = 0.9086 \times 10^{-3}$ m

$=$ **0.9086 mm. (Ans.)**

Example 3.5. *If 22.092 g of nickel is deposited by 110 A current flowing for 11 minutes, how much copper would be deposited by 55 A current in 7 minutes ? Atomic weights of nickel and copper are 58.6 and 63.18 respectively and valency of both is 2.* **(Gorakhpur University)**

Solution. *Given* : Nickel : $m_{Ni} = 22.092 \times 10^{-3}$ kg, $I = 110$ A, $t = 11$ minutes ; Atomic wt. = 58.6, Valency = 2.

Copper : $m_{Cu} = ?$, $I = 55$ A, $t = 7$ minutes, Atomic wt. = 63.18, Valency = 2.

E.C.E. of nickel, $Z_{Ni} = \frac{m}{I \times t} = \frac{22.092 \times 10^{-3}}{110 \times (11 \times 60)} = 30.43 \times 10^{-8}$ kg/C

E.C.E. of copper, $Z_{Cu} = Z_{Ni} \times \frac{\text{Chemical equivalent of copper}}{\text{Chemical equivalent of nickel}}$

$= 30.43 \times 10^{-8} \times \frac{63.18/2}{58.6/2} = 32.81 \times 10^{-8}$ kg/C

Mass of copper deposited, $m_{Cu} = Z_{Cu} \times I \times t$

$= 32.81 \times 10^{-8} \times 55 \times (7 \times 60) = 7.58 \times 10^{-3}$ kg $=$ **7.58 g. (Ans.)**

Example 3.6. *A copper refining plant, using 450 electrolytic cells, carries a current of 5500 A, voltage per cell being 0.25 V. If the plant were to work 45 hours/week, calculate the energy consumption per tonne. Assume E.C.E. of copper as 32.8 × 10⁻⁸ kg/C.*

Solution. No. of electrolytic cells used $= 450$

Voltage per cell $= 0.25$ V

Current carried $= 5500$ A

No. of working hours of the plant per week $= 45$ hours

E.C.E. of copper, $Z = 32.8 \times 10^{-8}$ kg/C

$= 32.8 \times 10^{-8} \times 3600 = 0.0011808$ kg/A-h

Total number of ampere-hours per annum $= 450 \times 5500 \times 45 \times 52 = 579 \times 10^7$ A-h

Annual output of plant $= 0.0011808 \times 579 \times 10^7$ kg

$= 6836832$ kg $= 6836.83$ tonnes.

Energy consumed per annum $=$ A-h per annum $\times$ voltage per cell

$= 579 \times 10^7 \times 0.25 = 144.75 \times 10^7$ W-h

$= 144.75 \times 10^4$ kWh

Energy consumption per tonne $= \frac{144.75 \times 10^4}{6836.83} =$ **211.72 kWh/tonne. (Ans.)**

Example 3.7. *Calculate the maximum voltage required for electrolysis of water if one kg of hydrogen on oxidation to water liberates 13.985×10^7 joules and E.C.E. of hydrogen is 1.0384×10^{-8} kg/C.*

Solution. The energy is required, during electrolysis of water, to decompose water into hydrogen and oxygen and this is equal to the energy expended in the circuit in forcing the quantity of electricity through the electrolyte.

Energy expended during electrolysis $= \frac{1}{Z} \times V$ watt-sec per kg

$= \frac{V}{Z}$ joules per kg

Energy liberated by 1 kg of hydrogen when it combines with oxygen

$= 13.985 \times 10^7$ joules

$\therefore \quad \frac{V}{Z} = 13.985 \times 10^7$

or $\quad V = Z \times 13.985 \times 10^7 = 1.0384 \times 10^{-8} \times 13.985 \times 10^7 = \mathbf{1.452\ V.}$ **(Ans.)**

Example 3.8. *Calculate the quantity of aluminium produced from aluminium oxide in 24 hours if the average current is 2800 A and current efficiency is 95 per cent. Aluminium is trivalent and atomic weight is 27. The chemical equivalent weight and E.C.E. of silver are 107.98 and 111 × 10⁻⁸ kg/C respectively.* **(Gorakhpur University)**

Solution. Average current $= 2800$ A

Current efficiency $= 95\%$

Valency of aluminium $= 3$

Atomic weight of aluminium $= 27$

Chemical equivalent weight of silver $= 107.98$

E.C.E. of silver $= 111 \times 10^{-8}$ kg/C

Chemical equivalent weight of aluminium $= \frac{\text{Atomic weight}}{\text{Valency}} = \frac{27}{3} = 9$

$\therefore$ E.C.E. of aluminium,

$$Z = \frac{\text{E.C.E. of silver} \times \text{Chemical equivalent weight of aluminium}}{\text{Chemical equivalent weight of silver}}$$

$$= \frac{111 \times 10^{-8} \times 9}{107.98} = 9.252 \times 10^{-8} \text{ kg/C}$$

Mass of aluminium produced, $m = ZIt \times$ current efficiency

$= 9.252 \times 10^{-8} \times 2800 \times (24 \times 60 \times 60) \times 0.95$

$= \mathbf{21.26\ kg.}$ **(Ans.)**

3.5. APPLICATIONS OF ELECTROLYSIS

The *major applications of electrolysis* are as under :

1. Electro-deposition
   - (*i*) Electroplating
   - (*ii*) Electro-deposition of rubber
   - (*iii*) Electro-metallisation
   - (*iv*) Electro-facing
   - (*v*) Electro-forming
   - (*vi*) Electro-typing.
2. Manufacture of chemicals
3. Anodizing
4. Electropolishing
5. Electro-cleaning or pickling

6. Electro-parting or electro-stripping
7. Electro-metallurgy
(i) Electro-extraction (ii) Electro-refining.

3.5.1. Electro-deposition

The process of depositing a coating of one metal over another metal or non-metal electrically is called the **electro-deposition.**

- It is used for protective, decorative and functional purposes and includes such processes as electro-plating, electro-forming, electro-typing, electro-facing, electro-metallisation etc.

As earlier discussed that, the compounds in the solution dissociate into positive and negative ions which when subjected to electric field travel towards respective electrodes then, one of the following *events* may take place :

(*i*) In case the ion, after giving off electric charge to electrode, has *stable existence* and *does not have chemical reaction* with electrode material, it will be *deposited on the electrode*. ... This is the *principle of electro-deposition and electro-extraction.*

(*ii*) The ion after giving off electric charge to electrode may *undergo chemical reaction* with electrode material, the product of reaction in turn is *soluble in the electrolyte* and the *electrode is gradually eaten away* .... This principle is employed in *Electro-refining.*

(*iii*) Ion if after giving off charge to electrode, does not react with the electrode material, or has any independent and stable existence, will react with the water of solution, thereby liberating oxygen or hydrogen. ...

This principle relates to *electrolysis of acidulated water.*

Factors on which quality of electrodeposition depends :

Following are the factors on which the quality of electro-deposition depends :

1. *Nature of electrolyte.* The electrolyte from which complex ions can be obtained (*e.g.*, cyanides) provides a smooth deposit.

2. *Current density.* The deposit of metal will be uniform and fine-grained if the current density is used at a rate higher than that at which the nuclei are formed. The deposit will be strong and porous if the rate of nuclei formation is very high due to very high current density.

3. *Temperature.* A low temperature of the solution favours formation of small crystals of metal ; and a high temperature, large crystals.

4. *Conductivity.* The solution of good conductivity provides economy in power consumption and also reduces the tendency to form trees and rough deposits.

5. *Electrolytic concentration.* By increasing the concentration of the electrolyte, higher current density can be achieved, which is necessary to obtain uniform and fine-grain deposit.

6. *Addition agents.* The addition of acid or other substances to the electrolyte reduces its resistance. Addition agents like glue, gums, dextrose, dextrin etc. influence the nature of deposit. The crystal nuclei absorb the addition agent added in the electrolyte ; this prevents it to have large growth and thus deposition will be fine-grained.

7. *"Throwing power".* It is defined as the *ability of the electrolyte to produce even irregular surfaces.* Due to irregular shape of the cathode the distance between the various portions of the cathode and anode will be different. Due to unequal distance, the resistance of the current path through the electrolyte for various portions of the cathode will be different but the potential difference between the anode and any point on the article to be plate (cathode) will be, of course, be the same and the result will be that the *current density will be more on the portion nearer to anode and it will cause uneven deposit of the metal.*

Throwing power can improved by the following *two ways* :

(*i*) By *increasing the distance between the anode and cathode* ;

(*ii*) By *reducing the voltage drop at the cathode surface.* In some cases decrease of current density causes a decrease in voltage drop at the cathode, leaving more voltage available for overcoming the resistance of the electrolyte, thus tending to counteract any change in current concentration.

— Copper cyanide bath is better suited for electroplating intricate articles.

— Zinc cyanide bath has better throwing power than zinc sulphate solution for zinc plating.

8. **Polarization.** With the increase in the electroplating current density, rate of metal deposition is also increased upto certain limit after which electrolyte surrounding the base metal becomes so much depleted of metal ions that rate of deposition does not increase with increase in current density. If current density more than this limit is employed, it will result in electrolysis of water and *hydrogen deposition on the cathode.* This hydrogen evolved, *blankets the base metal which diminishes the rate of metal deposition. This phenomenon is called* **polarization.** *Blanketing effect can be reduced by agitating the electrolyte.*

With "*reverse current electroplating*", in which at regular intervals plating current is reversed for a second or so, sufficient electron concentration is established around the base metal and the *polarization effect becomes negligible* even with very high overall speed of plating. The other advantages of reverse current plating are : (*i*) During reverse current period the unsound and inferior metal is depleted and the *flat level surfaces are produced.* (*ii*) Metal surface is *brightened* causing elimination of buffing or polishing operation.

3.5.1.1. Electroplating

"Electroplating" is an art of depositing a superior or a more noble metal on an inferior or a base metal by means of electrolysis of an aqueous solution of a suitable electrolyte.

Or

"Electroplating" is *defined as the electro-deposition of metal upon metallic surfaces.*

Electroplating is done to *accomplish the following* :

(*i*) To protect the metals against corrosion.

(*ii*) To give reflecting properties reflectors.

(*iii*) To give a shiny appearance to articles.

(*iv*) To replace worn out material.

- The electrolytic deposits are crystalline in nature. The crystals must be very fine in order to get firm, coherent and uniform deposits. For this purpose, suitable electrolytes should be used in the electrolytic bath and current density used should have an appropriate value. The temperature should also be maintained at a proper level.
- The articles to be coated with nobler metals should be in as high a state of purity as possible.

Operations involved in electroplating :

Various operations involved in electroplating are :

1. Cleaning operation.

2. Deposition of metal.

1. **Cleaning operation.** In case the object to be electroplated is not cleaned, polished and degreased, the deposit formed may not be well adherent to the base metal and is likely to peel off.

Cleaning operation includes the following :

- Removal of oil, grease, or other organic material. .... To accomplish this, soaps, hot alkali solutions, or organic solvents such as gasoline or carbon tetrachloride are used.

- Removal of rust, scale, oxides, or other inorganic coatings adhering to the base metal/ work piece .... To accomplish this various acids, alkali and salt solutions are employed.
- Mechanical preparation of the surface of the metal to remove the deposited metal, by polishing, buffing etc. ... To accomplish this mechanical abrasion and polishing are used.

2. **Deposition of metal.** In all types of metal deposition processes, *article to be electroplated is made cathode, solution is made up of salt of the metal to be deposited and anode is often of the same metal which is to be deposited.*

Details of preparation of solutions and current densities employed for deposition of various metals are given henceforth.

Copper plating. Copper plating baths are of the following two types :

- In *acid bath*, solution is made of 150 to 200 gm of copper sulphate, 25 to 37 gm of H_2SO_4 per 1000 c.c. of solution. Current density employed is 200 to 400 A/m^2 and temperature of 25° to 50°C. Anode is made of copper.

Deposit obtained is *thick and rough which requires polishing.*

- *Cyanide bath* consists of 25 gm of copper cyanide, 28 gm of sodium cyanide, 6 gm of sodium carbonate and 6 gm of sodium biphosphate per 1000 c.c. Current density used is 4 to 150 A/m^2, and temperature of 35° to 50°C. Anode employed is of copper.

This gives *thin and smooth deposit.*

Copper plating is used :

(*i*) for iron articles to prevent them from rusting ;

(*ii*) as undercoat for silver and nickel plating.

Nickel plating :

- For steel and brass articles, nickel bath consists of nickel sulphate 100 gm, ammonium chloride 12 gm, boric acid 12 gm per 100 c.c. Temperature employed is 20° to 30°C and current density of 10 to 20 A/m^2. Anode is of pure nickel.
- For steel, brass, copper and zinc, nickel plating bath consists of nickel sulphate 180 to 240 gm, nickel chloride 36 gm, boric acid 24 gm for 1000 c.c. Working temperature employed is 40° to 65°C with current density of 250 to 500 A/m^2. Anode is of pure nickel.

Chromium plating. Chromium plating bath consists of 180 to 300 gm of chromic acid with 2 to 3 gm of sulphuric acid per 1000 c.c. Working temperature of 40° to 70°C with current density of 600 to 5000 A/m^2 are employed. Current density is higher for hard chromium plating than for decorative plating. Anodes are of antinominal lead and chromic acid is added as may be necessary.

Vats employed for chromium plating are of steel with lead lining. Arrangement for removal of fumes must be made.

- Chromium plating *gives highly polished and extremely hard coating.*
- It is used where surface is required to be protected from atmospheric corrosion.

Silver plating. Plating solution consists of 24 gm of silver cyanide, 36 gm of potassium cyanide, 24 gm potassium carbonate per 1000 c.c. Temperature of 20° to 35°C with current density of 30 to 40 A/m^2 are employed. 25 c.c. of bisulphide are also added as brightening agent.

- In radio communication, high frequencies are employed. Due to skin effect current flows only on the surface layer. Conductor deposited with 0.0125 mm thick silver layer is as good as solid silver conductor.
- Silver plating is also used for plating of bearings. Silver has *high resistance to corrosion effect of lubricating oil, high melting point and good seizure resistance.* For this purpose, a layer of 0.05 mm is sufficient.

Gold plating. Plating solution consists of 18 gm of potassium gold cyanide, 12 gm of potassium cyanide, 12 gm of caustic potash, 6 gm of potassium sulphate per 1000 c.c. Temperature of 50° to 80°C and current density of 20 to 60 A/m^2 are employed. Anode is of stainless steel.

3.5.1.2. Electro-deposition of rubber

Rubber latix obtained from the tree consists of very fine colloidal particles of rubber suspended in water. Like other colloidal solutions, particles of rubber are negatively charged. On electrolysis of the solution the rubber particles migrate towards the anode and deposit on it. Current density of 100 A/m^2 (apprr.) is used.

3.5.1.3. Electro-metallisation

- It is the *process of depositing metal on conducting base for decoration and for protective purposes.*
- Nonconductive base is made conductive by a coating of graphite which is made the cathode.

3.5.1.4. Electro-facing

It *is a process of coating of metallic surface with a harder metal by electro-deposition in order to increase its durability.*

3.5.1.5. Electro-forming

"Electro-forming" is *the reproduction of objects by electro-deposition on some sort of a mould or form.*

- In the *reproduction of medals, coins, engravings etc.*, a mould is first made by impressing the object, say, in wax. The wax surface, which bears exact impressions of the object, is coated by powdered graphite in order to make it conducting. The mould is then dipped in an electro-forming cell as a cathode. After obtaining coating of desired thickness, the article is removed and the wax core is melted out of the metal shell.
- Other applications of electro-forming are :

(*i*) *Manufacture of gramophone records.*

(*ii*) *Production of seamless tube.*

3.5.1.6. Electro-typing

Electro-typing is a *special application of electro-forming* and it is used to reproduce printing, setup type, engraving and medals etc. The process is same as for electro-forming.

3.5.2. Manufacture of Chemicals

The industrial applications of electrolysis includes the manufacture of chemicals like chlorine, caustic soda, potassium permanganate, ammonium per sulphate, hydrogen and oxygen. Production of caustic soda and production of hydrogen and oxygen by electrolysis of water are described below.

1. **Production of caustic soda :**

- *Diaphragm process* of producing caustic soda is the oldest one. Here chlorine is formed at the anode, and most of it is evolved as a gas, a small part going into solution. Sodium is discharged at the cathode and reacts with the hydroxyl ions to form sodium hydroxide and hydrogen gas is liberated at the cathode. Usually, the brine is fed into the anode compartment *to resist the flow of hydroxyl ions towards the anode.*
- *Mercury cathode process* is another method of producing caustic soda by electrolytic process.
  - — Mercury cells may be built in small (1000 A) to very large sizes (3000 A) or even as high as 50000 A or more per unit. Purified saturated brine is fed to the cells and commonly circulated through them with a reduction in salt concentration. Brine is resaturated for another cycle.

— Each unit consists of two sections ; *one* which is closed and has a chlorine outlet, graphite anodes, and a mercury cathode ; a *second*, the denuder section, containing a mercury amalgam anode and iron cathodes. Circulation of mercury between two sections constitutes the electrical connection.

— Current densities in mercury cells are considerably higher than in diaphragm units, being of the order of 1.5 mA or more sq. mm. of anode surface.

— The voltage requirement is about 4 volts per cell. When graphite anodes are used, the anode consumption is of the order of 3-3.5 kg/tonne of chlorine.

Energy efficiencies are of the order of 50-60 percent.

2. **Production of hydrogen and oxygen by electrolysis of water :**

- The gases obtained by this process are of high purity and at a cheap cost due to *low energy consumption.*
- The electrodes are of iron and the electrolyte consists of 15-20 percent solution of caustic soda (or its equivalent caustic potash). The chemical reactions are :

At Cathode : $2Na + 2H_2O = 2NaOH + H_2$

At Anode : $2OH = H_2O + \frac{1}{2}O_2$

Thus hydrogen and oxygen are liberated at the cathode and anode respectively and water disappears, the quantity of NaOH remaining constant. It is, therefore, necessary to add water to the solution periodically.

Voltage required : During operation ...... 2 to 2.2 V

During starting period ...... 2.3 to 2.5 V

Energy consumption : ...... 6 kWh per m^3 of H_2, and $\frac{1}{2}m^3$ of O_2

Summary of manufacture of chemicals under various industrial applications of electrolysis is given in table 3.2.

Table 3.2. Manufacture of chemicals

S. No.	*Product*	*Solution*	*Energy consumption (kWh/tonne)*
1.	*Caustic soda*	Brine	3000 to 4000
2.	*Chlorine, compounds of sodium (Chlorates per-chlorates, hypochlorates)*	Brine	3000 to 7000
3.	*Potassium-permanganate*	Potassium-manganate	70 to 80
4.	*Hydrogen and oxygen*	Water	About 6 kWh per m^3 of H_2 and $\frac{1}{2}m^3$ of O_2
5.	*Ammonium per sulphate*	Ammonium-sulphate	2000 to 2500

3.5.3. Anodizing

An "*anodic coating*" means an oxide film deposited/created on a metal surface, with the help of an *anode* and *oxidation.*

The process of providing an oxide film is known as ***Anodizing.***

The surface of the metal can be anodized to achieve one or more of the following :

(*i*) To provide a protective coating on the surface.

(*ii*) To provide a decorative appearance to the surface.

(*iii*) To provide a specific colour base for subsequent painting on the surface.

(*iv*) To provide a bright and smooth surface on aluminium articles as a prior treatment to electroplating.

(*v*) To improve the corrosion resistance of aluminium and aluminium products.

The objectives mentioned above can be achieved by *using different electrolytes and varying their densities during the process, varying current and temperature of solution and similar other measures.*

In the ***anodizing process*** an electrolyte circuit is used in which the workpiece is made the *anode.* The *electrolyte should give up oxygen on electrolysis.* In *anodizing,* there is no deposition of a layer from outside. As the current is passed, a chemical reaction takes place and oxygen is liberated which reacts with the metal surface to convert it into an oxide. In this process the reaction progresses inwards into the metal *i.e.,* it is a part of the workpiece metal itself, all along its surface, that is converted into an oxide film. *Deeper the penetration of the chemical reaction the thicker will be the oxide film formed.*

Anodizing process is neither a purely electrical process nor a purely chemical process ; it is a combination of both *i.e.*, an *electrochemical process.* It is completed in the following *three stages :*

(*i*) The "*first stage*" consists of *cleaning and preparing the surface* through different cleaning methods.

(*ii*) The "*second stage*" consists of *anodizing i.e.*, converting the metal surface into an oxidized film.

(*iii*) The "*third stage*" consists of *providing desired colour* and *stability to the anodized film*, called *sealing* of oxide film.

For anodizing *aluminium and aluminium alloy products* only acid solutions are used as *electrolytes.*

— Sulphuric acid, chromic acid and oxalic acid are the three commonly used acids for making electrolyte solutions. The selection of a particular acid depends upon the type of colour shade we intend to provide to the anodized surface.

- Although *anodizing* is mainly employed for aluminium and aluminium alloys, yet with suitable modifications in the techniques and variations in the solutions used it is possible to anodize other metals such as *steel, galvanised iron and steel, magnesium, brass and bronze, zinc, copper and silver.*

3.5.4. Electro-polishing

This process, in principle, *consists of making the work as anode in a suitable position.* This produces insoluble compounds, which are broken down by more anodic action on the hills than on valleys of the surface. In this way smoothening of surface takes place.

- Aluminium workpiece which is to be polished after being properly buffed is subjected to two anodic treatments in series. In the first treatment job is made anode in fluoboric acid bath, this removes thin coating of metal from the surface uniformly. This surface is then given anodizing treatment in H_2SO_4 bath. This process produces an oxide film clear and transparent with reflectivity as much as 90 per cent.
- In case of silver plated article, polishing is done after plating is over by reversing the current through bath at 4 to 5 times the current strength used for plating and at intermittent intervals every few seconds.

3.5.5. Electro-cleaning or Pickling

Electrolytic solution of sodium phosphate is contained in iron tank which is made *anode.* Work is suspended as *cathode.* When the current is passed, it produces caustic soda on cathode which has *cleaning action.* Also large volume of *hydrogen* evolved at the cathode *quickly removes grease.* This is called ***cathodic cleansing.***

In ***anodic cleansing***, *work is made anode.* Dirt particles are positively charged in alkaline bath and by electrophoresis process they migrate to negative pole.

3.5.6. Electro-parting or Electro-stripping

Two or more metals may be *separated electrolytically.* As an example if copper is to be stripped off from steel, the workpiece is made anode in a solution of 75 gm of sodium cyanide, 25 gm of caustic soda in 1000 c.c. of water. Sheet of iron is made cathode and pressure used is 6 V.

3.5.7. Electro-metallurgy

Electro-metallurgy includes electro-extraction and electro-refining processes.

3.5.7.1. Electro-extraction

Extraction of metal is an *electro-chemical process employed for production of metal with commercially acceptable purity.*

Depending upon the physical state of the ore, following are the two methods of extraction of metals :

(*i*) In one of the methods the ore is treated with a strong acid to obtain a salt and the solution of such a *salt is electrolysed to liberate the metal.*

(*ii*) The second method is used when the ore is available in molten state or can be fused and in this method the ore, which is in a molten state is *electrolysed in a furnace.*

The methods used for extraction of some of the metals are described below :

1. **Extraction of zinc :**

The ore (consisting largely of zinc oxide) is treated with concentrated H_2SO_4, roasted and passed through various chemical processes in order to remove impurities (*e.g.,* cadmium, copper etc.) *by precipitation.* The zinc sulphate so obtained is then *electrolysed.* The electrolysis of zinc sulphate is accomplished in large lead-lined wooden boxes having a number of aluminium cathodes and lead anodes. Zinc gets deposited on the cathodes and is removed periodically, once or twice a day.

— The current density on the cathode is 1000 A/m^2 (appro.)

— The voltage per cell is 3.5 V (appro.). Usually 100 to 150 cells are employed in series giving a pressure of about 500 V.

— The energy consumption is 3000 to 5000 kWh/tonne.

2. **Extraction of aluminium :**

Aluminium is produced from *bauxite* containing aluminium oxide or alumina (70 per cent in high grade bauxite), silica (silicon oxide) and iron oxide as follows :

— The ore (bauxite) is first *reduced to aluminium oxide* by chemical treatment and then it is dissolved in *fused* cryolite (Cryolite is a solution of aluminium fluoride and fluoride of either of sodium, potassium or calcium). The mixture thus obtained is *electrolysed.*

— The fusion and electrolysis ore are accomplished in a large shallow rectangular steel bath lined with carbon ; carbon anodes projecting downwards into the bath and bottom of the bath forms the cathode. The charge is melted by the arc struck between the carbon anodes and cathode and is then maintained in a molten state by heating action of the electric current flowing through the charge. The *liquid metal deposits at the cathode and settles at the bath bottom and periodically siphoned out* into the large ladles from which it is poured into pig or ingot moulds.

— Fresh alumina is fed into the bath at short intervals to replace that which has been decomposed by the current and the process is, therefore, *continuous one*.

- The aluminium obtained by this process is 99.5%.
- A furnace having an area of about 15 m^2 will require a voltage of about 6 V and a current of about 40000 A.
- Energy consumption is 20000 to 25000 units per tonne.
- The high temperature (1000°C) necessary to keep the ores in a fused state is maintained by the ohmic losses due to the current flowing through the electrodes and electrolyte.

Almost the whole of the aluminium required in the present day industry is produced in this way. Since the electrolytic process requires a large amount of electric power and process is continuous, so such plants are installed near hydro-electric power stations.

3. **Extraction of magnesium.** Magnesium is obtained by *electrolysis of magnesium chloride*, employing current density of 160 to 350 A/m^2. Energy expenditure is 17 to 20 kWh/kg and operating temperature of 360 to 670°C.

4. **Extraction of sodium.** Sodium is obtained by the *electrolysis of sodium hydroxide, sodium nitrate or sodium chloride* requiring expenditure of 10 to 20 kWh/kg of sodium.

3.5.7.2. Electro-refining

Refining *is the process whereby a highly concentrated mixture of metals is subjected to electrochemical treatment for recovering not only the principal metal in pure form, but also the precious metals like gold, silver, bismuth etc., which may be present in the form of minute traces.*

By electro-refining, it is possible to get metal of *almost 100 % purity.* This is one of the most important prerequisite expected out of copper and aluminium in order *to have high electrical conductivity.* Electro-refining process in essence is same as electro-plating, *anode being made of impure metal and electrolyte being made of the salt of the metal to be refined. Pure metal is deposited at the cathode.*

- *Copper refining* requires copper sulphate solution with electric consumption of 150 to 300 kWh/tonne of copper refined.
- *Silver* is refined requiring solution of nitric acid and silver nitrate with electric consumption of 400 to 420 kWh/tonne.
- Iron is refined by using solution of iron-ammonium sulphate with electric consumption of 1000 to 1600 kWh/tonne.
- *Lead* is refined by using solution of lead flurosilicate with electric consumption of 100 to 120 kWh/tonne.
- *Gold refining* requires gold chloride solution with electric consumption of 300 to 350 kWh/tonne of gold refined.
- *Nickel refining* requires nickel-ammonium sulphate solution with electric consumption of 2500-4000 kWh/tonne of nickel refined.

3.6. POWER SUPPLY FOR ELECTROLYTIC PROCESSES

For electrolytic process, the power supply required is *D.C.* and at *very low voltage.*

- The power required for electro-deposition is usually very small (between 100 and 200 A at 10 or 12 V) and can be obtained either by employing a motor-generator set consisting of a standard induction motor driving a heavy current low voltage D.C. generator (preferably separately excited) or by employing the copper oxide rectifier. The latter is preferred since it has high operating efficiency, occupies less space and its maintenance cost is low.

Solid state rectifying devices employing germanium and silicon diodes have been developed. These devices occupy very small space even as compared to metal rectifiers.

- For extraction and refining of metals and large scale manufacture of chemicals a very large amount of power is required. Since most of the processes are continuous and as such have a 100 percent load factor, these plants are located near hydroelectric power stations or atomic power plants even if extra transportation or raw material is necessitated.

HIGHLIGHTS

1. The processes based on the fact that electrical energy can produce chemical changes are called **Electrolytic processes.**
2. *Faraday's First Law.* It states : "The mass of a substance liberated from an electrolyte in a given time is proportional to the quantity of electricity passing through the electrolyte".
   Faraday's Second Law. It states : "When the same quantity of electricity is passed through several electrolytes, the masses of the substances deposited are proportional to their respective chemical equivalents or equivalent weights".
3. *Current efficiency* is defined as the ratio of the actual quantity of substance liberated or deposited to the theoretical quantity, as calculated from Faraday's laws.
4. *Energy efficiency* is defined as the ratio of theoretical energy required to the actual energy required for depositing a given quantity of metal.
5. *Applications of electrolysis :*
   1. Electro-deposition
   2. Manufacture of chemicals
   3. Anodizing
   4. Electro-polishing
   5. Electro-cleaning or pickling
   6. Electro-parting of electro-stripping
   7. Electro-metallurgy :
      (*i*) Electro-extraction
      (*ii*) Electro-refining

OBJECTIVE TYPE QUESTIONS

Fill in the Blanks or Say 'Yes' or 'No'

1. The processes based on the fact that electrical energy can produce chemical changes are called ...... processes.
2. ...... bond is responsible for the formation of inorganic compound.
3. Faraday's ...... law states that the mass of a substance liberated from an electrolyte in a given time is proportional to the quantity of electricity passing through the electrolyte.
4. Chemical equivalent is the ratio of atomic weight to valency.
5. The ratio of the actual quantity of substance liberated or deposited to the theoretical quantity, as calculated from Faraday' laws is called energy efficiency.
6. The value of current efficiency lies between 90 to 98%.
7. The process of depositing a coating of one metal over another metal or non-metal electrically is called the ......
8. Throwing power is defined as the ability of the electrolyte to produce even irregular surfaces.
9. ...... is an art of depositing a superior or a more noble metal on an inferior or a base metal by means of electrolysis of an aqueous solution of a suitable electrolyte.
10. Electro-metallisation is the process of depositing metal on a conducting base for decoration and for protective surfaces.
11. ...... is the process of coating of a metallic surface with a harder metal by electro-deposition in order to increase its durability.

12. ...... is the reproduction of objects by electro-deposition on some sort of a mould or form.
13. The process of providing an oxide film is known as ....
14. Electro-polishing process, in principle, consists of making the work as anode in a suitable position.
15. By electro-refining, it is possible to get metal of almost 100% purity.

ANSWERS

1. Electrolytic	2. Electrovalent	3. First	4. Yes	5. No.
6. Yes	7. Electro-deposition	8. Yes	9. Electroplating	10. Yes
11. Electro-facing	12. Electro-forming	13. Anodizing	14. Yes	15. Yes.

THEORETICAL QUESTIONS

1. What are electrolytic processes ?
2. What is electrolysis ? Explain briefly.
3. State Faraday's laws of electrolysis and explain them clearly.
4. Explain the following terms used in electrolytic processes :
   (*i*) Current efficiency (*ii*) Energy efficiency
   (*iii*) Throwing power (*iv*) Electro-chemical equivalent.
5. What are the applications of electrolysis ?
6. What is electro-deposition ? Explain in detail various factors which have effect on the appearance and quality of the deposited surface.
7. Explain the basic laws which govern electro-deposition.
8. Give the advantages of reverse-current plating.
9. Explain the term 'polarisation, 'throwing power' and 'electro-deposition'. How are zinc and copper refined from their base metal electrically.
10. What is electroplating and what for is it done ? Describe the various operations involved in electroplating.
11. Describe in detail the process of nickel-electroplating in industry, giving the composition of electrolyte. What are the factors on which the quality of electroplating depends ?
12. Describe the process of copper plating on a piece of job. What arrangements are made to feed and control the supply of electric power to the electro-plating plant ?
13. Describe the equipment and process used for chromium plating. What is the composition of electrolyte used ?
14. What do you understand by 'reverse current process' in electro-plating ? Discuss its advantages.
15. What is meant by anodizing ? Explain the process of anodizing and describe the equipment used for it.
16. Explain briefly extraction and refining of metals by electrolysis.
17. Describe briefly the extraction of the following metals. Zinc, aluminium, magnesium and sodium.
18. What do you mean by 'Electro-refining' ?

UNSOLVED EXAMPLES

1. It is required to repair a worn out circular shaft 14 cm in diameter and 30 cm long by coating it with a layer of 1.5 mm of nickel. Determine the theoretical value of quantity of electricity required and the time taken if the current density used is 200 A/m^2. Electro-chemical equivalent of nickel is 30.4 $\times$ 10^{-8} kg/C of electricity and density of nickel is 8.9 $\times$ 10^3 kg/m^3. **[Ans.** 1610 A-h ; 61 hours]
2. In a copper sulphate voltameter, the copper cathode is increased in weight by 0.05 kg in two hours, when the current was maintained constant. Calculate the value of this current. Given Atomic weight of copper = 63.5 ; Atomic weight of hydrogen = 1 ; Atomic weight of silver = 108 ; Electro-chemical equivalent of silver = 111.8 $\times$ 10^{-8} kg/C. **[Ans.** 21.13 A]

3. **Calculate the ampere-hours required to deposit a coating of silver 0.05 mm thick on a sphere of 5 cm radius. Assume E.C.E. of silver = 111.8 × 10^{-8} kg/C and relative density of silver to be 10.5.** **[Ans.** 4.12]

4. **Estimate the thickness of copper deposited on a plate area of 2.5 cm^2 during electrolysis if a current of 1 A is passed for 100 minutes. E.C.E. of copper = 32.95 × 10^{-8} kg/C and density of copper = 8900 kg/m^3.** **[Ans.** 0.888 mm]

5. **A copper refining plant, using 500 electrolytic cells, carries a current of 6000 A, voltage per cell being 0.25 V. If the plant were to work 40 hours/week, calculate the energy consumption per tonne, assuming E.C.E. of copper as 32.81 × 10^{-8} kg/C.** **[Ans.** 211.66 kWh/tonne]

6. **Estimate the quantity of aluminium produced from aluminium oxide in 24 hours if the average current is 3000 A and current efficiency is 92 per cent. Aluminium is trivalent and atomic weight is 27. The chemical equivalent weight of silver is 107.98 and 0.00111 gm of silver is deposited by one coulomb.** **[Ans.** 22.062 kg]

4

Refrigeration and Air-conditioning

REFRIGERATION

4.1. INTRODUCTION TO "REFRIGERATION"

- **Refrigeration** *is the science of producing and maintaining temperatures below that of the surrounding atmosphere.* This means the removing of heat from a substance to be cooled. Heat always passes downhill, from a warm body to a cooler one, until both bodies are at the same temperature. Not only perishables, today many human work spaces in offices and factory buildings are air-conditioned and a refrigeration unit is the heart of the system.
- Before the advent of mechanical refrigeration water was kept cool by storing it in semi-porous jugs so that the water could seep through and evaporate. The evaporation carried away heat and cooled the water. This system was used by the Egyptians and by Indians in the Southwest. Natural ice from lakes and rivers was often cut during winter and stored in caves, straw-lined pits, and later in sawdust insulated buildings to be used as required. The Romans carried pack trains of snow from Alps to Rome for cooling the Emperor's drinks. Though these methods of cooling all make use of natural phenomena, they were used to maintain a lower temperature in a space or product and may properly be called refrigeration.
- In simple, **refrigeration** *means the cooling or removal of heat from a system.* The *equipment employed to maintain the system at a low temperature is termed as* **refrigerating system** and the system which is *kept at lower temperature is called* **refrigerated system.**

Refrigeration is generally produced in one on the following *three ways* :

(*i*) By melting of a solid,

(*ii*) By sublimation of a solid, and

(*iii*) By evaporation of a liquid.

- Most of the commercial refrigeration is produced by the evaporation of a liquid called *refrigerant*. **Mechanical refrigeration** depends upon the evaporation of liquid refrigerant and its circuit includes the equipments, naming evaporator, compressor, condenser and expansion valve. It is used for preservation of food, manufacture of ice, solid carbon-dioxide and control of air temperature and humidity in the air-conditioning system.

4.2. APPLICATIONS OF REFRIGERATION

Important **refrigeration applications** are given below :

1. Ice making
2. Transportation of foods above and below freezing
3. Industrial air-conditioning
4. Comfort air-conditioning
5. Chemical and related industries
6. Medical and surgical aids
7. Processing food products and beverages
8. Oil refining and synthetic rubber manufacturing
9. Manufacturing and treatment of metals
10. Freezing food products
11. Miscellaneous applications :
    (*i*) Extremely low temperatures
    (*ii*) Plumbing
    (*iii*) Building construction etc.

4.3. ELEMENTS OF REFRIGERATION SYSTEMS

All refrigeration systems must include atleast **four basic** units as given below :

(*i*) A low temperature thermal "sink" to which heat will flow from the space to be cooled.

(*ii*) Means of extracting energy from the sink, raising the temperature level of this energy, and delivering it to a heat receiver.

(*iii*) A receiver to which heat will be transferred from the high temperature high pressure refrigerant.

(*iv*) Means of reducing pressure and temperature of the refrigerant as it returns from the receiver to the "sink".

4.4. REFRIGERATION SYSTEMS

The various refrigeration systems may be enumerated as below :

1. Ice refrigeration
2. Air-refrigeration system
3. Vapour compression refrigeration system
4. Vapour absorption refrigeration system
5. Special refrigeration systems :
    (*i*) Absorption refrigeration system
    (*ii*) Cascade refrigeration system
    (*iii*) Mixed refrigeration system
    (*iv*) Vortex tube refrigeration system
    (*v*) Thermoelectric refrigeration
    (*vi*) Steam jet refrigeration system.

4.5. CO-EFFICIENT OF PERFORMANCE (C.O.P.)

The performance of a refrigeration system is expressed by a term known as the **Co-efficient of performance** *which is defined as the ratio of heat absorbed by the refrigerant while passing through the evaporator to the work input required to compress the refrigerant in the compressor ; in short it is the ratio between heat extracted and work done (in heat units).*

If, R_n = net refrigerating effect , and W = work expended in by the machine during the same interval of time, then,

$$\text{C.O.P.} = \frac{R_n}{W}$$

$$\textbf{Relative C.O.P.} = \frac{\text{Actual C.O.P.}}{\text{Theoretical C.O.P.}}$$

where, actual C.O.P. = ratio of R_n and W actually measured during a test and, theoretical C.O.P. = ratio of theoretical values of R_n and W obtained applying laws of thermodynamics to the refrigeration cycle.

4.6. STANDARD RATING OF A REFRIGERATION MACHINE

The rating of a refrigerating machine is obtained by refrigerating effect or amount of heat extracted in a given time from a body. The rating of the refrigeration machine is given by a unit of refrigeration known as **standard commercial tonne of refrigeration** *which is defined as the refrigerating effect (RE) produced by the melting of 1 tonne of ice from and at 0°C in 24 hours.* Since the latent heat of fusion of ice is 336 kJ/kg, the refrigerating effect of 336 × 1000 kJ/kg in 24 hours is rated as **one tonne,** *i.e.*,

$$\textbf{A tonne of refrigeration} = \frac{336 \times 1000}{24} = 14000 \text{ kJ/h.}$$

Note. Ton of refrigeration (TR). A ton of refrigeration is basically an American unit of *RE*. It originated from the rate at which heat is required to be removed to freeze *one ton of water from and at 0°C.* Using American unit this is equal to removal of 200 BTU of heat per minute, and in MKS units it is adopted as 50 kcal/min or 3000 kcal/hour. In S.I. units its conversion is rounded off to 3.5 kJ/s (kW) or 210 kJ/min.

(*1 ton* of refrigeration may be taken equal to *0.9 tonne* of refrigeration)

4.7. AIR-REFRIGERATION SYSTEM

4.7.1. Introduction

Air cycle refrigeration is one of the earliest methods of cooling developed. It became obsolete for several years because of its *low* coefficient of performance (C.O.P.). It has, however, found its application again in *aircraft refrigeration systems,* where with low equipment weight, it can utilise a portion of the cabin air according to the supercharger capacity.

The *main characteristic feature* of air-refrigeration system is that *throughout the cycle the refrigerant remains in gaseous state.*

The air-refrigeration system can be divided in *two systems :*

1. Closed system.

2. Open system.

1. Closed system (or Dense air machine). In this system the *air refrigerant is contained within the piping or components/parts of the system at all times and with refrigerator usually pressure is above atmospheric pressure.*

A closed system has the following *thermodynamic advantages :*

(*i*) It can work at suction pressure higher than atmospheric. This reduces the volume handled by the compressor and expander.

(*ii*) The operating pressure ratio can be reduced, resulting in higher C.O.P.

2. Open system. In this system the refrigerator is replaced by the actual space to be cooled with the air expanded to atmospheric pressure, circulated through the cold room and then compressed to the cooler pressure. The pressure of operation in this system is *inherently limited to operation at atmospheric pressure in the refrigerator.*

In this system, the refrigeration is obtained by three basic steps *viz. compression, cooling and expansion,* accompanied by extraction of work. The air after expansion is directly led to the conditioned space. It is therefore necessary that air be expanded to one atmosphere pressure. This requires larger volumes to be handled. Notwithstanding this, the sizes of the compressor and expander, now a days, are not affected significantly since both of them are turbo-type. This system, therefore, has an *advantage over closed system*, in respect that it *does not require a heat exchange for refrigeration process. This saves the weight and cost of the equipment.*

Advantages of closed system over open system :

1. In a *closed system* the suction to compressor may be at high pressure. The sizes of expander and compressor can be kept within reasonable limits by using dense air.

2. In *open air system* the *air picks up moisture from the products kept in the refrigerated chamber, the moisture may freeze during expansion and is likely to choke the valves. Thus a drier in the circuit is required whereas it does not happen in closed system.*

3. C.O.P. is higher.

4. In open system, the expansion of the refrigerant can be varied only upto atmospheric pressure prevailing in the cold chamber but for a *closed system there is no such restriction.*

4.7.2. Air-refrigeration System Working on Reversed Brayton Cycle

Fig 4.1 shows a schematic diagram of an air-refrigeration system working on the reversed Brayton (or Bell Coleman on Joule) cycle. Elements of this system are :

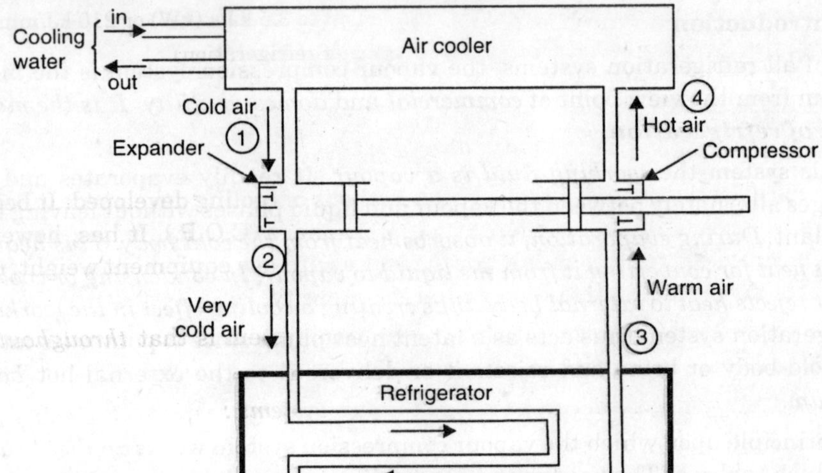

Fig. 4.1. Air-refrigeration system working on reversed Brayton or Bell-Coleman cycle.

1. Compressor 2. Cooler (heat exchanger)
3. Expander 4. Refrigerator.

In this system, work gained from expander is employed for compression of air, consequently less external work is needed for operation of the system. In practice it may not be done *e.g.*, in some aircraft refrigeration systems which employ air-refrigeration cycle the expansion work may be used for *driving other devices.*

Note. *The reversed Brayton cycle is same as the Bell-Coleman cycle. Conventionally Bell-Coleman cycle refers to closed cycle with expansion taking place in reciprocating expander and compressor respectively, and heat rejection and heat absorption taking place in condenser and evaporator respectively.*

With the development of efficient centrifugal compressors and gas turbines, the processes of compression and expansion can be carried out in centrifugal compressors and gas turbines respectively. Thus the shortcoming encountered with conventional reciprocating expander and compressor is overcome. Reversed Brayton cycle finds its application for air-conditioning of aeroplanes where air is used as refrigerant.

Merits and demerits of Air-refrigeration system :

Merits :

1. Since air is non-flammable, therefore there is no risk of fire as in the machine using NH_3 as the refrigerant.

2. It is cheaper as air is easily available as compared to the other refrigerants.

3. As compared to the other refrigeration systems the weight of *air-refrigeration system per tonne of refrigeration is quite low, because of this reason this system is employed in aircrafts.*

Demerits :

1. The C.O.P. of this system is *very low* in comparison to other systems.

2. The weight of air required to be circulated is more compared with refrigerants used in other systems. This is due to the fact that heat is carried by air in the form of *sensible heat.*

4.8. SIMPLE VAPOUR COMPRESSION REFRIGERATION SYSTEM

4.8.1. Introduction

- Out of all refrigeration systems, the vapour compression system is the most important system from the view point of *commercial* and *domestic utility. It is the* ***most practical form of refrigeration.***
- In this system the *working fluid is a vapour.* It readily evaporates and condenses or changes alternately between the vapour and liquid phases without leaving the refrigerating plant. *During evaporation, it absorbs heat from the cold body. This heat is used as its latent heat for converting it from the liquid to vapour. In condensing or cooling or liquifying, it rejects heat to external body, thus creating a cooling effect in the working fluid.* This refrigeration system thus acts as a latent heat pump since it pumps its latent heat from the cold body or brine and rejects it or delivers it to the external hot *body or cooling medium.*
- The principle upon which the vapour compression system works *applies to all the vapours* for which tables of Thermodynamic properties are available.

4.8.2. Simple Vapour Compression System

In a simple vapour compression system *fundamental processes are completed in one cycle.* These are :

1. Compression 2. Condensation
3. Expansion 4. Vapourisation

The flow diagram of such a cycle is shown in Fig. 4.2.

The *vapour at low temperature and pressure* (state '2') enters the **"compressor"** where it is compressed isentropically and subsequently its temperature and pressure increase considerably (state '3'). The vapour after leaving the compressor enters the **"condenser"** where it is condensed into *high pressure liquid* (state '4') and is collected in a *"receiver tank"*. From receiver tank it passes through the **"expansion valve",** here it is *throttled down to a lower pressure and has a low temperature* (state '1'). After finding its way through expansion valve it finally passes on to **"evaporator"** where it extracts *heat from the surroundings or circulating fluid being refrigerated and vaporises to low pressure vapour* (state '2').

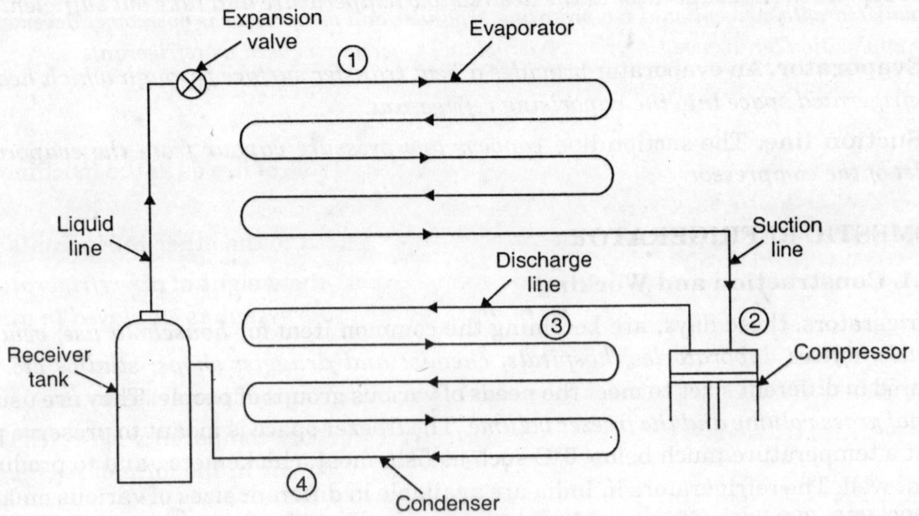

Fig. 4.2. Vapour compression system.

Merits and demerits of vapour compression system over air refrigeration system

Merits :

1. C.O.P. is quite high as the working of cycle is very near to that of reversed Carnot cycle.

2. When used on ground level the running cost of vapour-compression refrigeration system is only 1/5th of air refrigeration system.

3. For the same refrigeration effect the size of the evaporator is smaller.

4. The required temperature of evaporator can be achieved simply by adjusting the throttle valve of the same unit.

Demerits :

1. Initial cost is high.

2. The major disadvantages are inflammability, leakage of vapours and toxcity. These have been overcome to great extent by improvement in design.

4.8.3. Functions of Parts of a Simple Vapour Compression System

Here follows the brief description of various parts of a simple vapour compression system shown in Fig. 4.2.

1. Compressor. The function of a compressor is to remove the *vapour from the evaporator and to raise its temperature and pressure to a point such that it* (vapour) *can be condensed with available condensing media.*

2. Discharge line (or hot-gas line). A hot gas or discharge line *delivers the high-pressure, high temperature vapour from the discharge of the compressor to the condenser.*

3. Condenser. The function of a condenser is to *provide a heat transfer surface through which heat passes from the hot refrigerant vapour to the condensing medium.*

4. Receiver tank. A receiver tank is used to provide *storage for a condensed liquid* so that a constant supply of liquid is available to the evaporator as required.

5. Liquid line. A liquid line carries the liquid refrigerant from the receiver tank to the refrigerant flow control.

6. Expansion valve (refrigerant flow control). Its function is to *meter the proper amount of the refrigerant to the evaporator and to reduce the pressure of liquid entering the evaporator so that liquid will vaporize in the evaporator at the desired low temperature and take out sufficient amount of heat.*

7. Evaporator. An evaporator *provides a heat transfer surface through which heat can pass from the refrigerated space into the vaporising refrigerant.*

8. Suction line. The suction line *conveys low pressure vapour from the evaporator to the suction inlet of the compressor.*

4.9. DOMESTIC REFRIGERATOR

4.9.1. Construction and Working

Refrigerators, these days, are becoming the common item for *household use, vendor's shop, hotels, motels, offices, laboratories, hospitals, chemist and druggist shops, studios etc.* They are manufactured in different sizes to meet the needs of various groups of people. They are usually rated with *internal gross volume and the freezer volume.* The freezer space is meant to preserve perishable products at a temperature much below 0°C such as fish, meat, chicken etc., and to produce ice and ice-cream as well. The refrigerators in India are available in different sizes of various makes, *i.e.*, 90, 100, 140, 200, 250, 380 litres of gross volume. The freezers are usually provided at top portion of the refrigerator space occupying around one-tenth to one-third of the refrigerator volume. In some refrigerators, freezers are provided at the bottom.

A domestic refrigerator consists of the following *two* main parts :

1. The refrigeration system.
2. The insulated cabinet.

Fig. 4.3 shows a flow diagram of a typical refrigeration system used in a domestic refrigerator. A simple domestic refrigerator consists of a hermetic compressor placed in the cabinet base. The condenser is installed at the back and the evaporator is placed inside the cabinet at the top.

The working of the refrigerator is as follows :

- The low pressure and low temperature refrigerant vapour (usually R-12) is drawn through the suction line to the compressor. The *accumulator* provided between the *suction line* and the evaporator collects liquid refrigerant coming out of the evaporator due to incomplete evaporation, if any, prevents it from entering the compressor. The *compressor* then compresses the refrigerant vapour to a high pressure and high temperature. The compressed vapour flows through the *discharge line* into condenser (vertical natural draft, wire-tube type).
- In the *condenser* the vapour refrigerant at high pressure and at high temperature is condensed to the liquid refrigerant at high pressure and low temperature.

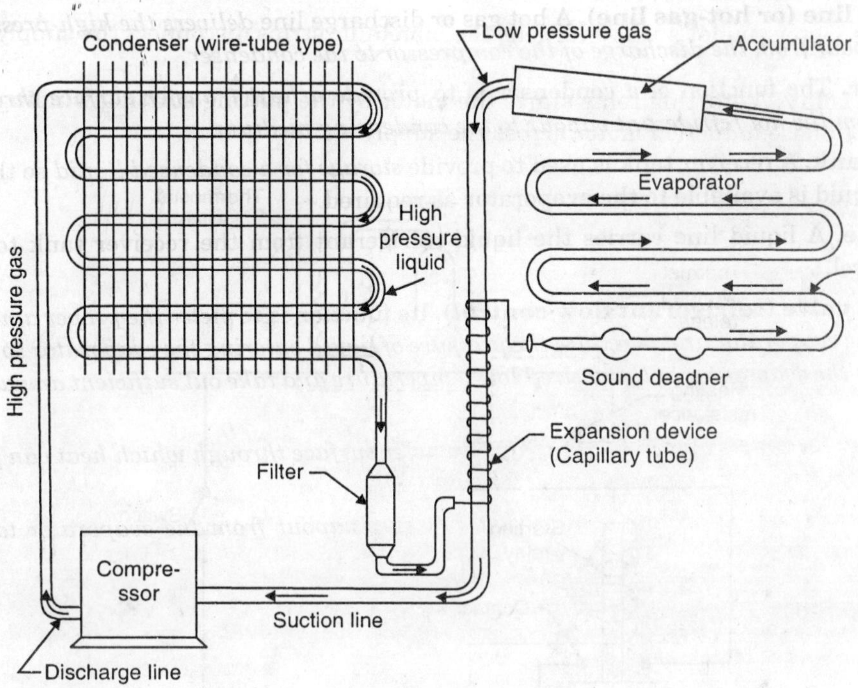

Fig. 4.3. Domestic refrigerator.

- The high pressure liquid refrigerant then flows through the *filter* and then enters the *capillary tube* (expansion device). The capillary tube is attached to the suction line as shown in Fig. 4.3. The warm refrigerant passing through the capillary tube gives some of its heat to cold suction line vapour. This increases the heat absorbing quality of the liquid refrigerant slightly and increases the superheat of vapour entering the compressor.

  The capillary tube expands the liquid refrigerant at high pressure to the liquid refrigerant at low pressure so that a measured quantity of liquid refrigerant is passed into the *evaporator*.
- In the *evaporator* the liquid refrigerant gets evaporated by absorbing heat from the container/articles placed in the evaporative chamber and is sucked back into the compressor and the cycle is repeated.

4.9.2. Electrical Circuit of a Refrigerator

Fig. 4.4 shows a schematic diagram of electric circuit of a refrigerator.

Components :

1. Lamp and switch. The arrangement is made in such a way that lamp remains *'off'* as the door is closed and becomes *'on'* whenever door is opened. When the lamp is *'on'* it is easy to trace the commodities placed in the refrigerator.

2. Thermostat switch. A thermostat switch maintains a requisite temperature in the refrigerator.

Freezer. Temperature remains between – 7°C to 5°C approx.

Remaining part of the refrigerator. Temperature remains between 7°C to 15°C.

3. Thermal overload release. This component is a protective device for compressor motor unit. It operates when temperature of the compressor rises beyond a certain value or excessive current flows in the motor ; under such conditions the bimetallic strips disconnect the supply to the motor.

4. Starting relay. A starting relay starts the motor by the putting starting winding/auxiliary winding of split phase induction motor across the supply.

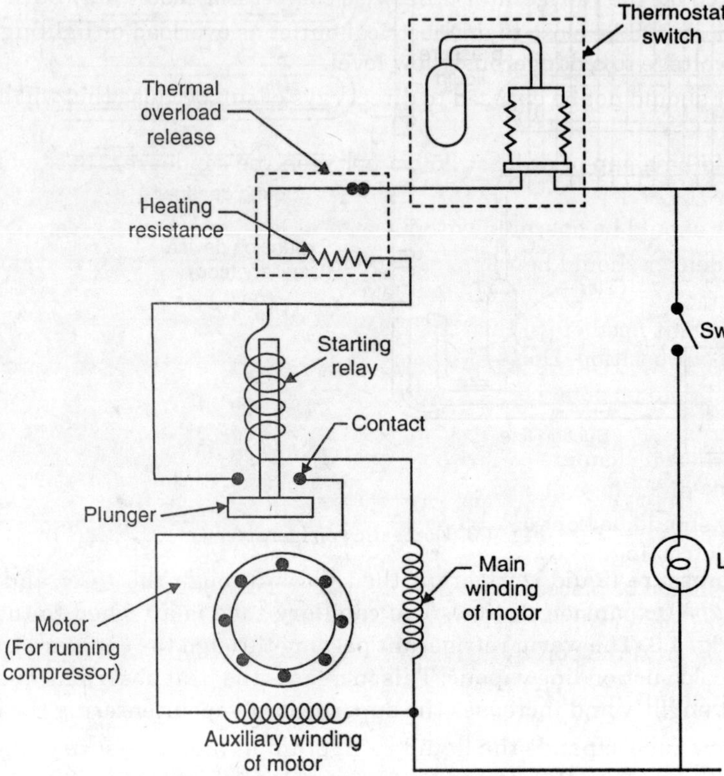

Fig. 4.4. Electric circuit of a refrigerator.

5. Electric motor. Electric motor used is single phase induction motor, split phase type. It is a fractional horse power induction motor ; its size depends upon the capacity of the refrigerator.

Working :

When electric supply is given to the refrigerator, current passes through the thermostat switch, thermal overload release, coil of starting relay and main winding of the motor. To start with, when the motor is at rest it draws a very heavy current. When this heavy current flows through the coil of the starting relay, the coil gets energised and it pulls up the plunger, short circuiting the contacts and putting auxiliary winding also in the circuit. Now since both the main winding and auxiliary winding are energised, motor starts running. When the motor gains normal speed, the current drawn by the main winding of motor becomes normal. At normal current plunger in the coil of starting relay cannot remain pulled and it is released down, opening the contacts, thus auxiliary winding gets out of circuit. The main function of starting relay is to put auxiliary winding in the circuit at the time of starting the motor and to disconnect it when the motor gains normal speed. In case starting relay fails to close, motor will not start (as explained above). But once it closes and it fails to open, then either thermal overload shall trip out or fuse shall be blown off.

4.9.3. Maintenance of Domestic Refrigerator

In order to achieve proper results in the maintenance of the refrigerator, the following points should be taken into consideration :

1. Refrigerator must be levelled properly.

2. Refrigerator should be operated on 230 volts, 50 Hz supply. If the voltage fluctuates widely it is advisable to switch off the refrigerator otherwise compressor motor may burn out.

3. Refrigerator should be close to an electrical outlet as overload or lighting extension cords may reduce the line voltage to a dangerously low level.

4. Refrigerator should not be exposed to direct rays of sun because this will add to the load of operation.

5. There should be a gap of at least 25 cm between the condenser (back of the refrigerator) and the wall.

6. Refrigerator should be operated without load at least two hours before loading it.

7. Dust of condenser should be removed periodically and made sure that no material covers the portion.

8. If there is urgent need of ice turn the knob of thermostat to 5 or more and turn the baffle tray deflection to the out position. Don't forget to change these to their normal position afterwards.

9. Defrosting should be done periodically as frost is a barrier which hinders cooling and adds to the load operation.

10. Never use harsh cleaners to hasten the defrosting process by striking at or chipping off the ice as it will result in leakage of gas.

11. Baffle tray should be kept out when it is desired to collect the dripping water from freezer chest at the time of defrosting.

12. Moist food should be placed in a closed container or wrapped in polythene or aluminium foil.

13. To safeguard against electric shock the refrigerator should be properly earthed with green wire and operated only on three point plugs. *Improper earthing may result in fatal accidents.*

4.9.4. Troubleshooting of Refrigerator

In a refrigerator, following troubles are often found :

S. No.	*Symptoms*	*Causes*
1.	*Motor fails to start on giving supply.*	• Thermostat contacts open. • Blown fuses in the main switch. • Overhead release open. • Open circuit in the main winding. • Open circuit in the auxiliary winding. • Grounded winding. • Burnt or shorted winding. • Worn out or tight bearing. • Bent rotor shaft.
2.	*Motor runs slower than normal speed.*	• Low voltage. • Overload. • Shorted main winding. • Defective electro-magnetic relay. • Worn out bearings.

3.	*Motor runs hot.*	• Low voltage. • Overload. • Shorted or grounded winding. • Worn out bearings.
4.	*Motor does not start and gives humming noise.*	• Low voltage. • Overload. • Open circuit in the auxiliary winding. • Defective electro-magnetic (starting) relay-contacts being not closed.
5.	*Motor runs with noise.*	• Shorted winding. • Improperly connected poles. • Loose rotor bar. • Worn out bearings. • Too much end play. • Foreign material in the rotor.
6.	*Motor keeps on running but* *(a) cooling is nil.* *(b) cooling is insufficient.*	• (*i*) No refrigerant gas in the system (indicated by the unit, tubing and condenser being at ambient temperature) • (*ii*) Complete choking of capillary. • (*i*) Less refrigerant gas in the system. • (*ii*) Partial choking of capillary.
7.	*Motor keeps on running even though it is very cold inside the refrigerator.*	• Wrong setting of thermostat. • Defective thermostat—shorted wiring or sticking contacts.
8.	*Motor starts very frequently.*	• Bad door seal. • Wrong setting of thermostat.
9.	*Motor operates normal with normal cooling in freezer but cooling in the rest portion is unsatisfactory.*	• Bad door seal. • Opening of door too frequent.
10.	*Motor operates normal with good cooling but defrosting starts all of a sudden. Cooling again starts after some time.*	• Presence of moisture in the refrigerant cycle.
11.	*Too much frosting around the freezer.*	• High atmospheric humidity or steaming hot liquids stored.

4.10. WATER COOLERS

Water is one of the most needed thing for a man. In summer season cold water gives life to a thirsty man. At 10°C water is most refreshing. Thus cooling of water in summer becomes necessary. *Water coolers* are used to produce cold water at about 7 to 13°C. The temperature of water is controlled with the help of a thermostatic switch.

Water cooler may be *classified* as follows :

1. *Instantaneous type water coolers :*
   (*i*) Bottle type cooler.
   (*ii*) Pressure type cooler.
   (*iii*) Self-controlled or remote type cooler.
2. *Storage type water coolers.*

1. Instantaneous type water coolers. In this type of coolers the cooling coil is *wrapped round the pipe line such that by the time water reaches the tank it is cooled to desired temperature.* The description of various types of instantaneous type water coolers is given below :

(*i*) ***Bottle type cooler.*** In this type water to be cooled is stored in a bottle or reservoir. For filling glass tumblers or containers faucet or similar means are provided. The dripping water from the faucet is collected in the waste water basin or water drip as shown in Fig. 4.5. Its usual *size is 25 litres* and is *suitable for places where plumbing installations is expensive and drains are available.*

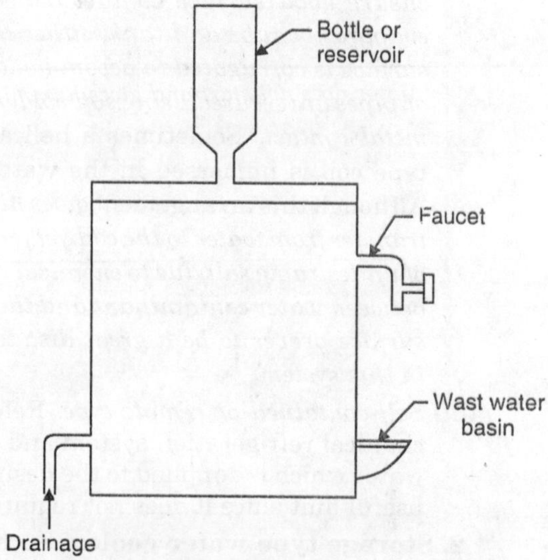

Fig. 4.5. Bottle type cooler.

(*ii*) ***Pressure type cooler.*** Refer to Fig. 4.6. Here water is supplied under pressure. For filling glass tumblers or containers faucets or similar means are provided. A valve is employed to control an appropriate flow of water or projected stream of water from a bubbler. An arrangement should be made to collect water and allow complete collection of water spreading from the bubbler. The temperature of waste water is low, it is used for cooling the supply water by passing through a pipe coil wrapped round the drainage line. By doing so, the cooling load for cooler is reduced. Since the water is supplied under pressure the cold water can be obtained from the top mounted at any height of the cooler. In case of bottle type, faucet has to be at a height upto which syphoned water can be obtained from the tank of the cooler. The *refrigeration system is usually mounted at the bottom*

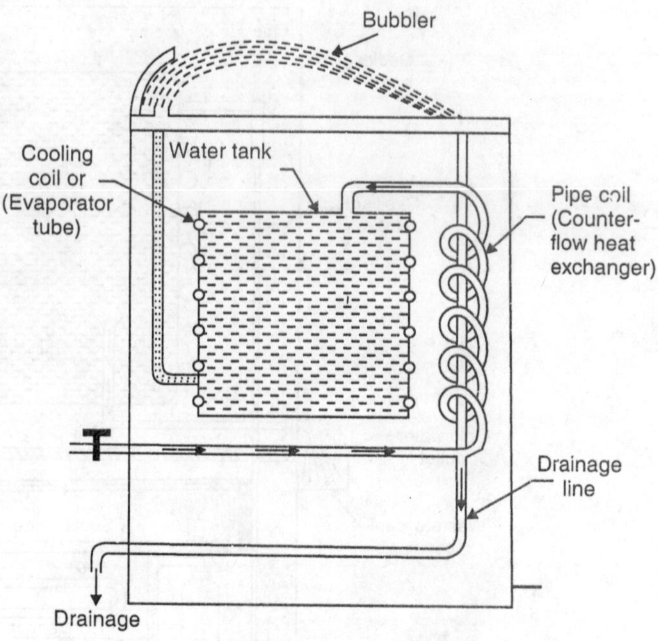

Fig. 4.6. Pressure type water cooler.

structure of the cooler body and a cooling coil is wrapped round the water tank, to ensure good surface contact between the evaporative tube and tank, either the tank surface is corrugated to accommodate pipe or pipes are secured using soft solder to give metal contact. Sometimes a helical or U-type coil is immersed in the water tank. Although this arrangement *gives high heat transfer from water to the coil yet formation of undesirable salt due to chemical reaction between water contaminant and the copper surface proves to be a great disadvantage in this system.*

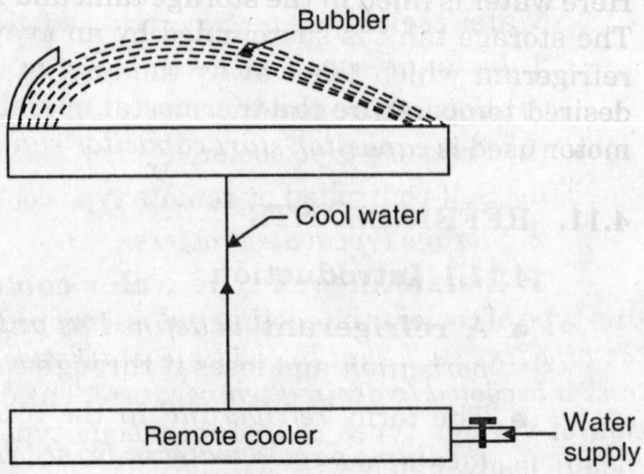

Fig. 4.7. Remote type cooler.

(*iii*) *Self-contained or remote type.* Refer to Fig. 4.7. This type of water cooler employs mechanical refrigeration system and is a factory assembled unit. A remote cooler cools the water which is supplied to the desired drinking place (away from the system). It is quite a useful unit since it does not require extra space near the place of work.

2. Storage type water cooler. Such type of coolers are used where continuous supply of water is not available. Fig. 4.8 shows a schematic storage type water cooler which is self explanatory.

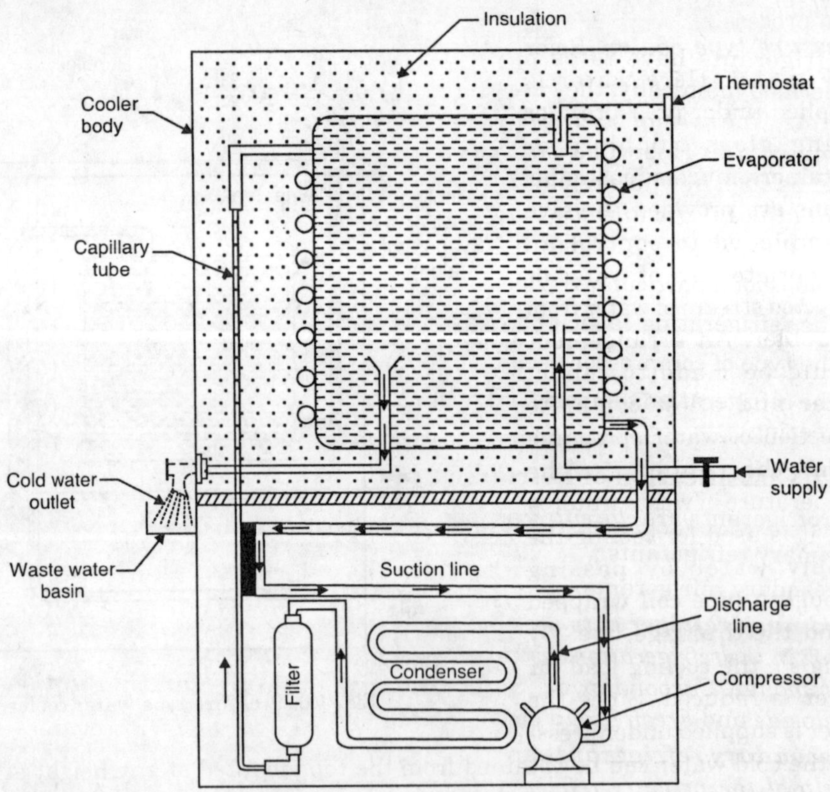

Fig. 4.8. Storage type water cooler.

Here water is filled in the storage tank and level of the water is kept same by the use of a float valve. The storage tank is surrounded by an evaporator coil through which flows a low pressure liquid refrigerant which takes away the heat of water and thus makes it cold. When the water attains desired temperature the thermostat operates and disconnects the power supply to the motor. The motor used is *capacitor-start capacitor run single-phase induction motor.*

4.11. REFRIGERANTS

4.11.1. Introduction

- A **'refrigerant'** *is defined as any substance that absorbs heat through expansion or vaporisation* and loses it through *condensation in a refrigeration system.*
- The term *'refrigerant'* in the broadest sense is also applied to such *secondary cooling mediums* as *cold water or brine, solutions.* Usually refrigerants include only those working mediums which pass through the cycle of *evaporation, recovery, compression, condensation and liquefaction. These substances absorb heat at one place at low temperature level and reject the same at some other place having higher temperature and pressure. The rejection of heat takes place at the cost of some mechanical work.* Thus circulating cold mediums and cooling mediums (such as ice and solid carbondioxide) are not primary refrigerants. In the early days only four refrigerants, *Air, Ammonia* (NH_3), *Carbondioxide* (CO_2), *Sulphur Dioxide* (SO_2), possessing chemical, physical and thermodynamic properties permitting their efficient application and service in the practical design of refrigeration equipment were used. All the refrigerants change from liquid state to vapour state during the process.
- Refrigerants have to be physiologically non-toxic and non-flammable substances. Thermodynamically, they are essentially characterised by their respective normal boiling points. The *normal boiling point of a refrigerant should be little lower than the evaporator temperature.* This will *ensure positive pressure both in the condenser and the evaporator,* and as such in whole refrigeration system. Positive pressure in the system is desirable *so that atmospheric air may not enter into the system* through gap across compressor shaft seal.

The *choice* of a particular refrigerant depends upon the application in *terms of :*

(*i*) The refrigerating capacity (very small, small, medium or large) ;

(*ii*) The type of compressor used (generally reciprocating or centrifugal or screw) ;

(*iii*) The refrigeration temperature required (whether for air-conditioning, cold storage or food freezing).

4.11.2. Classification of Refrigerants

The refrigerants are *classified* as follows :

1. Primary refrigerants ;
2. Secondary refrigerants.

- ***Primary refrigerants*** are those working mediums or heat carriers which *directly take part in the refrigeration system and cool the substance by the absorption of latent heat e.g., Ammonia, Carbondioxide, Sulphur dioxide, Methyl chloride, Methylene chloride, Ethyl chloride* and *Freon group* etc.
- ***Secondary refrigerants*** are those circulating substances *which are first cooled with the help of the primary refrigerants and are then employed for cooling purposes, e.g., ice, carbondioxide etc.* These refrigerants cool substances by *absorption of their sensible heat.*

4.11.3. Desirable Properties of an Ideal Refrigerant

An ideal refrigerant should possess the following *properties* :

A. Thermodynamic properties :

(*i*) Low boiling point
(*ii*) Low freezing point
(*iii*) Positive pressure (but nor very high) in evaporator and condenser
(*iv*) High saturation temperature
(*v*) High latent heat of vapourisation.

B. Chemical properties :

(*i*) Non-toxicity
(*ii*) Non-flammable and non-explosive
(*iii*) Non-corrosiveness
(*iv*) Chemical stability in reacting
(*v*) No effect on the quality of stored (food and other) products like flowers, with other materials, *i.e.*, furs and fabrics.
(*vi*) Non-irritating and odourless.

C. Physical properties :

(*i*) Low specific volume of vapour
(*ii*) Low specific heat
(*iii*) High thermal conductivity
(*iv*) Low viscosity
(*v*) High electrical insulation.

D. Other properties :

(*i*) Ease of leakage location
(*ii*) Availability and low cost
(*iii*) Ease of handling
(*iv*) High C.O.P.
(*v*) Low power consumption per tonne of refrigeration
(*vi*) Low pressure ratio and pressure difference.

4.12. REFRIGERATION COMPONENTS AND CONTROLS

In a vapour compression system the following are the *major refrigeration components* :

1. Compressors
2. Condensers
3. Evaporators

Refrigeration controls *are the devices used to control the flow of refrigerant at various points throughout the refrigeration cycle.* The following are the *six* basic types of *refrigerant flow controls* :

(*i*) Hand expansion valve
(*ii*) Automatic expansion valve
(*iii*) Thermostatic expansion valve
(*iv*) Capillary tube
(*v*) Low-side float
(*vi*) High-side float.

AIR-CONDITIONING

4.13. CONCEPT OF PSYCHROMETRY AND PSYCHROMETRICS

- Air consists of fixed gases principally, *nitrogen and oxygen with an admixture of water vapour in varying amounts.* In atmospheric air water is always present and its relative weight averages less than 1% of the weight of atmospheric air in temperate climates and less than 3% by weight under the most extreme natural climatic conditions, it is nevertheless one of most important factors in human comfort and has significant effects on many materials. Its effect on human activities is in fact altogether disproportionate to its relative weights.
- *The art of measuring the moisture content of air is termed* **"psychrometry".**
- The *science which investigates the thermal properties of moist air, considers the measurement and control of the moisture content of air, and studies the effect of atmospheric moisture on material and human comfort may properly be termed* **"psychrometrics".**

4.14. HUMAN COMFORT

4.14.1. Thermodynamics of Human Body

- A human body feels comfortable thermodynamically when the *heat produced by the metabolism of human body is equal to the sum of the heat dissipated to the surroundings and the heat stored in human body by raising the temperature of body tissues.* These phenomena can be represented by the following equation.

$$Q_M = \pm Q_S + Q_E \pm Q_R \pm Q_C \quad ...(14.1)$$

where, Q_M = *Metabolic heat* produced within the body (kJ/h.)

$\pm Q_S$ = *Stored energy* in temperature rise of body tissues ; + ve when the tissue temperature rises and – ve when it falls.

Q_E = *Evaporative* heat loss due to the water evaporating at skin surface and the lungs expressed in kJ/h.

$\pm Q_R$ = Heat loss and gain by *radiation* ; + ve when heat is lost to the surrounding, and –ve, when it is gained from the surroundings kJ/h.

Q_C = Heat loss or gain by *conduction and convection* ; +ve when heat is lost, and –ve, when heat is gained from the surroundings in kJ/h.

- *The rate at which the body produces heat is termed as* the **metabolic rate,** the value of which depends upon a number of factors such as *individual's health, his physical activity* and *his environments.* The metabolic heat produced depends on the rate of food energy consumption in body.
  - — A fasting man, the weak or sick, will have less metabolic heat production. If Q_E, Q_R and Q_C are high and +ve, Q_S will become –ve when Q_M is low and hence sick, old weak or fasting man feels more cold when exposed to low temperature, high draft air.
  - — A man gets fever, when internal body activities increase Q_M to such an extent that Q_S becomes +ve for given Q_E, Q_R and Q_C.
  - — The stored energy Q_S has the maximum and minimum limits which when exceeded brings death.
  - — Body temperature when exceeds 40.5°C and falls below 36.7°C is dangerous. The usual body temperature when $Q_S = 0$ is 37°C for a normal man. There is some kind of

thermo-static control in human body, which tries to maintain temperature of human body at the normal level of 37°C.

- Whenever the environment temperature changes, the human body reacts by trying many physiological adjustments to keep the deep temperature of the body constant. There are two types of adjustments generally adopted by the human control system.

(*i*) **Vasomotor control.** This control *regulates the blood supply to the skin.* It acts by causing vasodilation of the peripheral blood vessels, if environment temperature is increased. The increased circulation of the blood increases the convective transport of heat from the interior of the body to the surface. In other words, the thermal conductance of the skin is increased, thus minimising body heating. The vasomotor mechanism alone is sufficient to maintain the heat balance at a low level of heat load, and is therefore known as "*First Line of Defence*".

If environment temperature is decreased, the control causes vasocontraction of the peripheral blood vessels. This decreases blood circulation and reduces body cooling.

(*ii*) **Sudomotor control.** This control *regulates sweat production.* Whenever the heat loss by convection and radiation becomes negative due to high temperature of atmosphere compared to body temperature, the only mode of heat transfer to dissipate the heat is by evaporation.

The sudomotor control *acts by initiating sweat gland activity*. The sweating capacity differs according to persons degree of acclimatization to heat and work.

Note. *Effect of activities on the heat load.* The rate at which body produces heat is called the *metabolic rate.* The metabolic rate depends upon the type of activity of the human body. The heat produced (the heat load) by a normal healthy person while sleeping is of the order of 60 W. It increases when the human body is active, the value of heat load being dependent on the severity of activity. In extreme case it can be as high as 600 W, say for a person performing very heavy exercise.

The metabolic heat load is both, sensible and latent. For different types of activities the heat loads have been found experimentally, and are available in databook.

4.14.2. Factors Affecting Comfort

Air-conditioning of buildings mainly concerns the comfort of people and not the maintenance of exact conditions as required for products. The conditions conductive to comfort depend upon the following factors :

1. Temperature
2. Humidity
3. Air motion
4. Air purity.

Temperature. Proper control of temperature of the air medium surrounding the body removes a physiological stress of accommodation, thereby making for greater comfort and improved physical well-being and health.

Humidity. It is through evaporation from skin that a large proportion of body heat is lost. Since evaporation is promoted by a low relative humidity of the air and is related by a high humidity control has an important effect on the comfort. Extremes of humidity not only results undesirable physiological relations but also effect (usually adversely) the properties of many substances in the treated space, clothing and furniture in particular.

Air motion. Air movement over the body increases rate of heat and moisture dissipation above the still-air rate, thereby modifying the feeling of warmth or coldness. Depending on the amounts of motion there are pleasing and displeasing effects.

Air purity. People feel uncomfortable when breathing contaminated air even if it is within acceptable temperature and humidity ranges. Therefore proper filtration, cleaning and purification of air is necessary to keep it away from dust, dirt and other properties.

4.14.3. Effective Temperature

In addition to major four factors of *temperature, humidity, air movement, purity of air* there are several other factors such as *clothing, economical status of persons, climate* (a person living in the cold climates takes about 7 to 9 days to become acclimatized to the hot environments) to which a person is accustomed to, etc., which influence human comfort and deserve due consideration. Unfortunately it is not practically feasible to study the combined effects of all the factors on human comfort. Therefore, it is usual practice to consider the aforesaid four factors for most applications which are included in a term called the '*effective temperature*'.

- **Effective temperature is** *defined as a sensory index that combines into a single factor the effects of temperature, humidity and air movement on human comfort in a noise free pure air environment.* The effective temperature corresponds to the *dry-bulb temperature of the saturated air at which a given percentage of people feel comfortable.* For *example*, at 21°C (and RH = 100% and air movement 8 m/min.) most people feel comfortable.

Although a particular effective temperature may have a humidity ranging all the way from 0 to 100% and air motion from still air to high velocities, it does not follow that all these combinations are equally comfortable. *Each combination produces the same sensation of warmth or coolness but other affects may contribute a feeling of discomfort.*

— Too low a humidity produces a parched condition of skin, mouth, and nose.

— *Too high humidity* causes an accumulation of moisture in clothing and promotes the production of body odours.

— *High air motion* will usually cause annoyance because of its directional effects such as blowing of hair.

— When the ranges of humidity and air motion are limited to avoid these physiological or psychological regions of discomfort the effective temperature index may be said to measure comfort fairly closely. This is on the assumption that air purity and noise level is maintained at an acceptable standard.

4.15. AIR-CONDITIONING SYSTEMS

4.15.1. Introduction

An **air-conditioning system** *is defined as an assembly of different parts of the system used to produce a specified condition of air within a required space of building.*

The *basic elements of air-conditioning systems* (of whatever form) are :

1. **Fans :** For moving air.
2. **Filters :** For cleaning air, either fresh, recirculated or both.
3. **Refrigerating plant :** Connected to heat exchange surface, such as finned coils or chilled water sprays.
4. **Means for warming :** The air, such as hot water or steam heated coils or electrical elements.
5. **Means for humidification ; and/or dehumidification.**
6. **Control system :** To regulate automatically the amount of cooling or warming.

4.15.2. Air-conditioning Cycle

Refer to Fig. 4.9. An air-conditioning cycle comprises the *following steps :*

1. The fan forces air into duct-work which is connected to the openings in the room. These openings are commonly called *outlets or terminals.*
2. The duct-work directs the air to the room through the outlets.

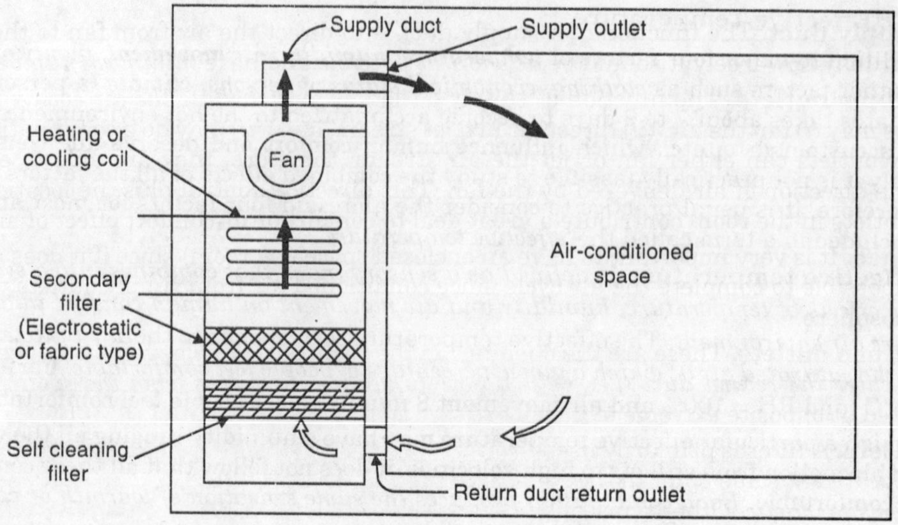

Fig. 4.9. Air-conditioning cycle.

3. The air enters the room and either heats or cools as required. Dust particles from the room enter the air stream and are carried along with it.
4. Air then flows from the room through a second outlet (sometimes called the return outlet) and enters the return duct-work, where dust particles are removed by a *filter*.
5. After the air is cleaned, it is either heated or cooled depending upon the condition in the room. If cool air is required, the air is passed over the surface of a *cooling coil* ; if warm air required, the air is passed through a *combustion chamber* or over the surface of a *heating coil.*
6. Finally the air flows back to the fan, and the cycle is completed.

The *main parts* of the equipment in the air-conditioning cycle are :

1. Fan ;
2. Supply ducts ;
3. Supply outlets ;
4. Space to be conditioned ;
5. Return outlets ;
6. Return ducts ;
7. Filter ;
8. Heating chamber or cooling coil.

1. **Fan.** The primary function of a fan is to move air *to and from the room.* The air that a fan moves in air-conditioning system is made up of :

(*i*) All outdoor air

(*ii*) All indoor room air (called recirculated air)

(*iii*) A combination of outdoor and indoor air.

The fan *pulls* air from outdoors or from the room but in most systems it *pulls air* from both sources at the same time. The amount of air supplied by the fan must be regulated since drafts in the room cause discomfort, and poor air movement slows the heat rejection processes. This can be achieved by (*i*) choosing a fan that can deliver the correct amount of air, and (*ii*) by controlling the speed of the fan so that air stream in the room provides good circulation but does not cause drafts. Of course, the fan is only one of the pieces of equipment that contributes body comfort.

2. **Supply duct.** The function of a supply duct is to direct the air from fan to the room. In order that air may flow freely it should be *as short as possible* and have *minimum number of turns.*

3. **Supply outlets.** The function of supply outlets is to *distribute the air evenly in a room.* These outlets may (*i*) fan the air, (*ii*) direct air in a jet stream, and (*iii*) may do a combination of both. Since supply outlets can either fan or jet the air stream, therefore, they are able to exert some control on the direction of air delivered by the fan. This direction control plus the location and the number of outlets in the room contribute a great deal to comfort or discomfort effect of air pattern.

4. **Space.** It is very important to have an enclosed space (*i.e.* room) since if it does not exist it would be impossible to complete the air cycle since conditioned air from supply outlets would flow into the atmosphere.

5. **Return outlets.** These are the openings in the room surface. They are employed to *allow room air to enter the return duct* (*i.e., return* outlets allow air to pass *from the* room). They are usually located at opposite extreme of a wall or room from the supply outlet.

6. **Filters.** A filter is primarily used *to clean the air by removing dust and dirt particles.* They are usually located at some point in the return air duct. They are made of any materials from *spun glass to composite plastic.* Other types operate on electrostatic principle.

7. **Cooling coil and heating coil or combustion chamber.** The cooling coil and heating coil or combustion chamber can be located either ahead or after the fan, but *should always be located after the filter.* A filter ahead of the coil is necessary to prevent the excessive dirt, dust and dirt particles from covering the coil surface.

8. **Summer operation.** The air-conditioning cycle *cools the air* during summer operation. Return air from the room passes over the surface of *cooling coil,* and the air cooled to the required temperature. If there is too much moisture present, it is removed automatically as the air is cooled by the coil.

9. **Winter operation.** The air-conditioning cycle *adds heat to the air during winter operation.* This is achieved by passing the return air from the room over the surface of a *heating coil*, etc.

4.15.3. Classification of Air-conditioning Systems

The air-conditioning systems may be *classified* as follows :

According to the arrangement of equipment :

(*i*) Central systems ;
(*ii*) Zoned systems ;
(*iii*) Unitary systems ;
(*iv*) Unitary-central systems.

According to the purpose :

(*i*) Comfort air-conditioning system ;
(*ii*) Industrial air-conditioning system.

According to season of the year :

(*i*) Winter air-conditioning system ;
(*ii*) Summer air-conditioning system ;
(*iii*) Year-round air-conditioning system.

Another method of classification of air-conditioning system is as follows :

(*i*) Single-air systems
(*ii*) Dual-air systems
(*iii*) Primary-air systems
(*iv*) Unit systems
(*v*) Panel systems.

4.15.4. Central System

This type of system is suitable for air-conditioning large space such as *theatres, cinemas, restaurants, exhibition halls, or big factory spaces where no sub-division exists.* The central systems are generally employed for the loads above 25 tonnes of refrigeration and 2500 m^3/min of conditioning air. The unitary systems can be more economically employed for low capacity (below 25 tonnes) units.

In central system, the equipment such as fans, coils, filters and their encasement are designed for assembly in the field.

- A *central system serves differing rooms, requires individual control of each room.* The condenser, compressor, dampers, heating, cooling and humidifying coils and fan are *located at one place say basement.*
- The conditioned air is carried to the different rooms by means of supply ducts and returned back to the control plant though return ducts. Part of the supply air to the rooms may be exhausted outdoors.
- Outdoor air enters from an intake which should be situated on that side of the building least exposed to solar heat. It should not be close to the ground or to dust collected roof.
- The air after passing through damper passes through *filters.* The filters may be of a mechanical cleaned type, replaceable-cell type or may be electrostatic.
- The cleaned air then passes to the conditioning equipment in the following order : *Temperature (or preheater) coil, cooling coil, humidifier (air washer), heating coils and finally fan.*

Fig. 4.10 shows a schematic diagram of complete (year round) air-conditioning system.

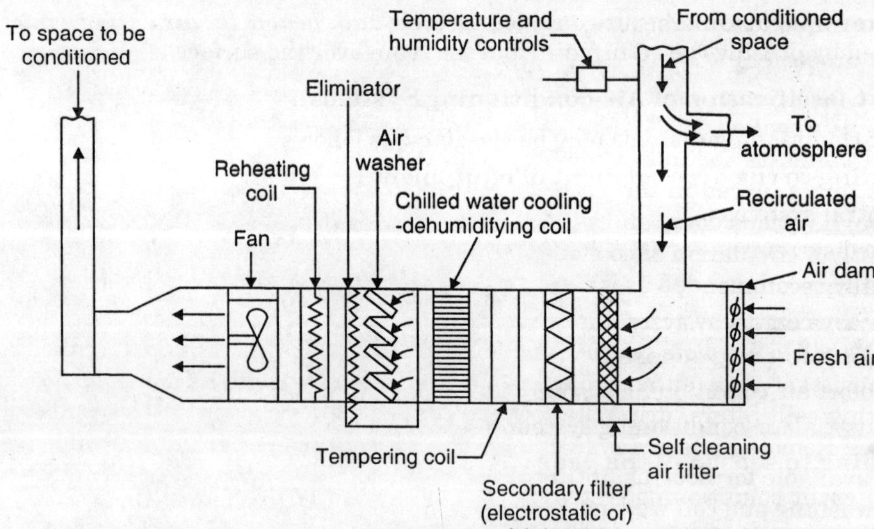

Fig. 4.10. Central system.

Advantages of central system :

1. Low investment cost as compared to total cost of separate units.
2. Space occupied is unimportant as compared to a room unit conditioner which must be placed in the room.

3. Better accessibility for maintenance.
4. The running cost is less per unit of refrigeration.
5. Noise and vibration troubles are less to the people in air-conditioned places as the air-conditioning plant is far away from the air-conditioned places.
6. The exhaust air can be returned and partly reused with obvious saving in heating and refrigeration.

Disadvantages of Central System:

The main disadvantage of the central system is that it results in *large size ducts which are costly and occupy large space.*

4.15.5. Unitary Systems

Unitary systems are packaged units in which all the equipments and controls, for fulfilling the requirements of air-conditioning, are *assembled in a single casing with decorative covering.* The *components are factory assembled, tested and balanced.* They can be directly installed in or near the vicinity to be air-conditioned. The only utility required is a source of electricity, and for water-cooled units, a supply of cooling water.

They are **used** for cooling capacities upto 100 tons (350 kW), with 10 to 50 tons (3.5 kW to 175 kW) installations a common present day practice. *Multiple units are used for zone control due to modular flexibility.*

Various **factory assembled units** available are :

(*i*) Attic (or exhaust) fans.
(*ii*) Remote units.
(*iii*) Self contained units.
(*iv*) Room air-conditioners.
(*v*) Unit air coolers.

(*i*) **Attic fans.** An attic or exhaust fan is a cooling unit without any heat transfer element such as a cooling coil. When the sun sets the temperature of outdoor air reduces to cool levels whereas the indoor temperatures are high. To reduce the inside temperature an attic fan is placed in the attic. It draws outdoor air into several rooms of the building through various doors and windows and finally discharges from attic to outdoors. Consequently a circulation of cool outdoor air is set up in the building. A propeller type fan is usually recommended as it can handle large volume of air at low pressure efficiently.

(*ii*) **Remote units.** A system in which air handling unit is separated from the condensing unit is called a *remote system.* The conditioning or air-handling unit is called *remote unit.* It consists of a fan (either propeller or centrifugal type) with its driving motor, cooling coil, heating coil, filters, drip pan, louvers etc. with or without the duct connections at the outlet. Remote units are available in capacities ranging from 2 to 100 tonnes. These units are available for floor mounting or for suspension from ceiling. Some remote units have air-washing and coil wetting features.

(*iii*) **Self contained units.** In a self-contained unit the condensing unit and other functional elements (such as coils and fans) are encased in the same cabinet. Fresh air can be introduced if required. The discharge from the casing may be free pressure type (*i.e.*, with or without duct-work). Proper means should be adopted to cool the compressor.

(*iv*) **Room air-conditioners.** Fig. 4.11 shows a unit air-conditioner for mounting in an window or wall bracket. Unit air-conditioners of small size generally have the condenser of the refrigerator air-cooled but in large sizes the condenser may be water-cooled, in which case piping connections are required. Apart from this the only services needed are an electric supply and a connection to drain to conduct away any moisture condensed out of the atmosphere during dehumidification. Compressors in most units are *hermetic and therefore quiet in running.* Units of considerable size are suitable for industrial applications, in which case ducting may be connected for distribution.

(*v*) **Unit air-coolers.** A unit air-cooler is a special application of remote units. It primarily reduces the temperature in insulated and sealed storage rooms. The rating of these coolers is the basic rating expressed in kJ per hour per C-degree temperature differential between the refrigerant and air. A defrosting coil is necessary when the temperature is below 2°C. It may be mounted on the floor or wall or suspended from the ceiling.

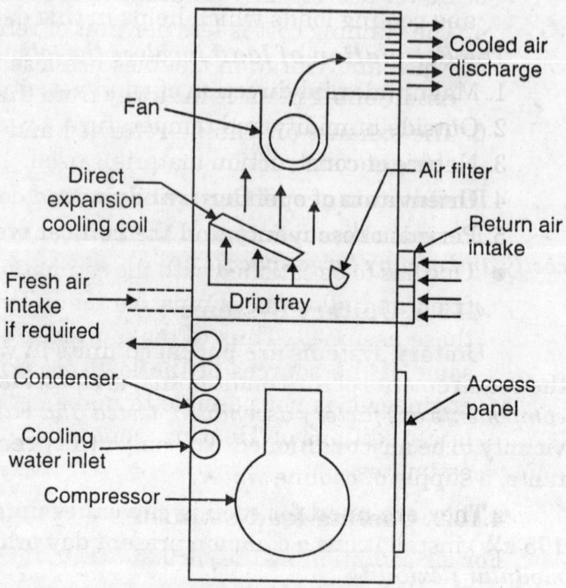

Fig. 4.11. Unit air-conditioner.

Advantages of unitary system. The unitary system commands following *advantages* over the central system.

1. There is saving in the installation and assembly labour charges.
2. Zoning and duct-work eliminated.
3. In unitary system exact requirement of each separate room is met where as in central system the individual needs of separate rooms cannot be met.
4. Failure of the unit puts off conditioning in only one room whereas the failure of the central plant off-set all the rooms to be served.
5. Only those rooms which need cooling will have their units running, whereas the central plant will have to run all the time for the sake of only a few rooms.
6. The specific feature of a unitary system is that there is *individual room-temperature control.*

4.16. LOAD ESTIMATION

4.16.1. Introduction

- *The primary requirement of cooling or heating equipment is that these must be able to remove or add heat at the rate at which it is produced, or removed and maintain the given comfort conditions in the room.*
- The proportion of sensible heat to latent heat decides the slope of '*sensible heat ratio*' or '*enthalpy humidity difference ratio*' and hence also the condition at which the conditioned air must enter the room *because it lies on the line drawn from the room condition at the slope of above ratios.*
- The cost of the equipment will be quite high if it is designed for maximum heating or cooling loads. The owner may be prepared to tolerate some uncomfort for a short time to save capital cost investment. If the equipment is designed for average loads, there may be long periods of uncomfort and the equipment may soon become unpopular. Hence the *capacity equipment must be estimated at a value, which wisely accounts for the physical*

and economical comfort of the owner. It is only the exact and detailed estimation of heating and cooling loads which helps in this decision.

The ***Estimation of load*** *involves the following variables :*

1. Magnitude and direction of wind velocity.
2. Outside humidity and temperature.
3. Nature of construction materials used.
4. Orientation of openings windows and doors.
5. Periods of occupancy and the number of persons in the room, activities of the persons etc.

- One has to be satisfied with the estimation under assumed conditions keeping the variables fixed and allowances must be made for, when the actual conditions are different from those assumed. Thus to make precise load calculation is unfortunately not simple and some of the sources of the load are difficult to predict and evaluate. For a practising engineer it is not possible to make detailed calculations and therefore he must consult latest editions of 'Heating, ventilation and air conditioning guide' and make quick easy estimates.

4.16.2. Cooling-load Estimate

For air-conditioning the cooling-load can be classified as follows :

1. **Room load**—which falls on the *room directly.*
2. **Total load**—which falls on the *air-conditioning apparatus.*

1. **Room load**

(*a*) Room Sensible Heat (RSH) :

(*i*) Solar and transmission heat gain through walls, roof, etc.
(*ii*) Solar and transmission heat gain through glass.
(*iii*) Transmission gain through partition walls, ceiling, floor etc.
(*iv*) Infiltration.
(*v*) Internal heat gain from people, power, lights, appliances etc.
(*vi*) Additional heat gain not accounted above, safety factor etc.
(*vii*) Supply duct heat gain, supply duct leakage loss and fan power.

(*b*) Room Latent Heat (RLH) :

(*i*) Infiltration.
(*ii*) Internal heat from people, steam, appliances etc.
(*iii*) Vapour transmission.
(*iv*) Additional heat gain not accounted above, safety factor etc.
(*v*) Supply duct leakage loss.

It may be noted that if we add '*by-passed air load*' to room sensible heat, we shall get '*effective room sensible heat*' (ERSH). Similarly adding 'by-passed air load' to room latent heat we shall get '*effective room latent heat*' (ERLH).

2. **Grand total load** (on air-conditioning apparatus)

(*a*) Sensible heat

(*i*) Effective room sensible heat.
(*ii*) Sensible heat of the outside air that is not by-passed.
(*iii*) Return duct heat gain, return duct leakage gain, dehumidifier pump power and dehumidifier and piping losses.

Total Sensible Heat (TSH) is obtained by adding items (*i*) to (*iii*).

(*b*) Latent heat

(*i*) Effective room latent heat.

(*ii*) Latent heat of outside air which is not by-passed.

(*iii*) Return duct leakage gain.

Total latent heat (TLH) is obtained by adding items (*i*) to (*iii*).

Grand total heat (GTH) = Total sensible heat (TSH) + Total latent heat (TLH)

i.e., $$\text{GTH} = \text{TSH} + \text{TLH} \quad \ldots(4.2)$$

4.16.3. Heating-load Estimate

Heating-load estimate is prepared *on the basis of 'maximum probable heat loss' of the room or space to be heated.* Following points should be considered while making heat-load calculations :

1. Transmission heat loss. The transmission heat loss from walls, roof, etc., is calculated *on the basis of just the outside and inside temperature difference.*

2. Solar radiation. Normally there is no solar radiation present and hence no solar heat gain at the time of the peak load which normally occurs in the early morning hours.

3. Internal heat gains. The heating requirement is reduced due to internal heat gains from occupants, lights, motors and machinery etc.

4.16.4. Solar Radiation

- The *solar radiation intensity* normal to the sun's rays incident upon a plane surface situated in the limits of the earth's atmosphere, varies with the time of the year as the distance of the earth from the sun changes. Its value when the earth is at its mean distance from the sun is called the *'solar constant'*. The normal value of the solar constant is assumed as 5045 kJ/m^2-h. Radiation received at the surface of the earth is much less because, much of it, while passing through earth's atmosphere, is scattered and absorbed by dust and vapour particles and the gases in the atmosphere.
- The solar heat reaches part of the earth's surface in the form of two radiations :

1. Beam or direct radiation. The part of the sun's radiation which travels through the atmosphere and reaches the earth's surface directly is called *Beam or direct radiation.* It *is maximum when the surface is normal to the sun's rays.* The intensity of the radiation can be increased or decreased by *changing the orientation of the surface.*

2. Diffuse or sky radiation. A large part of the sun's radiation is scattered, reflected back into space and absorbed by the earth's atmosphere. A part of this radiation is re-radiated and reaches the earth's surface uniformly from all directions. It is called diffuse or sky radiation. It *does not normally change with orientation of the surface.*

The total solar radiation reaching a surface is equal to the *sum of the direct and diffuse radiations.*

4.16.5. Solar Heat Gain Through Glass

- Glass which is major material of most buildings, provides the most direct route for entry of solar radiation. For these reasons, the proper estimation of heat gain through glass is necessary.
- *Heat transmitted through a glass surface depends on the wavelength of radiation and physical and chemical characteristics of the glass.* Part of the radiation is *absorbed,* part is *reflected* and the rest is *transmitted.* Glass is opaque to the radiant energy emitted

from sources below 200°C. Thus glass has high *transmittivity for short wavelength and low transmittivity for long wave length radiation.*

- *Direct solar heat gain can be reduced* by using different types of glass, glass construction and shades as given below :

(*i*) Double pan glass reduces the solar heat by 10% to 20%.
(*ii*) Special heat absorbing glass reduces the solar heat by 25%.
(*iii*) Stained glass can reduce it upto 65% depending upon its colour.
(*iv*) Shading devices installed on the outside of windows reduce sun load upto 15%.
(*v*) Ventilation blinds and curtain shades reduce it by 30 to 35%.

4.16.6. Internal Heat Gains

The sensible and latent heat gains due to occupants, lights, appliances, machines, piping etc. within the conditioned space form the components of *internal heat gains.*

4.17. AIR-CONDITIONING OF THEATRES

In all theatres central system of air-conditioning is employed. The cooling load mainly consists of *occupancy* and *amount of outdoor air used.*

The *outdoor air* required per occupant is 0.14 cmm without smoking and about 0.177 cmm with smoking.

The *occupancy load consists* of men, women and children at rest. On an average, the amount of heat released in kJ/h is 420 per male occupant, 370 per female occupant and 30 per child. The heat released per occupant may therefore be taken as 370 kJ/h.

The *design condition of air* inside is 25.5 to 26.5°C and 55% R.H. which may have apparatus dew point slightly in excess of 11.5°C.

The ventilation air per occupant is 0.42 and 0.51 cmm for air delivered horizontally from the near or side wall outlets and about 0.7 cmm for air delivered from the ceiling outlets with downward diffusion.

When the weather is mild arrangement should be provided for providing full outdoor air without recirculation and conditioning in order to reduce the operating costs.

On *an average one tonne of refrigeration is required per 15 to 20 seats.*

Types of air distribution. The following *two* types of air distribution is found in theatre application :

1. *Horizontal distribution from near wall outlets with ejector nozzles or wall diffusers.* Refer to Fig. 4.12. This arrangement is particularly applicable for *long narrow theatres with flat ceilings* without any projecting beams or obstructions. It is necessary that there should not be any obstructions in the path of high velocity air from ejector nozzles to the stage. High velocities can be used for large theatres. For small theatres the outlets may be located inside or on front walls and returns located in front walls at floor level.

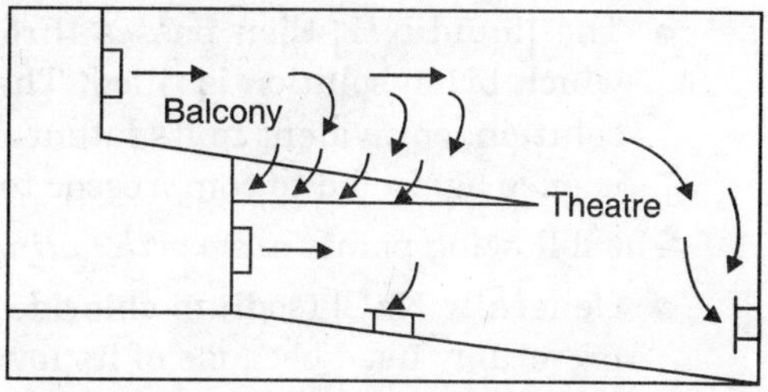

Fig. 4.12

2. *Overhead distribution from ceiling outlets with ceiling plaques on diffusers.* Refer to Fig. 4.13. In this system, the supply outlets are located under main ceiling and balcony. This arrangement is useful when there are obstructions under the ceiling or balcony. Such locations are unsuitable for heating in winter and will produce down cold drafts near windows unless ventilation is suitably enhanced.

Refrigerant and components :

The central plant *may use R–11 as refrigerant with centrifugal compressor.*

The *condenser* may be :

(*i*) Tube and shell type with cooling water supply cooled by cooling towers, or

(*ii*) Evaporative type, or

(*iii*) Air-cooled.

The air may be conditioned by :

(*i*) Direct expansion cooling coil, or

(*ii*) Cooling coil supplied with chilled water, or

(*iii*) Air washer using chilled water sprays.

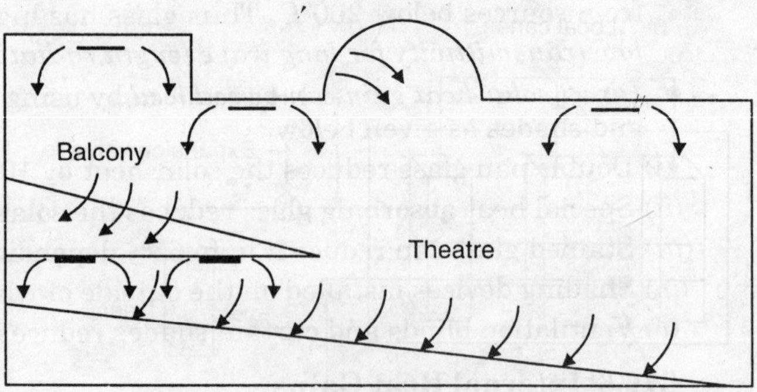

Fig. 4.13

It may be noted that the temperature of supply air from outlets in the theatre must be such as to maintain a temperature difference between room and supply air not exceeding 8 to 14°C otherwise it will create cold drafts causing shock to audience.

4.18. MANUFACTURE OF ICE

Fig. 4.14 shows the layout of an ice plant. The ice used for commercial purposes is produced by *freezing potable water* in standard cans placed in rectangular tanks which are filled with *chilled brine*. In order to enhance the heat transfer from the water agitators are employed to keep the brine solution in constant motion. The brine temperature is maintained at – 10°C to – 11°C by the refrigeration plant. The ammonia gas is used as refrigerant since it has excellent thermal properties, produces very high refrigerating effect per kg of refrigerant and has low specific volume of the refrigerant (in vapour state).

Working of ice plant :

- The high temperature, high pressure NH_3 vapours leaving the compressor are condensed in a condenser (may be of shell and tube type or evaporative type).
- The condensed liquid NH_3 is collected in the receiver and then expanded trough the expansion valve. Due to expansion, the pressure of the liquid NH_3 is considerably reduced.
- The liquid NH_3 then passes through the evaporator coils surrounding a brine tank in which brine solution is filled. The low pressure liquid NH_3 absorbs heat from the brine solution, equivalent to its latent heat of evaporation, gets converted to vapour state and is once again fed to compressor to complete the cycle.

The following points are *worth noting :*

- Generally NaCl (sodium chloride) or CaCl (calcium chloride) are used as brine. NaCl is preferably used because of its low cost and being less injurious.
- The freezing line is dependent upon the temperature and extent of brine and water agitation.
- In order to get clear transparent ice, water in the can is agitated using low pressure air through the tubes suspended from the top.
- The ice cans are fabricated from galvanised steel sheets and are given chromium treatment to prevent corrosion due to chemical reactions.

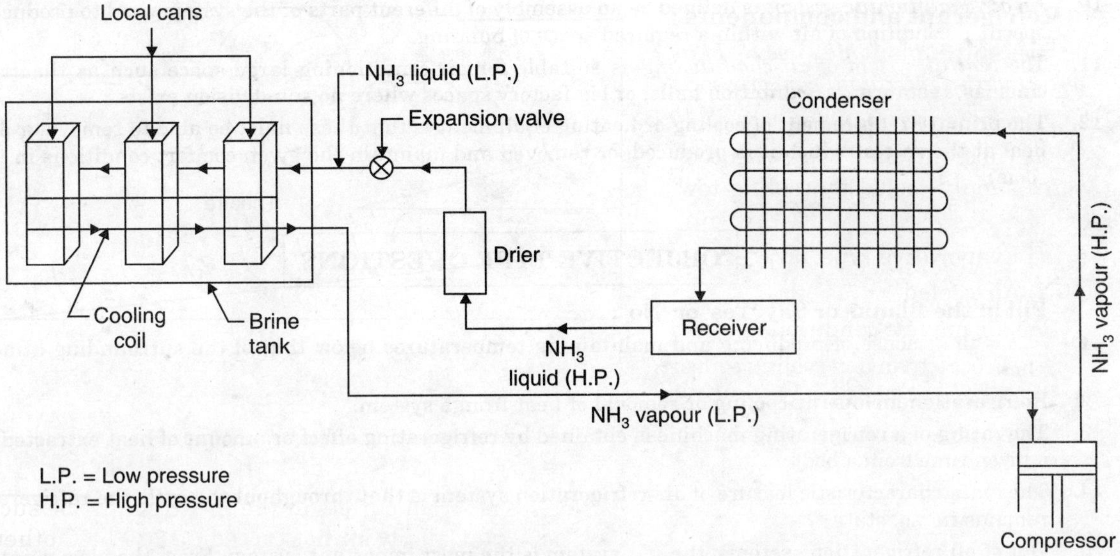

Fig. 4.14. Layout of an ice plant.

HIGHLIGHTS

1. *Refrigeration* is the science of producing and maintaining temperatures below that of the surrounding atmosphere.
2. The rating of a refrigeration machine is given by a unit of refrigeration known as "*standard commercial tonne* of *refrigeration*" which is defined as the *refrigerating effect* (*RE*) *produced by the melting of 1 tonne of ice from and at 0°C in 24 hours*. A *tonne of refrigeration* = 14000 kJ/h.
3. The main characteristic feature of air-refrigeration system is that throughout the cycle the *refrigerant remains in gaseous state.*
4. In a simple vapour compression system *fundamental processes are completed in one cycle.* These are :
   (*i*) Compression (*ii*) Condensation
   (*iii*) Expansion (*iv*) Vaporisation.
5. *Water coolers* are used to produce cold water at about 7 to 13°C. The temperature of water is controlled with the help of a thermostatic switch.
6. A *refrigerant* is defined as any substance that absorbs heat through expansion or vaporisation and loses it through condensation in a refrigeration system.
7. In a vapour compression system the following are the major refrigeration components :
   (*i*) Compressor (*ii*) Condensers
   (*iii*) Evaporators.
   The following are the *six basic types of refrigerant flow controls :*
   (*i*) Hand expansion valve
   (*ii*) Automatic expansion valve
   (*iii*) Thermostatic expansion valve
   (*iv*) Capillary tube
   (*v*) Low-side float
   (*vi*) High-side float.
8. The art of measuring the moisture content of air is termed "*psychrometry*".
9. *Effective temperature* is defined as a sensory index that combines into a single factor the effects of temperature, humidity, and air movement on human comfort in a noise free pure air environment.

10. An *air-conditioning system* is defined as an assembly of different parts of the system used to produce a specified condition of air within a required space of building.

11. The *central system of air-conditioning* is suitable for air-conditioning large space such as theatres, cinemas, restaurants, exhibition halls, or big factory spaces where no sub-division exists.

12. The primary requirement of cooling or heating equipment is that these must be able to remove or add heat at the rate at which it is produced, or removed and maintain the given comfort conditions in the room.

OBJECTIVE TYPE QUESTIONS

Fill in the Blanks or Say 'Yes' or 'No' :

1. ...... is the science of producing and maintaining temperatures below that of the surrounding atmosphere.
2. Refrigeration means the cooling or removal of heat from a system.
3. The rating of a refrigerating machine is obtained by refrigerating effect or amount of heat extracted in a given time from a body.
4. The main characteristic feature of air-refrigeration system is that throughout the cycle the refrigerant remains in ...... state
5. Out of all refrigeration systems, the ...... system is the most important system. From the view point of commercial and domestic utility.
6. Water coolers are used to produce cold water at about 7 to 13°C.
7. A ...... is any substance that absorbs heat through expansion or vaporisation and loses it through condensation in a refrigeration system.
8. The art of measuring the moisture content is termed ....
9. Sudomotor control does not regulate sweat production.
10. Vasomotor control regulates the blood supply to the skin.
11. The central systems are generally employed for loads above 25 tonnes of refrigeration and 2500 m^3/min of conditioning air.

ANSWERS

1. Refrigeration	**2.** Yes	**3.** Yes	**4.** gaseous
5. Vapour compression		**6.** Yes	**7.** Refrigerant
8. psychrometry	**9.** No	**10.** Yes	**11.** Yes

THEORETICAL QUESTIONS

1. Define the following :
   (*i*) Refrigeration (*ii*) Refrigeration system, and
   (*iii*) Refrigerated system.
2. Enumerate different ways of producing refrigeration.
3. Enumerate important refrigeration applications.
4. What is standard rating of a refrigeration machine ?
5. What is main characteristic feature of an air-refrigeration system ?
6. Differentiate clearly between open and closed air-refrigeration systems.
7. State merits and demerits of an air-refrigeration system.
8. Describe a simple vapour compression cycle giving clearly its flow diagram.
9. State the merits and demerits of 'Vapour compression system' over 'Vapour absorption system'.
10. Explain a refrigeration cycle by means of a neat-sketch.
11. Describe the working of a thermostat used in a domestic refrigerator.
12. Draw electric circuit of a refrigerator and explain its working. How can temperature inside the refrigerator be adjusted ?

13. Enlist the main requirements of a good refrigerant. What are the primary and secondary refrigerants ? Name the refrigerants generally used.

14. Define air conditioning. On what factors does the air-conditioning depend ? Explain in detail.

15. Discuss the role of air-conditioning in our day-to-day life.

16. What are different conditions which constitute Human comfort ?

17. Differentiate between comfort air-conditioning and industrial air-conditioning.

18. What is air-conditioning ? How is air from micro-organism, gaseous contaminant and odours purified ?

19. Draw a complete diagram showing there-in different components of an air-conditioning plant. What is the function of each component ?

20. Explain in brief the different processes involved in comfort air-conditioning. What makes a man comfortable in a conditioning space ?

21. What is 'effective temperature' ?

22. Discuss different methods of heating building.

23. Explain the working of a central air-conditioning system. Use a neat sketch.

5

Electrical Drives

5.1. INTRODUCTION

Electrical drive. A *form of machine equipment designed to convert electrical energy into mechanical energy and to provide electrical control of this process is known as* **electrical drive.**

An electrical drive has the following *major parts :*

(*i*) Load,
(*ii*) Motor,
(*iii*) Power modulator,
(*iv*) Control unit ; and
(*v*) Source.

Fig. 5.1 shows the block diagram of an electrical drive.

- *Load* comprises a machinery designed to accomplish a given task (*e.g.*, pumps, fans, machine tools, washing machines, drills, trains etc.)
- A *motor* having speed-torque characteristics and capabilities compatible to the load requirements is selected.

- *Power modulator* performs one or more of the following four functions :

(*i*) It modulates flow of power from the source, to the motor in such a way that motor is imparted speed-torque characteristics required by the load.

(*ii*) It restricts source and motor currents within permissible values, during transient operations (*e.g.*, starting, braking and speed reversal) ; excess current drawn from source may overload it or may cause a voltage dip.

(*iii*) It converts electrical energy of source in the form suitable to motor.

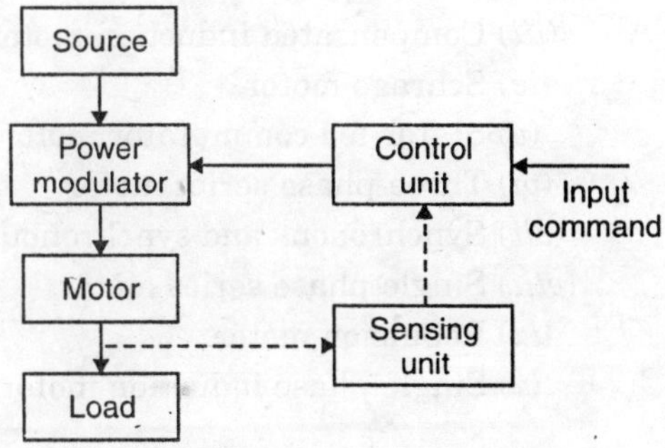

Fig. 5.1. Block diagram of an electrical drive.

(*iv*) It selects the mode of operation of the motor, *i.e.*, motoring or braking.

5.2. ADVANTAGES AND DISADVANTAGES OF ELECTRICAL DRIVES

Electrical drives entail the following *advantages* :

1. Available in wide range of torque, speed and power.
2. High efficiency.
3. Low no-load losses and considerable short time overloading capability.
4. Do not pollute environment.
5. Adaptable to almost any operating conditions such as explosive and radioactive environment, submerged in liquids, vertical mounting and so on.
6. They have flexible control characteristics.
7. Can be started instantly and can immediately be fully loaded.
8. They are powered by *electrical energy* which has a number of *advantages* (over other forms of energy) : (*i*) It can be generated and transported to the desired point economically and efficiently ; (*ii*) Conversion of electrical to mechanically energy and vice versa, and electrical energy from one form to another can also be done economically and efficiently.
9. Can operate in all the four quadrants of speed-torque plane.
   — Electric braking gives smooth deceleration and increases the life of equipment compared to other forms of braking.
   — When regenerative braking is possible, considerable saving of energy is achieved.
10. They have comparatively long life as compared to mechanical drive.
11. Being compact, require less space.
12. Speed control is easy and smooth.
13. Can be remotely controlled.

Disadvantages. The *two inherent disadvantages* of electrical drive system are :

1. It comes to stop as soon as there is failure of electric supply.
2. It cannot be used at far off places which are not served by electric supply.

However, the above two disadvantages can be overcome by installing diesel-driven D.C. generators and turbine-driven 3-phase alternators which can be used either in the absence of or on the failure of normal electric supply.

5.3. TYPES OF MOTORS USED FOR ELECTRICAL DRIVE

Following types of motors are generally used for electrical drive :

(*i*) D.C. shunt, series and compound motor.

(*ii*) Three-phase induction motor.

(*iii*) Compensated induction motor.
(*iv*) Schrage motor.
(*v*) Stator-fed commutator motor.
(*vi*) Three-phase series motor.
(*vii*) Synchronous and synchronous induction motor.
(*viii*) Single-phase series motor.
(*ix*) Repulsion motor.
(*x*) Single-phase induction motor.

5.4. SELECTION OF ELECTRICAL DRIVES

Some of the *important factors* on which the selection of electrical drives depend are :

1. *Steady state operation requirements :*
   - Nature of speed-torque characteristics.
   - Speed regulation.
   - Speed range.
   - Efficiency.
   - Duty cycle.
   - Quadrants of operation.
   - Speed fluctuations if any, ratings.
2. *Transient operation requirements :*
   - Values of acceleration and deceleration.
   - Starting, braking and reversing performance.
3. *Requirement related to the source :*
   - Type of source, and its capacity.
   - Magnitude of voltage, voltage fluctuations, power factor, harmonics and their effect on other loads.
   - Ability to accept regenerated power.
4. *Capital and running cost, maintenance needs, life.*
5. *Reliability.*
6. *Environment and location.*
7. *Space and weight restrictions, if any.*

5.5. STATUS OF D.C. AND A.C. DRIVES

Earlier induction and synchronous motor drives were employed for fixed speed applications ; D.C. motors dominated the variable speed applications.

The advent of thyristors in 1957 *led to the development of variable speed induction motor drives* in late sixties which were efficient and could match the performance of D.C. drives. Consequently it was predicted that induction motor will replace D.C. drives in variable speed applications, but the following problems/hurdles hindered the prediction to be realised :

1. Since the converter and control circuit of an induction motor was very expensive as compared to D.C. drive (although squirrel-cage induction motor was cheaper), the *total cost of an induction motor drive was much higher than that a D.C. drive.*

2. The technology of A.C. drive was new (in comparison to that of D.C. drive which was well established).

3. A.C. drives were less reliable than D.C. drives.

4. Although developments in linear and digital ICs, and VLSIs were helpful in improving the performance and reliability of A.C. drives, yet these developments also brought similar improvements in D.C. drives as well.

- The *following factors have resulted into reduction in cost, simple controllers, and improvement in performance and reliability of A.C. drives* :
  - — Improvement in thyristor capabilities, availability of power transistors in early seventies and that of GTDs and IGBTs in late seventies and late eighties respectively,
  - — Reduction in cost of thyristors, power transformers and GTDs ;
  - — Development of VLSIs and microprocessors ;
  - — Improvement in control techniques of converters.
- Although even now D.C. drives are invariably used in number of applications, the applications of A.C. drives are growing.
  - — "*Inducton motor drives*" are employed for *low to high power applications* ;
  - — "*Synchronous motor drives*" are used in *high power (MW) and medium power drives* ;
  - — The permanent magnet synchronous motor and brushless D.C. motors drives are being considered for replacing D.C. servo motors for fractional H.P. range.

5.6. CLASSIFICATION OF ELECTRICAL DRIVES

Electrical drives may be grouped into the following *three* categories :

1. Group drive.
2. Individual drive.
3. Multimotor drive.

1. **Group drive.** *A drive in which a single electric motor drives a line shaft by means of which an entire group of working machines may be operated is called* **group drive.** It is also sometimes call the *line shaft drive.* The line shaft is fitted with multi-stepped pulleys and belts that connect these pulleys and the shafts of the driven machines serve to vary their speed.

Advantages :

(*i*) Saving in initial cost (One 150 kW motor costs much less than ten 15 kW motors required to drive 10 separate machines.

(*ii*) The efficiency and power factor of a large group drive motor will be higher, provided it is operated fairly near its rated load.

(*iii*) If the machines are liable to short but sharp overloads, group drive is again advantageous, because 100 percent overload on an individual machine will cause hardly 10 percent overload when driven by group drive.

(*iv*) Group drive can be used with advantage in those industrial processes where there is a sequence of continuity in operation and where it is desirable to stop these processes simultaneously as in a flour mill.

Disadvantages. Group drive is seldom used these days due to the following *disadvantages* :

(*i*) Group drive does not give good appearance (owing to the use of line shafting pulley and belts) and is less safe to operate.

(*ii*) In group drive the speed control of individual machine is very cumbersome using stepped pulleys, belts etc.

(*iii*) Noise level at the working site is quite high.

(*iv*) This system is unreliable since any fault in driving motor renders all the driven equipment idle.

(*v*) Considerable amount of power is lost in energy transmitting mechanism.

(*vi*) If all the machines driven by the line shaft do not work together, the main motor runs at reduced load. Consequently it runs with low efficiency and with poor power factor.

(*vii*) Group drive cannot be used where constant speed is required as in paper and textile industry.

(*viii*) Flexibility of layout of different machines is lost since they have to be so located as to suit the position of the line shaft.

- Group drive is *adopted*, when existing factories are changed from engine drive to electric motor drive simply by replacing the oil or steam engine by an electric motor of corresponding output retaining all the old shafts and belts.

2. **Individual drive.** In "*individual drive*", *each machine is driven by its own separate motor with the help of gears, pulley etc.* ***Examples :*** Single-spindle drilling machines, various types of electrical hand tools and simple types of metal working machine tools and mechanisms.

Advantages :

(*i*) Flexibility in the installation of different machines.

(*ii*) Each operator has full control of the machine which can be quickly stopped if an accident occurs.

(*iii*) Since each machine is driven by a separate motor, it can be run and stopped as desired.

(*iv*) The motor and its control unit can be built as an integral part of the machine which results in a good appearance, cleanliness and safety.

(*v*) Machines not required can be shut down and also replaced with a minimum of dislocation.

(*vi*) In the case of motor fault, only its connected machine will stop whereas other will continue working undisturbed.

(*vii*) The maintenance of line shafts, bearings, pulleys and belts etc. is eliminated. Furthermore, there is no danger of oil falling on articles being manufactured, which is very important in textile industry.

Disadvantage. The only disadvantage of individual drive is its *high cost.*

- For driving heavy machines such as for lifts, cranes, shapers, lathes etc. and for the purposes where constancy of speed and flexibility of control is required, such as in paper mills and textile industry, *individual drive is essential.*
- Individual drive is *preferred for new factories*, as it causes some saving in the cost of superstructure which is much lighter and less expensive.

3. **Multimotor drive.** In "*multimotor drives*" *separate motors are provided for actuating differents parts of the driven mechanism.* ***Example :*** In *travelling cranes*, three motors are used : *one for hoisting, another for long travel motion and third for cross travel motion.*

- Such a drive is *essential in complicated metal-cutting machine tools, paper making machines, rolling mills etc.*
- The use of individual drives and multimotor drives has led to the introduction of automation in production processes which, apart from increasing the productivity of various undertakings, has enhanced the reliability and safety of operation.

Example 5.1. *Compare the total annual cost of a group drive with a motor costing Rs. 18000 with that of ten individual motors, each costing Rs. 5000. With group drive annual consumption is 80000 kWh. With separate drives annual consumption is 55000 kWh. Electrical energy costs*

20 paise per kWh. Depreciation, maintenance and other fixed charges amount to 10 percent in case of group drive and 15 percent in case of individual drive.

Solution. Capital cost on group drive = Rs. 18000.00

Annual depreciation, maintenance and

fixed charges $= 18000 \times \frac{10}{100}$ = Rs. 1800.00

Energy consumption per annum = 80000 kWh

Annual energy charges = Rs. 0.20 × 80000 = Rs. 16000.00

Total annual cost in case of group drive = Rs. 1800.00 + Rs. 16000.00 = Rs. 17800.00

Capital cost on individual drive = Rs. 5000 × 10 = Rs. 50000.00

Annual depreciation, maintenance and

fixed charges = Rs. $50000 \times \frac{15}{100}$ = Rs. 7500.00

Annual energy consumption = 55000 kWh

Annual energy charges = Rs. 0.20 × 55000 = Rs. 11000.00

Total annual cost in case of individual drive = Rs. 7500.00 + Rs. 11000.00 = Rs. 18500.00

Although capital and as well annual cost of *individual drive* is higher than that of the group drive yet because of its inherent advantages over group drive such as flexibility of control and convenience *it is usually preferred.*

5.7. TYPES OF LOADS/LOAD TORQUES

Load torques are *classified* into the following two categories :

1. Active or potential loads.
2. Passive loads.

1. **Active loads :**

- *The loads which are due to the forces of gravity, tension or compression in a spring or any elastic body are called* ***active loads.*** The term potential may be used for such loads/ torques as they are associated with the changes in the potential energy of various elements of an electric drive system.
- The active or potential torques may be +ve or –ve depending upon whether they *aid* or *resist* the motion of the drive. Such torques retain their sign irrespective of the direction of rotation of the drive. *Example :* When a load is moved upwards, the stored potential energy increases and the torque opposes the upward movement of the load. On the other hand, when a load is brought downward, the potential energy decreases but the active torque associated with it aids the downward motion of the load. Thus it may be observed that *active torques act in one direction only.*
- Active torques come to play for *hoists, lifts, elevators, railway locomotives on gradients* etc.

2. **Passive torques :**

- These torques are *due to friction, cutting and deformation of inelastic bodies.*
- They always *oppose* the motion of drive and change their sign when the direction of rotation of drive is changed. *Example :* The frictional torque always acts in a direction opposite to that of driving torque.

Nature of Mechanical Load :

A. Variation with speed. Depending upon the speed, load torques can be*classified* as follows :

(*i*) Constant load torque. (*ii*) Load torque ∝ speed.

(*iii*) Load torque ∝ speed^2. (*iv*) Load torque ∝ $\frac{1}{\text{speed}}$.

Fig. 5.2 shows load torque as a function of speed of cranes, hoists etc.

(*i*) *Constant load torque.* It is independent of the speed and occurs in case of cranes, hoists etc.

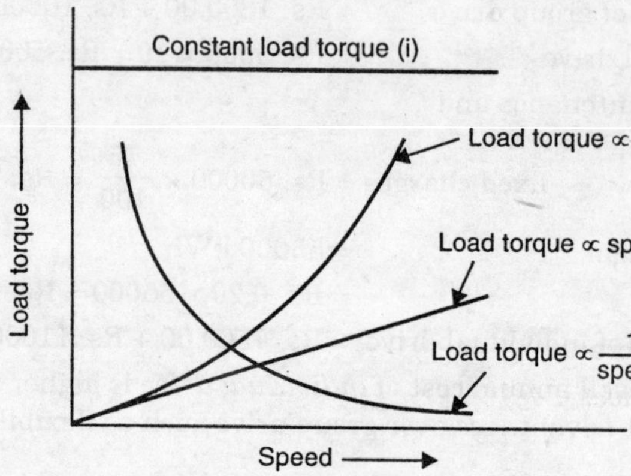

Fig. 5.2. Load torque as a function of speed of cranes, hoists, etc.

(*ii*) *Load torque ∝ speed.* The torque increases *linearly* with speed and occurs in case of fluid friction *where lubricant is used.*

(*iii*) *Load torque ∝ speed*^{2}. The torque increases as a square of speed and occurs in case of *air and fluid friction* such as in *fans, water wheels* etc.

(*iv*) *Load torque* ∝ $\frac{1}{speed}$. The torque vaires inversely as the speed and occurs where *deformation of material takes place e.g.,* in *grinding, metal drawing etc.*

B. Variation with time. The load torques, depending upon the time for which load exists on an electrical drive, can be *classified* as follows :

(*i*) *Continuous and constant load.*

Example : Centrifugal pumps operating under same condition for a long time.

(*ii*) *Continuous but variable loads.*

Example. Conveyors, hoisting winches etc.

(*iii*) *Pulsating loads.*

Example : Reciprocating pumps, textile looms etc.

(*iv*) *Impact loads.*

Example : Rolling mills, forging hammers, shearing machines etc. (Each of these machines has a flywheel).

(*v*) *Short time intermittent loads.*

Example : Excavators, crane hoists etc.

5.8. D.C. MOTOR DRIVES

5.8.1. General Aspects

- D.C. drives are widely used in aplications requiring :
  - — Adjustable speed ;
  - — Good speed regulation ;
  - — Frequent starting, braking and reversing.
- The applications of D.C. drives include :
  - — Rolling mills ;
  - — Paper mills ;
  - — Mine winders ;
  - — Hoists,
  - — Machine tools ;
  - — Traction ;
  - — Printing presses ;
  - — Textile mills,
  - — Excavators and cranes.
- *Fractional horse power D.C. motors* are widely used as *servomotors for positioning and tracking.*

5.8.2. Advantages and Disadvantages of D.C. Motors

Advantages. The D.C. motors possess the following *advantages* :

1. High starting torque.
2. Speed control over a wide range, both below and above the normal speed.
3. Accurate stepless speed control with constant torque.
4. Quick starting, stopping, reversing and accelerating.
5. High reliability.

Disadvantages. The disadvantages of D.C. motors are :

1. High initial cost.
2. Increased operating and maintenance costs because of the commutators and brushgear.

5.8.3. D.C. Motors and their Performance

Commonly used D.C. motors are :

1. Shunt and separately excited motor.
2. Series motor.
3. Compound motor.
4. Universal motor.
5. Permanent magnet motors.
6. D. C. servo motors.
7. Moving coil motors.
8. Torque motors.

Commonly used motors are shown in Fig. 5.3 [(*i*), (*ii*), (*iii*) and (*iv*)]

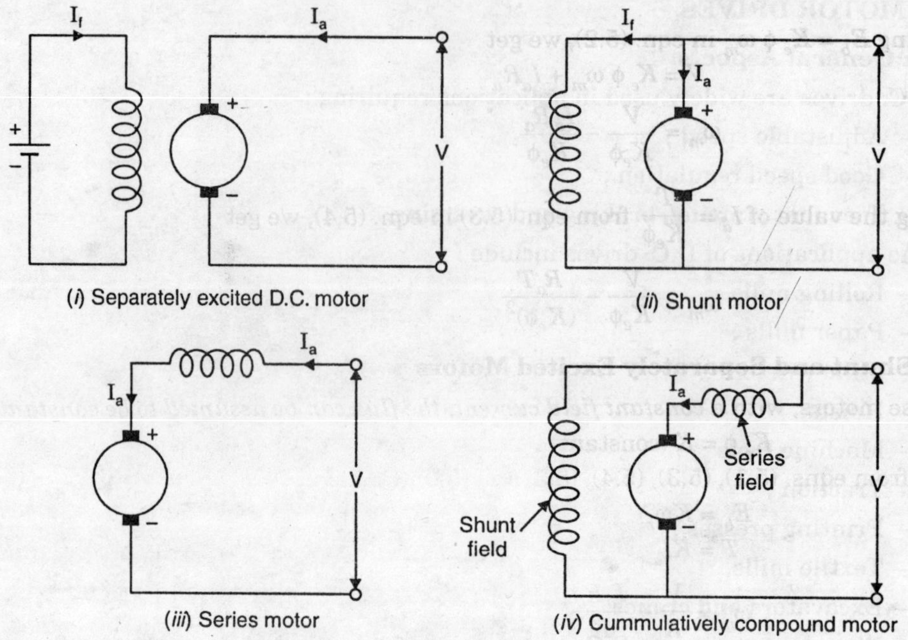

(*i*) Separately excited D.C. motor (*ii*) Shunt motor

(*iii*) Series motor (*iv*) Cummulatively compound motor

Fig. 5.3. Commonly used D.C. motors.

- *Refer Fig. 5.3* (*i*). In a *separately excited motor*, the field and armature voltages can be controlled independent of each other.
- *Refer Fig. 5.3* (*ii*). In a *shunt motor*, field and armature are connected to a common source.
- *Refer Fig. 5.3* (*iii*). In case of a *series motor*, field current is same as armature current, therefore, field flux is a function of armature current.
- *Refer Fig. 5.3* (*iv*). In a *cumulatively compound motor*, the m.m.f. (magneto-motive force) of the series field is a function of armature current and is in the *same direction* as m.m.f. of shunt field.

Steady state equivalent circuit of armature of a D.C. machine. The steady state equivalent circuit of armature of a D.C. machine is shown in Fig. 5.4.

Basic equations applicable to all D.C. motors are :

$$E_b = K_e \phi \omega_m \quad ....(5.1)$$

$$V = E_b + I_a R_a \quad ...(5.2)$$

$$T = K_e \phi I_a \quad ...(5.3)$$

where, K_e = Motor constant,

ϕ = Flux per pole,

ω_m = Armature speed, rad/s,

V = Supply voltage,

E_b = Back e.m.f.,

I_a = Armature current,

R_a = Resistance of armature circuit, and

T = Torque developed by the motor.

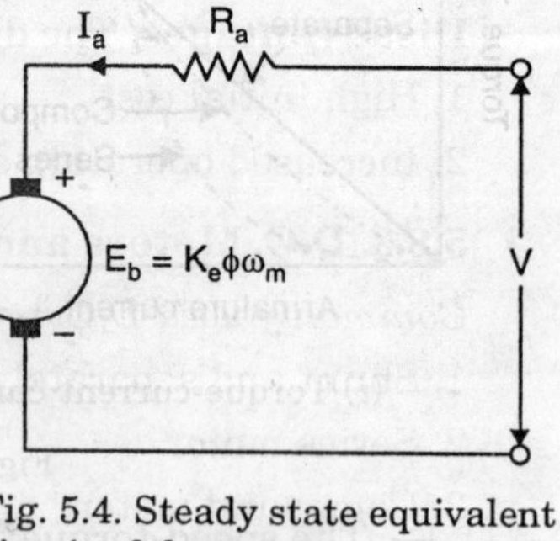

Fig. 5.4. Steady state equivalent circuit of the armature of a D.C. machine.

Inserting $E_b = K_e \phi \omega_m$ in eqn. (5.2), we get

$$V = K_e \phi \omega_m + I_a R_a$$

or,
$$\omega_m = \frac{V}{K_e\phi} - \frac{I_a R_a}{K_e\phi} \quad \text{...(5.4)}$$

Putting the value of $I_a = \frac{T}{K_e\phi}$ from eqn. (5.3) in eqn. (5.4), we get

$$\omega_m = \frac{V}{K_e\phi} - \frac{R_a T}{(K_e\phi)^2} \quad \text{...(5.5)}$$

5.8.4. Shunt and Separately Excited Motors

In these motors, with a *constant field current, the flux can be assumed to be constant.*

Let, $K_e \phi = K$ (constant) ...(5.6)

Then, from eqns. (5.1), (5.3), (5.4), (5.5) and (5.6), we have

$$E_b = K\omega_m \quad \text{...(5.7)}$$

$$T = KI_a \quad \text{...(5.8)}$$

$$\omega_m = \frac{V}{K} - \frac{I_a R_a}{K} \quad \text{...(5.9)}$$

Again,
$$\omega_m = \frac{V}{K} - \frac{R_a T}{K^2} \quad \text{...(5.10)}$$

- The torque-current (T/I_a) and speed-torque (ω_m/T) characteristics of a separately excited motor for rated terminal voltage and full field are shown in Fig. 5.5.

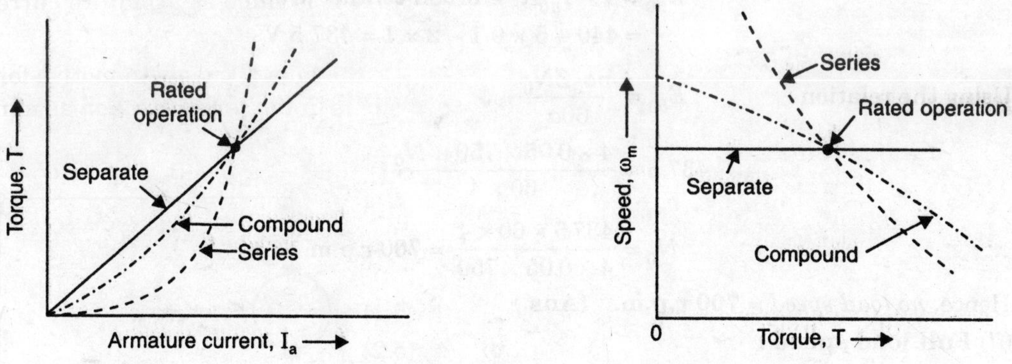

(*i*) Torque-current curves (*ii*) Speed-torque curves

Fig. 5.5. Performance curves of D.C. motors.

  — The speed-torque curve is a straight line. The no-load speed ω_{m0} is determined by the values of armature voltage and field excitation.
  — Speed decreases as torque increases and speed regulation depends on the armature circuit resistance [eqn. (5.10)].
  — In a medium size motor, the usual *drop in speed from no load to full load* is of the order of *5 percent.*

- Separately excited motors are employed in *applications requiring good speed regulation and adjustable speed.*

Example 5.2. *A 440 V, 4-pole, lap connected shunt motor has a no-load input current of 15 A and a shunt field current of 10 A. At full load it takes a current of 150 A. If armature resistance = 0.1 ohm, flux per pole on no-load = 0.05 weber, number of armature conductors = 750 and contact drop per brush = 1 V, calculate :*

(i) No-load speed *(ii) Full load speed*

(iii) Speed regulation.

Armature reaction weakens the field by 1.5% on full load.

Solution. Supply voltage, $V = 440$ Volts

Number of poles, $p = 4$

Number of parallel paths, $a = p = 4$ [Motor being lap-connected]

Armature resistance, $R_a = 0.1$ ohm

Flux per pole on no-load, $\phi_0 = 0.05$ weber

Number of armature conductors, $Z = 750$

Contact drop per brush, $= 1$ V

No-load input current, $I_0 = 15$ A

Shunt field current, $I_{sh} = 10$ A

Full load current, $I = 150$ A

No-load speed, N_0 :

Full load speed, N :

Speed regulation :

(i) **No load speed :**

Armature current, $I_{a0} = I_0 - I_{sh} = 15 - 10 = 5$ A

$E_{b0} = V - I_{a0}R_a$ – brush contact drop

$= 440 - 5 \times 0.1 - 2 \times 1 = 437.5$ V

Using the relation : $$E_{b0} = \frac{p\phi_0 Z N_0}{60a}$$

or $$437.5 = \frac{4 \times 0.05 \times 750 \times N_0}{60 \times 4}$$

$\therefore$ $$N_0 = \frac{437.5 \times 60 \times 4}{4 \times 0.05 \times 750} = 700 \text{ r.p.m.}$$

Hence, *no load speed* = **700 r.p.m. (Ans.)**

(ii) **Full load speed :**

Armature current, $I_a = I - I_{sh} = 150 - 10 = 140$ A

$E_b = V - I_aR_a$ – brush contact drop

$= 440 - 140 \times 0.1 - 2 \times 1 = 424$ V

$\phi = (1 - 0.015)\,\phi_0$

[Since armature reaction weakens the field by 1.5 per cent]

$= 0.985\,\phi_0$

Using the relation : $$\frac{N}{N_0} = \frac{E_b}{E_{b0}} \times \frac{\phi_0}{\phi}$$

or $$\frac{N}{700} = \frac{424}{437.5} \times \frac{\phi_0}{0.985\phi_0}$$

$$N = 700 \times \frac{424}{437.5} \times \frac{\phi_0}{0.985\,\phi_0} = 688.73 \text{ r.p.m.}$$

Hence, *full load speed* = **688.73 r.p.m. (Ans.)**

(*iii*) **Percentage speed regulation**

$$= \frac{\text{No load speed} - \text{Full load speed}}{\text{Full load speed}} \times 100$$

$$= \frac{700 - 688.73}{688.73} \times 100 = 1.637$$

Hence, *percentage speed regulation* = **1.637%. (Ans.)**

Example 5.3. *A 250 V shunt motor takes a line current of 60 A and runs at 800 r.p.m. Its armature and field resistances are 0.2 ohm and 125 ohms respectively. Contact drop/brush = 1 V. Calculate :*

(i) No-load speed if the no-load current is 6 A.

(ii) The percentage reduction in the flux per pole in order that the speed may be 1000 r.p.m. when the armature current is 40 A

Neglect the effects of armature reaction.

Solution. Supply voltage, $V = 250$ Volts

Load current, $I_1 = 60$ A

Load speed, $N_1 = 800$ r.p.m.

Armature resistance, $R_a = 0.2$ ohm

Shunt field resistance, $R_{sh} = 125$ ohms

Contact drop $= 2$ V

Armature current, $I_{a2} = 40$ A

Load speed, $N_2 = 1000$ r.p.m.

No load current, $I_0 = 6$ A

No load speed, N_0 :

Percentage reduction in flux/pole :

Shunt field current, $I_{sh} = \frac{250}{125} = 2$ A

At no load :

$$I_{a0} = I_0 - I_{sh} = 6 - 2 = 4 \text{ A}$$

$$E_{b0} = V - I_{a0}R_a - \text{brush drop}$$
$$= 250 - 4 \times 0.2 - 2 \times 1 = 247.2 \text{ V}$$

At load (I_1 = 60 A) :

$$I_{a1} = I_1 - I_{sh} = 60 - 2 = 58 \text{ A}$$

$$E_{b1} = V - I_{a1}R_a - \text{brush drop}$$
$$= 250 - 58 \times 0.2 - 2 \times 1 = 236.4 \text{ V}$$

Using the relation, $\frac{N_1}{N_0} = \frac{E_{b1}}{E_{b0}} \times \frac{\phi_0}{\phi_1}$

Since the motor is shunt wound and armature reaction effect is negligible so *flux may be assumed to be constant* and therefore speed is directly proportional to the back e.m.f. developed.

$$\therefore \quad \frac{N_1}{N_0} = \frac{E_{b1}}{E_{b0}}$$

$$\frac{800}{N_0} = \frac{236.4}{247.2}$$

$$N_0 = \frac{800 \times 247.2}{236.4} = 836.5 \text{ r.p.m.}$$

Hence, *no load speed* = **836.5 r.p.m. (Ans.)**

At armature current, I_{a2} = **40 A**

Let the flux be reduced to ϕ_2.

$$E_{b2} = V - I_{a2}R_a - \text{brush drop}$$
$$= 250 - 40 \times 0.2 - 2 \times 1 = 240 \text{ V}$$

Using the relation,
$$\frac{N_2}{N_1} = \frac{E_{b2}}{E_{b1}} \times \frac{\phi_1}{\phi_2}$$

or
$$\frac{1000}{800} = \frac{240}{236.4} \times \frac{\phi_1}{\phi_2}$$

$$\therefore \quad \frac{\phi_2}{\phi_1} = \frac{240 \times 800}{1000 \times 236.4} = 0.812$$

$\therefore$ Percentage reduction in flux

$$= \frac{\phi_1 - \phi_2}{\phi_1} \times 100 = \left(1 - \frac{\phi_2}{\phi_1}\right) \times 100 = (1 - 0.812) \times 100 = 18.8\%$$

Hence, *%age reduction in flux* = **18.8%. (Ans.)**

☞**Example 5.4.** *A 220 V shunt motor takes 60 A when running at 800 r.p.m. It has an armature resistance of 0.1 ohm. Determine the speed and armature current if the magnetic flux is weakened by 20%. Contact drop per brush = 1 V.*

Total torque developed remains constant.

Solution. Supply voltage = 220 V

Load current, $= I_1 = I_{a1} = 60$ A [Neglecting shunt field current]

Speed at 60 A, $N_1 = 800$ r.p.m.

Armature resistance, $R_a = 0.1$ ohm

Contact drop/brush = 1 V

$\mathbf{N_2} = ?, \quad \mathbf{I_{a2}} = ?$

$$\phi_2 = \left(1 - \frac{20}{100}\right)\phi_1 \quad \text{[Since magnetic flux is weakened by 20\%]}$$
$$= (1 - 0.20)\phi_1 = 0.8\phi_1$$

Since, $T \propto \phi I_a$

and torque remains constant, [*Given*]

$$\therefore \quad \phi_1 I_{a1} = \phi_2 I_{a2}$$
$$\phi_1 \times 60 = 0.8\phi_1 I_{a2}$$
$$\therefore \quad I_{a2} = \mathbf{75\ A.}$$

Back e.m.f., $E_{b1} = V - I_{a1}R_a - \text{brush drop}$
$$= 200 - 60 \times 0.1 - 2 \times 1 = 212 \text{ V}$$

Back e.m.f., $E_{b2} = V - I_{a2}R_a - \text{brush drop}$
$= 220 - 75 \times 0.1 - 2 \times 1 = 210.5$ V

Using the relation, $$\frac{N_2}{N_1} = \frac{E_{b2}}{E_{b1}} \times \frac{\phi_1}{\phi_2}$$

$$\frac{N_2}{800} = \frac{210.5}{212} \times \frac{\phi_1}{0.8\phi_1}$$

$\therefore$ $$N_2 = \frac{800 \times 210.5}{212 \times 0.8} = \mathbf{992.9\ r.p.m.}$$

Hence, *Speed* = **992.9 r.p.m. (Ans.)**

Armature current **= 75 A. (Ans.)**

☞**Example 5.5.** *A 440 V shunt motor takes 105 A (armature current) from the supply and runs at 1000 r.p.m. Its armature resistance is 0.15 ohm. If total torque developed is unchanged, calculate the speed and armature current if the magnetic field is reduced to 70% of the initial value.*

(Panjab University)

Solution. Supply voltage, $V = 440$ Volts
Armature current, $I_{a1} = 105$ A
Speed, $N_1 = 1000$ r.p.m.
Armature resistance, $R_a = 0.15$ ohm
$\phi_2 = 0.7\ \phi_1.$

(*i*) **Armature current, I_{a2} :**

As the torque developed is same

$\therefore$ $$\phi_1 I_{a1} = \phi_2 I_{a2}$$

$$I_{a2} = \frac{\phi_1}{\phi_2} \times I_{a1} = \frac{\phi_1}{0.7\phi_1} \times 105 = 150 \text{ A}$$

Hence, *armature current* = **150 A. (Ans.)**

Now, $E_{b1} = V - I_{a1}R_a = 440 - 105 \times 0.15 = 424.25$
and, $E_{b2} = V - I_{a2}R_a = 440 - 150 \times 0.15 = 417.5$ V.

(*ii*) **Speed, N_2 :**

Using the relation, $$\frac{N_2}{N_1} = \frac{E_{b2}}{E_{b1}} \times \frac{\phi_1}{\phi_2}$$

$$\frac{N_2}{1000} = \frac{417.5}{424.25} \times \frac{\phi_1}{0.7\phi_1}$$

$\therefore$ $$N_2 = 1000 \times \frac{417.5}{424.25} \times \frac{1}{0.7} = 1405 \text{ r.p.m.}$$

Hence, *speed of the motor when magnetic field is reduced to 70% of initial value* = **1405 r.p.m. (Ans.)**

Example 5.6. *A 250 V D.C. shunt motor has an armature resistance of 0.5 Ω and a field resistance of 250 Ω. When driving a constant torque load at 600 r.p.m., the motor draws 21 A. What will be the new speed of the motor if an additional 250 Ω resistance is inserted in the field circuit.*

(GATE, 2000)

Solution. *Given* : $V = 250$ volts ; $R_a = 0.5\ \Omega$, $R_{sh} = 250\ \Omega$, $N_1 = 600$ r.p.m., $I = 21$ A

New speed, N_2 :

Shunt field current, $$I_{sh1} = \frac{V}{R_{sh}} = \frac{250}{250} = 1\text{ A}$$

Armature current, $$I_{a1} = 21 - 1 = 20\text{ A}$$

Back e.m.f., $$E_{b1} = V - I_a R_a = 250 - 20 \times 0.5 = 240\text{ V}$$

Shunt field current when an additional 250 Ω resistance is inserted in the field circuit,

$$I_{sh2} = \frac{250}{250 + 250} = 0.5\text{ A}$$

Neglecting magnetic saturation,

$$\phi_1 \propto I_{sh1}$$

or, $$\frac{\phi_1}{\phi_2} = \frac{I_{sh1}}{I_{sh2}} \quad \text{and} \quad \phi_2 \propto I_{sh2}$$

For constant load torque, $$T_1 = T_2$$

or, $$\phi_1 I_{a1} = \phi_2 I_{a2}$$

or, $$I_{a2} = I_{a1} \times \frac{\phi_1}{\phi_2} = I_{a1} \times \frac{I_{sh1}}{I_{sh2}} = 20 \times \frac{1}{0.5} = 40\text{ A}$$

Back e.m.f., $$E_{b2} = 250 - 40 \times 0.5 = 230\text{ V}$$

Also, $$\frac{N_2}{N_1} = \frac{E_{b2}}{E_{b1}} \times \frac{\phi_1}{\phi_2}$$

$$\frac{N_2}{600} = \frac{230}{240} \times \frac{I_{sh1}}{I_{sh2}} = \frac{230}{240} \times \frac{1}{0.5}$$

or, $$\mathbf{N_2} = 600 \times \frac{230}{240} \times \frac{1}{0.5} = \mathbf{1150\ r.p.m.} \quad \textbf{(Ans.)}$$

Example 5.7. *A 15 kW, 230 V, 1150 r.p.m., 4-pole, D.C. shunt motor has a total of 882 armature conductors arranged in four parallel paths and yielding an armature ckt resistance of 0.2 ohm. When it delivers rated power at rated speed, the motor draws an armature current of 73 A at a field current of 1.6 A. Calculate the developed torque. Also find new operating speed if the field flux is reduced by 80% of its original value for the same developed torque.* **(A.M.I.E. Summer, 2000)**

Solution. *Given :* Power output = 15 kW ; $V = 230$ volts ; $N = 1150$ r.p.m.,

$Z = 882$, $R_a = 0.2\ \Omega$, $a = 4$; $I_a = 73$ A ; $I_{sh} = 1.6$ A ; $\phi_2 = 0.8\ \phi_1$.

Torque developed, T_a :

We know that, $$T_a = 0.159\, Z\, \phi\, p \,.\, \frac{I_a}{a}\ \text{N-m}$$

and, $$E_b = \frac{p\phi ZN}{60a} = V - I_a R_a$$

$$\therefore \quad \phi = \frac{(V - I_a R_a) \times 60a}{pZN}$$

Putting the given values, we have

$$\phi = \frac{(230 - 73 \times 0.2) \times 60 \times 4}{4 \times 882 \times 1150} = 0.01274 \text{ Wb}$$

$$\therefore \quad \mathbf{T_a} = 0.159 \times 882 \times 0.01274 \times 4 \times \frac{73}{4}$$

$$= \mathbf{130.4 \text{ N-m.}} \quad \textbf{(Ans.)}$$

Also, $T_a \propto \phi I_a$

$$\therefore \quad \frac{T_1}{T_2} = \frac{\phi_1 I_{a1}}{\phi_2 I_{a2}}$$

Again, $N \propto \frac{E_b}{\phi}$ or $\phi \propto \frac{E_b}{N}$

$$\therefore \quad \frac{T_1}{T_2} = \frac{E_{b1}}{N_1} \times \frac{N_2}{E_{b2}} \times \frac{I_{a1}}{I_{a2}} \quad \text{...}(i)$$

When $T_1 = T_2$ (Given), we have

$$\therefore \quad \phi_1 I_{a1} = \phi_2 I_{a2}$$

$$I_{a2} = I_{a1} \times \frac{\phi_1}{\phi_2} = 73 \times \frac{1}{0.8} = 91.25 \text{ A}$$

Now
$$E_{b1} = V - I_{a1}R_a = 230 - 73 \times 0.2 = 215.4 \text{ V}$$
$$E_{b2} = V - I_{a2}R_a = 230 - 91.25 \times 0.2 = 211.75 \text{ V}$$

Substituting the values in (*i*), we have

$$1 = \frac{215.4}{1150} \times \frac{N_2}{211.75} \times \frac{73}{91.25}$$

$$\therefore \quad \mathbf{N_2} = \frac{1150 \times 211.75 \times 91.25}{215.4 \times 73} = \mathbf{1413 \text{ r.p.m.}} \quad \textbf{(Ans.)}$$

Example 5.8. *A 230 V D.C. shunt motor with constant field drives a load whose torque is proportional to the speed. When running at 750 r.p.m. it takes 30 A. Find the speed at which it will run if a 10 Ω resistance is connected in series with its armature. The resistance of armature may be neglected.* **(A.M.I.E. Electrical Drives and their Control Winter, 2001)**

Solution. *Given* : $V = 230$ volts ; $I_1 (= I_{a1}) = 30$ A ; $N_1 = 750$ r.p.m.

Speed N_2 :

Back e.m.f., $E_{b1} = V = 230$ Volts [∵ Armature resistance is negligible]

Let the armature current be I_{a2} and speed N_2 on insertion of a 10 Ω resistor in the armature circuit.

$$E_{b2} = V - I_{a2}R = 230 - 10I_{a2}$$

Also
$$N \propto \frac{E_b}{\phi}$$

or,
$$N \propto E_b \quad \text{[∵ Flux remains constant.]}$$

or,
$$\frac{N_2}{N_1} = \frac{E_{b2}}{E_{b1}} \quad \text{...}(i)$$

Also, $T \propto N$ [Given]

$$\therefore \quad \frac{T_2}{T_1} = \frac{N_2}{N_1} \quad \text{...(ii)}$$

Since, $T \propto \dfrac{E_b I_a}{N}$

$$\therefore \quad \frac{T_2}{T_1} = \frac{E_{b2} I_{a2}}{E_{b1} I_{a1}} \times \frac{N_1}{N_2} \quad \text{...(iii)}$$

From (*ii*) and (*iii*), we have

$$\frac{E_{b2} I_{a2}}{E_{b1} I_{a1}} \times \frac{N_1}{N_2} = \frac{N_2}{N_1}$$

or, $$\frac{E_{b2} I_{a2}}{E_{b1} I_{a1}} = \left(\frac{N_2}{N_1}\right)^2 = \left(\frac{E_{b2}}{E_{b1}}\right)^2 \qquad \left[\because \text{ From (i), } \frac{N_2}{N_1} = \frac{E_{b2}}{E_{b1}}\right]$$

or, $$\frac{I_{a2}}{I_{a1}} = \frac{E_{b2}}{E_{b1}}$$

or, $$\frac{I_{a2}}{30} = \frac{230 - 10I_{a2}}{230}$$

or, $$230I_{a2} = 6900 - 300I_{a2}$$

or, $$530I_{a2} = 6900$$

$$\therefore \quad I_{a2} = 13.02 \text{ A}$$

From (*i*), we have $$\frac{N_2}{N_1} = \frac{E_{b2}}{E_{b1}} = \frac{230 - 10 \times 13.02}{230}$$

$$\therefore \quad \mathbf{N_2} = 750 \times \left(\frac{230 - 10 \times 13.02}{230}\right) = \mathbf{325.4 \text{ r.p.m.}} \quad \textbf{(Ans.)}$$

Example 5.9. *A 440 V D.C. shunt motor has a no-load (I_a = 0) speed of 2000 r.p.m. It is running at 1000 r.p.m. at full load torque, reduced armature voltage and full field. If load torque is reduced to 50% of rated value with armature voltage and field voltage held constant at previous values, the speed increases to 1050 r.p.m. Find the armature voltage drop at full load. Neglect effect of armature reaction.*

(GATE, 1998)

Solution. *Given :* $V = 440$ volts ; $N_0 = 2000$ r.p.m. (at $I_a = 0$) ;
$N_1 = 1000$ r.p.m. ; $T_2 = 0.5T_1$; $N_2 = 1050$ r.p.m.

Armature voltage drop at full load :

Let I_{a1} be the armature current at full-load torque.

Back e.m.f. at full-load torque, $E_{b1} = (440 - I_{a1}R_a) \propto N_1\phi_1$ ...(*i*)

Motor torque, $T_1 \propto \phi_1 I_{a1}$

Also, $T_2 \propto \phi_2 I_{a2}$

or, $$\frac{T_1}{2} \propto \phi_2 I_{a2} \qquad \left[\because T_2 = \frac{T_1}{2} \text{ ...... Given}\right]$$

As the field voltage is held constant, $\phi_2 = \phi_1$

$$\therefore \quad \frac{T_1}{(T_1/2)} = \frac{\phi_1 I_{a1}}{\phi_2 I_{a2}} = 2$$

$$\therefore \qquad I_{a2} = \frac{I_{a1}}{2}$$

Back e.m.f. at half rated torque, $E_{b2} = \left(440 - \frac{I_{a1}}{2} . R_a\right) \propto N_2\phi_2$ ...(*ii*)

From (*i*) and (*ii*), we have $\frac{E_{b1}}{E_{b2}} = \frac{N_1 . \phi_1}{N_2\phi_2}$

$$\frac{440 - I_{a1}R_a}{440 - \frac{I_{a1}}{2} R_a} = \frac{1000 \times \phi_1}{1050 \times \phi_1} = \frac{20}{21}$$

$$21(440 - I_{a1}R_a) = 20\left(440 - \frac{I_{a1}}{2} R_a\right)$$

or, $9240 - 21I_{a1}R_a = 8800 - 10I_{a1}R_a$

or, $11I_{a1}R_a = 440$

or, $I_{a1}R_a = 40$ V

$\therefore$ *Armature drop at full-load* $= I_{a1}R_a =$ **40 V. (Ans.)**

Example 5.10. *A 230 V, 1000 r.p.m. D.C. shunt motor has field resistance of 115 Ω and armature circuit resistance of 0.5 Ω. At no load, the motor runs at 1000 r.p.m. with armature current of 4 A and with full field flux.*

(*i*) *For a developed torque of 80 Nm. compute armature current and speed of the motor.*

(*ii*) *If it is desired that motor develops 8 kW at 1250 r.p.m., determine the value of external resistance that must be inserted in series with the field winding. Saturation and armature reaction are neglected.* **(A.M.I.E. Winter, 2002)**

Solution. *Given :* $V = 230$ volts ; $N_1 (= N_0) = 1000$ r.p.m., $I_{ao} = 4$ A ;

$R_{sh} = 115\ \Omega$, $R_a = 0.5\ \Omega$, $T_a = 80$ Nm

(*i*) $\mathbf{I_a ; N_2}$ **:**

$$I_{sh} = \frac{V}{R_{sh}} = \frac{230}{115} = 2 \text{ A}$$

At no load :

$$E_{bo} = V - I_{ao} R_a$$
$$= 230 - 4 \times 0.5 = 228 \text{ V}$$

Also, $E_{bo} = \frac{p\phi ZN}{60\, a}$

or, $228 = \left(\frac{p\phi Z}{a}\right) \times \frac{1000}{60}$ or $\frac{p\phi Z}{a} = \frac{228 \times 60}{1000} = 13.68$

Now torque in Nm, $T_a = \frac{1}{2\pi}\left(\frac{p\phi Z}{a}\right) . I_a$

or, $80 = \frac{13.68}{2\pi} \times I_a$

$$\therefore \qquad \mathbf{I_a} = \frac{80 \times 2\pi}{13.68} = \mathbf{36.74\ A.\ (Ans.)}$$

Now $E_b = 230 - 36.74 \times 0.5 = 211.63$ V

Also, $$\frac{N_2}{N_1} = \frac{E_{b2}}{E_{b1}} \times \frac{\phi_1}{\phi_2}$$

or, $$\frac{N_2}{N_1} = \frac{E_{b2}}{E_{b1}} \qquad [\because \phi_1 = \phi_2 = \phi]$$

or, $$N_2 = N_1 \times \frac{E_{b2}}{E_{b1}} \qquad [E_{b1} = E_{b0} = 228 \text{ V} ; E_{b2} = E_b = 211.63 \text{ V}]$$

or, $$\mathbf{N_2 = 1000 \times \frac{211.63}{228} = 928.2 \text{ r.p.m.}} \textbf{ (Ans.)}$$

(*ii*) **Additional resistance to be inserted externally :**

For 8 kW power to be developed at 1250 r.p.m.

$$8000 = E_b \times I_a$$
$$= (V - I_a R_a) I_a = V I_a - I_a^2 R_a$$

or, $$8000 = 230 I_a - 0.5\, I_a^2$$

or, $$I_a^2 - 460 I_a + 16000 = 0$$

$\therefore$ $$I_a = \frac{460 \pm \sqrt{460^2 - 4 \times 16000}}{2} = \frac{460 \pm 384.2}{2} = 37.9 \text{ A}$$

[Taking only – ve sign]

$$E_b = V - I_a R_a = 230 - 37.9 \times 0.5 = 211.05 \text{ V}$$

When flux does not remain constant, we have

$$\frac{N_2}{N_1} = \frac{E_{b2}}{E_{b1}} \times \frac{\phi_1}{\phi_2}$$

or, $$\frac{N_2}{N_1} = \frac{E_{b2}}{E_{b1}} \times \frac{I_{sh1}}{I_{sh2}}$$

or, $$I_{sh2} = \frac{E_{b2}}{E_{b1}} \times \frac{N_1}{N_2} \times I_{sh1}$$

$$= \frac{211.05}{228} \times \frac{1000}{1250} \times 2 = 1.48 \text{ A}$$

$$R_{sh2} = \frac{V}{I_{sh2}} = \frac{230}{1.48} = 155.4\ \Omega$$

$\therefore$ Additional resistance to be inserted in series with the field winding

$$= 155.4 - 115 = \mathbf{40.4\ \Omega.} \quad \textbf{(Ans.)}$$

☞**Example 5.11.** *A 4-pole, 250 V, wave-connected shunt motor gives 10 kW when running at 1000 r.p.m. and drawing armature and field currents of 60 A and 1 A respectively. It has 560 conductors. Its armature resistance is 0.2 ohm. Assuming a drop of 1 volt per brush, determine :*

(*i*) *Total torque* (*ii*) *Useful torque*

(*iii*) *Useful flux per pole* (*iv*) *Rotational losses*

(*v*) *Efficiency.*

Solution. Number of poles, $p = 4$

Number of parallel paths, $a = 2$ [Motor being wave connected.]

Supply voltage, $V = 250$ Volts

Net power output $= 10$ kW or 10000 W

Load speed, $N = 1000$ r.p.m.

Armature current, $I_a = 60$ A

Shunt field current, $I_{sh} = 1$ A

Number of conductors, $Z = 560$

Armature resistance, $R_a = 0.2$ ohm,

Drop per brush $= 1$ V

We know that, $E_b = V - I_aR_a -$ brush drop

$= 250 - 60 \times 0.2 - 2 \times 1 = 236$ V.

(*i*) **Total or armature torque, T_a :**

We know that, $$T_a \times \frac{2\pi N}{60} = E_bI_a$$

$$\therefore \quad T_a = \frac{E_bI_a \times 60}{2\pi N} = \frac{236 \times 60 \times 60}{2\pi \times 1000} = 135.2 \text{ N-m}$$

Hence, *total or armature torque* = **135.2 N-m. (Ans.)**

(*ii*) **Useful torque, T_{useful} :**

We know that, $$T_{useful} \times \left(\frac{2\pi N}{60}\right) = 10000$$

$$\therefore \quad T_{useful} = \frac{10000 \times 60}{2\pi N} = \frac{10000 \times 60}{2\pi \times 1000} = 95.48 \text{ N-m.}$$

Hence, *useful torque* = **95.48 Nm. (Ans.)**

(*iii*) **Useful flux per pole, ϕ :**

Using the relation, $$E_b = \frac{p\phi ZN}{60\,a}$$

$$\phi = \frac{E_b \times 60\,a}{pZN} = \frac{236 \times 60 \times 2}{4 \times 560 \times 1000} = 0.0126 \text{ Wb} \quad \text{or} \quad 12.6 \text{ mWb}$$

Hence, *useful flux per pole* = **12.6 mWb. (Ans.)**

(*iv*) **Rotations losses :**

Armature input $= VI_a = 250 \times 60 = 15000$ W

Armature copper loss $= I_2^2R_a = 60^2 \times 0.2 = 720$ W

Brush contact loss $= 60 \times 2 = 120$ W

$\therefore$ Power developed $= VI_a -$ armature copper loss $-$ brush contact loss

$= 15000 - 720 - 120 = 14160$ W

But, net power output $= 10000$ W

$\therefore$ Rotational losses $=$ Power developed $-$ net power output

$= 14160 - 10000 = 4160$ W

Hence, *rotational losses* = **4160 W. (Ans.)**

(*v*) **Efficient of motor, η :**

Total motor input, $= VI = V(I_a + I_{sh}) = 250(60 + 1) = 15250$ W

Net power output $= 10000$ W

$\therefore$ Efficiency, $\eta = \dfrac{\text{Output}}{\text{Input}} = \dfrac{10000}{15250} = 0.656$ or 65.6%

Hence, *efficiency of motor* = **65.6%. (Ans.)**

5.8.5. Series Motors

In **series motors**, the *flux is a function of armature current.* In unsaturated region of magnetization **characteristic**, ϕ **can** be assumed to be proportional to I_a. Thus,

$$\phi = K_f I_a \quad \text{...(5.11)}$$

Substituting $\phi = I_f I_a$ is eqns. (5.3), (5.4) and (5.5), we get

$$T = K_e K_f I_a^2 \quad \text{...(5.12)}$$

$$\omega_m = \frac{V}{K_e K_f I_a} - \frac{R_a}{K_e K_f} \quad \text{...(5.13)}$$

Again,
$$\omega_m = \frac{V}{K_e K_f I_a} - \frac{R_a T}{(K_e K_f I_a)^2} \quad \text{...[15.13 (a)]}$$

or,
$$\omega_m = \frac{V}{\sqrt{K_e K_f}} \cdot \frac{1}{\sqrt{T}} - \frac{R_a}{K_e K_f} \quad \text{...(15.14)}$$

...[Substituting for I_a and I_a^2 from eqn. (5.12)]

Here, armature circuit resistance R_a is now the *sum of armature and field winding resistances.*

The *torque-current* (T/I_a) and *speed-torque* (ω_m/T) characteristics of a series motor at rated terminal voltage and full field are shown in Fig. 5.5.

***Series motors** are suitable to applications requiring high starting torque and heavy torque overloads.*

— Since torque is proportional to the *armature current squared, (i.e.,* $T \propto I_a^2$), for the same increase in torque, increase in motor current is *less* compared to that in a separately excited motor where torque is proportional to armature current. Thus, during heavy torque overloads and starting, power overload on the source and thermal overloading of the motor are kept limited to reasonable values. According to eqn. (5.14), as *speed varies inversely as the square root of torque, machine runs at a large speed at light load.* Generally, *mechanical strength of D.C. motor permits it to operate upto about twice rated speed. Hence,* the series motor should not be used in those drives where there is a possibility of the load torque being dropped to the extent that the speed may exceed twice rated value.

Example 5.12. *A 230 V series motor develops torque of 310 N-m at 800 r.p.m. The torque lost due to iron and friction loss is 10 N-m. If the efficiency is 85% determine the current taken by the motor at 800 r.p.m.*

Solution. Supply voltage $= 230$ V

Torque developed, $T_a = 310$ N-m

Torque lost due to iron and friction, $T_{lost} = 10$ N-m

Efficiency $= 85\%$

Speed of the motor, $N = 800$ r.p.m.

Current taken by motor at 800 r.p.m. :

$$T_{useful} = T_a - T_{lost} = 310 - 10 = 300 \text{ N-m}$$

Watts corresponding to useful torque

$$= \frac{T_{\text{useful}} \times 2\pi N}{60} = \frac{300 \times 2\pi \times 800}{60} = 25136 \text{ W}$$

Power in watts taken from mains $= \dfrac{25136}{0.85} = 29571.7$ W

Current taken by motor $= \dfrac{29571.7}{230} = 128.57$ A.

Example 5.13. *A 8-pole 240 V lap-wound, series motor has armature and series field resistances of 0.2 ohm and 0.02 ohm respectively. There are 660 armature conductors. If the flux/pole is 0.3 Wb and the total torque developed in the armature is 320 N-m, find the current taken by the motor and its speed.*

Solution. Number of poles, $p = 4$

Number of parallel paths, $a = p = 4$ [Motor being lap-wound]

Armature resistance, $R_a = 0.2$ ohm

Series field resistance, $R_{se} = 0.02$ ohm

Number of armature conductors, $Z = 660$

Flux/pole, $\phi = 0.03$ Wb

Total torque developed, $I_a = 320$ N-m

(*i*) **Current taken by motor, I_a :**

$$E_b = \frac{p\phi ZN}{60a}$$

$$\frac{E_b}{N} = \frac{p\phi Z}{60a} = \frac{4 \times 0.03 \times 660}{60 \times 4} = 0.33$$

Also, $$T_a \times \left(\frac{2\pi N}{60}\right) = E_b I_a$$

or, $$T_a = \frac{E_b}{N} \cdot \frac{60 I_a}{2\pi} = 0.33 \times \frac{60 \times I_a}{2\pi}$$

$\therefore$ $$320 = \frac{0.33 \times 60 \times I_a}{2\pi}$$

or, $$I_a = \frac{320 \times 2\pi}{0.33 \times 60} = 101.56 \text{ A}$$

Hence, current taken by motor = **101.56 A. (Ans.)**

(*ii*) **Speed of motor, N :**

$$E_b = V - I_a(R_a + R_{se})$$
$$= 240 - 101.56(0.2 + 0.02) = 217.66 \text{ V}$$

But, $$\frac{E_b}{N} = 0.33$$ [Already calculated above]

$\therefore$ $$N = \frac{E_b}{0.33} = \frac{217.66}{0.33} = 659.6 \text{ r.p.m.}$$

Example 5.14. *A six-pole, lap-wound 400 V series motor has the following data : Number of armature conductors = 920, flux/pole = 0.045 Wb, total motor resistance = 0.6 ohm, iron and friction losses = 2 kW. If the current taken by the motor is 90 A, find :*

(*i*) *Total torque ;* (*ii*) *Useful torque at the shaft ;*

(*iii*) *Power output ;*

(*iv*) *Pull at the rim of a pulley of 40 cm diameter connected to the shaft.*

Solution. Number of poles, $p = 6$

Supply voltage, $V = 400$ Volts

Number of parallel paths, $a = p = 6$ [Motor being lap-wound]

Number of armature conductors, $Z = 920$

Flux/pole, $\phi = 0.045$ Wb

Motor resistance, $R_m = 0.6$ ohm

Iron and friction losses = 2 kW or 2000 W

Current taken by the motor, $I_a = 90$ A

Radius of the pulley, = 40/2 = 20 cm or 0.2 m

Using the relation, $E_b = V - I_a R_m = 400 - 90 \times 0.6$

$= 346$ V

Also, $$E_b = \frac{p\phi ZN}{60a}$$

$$346 = \frac{6 \times 0.045 \times 920 \times N}{60 \times 6}$$

$$\therefore \quad N = \frac{346 \times 60 \times 6}{6 \times 0.045 \times 920} = 501 \text{ r.p.m.}$$

(*i*) **Total torque, T_a :**

We know that, $$T_a = 0.159 \times Z\phi p \times \left(\frac{I_a}{a}\right)$$

$$= 0.159 \times 920 \times 0.045 \times 6 \times \frac{90}{6} = 592.4 \text{ Nm}$$

Hence, *Total torque* = **592.04 N-m. (Ans.)**

(*ii*) **Useful torque, T_{useful} :**

$$T_{lost} \times \left(\frac{2\pi N}{60}\right) = \text{Iron and friction loss} = 2000 \text{ W}$$

$$\therefore \quad T_{lost} = \frac{2000 \times 60}{2\pi N} = \frac{2000 \times 60}{2\pi \times 501} = 38.11 \text{ Nm}$$

$$\therefore \quad T_{useful} = T_a - T_{lost} = 592.4 - 38.11$$

$$= 554.29 \text{ Nm}$$

Hence, *useful torque* = **554.29 Nm. (Ans.)**

(*iii*) **Power output :**

Power output $$= T_{useful}\left(\frac{2\pi N}{60}\right) = \frac{554.29 \times 2\pi \times 501}{60}$$

$$= 29084.4 \text{ or } 29.08 \text{ kW}$$

Hence, *power output* = **29.08 kW. (Ans.)**

(iv) If F is the pull at the rim of the pulley and r is the radius,

Torque at the shaft, $T_{useful} = F \times r$

i.e., $F \times 0.2 = 554.29$

$\therefore$ $F = 2771.45$ N

Hence, *pull at the rim of the pulley* = **2771.45 N. (Ans.)**

Example 5.15. *A 240 V series motor takes 40 A when giving its rated output at 1500 r.p.m. Its resistance is 0.3 Ω. Find what resistance must be added to obtain rated torque (i) at starting and (ii) at 1000 r.p.m.* **(A.M.I.E. Winter, 2002)**

Solution. *Given :* $V = 240$ volts ; $I (= I_a) = 40$ A, $N = 1500$ r.p.m., $R = 0.3\ \Omega$.

(i) **Resistance to be added to obtain rated torque at starting, R_{add} :**

Since the torque remains the same in both the cases, it is obvious that the current drawn by the motor remains constant at 40 A.

$$\therefore \quad 40 = \frac{240}{R_{add} + 0.3}$$

or, $$R_{add} = \frac{240}{40} - 0.3 = 5.7\ \Omega.$$

(ii) **Resistance to be added to obtain rated torque at 1000 r.p.m., R_{add} :**

We know that, $$\frac{N_2}{N_1} = \frac{E_{b2}}{E_{b1}}$$

$$E_{b1} = 240 - 40 \times 0.3 = 228 \text{ V}$$

$$\therefore \quad \frac{1000}{1500} = \frac{E_{b2}}{228}$$

or, $$E_{b2} = \frac{1000 \times 228}{1500} = 152 \text{ V}$$

Now, $E_{b2} = V - 40(R_{add} + 0.3)$ or $152 = 240 - 40(R_{add} + 0.3)$

$$\therefore \quad \mathbf{R_{add}} = \frac{240 - 152}{40} - 0.3 = \mathbf{1.9\ \Omega}. \quad \textbf{(Ans.)}$$

Example 5.16. *A 200 V D.C. series motor runs at 700 r.p.m. when operating at its full load current of 20 A. The motor resistance is 0.5 Ω and the magnetic circuit can be assumed unsaturated. What will be the speed if (i) the load torque is increased by 44% (ii) the motor current is 10 A.*

(A.M.I.E. Elec. Drives and their Control, Summer, 1999)

Solution. *Given :* $V = 200$ volts ; $N_1 = 700$ r.p.m., $I_1 = 20$ A, $R_m = (R_a + R_{sh}) = 0.5\ \Omega$

Speeds, (N_2, N_3) :

(i) **When the load torque is increased by 44% :**

$$T_2 = 1.44 T_1$$

$$\phi_2 I_2 = 1.44\ \phi_1 I_1 \qquad (\because\ T \propto \phi I_a)$$

$$I_2^2 = 1.44\ I_1^2 \qquad (\because\ \phi \propto I)$$

or, $$I_2 = I_1\sqrt{1.44} = 20 \times 1.2 = 24 \text{ A}.$$

Back e.m.f. $E_{b1} = V - I_1(R_a + R_{se}) = 200 - 20 \times 0.5 = 190$ V

Back e.m.f. $E_{b2} = V - I_2(R_a + R_{se}) = 200 - 24 \times 0.5 = 188$ V

Also $$\frac{N_2}{N_1} = \frac{E_{b2}}{E_{b1}} \times \frac{\phi_1}{\phi_2}$$

or, $$\frac{N_2}{700} = \frac{188}{190} \times \frac{20}{24}$$

or, $$N_2 = 577.2 \text{ r.p.m.} \quad \textbf{(Ans.)}$$

(ii) **When the motor current is 10 A :**

Back e.m.f., $$E_{b3} = V - I_3(R_a + R_{se})$$
$$= 200 - 10 \times 0.5 = 195 \text{ V}$$

Also, $$\frac{N_3}{N_1} = \frac{E_{b3}}{E_{b1}} \times \frac{\phi_1}{\phi_3}$$
$$\frac{N_3}{700} = \frac{195}{190} \times \frac{20}{10}$$

or, $$N_3 \simeq 1437 \text{ r.p.m.} \quad \textbf{(Ans.)}$$

Example 5.17. *A 220 V D.C. series motor draws full load line current of 38 A at the rated speed of 600 r.p.m. The motor has armature resistance of 0.4 Ω and the series field resistance is 0.2 Ω. The brush voltage drop irrespective of load is 3.0 volts, find :*

(i) The speed of the motor when the load current drops to 19 A.

(ii) The speed on removal of load when the motor takes only 1 A from supply.

(iii) The internal horse power developed in each of the above cases. Neglect the effect of armature reaction and saturation. **(A.M.I.E. Summer, 2001)**

Solution. *Given :* $V = 220$ volts ; $I_1 = 38$ A, $N_1 = 600$ r.p.m., $R_a = 0.4\ \Omega$, $R_{se} = 0.2\ \Omega$, Brush drop = 3 V ; $I_2 = 19$ A, $I_0 = 1$ A.

(i) **Speed of the motor N_2 (at 19 A)**

On full-load :

Back e.m.f., $$E_{b1} = V - I_1(R_a + R_{se}) - \text{brush voltage drop}$$
$$= 220 - 38(0.4 + 0.2) - 3 = 194.2 \text{ V.}$$

On load current of 19 A :

Back e.m.f., $$E_{b2} = 220 - 19(0.4 + 0.2) - 3 = 205.6 \text{ V}$$

No-load, back e.m.f., $$E_{b0} = 220 - 1(0.4 + 0.2) - 3 = 216.4 \text{ V}$$

We know that $$\frac{N_2}{N_1} = \frac{E_{b2}}{E_{b1}} \times \frac{\phi_1}{\phi_2}$$

$$\frac{N_2}{600} = \frac{205.6}{194.2} \times \frac{I_1}{I_2} \quad \left[\text{Assuming unsaturated field } i.e., \frac{\phi_1}{\phi_2} = \frac{I_1}{I_2}\right]$$

$$N_2 = 600 \times \frac{205.6}{194.2} \times \frac{38}{19} = 1270 \text{ r.p.m.} \quad \textbf{(Ans.)}$$

(ii) **N_0 (at 1 A) :**

Now, $$\frac{N_0}{N_1} = \frac{E_{b0}}{E_{b1}} \times \frac{\phi_1}{\phi_0}$$
$$= \frac{E_{b0}}{E_{b1}} \times \frac{I_1}{I_0}$$

∴ $$N_0 = N_1 \times \frac{E_{b0}}{E_{b1}} \times \frac{I_1}{I_0}$$
$$= 600 \times \frac{216.4}{194.2} \times \frac{38}{1} = 25406 \text{ r.p.m.} \quad \textbf{(Ans.)}$$

(*iii*) **Internal H.P. developed (in the above cases) :**

Internal H.P. developed at a load current 38 A

$$= \frac{194.2 \times 3.8}{735.5} = \mathbf{10.} \quad \textbf{(Ans.)}$$

Internal H.P. developed at a load current of 19 A

$$= \frac{205.6 \times 19}{735.5} = \mathbf{5.31.} \quad \textbf{(Ans.)}$$

Internal H.P. developed at no load current of 1 A

$$= \frac{216.4 \times 1}{735.5} = \mathbf{0.294.} \quad \textbf{(Ans.)}$$

Example 5.18. *A 220 V series motor runs at 800 r.p.m. when taking a current of 15 A. The motor has an armature resistance of 0.3 ohm and series field resistance of 0.2 ohm. Find the resistance to be connected in series with the armature if it has to take the same current at the same voltage at 600 r.p.m. Assume flux is proportional to current.*

Solution. Supply voltage $= 220$ V

Load current, $I = I_a = 15$ A

Armature resistance, $R_a = 0.3$ ohm

Series field resistance, $R_{se} = 0.2$ ohm

$N_1 = 800$ r.p.m., $N_2 = 600$ r.p.m.

Additional resistance to be added, R :

$$E_{b1} = V - I_{a1}(R_a + R_{se})$$
$$= 220 - 15(0.3 + 0.2) = 212.5 \text{ V}$$

Using the relation, $\dfrac{N_2}{N_1} = \dfrac{E_{b2}}{E_{b1}} \times \dfrac{\phi_1}{\phi_2}$

Since current remains same,

Flux at 800 r.p.m, ϕ_1 = Flux at 600 r.p.m., ϕ_2

$$\therefore \quad \frac{N_2}{N_1} = \frac{E_{b2}}{E_{b1}}$$

or,
$$E_{b2} = \frac{N_2}{N_1} \times E_{b1} = \frac{600}{800} \times 212.5 = 159.37 \text{ V}$$

But,
$$E_{b2} = V - I_{a2}(R_a + R_{se} + R)$$
$$\therefore \quad 159.37 = 220 - 15(0.3 + 0.2 + R)$$

or,
$$(0.3 + 0.2 + R) = \frac{220 - 159.37}{15} = 4.04$$

$$\therefore \quad R = 3.54 \text{ ohms}$$

Hence, *additional resistance to be added* = **3.54 ohms. (Ans.)**

Example 5.19. *A 440 V series motor has a load resistance of 0.8 ohm. It takes 110 A to give the rated output at 800 r.p.m. Determine the resistance of the resistor to be added to obtain rated torque :*

(*i*) *At starting* (*ii*) *At 600 r.p.m.*

Assume flux is proportional to the current.

Solution. Supply voltage $= 440$ V

Load resistance, $R_L = 0.8$ ohm

Motor current, $I = I_a = 110$ A

$N_1 = 800$ r.p.m., $N_2 = 600$ r.p.m.

(*i*) **Additional resistance to be added at starting :**

We know that $T \propto \phi I_a$

Since flux is proportional to current

$$T \propto I_a^2.$$

Since the *torque remains constant and is equal to the rated output in these cases,*

$\therefore$ I_a will be the same in all these cases

$$I_a = 110 \text{ A}$$

Let R_1 be the resistance of the resistor to be added

Then, $E_b = V - I_a (R_L + R_1)$

At starting, $E_b = 0$

$\therefore$ $V - I_a(R_L + R_1) = 0$

or, $$R_L + R_1 = \frac{V}{I_a} = \frac{440}{110} = 4 \text{ ohms}$$

or, $0.8 + R_1 = 4$

or, $R_1 = 4 - 0.8 = 3.2$ ohms

Hence, *additional resistance* = **3.2 ohms. (Ans.)**

(*ii*) **Additional resistance to be added at 600 r.p.m. :**

$$E_{b1} = V - I_{a1}R_L = 440 - 110 \times 0.8 = 352 \text{ V}.$$

Since the *armature current remains the same in all cases flux will also remain unaltered in all these cases.*

Therefore if E_{b2} is the back e.m.f. at $N_2 = 600$ r.p.m., then

$$\frac{N_2}{N_1} = \frac{E_{b2}}{E_{b1}}$$

$\therefore$ $$E_{b2} = \frac{N_2}{N_1} \times E_{b1} = \frac{600}{800} \times 352 = 264 \text{ V}$$

If R_2 is the additional resistance to be added

$$E_{b2} = V - I_{a2}(R_L + R_2)$$

or, $264 = 440 - 110(0.8 + R_2)$

or, $264 - 440 = -110(0.8 + R_2)$

or, $0.8 + R_2 = 1.6$

$\therefore$ $R_2 = 0.8$ ohm

Hence, *additional resistance to be added* = **0.8 ohm. (Ans.)**

Example 5.20. *A 250 V series motor in which the total armature and field resistance is 0.15 ohm is working with unsaturated field, taking 120 A and running at 750 r.p.m. Calculate at what speed the motor will run when developing half the torque.*

Solution. Supply voltage $V = 250$ V

Armature and field resistance, $R_m = 0.15$ ohm

Load current, $I_1 = 120$ A

Speed when current is 120 A, $N_1 = 750$ r.p.m.

Speed at half torque, N_2 :

Back e.m.f., $E_{b1} = V - I_1 R_m$
$= 250 - 120 \times 0.15 = 232$ V

When developing half the torque,

i.e. $$T_2 = \frac{1}{2} T_1$$

or, $$\phi_2 I_2 = \frac{1}{2} \phi_1 I_1 \qquad [\because \; T \propto \phi I_a \propto \phi I]$$

or, $$I_2^2 = \frac{1}{2} I_1^2 \qquad [\because \; \text{The field is unsaturated, } \phi \propto I_a \propto I]$$

or, $$I_2^2 = \frac{1}{2} (120)^2$$

or, $$I_2 = 84.86 \text{ A}$$

Back e.m.f., $E_{b2} = V - I_2 R_m = 250 - 84.86 \times 0.15 = 237.27$ V

Using the relation, $$\frac{N_2}{N_1} = \frac{E_{b2}}{E_{b1}} \times \frac{\phi_1}{\phi_2}$$

or, $$\frac{N_2}{N_1} = \frac{E_{b2}}{E_{b1}} \times \frac{I_1}{I_2} \qquad [\because \; \phi \propto I]$$

or, $$\frac{N_2}{750} = \frac{237.72}{232} \times \frac{120}{84.8}$$

$\therefore$ $$N_2 = 750 \times \frac{237.27}{232} \times \frac{120}{84.86} = 1084.6 \text{ r.p.m.}$$

Hence, *speed of motor when developing half the torque*
= 1084.6 r.p.m. (Ans.)

Example 5.21. *A 460 V D.C. series motor runs at 500 r.p.m. taking a current of 40 A. Calculate the speed and percentage change in torque if the load is reduced so that the motor is taking 30 A. Total resistance of the armature and field circuits is 0.8 Ω. Assume flux is proportional to the field current.*

Solution. *Given :* $V = 460$ volts ; $N = 500$ r.p.m. ; $I_1 (= I_{a1}) = 40$ A,
$I_2 (= I_{a2}) = 30$ A ; $R_t = (R_a + R_{se}) = 0.8\ \Omega$
$\phi \propto I_f$ (*i.e.*, I_a in this case)

Speed N_2 :

Armature current, $I_1 = I_{a1} = 40$ A

Back, e.m.f., $E_{b1} = V - I_{a1} R_t - 460 - 40 \times 0.8 = 428$ V

Also, $\phi_1 \propto I_{a1}$ and $\phi_2 \propto I_{a2}$

$\therefore$ $$\frac{\phi_2}{\phi_1} = \frac{I_{a2}}{I_{a1}} = \frac{30}{40} = 0.75$$

$$E_{b2} = 460 - 30 \times 0.8 = 436 \text{ V}$$

We know that, $$\frac{N_2}{N_1} = \frac{E_{b2}}{E_{b1}} \times \frac{\phi_1}{\phi_2}$$

$\therefore$ $$\mathbf{N_2} = 500 \times \frac{436}{428} \times \frac{1}{0.75} \simeq \mathbf{680 \text{ r.p.m.}} \quad \textbf{(Ans.)}$$

Percentage change in torque :

Since torque developed, $T \propto \phi I_a$

$$\therefore \quad \frac{T_2}{T_1} = \frac{\phi_2 I_{a2}}{\phi_1 I_{a1}} = 0.75 \times \frac{30}{40} = \frac{9}{16}$$

$\therefore$ *Percentage change in torque*

$$= \frac{T_1 - T_2}{T_1} \times 100$$

$$= \left(1 - \frac{T_2}{T_1}\right) \times 100 = \left(1 - \frac{9}{16}\right) \times 100 = \mathbf{43.75\%.} \quad \textbf{(Ans.)}$$

☞**Example 5.22.** *A 240 V series motor runs at 800 r.p.m. to give a total torque of 110 Nm. The current taken by the motor at this torque is 90 A. If the total resistance of the motor is 0.6 ohm find the torque developed when motor runs at 1200 r.p.m.*

Assume that flux is proportional to current.

Solution. Supply voltage $= 240$ V

Torque developed at 800 r.p.m., $T_{a1} = 110$ Nm

Current drawn at the above torque, $I_{a1} = 90$ A

Total resistance of motor, $R_m = 0.6$ ohm

$N_1 = 800$ r.p.m.

$N_2 = 1200$ r.p.m.

Torque developed at 1200 r.p.m. :

Back e.m.f. at speed 800 r.p.m.,

$$E_{b1} = V - I_{a1}R_m = 240 - 90 \times 0.6 = 186 \text{ V}$$

Using the relation, $\dfrac{N_2}{N_1} = \dfrac{E_{b2}}{E_{b1}} \times \dfrac{\phi_1}{\phi_2}$

But $\phi \propto I_a$

$$\therefore \quad \frac{N_2}{N_1} = \frac{E_{b2}}{E_{b1}} \times \frac{I_{a1}}{I_{a2}}$$

or, $$E_{b2} = E_{b1} \times \frac{N_2}{N_1} = \frac{I_{a2}}{I_{a1}} = 186 \times \frac{1200}{800} \times \frac{I_{a2}}{90} = 3.1 I_{a2}$$

$$\therefore \quad E_{b2} = 3.1 I_{a2} \quad \ldots(i)$$

where E_{b1} and I_{a2} are the back e.m.f. and armature current respectively at 1200 r.p.m.

But, $$E_{b2} = V - I_{a2}R_m = 240 - 0.6 I_{a2} \quad \ldots(ii)$$

Equating (*i*) and (*ii*), we get

$$3.1 I_{a2} = 240 - 0.6 I_{a2}$$

$$3.7 I_{a2} = 240$$

$$\therefore \quad I_{a2} = 64.86 \text{ A}$$

Torque at 800 r.p.m., $T_{a1} = 110$ Nm (Given)

Let T_{a2} be the torque at 1200 r.p.m

$$T_{a1} \propto \phi_1 I_{a1}$$

$$\propto I_{a1}^2 \quad [\because \ \phi_1 \propto I_{a1}]$$

$$T_{a2} \propto \phi_2 I_{a2}$$

$$\propto I_{a2}^2 \qquad [\because \phi_2 \propto I_{a2}]$$

$$\therefore \quad \frac{T_{a2}}{T_{a1}} = \frac{I_{a2}^2}{I_{a1}^2}$$

or,
$$\frac{T_{a2}}{110} = \frac{64.86^2}{90^2}$$

$$\therefore \quad T_{a2} = 110 \times \left(\frac{64.86}{90}\right)^2 = 57.13 \text{ Nm}$$

Hence, *torque developed at 1200 r.p.m.* = **57.13 N-m. (Ans.)**

Example 5.23. *A 3 kW series motor runs normally at 800 r.p.m. on a 200 V supply taking 16 A ; when the field coils are all connected in series. Estimate the speed and current taken by the motor if the coils are reconnected in two parallel groups of two each in series. Load torque increases as square of the speed.*

Assume that the flux is directly proportional to the current and ignore losses.

(A.M.I.E. Summer, 1999)

Solution. *Given* : N_1 = 800 r.p.m., $I_1 = I_{a1}$ = 16 A

I_{a2}, N_2 :

When the coils are connected in two parallel groups, current through each is $(I_{a2}/2)$, where I_{a2} is the new armature current.

Hence $\phi_2 \propto (I_{a2}/2)$

Now, $T \propto \phi I_a \propto N^2$ (Given)

$\therefore \quad \phi_1 I_{a1} \propto N_1^2$, and $\phi_2 I_{a2} \propto N_1^2$

$$\therefore \quad \left(\frac{N_2}{N_1}\right)^2 = \frac{\phi_2 I_{a2}}{\phi_1 I_{a1}} \qquad ...(i)$$

Since losses are negligible, field coil resistance as well as armature resistance are negligible. It means that armature and series field voltage drops are negligible. Hence back e.m.f. in each case equals to supply voltage.

$$\therefore \quad \frac{N_2}{N_1} = \frac{E_{b2}}{E_{b1}} \times \frac{\phi_1}{\phi_2} \text{ becomes } \frac{N_2}{N_1} = \frac{\phi_1}{\phi_2} \qquad ...(ii)$$

From (*i*) and (*ii*), we have $\left(\frac{\phi_1}{\phi_2}\right)^2 = \frac{\phi_2 I_{a2}}{\phi_1 I_{a1}}$ or $\frac{I_{a2}}{I_{a1}} = \left(\frac{\phi_1}{\phi_2}\right)^3$

Now $\phi_1 \propto 16$ and $\phi_2 \propto I_{a2}/2$

$$\therefore \quad \frac{I_{a2}}{16} = \left[\frac{16}{(I_{a2}/2)}\right]^3$$

or
$$\frac{I_{a2}}{16} = \frac{32^3}{I_{a2}^3} \quad \text{or} \quad I_{a2}^4 = 16 \times 32^3 = 524286$$

$$\therefore \quad I_{a2} = 26.9 \text{ A}$$

Again from (*ii*), we have $\frac{N_2}{N_1} = \frac{\phi_1}{\phi_2} = \frac{I_{a1}}{(I_{a2}/2)} = \frac{2I_{a1}}{I_{a2}}$

$$\therefore \quad N_2 = N_1 \times \frac{2I_{a1}}{I_{a2}} = 800 \times \frac{2 \times 16}{26.9} \simeq \mathbf{952\ r.p.m.} \quad \textbf{(Ans.)}$$

☞**Example 5.24.** *A series motor runs at 1000 r.p.m. taking 90 A with 110 V. What resistance would be connected in parallel with field circuit to get 1500 r.p.m. speed for delivering same load torque ? Armature resistance is 0.08 Ω and series field resistance is 0.06 Ω. Assume magnetic circuit is unsaturated.* **(A.M.I.E. Winter, 2002)**

Solution. *Given* : $N_1 = 1000$ r.p.m. ; $I_1 (= I_{a1}) = 90$ A, $V = 110$ volts ;
$N_2 = 1500$ r.p.m. ; $R_a = 0.08\ \Omega$; $R_{se} = 0.06\ \Omega$.

Shunt resistance :

We know that for a series motor

$$T \propto \phi I_a$$

In the *first case* :

$$\phi \propto I_a$$

$$\therefore \quad T_1 \propto I_{a1}^{\ 2} \propto 90^2$$

In the *second case* let us take the armature current as I_{a2} and a resistance to be connected *across the series field* (shunt resistance) to reduce the field current to KI_{a2}.

$$\therefore \quad T \propto \phi I_a \propto KI_{a2} \cdot I_{a2} \propto KI_{a2}^{\ 2}$$

As the load torque remains unchanged we have

$$T_1 = T_2$$

$$\therefore \quad 90^2 = KI_{a2}^{\ 2}$$

or, $$K = \frac{8100}{I_{a2}^{\ 2}} \quad ...(i)$$

Also, $$\frac{N_2}{N_1} = \frac{E_{b2}}{E_{b1}} \times \frac{\phi_1}{\phi_2}$$

or, $$\frac{E_{b2}}{E_{b1}} = \frac{N_2}{N_1} \times \frac{\phi_2}{\phi_1} = \frac{1500}{1000} \times \frac{KI_{a2}}{I_{a1}}$$

or, $$\frac{E_{b2}}{E_{b1}} = \frac{1500}{1000} \times \frac{KI_{a2}}{90} = \frac{1500 \times KI_{a2}}{1000 \times 90} \quad ...(ii)$$

Now, $$E_{b1} = V - I_{a1}(R_a + R_{se}) = 110 - 90(0.08 + 0.06) = 97.4 \text{ V}$$

$$E_{b2} = V - I_{a2}(R_a + R_{se2}) = 110 - I_{a2}\,(0.08 + R_{se2})$$

R_{se2} = Parallel combination of field and shunt resistances $= 0.06$ K

Putting the values of E_{b1} and E_{b2} in (*ii*), we get

$$\frac{110 - I_{a2}(0.08 + 0.06K)}{97.4} = \frac{1500 \times KI_{a2}}{1000 \times 90}$$

or, $$110 - I_{a2}\,(0.08 + 0.06K) = \frac{97.4 \times 1500}{1000 \times 90} \times KI_{a2} = 1.476\, KI_{a2}$$

$$\therefore \quad KI_{a2} = 0.616[110 - I_{a2}\,(0.08 + R_{se2})]$$

Putting the value of K from (*i*), we have

$$\frac{8100}{I_{a2}} = 0.616\left[110 - I_{a2}\left\{0.08 + 0.06 \times \frac{8100}{I_a^{\ 2}}\right\}\right]$$

or, $$\frac{8100}{I_{a2}} = 67.76 - 0.05I_{a2} - \frac{299.4}{I_{a2}}$$

or, $$8100 = 67.76I_a - 0.05I_{a2}^2 - 299.4$$

or, $$0.05I_{a2}^2 - 67.76I_a + 8399.4 = 0$$

or, $$I_{a2} = \frac{67.76 \pm \sqrt{(67.76)^2 - 4 \times 0.05 \times 8399.4}}{2 \times 0.05}$$

$$= \frac{67.76 - 53.96}{0.1} = 138 \text{ A} \qquad \text{(Taking only – ve sign)}$$

and, $$K = \frac{8100}{(138)^2} = 0.4253$$

Hence, *shunt resistance* $$= \frac{{}^*0.06K}{1-K}$$

$$= \frac{0.06 \times 0.4253}{1 - 0.4253} = \mathbf{0.044\ \Omega.} \quad \textbf{(Ans.)}$$

$$\left[\begin{array}{l} \dfrac{1}{R_{se2}} = \dfrac{1}{R_{se1}} + \dfrac{1}{R_{sh}} \\ \text{or,} \quad R_{sh2} = KR_{se1} = \dfrac{R_{se1} \times R_{sh}}{R_{se1} + R_{sh}} \\ \text{where,} \quad K = \dfrac{R_{sh}}{R_{se1} + R_{sh}} \\ \text{or,} \quad KR_{se1} + KR_{sh} = R_{sh} \\ \text{or,} \quad R_{sh}(1 - K) = KR_{se1} \\ \text{or,} \quad {}^*R_{sh} = \dfrac{KR_{se1}}{1-K} \end{array}\right]$$

Example 5.25. *A series motor takes 20 A at 400 V to drive a fan at 200 r.p.m. Its resistance is 1 Ω (field and armature). If the torque required to drive the fan varies as the square of the speed, find the necessary applied voltage and current to drive the fan at 300 r.p.m.*

(S.C.E. & T.E.W.B. Elec. Machines 2001)

Solution. *Given :* $V = 400$ volts ; $I_1 (= I_{a1}) = 20$ A ; $N = 200$ r.p.m., $(R_a + R_{se}) = 1\ \Omega$; $T \propto N^2$, $N_2 = 300$ r.p.m.

I_2 ; V_2 :

Back e.m.f., $$E_{b1} = V - I_1 (R_a + R_{se})$$
$$= 400 - 20 \times 1 = 380 \text{ V}$$

Since $$T \propto N^2 \qquad \text{...(Given)}$$

$$\therefore \quad \frac{T_2}{T_1} = \frac{N_2^2}{N_1^2} \quad \text{or} \quad \frac{\phi_2 I_2}{\phi_1 I_1} = \frac{N_2^2}{N_1^2}$$

or, $$\frac{I_2^2}{I_1^2} = \left(\frac{300}{200}\right)^2 \qquad (\because \ \phi \propto I, \text{ neglecting saturation})$$

or, $$\frac{I_2}{I_1} = 1.5$$

or, $$\mathbf{I_2} = 1.5I_1 = 1.5 \times 20 = \mathbf{30\ A.} \quad \textbf{(Ans.)}$$

Also, $$\frac{N_2}{N_1} = \frac{E_{b2}}{E_{b1}} \times \frac{\phi_1}{\phi_2}$$

$$\frac{300}{200} = \frac{E_{b2}}{380} \times \frac{I_1}{I_2} = \frac{E_{b2}}{380} \times \frac{1}{1.5}$$

∴ $$E_{b2} = \frac{300}{200} \times 380 \times 1.5 = 85\ V$$

∴ $$\mathbf{V_2} = E_{b2} + I_2(R_a + R_{se})$$
$$= 855 + 30 \times 1 = \mathbf{885\ V.} \quad \textbf{(Ans.)}$$

Example 5.26. *A D.C. series motor drives a load, the torque of which varies as the square of speed. The motor takes a current of 15 A when the speed is 600 r.p.m. Calculate the speed and current when the motor field winding is shunted by a divertor of the same resistance as that of field winding. Mention the assumptions made if any.*

Solution. *Given* : $T \propto N^2$; $I_1 (= I_{a1}) = 15$ A, $N_1 = 600$ r.p.m.

$\mathbf{N_2}$; $\mathbf{I_2 (= I_{a2})}$:

Series field current, $I_{se1} = I_{a1} = 15$ A

Let the speed be N_2 and armature current I_{a2} when the field winding is shunted by a *"divertor"* of the *same resistance as that of field winding.*

∴ Series field current, $$I_{se2} = \frac{1}{2} I_{a2}$$

(∵ Current I_{a2} divided equally between series field and divertor)

Since, $$T \propto \phi I_a$$

∴ $$\frac{T_2}{T_1} = \frac{\phi_2 I_{a2}}{\phi_1 I_{a1}} = \frac{\frac{1}{2} I_{a2}^2}{I_{a1}^2}$$ (∵ $\phi \propto I_{se}$, neglecting magnetic saturation)

or, $$\frac{T_2}{T_1} = \frac{I_{a2}^2}{2I_{a1}^2} \quad ...(i)$$

Also, $$\frac{T_2}{T_1} = \frac{N_2^2}{N_1^2} \quad ...(ii)$$

From (*i*) and (*ii*), we have $$\frac{N_2^2}{N_1^2} = \frac{I_{a2}^2}{2I_{a1}^2}$$

or, $$\frac{N_2}{N_1} = \frac{1}{\sqrt{2}} \frac{I_{a2}}{I_{a1}} \quad ...(iii)$$

Neglecting voltage drop in armature and assuming applied voltage constant

$$N \propto \frac{1}{\phi}$$

$$\frac{N_2}{N_1} = \frac{\phi_1}{\phi_2} = \frac{I_{a1}}{\frac{1}{2} I_{a2}} = \frac{2I_{a1}}{I_{a2}} \quad ...(iv)$$

From (*iii*) and (*iv*), we have $\frac{1}{\sqrt{2}} \frac{I_{a2}}{I_{a1}} = \frac{2I_{a1}}{I_{a2}}$

or, $$I_{a2}^2 = 2\sqrt{2} I_{a1}^2 = 2\sqrt{2} \times (15)^2 = 636.4 \text{ A}$$

or, $$I_{a2} = 25.227 \text{ A}$$

Substituting $I_{a2} = 25.227$ A in (*iv*), we get

$$\frac{N_2}{N_1} = \frac{2 \times 15}{25.227}$$

or, $$\mathbf{N_2} = 600 \times \frac{2 \times 15}{25.227} = \mathbf{713.5 \text{ r.p.m.}} \quad \textbf{(Ans.)}$$

☞**Example 5.27.** *A 500 V D.C. series motor running at 500 r.p.m. has the following magnetisation characteristic :*

E.m.f., V :	*290*	*354*	*390*	*420*	*470*
Field current, A :	*75*	*100*	*125*	*150*	*200*

The armature resistance is 0.3 ohm and the field resistance is 0.4 ohm. Find the speed when the input current is 180 A with the rated applied voltage and a divertor of 0.6 ohm is connected across the field. If the sum of windage, friction and iron losses at this load is 5 kW, find the output torque.

Solution. Supply voltage, $V = 500$ Volts

Armature resistance, $R_a = 0.3$ ohm

Series field resistance, $R_{se} = 0.4$ ohm

Input current $= 180$ A

Resistance of divertor $R_{div.} = 0.6$ ohm

Windage, friction and iron losses $= 5$ kW

$\mathbf{T_{useful}}$ **(or $\mathbf{T_{sh}}$) :**

Refer to Fig. 5.6

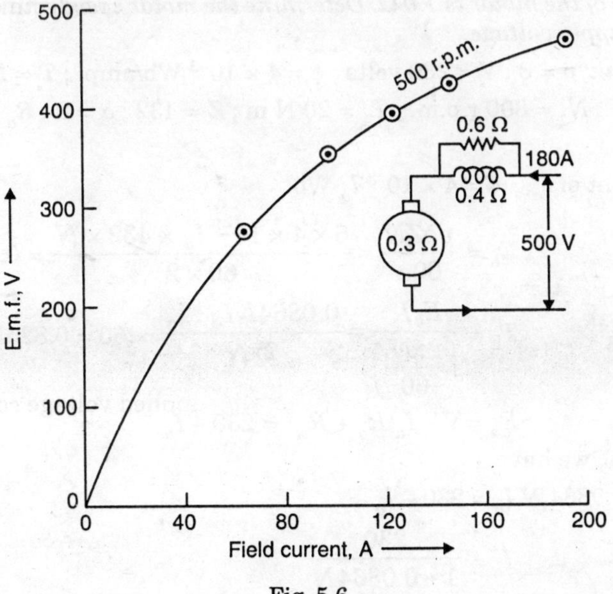

Fig. 5.6

When divertor is used, the field current is given by

$$I_f = 180 \times \frac{R_{div.}}{(R_{div.} + R_{se})} = 180 \times \frac{0.6}{0.6 + 0.4} = 108 \text{ A}$$

From the magnetisation curve (at 500 r.p.m.) it is seen that the corresponding e.m.f. is 365 V. This acts as the back e.m.f. E_{b1}. The new back e.m.f. is given by

$$E_{b2} = 500 - (108 \times 0.4) - (180 \times 0.3)$$
$$= 500 - 43.2 - 54 = 402.8 \text{ V.}$$

Using the relation, $\frac{N_2}{N_1} = \frac{E_{b2}}{E_{b1}}$

$$N_2 = N_1 \times \frac{E_{b2}}{E_{b1}} = 500 \times \frac{402.8}{365} = 551.8 \text{ r.p.m.}$$

Electrical power developed $= E_b I_a = 402.8 \times 180 = 72504$ W or 72.504 kW

Stray losses = 5 kW

∴ Output = 72.504 – 5 = 67.504 kW

Now using the relation,

$$T_{useful} \times \frac{2\pi N}{60} = \text{output}$$

$$\therefore \quad T_{useful} \times \frac{2\pi \times 551.8}{60} = 67.504 \times 1000$$

or,
$$T_{useful} = \frac{60 \times 67.504 \times 1000}{2\pi \times 551.8} = 1168 \text{ Nm}$$

Hence, *useful torque available* = **1168 Nm. (Ans.)**

Example 5.28. *A 6-pole, 230 V D.C. series motor has a flux per pole of 4 m Wb / amp. over the working range of the magnetization curve which is assumed to be linear. The load torque is proportional to speed squared and its value is 20 Nm at 800 r.p.m. There are 432 wave-connected conductors and the total resistance of the motor is 1.0 Ω. Determine the motor speed and current when this motor is connected to rated supply voltage.* **(Nagpur University)**

Solution. *Given :* $p = 6$; $V = 230$ volts ; $\phi = 4 \times 10^{-3}$ Wb/amp. ; $T \propto N^2$;

$N_1 = 800$ r.p.m. ; $T_1 = 20$ N m ; $Z = 432$; $a = 2$; $R_a + R_{se} = 1.0\ \Omega$.

N_2 : ; I_a :

For motor current of I_a, $\phi = 4 \times 10^{-3} I_a$ Wb

$$\therefore \quad E_b = \frac{p\phi ZN}{60a} = \frac{6 \times 4 \times 10^{-3} I_a \times 432 \times N}{60 \times 2} = 0.0864\, N\, I_a \quad ...(i)$$

$$T_a = \frac{E_b I_a}{\left(\frac{2\pi N}{60}\right)} = \frac{0.0864 N I_a \,.\, I_a}{2\pi N} \times 60 = 0.825 I_a^2 \quad ...(ii)$$

But,
$$E_b = V - I_a(R_a + R_{se}) = 230 - I_a \quad ...(iii)$$

From (*i*) and (*iii*), we have

$$0.0864\, N\, I_a = 230 - I_a$$

or,
$$I_a = \frac{230}{1 + 0.0864 N} \quad ...(iv)$$

Substituting the value of I_a in (*ii*), we have

$$T_a = 0.825 \left[\frac{230}{1+0.0864N}\right]^2$$

Also, $T \propto N^2$ (*Given*)

or, $T = KN^2$

or, $20 = K(800)^2$

$\therefore$ $K = \frac{20}{(800)^2} = 3.125 \times 10^{-5}$ Nm/r.p.m.

Under steady state conditions, $T = T_a$

or, $$3.125 \times 10^{-5} N^2 = 0.825 \left[\frac{230}{1+0.0864N}\right]^2$$

From which, $N = 651.888 \simeq$ **652 r.p.m.** **(Ans.)**

Substituting the value of N = 652 r.p.m. in (*iv*), we get

$$I_a = \frac{230}{1+0.0864 \times 652} = \textbf{4.012 A.} \quad \textbf{(Ans.)}$$

Example 5.29. *The torque-armature current characteristics of a series traction motor are given as :*

Armature current (A)	*5*	*10*	*15*	*20*	*25*	*30*	*35*	*40*
Torque (N-m)	*20*	*50*	*100*	*155*	*215*	*290*	*360*	*430*

The motor resistance is 0.3 ohm.

If this motor is connected across 230 V deduce the speed-armature current characteristics.

(A.M.I.E.)

Solution. *Given* : Supply voltage = 230 V ; Motor resistance, $R_m = R_a + R_{se} = 0.3\ \Omega$.

Armature current (I_a) amps. (A)	5	10	15	20	25	30	35	40
Torque (T), N-m	20	50	100	155	215	290	360	430
Back e.m.f. (E_b); $= V - I_a R_m$ *volts (V)*	228.5	227.0	225.5	224.0	222.5	221.0	219.5	218.0
Speed (N) $= \frac{9.55\,(V - I_a R_m)}{T} \cdot I_a$, *r.p.m.*	545	434	323	276	247	218	204	194

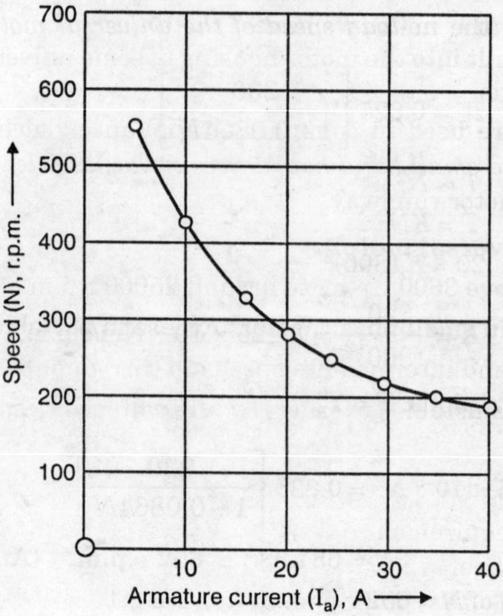

Fig. 5.7

The deduced speed-armature current characteristic is shown in Fig. 5.7.

5.8.6. Compound Motor

The torque-current (T/I_a) and speed-torque (ω_m/T) characteristics of a cumulative compound motor are shown in Fig. 5.5.

— These characteristics which are obtained at rated terminal voltage and full field are known as *natural speed-torque characteristics* ; rated (or full load) speed is known as the *base speed.*

— The no-load speed depends on the strength of shunt field and slope of the characteristic on the strength of series field.

Applications. (*i*) *"Cumulative compound motors"* are used in those applications where a drooping characteristic similar to that of a series motor is required and at the same time the no-load speed must be limited to a safe value ; typical *examples* are *"lifts and winches"*

(*ii*) A cumulative motor is also used in *intermittent load applications*, where the *load varies from almost no load to very heavy loads.* In these applications a flywheel may be mounted on the motor shaft for load equilisation. This apart from equilising load on the supply, permits the use, of a smaller size motor ; typical *example* of this type of application is *"pressing machine"*.

5.8.7. Universal Motor

- Fractional-horsepower *series motors* that are adapted for use on either D.C. or A.C. circuits of a given voltage are called *universal motors*.
- The universal motor is designed for commercial frequencies from 60 cycles down to D.C. (zero frequency), and for voltage from 250 V to 1.5 V. A commercial universal motor may have a somewhat weaker series field and more armature conductors than a D.C. series motor of equivalent horsepower. *It is manufactured in ratings up to 3/4 H.P., particularly for vacuum cleaners and industrial sewing machines.* In smaller sizes of $\frac{1}{4}$ H.P. or less, it is used in *electric hand drills*.

Like all series motors, the *no-load speed of the universal motor is universally high.* Quite frequently, *gears trains* are built into the motor housing of some universal motors to provide exceedingly high torque at low speeds.

When these motors are used in commercial appliances such as *electric shavers, sewing machines, office machines, and small hand hair dryers or vacuum cleaners,* they are always *directly loaded* with little danger of motor runaway.

Advantages of a universal motor :

1. High speed from above 3600 r.p.m. to around 25000 r.p.m.
2. High power output in small physical sizes for use in portable tools.
3. High torque at low and intermediate speeds to carry a particularly severe load.
4. Variable speed by adjustable governor, by line voltage or especially by modern pulse techniques.

Disadvantages :

1. Increased service requirement due to use of brushes and commutators. The life of these parts is limited in severe service.
2. Relatively high noise level at high speeds.
3. Moderate to severe radio and television interference due to brush sparking.
4. Requirement for careful balancing to avoid vibration.
5. Requirement for reduction gearing in most portable tools.

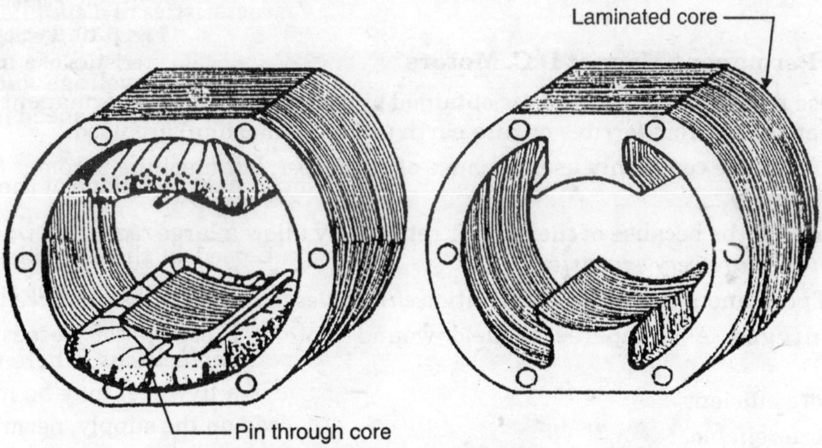

Fig. 5.8. Field core of a two pole universal motor.

Universal motors are manufactured in two types :

1. Concentrated-pole, non-compensated type (low H.P. rating).
2. Distributed field compensated type (high H.P. rating).

Fig. 5.8 shows the laminated field structure of a typical concentrated field universal motor.

Operation of a universal motor. Such motors develop unidirectional torque regardless of whether they operate on D.C. or A.C. supply. The production of unidirectional torque when the motor runs on A.C. supply can be easily understood from Fig. 5.9. The motor works on the same principle as a D.C. motor *i.e.*, the force between the main pole flux and the current carrying armature conductors. This is true regardless of whether current is alternating or direct.

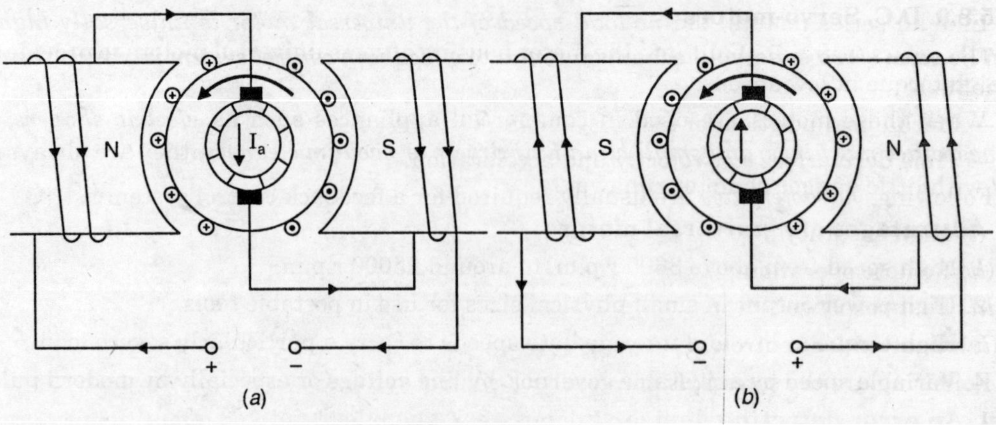

Fig. 5.9. Universal motor operation.

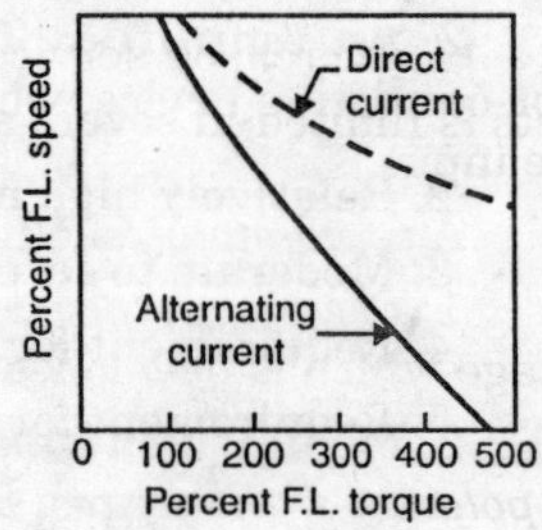

Fig. 5.10. Typical torque characteristics of a universal motor.

- Fig. 5.10 shows the typical torque characteristics of a universal motor both for D.C. and A.C. supply.
- The speed of a universal motor may be *controlled* by the following methods :

(*i*) Reactance method.

(*ii*) Tapped-field method.

(*iii*) Centrifugal mechanism.

5.8.8. Permanent Magnet D.C. Motors

In these motors, *field excitation* is obtained by suitably mounting permanent magnets on the stator. Magnets made from ferrites or rare earth (cobalt samarium) are used.

— *Ferrites* are commonly used because of *lower cost*, but machine becomes *bulky due to low retentivity*.

— Rare earths because of their hight retentivity allow a large *reduction in weight and size, but they are very expensive.*

Use of permanent magnets for excitation *eliminates field copper loss and need for field supply.*

Advantages. As compared to field wound motors, these motor possess the following *advantage :*

(*i*) More efficient.

(*ii*) More reliable.

(*iii*) More study and compact.

(*iv*) The field flux remains constant for all loads giving a more linear speed-torque characteristic.

(*v*) In a separately excited motor, failure of field can lead to runaway condition. This does not happen in permanent magnet motors.

Limitation. As the flux is constant in these motors, speed *cannot be controlled above base speed.*

Applications :

- These motors are mainly employed in fractional H.P. range, but they are available upto 5 kW rating.
- They find applications in electric vehicles like *wheel chairs, mopeds, forklift trucks etc.*

5.8.9. D.C. Servo-motors

The term *servo* or servo-mechanism refers to a *feedback control system* in which the controlled variable is :

— Mechanical position, or

— Time derivatives *e.g.* velocity and acceleration.

Following *characteristics* are usually required for a feedback control system :

(*i*) High accuracy ;

(*ii*) Remote operation ;

(*iii*) Fast-response ;

(*iv*) Unattended control.

Following are the *essentials* of a feedback control system :

1. **An error detecting device.** It determines when the regulated quantity is *different* from the reference quantity and sends out the error signal to the *amplifier*.

2. **An amplifier.** The amplifier receives the error signal and then supplies power to the error-correcting device, which in turn changes the regulated quantity so that it matches the reference input.

A *servo-motor* should entail the following *characteristics* :

1. *The output torque of the motor should be proportional to the voltage applied* (*i.e.*, the control voltage which is developed by the amplifier in response to error signal).

2. *The direction of the torque developed by the servo-motor should depend upon the instantaneous polarity of the control voltage.*

Types of servo-motors :

The servo-motors are of the following *two* types :

1. *D.C. servo-motors.*

2. *A.C. servo-motors.*

We shall discuss here only D.C. servo-motors.

D.C. servo-motors :

These motors are preferred for very high power systems since they operate more efficiently (as compared to A.C. servo-motors).

These motors may be of the following types :

— Series motors ;

— Split series motors ;

— Shunt control motors ;

— Permanent magnet (*fixed excitation*) shunt motor.

(*i*) **Series motor :**

- This motor has a high starting torque.
- It draws large current.
- The speed regulation is poor.
- Reversal can be obtained by reversing field voltage polarity with split series field winding.

(*ii*) **Split series motor :**

- The D.C. series motor with split field (small fractional kW) may be operated as a separately excited field-controlled motor (Fig. 5.11).

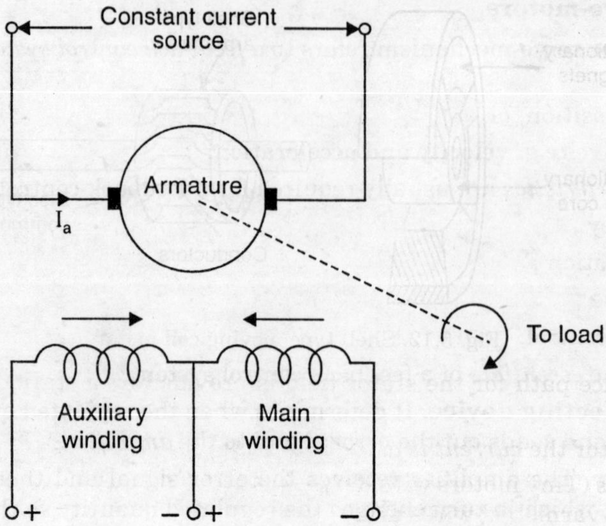

Fig. 5.11. From D.C. amplifier.

The armature may be supplied from a constant current source.

- A typical torque curve shows the following :
  - — High stall torque ;
  - — Rapid reduction in torque with increase in speed.

(*iii*) **Shunt control motor :**

- This type of motor has *two separate windings : Field winding placed on the stator and the armature winding placed on the rotor of the machine.* Both the windings are connected to a D.C. supply source.
- Whereas in a conventional D.C. shunt motor, the two windings are connected in parallel across the D.C. supply mains, but in a *servo-application* the windings are driven by *separate D.C. supplies.*

(*iv*) **Permanent magnet shunt motor :**

- It is a fixed excitation shunt motor where the field is actually supplied by a permanent magnet.
- Its performance is similar to that of armature controlled fixed field motor.

5.8.10. Moving Coil Motors

There are certain applications which require acceleration much higher than what can be achieved in a conventional D.C. servo-motor. The armatures of moving coil D.C. motors have special constructions which allow a substantial reduction in armature inertia and inductance, permitting very high accelerations.

Moving coil motors are of the following two types :

1. Shell type
2. Disc or Pancake type.

1. Shell type moving coil motor. Refer to Fig. 5.12.

- In this type of motor, the rotor consists of *only armature winding* due to which it has very low inertia ; consequently high acceleration is obtained. Armature winding consists of conductors assembled to form a thin walled cylinder. The *commutator* may have a cylindrical construction as in conventional D.C. motors or disc type construction.

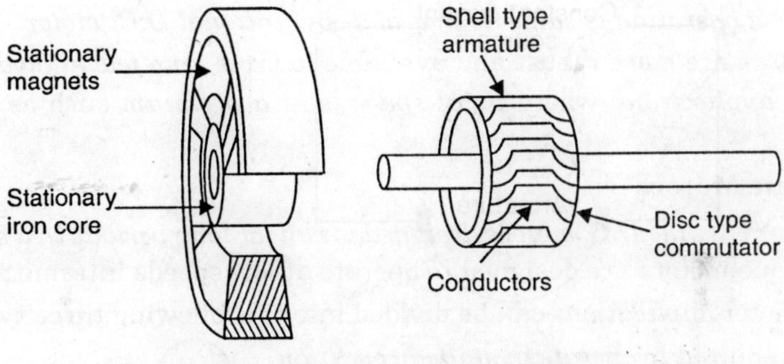

Fig. 5.12. Shell type moving coil motor.

Low reluctance path for the stator field is provided by a stationary magnetic material cylinder.

In such a motor the *current is axial and flux is radial*.

- ***Micromotors*** (Tiny motors with diameters around 1 cm) have armature winding consisting of simply varnished wires arranged in cylindrical form and a disc type commutator.

Such motors are find wide applications in *card readers, video systems, cameras* etc.

In *bigger size motors* the armature winding is made by bonding conductors together using polymer resins and fibreglass to provide adequate mechanical strength.

2. Disc or Pancake type moving coil motor. Refer to Fig. 5.13.

In this motor armature is made in disc or pancake form, and armature conductors resemble spokes on a wheel. The armature winding is formed by stamping conductors from a sheet of copper, welding them together and placing them on a light weight disc. Conductor segments are then joined with a *commutator* at the centre of the disc.

Here the direction of *flux is axial* and *armature current is radial* (just opposite to shell type conventional motors).

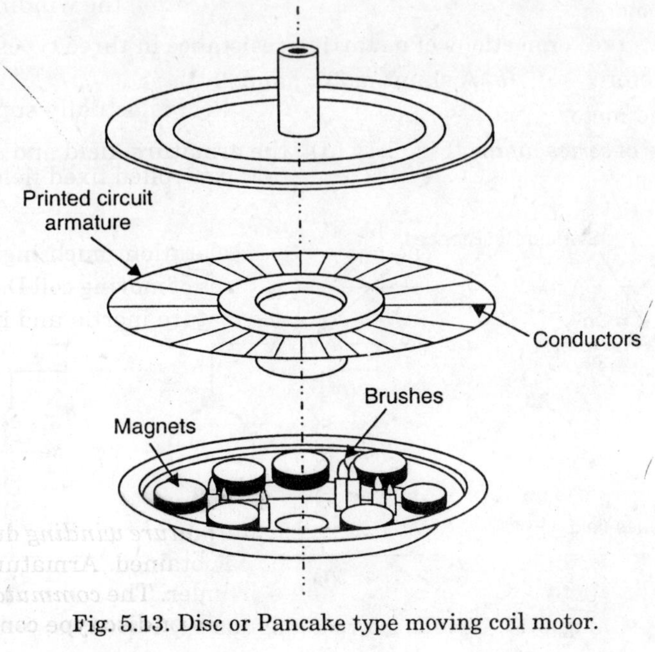

Fig. 5.13. Disc or Pancake type moving coil motor.

The *principle of operation is same as that of a conventional D.C. motor.*

- These motors are more robust and available in sizes *upto few kilowatts.*
- They find *applications* where *axial space is at a premium* such as *machine tools, disc drives* etc.

5.8.11. Torque Motors

"Torque motors" are *the D.C. motors designed to run for long periods in a stalled or a low speed condition.* Some torque motors are designed to operate at low speeds intermittently.

The torque motor *applications* can be divided into the following three types :

(*i*) Motor is required to *operate in stalled condition.*

— The purpose of the motor is develop required tension or pressure on a material, similar to spring.

(*ii*) Motor is required to *move through only a few revolutions* or *degrees of revolution.*

Examples. Opening of valves, switches, clamping devices etc.

(*iii*) This category involves *continuous movement of the motor at low speed.*

Example. Reel drive.

5.9. STARTING OF D.C. MOTORS

5.9.1. Need for Starters

A motor at *rest* has no back or counter e.m.f. At starting therefore, the *armature current is limited only by the resistance of the armature circuit.* The *armature resistance is very low*, however, and if full voltage were impressed upon the motor terminals at stand still, the resulting armature current would be *many times full-load* value—*usually sufficient to damage the machine.* For this reason, *additional resistance* is introduced into the armature circuit at starting. As the motor gains speed, its back e.m.f. builds up and the starting resistance is cut out.

Note. Very small D.C. motors, either shunt, series or compound wound, have sufficient armature resistance so that they may be started directly from the line without the use of a starting resistance and without injury to the motor.

Fig. 5.14 shows the connections of a starting resistance in three types of D.C. motors :

(*a*) A series motor ; (*b*) A shunt motor ; and

(*c*) A compound motor.

- In the case of *series motor* [Fig. 5.14 (*a*)], the armature, field and starting resistance are all in series.

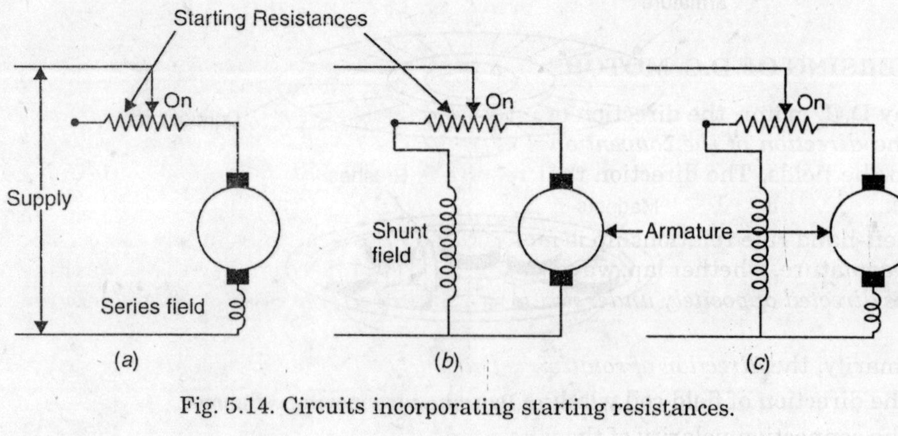

Fig. 5.14. Circuits incorporating starting resistances.

- In the case of *shunt motor* [Fig. 5.14 (*b*)], it will be seen that the *top end of shunt field is connected to the first contact* on the starting resistance. This is to *ensure* that the *field winding receives the full supply at the moment of switching on.* If the fields were connected to the *last stud* of the starting resistance, then, on starting, the field would receive only a *proportion of the supply voltage, the field current would be correspondingly weak and the torque might be too small to start the motor against the friction of the moving parts.*
- The connections for the *compound motor* are seen from [Fig. 5.14 (*c*)] to be a combination of those of the series and the shunt connections.

5.9.2. Starters for Shunt and Compound Motors

- The starters of D.C. motors are generally manufactured in convenient sizes and styles for use as auxiliaries with D.C. shunt and compound motors. Their *primary function is to limit the current in the armature circuit during the starting accelerating period.*
- The motor starters are always rated on the basis of output power and voltage of the motors with which they are to be used.
- There are two standard types of motor starters for shunt and compound motors. These are :

  (*i*) Three-point type ; and (*ii*) Four-point type.

Three-point starters are not completely satisfactory when used with motors whose speeds must be controlled by inserting resistance in the shunt field circuit. However, when applications require *little or no speed control,* either may be employed.

5.9.3. Automatic Starters

There is a frequent use of automatic starters in industry. They offer the following *advantages* :

(*i*) Motor can be started in the minimum of time.

(*ii*) Inexperienced operators can start motors without the possibility of blowing fuses and causing delays.

Because of the ease of starting and stopping motors by this means, operators will conveniently shut down equipment when not in use. This *will prolong the life of equipment and reduce power consumption.*

The primary requirement of a device which is to work automatically is that must be *capable of sensing a change and then responding to the change*. The changes which can be sensed during the starting process are the current, voltage and time. These changes form the basis for most automatic starters.

5.10. REVERSING OF D.C. MOTORS

In any D.C. motor, the direction of rotation depends upon the *magnetic polarity of its main fields* and the *direction of the conventional current that flows in the armature windings* that are immersed in the fields. The direction that results is then in *accordance with the left-hand rule of motor action.*

The left-hand rule relationship is met under all the field poles in any motor. The methods of winding the armature, whether lap, wave, or frog leg coils, all accomplish the same thing in a motor : *the current is directed oppositely under north and south poles, so each region pulls in the same direction.*

Summarily, the *direction of rotation is determined by the following factors* :

(*i*) The direction of field coil winding, which is built in.

(*ii*) The connection polarity of the whole field which may be changed or switched.

(*iii*) The direction of armature coil winding, which is also built in.

(*iv*) The connection polarity of the brush group, which fixes the access point to the armature winding and which may be changed or switched.

Factors (*ii*) and (*iv*) are *accessible modifications* and may be manipulated.

In a D.C. motor if *the overall terminal connection polarities are changed, the current direction of both the armature and the field are changed.* Since this, in effect, does the same thing as moving from one pole to another locally, the *rotating torque direction is not changed.* The exception to this is a permanent magnet field motor where only the armature has coil windings. In this case the reversal can be produced only by reversing the line connections.

Reversal is accomplished by changing the polarity of *either* the armature or the field, *but not by changing both.*

Reversing circuit connections of D.C. motors :

Shunt motor. Fig. 5.15 shows the motor reversing connection schematically. In a shunt motor the field normally has *much less current* than the armature, and so it would seem the *field is the logical place to switch.* This is *not* usually the *best practice* since the field is *highly inductive circuit.* Thus, if any switching is performed before the field current is fully decayed, the switch points will arc viciously or even dangerously in large size. Further more, since any switching point is usually the most undesirable part of an electric circuit, it is unwise to switch where the basic action deserved contributes to further unreliability. Finally, upon failure of a field reverse switch contact, the next start will take place with little or no field, which causes *high-speed run away*. Owing to these reasons usually *armature circuit is reversed.*

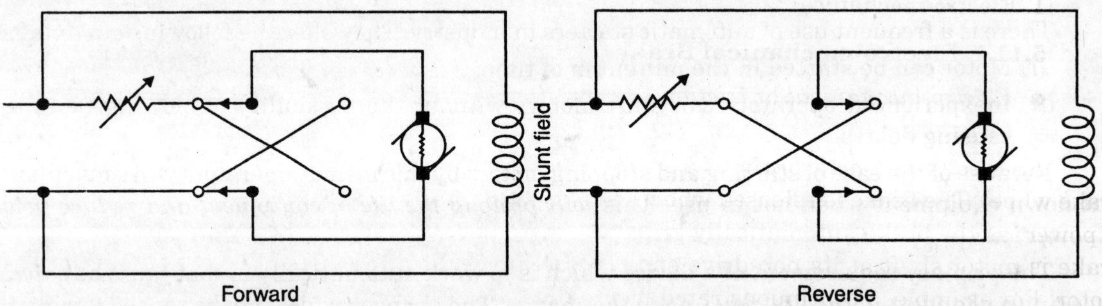

Fig. 5.15. Shunt motor reversing.

Series motor. Fig. 5.16 shows the series motor reversing connections. In a series motor (as discussed in shunt motor) it does not make much difference since the field coil is *much less inductive* with its *fewer winding turns.* The field also has the same current as the armature.

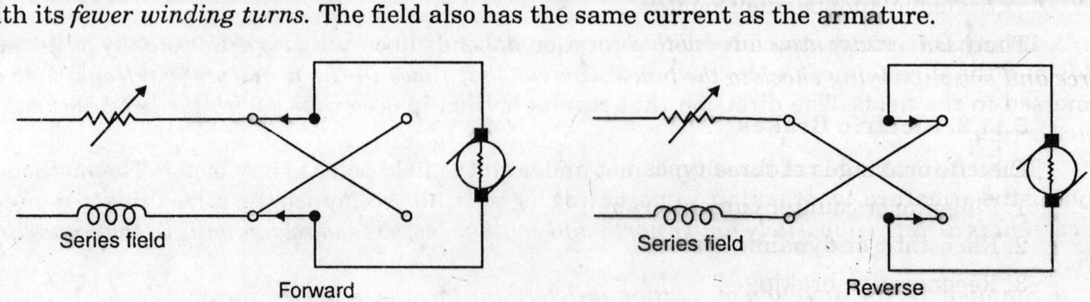

Fig. 5.16. Series motor reversing.

Compound motor. Fig. 5.17 shows the reversing connections for a compound motor. A compound motor *must have both fields changed* if field reversing is used, so *armature switching is less complex.* If *only one field were reversed, a compound motor would be changed from cumulative to differential* (which changes the whole character of the motor).

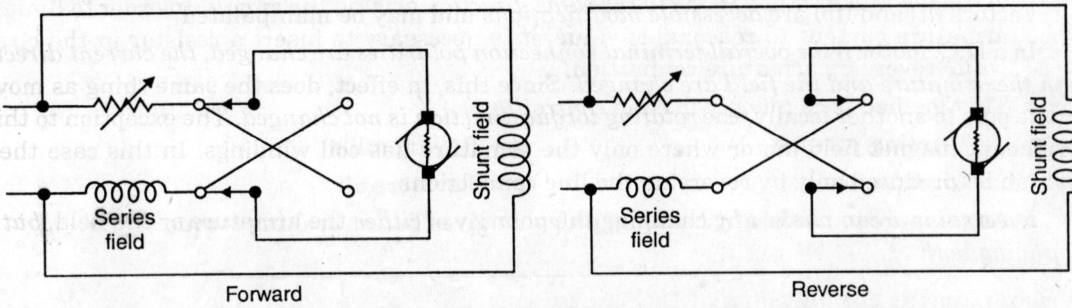

Fig. 5.17. Compound motor reversing.

Note. The commutating fields and compensating windings, if present, *must be left in their original relationship* with the armature, so the reversal must take the *armature circuit as a whole*. If the commutating field or the compensating winding are inadvertently switched in relation to the armature, the commutation quality would be destroyed. Taking either of these windings out of their proper relationship is worse than if they were not there at all.

5.11. ELECTRIC BRAKING OF D.C. MOTORS

Electric braking of motors can be broadly *classified* as

1. Electro-mechanical ; 2. Electrical.

5.11.1. Electro-mechanical Brakes

- *Electro-mechanical* or *friction brakes* are operated by *electromagnets* or *electric-operated thrustors.*

An *'electro-magnetic brake'* is a spring-loaded brake with friction liners pressing against the brake wheel. The brake is released when magnet is energized. This ensures safety in case of failure of power, as the brake automatically grips the wheel in that case. The best place for mounting the brake is motor shaft at its non-drive end. The rating of the brake should be the same that of the motor. For example, a continuous rated motor should have a continuous rated brake.

- In *thrustor* or *hydraulic type brake*, the electro-magnet is replaced by a pump, and fluid under pressure displaces the piston which, in turn, releases the brake. The electromagnets or thrustors can be connected directly across the terminals of motors or can be controlled by independent motors.

The *disadvantage associated with electro-mechanical brakes is the sudden application of braking force and accompanying shock to the machine.*

5.11.2. Electric Brakes

Electric braking is of three types :

1. Plugging or counter current braking
2. Rheostatic or dynamic braking
3. Regenerative braking.

5.11.2.1. Electric braking of shunt motor :

(*i*) ***Plugging or counter current braking :***

- In this method, *connection to the armature terminals are reversed so that motor tends to run in the opposite direction* (Fig. 5.18). Due to reversal of armature connections, applied voltage V and E_b start acting *in the same direction* around the circuit. In order to limit the armature current to reasonable value, it is necessary to insert a resistor in the circuit while reversing armature connection.
- This method is commonly used in *controlling :*

  (*i*) Printing presses (*ii*) Rolling mills

  (*iii*) Machine tools (*iv*) Elevators etc.
- As compare to rheostatic braking, plugging gives better braking torque.

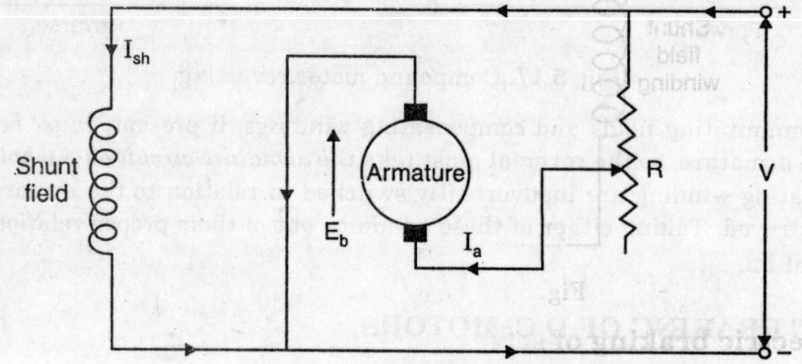

Fig. 5.18. Plugging or counter-current braking.

(*ii*) ***Rheostatic or dynamic braking.*** In this method of electric braking of shunt motors, the armature of the shunt motor is disconnected from the supply and is connected across a variable resistance R as shown in Fig. 5.19 (*b*). The field winding is, however, left connected across the supply undisturbed. The braking effect is *controlled by varying the series resistance R.*

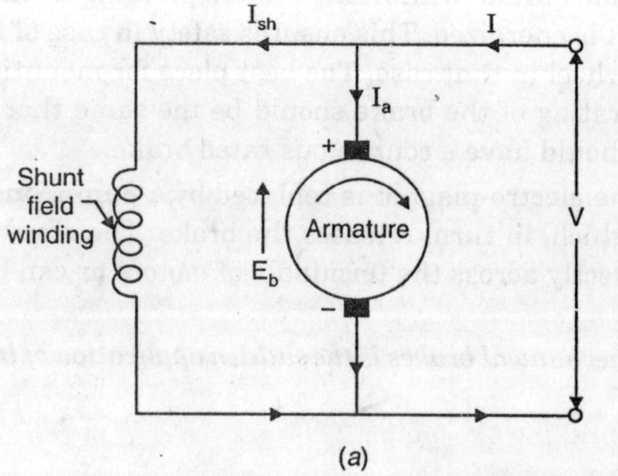

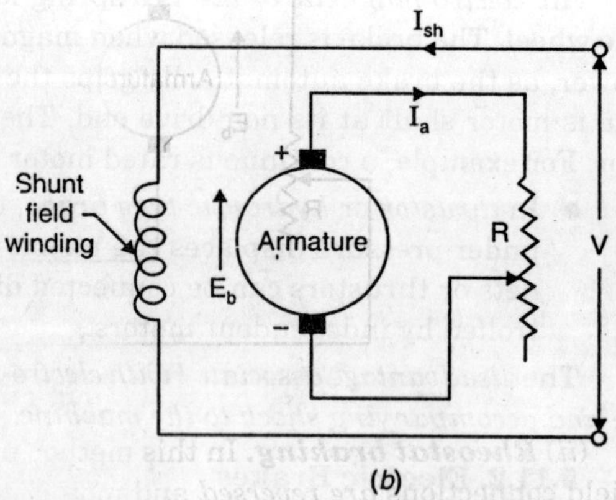

Fig. 5.19. Rheostatic or dynamic braking.

(*iii*) ***Regenerative braking.*** Refer to Fig. 5.20. Regenerative braking method is used *when the load on the motor has overhauling characteristic as in the lowering of the case of a hoist or downgrade motion of an electric train. Regeneration takes place when* E_b *becomes greater than* V. This happens when the overhauling load acts as a prime mover and so *drives the machine as a generator*. Consequently, direction of I_a and hence of armature torque is reversed and speed falls until E_b becomes less than V. It is obvious that during slowing down of the motor, power is returned to the line which may be used for supplying another train on an upgrade thereby relieving the power house of part of its load.

As a protective measure, it is necessary to have some type of mechanical brake in order to hold the load in the event of a power failure.

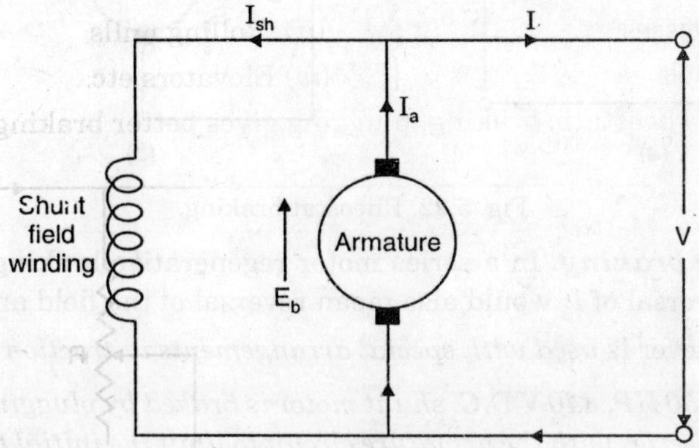

Fig. 5.20. Regenerative braking.

5.11.2.2. Electric braking of series motor :

(*i*) ***Plugging.*** In this method (as in the case of shunt motors) the connections of the armature are reversed and a variable resistance R is put in series with the armature as shown in Fig. 5.21.

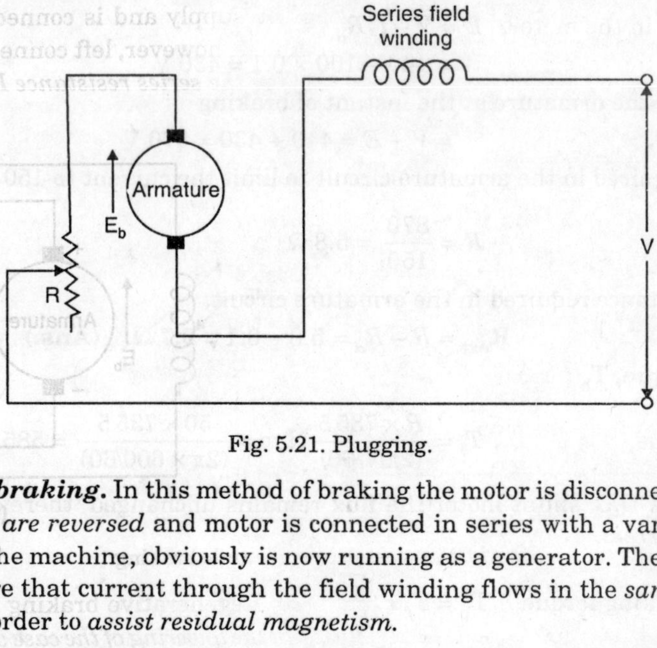

Fig. 5.21. Plugging.

(*ii*) ***Rheostat braking.*** In this method of braking the motor is disconnected from the supply, the field connections *are reversed* and motor is connected in series with a variable resistance R as shown in Fig. 5.22. The machine, obviously is now running as a generator. The field connections are reversed to make sure that current through the field winding flows in the *same* direction as before (*i.e.*, from A to B) in order to *assist residual magnetism.*

In practice, the variable resistance employed for starting purpose is itself used for braking purposes.

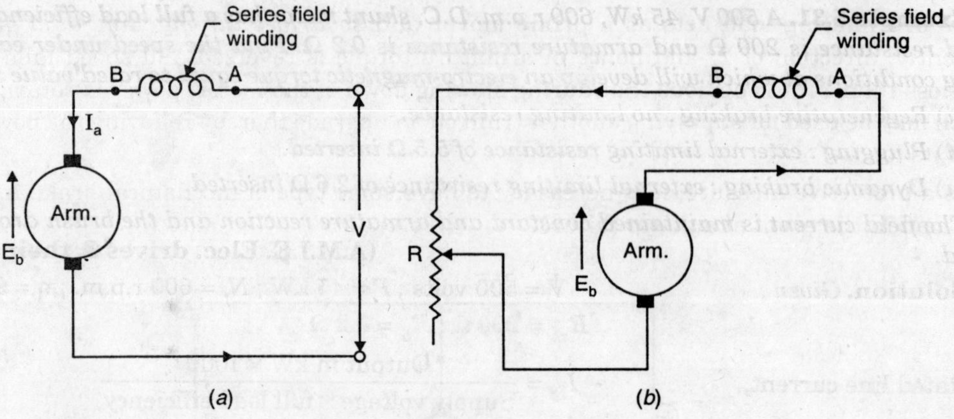

Fig. 5.22. Rheostat braking.

(*iii*) ***Regenerative braking.*** In a series motor regenerative braking is ***not possible without modification*** because reversal of I_a would also mean reversal of the field and hence of E_b.

This method, however is *used with special arrangements in traction motors.*

Example 5.30. *A 50 HP, 440 V D.C. shunt motor is braked by plugging. Calculate the value of resistance to be placed in series with the armature circuit to limit the initial braking current to 150 A. Calculate the braking torque so obtained. Assume armature resistance as 0.1 Ω ; full load armature current = 100 A, full load speed = 600 r.p.m.* **(B.T.E. U.P., 2000)**

Solution. *Given* : Power, $P = 50$ H.P. ; $V = 440$ volts ; $I_b = 150$ A ;

$R_a = 0.1\ \Omega$; $I_f = 100$ A ; $N = 600$ r.p.m.

External resistance to be connected in series, $\mathbf{R_{ext}}$:

Induced e.m.f. in the motor, $E = V - I_f R_a$

$= 440 - 100 \times 0.1 = 430$ V

Voltage across the armature at the instant of braking

$= V + E = 440 + 430 = 870$ V

Resistance required in the armature circuit to limit the current to 150 A,

$$R = \frac{870}{150} = 5.8\ \Omega$$

External resistance required in the armature circuit,

$$\mathbf{R_{ext}} = R - R_a = 5.8 - 0.1 = \mathbf{5.7\ \Omega.} \quad \textbf{(Ans.)}$$

Braking torque, $\mathbf{T_b}$:

Full load torque,
$$T_f = \frac{P \times 735.5}{(2\pi N/60)} \text{ N m} = \frac{50 \times 735.5}{(2\pi \times 600/60)} = 585.3 \text{ N m}$$

As in case of a D.C. shunt motor the flux remains unchanged, therefore, *torque is proportional to armature current*.

∴ Initial braking torque, $\mathbf{T_b} = T_f \times \dfrac{I_b}{I_f}$

$$= 585.3 \times \frac{150}{100} = \mathbf{877.95\ N\ m.\ (Ans.)}$$

Example 5.31. *A 500 V, 45 kW, 600 r.p.m. D.C. shunt motor has a full load efficiency of 90%. The field resistance is 200 Ω and armature resistance is 0.2 Ω. Find the speed under each of the following conditions at which will develop an electro-magnetic torque equal to rated value :*

(*i*) *Regenerative braking : no limiting resistance.*

(*ii*) *Plugging : external limiting resistance of 5.5 Ω inserted.*

(*iii*) *Dynamic braking : external limiting resistance of 2.6 Ω inserted.*

The field current is maintained constant and armature reaction and the brush drop may be neglected. **(A.M.I.E. Elec. drives & their control)**

Solution. *Given :* $V = 500$ volts ; $P = 45$ kW ; $N_f = 600$ r.p.m. ; $\eta = 90\%$;

$R_{sh} = 200\ \Omega$; $R_a = 0.2\ \Omega$

Rated line current,
$$I_{Lf} = \frac{\text{Output in kW} \times 1000}{\text{Supply voltage} \times \text{full load efficiency}}$$
$$= \frac{45 \times 1000}{500 \times 0.9} = 100 \text{ A}$$

Shunt field current, $I_{sh} = \dfrac{V}{R_{sh}} = \dfrac{500}{200} = 2.5$ A

Armature current on full load, $I_{af} = I_{Lf} - I_{sh} = 100 - 2.5 = 97.5$ A

Induced e.m.f. on full load, $E_f = V - I_{af}R_a = 500 - 97.5 \times 0.2 = 480.5$ V.

Since field current remains constant and the armature reaction and brush drop are negligible so *electro-magnetic torque developed is directly proportional to armature current and speed is directly proportional to induced e.m.f.*

As torque to be developed in each case is equal to rated torque so armature current in each case is equal to armature current at full load *i.e.*, 97.5 A.

Speeds under various conditions, N_1, N_2, N_3 :

(*i*) Induced e.m.f. to give armature current of 97.5 A for *regenerative braking,*

$$E_1 = V + I_{af}R = 500 + 97.5 \times 0.2 = 519.5 \text{ V}$$

and speed,
$$N_1 = N_f \times \frac{E_1}{E_f} = 600 \times \frac{519.5}{480.5} = \mathbf{648.7\ r.p.m.\ (Ans.)}$$

(*ii*) Induced e.m.f. to give armature current of 97.5 A for *plugging* with external limiting resistance of 5.5 Ω,

$$E_2 = I_{af}(R + R_a) - V = 97.5(5.5 + 0.2) - 500 = 55.75 \text{ V}$$

and speed,
$$N_2 = N_f \times \frac{E_2}{E_f} = 600 \times \frac{5575}{480.5} = \mathbf{69.6\ r.p.m.\ (Ans.)}$$

(*iii*) Induced e.m.f. to give armature current of 97.5 A for *dynamic braking* with external limiting resistance of 2.6 Ω,

$$E_3 = I_{af}(R + R_a) = 97.5(2.6 + 0.2) = 273 \text{ V}$$

and speed,
$$N_3 = N_f \times \frac{E_3}{E_f} = 600 \times \frac{273}{480.5} = \mathbf{340.9\ r.p.m.\ (Ans.)}$$

5.12. SPEED CONTROL OF D.C. MOTORS

5.12.1. Factors Controlling the Speed

D.C. machines are generally much more adaptable to adjustable speed service. *The ready availability of D.C. motors to adjustment of their operating speed over wide ranges and by a variety of methods is one of the important reasons for the strong competitive position of D.C. machinery in modern industrial applications.*

The speed of a D.C. motor can be expressed by the following relationship.

$$N \propto \frac{V - I_a R_a}{\phi}.$$

Therefore, the speed of D.C. motor can be regulated by changing ϕ, R or V.

The speed of D.C. motors can be controlled by the following *methods* :

1. Field control.
2. Rheostatic control.
3. Voltage control.
4. Thyristor control.

5.12.2. Field Control Method

- **Field control** is the most common method and forms one of the *outstanding advantages of shunt motors*. The method is, of course, also applicable to compound motors. Adjustment of field current and hence the flux and speed by adjustment of the shunt field circuit resistance or with a solid-state control when the field is *separately excited* is accomplished *simply*, *inexpensively*, and *without much change in motor losses.*

The *speed is inversely proportional to the field current*

i.e.,
$$N \propto \frac{1}{I_f} \propto \frac{1}{\phi}.$$

- The *lowest speed obtainable is that corresponding to maximum field current* ; the *highest speed is limited electrically by the effects of armature reaction under weak-field conditions in causing motor instability and poor commutation.*
- *Since, voltage across the motor remains constant, it continues to deliver constant output.* This characteristic makes this method suitable for fixed output loads. The performance curve of a D.C. motor with voltage and field control is shown in Fig. 5.23.

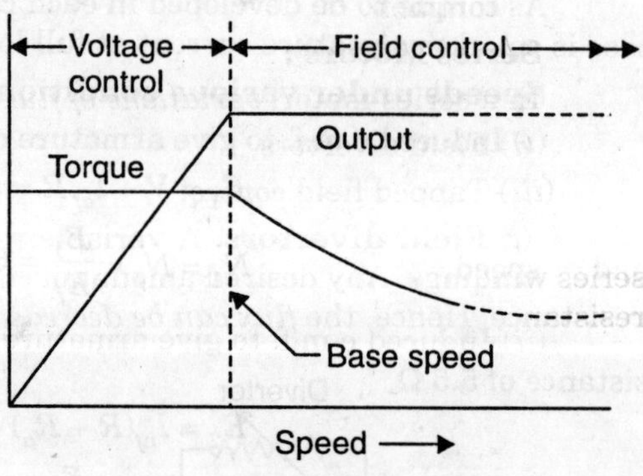

Fig. 5.23

Merits. The *merits* of this method are :

1. Good working efficiency.
2. Compact controlling equipment.
3. Capability of minute speed control.
4. The speed is not effected by load, and speed control can be performed effectively even at light loads.
5. Relatively inexpensive and simple to accomplish, both manually and automatically.
6. Within limits, field control does not affect speed regulation in the cases of shunt, compound, and series motors.
7. Provides relatively smooth and stepless control of speed.

Demerits. The *demerits* of field control as a method of speed control are :

1. Inability to obtain speeds below the basic speed.
2. Instability at high speeds because of armature reaction.
3. Commutation difficulties and possible commutator damage at high speeds.

Shunt Motors :

- The flux of a D.C. shunt motor can be changed by changing shunt field current (I_{sh}) with the help of a shunt field rheostat as shown in Fig. 5.24. Since the field current is very small, the power wasted in the controlling resistance is very small.

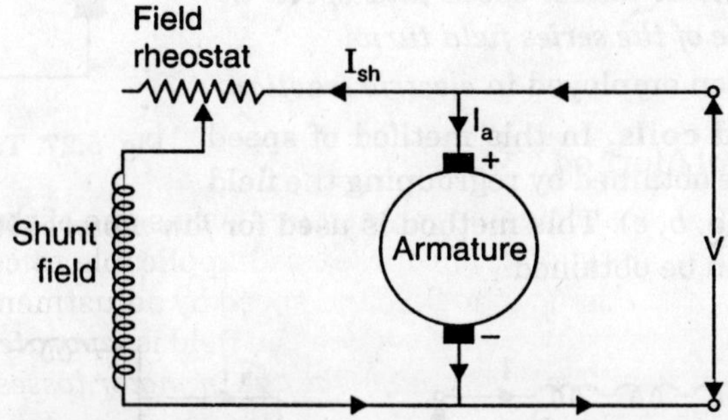

Fig. 5.24. Field rheostatic control for a D.C. shunt motor.

- In *non-interpolar machines* the speed can be increased by this method in the ratio **2 : 1**. In machines fitted *with interpoles* a ratio of maximum to minimum speeds of **6 : 1** is fairly common.

Series Motors :

In a series motor, *variations of flux* can be brought about in any one of the following ways :

(*i*) Field divertors (*ii*) Armature divertor
(*iii*) Tapped field control (*iv*) Paralleling field coils.

(*i*) **Field divertors.** A variable resistance, known as field divertor (Fig. 5.25) *shunts* the series windings. Any desired amount of current can be passed through the divertor by adjusting its resistance. Hence, the *flux can be decreased and consequently the speed of the motor increased.*

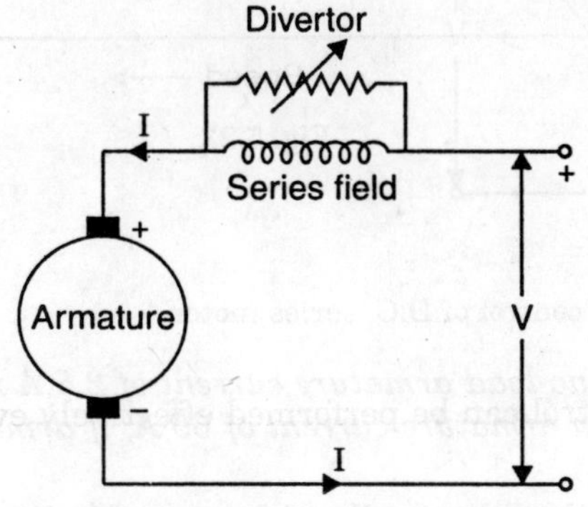

Fig. 5.25. Field divertor method of speed control for D.C. series motor.

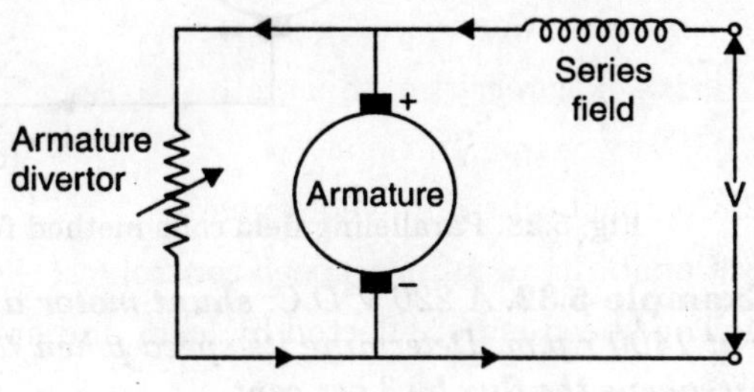

Fig. 5.26. Armature divertor method of speed control of D.C. series motor.

(*ii*) **Armature divertor.** In order to *get speeds lower than the normal speed* a divertor across the armature can be used (Fig. 5.26). For a *given constant load torque,* if I_a is reduced due to armature divertor, then ϕ must increase ($\because T_a \propto I_a$). This results in an increase in current taken from the supply which increases the flux and a fall in speed $\left(\because N \propto \frac{1}{\phi}\right)$. The variations in speed can be controlled by varying the divertor resistance.

(*iii*) **Tapped field control.** In this method a number of tappings from the field winding are brought outside, as shown in Fig. 5.27. A number of series field turns can be short-circuited according to the requirement. *When all field turns are in circuit, the motor runs at lowest speed and speed increases with cutting out some of the series field turns.*

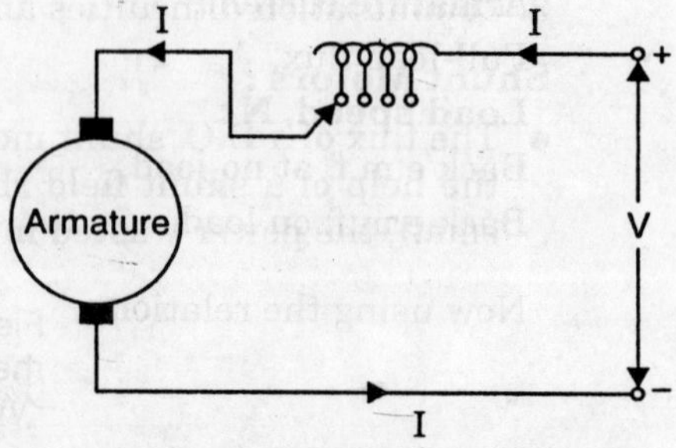

Fig. 5.27. Tapped field control for D.C. series motor.

- This method is often employed in *electric traction.*

(*iv*) **Paralleling field coils.** In this method of speed control several speeds can be obtained by regrouping the field coils as shown in Fig. 5.28 (*a*, *b*, *c*). This method is used for *fan motors*. It is seen that for a 4-pole motor, three fixed speeds can be obtained.

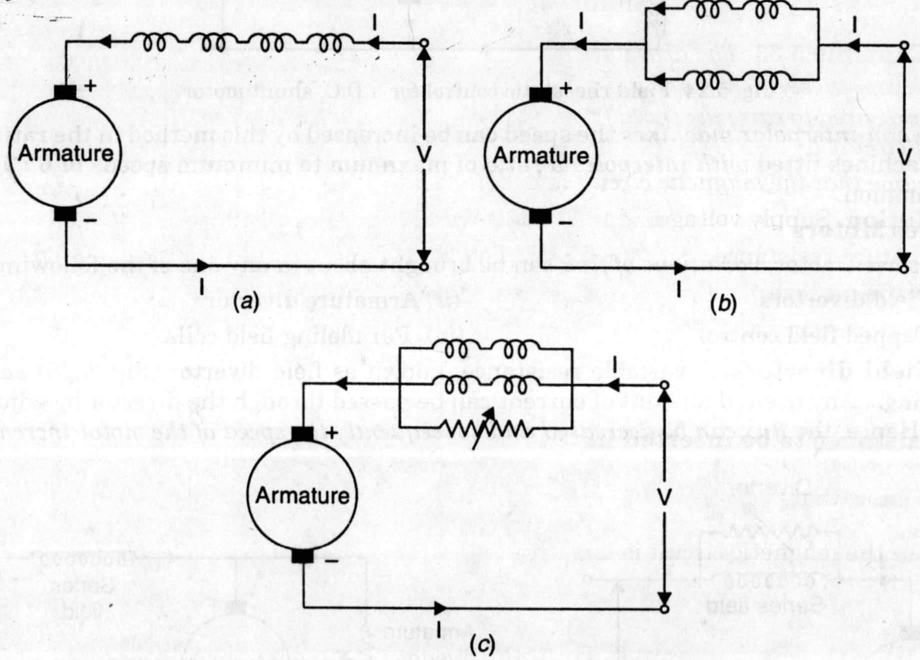

Fig. 5.28. Paralleling field coils method for speed control of D.C. series motor.

Example 5.32. *A 220 V D.C. shunt motor draws a no-load armature current of 2.5 A when running at 1400 r.p.m. Determine its speed when taking an armature current of 60 A, if armature reaction weakens the flux by 3 per cent.*

Take armature resistance = 0.2 Ω.

Solution. Supply voltage, $V = 220$ Volts

No-load current, $I_{a0} = 2.5$ A

No-load speed, $N_0 = 1400$ r.p.m.

Armature resistance, $R_a = 0.2\ \Omega$

Armature current, $I_a = 60$ A

Full-load flux, $\phi = 0.97\ \phi_0$

Load speed, N :

Back e.m.f. at no-load, $E_{b0} = V - I_{a0} R_a = 200 - 2.5 \times 0.2 = 219.5$ V

Back e.m.f. on load, $E_b = V - I_a R_a = 220 - 60 \times 0.2 = 208$ V.

Now using the relation, $\frac{N}{N_0} = \frac{E_b}{E_{b0}} \times \frac{\phi_0}{\phi}$

$$\frac{N}{1400} = \frac{208}{219.5} \times \frac{\phi_0}{0.97\ \phi_0}$$

$\therefore$ $N = \frac{1400 \times 208}{219.5 \times 0.97} = 1367.7$ r.p.m.

Hence, *Load speed* = **1367.7 r.p.m. (Ans.)**

Shunt Motors (Field Control)

☞**Example 5.33.** *The armature and field resistances of a 250 V D.C. shunt motor are 0.5 Ω and 250 Ω respectively. When driving a load of constant torque at 600 r.p.m., the armature current is 20 A. If it is desired to raise the speed from 600 to 800 r.p.m., what resistance should be inserted in the shunt field circuit ?*

Assume that the magnetic circuit is unsaturated.

Solution. Supply voltage, $V = 250$ Volts

Armature resistance, $R_a = 0.5$ ohm

Field resistance, $R_{sh} = 250$ ohms

Armature current, $I_a = 20$ A

Speed, $N_1 = 600$ r.p.m.

Speed, $N_2 = 800$ r.p.m.

Resistance to be inserted in the shunt field circuit, R :

We know that, $\frac{N_2}{N_1} = \frac{E_{b2}}{E_{b1}} \times \frac{\phi_1}{\phi_2}$ ...(*i*)

Since the magnetic circuit is unsaturated,

$\therefore$ $\phi \propto I_{sh}$

$\therefore$ $\frac{N_2}{N_1} = \frac{E_{b2}}{E_{b1}} \times \frac{I_{sh1}}{I_{sh2}}$

Since *torque remains constant,*

$\therefore$ $\phi_1 I_{a1} = \phi_2 I_{a2}$ $[\because\ T \propto \phi I_a]$

$$I_{a2} = \frac{\phi_1 I_{a1}}{\phi_2} = I_{a1} \times \frac{I_{sh1}}{I_{sh2}}$$

Now, $I_{sh1} = \frac{250}{250} = 1$ A

and $I_{sh2} = \frac{250}{R_t}$

where, R_t is the total resistance of the shunt field circuit.

$\therefore \quad I_{a2} = 20 \times \dfrac{1}{250/R_t} = 0.08\, R_t$

Also, $\quad E_{b1} = V - I_{a1} R_a = 250 - 20 \times 0.5 = 240$ V

and, $\quad E_{b2} = V - I_{a2} R_a = 250 - 0.08\, R_t \times 0.5 = 250 - 0.04\, R_t$

Substituting these values in eqn. (*i*), we get

$$\frac{800}{600} = \frac{250 - 0.04\, R_t}{240} \times \frac{1}{250/R_t}$$

$$\frac{4}{3} = \frac{250 - 0.04 R_t}{240} \times \frac{R_t}{250}$$

or, $\quad R_t (250 - 0.04 R_t) = \dfrac{4}{3} \times 240 \times 250$

or, $\quad 250 R_t - 0.04 R_t^2 = 80000$

or, $\quad 0.04 R_t^2 - 250 R_t + 80000 = 0$

or, $\quad R_t = \dfrac{250 \pm \sqrt{(250)^2 - 4 \times 0.04 \times 80000}}{2 \times 0.04}$

$$= \frac{250 \pm 222.9}{0.08} = \frac{27.1}{0.08} = 338.75 \text{ ohms.} \quad \text{[Neglecting +ve sign]}$$

Additional resistance required in the shunt field circuit,

$$R = 338.75 - 250 = 88.75 \text{ ohms}$$

Additional resistance = **88.75 ohms. (Ans.)**

Example 5.34. *A 220 V shunt motor has an armature resistance of 0.5 Ω and takes an armature current of 40 A on a certain load. By how much must the main flux be reduced to raise the speed by 50 percent if the developed torque is constant ? Neglect saturation and armature reaction.*

(A.M.I.E. Summer, 2002)

Solution. *Given* : $V = 220$ volts ; $R_a = 0.5\ \Omega$, $I_{a1} = 40$ A ; $N_2 = 1.5\, N_1$,

Percentage reduction in main flux :

We know that, $\quad \dfrac{N_2}{N_1} = \dfrac{E_{b2}}{E_{b1}} \times \dfrac{\phi_1}{\phi_2} \quad$ ...(*i*)

Now $T \propto \phi I_a$, hence $T_1 \propto \phi_1 I_{a1}$ and $T_2 \propto \phi_2 I_{a2}$

Also, $\quad T_1 = T_2 \quad$ (Given)

$\therefore \quad \phi_1 I_{a1} = \phi_2 I_{a2}$

Now $\quad \phi_2 = x\phi_1$

$\therefore \quad I_{a2} = \dfrac{\phi_1 I_{a1}}{x\phi_1} = \dfrac{40}{x}$

Now, $\quad E_{b1} = V - I_{a1} R_a$

$$= 220 - 40 \times 0.5 = 200 \text{ V}$$

Putting this value of E_{b1} (= 200 V) in (*i*), we get

$$\frac{1.5 N_1}{N_1} = \frac{E_{b2}}{200} \times \frac{\phi_1}{x\phi_1} \quad \text{or} \quad E_{b2} = 300x$$

Also, $$E_{b2} = V - I_{a2} R_a = 220 - \frac{40}{x} \times 0.5 = 220 - \frac{20}{x}$$

or, $$300x = 220 - \frac{20}{x}$$

or, $$300x^2 = 220x - 20$$

or, $$300x^2 - 220x + 20 = 0$$

or, $$x^2 - 0.733x + 0.066 = 0$$

or, $$x = \frac{0.733 \pm \sqrt{0.733^2 - 4 \times 0.066}}{2}$$

$$= \frac{0.733 \pm 0.5228}{2} = 0.6279 \text{ or } 0.1051$$

As the expected value of ϕ_2 should be much less than the original value hence acceptible value of x taken is 0.6279.

Hence the *main flux be reduced to* **62.79%. (Ans.)**

☞**Example 5.35.** *A 250 V shunt motor, having an armature resistance of 0.2 ohm draws from the mains a current of 50 A on half full-load. The speed is to be increased to twice half full-load speed. If the torque of the motor is of constant magnitude, determine the percentage change in flux required.*

Solution. Supply voltage, $V = 250$ Volts

Armature resistance, $R_a = 0.2$ ohm

Armature current, $I_{a1} = 50$ A [Considering shunt field current to be negligible.]

$$\frac{\text{Speed, } N_2}{\text{Speed, } N_1} = 2$$

Percentage change in flux, $\frac{\phi_1 - \phi_2}{\phi_1}$:

We know that, $$\frac{N_2}{N_1} = \frac{E_{b2}}{E_{b1}} \times \frac{\phi_1}{\phi_2} \quad ...(i)$$

Now, $$E_{b1} = V - I_{a1} R_a = 250 - 50 \times 0.2 = 240 \text{ V}$$

and $$E_{b2} = V - I_{a2} R_a = 250 - 0.2 I_a$$

Substituting these in eqn. (*i*), we get

$$\frac{2}{1} = \frac{250 - 0.2 I_{a2}}{240} \times \frac{\phi_1}{\phi_2}$$

$$480 \times \frac{\phi_2}{\phi_1} = 250 - 0.2 I_{a2} \quad ...(ii)$$

Since the torque remains constant

$$\phi_1 I_{a1} = \phi_2 I_{a2} \quad (\because \; T \times \phi I_a)$$

$$\therefore \quad I_{a2} = \frac{\phi_1}{\phi_2} I_{a1} = \frac{\phi_1}{\phi_2} \times 50 \quad ...(iii)$$

Substituting (*iii*) in (*ii*), we get

$$480 \times \frac{\phi_2}{\phi_1} = 250 - 0.2 \times 50 \times \frac{\phi_1}{\phi_2}$$

$$480 \times \frac{\phi_2}{\phi_1} = 250 - 10 \times \frac{\phi_1}{\phi_2}$$

Let, $\frac{\phi_2}{\phi_1} = x$

Then $480x = 250 - \frac{10}{x}$

or, $480x^2 - 250x + 10 = 0$

or, $x = \frac{250 \pm \sqrt{(250)^2 - 4 \times 480 \times 10}}{2 \times 480}$

$= \frac{250 \pm 208.08}{960} = 0.4771$ [Neglecting – ve sign]

i.e., $x = \frac{\phi_2}{\phi_1} = 0.4771$

$\therefore$ $\phi_2 = 0.4771\ \phi_1$

$\therefore$ Percentage change in flux $= \frac{\phi_1 - \phi_2}{\phi_1} \times 100$

$= \left(\frac{\phi_1 - 0.4771\phi_1}{\phi_1}\right) \times 100 = 52.29\%$

Hence, *percentage change in flux* = **52.29%. (Ans.)**

☞**Example 5.36.** *A 220 V shunt motor develops a total torque of 100 N-m and takes 31 A at 600 r.p.m. The armature and shunt field resistances are 0.3 ohm and 220 ohms respectively. If the speed is to be increased to 800 r.p.m. determine the percentage reduction of the field and additional resistance to be inserted in the field circuit. Total torque developed at 800 r.p.m. is 70 Nm.*

Neglect armature reaction and assume that magnetization characteristic is a straight line.

Solution. Supply voltage, $V = 220$ Volts

Armature resistance, $R_a = 0.3$ ohm

Shunt field resistance, $R_{sh} = 220$ ohms

$T_1 = 100$ Nm at $N_1 = 600$ r.p.m.

$T_2 = 70$ Nm at $N_2 = 800$ r.p.m.

Load current at 600 r.p.m., $I_1 = 31$ A

% reduction in flux :

Additional resistance to be inserted :

Let, ϕ_1 = Flux at 600 r.p.m. (N_1)

ϕ_2 = Flux at 800 r.p.m. (N_2)

$$I_{sh1} = \frac{V}{R_{sh}} = \frac{220}{220} = 1 \text{ A}$$

$$I_{a1} = I_1 - I_{sh1} = 31 - 1 = 30 \text{ A}$$

$$E_{b1} = V - I_{a1}R_a = 220 - 30 \times 0.3 = 211 \text{ V}$$

Also, $T \propto \phi I_a$

$\therefore$ $\frac{T_2}{T_1} = \frac{\phi_2 I_{a2}}{\phi_1 I_{a1}}$

Let, $\frac{\phi_2}{\phi_1} = x$

Then, $$\frac{T_2}{T_1} = x \cdot \frac{I_{a2}}{I_{a1}}$$

and, $$I_{a2} = \frac{T_2}{T_1} \times \frac{I_{a1}}{x} = \frac{70}{100} \times \frac{30}{x} = \frac{21}{x} \quad ...(i)$$

Also $$\frac{N_2}{N_1} = \frac{E_{b2}}{E_{b1}} \times \frac{\phi_1}{\phi_2}$$

or, $$\frac{E_{b2}}{E_{b1}} = \frac{N_2}{N_1} \times \frac{\phi_2}{\phi_1}$$

or, $$E_{b2} = \frac{800}{600} \times x \times E_{b1}$$

or, $$E_{b2} = \frac{800}{600} \times x \times 211 = 281.33x \quad ...(ii)$$

But, $$E_{b2} = V - I_{a2} R_a$$

$$281.33x = 220 - \frac{21}{x} \times 0.3 = 220 - \frac{6.3}{x}$$

or, $$281.33x^2 - 220x + 6.3 = 0$$

$$x = \frac{220 \pm \sqrt{(220)^2 - 4 \times 281.33 \times 6.3}}{2 \times 281.33}$$

$$= \frac{220 \pm \sqrt{48400 - 7089.52}}{562.66}$$

$$= \frac{220 \pm 203.25}{562.66} = 0.72 \quad \text{[Neglecting – ve sign]}$$

or, $$x = \frac{\phi_2}{\phi_1} = 0.72$$

i.e., $$\phi_2 = 0.72\phi_1$$

Reduction in flux $= \phi_1 - \phi_2 = \phi_1 - 0.72\phi_1 = 0.28\phi_1$

% reduction in flux $= \dfrac{\phi_1 - \phi_2}{\phi_1} = \dfrac{0.28\phi_1}{\phi_1} \times 100 = 28\%.$

Hence, *% reduction in flux* = **28%. (Ans.)**

Since the magnetization characteristic is a straight line,

$$\frac{\phi_2}{\phi_1} = \frac{I_{sh2}}{I_{sh1}}$$

$$\therefore \quad \frac{I_{sh2}}{I_{sh1}} = 0.72$$

or, $$I_{sh2} = 0.72 \times I_{sh1} = 0.72 \times 1.0 = 0.72 \text{ A}$$

If R is the additional resistance to be inserted

$$I_{sh2} = \frac{V}{R_{sh} + R}$$

$$0.72 = \frac{220}{220 + R}$$

or, $$220 + R = \frac{220}{0.72}$$

or, $$R = 85.55 \text{ ohms}$$

Hence, *additional resistance to be inserted* = **85.55 ohms. (Ans.)**

Series Motors (Field Control)

☞ **Example 5.37.** *A 6-pole 230 V series motor takes 20 A when running at 750 r.p.m. with the field coils connected in series. If the series fields are connected in two parallel groups of three in series find the speed and current taken by the motor. The load torque remains unaltered. Assume that the flux is proportional to the current when the field coils are connected in series.*

Armature resistance = 0.2 ohm and series field resistance = 0.1 ohm.

Solution. Supply voltage, $V = 230$ Volts

$I_{a1} = 20$ A, $N_1 = 750$ r.p.m.

Armature resistance, $R_a = 0.2$ ohm

Series field resistance, $R_{se} = 0.1$ ohm

Speed and current :

Back e.m.f., $$E_{b1} = V - I_{a1}(R_a + R_{se}) = 230 - 20\,(0.2 + 0.1) = 224 \text{ V}$$

$$\phi_1 \propto I_{a1}$$

When the field coils are connected in two parallel groups,

Current through each parallel group $= \dfrac{I_{a2}}{2}$

where, I_{a2} is the *new intake current.*

Then, $$\phi_2 \propto \frac{I_{a2}}{2}$$

Since the *torque remains constant*

$$\phi_1 I_{a1} = \phi_2 I_{a2}$$

$$I_{a1}^2 = \frac{I_{a2}}{2} \times I_{a2}$$

$$I_{a2}^2 = 2I_{a1}^2 = 2 \times 20^2 = 800$$

$$I_{a2} = 28.28 \text{ A}$$

Hence, *current taken by the motor* = **28.28 A. (Ans.)**

Resistance of each parallel group

$$= \frac{R_{se}}{2} = \frac{0.1}{2} = 0.05 \text{ ohm}$$

Equivalent resistance of two parallel groups

$$= \frac{0.05 \times 0.05}{0.05 + 0.05} = 0.025 \text{ ohm}$$

Back e.m.f., $$E_{b2} = V - I_{a2}(R_a + 0.025)$$

$$= 230 - 28.28(0.2 + 0.025) = 223.6 \text{ V}$$

Let N_2 be the *speed* of the motor when the *field coils are connected in parallel*,

then,
$$\frac{N_2}{N_1} = \frac{E_{b2}}{E_{b1}} \times \frac{\phi_1}{\phi_2}$$

$$\frac{N_2}{750} = \frac{223.6}{224} \times \frac{20}{28.28/2} \qquad [\because \ \phi \propto I_a]$$

$$\therefore \quad N_2 = 750 \times \frac{223.6}{224} \times \frac{20 \times 2}{28.28} = 1058.9 \text{ r.p.m.}$$

Hence, *the speed of the motor* = **1058.9 r.p.m. (Ans.)**

Example 5.38. *A 4-pole series wound fan motor runs normally at 600 r.p.m. on a 250 V supply taking 20 A. The field coils are connected all in series. Estimate the speed and current taken by the motor if the coils are reconnected in two parallel groups of two in series. Assumed that the flux is directly proportional to the current and ignore losses.*

The load torque increases as square of the speed.

Solution. Number of poles, $p = 4$

Supply voltage, $V = 250$ Volts

$I_{a1} = 20$ A, $N_1 = 600$ r.p.m.

Speed and current :

When coils are connected in two parallel groups, current through each becomes $\frac{I_{a2}}{2}$, where I_{a2} is the new armature current.

Hence
$$\phi_2 \propto \frac{I_{a2}}{2}$$

Now
$$T_a \propto \phi I_a$$
$$\propto N^2 \qquad \textit{(Given)}$$

$$\therefore \quad \phi_1 I_{a1} \propto N_1^2$$

and
$$\phi_2 I_{a2} \propto N_2^2$$

$$\therefore \quad \left(\frac{N_2}{N_1}\right)^2 = \frac{\phi_2 I_{a2}}{\phi_1 I_{a1}} \qquad ...(i)$$

Since losses are negligible, field coil resistance as well armature resistance are negligible. It means armature and series field voltage drops are negligible. Hence, back e.m.f., in each case equals the supply voltage.

$$\therefore \quad \frac{N_2}{N_1} = \frac{E_{b2}}{E_{b1}} \times \frac{\phi_1}{\phi_2} \text{ becomes}$$

$$\frac{N_2}{N_1} = \frac{\phi_1}{\phi_2} \qquad ...(ii)$$

Putting this value in (*i*), we get

$$\left(\frac{\phi_1}{\phi_2}\right)^2 = \frac{\phi_2 I_{a2}}{\phi_1 I_{a1}}$$

$$\therefore \quad \frac{I_{a2}}{I_{a1}} = \left(\frac{\phi_1}{\phi_2}\right)^3$$

Now, $\phi_1 \propto 20$

and, $\phi_2 \propto \dfrac{I_a}{2}$

$$\therefore \quad \frac{I_{a2}}{20} = \left(\frac{20}{I_{a2}/2}\right)^3$$

$$\frac{I_{a2}}{20} = \left(\frac{40}{I_{a2}}\right)^3$$

$$I_{a2}^{\ 4} = 20 \times 40^3$$

$$\therefore \quad \mathbf{I_{a2} = 33.64. \ (Ans.)}$$

From (*ii*) above, we get

$$\frac{N_2}{N_a} = \frac{\phi_1}{\phi_2} = \frac{I_{a1}}{I_{a2}/2} = \frac{2I_{a1}}{I_{a2}}$$

or, $$\frac{N_2}{600} = \frac{2 \times 20}{33.64}$$

$$\therefore \quad \mathbf{N_2 = 713.4 \ r.p.m. \quad (Ans.)}$$

☞**Example 5.39.** *A 500 V series motor has an armature resistance of 0.4 ohm and series field resistance of 0.3 ohm. It takes a current of 100 A at a speed of 600 r.p.m. Find the speed of the motor if a divertor of resistance 0.6 ohm is connected across the field, the load torque being kept constant.*

Neglect armature reaction and assume that flux is proportional to the current.

Solution. Supply voltage, $V = 500$ Volts

Armature resistance, $R_a = 0.4$ ohm

Series field resistance, $R_{se} = 0.3$ ohm

Divertor resistance, $R_{div.} = 0.6$ ohm

$I_{a1} = 100$ A, $N_1 = 600$ r.p.m.

Speed, N_2 :

Back e.m.f., $E_{b1} = V - I_{a1}(R_a + R_{se})$

$= 500 - 100(0.4 + 0.3) = 430$ V.

Let I_{a2} be the current taken and ϕ_2 be the flux produced when a divertor is connected across the series field (Fig. 5.29)

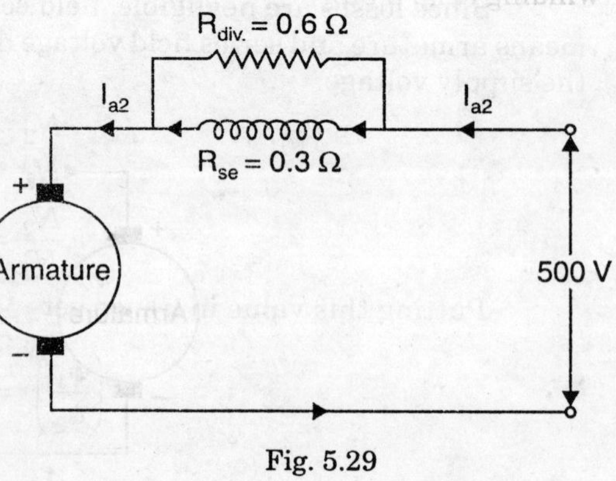

Fig. 5.29

Since the *torque remains constant,*

$$\therefore \quad \phi_1 I_{a1} = \phi_2 I_{a2} \quad \ldots(i)$$

But $\phi \propto$ current through series field,

$$\therefore \quad \phi_1 \propto I_{a1}$$

Current through the series field when a divertor is connected

$$= I_{a2} \times \frac{R_{div.}}{R_{div.} + R_{se}}$$

$$= I_{a2} \times \frac{0.6}{0.6 + 0.3} = 0.667 I_{a2}$$

Flux in this case $\phi_2 \propto 0.667 I_{a2}$

Substituting in (*i*), we get

$$I_{a1}^2 = 0.667 I_{a2}^2$$

$$I_{a2}^2 = \frac{I_{a1}^2}{0.667}$$

$$\therefore \quad I_{a2} = \frac{I_{a1}}{\sqrt{0.667}} = \frac{100}{\sqrt{0.667}} = 122.44 \text{ A}$$

Series field current, $I_{se} = 0.667 I_{a2}$

$= 0.667 \times 122.44 = 81.64$ A

Back e.m.f., $E_{b2} = V - I_{a2}R_a - I_{se}R_{se}$

$= 500 - 122.44 \times 0.4 - 81.64 \times 0.3$

$= 500 - 48.97 - 24.49 = 426.54$ V

Using the relation, $\dfrac{N_2}{N_1} = \dfrac{E_{b2}}{E_{b1}} \times \dfrac{\phi_1}{\phi_2}$

$$\frac{N_2}{600} = \frac{426.54}{430} \times \frac{100}{81.64}$$

$$\therefore \quad N_2 = 729 \text{ r.p.m.}$$

Hence, *speed of the motor* = **729 r.p.m. (Ans.)**

☞**Example 5.40.** *A D.C. series motor drives a load the torque of which varies as square of the speed. The motor takes a current of 20 A when the speed is 800 r.p.m. Calculate the speed and current when the motor field winding is shunted by a divertor of the same resistance as that of the field winding.*

Neglect all motor losses and assume that the magnetic circuit is unsaturated.

Solution. $I_{a1} = 20$ A, $N_1 = 800$ r.p.m.

$\mathbf{I_{a2} = ?}$, $\mathbf{N_2 = ?}$

When the field winding is shunted by a divertor of equal resistance, then current through either is half the armature current. If I_{a2} is the new armature current, then $\dfrac{I_{a2}}{2}$ passes through the winding (Fig. 5.30).

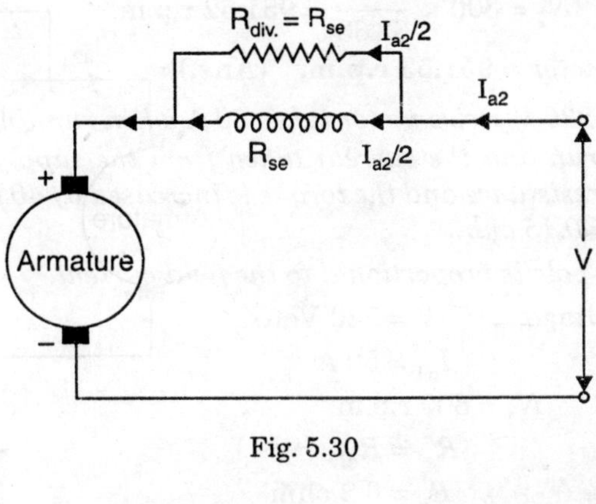

Fig. 5.30

$\therefore \qquad \phi_2 = \dfrac{I_{a2}}{2}$

Now, $\qquad T_1 \propto \phi_1 I_{a1} \propto N_1^2$ *(Given)*

and, $\qquad T_2 \propto \phi_2 I_{a2} \propto N_2^2$ *(Given)*

From (*i*) and (*ii*), we get

$$\left(\frac{N_2}{N_1}\right)^2 = \frac{\phi_2 I_{a2}}{\phi_1 I_{a1}} \qquad ...(i)$$

Because all losses are negligible, hence the armature and series field resistances are negligible. This means that back e.m.f. in both cases is the same as the applied voltage.

$$\therefore \qquad \frac{N_2}{N_1} = \frac{E_{b2}}{E_{b1}} \times \frac{\phi_1}{\phi_2} \text{ becomes}$$

$$\frac{N_2}{N_1} = \frac{\phi_1}{\phi_2} \qquad ...(ii)$$

$[\because \; E_{b2} = E_{b1} = \text{applied voltage}]$

Putting this value in (*i*) above, we get

$$\left(\frac{\phi_1}{\phi_2}\right)^2 = \frac{\phi_2 I_{a2}}{\phi_1 I_{a1}}$$

or, $$\frac{I_{a2}}{I_{a1}} = \left(\frac{\phi_1}{\phi_2}\right)^3$$

$$\frac{I_{a2}}{20} = \left(\frac{20}{I_{a2}/2}\right)^3 \qquad [\because \; \phi_2 \propto I_a/2]$$

$$\therefore \qquad I_{a2}^4 = 20 \times 40^3$$

or, $$I_{a2} = 33.63 \text{ A}$$

Hence, *current* = **33.63 A. (Ans.)**

From (*ii*), we get $\qquad \dfrac{N_2}{800} = \dfrac{20}{(33.63)/2}$

$$N_2 = 800 \times \frac{40}{33.63} = 951.52 \text{ r.p.m.}$$

Hence, *speed of the motor* = **951.53 r.p.m. (Ans.)**

☞**Example 5.41.** *A 230 V series motor takes 50 A when running at 800 r.p.m. Calculate the speed at which motor will run and the current taken from the supply if the field is shunted by a resistance equal to the field resistance and the torque is increased by 40 per cent. Armature resistance = 0.2 ohm, field resistance = 0.15 ohm.*

Assume that flux per pole is proportional to the field current.

Solution. Supply voltage, $\qquad V = 230$ Volts

Current, $\qquad I_{a1} = 50$ A

Speed $\qquad N_1 = 800$ r.p.m.

$$R_{se} = R_{div.}$$

Armature resistance, $\qquad R_a = 0.2$ ohm

Field resistance, $R_{se} = 0.15$ ohm

Speed N_2 :

In a series motor, prior to magnetic saturation

$$T \propto \phi I_a \propto I_a^{\,2}$$

$$\therefore \quad T_1 \propto I_{a1}^{\,2} \propto 50^2 \quad ...(i)$$

If I_{a2} be the armature or motor current in the second case when the divertor is used, then only $\frac{I_{a2}}{2}$ passes through the series field winding.

$$\therefore \quad \phi_2 \propto \frac{I_{a2}}{2}$$

$$\therefore \quad T_2 \propto \phi_2 I_{a2}$$

$$\propto \frac{I_{a2}}{2} \times I_{a2} = \frac{I_{a2}^{\,2}}{2} \quad ...(ii)$$

From (*i*) and (*ii*), we get $\frac{T_2}{T_1} = \frac{I_{a2}^{\,2}}{2 \times 50^2}$

Also, $\frac{T_2}{T_1} = 1.4$ [Given]

$$\therefore \quad 1.4 = \frac{I_{a2}^{\,2}}{2 \times 50^2}$$

$$\therefore \quad I_{a2} = (1.4 \times 2 \times 50^2)^{1/2} = 83.67 \text{ A}$$

Now, back e.m.f., $E_{b1} = V - I_{a1}(R_a + R_{se})$

$$= 230 - 50(0.2 + 0.15) = 212.5 \text{ V}$$

Combined resistance of series field winding and divertor

$$= \frac{0.15}{2} = 0.075 \text{ ohm}$$

$\therefore$ Back e.m.f., $E_{b2} = 230 - 83.67(0.2 + 0.075) = 207$ V (app.)

Using the relation, $\frac{N_2}{N_1} = \frac{E_{b2}}{E_{a1}} \times \frac{\phi_1}{\phi_2}$

$$\frac{N_2}{800} = \frac{207}{212.5} \times \frac{50}{\frac{83.67}{2}} \quad [\because \ \phi \propto I_{se}]$$

$$N_2 = 800 \times \frac{207}{212.5} \times \frac{50 \times 2}{83.67} = 931.4 \text{ r.p.m.}$$

Hence, *speed of the motor* = **931.4 r.p.m. (Ans.)**

☞**Example 5.42.** *The shaft torque, speed and efficiency of a 500 V series motor on full-load are 1800 Nm, 800 r.p.m. and 90 per cent respectively. When the speed is increased to 1200 r.p.m. by connecting a resistance across the series field it develops a shaft torque of 900 Nm and has an efficiency of 80 per cent. Find the value of resistance connected across the series field.*

Armature resistance = *0.1 ohm*

Series field resistance = *0.05 ohm.*

Solution. Full-load torque, $T_1 = 1800$ Nm

Full-load speed, $N_1 = 800$ r.p.m.

Full-load efficiency, $\eta_1 = 90$ per cent

$N_2 = 1200$ r.p.m., $T_2 = 900$ Nm, $\eta_2 = 80$ per cent

Resistance to be connected across series field, R_{div} :

At $N_1 = 800$ r.p.m. :

Full-load output $= \dfrac{2\pi N_1 T_1}{60} = \dfrac{2\pi \times 800 \times 1800}{60} = \mathbf{150796.45\ W}$

Full load input $= \dfrac{150796.45}{0.9} = \mathbf{167551.61\ W}$

Full-load input current, $I_{a1} = \dfrac{167551.61}{500} = 335.1$ A

Back e.m.f., $E_{b1} = V - I_{a1}(R_a + R_{se})$

$= 500 - 335.1(0.1 + 0.05) = 449.7$ V

At $N_2 = 1200$ r.p.m. :

Output $= \dfrac{2\pi N_2 T_2}{60} = \dfrac{2\pi \times 1200 \times 900}{60}$

Input $= \dfrac{2\pi \times 1200 \times 900}{60 \times 0.8} = 141371.67$ W

Input current, $I_{a2} = \dfrac{141371.67}{500} = 282.7$ A

Let I_{se} be the current through the series field winding when a resistance $R_{div.}$ is connected across it.

Back e.m.f., $E_{b2} = V - I_{a2} R_a - I_{se} R_{se}$

$= 500 - 282.7 \times 0.1 - 0.05 I_{se} = 471.73 - 0.05 I_{se}$

Flux in the first case, $\phi \propto I_{a1}$

Flux in the second case, $\phi_2 \propto I_{se}$

We know that, $\dfrac{N_2}{N_1} = \dfrac{E_{b2}}{E_{b1}} \times \dfrac{\phi_1}{\phi_2}$

or, $$\frac{1200}{800} = \frac{471.73 - 0.05 I_{se}}{449.7} \times \frac{335.1}{I_{se}}$$

or, $$\frac{471.73 - 0.05 I_{se}}{I_{se}} = \frac{1200}{800} \times \frac{449.7}{335.1} = 2.013$$

$\therefore$ $471.73 - 0.05 I_{se} = 2.013 I_{se}$

or, $I_{se} = 228.66$ A

But, $$I_{se} = \left(\frac{R_{div.}}{R_{se} + R_{div.}}\right) I_{a2}$$

$$228.66 = \left(\frac{R_{div.}}{0.05 + R_{div.}}\right) 282.7$$

$$\frac{R_{div.}}{0.05 + R_{div.}} = \frac{228.66}{282.7} = 0.8088$$

$$R_{div.} = (0.05 + R_{div.})\ 0.8088 = 0.0404 + 0.8088\ R_{div.}$$

$$R_{div.} = \frac{0.0404}{0.1912} = 0.2112 \text{ ohm.}$$

Hence, *resistance connected across the series field*

$$= \mathbf{0.2112\ ohm.}\quad \textbf{(Ans.)}$$

☞**Example 5.43. *(Armature divertor).*** *A 500 V series motor takes 70 A, when running at a speed of 800 r.p.m. The armature and series field resistances are 0.8 ohm and 0.6 ohm respectively. If a resistance of 10 ohms is connected across the armature and torque remains constant determine the speed of the motor.*

Assume straight line magnetisation.

Solution. Supply voltage, $V = 500$ Volts

Armature resistance, $R_a = 0.8$ ohm

Series field resistance, $R_{se} = 0.6$ ohm

$I_1 = I_{a1} = 70$ A, $N_1 = 800$ r.p.m.

Resistance of armature divertor, $R_{div.} = 10$ ohms

Speed of motor, N_2 :

Back e.m.f., $E_{b1} = V - I_{a1}(R_a + R_{se})$

$= 500 - 70(0.8 + 0.6)$ $\quad [\because\ I_1 = I_{a1} = 70 \text{ A}]$

$= 402$ V

Let I_2 be the input current with an armature divertor of 10 ohms and I_{a2} be the armature current (Fig. 5.31).

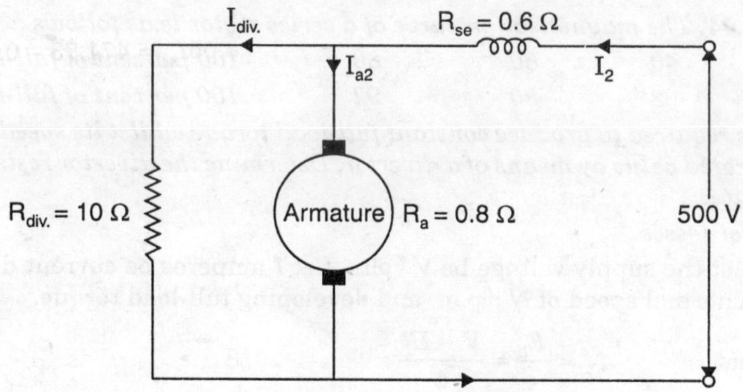

Fig. 5.31

Potential difference across the divertor

$$= V - I_2 R_{se} = 500 - 0.6I_2$$

Divertor current, $$I_{div.} = \frac{\text{P.D. across divertor}}{R_{div.}} = \frac{500 - 0.6I_2}{10} = 50 - 0.06I_2$$

$$I_{a2} = I_2 - I_{div.}$$

$$= I_2 - (50 - 0.06I_2) = 1.06I_2 - 50$$

Since *torque is the same in both cases,*

$$\therefore \quad \phi_1 I_1 = \phi_2 I_2$$

But, $\phi_1 \propto I_1$

and, $\phi_2 \propto I_{a2}$

$$I_1^2 = I_{a2} I_2$$

$$70^2 = I_2(1.06I_2 - 50)$$

or, $1.06I_2^2 - 50I_2 - 4900 = 0$

$$I_2 = \frac{50 \pm \sqrt{50^2 + 4 \times 1.06 \times 4900}}{2 \times 1.06}$$

$$= \frac{50 \pm 152.56}{2.12} = 95.55 \text{ A} \qquad \text{[Neglecting – ve sign]}$$

$$I_{a2} = 1.06 \times 95.55 - 50 = 51.28 \text{ A}$$

Back e.m.f., $E_{b2} = V - I_2 R_{se} - I_{a2} R_a$

$$= 500 - 95.55 \times 0.6 - 51.28 \times 0.8$$

$$= 500 - 57.33 - 41.024 = 401.6 \text{ V}$$

Using the relation, $\dfrac{N_2}{N_1} = \dfrac{E_{b2}}{E_{b1}} \times \dfrac{\phi_1}{\phi_2}$

$$\frac{N_2}{800} = \frac{401.6}{402} \times \frac{I_1}{I_2}$$

$\therefore$ $N_2 = \dfrac{401.6}{402} \times \dfrac{70}{95.55} \times 800 = 585.5$ r.p.m.

Speed of motor with armature divertor of 10 ohms

= 585.5 r.p.m. (Ans.)

Example 5.44. *The magnetisation curve of a series motor is as follows :*

Current	*40*	*60*	*80*	*100 per cent of full-load value*
Flux/pole	*62*	*80*	*92*	*100 per cent of full-load value*

The motor is required to produce constant full-load torque whilst its speed is raised by 20 per cent above the full rated value by means of a divertor. Determine the divertor resistance as a percentage of field resistance.

Neglect motor losses.

Solution. Let the supply voltage be V volts. Let I amperes be current drawn by the motor when running at a normal speed of N r.p.m. and developing full-load torque.

We know that, $N \propto \dfrac{E_b}{\phi} = \dfrac{V - IR}{\phi}$

$$\propto \frac{V}{\phi} \qquad [\because \text{ Losses are negligible}]$$

After connecting a divertor, the speed $N_2 = 1.2$ N

Now $\dfrac{N_2}{N_1} = \dfrac{\frac{V}{\phi_2}}{\frac{V}{\phi_1}}$ or $\dfrac{1.2N}{N} = \dfrac{\phi_1}{\phi_2}$

or, $\dfrac{\phi_1}{\phi_2} = 1.2$ or $\phi_2 = \dfrac{\phi_1}{1.2} = 83.3\%$ of full-load value of flux.

From saturation curve (Fig. 5.32) the field current corresponding to 83.3% of full-load value of flux is 65% of full-load value of field current.

i.e., $$I_{se2} = 0.65 I_{se1} = 0.65 I_{a1}$$

Since load torque is constant

$$\therefore \quad \phi_2 I_{a2} = \phi_1 I_{a1}$$

$$\therefore \quad I_{a2} = \frac{\phi_1 I_{a1}}{\phi_2} = \frac{\phi_1 I_{a1}}{0.833\,\phi_1} = 1.2 I_{a1}$$

Also, $$I_{se2} = \frac{R_{div.}}{R_{div.} + R_{se}} \times I_{a2}$$

or, $$0.65 I_{a1} = 1.2 I_{a1} \times \frac{R_{div.}}{R_{div.} + R_{se}}$$

or, $$R_{div.} = 0.54\,(R_{div.} + R_{se})$$

or, $$0.46 R_{div.} = 0.54 R_{se}$$

or, $$R_{div.} = 1.17 R_{se}.$$

Hence, *resistance of divertor is 117 per cent of series field resistance.* **(Ans.)**

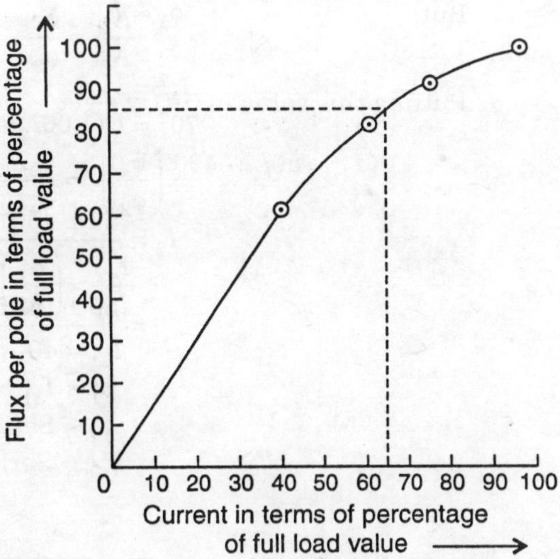

Fig. 5.32

Example 5.45. *A 200 V D.C. series motor takes a current of 40 A at 800 r.p.m. Determine the speed and current when a divertor of resistance equal to double that of series field winding is connected across it. The load torque varies as the cube of the speed. Neglect all losses.*

Assume the flux is proportional to current.

Solution. Supply voltage, $V = 220$ Volts ;

$$I_{a1} = 40\text{ A}; \quad N_1 = 800\text{ r.p.m.}; \quad R_{div.} = 2R_{se}$$

$\mathbf{I_{a2}}$: ; $\mathbf{N_2}$:

Let ϕ_1, I_{a1} and T_1 be the flux, current and torque at a speed

$$N_1 = 800\text{ r.p.m.}$$

and ϕ_2, I_{a2} and T_2 be the flux, current and torque at a speed when a resistance is connected across the series field.

$$T_1 \propto \phi_1 I_{a1} \qquad ...(i)$$

Also, $$T_1 \propto N_1^3 \quad (Given) \qquad ...(ii)$$

From (*i*) and (*ii*), we get $\phi_1 I_{a1} \propto N_1^3$

Similarly, $$\phi_2 I_a \propto N_2^3$$

$$\frac{N_2^3}{N_1^3} = \frac{\phi_2 I_{a2}}{\phi_1 I_{a1}}$$

$$\left(\frac{N_2}{N_1}\right)^3 = \frac{\phi_2 I_{a2}}{\phi_1 I_{a1}} \qquad ...(iii)$$

Also, $$\frac{N_2}{N_1} = \frac{E_{b2}}{E_{b1}} \times \frac{\phi_1}{\phi_2} \qquad ...(iv)$$

Since losses are negligible

$$E_{b2} = E_{b1} = \text{Supply voltage}$$

$$\therefore \qquad \frac{N_2}{N_1} = \frac{\phi_1}{\phi_2} \qquad \text{...(v)}$$

Putting this value in (*iii*), we get

$$\left(\frac{\phi_1}{\phi_2}\right)^3 = \frac{\phi_2 I_{a2}}{\phi_1 I_{a1}}$$

i.e.,
$$\frac{I_{a2}}{I_{a1}} = \left(\frac{\phi_1}{\phi_2}\right)^4 \qquad \text{...(vi)}$$

$$I_{a1} = 40 \text{ A}$$

$$\phi_1 \propto I_{a1} \propto 40$$

$$\phi_2 \propto \text{Series full current}$$

Let, R_{se} = Series field resistance

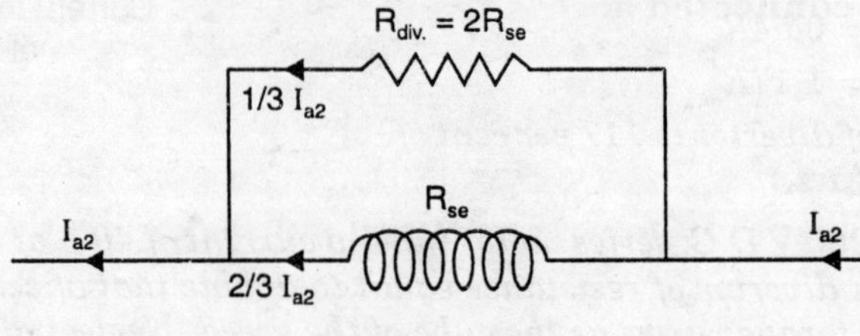

Fig. 5.33

Then, resistance of the divertor= $2R_{se}$ (*Given*)

Current through the series field winding (Fig. 5.33)

$$= I_{a2} \times \frac{2R_{se}}{R_{se} + 2R_{se}} = \frac{2}{3} I_{a2}$$

$$\therefore \qquad \phi_2 \propto \frac{2}{3} I_{a2}$$

$$\frac{\phi_1}{\phi_2} = \frac{40}{\frac{2}{3} I_{a2}} = \frac{60}{I_{a2}}$$

Putting this value in (*vi*), we get

$$\frac{I_{a2}}{40} = \left(\frac{60}{I_{a2}}\right)^4$$

$$I_{a2}{}^5 = 60^4 \times 40$$

i.e.,
$$\mathbf{I_{a2} = 55.33\ A}$$

$$\frac{\phi_1}{\phi_2} = \frac{60}{I_{a2}} = \frac{60}{55.33} = 1.084$$

Putting this value in (*v*), we get

$$\frac{N_2}{800} = \frac{\phi_1}{\phi_2} = 1.084$$

$$\therefore \qquad \mathbf{N_2 = 867.2\ r.p.m.}$$

Hence, when a divertor of resistance $2R_{se}$ is connected across the series field,

Speed = 867.2 r.p.m. (Ans.)

Current = 55.33 A. (Ans.)

Example 5.46. *The shaft torque, speed and efficiency of a 500 V series motor on full-load are 1800 N-m, 800 r.p.m. and 90 per cent respectively. When the speed is increased to 1200 r.p.m. by connecting a resistance across the series field it develops a shaft torque of 900 N-m and has an efficiency of 80 per cent. Find the value of resistance connected across the series field.*

Armature resistance = 0.1 ohm

Series field resistance = 0.05 ohm.

Solution. Full-load torque, $T_1 = 1800$ N-m

Full-load speed, $N_1 = 800$ r.p.m.

Full-load efficiency, $\eta_1 = 90$ per cent

$N_2 = 1200$ r.p.m., $T_2 = 900$ N-m, $\eta_2 = 80$ per cent

Resistance to be connected across series field, $R_{div.}$:

At $N_1 = 800$ r.p.m.

Full-load output $= \dfrac{2\pi N_1 T_1}{60} = \dfrac{2\pi \times 800 \times 1800}{60} = 150796.45$ W

Full-load input $= \dfrac{150796.45}{0.9} = 167551.61$ W

Full-load input current, $I_{a1} = \dfrac{167551.61}{500} = 335.1$ A

Back e.m.f., $E_{b1} = V - I_{a1}(R_a + R_{se})$

$= 500 - 335.1(0.1 + 0.05) = 449.7$ V

At $N_2 = 1200$ r.p.m.

Output $= \dfrac{2\pi N_2 T_2}{60} = \dfrac{2\pi \times 1200 \times 900}{60}$

Input $= \dfrac{2\pi \times 1200 \times 900}{60 \times 0.8} = 141371.67$ W

Input current, $I_{a2} = \dfrac{141371.67}{500} = 282.7$ A

Let I_{se} be the current through the series field winding when a resistance $R_{div.}$ is connected across it.

Back e.m.f., $E_{b2} = V - I_{a2}R_a - I_{se}R_{se}$

$= 500 - 282.7 \times 0.1 - 0.05I_{se} = 471.73 - 0.05I_{se}$

Flux in the first case, $\phi_1 \propto I_{a1}$

Flux in the second case, $\phi_2 \propto I_{se}$

We know that, $\dfrac{N_2}{N_1} = \dfrac{E_{b2}}{E_{b1}} \times \dfrac{\phi_1}{\phi_2}$

or, $\dfrac{1200}{800} = \dfrac{471.73 - 0.05I_{se}}{449.7} \times \dfrac{335.1}{I_{se}}$

or, $\dfrac{471.73 - 0.05I_{se}}{I_{se}} = \dfrac{1200}{800} \times \dfrac{449.7}{335.1} = 2.013$

$\therefore \quad 471.73 - 0.05\, I_{se} = 2.013\, I_{se}$

or, $\quad I_{se} = 228.66$ A

But, $\quad I_{se} = \left(\dfrac{R_{div.}}{R_{se} + R_{div.}}\right) I_{a2}$

$$228.66 = \left(\frac{R_{div.}}{0.05 + R_{div.}}\right) \times 282.7$$

$$\frac{R_{div.}}{0.05 + R_{div.}} = \frac{228.66}{282.7} = 0.8088$$

$$R_{div.} = (0.05 + R_{div.})\, 0.8088$$
$$= 0.0404 + 0.8088\, R_{div.}$$

$$R_{div.} = \frac{0.0404}{0.1912} = 0.2112 \text{ ohm.}$$

Hence, *resistance connected across the series field*

= 0.2112 ohm. (Ans.)

Example 5.47. *A D.C. series motor drives a load, the torque of which varies as the square of the speed. The motor takes a current of 15 A when the speed is 600 r.p.m. Calculate the speed and current when the motor field winding is shunted by a divertor of the same resistance as that of field winding. Mention the assumptions made if any.*

Solution. $\quad I_{a1} = (I_1) = 15$ A ; $N_1 = 600$ r.p.m.

Speed and current ; N_2, I_{a2} :

Since $\quad T \propto N^2$ ...[*Given*]

$$\therefore \quad \frac{T_2}{T_1} = \left(\frac{N_2}{N_1}\right)^2 \quad ...(i)$$

Also, $\quad T \propto \phi I_a$

$$\therefore \quad \frac{T_2}{T_1} = \frac{\phi_2 I_{a2}}{\phi_1 I_{a1}} = \frac{I_{se2} I_{a2}}{I_{se} \cdot I_{a1}}, \text{ assuming unsaturated magnetic field.}$$

Since in normal conditions the same current flows through series field winding and armature so $I_{se1} = I_{a1}$, but when the field winding is shunted by a divertor of the same resistance, the armature current will equally divide between series field and divertor so $I_{se2} = \dfrac{1}{2} I_{a2}$

and, $$\frac{T_2}{T_1} = \frac{I_{a2}^2}{2 I_{a1}^2} \quad ...(ii)$$

Comparing (*i*) and (*ii*), we have

$$\frac{N_2}{N_1} = \frac{I_{a2}}{\sqrt{2}\, I_{a1}} \quad ...(iii)$$

Neglecting armature and series field voltage drops, the back e.m.f. is nearly equal to applied voltage and so,

$$\frac{N_2}{N_1} = \frac{E_{b2}}{E_{b1}} \times \frac{\phi_1}{\phi_2}$$
$$= \frac{V}{V} \times \frac{I_{se1}}{I_{se2}} = \frac{I_{a1}}{I_{a2}/2} = \frac{2 I_{a1}}{I_{a2}} \quad ...(iv)$$

$$\frac{I_{a2}}{\sqrt{2}\, I_{a1}} = \frac{2I_{a1}}{I_{a2}}$$

$\therefore$ $$\mathbf{I_{a2}} = (2\sqrt{2})^{1/2} I_{a1} = (2\sqrt{2})^{1/2} \times 15 = \mathbf{25.23\ A.} \quad \textbf{(Ans.)}$$

From eqn. (*iv*), $$\mathbf{N_2} = N_1 \times \frac{2I_{a1}}{I_{a2}}$$

$$= 600 \times \frac{2 \times 15}{25.23} = \mathbf{713.4\ r.p.m.\ (Ans.)}$$

Example 5.48. *A series motor runs at 1,100 r.p.m. taking 90 A with 110 V. What resistance would be connected in parallel with field circuit to get 1,500 r.p.m. speed for delivering same load torque ? Armature resistance is 0.08 Ω and series field resistance is 0.06 Ω. Assume magnetic circuit as unsaturated.*

Solution. *Given* : N_1 = 1100 r.p.m. ; I_{a1} (= I_1) = 90 A ; V = 110 volts ; N_2 = 1500 r.p.m. ; R_a = 0.08 Ω ; R_{se} = 0.06 Ω.

Field divertor resistance, $\mathbf{R_{div.}}$:

Under normal conditions, back e.m.f.,

$$E_{b1} = V - I_{a1}(R_a + R_{se})$$
$$= 110 - 90(0.08 + 0.06) = 97.4 \text{ V}$$

Let, K = Ratio of divertor resistance ($R_{div.}$) to series field resistance (R_{se}) to give a speed of 1500 r.p.m., and

I_{a2} = New armature current.

Since, $$T_2 = T_1 \qquad \text{[Given]}$$

$\therefore$ $$\phi_2 I_{a2} = \phi_1 I_{a1}$$

or, $$I_{se2} \,.\, I_{a2} = I_{se1} \,.\, I_{a1}$$

[$\because$ Magnetic circuit is assumed to be unsaturated]

or, $$\frac{K}{1+K} I_{a2} \,.\, I_{a2} = I_{a1}^{\;2}$$

or, $$I_{a2} = \sqrt{\frac{1+K}{K}}\, I_{a1} = \sqrt{\left(1+\frac{1}{K}\right)} \times 90 \text{ A}$$

and, $$I_{se2} = \frac{K}{1+K} I_{a2} = \frac{K}{1+K} \times \sqrt{\frac{1+K}{K}}\, I_{a1} = \sqrt{\frac{K}{1+K}} \times 90 \text{ A}$$

Back e.m.f., $$E_{b2} = V - I_{a2}R_a - I_{se2}R_{se}$$

$$= 110 - \sqrt{\frac{K+1}{K}} \times 90 \times 0.08 - \sqrt{\frac{K}{1+K}} \times 90 \times 0.06$$

Also, $$\frac{N_2}{N_1} = \frac{E_{b2}}{E_{b1}} \times \frac{\phi_1}{\phi_2} = \frac{E_{b2}}{E_{b1}} \times \frac{I_{se1}}{I_{se2}}$$

$$\frac{1500}{1100} = \frac{110 - 7.2\sqrt{\dfrac{1+K}{K}} - 5.4\sqrt{\dfrac{K}{1+K}}}{97.4} \times \frac{90}{90\sqrt{\dfrac{K}{1+K}}}$$

Substituting $\sqrt{\dfrac{1+K}{K}} = x$ in the above equation, we have

$$\frac{1500}{1100} = \frac{\left[110 - 7.2x - \dfrac{5.4}{x}\right] \times x}{97.4}$$

or, $$\frac{1500}{1100} \times 97.4 = 110x - 7.2x^2 - 5.4$$

or, $$7.2x^2 - 110x + 138.22 = 0$$

or, $$x = \frac{110 \pm \sqrt{(110)^2 - 4 \times 7.2 \times 138.22}}{2 \times 7.2}$$

$$= \frac{110 \pm 90.11}{14.4} = 13.9 \text{ or } 1.38$$

$\therefore$ $$\sqrt{\frac{1+K}{K}} = 1.38, \text{ since the valve 13.9 is not practicable.}$$

or, $$\frac{1+K}{K} = 1.9 \text{ or } 1 + K = 1.9K$$

or, $$0.9K = 1$$

or, $$K = \frac{1}{0.9} = \frac{R_{div.}}{R_{se}} = 1.111$$

$\therefore$ Field divertor resistance, $R_{div.} = 1.111 \times 0.06 =$ **0.0667 Ω. (Ans.)**

Example 5.49. *The armature resistance of a 25 HP, 250 V series motor is 0.1 Ω, the brush drop is 3 V and the resistance of the series field is 0.05 Ω. When the series motor takes 85 A, the speed is 600 r.p.m. Calculate (i) the speed when the current is 100 A (ii) the speed when the current is 40 A (iii) the speed in (i) and (ii) if a divertor of 0.05 Ω is used. Neglect armature reaction and use linear-portion of saturation curve.*

Solution. *Given :* $V = 250$ volts ; $R_a = 0.1$ Ω ; brush drop = 3 V ; $R_{se} = 0.05$ Ω ;

$I_{a1} (= I_1) = 85$ A ; $N_1 = 600$ r.p.m.

Back e.m.f., $$E_{b1} = V - I_{a1}(R_a + R_{se}) - \text{brush drop}$$

$$= 250 - 85(0.1 + 0.05) - 3 = 234.25 \text{ V}$$

(i) **Speed when current is 100 A, N_2 :**

In this case, $I_{a2} (= I_2) = 100$ A

Back e.m.f., $$E_{b2} = V - I_{a2}(R_a + R_{se}) - \text{brush drop}$$

$$= 250 - 100(0.1 + 0.05) - 3 = 232 \text{ V}$$

Also, $$\frac{N_2}{N_1} = \frac{E_{b2}}{E_{b1}} \times \frac{\phi_1}{\phi_2}$$

or, $$\frac{N_2}{600} = \frac{232}{234.25} \times \frac{85}{100} \qquad [\because \ \phi \propto I_a]$$

$\therefore$ $N_2 =$ **505 r.p.m. (Ans.)**

(ii) **Speed when current is 40 A, N_3 :**

Here, $I_{a3} (= I_3) = 40$ A

Back e.m.f., $$E_{b3} = V - I_{a3}(R_a + R_{se}) - \text{brush drop}$$

$$= 250 - 40(0.1 + 0.05) - 3 = 241 \text{ V}$$

Also, $\frac{N_3}{N_1} = \frac{E_{b3}}{E_{b1}} \times \frac{\phi_1}{\phi_3}$

or, $\frac{N_3}{600} = \frac{241}{234.25} \times \frac{85}{40}$ $[\because \phi \propto I_a]$

$\therefore$ $\mathbf{N_3 = 1312}$ **r.p.m. (Ans.)**

(*iii*) The speed in (*i*) and (*ii*) when a divertor of 0.05 Ω is used ; $\mathbf{N_4, N_5}$:

When line current is 100 A (= I_4) and a divertor of 0.05 Ω is used :

Series field current, $I_{se4} = I_4 \times \frac{R_{div.}}{R_{se} + R_{div.}}$

$$= 100 \times \frac{0.05}{0.05 + 0.05} = 50 \text{ A}$$

Back e.m.f., $E_{b4} = V - I_{a4}R_a - I_{se4}R_{se}$ – brush drop

$$= 250 - 100 \times 0.1 - 50 \times 0.05 - 3 = 234.5 \text{ V}$$

Also, $\frac{N_4}{N_1} = \frac{E_{b4}}{E_{b1}} \times \frac{\phi_1}{\phi_4}$

$\frac{N_4}{600} = \frac{234.5}{234.25} \times \frac{85}{50}$ $[\because \phi_4 \propto I_{se4}]$

$\therefore$ $\mathbf{N_4 = 1021}$ **r.p.m. (Ans.)**

When line current is 40 A (= I_5) and a divertor of 0.05 Ω is used :

Series field current, $I_{se5} = I_5 \times \frac{R_{div.}}{R_{se} + R_{div.}}$

$$= 40 \times \frac{0.05}{0.05 + 0.05} = 20 \text{ A}$$

Back e.m.f., $E_{b5} = V - (I_{a5}R_a - I_{se5}R_{se}$ – brush drop)

$$= 250 - 40 \times 0.1 - 20 \times 0.05 - 3 = 242 \text{ V}$$

Also, $\frac{N_5}{N_1} = \frac{E_{b5}}{E_{b1}} \times \frac{\phi_1}{\phi_5}$

$\frac{N_5}{600} = \frac{242}{234.25} \times \frac{85}{20}$

$\mathbf{N_5} = 600 \times \frac{242}{234.25} \times \frac{85}{20} = \mathbf{2634}$ **r.p.m. (Ans.)**

5.12.3. Rheostatic Control

- This method consists of obtaining reduced speeds by the *insertion of external series resistance in the armature circuit*. It can be used with series, shunt and compound motors ; for the last two types, the *series resistors must be connected between the shunt field and the armature, not between line and the motor.*
- It is *common method of speed control for series motors* and is generally analogous in action to wound-rotor induction-motor control by series rotor resistance.
- This method is *used when speeds below the no-load speed is required.*

Advantages :

1. The ability to achieve speeds below the basic speed.
2. Simplicity and ease of connection.
3. The possibility of combining the functions of motor starting with speed control.

Disadvantages :

1. The relatively high cost of large, continuously rated, variable resistors capable of dissipating large amounts of power (particularly at higher power ratings).
2. Poor speed regulation for any given no-load speed setting.
3. Low efficiency resulting in high operating cost.
4. Difficulty in obtaining stepless control of speed in higher power ratings.

Shunt motors :

- In armature or rheostatic control method of speed the voltage across the armature (which is normally constant) is varied by inserting a variable rheostat or resistance, called *controller resistance,* in *series* with the armature circuit. As the controller resistance is increased, the potential difference across the armature is decreased thereby decreasing the armature speed. For a load of constant torque, speed is approximately proportional to the potential difference across the armature.

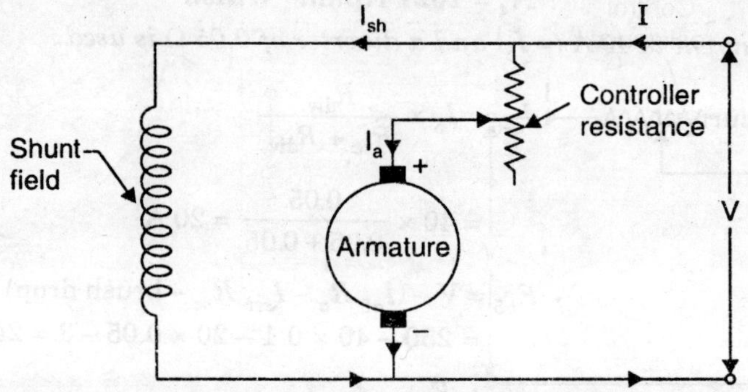

Fig. 5.34. Armature resistance control for D.C. shunt motor.

From the speed/armature current characteristic (Fig. 5.35) it is seen that *greater the resistance in armature, greater is the fall in speed.*

There is a particular load current for which the *speed would be zero. This is the maximum current and is known as 'stalling current'.*

- This method is *very wasteful, expensive and unsuitable for rapidly changing load*, because for a given value of R_t the speed will change with load. A more stable operation can be obtained by using a *divertor across the armature* (Fig. 5.36) in addition to armature control resistance. Now, the changes in armature current due to changes in the load torque *will not be so effective* in changing the potential difference across the armature and hence the speed of the armature.

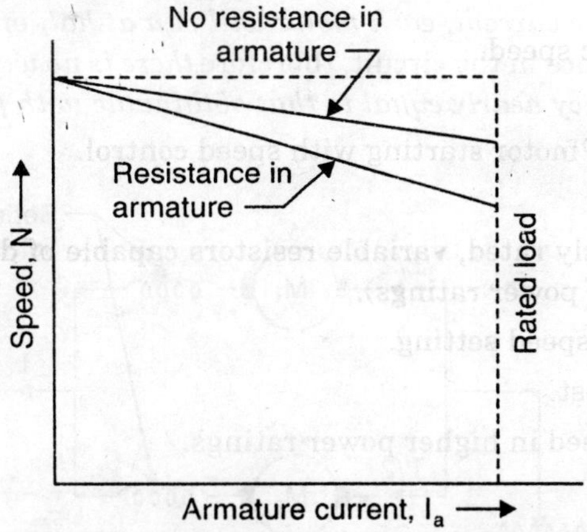

Fig. 5.35. Speed-current characteristic of D.C. shunt motor.

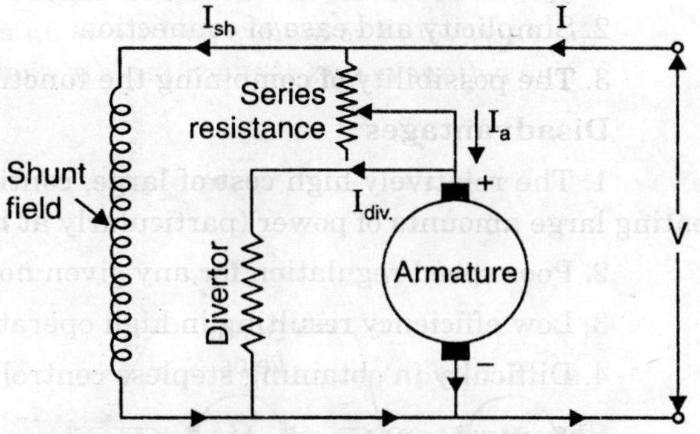

Fig. 5.36. Use of divertor across the armature for speed control of D.C. shunt motor.

Series motors :

Armature resistance control is the most common method employed for D.C. series motors (Figs. 5.37 and 5.38).

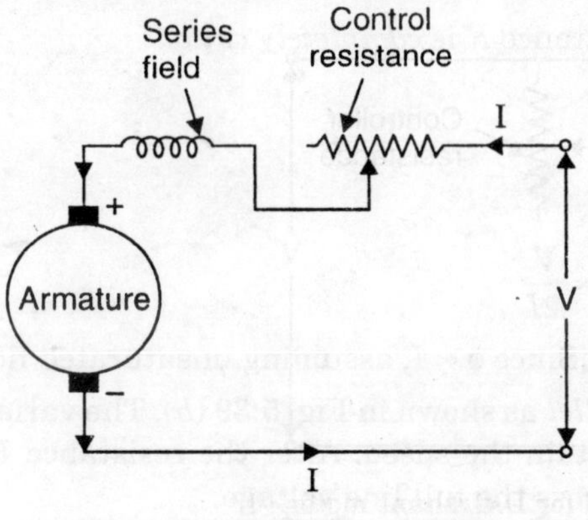

Fig. 5.37. Armature resistance control for D.C. series motor.

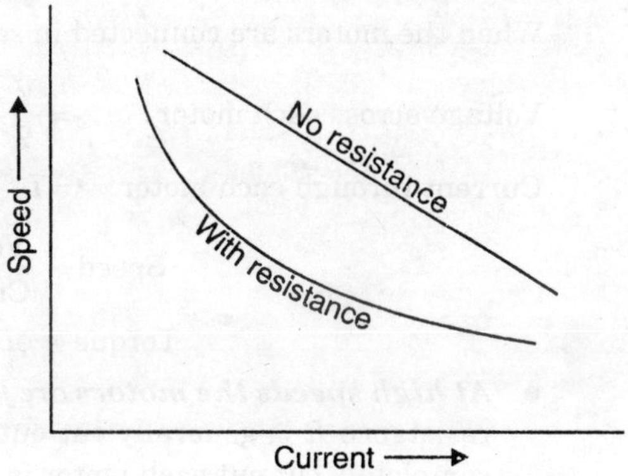

Fig. 5.38

By increasing the resistance in series with the armature the voltage applied across the armature terminals can be decreased. With the reduced voltage across the armature, the speed is reduced.

Since full motor current passes through the resistance, the loss of power is considerable.

Although terminal-voltage control by means of a variable voltage supply would effectively control the speed of a D.C. series motor, the high cost of the control equipment is seldom warranted.

Series-parallel control :

- This system is widely used in *electric traction.* Here two or more similar mechanically-coupled motors are employed.
- ***At low speeds*** the *motors are joined in series* as shown in Fig. 5.39 (*a*). The additional resistance is gradually cut out by controller as the motors attain the speed, and finally the resistance is totally removed, then *each motor has half of line voltage.* In this

arrangement, for *any given value of armature current, each motor will run at half of its normal speed*. As there is no external resistance in the circuit, therefore *there is no waste of energy and so motors operate at an efficiency nearly equal to that obtainable with full line voltage across the terminals of each motor.*

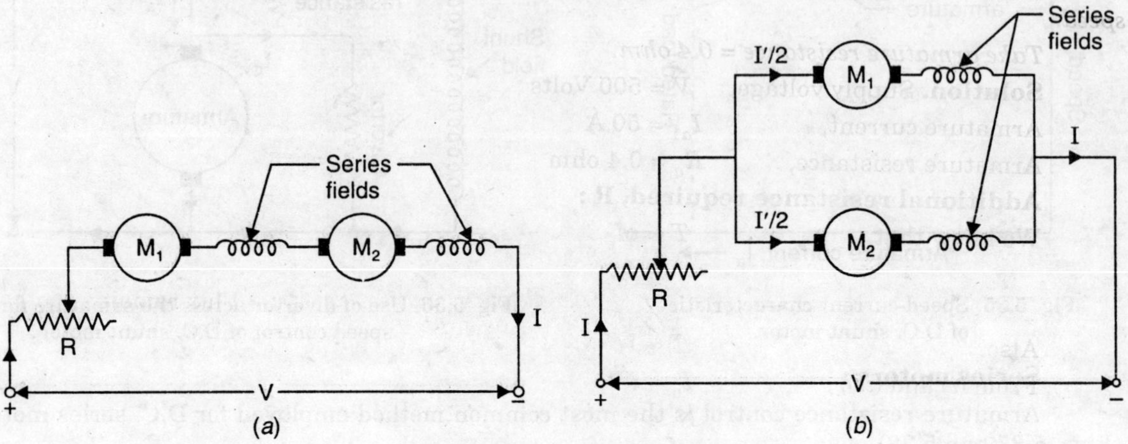

Fig. 5.39. Series-parallel control.

When the motors are connected in *series* and resistance R is *completely cut out* :

Voltage across each motor $= \dfrac{V}{2}$

Current through each motor $= I$

$$\text{Speed} \propto \frac{\text{Voltage}}{\text{Current}} \propto \frac{V/2}{I} \propto \frac{V}{2I}$$

$$\text{Torque} \propto \phi I \propto I^2 \qquad [\text{Since } \phi \propto I, \text{ assuming unsaturated field}]$$

- ***At high speeds the motors** are joined in parallel* as shown in Fig. 5.39 (*b*). The variable resistance R is generally cut out as motors attain the speed. After the resistance R is completely cut out each motor is connected across the full line voltage.

When the motors are connected in *parallel* and resistance R is completely cut out :

Voltage across each motor $= V$

Current through each motor $= \dfrac{I}{2}$ $\left[= \dfrac{I'}{2} \text{ when resistance } R \text{ is } \textit{not} \text{ completely cut out}\right]$

$$\text{Speed} \propto \frac{\text{Voltage}}{\text{Current}} \propto \frac{V}{I/2} \propto \frac{2V}{I}$$

Also, $\text{Torque} \propto \phi I \propto I^2 \qquad [\because \phi \propto I]$

$$\therefore \qquad T \propto \left(\frac{I}{2}\right)^2 \propto \frac{I^2}{4}$$

The *torque is $\frac{1}{4}$ times that produced by motors when in series.*

Shunt Motors (Rheostatic Control)

☞**Example 5.50.** *The torque of the load driven by a 500 V shunt motor varies as the cube of the speed. At certain speed the current taken by it is 50 A. Find the additional resistance required to be connected in series with the armature circuit to reduce the speed to 50 per cent of the original speed.*

Take armature resistance = 0.4 ohm.

Solution. Supply voltage, $V = 500$ Volts

Armature current, $I_{a1} = 50$ A

Armature resistance, $R_a = 0.4$ ohm

Additional resistance required, R :

We know that, $T \propto \phi I_a$

$\propto I_a$ ...(*i*)

[∵ Flux remains constant]

Also, $T \propto N^3$ [*Given*] ...(*ii*)

From (*i*) and (*ii*), $I_a \propto N^3$

$N_2 = 50\%$ of $N_1 = 0.5\, N_1$

$$\therefore \quad \frac{N_2}{N_1} = 0.5$$

Also,
$$\frac{I_{a2}}{I_{a1}} = \frac{N_2^3}{N_1^3} = \left(\frac{N_2}{N_1}\right)^3 = (0.5)^3 = 0.125$$

$I_{a2} = 0.125 I_{a1} = 0.125 \times 50 = 6.25$ A

Back e.m.f., $E_{b1} = V - I_{a1}R_a = 500 - 50 \times 0.4$ V $= 480$ V

Back e.m.f., $E_{b2} = V - I_{a2}(R_a + R)$

$= 500 - 6.25(0.4 + R) = 497.5 - 6.25R$

Using the relation,
$$\frac{N_2}{N_1} = \frac{E_{b2}}{E_{b1}}$$

$$0.5 = \frac{497.5 - 6.35R}{480}$$

or, $6.25R = 497.5 - 0.5 \times 480$

∴ $R = 41.2$ ohms

Hence, *additional resistance* = **41.2 ohms. (Ans.)**

Example 5.51. *A 440 V shunt motor while running at 1500 r.p.m. takes an armature current of 30 A and delivers an output of 15 H.P., the load torque varies as the square of speed. Calculate the value of resistance to be connected in series with the armature for reducing the motor speed to 1300 r.p.m.* **(B.T.E.A.P., 1998)**

Solution. *Given* : $V = 440$ volts, $N_1 = 1500$ r.p.m., $I_{a1} = 30$ A ; $N_2 = 1300$ r.p.m.

Resistance to be connected in series with the armature, R :

$E_{b1} = V = 440$ volts ...... neglecting armature resistance.

We know that, $T \propto N^2$ [*Given*]

$$\therefore \quad \frac{T_2}{T_1} = \left(\frac{N_2}{N_1}\right)^2$$

or, $$T_2 = T_1 \times \left(\frac{N_2}{N_1}\right)^2 = T_1 \times \left(\frac{1300}{1500}\right)^2 = 0.7511\, T_1$$

or, $$\phi_2 I_{a2} = 0.7511\, \phi_1 I_{a1} \qquad [\because \; T \propto \phi I_a]$$

or, $$I_{a2} = 0.7511\, I_{a1} \qquad [\because \; \phi_2 = \phi_1 = \phi]$$

or, $$I_{a2} = 0.7511 \times 30 = 22.53 \text{ A}$$

Now, $$E_{b2} = V - I_a R$$
$$= 440 - 22.53R$$

Also, $$\frac{N_2}{N_1} = \frac{E_{b2}}{E_{b1}} \times \frac{\phi_1}{\phi_2} = \frac{E_{b2}}{E_{b1}} \qquad [\because \; \phi_1 = \phi_2 = \phi]$$

$\therefore$ $$\frac{1300}{1500} = \frac{440 - 22.53R}{440}$$

$$22.53R = 440 - \frac{1300}{1500} \times 440 = 58.67$$

$\therefore$ $$\mathbf{R = 2.6\ \Omega.} \quad \textbf{(Ans.)}$$

☞**Example 5.52.** *A 500 V shunt motor drives a load the torque of which varies as the square of the speed. When running at certain speed it takes a current of 30 A. If a resistance of 10 ohms is included in series with the armature determine the new speed as a percentage of the original speed.*

Shunt field resistance of the motor = 250 ohms

Neglect armature resistance.

Solution. Supply voltage, $V = 500$ Volts

Shunt field resistance, $R_{sh} = 250$ ohms

Load current, $I_1 = 30$ A

Resistance in series with armature, $R = 10$ ohms

New speed, N_2 :

Shunt field current, $$I_{sh} = \frac{V}{R_{sh}} = \frac{500}{250} = 2 \text{ A}$$

Armature current $$I_{a1} = I_1 - I_{sh} = 30 - 2 = 28 \text{ A}$$

$$T \propto \phi I_a$$
$$\propto I_a \qquad \text{[Since flux remains constant]}$$

Also, $$T \propto N^2 \qquad ...[\textit{Given}]$$
$$I_a \propto N^2$$

Let I_{a2} be the armature current at the new speed N_2.

Then $$\frac{I_{a2}}{I_{a1}} = \frac{N_2^2}{N_1^2} = \left(\frac{N_2}{N_1}\right)^2$$

Let $$\frac{N_2}{N_1} = x$$

Then $$\frac{I_{a2}}{I_{a1}} = x^2$$

$\therefore$ $$I_{a2} = x^2 I_{a1} = 28x^2$$

Back e.m.f., $E_{b1} = 500$ volts [Neglecting armature resistance]

Back e.m.f., $E_{b2} = V - I_{a2} R$

$= 500 - 10 I_{a2}$ ...(*i*)

Also, since the flux remains constant,

$$\frac{N_2}{N_1} = \frac{E_{b2}}{E_{b1}} = x$$

$\therefore$ $E_{b2} = xE_{b1} = 500x$ ...(*ii*)

$I_{a2} = 28x^2$ (already calculated) ...(*iii*)

Substituting (*ii*) and (*iii*) in (*i*), we get

$$500x = 500 - 10 \times 28x^2$$

$$280x^2 + 500x - 500 = 0$$

or, $28x^2 + 50x - 50 = 0$

or, $$x = \frac{-50 \pm \sqrt{50^2 + 4 \times 28 \times 50}}{56} = \frac{-50 \pm 90}{56}$$

$= 0.714$ [Neglecting – ve sign]

i.e., $x = \frac{N_2}{N_1} = 0.714$

or, $N_2 = 0.714 N_1 = 71.4\%$ of N_1

Hence, *new speed* **= 71.4% *of the original speed.* (Ans.)**

Example 5.53. *A 230 V, 30 kW D.C. shunt motor has an efficiency of 88 per cent, when running at 1000 r.p.m. on full-load. The armature and field resistances are 0.1 ohm and 115 ohms respectively. Find :*

(*i*) *The net and developed torque on full-load.*

(*ii*) *The starting resistance to have the line starting current equal to 1.5 times the full-load current.*

(*iii*) *The torque developed at the starting.*

Solution. Supply voltage, $V = 230$ Volts

Output of the motor $= 30$ kW

Efficiency, $\eta = 88$ per cent

Full-load speed, $N = 1000$ r.p.m.

Armature resistance, $R_a = 0.1$ ohm

Shunt field resistance, $R_{sh} = 115$ ohms

(*i*) **Net and developed torque on full load ; T_{sh}, T_a :**

Net torque, $$T_{sh} = \frac{\text{Output}}{2\pi(N/60)}$$

$$= \frac{30 \times 1000}{2\pi \times (1000/60)} = 286.48 \text{ N-m}$$

Hence, T_{sh} **= 286.48 N-m. (Ans.)**

Torque developed (T_a) is given by the relation,

$$T_a = 0.159 \frac{E_b I_a}{(N/60)} \text{ N-m}$$

Input current, $I = \dfrac{\text{Output}}{\eta \times 230} = \dfrac{30 \times 1000}{0.88 \times 230} = 148.2 \text{ A}$

Shunt field current, $I_{sh} = \dfrac{V}{R_{sh}} = \dfrac{230}{115} = 2 \text{ A}$

Armature current, $I_a = I - I_{sh}$
$= 148.2 - 2 = 146.2 \text{ A}$

Back e.m.f., $E_b = V - I_a R_a = 230 - 146.2 \times 0.1 = 215.76 \text{ V}$

Putting these values in the above equation, we get

$$\mathbf{T_a} = 0.159 \times \frac{215.76 \times 146.2}{(1000/60)} = \mathbf{300.9 \text{ N-m.}} \quad \textbf{(Ans.)}$$

(*ii*) **Starting resistance required :**

Input line current (full-load) $= 148.2 \text{ A}$

Permissible input current $= 148.2 \times 1.5 = 222.3 \text{ A}$

Permissible armature current $= 222.3 - 2 = 220.3 \text{ A}$

Total armature resistance $= \dfrac{230}{220.3} = 1.044 \text{ ohm}$

∴ *Starting resistance required* = 1.044 – 0.1 = **0.944 ohm. (Ans.)**

(*iii*) **Torque developed at the starting :**

Torque developed at starting = 1.5 × full-load torque
= 1.5 × 300.9
= **451.35 N-m. (Ans.)**

☞**Example 5.54.** *A 230 V, 10 kW shunt motor takes a current of 50 A on full-load, the full-load speed being 1000 r.p.m. The armature and shunt field resistances are 0.1 ohm and 230 ohms respectively. If a resistor is to be connected in series with the armature to reduce the speed to 0.7 times the full-load speed find the resistance of the resistor and efficiency at this speed.*

Assume that the field and armature currents remain constant and the constant losses vary as (speed)$^{1.2}$.

Solution. Supply voltage, $V = 230$ Volts

Output of the motor = 10 kW

Full-load current, $I_1 = 50$ A

Speed, $N_1 = 1000$ r.p.m.

Armature resistance, $R_a = 0.1$ ohm

Shunt field resistance, $R_{sh} = 230$ ohms

Additional resistance to be connected, R :

Percentage efficiency :

Full-load output = 10 kW = 10000 W

Full-load input = $VI = 230 \times 50 = 11500$ W

Total losses on full-load = Input-output
= 11500 – 10000 = 1500 W

Shunt field current $I_{sh} = \dfrac{230}{230} = 1 \text{ A}$

Armature current, $I_{a1} = I_1 - I_{sh} = 50 - 1 = 49 \text{ A}$

Back e.m.f., $E_{b1} = V - I_{a1} R_a = 230 - 49 \times 0.1 = 225.1 \text{ V}$

Back e.m.f., $E_{b2} = V - I_{a2}(R_a + R)$

$= 230 - 49 \times (0.1 + R) = 225.1 - 49\ R$

Using the relation, $\dfrac{N_2}{N_1} = \dfrac{E_{b2}}{E_{b1}} \times \dfrac{\phi_1}{\phi_2}$

$$\frac{0.7N_1}{N_1} = \left(\frac{225.1 - 49R}{225.1}\right) \quad [\because \ N_2 = 0.7N_1 \ \ (Given) \text{ and } \phi_1 = \phi_2]$$

$\therefore$ $49R = 225.1 - 0.7 \times 225.1$

$R =$ **1.378 ohm. (Ans.)**

Total losses $= 1500$ W

Full-load armature copper loss $= I_a^2 R_a = 49^2 \times 0.1 = 240.1$ W

Shunt field copper loss $= VI_{sh} = 230 \times 1 = 230$ W

Total copper loss on full-load $= 240.1 + 230 = 470.1$ W

Constant losses at full-load speed

$= 1500 - 470.1 = 1029.9$ W

At 0.7 times full-load speed, constant losses

$= (0.7)^{1.2} \times 1029.9 = 671.29$ W

Copper losses in the armature circuit

$= I_a^2(R_a + R) = 49^2(0.1 + 1.378) = 3548.68$ W

Shunt field copper loss $= 230$ W

Total losses $= 671.29 + 3548.68 + 230 = 4449.97$ W

Output $=$ Input – losses

$= 11500 - 4449.97 = 7050.03$ W

$\therefore$ *Percentage efficiency* $= \dfrac{\text{Output}}{\text{Input}} \times 100 = \dfrac{7050.03}{11500} \times 100 =$ **61.3 per cent. (Ans.)**

Example 5.55. *The armature and shunt field resistances of a 500 V shunt motor are 0.1 ohm and 125 ohms respectively. A resistance of 1.5 ohm is connected in series with the armature and it runs at 800 r.p.m. taking a current of 60 A. What must be the resistance of a divertor connected across the armature to reduce the speed to 600 r.p.m.*

Assume the armature current to be constant.

Solution. Supply voltage, $V = 500$ Volts

Armature resistance, $R_a = 0.1$ ohm

Shunt field resistance, $R_{sh} = 125$ ohms

Resistance connected in series with armature,

$R = 1.5$ ohm

$I_1 = 60$ A, $N_1 = 800$ r.p.m.

$N_2 = 600$ r.p.m.

$\mathbf{R_{div.}}$:

Shunt field current, $I_{sh} = \dfrac{V}{R_{sh}} = \dfrac{500}{125} = 4$ A

Armature current, $I_{a1} = I_1 - I_{sh} = 60 - 4 = 56$ A

Back e.m.f., $E_{b1} = V - I_{a1}(R_a + R)$

$= 500 - 56(0.1 + 1.5) = 410.4$ V.

Let $R_{div.}$ be the resistance of the divertor and $I_{div.}$ be the divertor current (Fig. 5.40). Let I_2 be the current through the 1.5 ohm resistor. Since armature current remains constant,

$$I_{a2} = I_{a1} = 56 \text{ A}$$

Speed of the motor when the divertor is connected,

$$N_2 = 600 \text{ r.p.m.}$$

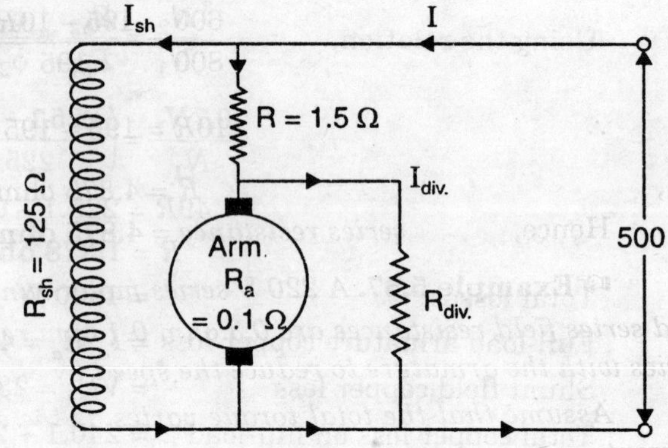

Fig. 5.40

Back e.m.f., $E_{b2} = V - I_2R - I_{a2}R_a$

$= 500 - 1.5I_2 - 56 \times 0.1$

$= 494.4 - 1.5I_2$

Using the relation, $\dfrac{N_2}{N_1} = \dfrac{E_{b2}}{E_{b1}} \times \dfrac{\phi_1}{\phi_2}$

or, $\dfrac{600}{800} = \dfrac{494.4 - 1.5I_2}{410.4}$ $[\because \phi_1 = \phi_2]$

$$1.5I_2 = 494.4 - 410.4 \times \frac{600}{800} = 186.6$$

or, $I_2 = 124.4$ A

$$I_{div.} = I_2 - I_{a2} = 124.4 - 56 = 68.4 \text{ A}$$

$$= V - I_2R = 500 - 124.4 \times 1.5 = 313.4 \text{ V}$$

$$R_{div.} = \frac{313.4}{I_{div.}} = \frac{313.4}{68.4} = 4.58 \text{ ohms}$$

Hence, *resistance of divertor connected across the armature*

= 4.58 ohms. (Ans.)

Series Motors (Rheostatic Control)

Example 5.56. *A 200 V series motor has a total resistance of 0.5 ohm. It runs at 800 r.p.m. taking an input current of 10 A. Find the series resistance required to reduce the speed to 600 r.p.m., the input current being kept constant.*

Assume the magnetisation characteristic to be a straight line.

Solution. Supply voltage, $V = 200$ Volts

Total resistance, $(R_a + R_{se}) = 0.5$ ohm

$I_1 = I_{a1} = 10$ A, $N_1 = 800$ r.p.m.

$N_2 = 600$ r.p.m.

Series resistance, R :

Back e.m.f., $E_{b1} = V - I_{a1}(R_a + R_{se}) = 200 - 10 \times 0.5 = 195$ V

Back e.m.f., $E_{b2} = V - I_{a2}(R_a + R_{se} + R)$

$= 200 - 10(0.5 + R)$ $[\because I_{a1} = I_{a2}]$

$= 195 - 10R$

Using the relation, $\frac{N_2}{N_1} = \frac{E_{b2}}{E_{b1}} \times \frac{\phi_1}{\phi_2}$

$$\frac{600}{800} = \frac{195 - 10R}{195} \qquad [\because \phi_1 = \phi_2]$$

$\therefore$ $10R = 195 - 195 \times \frac{600}{800}$

i.e., $R = 4.875$ ohms

Hence, *series resistance* = **4.875 ohms. (Ans.)**

☞**Example 5.57.** *A 220 V series motor running at a certain speed takes 25 A. Its armature and series field resistances are 0.3 ohm 0.1 ohm respectively. Find the resistance to be connected is series with the armature to reduce the speed by 30 per cent.*

Assume that the total torque varies as the cube of the speed and flux is proportional to the current.

Solution. Supply voltage, $V = 220$ Volts

$I_1 = I_{a1} = 25$ A

Armature resistance, $R_a = 0.3$ ohm

Series field resistance, $R_{se} = 0.1$ ohm

$N_2 = 0.7N_1$

Resistance to be connected in series, R :

$\phi \propto I_a$ [*Given*]

We know that, $T \propto \phi I_a$

$\propto I_a^2$ ...(*i*)

But, $T \propto N^3$ (*Given*) ...(*ii*)

From (*i*) and (*ii*), we get $I_a^2 \propto N^3$

$$\left(\frac{I_{a2}}{I_{a1}}\right)^2 = \left(\frac{N_2}{N_1}\right)^3 = (0.7)^3$$

$\therefore$ $\frac{I_{a2}}{I_{a1}} = 0.586$

$\therefore$ $I_{a2} = 0.586 I_{a1} = 0.586 \times 25 = 14.65$ A

Back e.m.f., $E_{b1} = V - I_{a1}(R_a + R_{se})$

$= 220 - 25(0.3 + 0.1) = 210$ V

Back e.m.f., $E_{b2} = V - I_{a2}(R_a + R_{se} + R)$

$= 220 - 14.65(0.3 + 0.1 + R) = 214.14 - 14.65\,R$

Using the relation, $\frac{N_2}{N_1} = \frac{E_{b2}}{E_{b1}} \times \frac{\phi_1}{\phi_2}$

or, $\frac{N_2}{N_1} = \frac{E_{b2}}{E_{b1}} \times \frac{I_{a1}}{I_{a2}}$

$$0.7 = \frac{214.14 - 14.65R}{210} \times \frac{25}{14.65}$$

$$\therefore \quad 214.14 - 14.65R = \frac{0.7 \times 210 \times 14.65}{25} = 86.142$$

$$\therefore \quad R = 8.737 \text{ ohms}$$

Hence, *the resistance to be connected in series*

$$= \mathbf{8.737 \text{ ohms.}} \quad \textbf{(Ans.)}$$

Example 5.58. *A D.C. series motor, with unsaturated magnetic circuit and negligible resistance, when running at a certain speed on a given load, takes 50 A at 500 V. If the load torque varies as the cube of the speed, find the resistance to be inserted to reduce the speed by 50%.*

(A.M.I.E. Elec. Drives & their Control, Summer, 2001)

Solution. *Given :* $I_1 = I_{a1} = 50 \text{ A} \; ; V = 500 \text{ Volts} \; ; \frac{N_2}{N_1} = 0.5$

Resistance to be inserted, R :

Back e.m.f., $E_{b1} = V = 500$ volts, since motor resistance is negligible.

Since, $T \propto N^3$ ...[*Given*]

$$\therefore \quad \frac{T_2}{T_1} = \left(\frac{N_2}{N_1}\right)^3 = (0.5)^3 = 0.125$$

or, $$\frac{\phi_2 I_{a2}}{\phi_1 I_{a1}} = 0.125 \qquad [\because \; T \propto \phi I_a]$$

or, $$\frac{I_{a2}^{\;2}}{I_{a1}^{\;2}} = 0.125 \qquad [\because \; \phi \propto I_a, \text{ magnetic circuit being unsaturated}]$$

or, $$I_{a2} = I_{a1} \times \sqrt{0.125} = 50\sqrt{0.125} = 17.68 \text{ A}$$

Back e.m.f., $$E_{b2} = V - I_{a2}R = 500 - 17.68 \times R$$

Also, $$\frac{N_2}{N_1} = \frac{E_{b2}}{E_{b1}} \times \frac{\phi_1}{\phi_2}$$

or, $$0.5 = \frac{500 - 17.68R}{500} \times \frac{50}{17.68} \qquad [\because \; \phi \propto I_a]$$

or, $$500 - 17.68 \text{ R} = 0.5 \times 500 \times \frac{17.68}{50} = 88.4$$

$$\therefore \quad \text{R} = \frac{500 - 88.4}{17.68} = \mathbf{23.28 \; \Omega.} \quad \textbf{(Ans.)}$$

Example 5.59. *A D.C. series motor runs at 1,000 r.p.m. and takes 25 A from 250 V mains. The armature current is then reduced to 15 A by inserting a series resistance. Find the new speed and the value of the resistance inserted if the load torque varies as the square of the speed and the field flux is reduced by 15% for above change in the armature current. Assume the combined resistance of the armature and series field to be 1.0 Ω.* **(B.T.E.U.P., 1997)**

Solution. *Given :* $N_1 = 1000 \text{ r.p.m.} \; ; I_{a1} (= I_1) = 25 \text{ A} \; ; V = 250 \text{ Volts} \; ;$

$I_{a2} = 15 \text{ A} \; ; \phi_2 = 0.85 \, \phi_1 \; ; (R_a + R_{se}) = 1.0 \; \Omega$

New speed, N_2 :

Resistance to be inserted in series, R :

Back e.m.f., $$E_{b1} = V - I_{a1}(R_a + R_{se})$$
$$= 250 - 25 \times 1 = 225 \text{ V}$$

Since, $T \propto N^2$ ...[*Given*]

$$\therefore \quad \frac{T_2}{T_1} = \left(\frac{N_2}{N_1}\right)^2$$

$$\text{or,} \quad \frac{\phi_2 I_{a2}}{\phi_1 I_{a1}} = \left(\frac{N_2}{100}\right)^2 \text{ or } 0.85 \times \frac{15}{25} = \left(\frac{N_2}{100}\right)^2$$

$$\therefore \quad \mathbf{N_2} = \left(0.85 \times \frac{15}{25}\right)^{1/2} \times 1000 = \mathbf{714\ r.p.m.} \quad \textbf{(Ans.)}$$

Back e.m.f.,

$$E_{b2} = V - I_{a2}(R_a + R_{se} + R)$$
$$= 250 - 15(1 + R) = 235 - 15R$$

Also,

$$\frac{N_2}{N_1} = \frac{E_{b2}}{E_{b1}} \times \frac{\phi_1}{\phi_2}$$

$$\frac{714}{1000} = \frac{235 - 15R}{225} \times \frac{1}{0.85}$$

$$\therefore \quad 235 - 15R = \frac{714}{1000} \times 225 \times 0.85 = 136.55$$

or, $\mathbf{R = 6.56\ \Omega}$. **(Ans.)**

Example 5.60. *A D.C. series motor runs at 500 r.p.m. drawing 40 A from 600 V supply. Determine the value of the external resistance to be added in series with the armature for the motor to run at 450 r.p.m. The load torque varies as the square of the speed. Assume linear magnetisation and take armature resistance as 0.3 Ω and series field resistance 0.2 Ω.* **(A.M.I.E. Summer, 2002)**

Solution. *Given* : $N_1 = 500$ r.p.m. ; $I_{a1} = 40$ A ; $V = 600$ Volts ;
$N_2 = 450$ r.p.m. ; $R_a = 0.3\ \Omega$; $R_{se} = 0.2\ \Omega$.

External resistance to be added in series with the armature, R :

Back e.m.f.,

$$E_{b1} = V - I_{a1}(R_a + R_{se})$$
$$= 600 - 40(0.3 + 0.2) = 580 \text{ V}$$

Since, $T \propto N^2$ [*Given*]

$$\therefore \quad \frac{T_2}{T_1} = \left(\frac{N_2}{N_1}\right)^2 = \left(\frac{450}{500}\right)^2 = 0.81$$

$$\text{or,} \quad \frac{\phi_2 I_{a2}}{\phi_1 I_{a1}} = 0.81 \qquad [\because \ T \propto \phi I_a]$$

$$\text{or,} \quad \left(\frac{I_{a2}}{I_{a1}}\right)^2 = 0.81 \qquad [\because \ \phi \propto I \text{ ... Given}]$$

$$\therefore \quad I_{a2} = 0.9 I_{a1} = 0.9 \times 40 = 36 \text{ A}$$

Back e.m.f.,

$$E_{b2} = V - I_{a2}(R + R_a + R_{se})$$
$$= 600 - 36(R + 0.3 + 0.2) = 582 - 36\ R$$

Also,

$$\frac{N_2}{N_1} = \frac{E_{b2}}{E_{b2}} \times \frac{\phi_1}{\phi_2} = \frac{E_{b2}}{E_{b1}} \times \frac{I_{a1}}{I_{a2}} \qquad [\because \ \phi \propto I_a]$$

$$\therefore \quad \frac{450}{500} = \frac{582 - 36R}{580} \times \frac{40}{36}$$

or, $$(582 - 36\text{ R}) = \frac{450}{500} \times 580 \times \frac{40}{36} = 469.8$$

$$\therefore \quad \mathbf{R} = \frac{582 - 469.8}{36} = \mathbf{3.12\ \Omega.} \quad \textbf{(Ans.)}$$

Example 5.61. *The resistance of a 220 V series lift motor is 0.15 ohm. At a speed of 1500 r.p.m. it takes 35 A. Find the resistance to be added in series with the motor to limit the speed to 3000 r.p.m. when the current is 10 A.*

Assume the flux to be proportional to the current.

Solution. Supply voltage, $V = 220$ Volts

Motor resistance $= 0.15$ ohm

$$I_1 = I_{a1} = 35\text{ A}, N_1 = 1500\text{ r.p.m.}, I_2 = I_{a2} = 10\text{ A}, N_2 = 3000\text{ r.p.m.}$$

Resistance to be added in series with the motor, R :

Back e.m.f., $$E_{b1} = V - I_{a1}(R_a + R_{se}) = 220 - 35 \times 0.15 = 214.75\text{ V}$$

Back e.m.f., $$E_{b2} = V - I_{a2}(R_a + R_{se} + R) = 220 - 10(0.15 + R) = 218.5 - 10R$$

Since $\phi \propto I$ ...*[Given]*

$$\frac{\phi_1}{I_1} = \frac{\phi_2}{I_2}$$

$$\therefore \quad \phi_2 = \frac{I_2}{I_1} \times \phi_2 = \frac{10}{35} \times \phi_1 = 0.2857\phi_1$$

Using the relation, $$\frac{N_2}{N_1} = \frac{E_{b2}}{E_{b1}} \times \frac{\phi_1}{\phi_2}$$

i.e., $$\frac{3000}{1500} = \frac{218.5 - 10R}{214.75} \times \frac{\phi_1}{0.2857\phi_1}$$

or, $$\frac{218.5 - 10R}{214.75} = 0.2857 \times \frac{3000}{1500} = 0.5714$$

$$\therefore \quad R = \frac{218.5 - 214.75 \times 0.5714}{10} = 9.58\text{ ohms.}$$

Hence, *resistance to be added in series* = **9.58 ohms. (Ans.)**

Example 5.62. *A 220 V D.C. series motor runs at 1500 r.p.m. when loaded to 45 A. The motor resistance is 0.15 ohm. Calculate the value of resistance to be inserted in series to bring down the speed to 1000 r.p.m. at which the load torque is half the previous value.*

The flux may be taken as proportional to the current.

Solution. Supply voltage, $V = 220$ Volts

$$I_1 = I_{a1} = 45\text{ A}, N_1 = 1500\text{ r.p.m.}$$

Motor resistance, $(R_a + R_{se}) = 0.15$ ohm

Speed, $N_2 = 1000$ r.p.m.

Resistance to be connected in series, R :

Since, $\phi \propto I_a$ (or I) ...[*Given*]

$\therefore$ $T_1 \propto \phi_1 I_{a1} \propto I_{a1}^2$

$T_2 \propto I_{a2}^2$

or, $$\frac{T_2}{T_1} = \frac{I_{a2}^2}{I_{a1}^2}$$

$$\frac{1}{2} = \frac{I_{a2}^2}{45^2} \qquad \left[\because T_2 = \frac{T_1}{2} \text{ (Given)}\right]$$

$\therefore$ $$I_a^2 = \frac{45^2}{2} = 1012.5 \text{ or } I_{a2} = 31.82 \text{ A}$$

Back e.m.f., $$E_{b1} = V - I_{a1}(R_a + R_{se}) = 220 - 45 \times 0.15 = 213.25 \text{ V}$$

Back e.m.f., $$E_{b2} = V - I_{a2}(R_a + R_{se}) = 220 - 31.82(0.15 + R) = 215.22 - 31.82R$$

Using the relation, $$\frac{N_2}{N_1} = \frac{E_{b2}}{E_{b1}} \times \frac{\phi_1}{\phi_2}$$

$$\frac{1000}{1500} = \frac{215.22 \times 31.82R}{213.25} \times \frac{I_{a1}}{I_{a2}}$$

or, $$\frac{1000}{1500} = \frac{215.22 - 31.82R}{213.25} \times \frac{45}{31.82}$$

or, $$215.22 - 31.82R = \frac{1000}{1500} \times 213.25 \times \frac{31.82}{45} = 100.46$$

or, $$R = \frac{215.22 - 100.46}{31.82} = 3.606\ \Omega$$

Resistance to be connected in series = **3.606 Ω. (Ans.)**

5.12.4. Voltage Control

When the speed is controlled by regulating the motor terminal voltage while maintaining constant field current, it is called **voltage control.**

With voltage control, the change in speed is almost proportional to the change in voltage as shown in Fig. 5.23. The *output varies directly with speed and the torque remains constant.* Since the voltage has to be regulated without affecting the field, the application of voltage control is limited to *separately excited motors* (Fig. 5.41) *only.*

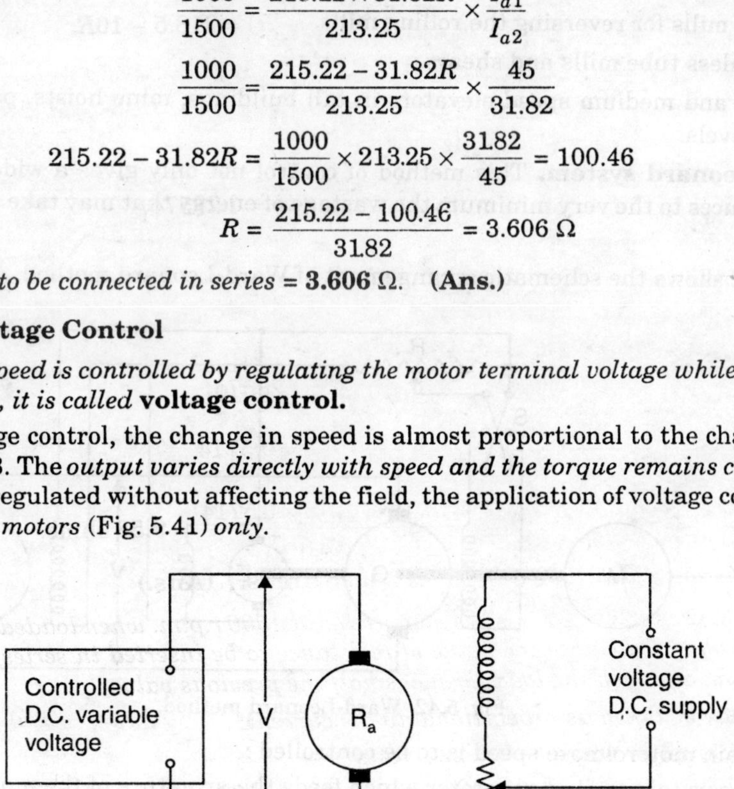

Fig. 5.41. Voltage control method.

- For D.C. motors of *fractional* and *relatively low power* rating, the variable D.C. voltage source may be a D.C. vacuum tube, a gas or thyraton tube, or a semiconductor (silicon controlled rectifier) amplifier, operating from a three-phase or single phase A.C. supply.
- *Motors of moderate rating up to* 75 kW may be controlled by this method, using Rototrol or Regulex or magnetic amplifiers as the adjustable D.C. voltage source.
- *Large D.C. motors* are controlled in this manner by means of rotary amplifiers such as the amplidyne or the Ward-Leonard control system.

Advantages :

1. Speed control over a wide range is possible.
2. This method eliminates the need for series armature starting resistance.
3. Uniform acceleration can be obtained.
4. Speed regulation is good.

Disadvantages :

1. Arrangement is costly as two extra machines are required.
2. The overall efficiency of the system is low, especially at light loads.

Applications :

Inspite of the high capital cost, this method finds wide applications in :

(*i*) Steel mills for reversing the rolling mills.

(*ii*) Seamless tube mills and shears.

(*iii*) High and medium speed elevators in tall buildings, mine hoists, paper machine drives and electric shovels.

Ward-Leonard system. This method of control not only gives a wide range of operating speeds, but reduces to the very minimum the wastage of energy that may take place at starting and stopping.

Fig. 5.42 shows the schematic arrangement of Ward-Leonard method.

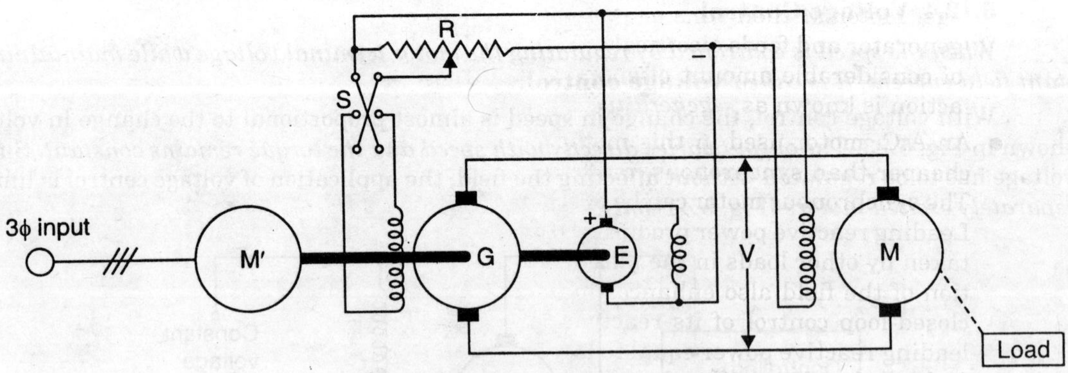

Fig. 5.42. Ward-Leonard method.

M = Main motor whose speed is to be controlled ;

G = Separately excited generator which feeds the armature of the motor M ;

E = An exciter (a small shunt generator) which provides field excitation to the generator G ; and motor M ;

M' = Driving motor—a constant speed motor which drives G and E ;

[If the system is to work on A.C. supply, the driving motor M' is a 3-phase induction motor. If the system is to work on D.C. supply, the motor M' is a shunt motor. In the latter case the exciter E is not necessary because the excitation for the generator G and motor M can be obtained from D.C. mains. A diesel engine can also be used in place of motor M'].

R = A potentiometer rheostat ;

S = A double throw switch.

The ***working*** of *this system* is as follows :

— The motor M' drives the generator G and exciter E at constant speed. The voltage fed to motor M can be controlled by varying the setting of R. A change in voltage applied to motor M changes its speed. The speed can be adjusted to any value from zero to maximum in either direction by means of a rheostat R and switch S.

— When the sliding contact of R is at extreme right, the motor is running at full speed in one direction. To decrease the speed the sliding contact is moved to the left. When the sliding contact is at the extreme left position, the speed of motor M is zero. In order to reverse the speed of the motor, the sliding contact is shifted to the extreme left, the switch S is reversed and the sliding contact shifted to right again.

— A modification of the Ward-Leonard system is known as *Ward-Leonard Ilgner system,* which uses a small motor-generator set with the *addition of a flywheel whose function is to reduce fluctuations in the power demand from the supply circuit*. When the main motor M becomes suddenly overloaded, the driving motor M' slows down, thus allowing the inertia of the flywheel to supply a part of the overload. However, when the load is suddenly thrown off the main motor M, then M' speeds up thereby again storing energy in the flywheel. When Ilgner system is driven by means of an A.C. motor (whether induction or synchronous) another refinement in the form of a 'slip regulator' can be usefully employed thus giving an additional control.

— One important feature of the Ward-Leonard system is its *regenerative action.* When a locomotive, fitted with this system, is descending a slope, it speeds up due to the action of gravity. The speed of motor M increases until its back e.m.f. exceeds the applied voltage. Motor M then runs as generator and feeds the machine G which now works as a generator and feeds electrical energy back into the trolley wire. This results in salvaging of considerable amount of energy and a superior and smooth braking action. Such an action is known as *regenerative braking*.

- An A.C. motor used in this method can be induction or a synchronous motor. Though cheaper than synchronous, induction motor always operates at a lagging power factor. The synchronous motor can be operated at a leading power factor by overexciting its field. Leading reactive power produced by the motor compensates for the lagging reactive power taken by other loads in the plant, thus improving power factor of the plant. Over excitation of the field also enhances maximum torque capability of the motor. By employing closed-loop control of its reactive power, synchronous motor can be made to generate leading reactive power equal to lagging reactive power of the plant caused by other loads, making the plant power factor unity.
- When the load is very heavy and intermittent, a slip-ring induction motor is employed and a flywheel is mounted on the shaft. This is called the *Ward-Leonard-Ilgener* scheme.

Advantages of Ward-Leonard system :

1. A wide range of speed from standstill to high speeds in either direction.
2. Rapid and instant reversal without excessively high armature currents.
3. Starting without the necessity of series armature resistances.

4. Stepless control from stand still to maximum speed in either direction.

5. Larger units employing generator field reversal eliminate the need for heavy armature conductors for reversing, and at the same time prevent motor runaway since the motor field is always excited.

6. The method lends itself to adaptation of intermediate electronic, semiconductor, and magnetic amplifiers to provide stages of amplification for an extremely large motor. Thus the power in the control circuit may be extremely small.

7. Extremely good speed regulation at any speed.

Disadvantages :

1. High initial cost.

2. Since the efficiency, neglecting the exciter efficiency, is essentially the product of the individual efficiencies of the two larger machines, the efficiency of this method is *not as high as rheostat speed control or the field control method.*

5.12.5. Thyristor Control of D.C. Motors

5.12.5.1. General aspects

- The direct current machines, even these days when power available is A.C. and A.C. machines have been developed which are simpler and rugged is construction and cheaper in initial as well as in maintenance cost, are finding extensive use in industrial and traction services with large speed range, owing to the following ***reasons/advantages :***
  - (*i*) The D.C. machines can be operated under *variable or constant torque conditions*, and in *closed-loop control systems to provide accurate speed or position control*.
  - (*ii*) In most cases, the control methods are *simpler* and *less costly* than the methods of control of A.C. motors to provide the same performance.
  - (*iii*) They can be controlled easily and rapidly accelerated, decelerated or reversed.
  - (*iv*) They can also be *operated under regenerative conditions*.
- Ample improvement in D.C. drive system took place in 1890's when the "Ward-Leonard Control System" was introduced.
- With the advent of electronic control system in 1950's a remarkable improvement in speed control system took place. The open-loop manual control was replaced by closed-loop feedback control, and this resulted in improved response and better accuracy. Initially, low power gas diodes and thyratrons were used to control the field current of the D.C. generator of the Ward-Leonard system but later high power gas diodes and ignitrons were developed and A.C. to D.C. convertors were used for D.C. control.
- The field of electric power control has been revolutionised due to the advent of *thyristor* which is capable of handling large currents. Now the solid-state circuits employing semiconductor diodes and thyristors have completely replaced thyratrons, ignitrons, mercury arc rectifiers, motor-generator sets etc.
- *Thyristor controlled drives employing both D.C. and A.C. motors find wide applications in industry as variable speed devices.*
- These days thyristors are used extensively for A.C. to D.C. conversion.

5.12.5.2. Advantages of thyristor control over Ward-Leonard system of speed control

Thyristor control entails the following *advantages* over the Ward-Leonard system of speed control :

1. Highly-reliable.
2. Easy maintenance.

3. Cost of installation is low.
4. Floor-space requirement is low.
5. Operation at a wide range of temperature.
6. Operation accuracy is higher.
7. Owing to smaller overall time constant of the control equipment, response is quick.
8. Owing to absence of moving parts and I^2R losses operational efficiency is high.

5.12.5.3. Speed control of D.C. motor with thyristor

The speed of a D.C. motor is controlled with *thyristor* as follows :

(*i*) By adjusting the voltage applied to the armature ;

(*ii*) By adjusting the field current ; or

(*iii*) By adjusting both the voltage and the field current.

With combined armature and field control, the speed can be controlled from zero to maximum value with automatic change-over from armature control to field control and vice-versa.

Speed from zero to rated value is obtained as usual from armature voltage control.

Adjustable armature voltage can either be obtained from controlled rectifier circuits, often called the **converters**, or from **chopper circuits**, the latter are, normally employed when D.C. supply is easily available.

- In *controlled rectifier circuit*, adjustable voltage to be applied to the D.C. motor is achieved by ***varying the phase angle*** *at which the thyristors are fixed relative to the applied alternating voltage waveform.*
- In **chopper circuits**, adjustable voltage to be applied to the D.C. motor is obtained by ***changing the on-to-off time ratio*** for which the supply voltage is applied to the motor.

Instead of controlled rectifier circuits, it is possible to use an *uncontrolled rectifier*, which provides a constant direct voltage, *followed by a chopper to give a variable mean direct voltage output*.

5.12.5.4. Uncontrolled rectifiers

When A.C. power supply is available, then, D.C. power can be supplied to D.C. motors by the following methods (as mentioned above) :

(*i*) By controlled rectifier circuits using thyristors, or

(*ii*) By *uncontrolled rectifiers* (using *only diodes* and *not thyristors*), *in conjunction with thyristor chopper circuits*.

A *rectifier* is a device which converts alternating voltage or current into unidirectional voltage or current. In a rectifier the conduction takes place in one direction only. *P-N* junction diode which conducts when forward biased and practically does not conduct when reverse biased, can be used for rectification. Such rectifiers may be either *half wave* or *full wave. Since unidirectional voltage available from such a rectifier is of fixed value depending upon the magnitude of A.C. input voltage so such a rectifier is called the uncontrolled rectifier.*

5.12.5.5. Controlled rectifiers

The supply of average current to a load or motor may be controlled by the use of silicon controlled rectifier (SCR), which performs most of the duties of a rheostat. The characteristics/working of an SCR are given below :

- An SCR is a three-terminal device used to control rather large currents to a load. *An SCR or any power thyristor does not have the drawbacks of high power rheostats.*
- SCRs are *small, inexpensive* and *energy efficient*.

- An SCR acts very much *like a switch*.
  - — When it is *turned-on* there is a low resistance current flow path from anode to cathode ; then it acts like a *"closed switch"*.
  - — When it *turned-off*, no current can flow from anode to cathode, then it acts like an *"open-switch"*.

Because an SCR is a solid-state device, its *switching action is very fast*.

The switching action of gate takes place only when :

(*i*) SCR is *forward biased* (*i.e.*, anode positive with respect to cathode) ;

(*ii*) A suitable *positive voltage* is applied between the gate and cathode.

— *Once the SCR has switched on it has no control on the magnitude of current flowing through it. The current through the SCR is entirely controlled by the external impedance connected in the circuit and the applied voltage.*

— The forward current through the SCR can be reduced by reducing the voltage or by increasing the circuit impedance. There is, however, a minimum forward current that must be maintained to keep the SCR in conducting state. This is ***called the holding current rating of SCR.***

— The SCR can be *switched off by reducing the forward current below the level of holding current which may be done either by reducing the applied voltage or by increasing the circuit impedance.*

— The *gate can only trigger or switch on the SCR, it cannot switch off.*

- In SCRs, output voltage or current can be varied by controlling the point in the input A.C. cycle at which thyristor is turned-on with the application of a suitable low-power gate pulse. Once triggered or fired into conduction, the thyristor remains in the conducting state for the rest of the half cycle *i.e.*, upto 180°. The firms angle α *can be adjusted with the help of a control circuit. Firing delay angle and conduction angle always total 180°.*

Controlled rectifiers may be ***either half-wave or full wave.***

5.12.5.6. Thyristor choppers

If a high-speed switch is inserted between the D.C. source and the load the fixed voltage of a D.C. source can be converted into an adjustable average voltage across the load. This high-speed switch is called the **chopper**. A chopper is a D.C. to D.C. converter and its basic circuit is shown in Fig. 5.43 (*a*) and (*b*).

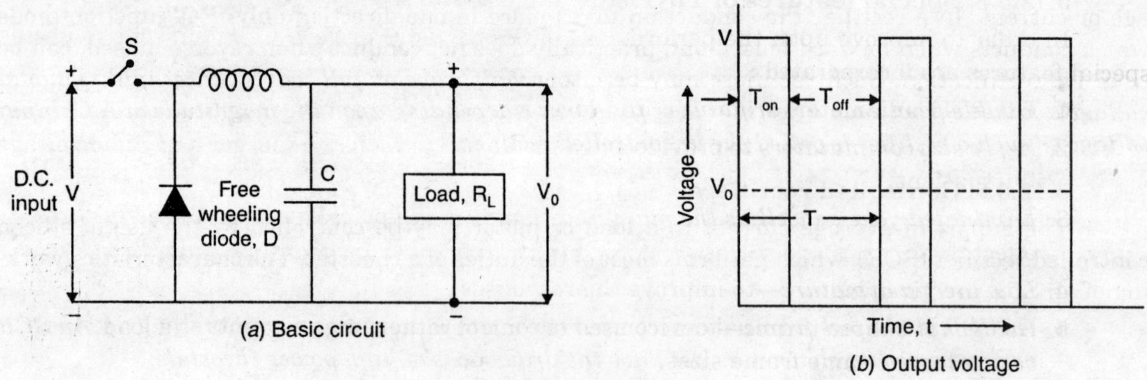

Fig. 5.43. Voltage chopping.

The average value of load voltage or output voltage V_0 for a resistive load,

$$V_0 = V \times \frac{T_{on}}{T_{on} + T_{off}} = \frac{V}{T} \times T_{on} = f V \times T_{on}$$

where, V = Input D.C. voltage,

f = Switching frequency,

T_{on} = Switching-on period,

T_{off} = Switching-off period, and

T = Total time period.

From the above equation it can be concluded that following are the *three ways* of obtaining the variable *mark-space ratio* or time-on to time-off ratio (*time ratio control TRC*) for voltage control :

(*i*) By varying the duration of on-time with respect to off-time keeping the total time period T constant ;

(*ii*) By keeping the on-time constant and varying the frequency ;

(*iii*) By adjustment of both.

5.12.5.7. Effects of Thyristor power supply on the D.C. motor performance

The performance of a D.C. motor may be affected by the thyristor power supply as follows :

(*i*) The output voltage may change very rapidly in comparison to that of a motor-generator set owing to the absence of field time constants associated with the generator.

(*ii*) The output voltage from thyristor convertor consists of a *D.C. component and A.C. harmonic components. Torque is developed by the D.C. component of the current whereas heating is developed by the r.m.s. (effective) value of current.*

(*iii*) In the event of thyristor fault (when operating in the invertor mode) the armature current may rise to an abnormally high value.

(*iv*) The commutating ability is seriously affected by the presence of harmonic currents.

(*v*) The *other effects of thyristor supply on D.C. motor performance are :*

- Heating of interpole winding ;
- Saturation of interpole magnetic circuit ;
- Transformer voltage at the brushes ;
- Increase in voltage per commutator segment.

5.12.5.8. Special features of Thyristor drive motors

In order to improve upon the performance of the thyristor drive D.C. motors, the following special features are incorporated :

1. *Large size* (diameter) *armatures and large size poles of reduced weight.*
2. *Large size commutators* to provide extra insulation to withstand larger and rapid voltage fluctuations.
3. *Laminated yoke as well as the main and commutating poles* to reduce the eddy currents effect.
4. *Low inertia armature*—to improve the responses.
5. *Octagonal shaped frame*—to accommodate more material and eventually give a larger rating for the same frame sizes.

6. *Use of better class of insulation* (class F materials)—to allow higher temperature rise and dissipate more losses from a given frame.
7. *Reduced pole arc / pole pitch ratio* to reduce the ratio of commutating zone to neutral zone.
8. *Forced cooling by an auxiliary motor* to improve cooling of the motor at reduced speeds.

5.12.5.9. Types of thyristor drives

The following types of *thyristor drives* are employed :

1. Single phase half-wave controlled rectifier circuits for D.C. motors upto 1 kW rating.
2. Single phase half bridge circuits for D.C. motors upto 5 kW rating.
3. Three phase full bridge circuits for D.C. motors of 5 to 75 kW rating.
4. Three phase full bridge circuits for D.C. motors of 75 to 400 kW rating.
5. Twelve pulse convertors for D.C. motors of rating exceeding 400 kW.

For various applications D.C. motors require speed control in a *forward direction, reverse direction* and *regenerative braking*. In *all thyristor drives, closed loop control is invariably used.*

For more details refer to Art. 5.12.6.

Voltage Control Method

Example 5.63. *A series motor drives a fan for which the torque varies as square of the speed. Its resistance between terminals is 1.2 ohm. On 220 V, it runs at 350 r.p.m. and takes 30 A. The speed is to be raised to 450 r.p.m. by increasing the voltage. Find the voltage.*

Assume that flux varies directly as current.

Solution. Resistance between terminals = 1.2 ohm

$$I_{a1} = 30 \text{ A}, N_1 = 350 \text{ r.p.m.}, N_2 = 450 \text{ r.p.m.}$$

Since, $\phi \propto I_a$

$\therefore$ $T \propto \phi I_a \propto I_a^2$ ...(*i*)

Also, $T \propto N^2$ ...[*Given*] ...(*ii*)

From (*i*) and (*ii*), we get $I_a^2 \propto N^2$ or $I_a \propto N$

or, $$\frac{I_{a2}}{I_{a1}} = \frac{N_2}{N_1} = \frac{450}{350}$$

$\therefore$ $$I_{a2} = I_{a1} \times \frac{450}{350} = 30 \times \frac{450}{350} = 38.57 \text{ A}$$

Back e.m.f., $E_{b1} = 220 - 30 \times 1.2 = 184$ V

Back e.m.f., $E_{b2} = V - 38.57 \times 1.2 = V - 46.28$

$$\frac{\phi_1}{\phi_2} = \frac{I_{a1}}{I_{a2}} = \frac{30}{38.57}$$

Now using the relation, $$\frac{N_2}{N_1} = \frac{E_{b2}}{E_{b1}} \times \frac{\phi_1}{\phi_2}$$

$$\frac{450}{350} = \frac{V - 46.28}{184} \times \frac{30}{38.57}$$

$\therefore$ $$V - 46.28 = \frac{450}{350} \times 184 \times \frac{38.57}{30} = 304.15$$

$\therefore$ **V = 304.15 volts. (Ans.)**

Example 5.64. *A series motor takes 20 A at 400 V and runs at 250 r.p.m. The armature resistance is 0.6 Ω and field resistance 0.4 Ω. The torque required by the device varies as the square of the speed. Find the applied voltage and current to run the device at 350 r.p.m.*

(B.T.E. U.P. Elec. Machines-I, 2002)

Solution. *Given :* $I_1 = I_{a1} = 20$ A ; $V = 400$ volts ; $N_1 = 250$ r.p.m. ; $R_a = 0.6\ \Omega$; $R_{se} = 0.4\ \Omega$; $N_2 = 350$ r.p.m.

Applied voltage, V_2 and current, I_2 (= I_{a2}) :

Back e.m.f., $E_{b1} = V - I_{a1}(R_a + R_{se})$
$= 400 - 20(0.6 + 0.4) = 380$ V

Since, $T \propto N^2$ ...[*Given*]

$$\therefore \quad \frac{T_2}{T_1} = \left(\frac{N_2}{N_1}\right)^2$$

$$\text{or,} \quad \frac{T_2}{T_1} = \left(\frac{350}{250}\right)^2 = 1.96$$

$$\text{or,} \quad \frac{\phi_2 I_{a2}}{\phi_1 I_{a1}} = 1.96 \quad \text{or} \quad \left(\frac{I_{a2}}{I_{a1}}\right)^2 = 1.96$$

[Assuming magnetic field *unsaturated*]

$\therefore$ $\mathbf{I_{a2}}$ $(= I_2) = I_{a1} \times \sqrt{1.96} = 20 \times \sqrt{1.96}$ = **28 A. (Ans.)**

Back e.m.f., $E_{b2} = V_2 - I_{a2}(R_a + R_{se})$
$= V_2 - 28(0.6 + 0.4) = V_2 - 28$

$$\text{Also,} \quad \frac{N_2}{N_1} = \frac{E_{b2}}{E_{b1}} \times \frac{\phi_1}{\phi_2}$$

$$\text{or,} \quad \frac{350}{250} = \frac{V_2 - 28}{380} \times \frac{20}{28} \qquad [\because \phi \propto I_a]$$

$$\text{or,} \quad V_2 - 28 = \frac{350}{250} \times 380 \times \frac{28}{20} = 744.8$$

$\therefore$ $\mathbf{V_2 = 772.8}$ **volts. (Ans.)**

Example 5.65. *A series motor when running at 600 r.p.m. takes a current of 25 A at 400 V. Its armature and series field resistances are 0.2 ohm and 0.1 ohm respectively. If the speed is to be increased to 800 r.p.m. by increasing the supply voltage, determine the new supply voltage and current if the field is*

(i) saturated *(ii) unsaturated.*

Assume that the torque varies as the cube of the speed.

Solution. Armature resistance, $R_a = 0.2$ ohm

Series field resistance, $R_{se} = 0.1$ ohm

$I_{a1} = 25$ A, $N_1 = 600$ r.p.m., $N_2 = 800$ r.p.m.

V : ; $\mathbf{I_{a2}}$:

(*i*) ***When the field is saturated :***

When the field is saturated the flux remain constant

$$T \propto \phi I_a$$
$$\propto I_a$$

Also, $T \propto N^3$ ...[*Given*]

Hence, $I_a \propto N^3$

$$\therefore \quad \frac{I_{a2}}{I_{a1}} = \left(\frac{N_2}{N_1}\right)^3 \quad ...(i)$$

Back e.m.f.,
$$E_{b1} = V - I_{a1}(R_a + R_{se})$$
$$= 400 - 25(0.2 + 0.1) = 392.5 \text{ V}$$

From (*i*), we get
$$\frac{I_{a2}}{25} = \left(\frac{800}{600}\right)^3$$
$$\therefore \quad I_{a2} = 59.26 \text{ A}$$

Hence **current taken by motor at 800 r.p.m. = 59.26 A. (Ans.)**

Back e.m.f.,
$$E_{b2} = V - I_{a2}(R_a + R_{se}) = V - 59.26(0.2 + 0.1) = V - 17.78$$

Using the relation,
$$\frac{N_2}{N_1} = \frac{E_{b2}}{E_{b1}} \times \frac{\phi_1}{\phi_2}$$
$$\frac{800}{600} = \frac{V - 17.78}{392.5} \quad [\because \ \phi_1 = \phi_2]$$
$$\therefore \quad V - 17.78 = \frac{800}{600} \times 392.5 = 523.33 \text{ or } V = 541.1 \text{ volts}$$

Hence, **supply voltage at 800 r.p.m. = 541.1 volts. (Ans.)**

(*i*) ***When the field is unsaturated :***

When the field is unsaturated $\phi \propto I_a$

$$T \propto \phi I_a \quad \text{or} \quad T \propto I_a^2$$

Also, $T \propto N^3$ ...[*Given*]

$$\therefore \quad I_a^2 \propto N^3$$
$$\left(\frac{I_{a2}}{I_{a1}}\right)^2 = \left(\frac{N_2}{N_1}\right)^3$$
$$\frac{I_{a2}}{I_{a1}} = \left(\frac{N_2}{N_1}\right)^{1.5}$$
$$\therefore \quad I_{a2} = 25 \times \left(\frac{800}{600}\right)^{1.5} = 38.49 \text{ A}$$

Hence, **current taken at 800 r.p.m. = 38.49 A. (Ans.)**

Back e.m.f.,
$$E_{b2} = V - I_{a2}(R_a + R_{se})$$
$$= V - 38.49(0.2 + 0.1) = V - 11.547$$

Using the relation,
$$\frac{N_2}{N_1} = \frac{E_{b2}}{E_{b1}} \times \frac{\phi_1}{\phi_2}$$
$$\frac{800}{600} = \frac{V - 11.547}{392.5} \times \frac{I_{a1}}{I_{a2}}$$
$$\frac{800}{600} = \frac{V - 11.547}{392.5} \times \frac{25}{38.49}$$

$$\therefore \qquad V - 11.547 = \frac{800}{600} \times 392.5 \times \frac{38.49}{25} = 805.724$$

i.e., $$V = 817.27 \text{ volts}$$

Hence, **supply voltage at 800 r.p.m. = 817.27 V. (Ans.)**

Example 5.66. *A 10 kW, 250 V, 1200 r.p.m. D.C. shunt motor has a full load efficiency of 80%. The field and armature resistances are 125 ohms and 0.2 ohm respectively. The speed of the motor is to be reduced to 75% with load torque remaining constant.*

(i) What resistance should be inserted in the armature circuit ?

(ii) With field current at its normal value, what voltage should be applied to the armature ?

(Panjab University)

Solution. *Given :* Motor input at full load = 10 kW ; $V = 250$ volts ;

$N_1 = 1200$ r.p.m. ; $\eta = 80\%$,

$R_{sh} = 125\ \Omega$; $R_a = 0.2\ \Omega$.

On full load :

Total input to the motor $= \dfrac{\text{Motor output at full load}}{\text{Full load efficiency}}$

$$= \frac{10}{0.8} = 12.5 \text{ kW}$$

Line current, $$I = \frac{\text{Motor input}}{\text{Line voltage}}$$

$$= \frac{12.5 \times 1000}{250} = 50 \text{ A}$$

Shunt field current, $$I_{sh} = \frac{V}{R_{sh}} = \frac{210}{125} = 2 \text{ A}$$

Armature current, $$I_{a1} = I - I_{sh} = 50 - 2 = 48 \text{ A}$$

Back e.m.f., $$E_{b1} = V - I_{a1}R_a$$

$$= 250 - 48 \times 0.2 = 240.4 \text{ V}$$

(i) **Resistance to be inserted in the armature circuit, R :**

Since, $$T_2 = T_1$$

$\therefore$ $$\phi_2 I_{a2} = \phi_1 I_{a1} \qquad [\because \ T \propto \phi I_a]$$

or, $$I_{a2} = I_{a1} = 48 \text{ A} \qquad [\phi \text{ remains constant}]$$

Back e.m.f., $$E_{b2} = V - I_{a2}(R + R_a)$$

$$= 250 - 48(R + 0.2) = 240.4 - 48\,R$$

Also, $$\frac{N_2}{N_1} = \frac{E_{b2}}{E_{b1}} \qquad (\because \ \phi_1 = \phi_2 = \phi)$$

or, $$0.75 = \frac{240.4 - 48R}{240.4} \qquad \left[\because \ \frac{N_2}{N_1} = 0.75 \ldots (Given)\right]$$

$$\mathbf{R} = \frac{240.4 - 0.75 \times 240.4}{48} = \mathbf{1.25\ \Omega.\ (Ans.)}$$

(ii) **Voltage to be applied, V′ :**

$$I_{a2} = I_{a1} = 48 \text{ A}$$

[∵ Load torque and flux remain constant]

Back e.m.f., $E_{b2} = V' - I_{a2} R_a$

$= V' - 48 \times 0.2 = (V' - 9.6)$ volts

Also, $$\frac{N_2}{N_1} = \frac{E_{b2}}{E_{b1}}$$

$$0.75 = \frac{V' - 9.6}{240.4}$$

$\therefore$ $\mathbf{V' = 9.6 + 0.75 \times 240.4 = 189.9\ V.}$ **(Ans.)**

Series-Parallel Control

Example 5.67. *Two series traction motor when working on 500 V supply and taking a current of 50 A run at 600 r.p.m. and 700 r.p.m. respectively. Total resistance of each motor is 0.2 ohm. If the two motors are mechanically coupled and put in series across 500 V, at what speed will they run when still taking a current of 50 A ?*

Solution. Supply voltage, $V = 500$ volts

$I_{a1} = I_{a2} = 50$ A

Total resistance of each motor = 0.2 ohm

$N_1 = 600$ r.p.m., $N_2 = 700$ r.p.m.

Common speed, N :

First motor :

Back e.m.f., $E_{b1} = 500 - 50 \times 0.2 = 490$ V

Now, $$N_1 \propto \frac{E_{b1}}{\phi_1}$$

or, $E_{b1} \propto N_1 \phi_1$ or $E_{b1} = kN_1\phi_1$

$\therefore$ $490 = k \times 600 \times \phi_1$

$\therefore$ $$k\phi_1 = \frac{490}{600} = 0.8167$$

Second motor :

$E_{b2} = 500 - 50 \times 0.2 = 490$ V

Similarly, $$k\phi_2 = \frac{490}{700} = 0.7$$

When both motors are in series

$E_b' = 500 - 50 \times 0.4 = 480$ V

Now, $E_b' = E_{b1} + E_{b2}$

where, $E_{b1} = k\phi_1 N$ and $E_{b2} = k\phi_2 N$

where, N is the common speed when joined in series

$\therefore$ $480 = 0.8167N + 0.7\,N$

or, $$N = \frac{480}{(0.8167 + 0.7)} = 316.5 \text{ r.p.m.}$$

Hence, **common speed = 316.5 r.p.m. (Ans.)**

Example 5.68. *A 3 kW series motor runs normally at 800 r.p.m. on a 200 V supply taking 16 A, when the field coils are all connected in series.*

Estimate the speed and current taken by the motor if the coils are re-connected in two parallel groups of two each in series. Load torque increases as square of the speed.

Assume that the flux is directly proportional to the current and ignore losses.

(A.M.I.E. Summer, 2001)

Solution. *Given :* $N_1 = 800$ r.p.m. ; $V = 200$ volts ; $I_{a1} (= I_1) = 16$ A.

Back e.m.f., $E_{b1} = V = 200$ volts, since losses are negligible.

When the field coils are re-connected in parallel groups of two each in series, the current through each coil is half the new input current I_2. Let the new speed be N_2.

Speed, N_2 ; Current, I_2 :

New series field current, $I_{se2} = \frac{1}{2} I_2$

Back e.m.f., $E_{b2} = V$ [∵ Losses are negligible]

Also, $$\frac{N_2}{N_1} = \frac{E_{b2}}{E_{b1}} \times \frac{\phi_1}{\phi_2}$$

or, $$\frac{N_2}{N_1} = \frac{\phi_1}{\phi_2} \qquad [\because E_{b2} = E_{b1} = V]$$

or, $$\frac{N_2}{N_1} = \frac{I_{se1}}{I_{se2}} = \frac{I_1}{I_2/2} = \frac{2I_1}{I_2} = \frac{2 \times 16}{I_2} = \frac{32}{I_2} \qquad [\because \phi \propto I_{se}]$$

Since, $T \propto N^2$ ...[*Given*]

∴ $$\frac{T_2}{T_1} = \left(\frac{N_2}{N_1}\right)^2$$

or, $$\frac{\phi_2 I_{a2}}{\phi_1 I_{a1}} = \left(\frac{32}{I_2}\right)^2 \quad \text{or} \quad \frac{I_2/2}{I_1} \times \frac{I_2}{I_1} = \left(\frac{32}{I_2}\right)^2$$

or, $$\frac{I_2^2}{2I_1^2} = \frac{(32)^2}{I_2^2}$$

or, $$I_2^4 = 2 \times (32)^2 \times I_1^2 = 2 \times (32)^2 \times (16)^2$$

∴ $$I_2 = 26.91 \text{ A}$$

and, $$\mathbf{N_2} = N_1 \times \frac{32}{I_2} = 800 \times \frac{32}{26.9} = \mathbf{951.7 \text{ r.p.m.}} \quad \textbf{(Ans.)}$$

Example 5.69. *Two series motors run at 650 r.p.m. and 750 r.p.m. respectively when taking a current of 60 A from a 600 V supply. Total resistance of each motor is 0.3 Ω. If the two motors are mechanically coupled and put in series across 600 V, at what speed will they run when still taking a current of 60 A ?* **(A.M.I.E. Winter, 2002)**

Solution. *Given :* $N_1 = 650$ r.p.m. ; $N_2 = 750$ r.p.m. ; $I_1 = 60$ A ;
$R_{m1} = R_{m2} = 0.3$ Ω ; $V = 600$ volts.

Speed, N :

First motor :

Back e.m.f., $$E_{b1} = V - I_1 R_{m1} = 600 - 60 \times 0.3 = 582 \text{ V}$$

Since back e.m.f., $N \propto \dfrac{E_b}{\phi}$, or $E_b \propto N\phi$

or, $E_{b1} = K_1 N_1 \phi_1$ or $K_1\phi_1 = \dfrac{E_{b1}}{N_1} = \dfrac{582}{650}$

Second motor :

Back e.m.f., $E_{b2} = V - I_2 R_{m2}$

$= 600 - 60 \times 0.3 = 582$ V

and, $K_2\phi_2 = \dfrac{E_{b2}}{N_2} = \dfrac{582}{750}$

When two motors are mechanically coupled and connected in series to 600 V supply they draw current of 60 A. Let the *common speed be N r.p.m.*

Sum of back e.m.f.s of two motors

$$E_{b1}' + E_{b2}' = 600 - I(R_{m1} + R_{m2})$$
$$= 600 - 60(0.3 + 0.3) = 564 \text{ V}$$

Also, $E_{b1}' = K_1\phi_1 N = \dfrac{582}{650}$ N and $E_{b2}' = K_2\phi_2 N = \dfrac{582}{750}$ N

$$\therefore \quad \frac{582}{650}N + \frac{582}{750}N = 564 \quad \text{or} \quad N\left(\frac{582}{650} + \frac{582}{750}\right) = 564$$

N = 337.4 r.p.m. (Ans.)

Example 5.70. *Four identical series motors are first connected in series and are supplied at a voltage V. If the motors are then connected in parallel and are supplied with the same current at the same supply voltage, determine the percentage increase in speed.*

Solution. ***Series connection :***

Voltage across each motor $= \dfrac{V}{4}$

Let, I = Current taken,

Then, $\phi \propto I$

Speed of the series set $N_{se} \propto \dfrac{E_b}{\phi}$

$$\propto \frac{V}{4I} \quad \text{...(i)}$$

Parallel connection :

Voltage across each motor $= V$

Current taken by each motor $= \dfrac{I}{4}$

$$\phi \propto \frac{I}{4}$$

Speed of the set, $N_p \propto \dfrac{E_b}{\phi} \propto \dfrac{V}{I/4} \propto \dfrac{4V}{I}$

$$\frac{N_p}{N_{se}} = \frac{4V/I}{V/4I} = 16$$

$\therefore$ $N_p = 16N_{se}$

Increase in speed $= 16N_{se} - N_{se} = 15N_{se}$

Percentage increase in speed $= \dfrac{15N_{se}}{N_{se}} \times 100 = 1500\%$

Hence, **percentage increase in speed = 1500%. (Ans.)**

5.12.6. Electronic Control of D.C. Motors

5.12.6.1. Introduction

Normally, it is essential to vary the speed of electrical drives in different fields of application. Usually, in all process industries, it is desired that the system be set at slow speed in the beginning and then gradually increased to meet the maximum production rate, *e.g. newspaper printing press.*

One of the major achievements of *thyristor technology* in the field of control is the control of D.C. and A.C. motor drives. *Thyristor controlled schemes* have totally dominated the field of control of D.C. as well as A.C. motors because of the following *advantages :*

(*i*) *Compactness ;* (*ii*) *Fast response ;*
(*iii*) *More efficiency ;* (*iv*) *More control capabilities ;*
(*v*) *More reliability ;* (*vi*) *Less cost etc.*

5.12.6.2. Advantages of electronic control systems

The electronic control system claim the following *advantages over conventional methods :*

1. Very compact and small in size.
2. Consumes very less power.
3. Very fast in response.
4. Much more accurate and efficient than a conventional system.
5. Control ranges are much more than any other systems.
6. High reliability comparatively.
7. Economical, since maintenance cost is minimum.
8. Highly protective.
9. In electronic systems more automation, as required for highly sophisticated machines, is possible.

5.12.6.3. D.C. Motor speed control

There are several methods by which the speed of a D.C. shunt motor can be controlled *using thyristors* ; some of the commonly used methods are discussed here.

1. **Armature voltage control method :**

This is also called the the *phase control method of speed control.* The complete diagram for this scheme is shown in Fig. 5.44.

- The field of motor is excited by a constant D.C. obtained from the *full-wave rectifier.*
- The armature voltage is *varied by varying the firing angle of the SCRs of the thyristor bridge.* Voltage across the armature terminals will be variable D.C. obtained from the full-wave half-controlled thyristor bridge. In the positive half-cycle SCR_1 and diode D_1 will conduct whereas in the negative half cycle SCR_2 and diode D_2 will conduct. Gates of SCRs will be given signal from the triggering circuit (not shown in the Fig. 5.44)

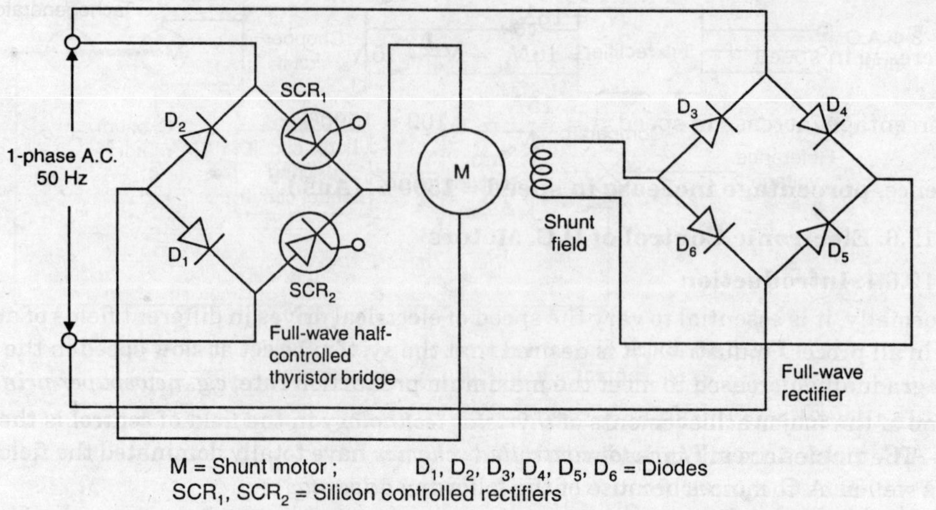

Fig. 5.44. Complete circuit diagram for the armature voltage control method for speed control of D.C. shunt motor.

- The wave shapes for the A.C. input voltage and controlled D.C. armature voltage are shown in Fig. 5.45.

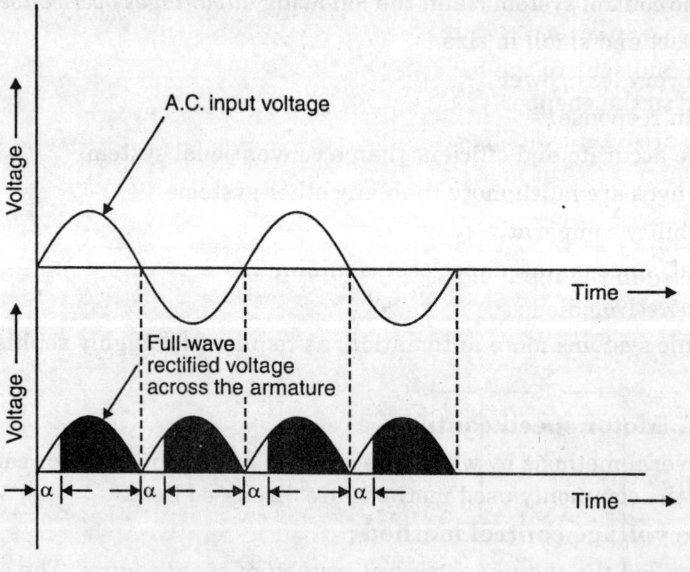

Fig. 5.45. Wave shapes for A.C. input voltage and controlled D.C. armature voltage.

2. D.C. chopper speed control :

A D.C. chopper can give variable D.C. at its output.

This variable can be utilised for the purpose of speed control of D.C. shunt motors. This method of speed control has gained popularity since the introduction of semiconductor devices.

Fig. 5.46 shows the block diagram representation of a D.C. chopper speed control scheme for D.C. shunt motors.

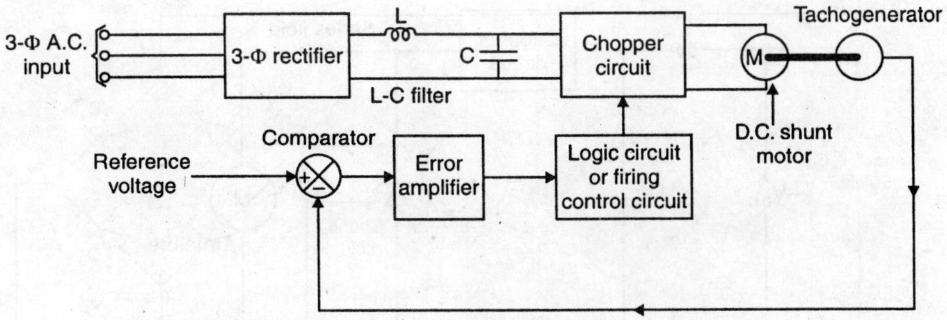

Fig. 5.46. Block diagram representation of a D.C. chopper speed control scheme for D.C. shunt motors.

— In this scheme the 3-phase A.C. is rectified into D.C. by means of a *3-phase rectifier.*

— The ripples are minimised with the help of a proper '*L-C filter*'. This filtered rectified D.C. serves as the *input* for the chopper circuit. There is a '*logic circuit*' which decides the firing of the thyristors used in the chopper. The ON, OFF durations for the thyristors used are decided by this unit. The input signal to this logic circuit is obtained from a '*comparator*' through an '*error amplifier*'.

— The speed feedback from the D.C. shunt motor is converted into equivalent voltage signal by means of a '*tachogenerator*'. The speed feedback in the form of voltage signal is given to the comparator where it is compared with the set reference voltage. If there is a difference between the two, it will generate an error signal which is amplified by the error amplifier and sent to the logic circuit to decide the ON, OFF duration of the thyristors connected in the chopper.

— Choppers are built by using one or two SCRs depending upon the type and circuitry used. *This is a very efficient method and is widely used in industries these days because of its fast response.*

- Fig 5.47 shows a simple circuit diagram for speed control of a D.C. shunt motor. An L-C filter is used in the input side of the chopper to *reduce ripples* in the D.C. input. Diode is the free wheeling diode.

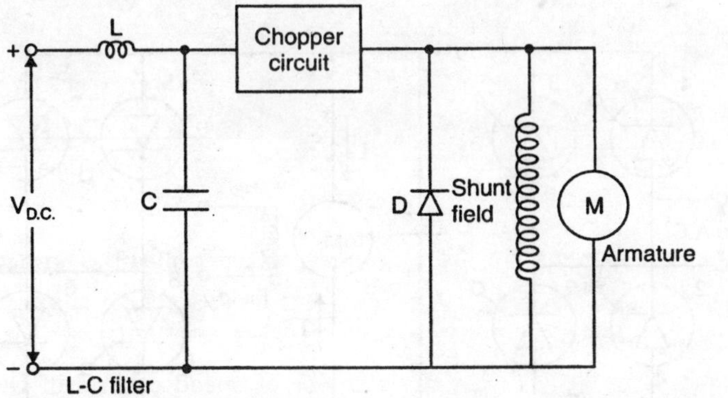

Fig. 5.47. Circuit diagram of a D.C. chopper for speed control of a D.C. shunt motor.

- Fig 5.48 shows a simple chopper circuit which may be used for controlling the speed of a D.C. series motor.

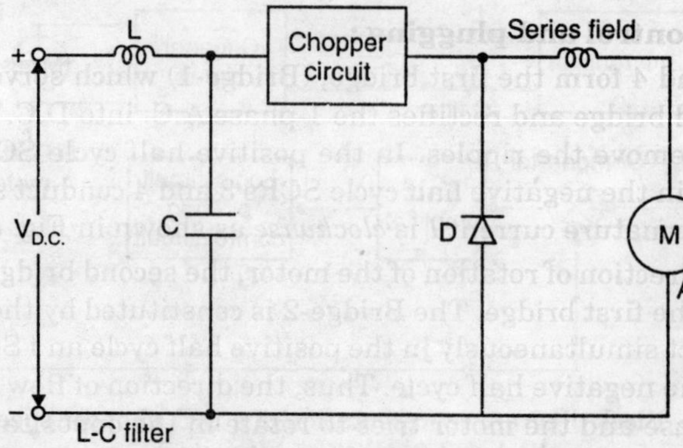

Fig. 5.48. D.C. chopper application for speed control of a D.C. series motor.

— Variation of T_{ON} and T_{OFF} will vary the load voltage at the output of the chopper which will change the speed of the motor accordingly.

— Diode D has been used as a *freewheeling diode to provide low resistance path for the current which will flow even at the OFF period of the thyristors.* This current flows for a little time due to the stored energy in the winding which is inductive in nature.

— An *L-C* filter has been used in the input side of the chopper to reduce the ripples in the D.C. input voltage.

3. Speed control by using a dual converter :

A dual converter, as the name indicates uses two converters a *rectifier* and an *inverter*. Both the bridges are built by using SCRs. A dual converter may be used to obtain the following controls of a D.C. motor :

- Reversible speed control ;
- Plugging ;
- Regenerative braking.

The above controls are discussed below.

Fig. 5.49 shows the circuit diagram for speed control of a D.C. shunt motor using a dual converter.

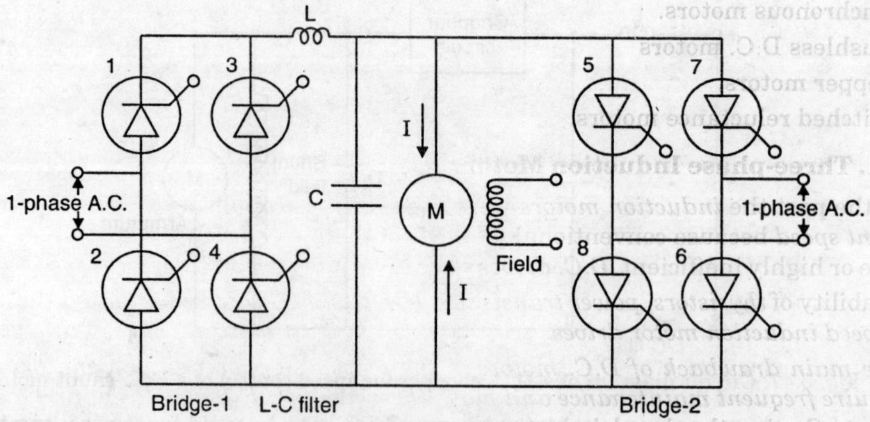

Fig. 5.49. Circuit diagram for speed control of a D.C. shunt motor using a dual converter.

1. **Reversible speed control and plugging :**

- Four SCRs 1, 2, 3 and 4 form the first bridge, (Bridge-1) which serves as a 1-phase full-wave fully-controlled bridge and rectifies the 1-phase A.C. into D.C. This D.C. is filtered by an *L-C filler* to remove the ripples. In the positive half cycle SCRs 1 and 2 conduct simultaneously and in the negative half cycle SCRs 3 and 4 conduct simultaneously. The direction of flow of armature current *I* is *clockwise* as shown in Fig. 5.49.
- For ***reversing*** the direction of rotation of the motor, the second bridge (Bridge-2) is gated after commutating the first bridge. The Bridge-2 is constituted by the SCRs 5, 6, 7 and 8. SCRs 5 and 6 conduct simultaneously in the positive half cycle and SCRs 7 and 8 conduct simultaneously in the negative half cycle. Thus, the direction of flow of armature current is reversed in this case and the motor tries to rotate in the opposite direction *i.e.,* in the *anticlockwise* direction.
- Because the motor was originally running in the clockwise direction, the inertia would oppose the torque developed in the anticlockwise direction. When the *two torques become equal, the motor becomes stationary provided bridge-2 is commutated. This process of stopping the motor is called* ***plugging.*** If the bridge-2 further continues to conduct, the motor would start running in the opposite direction resulting in speed reversal. In the opposite direction of rotation of the motor, the speed can be controlled by varying the *firing angle* of the second bridge.

2. **Regenerative braking :**

In this case, after bridge-1 is commutated and bridge-2 is triggered, the counter e.m.f. generated in the armature of the motor acts as input for bridge-2 which is connected in the *inverter mode.* The output of bridge-2 which is 1-phase A.C. may be fedback to the mains supply. Thus, we see that bridge-1 acts as a rectifier and bridge-2 acts as inverter. Therefore, in regenerative braking the K.E. of the motor is converted into electrical energy and fedback to the supply system thereby saving energy.

5.13. A.C. MOTORS

Under this topic the following motors will be discussed :

1. Three-phase induction motors.
2. Single-phase induction motors.
3. Synchronous motors.
4. Brushless D.C. motors
5. Stepper motors.
6. Switched reluctance motors.

5.13.1. Three-phase Induction Motors

- In the past the *induction motors* were used for those applications which required a *constant speed* because conventional methods of their speed control have either been expensive or highly inefficient. *D.C. drives* were invariably used for *variable speed applications.*

Availability of *thyristors, power transistors, IGBT and GTO* have permitted the *development of variable speed induction motor drives.*

- *The main drawback of D.C. motors is the presence of commutator and brushes, which require frequent maintenance and make them unsuitable for explosive and dirty environments.* On the other hand, induction motors, particularly squirrel-cage are *rugged, cheaper, lighter, smaller, more efficient require lower maintenance and can operate in dirty and explosive environments.*

Although variable speed induction motor drives are generally costlier than D.C. drives, they are used (because of advantages of induction motors) in a number of *applications* such as :

(*i*) Fans (*ii*) Blowers
(*iii*) Mill run-out tables (*iv*) Cranes
(*v*) Conveyors (*vi*) Traction
(*vii*) Underground and underwater installations
(*viii*) Explosive and dirty environments.

5.13.1.1. Theory of operation of an induction motor

When a three-phase is given to the stator winding a rotating field is set-up. This field sweeps past the rotor (conductors) and by *virtue of relative motion,* an e.m.f. is induced in the conductors which form the rotor winding. Since this winding is in the form of a closed circuit, a current flows, the direction of which is, by Lenz's law, such as to oppose the change causing it.

Now, *the change is the relative motion of the rotating field and the rotor, so that, to oppose this, the rotor runs in the same direction as the field and* attempts to catch up with it. It is clear that torque must be produced to cause rotation, and *this torque is due to the fact that currents flow in the rotor conductors* which are situated in, and at right angles to, a magnetic field.

Fig. 5.50 shows the induction motor action.

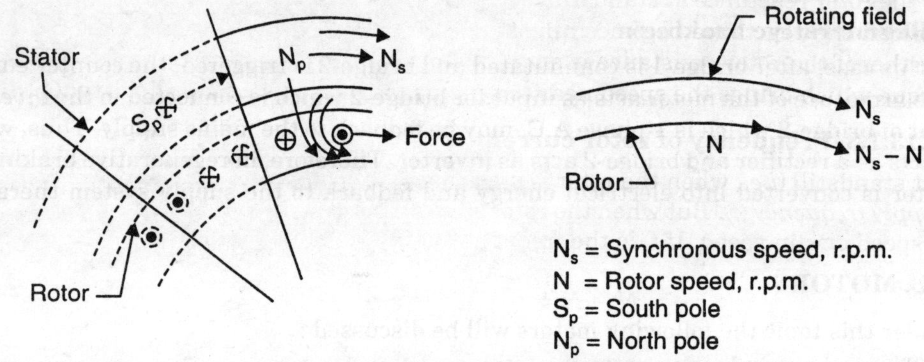

Fig. 5.50. Induction motor action.

- When the motor shaft is *not loaded*, the machine has only to rotate itself against the mechanical losses and the rotor speed is *very close to the synchronous speed.* However, the rotor speed cannot become equal to the synchronous speed because if it does so, *the e.m.f. induced in the rotor winding would become zero and there will be no torque.* Hence the speed remains slightly less than the synchronous speed. If the motor shaft is *loaded,* the rotor will slow down and the relative speed of the rotor with respect to the stator rotating field will increase. The e.m.f. induced in the rotor winding will increase and will produce more rotor current which will increase the electromagnetic torque produced by the motor. Conditions of equilibrium are attained when the rotor speed has adjusted to a new value so that the electromagnetic torque is sufficient to balance the mechanical or load torque applied to the shaft. The speed of the motor *when running under full load conditions is somewhat less than the no-load speed.*

5.13.1.2. Slip

- As earlier stated, *the rotor speed must always remain less than the synchronous speed. The difference between the synchronous speed and the rotor speed is known as* **'slip'**. It is usually expressed as a *fraction* of the synchronous speed. Thus slip *s* is,

$$s = \frac{N_s - N}{N_s} \quad ...(5.11)$$

or, $$N = N_s(1 - s) \quad ...[5.11\ (a)]$$

where, N_s = Synchronous speed (r.p.m.), and
N = Motor speed (r.p.m.).

In practice the value of slip is very small. At no-load, slip is around 1% or so and at full-load it is around 3%. For large efficient machines the slip at full-load may be around 1% only. The induction motor, is therefore, a motor with substantially constant speed and fills the same role as *D.C. shunt motor*.

- When the rotor is stationary (standstill) its speed is zero and $s = 1$. The rotor cannot run at synchronous speed because then there will be no rotor e.m.f. and no rotor current and torque. If the rotor is to run at synchronous speed an external torque is necessary. *If the rotor is driven such that $N > N_s$, the slip becomes negative, the rotor torque opposes the external driving torque and the* **machine acts as induction generator.**
- The induction motor derives its name from the fact that the *current in the rotor circuit is induced from the stator.* There is no external connection to the rotor except for some special purposes.

If the rotor reactance at standstill is X_2 its value at slip 's' becomes sX_2. *This is very desirable,* for at no-load the reactance becomes almost negligible and the rotor impedance is now all resistance. Further if the rotor resistance is small the rotor current is large, so that motor works with a large torque which brings the speed near to synchronous speed, *i.e.*, the slip is reduced.

5.13.1.3. Frequency of rotor current

At standstill (*i.e.*, when the rotor is stationary), the *frequency of the rotor current is the same as the supply frequency (f).* But when the rotor starts revolving, then the frequency depends upon the relative speed or slip-speed. If f_r is the frequency of the rotor current, then

$$N_s - N = \frac{120\, f_r}{p} \quad ...(i)$$

Also, $$N_s = \frac{120\, f}{p} \quad ...(ii)$$

Dividing (*i*) by (*ii*), we get

$$\frac{N_s - N}{N_s} = \frac{f_r}{f} \quad \text{or} \quad s = \frac{f_r}{f}$$

or, $$f_r = sf \quad ...(5.12)$$

5.13.1.4. Rotor e.m.f. and rotor current

Rotor e.m.f. :

When the rotor is stationary, an induction motor is equivalent to a 3-phase transformer with secondary short-circuited. Therefore, the induced e.m.f. per phase E_2 in the rotor at the instant of starting is given as :

$$E_2 = E_1 \times \frac{N_2}{N_1} \quad ...(5.13)$$

where, E_1 = Applied voltage per phase to primary *i.e.*, stator winding,
N_1 = Number of stator turns, and
N_2 = Number of rotor turns.

When the rotor starts gaining speed, the relative speed of the rotor with respect to stator flux *i.e.,* slip is decreased. Hence induced e.m.f. in the rotor, which is *directly proportional to the relative speed i.e., slip* is also decreased and is given by sE_2. Hence *for slip 's', the induced e.m.f. in the rotor is 's' times the induced e.m.f. in the rotor at standstill.*

Rotor current :

Let, R_2 = Rotor resistance/phase,

L_2 = Rotor inductance/phase, and

E_2 = Induced e.m.f. of rotor/phase at standstill.

At standstill :

Induced e.m.f. of rotor/phase $= E_2$

Rotor winding resistance/phase $= R_2$

Rotor winding reactance/phase, $X_2 = 2\pi f L_2$ where f is the supply frequency

Rotor impedance/phase, $Z_2 = \sqrt{R_2^2 + X_2^2}$

$\therefore$ Rotor current/phase $= \dfrac{E_2}{Z_2} = \dfrac{E_2}{\sqrt{R_2^2 + X_2^2}}$.

At slip 's' :

Induced e.m.f. of rotor/phase $= sE_2$

Rotor winding resistance $= R_2$

Rotor winding reactance $= 2\pi f_r L_2 = 2\pi s f L_2 = s(2\pi f L_2) = sX_2$

Rotor winding impedance/phase $= \sqrt{R_2^2 + (sX_2)^2}$

$\therefore$ Rotor current/phase, $I_2 = \dfrac{sE_2}{\sqrt{R_2^2 + (sX_2)^2}}$

$$= \frac{sE_2}{\sqrt{R_2^2 + s^2 X_2^2}} = \frac{E_2}{\sqrt{(R_2/s)^2 + X_2^2}} \quad \text{...(5.14)}$$

The rotor current I_2 lags the rotor voltage E_2 by rotor power factor angle ϕ_2 given by

$$\phi_2 = \tan^{-1}\left(\frac{sX_2}{R}\right)$$

Power factor of rotor current, $\cos \phi_2 = \dfrac{R_2}{\sqrt{R_2^2 + s^2 X_2^2}} \dfrac{R_2/s}{\sqrt{(R_2/s)^2 + X_2^2}}$ ...(5.15)

5.13.1.5. Torque and power

The torque of an induction motor (being due to interaction of a rotor and stator fields),

$$T \propto \phi I_2 \cos \phi_2$$

where, ϕ = Flux of rotating stator,

I_2 = Rotor current/phase, and

$\cos \phi_2$ = Rotor power factor.

Since rotor e.m.f./phase at standstill, $E_2 \propto \phi$

$\therefore$ $T \propto E_2 I_2 \cos \phi_2$

or, $T = kE_2 I_2 \cos \phi_2$ where, k is any constant ...(5.16)

Substituting the value of I_2 and $\cos \phi_2$ from eqns. (5.14) and (5.15) in eqn. (5.16), we get

$$T = kE_2 \frac{sE_2}{\sqrt{R^2 + s^2X_2^2}} \times \frac{R_2}{\sqrt{R^2 + s^2X_2^2}}$$

i.e.,

$$T = \frac{ksR_2E_2^2}{R_2^2 + s^2X_2^2} \quad ...(5.17)$$

Starting torque. At start slip 's' = 1. Therefore, expression for starting torque may be obtained by putting $s = 1$ in eqn. (5.17).

$$\text{Starting torque,} \quad T_{st} = \frac{kR_2E_2^2}{R_2^2 + X_2^2} \quad ...(5.18)$$

Condition for maximum torque. The value of torque when motor is running is given by

$$T = \frac{ksR_2E_2^2}{R_2^2 + s^2X_2^2}$$

Torque will be maximum when,

$$\frac{sR_2}{R_2^2 + s^2X_2^2} \quad \text{or} \quad \frac{R_2}{\frac{R_2^2}{s} + sX_2^2}$$

or,

$$\frac{R_2}{\left(\frac{R_2}{\sqrt{s}} - X_2\sqrt{s}\right)^2 + 2R_2X_2} \text{ is maximum, } viz., \frac{R_2}{\sqrt{s}} - X_2\sqrt{s} = 0$$

or,

$$s (= s_{mT}) = \frac{R_2}{X_2} \quad ...(5.19)$$

(where, s_{mT} = slip corresponding to maximum torque)

$$\therefore \quad \text{Maximum torque, } T_{max} = \frac{kE_2^2}{2X_2} \quad ...(5.20)$$

From the above expression, the following *conclusions* can be drawn :

- Maximum torque is *independent* of rotor circuit resistance.
- Maximum torque varies *inversely as standstill reactance* of the rotor. Therefore to have maximum torque, standstill reactance (*i.e.,* inductance) of the rotor should be kept as small as possible.
- The slip at which the maximum torque occurs depends upon the resistance of the rotor.

The condition for getting maximum torque at starting can be obtained by putting $s = 1$ in eqn. (5.19).

Thus, starting torque will be maximum if

$$\frac{R_2}{X_2} = s = 1 \quad \text{or} \quad R_2 = X_2.$$

Starting torque of a squirrel-cage motor :

The squirrel-cage rotor resistance is fixed and *small* as compared to its reactance which is very large especially *at start* (because at standstill the frequency of rotor current is equal to that of supply frequency). Hence, the starting current I_2 of the rotor, though very large in magnitude, *lags by a very large angle* behind E_2 ; consequently the *starting torque per ampere is very poor.* It is roughly 1.5 times the full-load torque although the starting current is 5 to 7 times the full-load current. Thus such motors are *not suitable* for applications where these have to be *started against heavy loads.*

Starting torque of a slip ring motor :

In a slip ring motor the torque is increased by *improving its power factor by adding external resistance in the rotor circuit from the star-connected rheostat* ; as the motor gains speed the rheostat resistance is gradually cut out. This additional resistance, however, increases the rotor impedance and so reduces the rotor current. At first, the effect of improved power factor *predominates the current-decreasing effect* of impedance, hence starting torque is increased. But after a certain point, the effect of increased impedance *predominates the effect of improved power factor and so the torque starts decreasing.*

5.13.1.6. Power

Eqn. (5.14) can be represented by a simple series circuit as shown in Fig. 5.51.

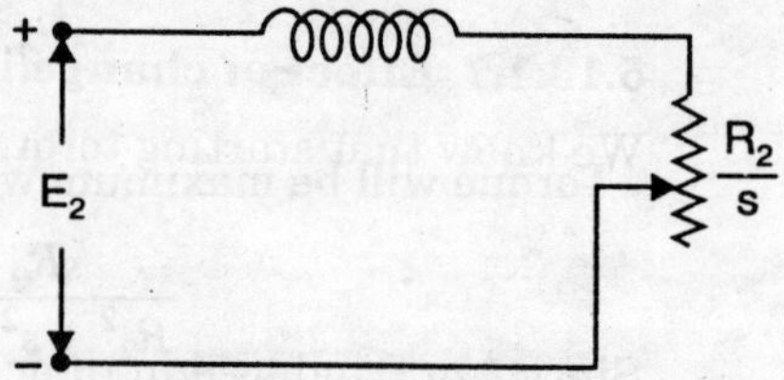

Fig. 5.51. Rotor equivalent circuit of an induction motor.

It is seen from this circuit that ***per phase** power input (gross) to rotor,*

$$P_g = E_2 I_2 \cos \phi_2$$

where, $\cos \phi_2 = \dfrac{R_2/s}{\sqrt{(R_2/s)^2 + X_2^2}}$

$$\therefore \quad P_g = \frac{E_2}{\sqrt{(R_2/s)^2 + X_2^2}} \cdot I_2 \frac{R_2}{s} = I_2^2 \frac{R_2}{s} \qquad \text{...(5.21)}$$

An examination of Fig. 5.51 also shows that per phase power input to rotor is equal to $I_2^2 \dfrac{R_2}{s}$ as the reactance X_2 consumes no power.

P_g is the power transferred from stator to rotor across the air gap. In view of this, P_g *is called the* **air-gap power**. The expression for P_g may be written as

$$P_g = I_2^2 \frac{R_2}{s} = I_2^2 R_2 + I_2^2 R_2 \left(\frac{1-s}{s}\right)$$

$$= \text{Rotor ohmic loss + internal mechanical power developed in rotor } (P_{mech.})$$

$$= sP_g + (1-s)P_g \qquad \text{...(5.22 a)}$$

$$\therefore \quad P_{mech.} = (1-s)P_g = I_2^2 R_2 \left(\frac{1-s}{s}\right) \qquad \text{...(5.22 b)}$$

$$\text{Rotor ohmic loss} = \left(\frac{s}{I-s}\right) P_{mech.} = sP_g \qquad \text{...(5.22 c)}$$

Eqn. (5.21) reveals that ohmic loss

$$= I_2^2 R_2 = sP_g = s(\text{power input to rotor}) \qquad \text{...(5.23)}$$

Internal (or gross) torque developed per phase is given by

$$T_g = \frac{\text{Internal mechanical power developed in rotor}}{\text{Rotor speed in mechanical rad. per sec.}}$$

$$= \frac{P_{mech.}}{\omega_r} = \frac{(1-s)P_g}{(1-s)\omega_s} = \frac{P_g}{\omega_s} \qquad \text{...(5.24)}$$

Here $\omega_s \left(= \frac{2\pi N_s}{60} \right)$ is the *synchronous speed in mechanical radians per second.*

Also, $$T = \frac{P_g}{\omega_s} = \frac{1}{\omega_s} \times \frac{I_2^2 R_2}{s} = \frac{\text{Rotor ohmic loss}}{(\omega_s)\text{ slip}} = \frac{1}{2\pi(N_s/60)} I_2^2 \frac{R_2}{s} \quad ...(5.25)$$

The power available at the shaft can be obtained from P_g as follows :

Output or shaft power, $P_{sh} = P_m$ – mechanical losses (friction, and windage losses)

or, $P_{sh} = P_g$ – rotor ohmic loss – friction and windage losses.

Output or shaft torque,

$$T_{sh} = \frac{P_{sh}}{\text{Rotor speed}} = \frac{P_{sh}}{(1-s)\omega_s}$$

If stator input is known, then air-gap power P_g is given by

P_g = Stator power input – stator I^2R loss – stator core loss.

5.13.1.7. Effect of change in supply voltage on starting torque

We know that starting torque,

$$T_{st} = \frac{kR_2E_2^2}{R_2^2 + X_2^2}$$

Since e.m.f. induced in rotor (at standstill) $E_2 \propto \phi \propto V$

$\therefore$ Starting torque, $$T_{st} = \frac{k' R_2 V_2^2}{R_2^2 + X_2^2}.$$

where k' is another constant

or, $$T_{st} \propto V^2$$

i.e., Starting torque is proportional to the square of the applied voltage.

5.13.1.8. Effect of change in supply voltage on torque and slip

The torque acting on the rotor when the motor is running with slip 's' is given by

$$T = \frac{ksR_2E_2^2}{R_2^2 + s^2X_2^2}$$

Since e.m.f. induced in rotor (at standstill), $E_2 \propto \phi \propto V$

$\therefore$ $$T = \frac{k' sR_2V^2}{R_2^2 + s^2X_2^2}, \text{ where } k' \text{ is another constant}$$

Since the slip 's' at full-load is *very low*, therefore $s^2X_2^2$ can be *neglected* in comparison with R_2^2.

$\therefore$ $$T = \frac{k' sR_2V^2}{R_2^2} = \frac{k' sV^2}{R_2} \quad \text{or} \quad T \propto sV^2$$

When the supply voltage is changed, it changes the torque under running condition also. With the decrease in supply voltage torque under running condition decreases, therefore, in order to *maintain the same torque, slip increases or speed decreases.*

5.13.1.9. Full-load torque and maximum torque

Let, s_f = Full-load slip of the motor, and

s_{mT} = Slip corresponding to maximum torque = $\frac{R_2}{X_2}$.

We know that,

Full-load torque, $$T_f = \frac{ks_f R_2 E_2^2}{R_2^2 + s_f^2 X^2}$$ [From Art. 5.13.3.5]

Maximum torque, $$T_m = \frac{kE_2^2}{2X_2}$$

$$\therefore \quad \frac{T_f}{T_m} = \frac{ks_f R_2 E_2^2/(R_2^2 + s_f^2 X_2^2)}{kE_2^2/(2X_2)} = \frac{2s_f R_2 X_2}{R_2^2 + s_f^2 X_2^2}$$

$$= \frac{2s_f R_2 X_2/(X_2^2)}{(R_2^2 + s_f^2 X_2^2)/(X_2^2)}$$

[Dividing numerator and denominator by X_2^2]

$$= \frac{2s_f (R_2/X_2)}{(R_2/X_2)^2 + s_f^2} = \frac{2s_f s_{mT}}{s_{mT}^2 + s_f^2}$$

i.e., $$\frac{T_f}{T_m} = \frac{2s_f s_{mT}}{s_f^2 + s_{mT}^2} \quad ...(5.26)$$

$$= \frac{2}{\frac{s_f}{s_{mT}} + \frac{s_{mT}}{s_f}} \quad ...(5.26\,a)$$

[Dividing numerator and denominator by $s_f s_{mT}$]

5.13.1.10. Starting torque and maximum torque

$$\frac{T_{st}}{T_m} = \frac{kR_2 E_2^2/(R_2^2 + X_2^2)}{kE_2^2/(2X_2)} = \frac{2R_2 X_2}{R_2^2 + X_2^2} = \frac{2(R_2/X_2)}{(R_2/X_2)^2 + 1}$$

(Dividing numerator and denominator by X_2^2)

or $$\frac{T_{st}}{T_m} = \frac{2s_{mT}}{s^2_{mT} + 1} \quad ...(5.27)$$

5.13.1.11. Torque-slip and torque-speed curves

The expression for torque is as follows :

$$T = \frac{ksR_2 E_2^2}{R_2^2 + s^2 X_2^2}$$

From the above expression, it is evident, that

- Torque is zero when slip $s = 0$ (*i.e.*, speed is synchronous).
- When slip '*s*' *is very low* the value of the term sX_2 is very small and is negligible in comparison with R_2, therefore torque T *is approximately proportional* to slip '*s*' if rotor resistance R_2 is constant. This means that at speeds near to synchronous speed the *torque-speed* and *torque-slip* curves are approximately straight lines (Figs. 5.52 and 5.53).
- When the slip '*s*' increases (*i.e.*, as the speed decreases with increase in load) torque increases and reaches its maximum value when $s = \frac{R_2}{X_2}$. The *maximum torque is also known as* ***'pull-out'*** or ***'break-down' torque.***

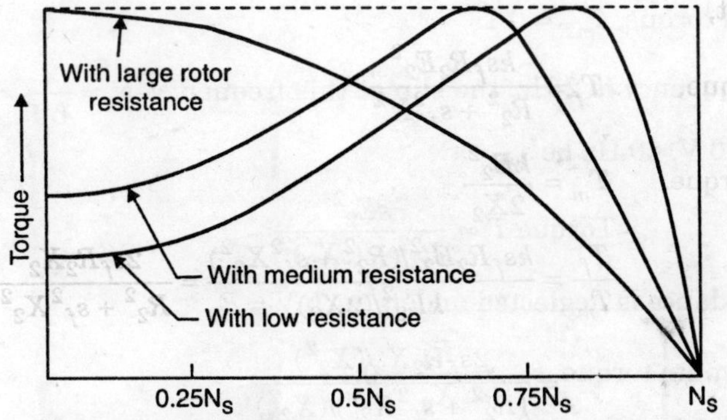

Fig. 5.52. Torque-speed curves.

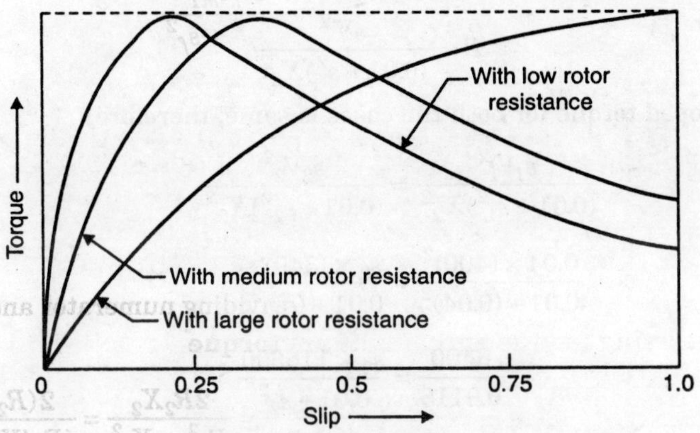

Fig. 5.53. Torque-slip curves.

- When the slip is further increased the torque decreases. The result is that *motor slows down* and *eventually stops. The motor operates for the value of slip between zero and that corresponding to maximum torque.*

With higher slip, R_2 becomes negligible as compared to sX_2 and torque varies as follows :

$$T \propto \frac{s}{s^2 X_2^{\ 2}} \propto \frac{1}{s} \text{ if standstill reactance is constant.}$$

This means that *speed-torque* or *slip-torque* curves are rectangular hyperbola with the speed or slip beyond that corresponding to maximum torque. Figs. 5.52 and 5.53 show the *torque-speed* and *torque-slip* curves for different values of rotor resistance. It is observed that although maximum torque is independent of rotor resistance R_2, yet the exact location of T_m is dependent on it. *Greater the R_2, greater is the value of slip at which maximum torque occurs.*

Example 5.71. *An induction motor runs at a slip frequency of 2 Hz when supplied from a three-phase 400 V, 50 Hz supply. For the same developed torque, find the slip frequency at which motor will run when supplied from a three-phase 340 V, 40 Hz system. Slip at which the machine develops maximum torque using 50 Hz supply is 0.1. Neglect the stator impedance and assume linear torque-slip characteristic between zero torque and maximum torque in the working region.*

(GATE, 1998)

Solution. Rating of induction motor = 400 V, 50 Hz

Slip frequency = 2 Hz

Slip at maximum torque, $s_{mT} = 0.1$

When the slip frequency is 2 Hz, the slip at this frequency, $s_1 = \frac{2}{50} = 0.04$

Let the slip at 340 V, 40 Hz be s_2.

$$\text{Torque } T \propto \frac{sE_2^2}{R_2^2 + (sX_2)^2} \quad ...(i)$$

Here stator impedance is neglected and as such $V = E_2$

Also slip at maximum torque, $s_{mT} = \frac{R_2}{X_2} = 0.1$

$$\therefore \quad R_2 = 0.1\,X_2$$

Substituting this value in (*i*), we get

$$T \propto \frac{sV^2}{(0.01 + s^2)X_2^2}$$

Since the developed torque for both the cases is same, therefore,

$$\frac{s_1 V_1^2}{(0.01 + s_1^2)X_2^2} = \frac{s_2 V_2^2}{(0.01 + s_2^2)X_2^2}$$

or, $$\frac{0.04 \times (400)^2}{0.01 + (0.04)^2} = \frac{s_2 \times (340)^2}{0.01 + (s_2)^2}$$

or, $$\frac{6400}{0.0116} = \frac{s_2 \times 115600}{0.01 + s_2^2}$$

or, $$6400(0.01 + s_2^2) = 0.0116 \times s_2 \times 115600$$

or, $$64 + 6400 s_2^2 = 1340.96 s_2$$

or, $$6400 s_2^2 - 1340.96 s_2 + 64 = 0$$

or, $$s_2 = \frac{1340.96 \pm \sqrt{(1340.96)^2 - 4 \times 6400 \times 64}}{2 \times 6400}$$

$$= \frac{1340.96 \pm 399.7}{2 \times 6400} = 0.136 \text{ or } 0.0735$$

As the slip cannot be high, thus select the value of slip as 0.0735.

$\therefore$ Slip at 40 Hz = 0.0735

Hence, slip frequency = 0.0735 × 40 = **2.94 Hz. (Ans.)**

Example 5.72. *A 3-phase star-connected 6.6 kV, 20 pole 50 Hz induction motor has rotor resistance of 0.12 Ω and standstill reactance of 1.12 Ω.*

The motor has speed of 292.5 r.p.m. at full load.

Calculate slip at maximum torque and ratio of maximum torque to full load torque.

(AMIE Summer, 2001)

Solution. *Given* : $p = 20$; $f = 50$ Hz ; $R_2 = 0.12\ \Omega$; $X_2 = 1.12\ \Omega$; $N = 292.5$ r.p.m.

s_{mT} ; $\frac{T_m}{T_f}$:

Slip corresponding to maximum torque is given by

$$s_{mT} = \frac{R_2}{X_2} = \frac{0.12}{1.12} = 0.107 \text{ or } \mathbf{10.7\%.\ (Ans.)}$$

We know that,
$$\frac{T_f}{T_m} = \frac{2s_f s_{mT}}{s_f^{\ 2} + s_{mT}^{\ 2}} \quad \text{[Eqn. (5.26)]}$$

where $s_{mT} = \frac{R_2}{X_2}$, and s_f = full-load slip of the motor.

Now,
$$s_{mT} = \frac{0.12}{1.12} = 0.107$$

$$N_s = \frac{120f}{p} = \frac{120 \times 50}{20} = 300 \text{ r.p.m.}$$

$$\therefore \quad s_f = \frac{N_s - N}{N_s} = \frac{300 - 292.5}{300} = 0.025$$

Substituting the values in the above equation, we get

$$\therefore \quad \frac{T_f}{T_m} = \frac{2 \times 0.107 \times 0.025}{(0.107)^2 + (0.025)^2} = 0.443$$

Hence,
$$\frac{T_m}{T_f} = \frac{1}{0.443} = \mathbf{2.257.\ (Ans.)}$$

Example 5.73. *A 3-phase induction motor has starting torque of 100% and a maximum torque of 200% of the full-load torque. Find slip at maximum torque.* **(UPSC, 1994)**

Solution. *Given* : Starting torque, T_{st} = 100% of T_f or = T_f

Maximum torque, T_m = 200% of T_f or = $2T_f$

Slip at maximum torque, s_{mT} :

We know that,
$$\frac{T_{st}}{T_m} = \frac{2s_{mT}}{s_{mT}^{\ 2} + 1} \quad \text{[Eqn. (5.27)]}$$

$$\frac{T_f}{2T_f} = \frac{2s_{mT}}{s_{mT}^{\ 2} + 1}$$

or,
$$\frac{1}{2} = \frac{2s_{mT}}{s_{mT}^{\ 2} + 1} \quad \text{or} \quad s_{mT}^{\ 2} - 4s_{mT} + 1 = 0$$

or,
$$s_{mT} = \frac{4 \pm \sqrt{4^2 - 4 \times 1}}{2} = 0.268 \text{ (rejecting higher value)}$$

$\therefore$ Slip at maximum torque, s_{mT} = **26.8%. (Ans.)**

Example 5.74. *In a 3-phase induction motor :*

Maximum torque = 2 × full-load torque

Starting torque = full-load torque.

Calculate :

(i) Full-load speed ; and *(ii) Slip at which maximum torque occurs.*

Solution. Maximum torque, $T_m = 2 \times T_f$

Starting torque, $T_{st} = T_f$

(*i*) Now, ratio of starting torque to maximum torque,

$$\frac{T_{st}}{T_m} = \frac{T_f}{2T_f} = 0.5$$

But, $$\frac{T_{st}}{T_m} = \frac{2s_{mT}}{s^2_{mT} + 1}, \text{ where } s_{mT} = \frac{R_2}{X_2}$$

$$\therefore \quad 0.5 = \frac{2s_{mT}}{s^2_{mT} + 1}$$

or, $$0.5(s^2_{mT} + 1) = 2s_{mT} \quad \text{or} \quad s^2_{mT} - 4s_{mT} + 1 = 0$$

or, $$s_{mT} = \frac{4 \pm \sqrt{4^2 - 4}}{2} = \frac{4 \pm \sqrt{12}}{2} = \frac{4 \pm 3.464}{2}$$

$$= 0.268 \text{ neglecting higher value}$$

Let, s_f = Full-load slip

Also, $$\frac{T_f}{T_m} = \frac{2s_f s_{mT}}{s_f^2 + s_{mT}^2}$$

$$\frac{T_f}{2T_f} = \frac{2 \times s_f \times 0.268}{s_f^2 + (0.268)^2}$$

$$0.5 = \frac{0.536 s_f}{s_f^2 + 0.0718}$$

or, $$0.5\,(s_f^2 + 0.0718) = 0.536 s_f \quad \text{or} \quad s_f^2 + 0.0718 = 1.072 s_f$$

or, $$s_f^2 - 1.072 s_f + 0.0718 = 0$$

or, $$s_f = \frac{1.072 \pm \sqrt{(1.072)^2 - 4 \times 0.0718}}{2} = \frac{1.072 \pm 0.928}{2}$$

$$= 0.072, \text{ rejecting the higher value}$$

Full-load speed, $N_f = N_s(1 - s) = N_s(1 - 0.072) = \mathbf{0.928\ N_s}$.

i.e., 0.928 times the synchronous speed. **(Ans.)**

(*ii*) **Slip corresponding to maximum torque,**

$$s_{mT} = \left(= \frac{R_2}{X_2}\right) = \mathbf{0.268\ (or\ 26.8\%)}. \quad \textbf{(Ans.)}$$

Example 5.75. *A 3-phase, 50 Hz induction motor has a starting torque which is 1.25 times full-load torque and a maximum torque which is 2.5 times the full-load torque. Neglecting stator resistance and rotational losses and assuming constant rotor resistance, find :*

(*i*) *Slip at maximum torque ;* (*ii*) *The slip at full-load ;*

(*iii*) *The current at starting in per unit of full-load current.* **(Panjab University)**

Solution. *Given* : f = 50 Hz ; $T_{st} = 1.25\,T_f$; $T_m = 2.5 T_f$

(*i*) **The slip at maximum torque, s_{mT} :**

$$\frac{T_{st}}{T_m} = \frac{2s_{mT}}{s^2_{mT} + 1} = \frac{1.25}{2.5} = 0.5 \quad \text{or} \quad s^2_{mT} - 4s_{mT} + 1 = 0$$

or, $$s_{mT} = \frac{4 \pm \sqrt{4^2 - 4}}{2} = 0.268, \text{ rejecting higher value.}$$

Hence slip at maximum torque = 0.268 or **26.8%.** **(Ans.)**

(*ii*) **The slip at full-load, s_f :**

$$\frac{T_f}{T_m} = \frac{2s_f s_{mT}}{s_f^{\,2} + s_{mT}^{\,2}} \quad \text{or} \quad \frac{1}{2.5} = \frac{2s_f \times 0.268}{s_f^{\,2} + 0.268^2}$$

or, $$s_f^2 - 1.34 s_f + 0.0718 = 0$$

or, $$s_f = \frac{1.34 \pm \sqrt{(1.34)^2 - 4 \times 0.0718}}{2}$$

$$= \frac{1.34 \pm 1.228}{2} = 0.056, \text{ rejecting higher value}$$

$\therefore$ $$\mathbf{s_f = 0.056 \text{ or } 5.6\%.} \quad \textbf{(Ans.)}$$

(*iii*) $\dfrac{\mathbf{I_{st}}}{\mathbf{I_f}}$:

Rotor current at start, $I_{st} = \dfrac{E_2}{\sqrt{R_2^{\,2} + X_2^{\,2}}}$

Rotor current at full load, $I_f = \dfrac{E_2}{\sqrt{\left(\dfrac{R_2}{s_f}\right)^2 + X_2^{\,2}}}$

$\therefore$ $$\frac{I_{st}}{I_f} = \frac{\sqrt{\left(\dfrac{R_2}{s_f}\right)^2 + X_2^{\,2}}}{\sqrt{R_2^{\,2} + X_2^{\,2}}} = \frac{\sqrt{\left(\dfrac{R_2/X_2}{s_f}\right)^2 + 1}}{\sqrt{\left(\dfrac{R_2}{X_2}\right)^2 + 1}}$$

$$= \frac{\sqrt{\left(\dfrac{0.268}{0.056}\right)^2 + 1}}{\sqrt{(0.268)^2 + 1}} = \frac{4.889}{1.0353} = 4.722$$

$$\left(\because \frac{R_2}{X_2} = s_{mT} = 0.268, \text{ as above}\right)$$

$\therefore$ *Starting current* is **4.722 times** *full-load current.* **(Ans.)**

Example 5.76. *The starting and maximum torques of a 3-phase induction motor are 1.5 times and 2.5 times its full-load torque. Determine the percentage change in rotor circuit resistance to obtain a full-load slip of 0.03. Neglect stator impedance.* **(Pb. Univ., 1998)**

Solution. *Given* : $T_{st} = 1.5\ T_f$; $T_m = 2.5\ T_f$; $s_f = 0.03$.

Percentage change in rotor circuit resistance :

$$\frac{T_{st}}{T_m} = \frac{2s_{mT}}{s^2_{mT} + 1} = \frac{1.5}{2.5} = 0.6$$

$$2s_{mT} = 0.6(s^2_{mT} + 1)$$

or, $$s^2_{mT} - 3.333s_{mT} + 1 = 0$$

or, $$s_{mT} = \frac{3.333 \pm \sqrt{3.333^2 - 4}}{2}$$

$$= \frac{3.333 \pm 2.666}{2} = 0.333 \text{ (rejecting higher value)}$$

or, $$s_{mT} = \frac{R_2}{X_2} = 0.333$$

or, Rotor resistance, $$R_2 = 0.333\, X_2$$

Also, $$\frac{T_f}{T_m} = \frac{2s_f s_{mT}}{s_f^{\,2} + s_{mT}^2}$$

or, $$\frac{1}{2.5} = \frac{2 \times 0.03 \times s_{mT}}{0.03^2 + s^2_{mT}}$$

or, $$0.03^2 + s^2_{mT} = 2.5 \times 2 \times 0.03 \times s_{mT} = 0.15s_{mT}$$

or, $$s^2_{mT} - 0.15s_{mT} + 0.0009 = 0$$

or, $$s_{mT} = \frac{0.15 \pm \sqrt{0.15^2 - 4 \times 0.0009}}{2} = \frac{0.15 \pm 0.1375}{2} = 0.1437$$

(other value is not feasible)

∴ New rotor resistance, $R_2' = s_{mT} X_2 = 0.1437X_2$

∴ Percentage reduction in rotor resistance

$$= \frac{R_2 - R_2'}{R_2} \times 100$$

$$= \frac{0.333X_2 - 0.1437X_2}{0.333X_2} = \textbf{56.85\%}. \quad \textbf{(Ans.)}$$

5.13.1.12. Induction motor as transformer. An induction motor is essentially a transformer with stator forming the primary and rotor forming (the short-circuited) rotating secondary (Fig. 5.54). This is so because the transfer of energy from stator to the rotor of an induction motor takes place entirely *inductively* with the help of flux mutually linking the two.

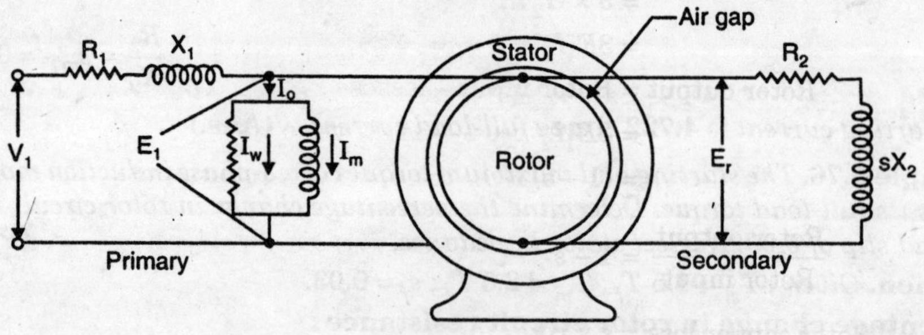

Fig. 5.54. Induction motor as transformer.

The vector diagram (Fig. 5.55) is similar to that of a transformer.

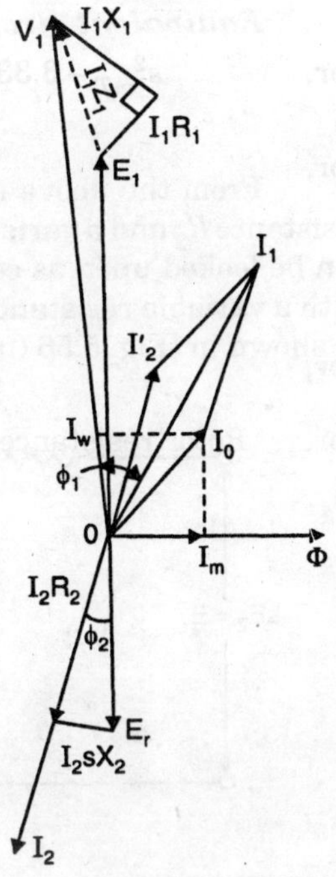

Fig. 5.55. Vector diagram of induction motor.

$$V_1 = E_1 + I_1 (R_1 + jX_1)$$

and, $$E_r = I_2 (R_2 + jsX_2)$$

However, the *important differences* between a transformer and an induction motor are :

1. In induction motor the magnetic leakage and leakage reactance of rotor and stator are *higher* than in a transformer.

2. The magnetic circuit of an induction motor has an air gap and this makes the per unit value of magnetising current *much higher* than that of a transformer.

3. Because of the distributed windings in an induction motor the *ratio of stator and rotor currents is not equal* to the ratio of turns per phase in the rotor and stator windings.

4. *The losses in an induction motor are higher and, therefore, the efficiency is lower than in a transformer.*

Rotor output. Primary current I_1 consists of two parts I_0 and I_2'. It is the latter which is transferred to the rotor, because I_0 is used in meeting the copper and iron losses in the stator itself. Out of the primary voltage V_1, some is absorbed in the primary itself $(= I_1 Z_1)$ and the remaining E_1 is transferred to the rotor. If the angle between E_2 and I_2' is ϕ, then

Rotor input/phase $= E_1 I_2' \cos\phi$

Total rotor input $= 3E_1 I_2' \cos\phi$

The electrical input to the rotor which is wasted in the form of the heat

$$= 3I_2 E_r \cos\phi \text{ (or } 3I_2^2 R_2)$$

Now $$I_2' = KI_2 \text{ or } I_2 = \frac{I_2'}{K}$$

$$E_r = sE_2 \text{ and } E_2 = KE_1$$

$$\therefore \quad E_r = sKE_1$$

$\therefore$ Electrical input wasted as heat

$$= 3 \times (I_2'/K) \times sKE_1 \cos\phi$$

$$= 3E_1 I_2' \cos\phi \times s = \text{rotor input} \times s$$

Now, Rotor output = Rotor input – losses

$$= 3E_1 I_2' \cos\phi - 3E_1 I_2' \cos\phi \times s$$

$$= (1 - s) E_1 I_2' \cos\phi = (1 - s) \text{ rotor input}$$

$$\therefore \quad \frac{\text{Rotor output}}{\text{Rotor input}} = 1 - s$$

$\therefore$ Rotor copper loss = $s \times$ rotor input

Rotor efficiency $$= 1 - s = \frac{N}{N_s} = \frac{\text{Actual speed}}{\text{Synchronous speed}}.$$

Equivalent circuit of the rotor. The rotor current I_2, when motor is loaded, is given by

$$I_2 = \frac{sE_2}{\sqrt{R_2^2 + (sX_2)^2}} = \frac{E_2}{\sqrt{(R_2/s)^2 + X_2^2}}$$

From the above relation it appears that the rotor circuit which actually consists of a fixed resistance R_2 and a variable reactance sX_2 (proportional to slip) connected across $E_r = sE_2$ [Fig. 5.56 (*i*)] can be looked upon as equivalent to a rotor circuit having a fixed reactance X_2 connected in series with a variable resistance R_2/s (inversely proportional to slip) and supplied with constant voltage E_2 as shown in [Fig. 5.56 (*ii*)].

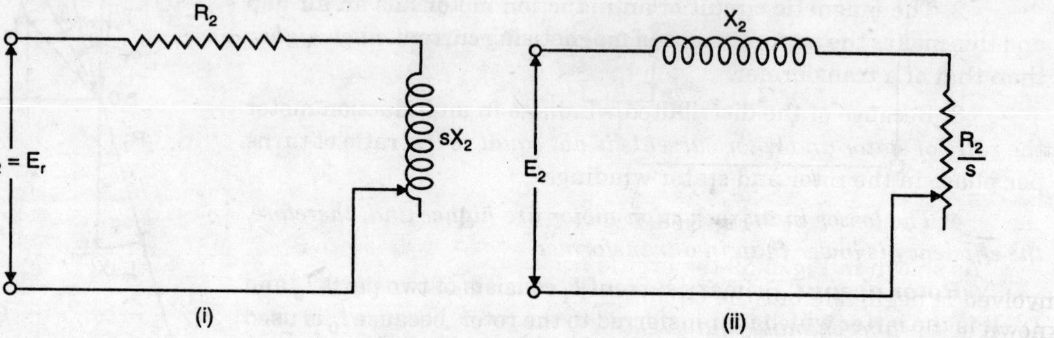

Fig. 5.56. Equivalent circuit of a rotor.

Also, the resistance R_2/s can be written as,

$$\frac{R_2}{s} = R_2 + R_2\left(\frac{1}{s} - 1\right).$$

It consists of two parts : (*i*) The part R_2 is the *rotor resistance* itself and *represents the rotor copper loss* (*ii*) the second part is $R_2\left(\frac{1}{s} - 1\right)$. This is known as *load resistance* R_L and is *the electrical equivalent of the mechanical load on the motor*. In other words the mechanical load on an induction motor can be represented by a *non-inductive resistance of the value* $R_2\left(\frac{1}{s} - 1\right)$.

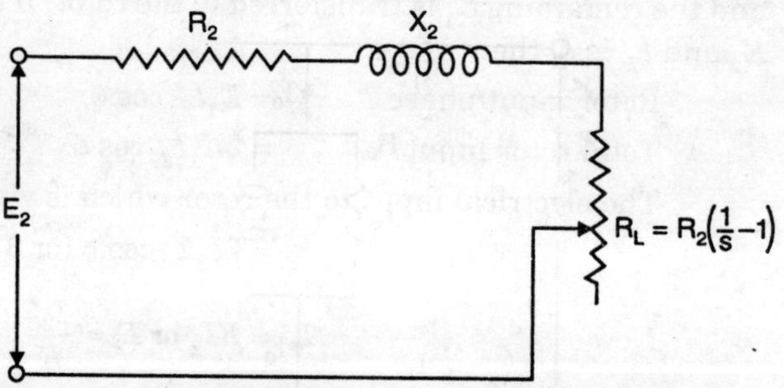

Fig. 5.57. Equivalent circuit of a rotor with load resistance R_L.

In Fig. 5.57 is shown the equivalent rotor circuit along with the load resistance R_L.

5.13.1.13. Equivalent circuit of an induction motor

The equivalent circuit for a polyphase induction motor is shown in Fig. 5.58, where

V_1 = Applied voltage per phase,
R_1 = Stator resistance/phase,
R_2 = Rotor resistance/phase,
X_1 = Stator leakage reactance/phase,
X_2 = Rotor standstill leakage reactance/phase,

K = Turn-ratio of secondary to primary, and

R_0 = No-load resistance/phase ⎤ Rotor being driven at synchronous speed.
X_0 = No-load reactance/phase ⎦

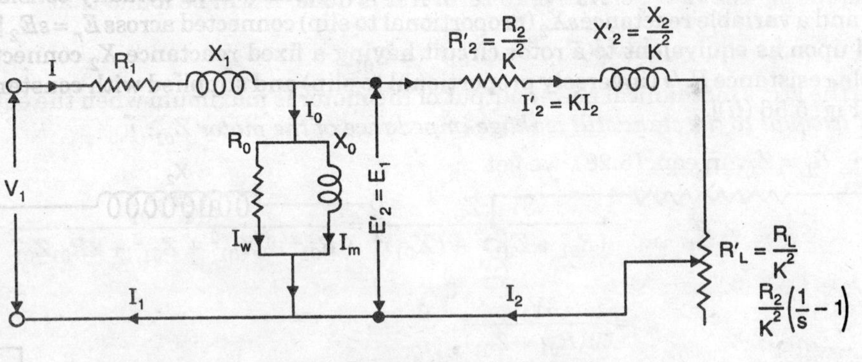

Fig. 5.58. Equivelent circuit of an induction motor.

As shown in Fig. 5.59 the exciting circuit may be transferred to the left, because inaccuracy involved is negligible but the circuit and hence the calculations are very much simplified. This is known as the *approximate equivalent* circuit of the induction motor.

Maximum power output Refer to Fig. 5.59.

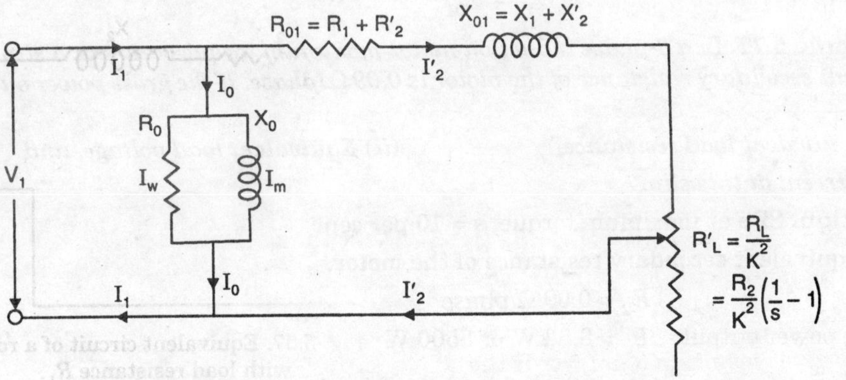

Fig. 5.59. Approximate equivalent circuit of an induction motor.

Here, $$R_{01} = R_1 + R_2' = R_1 + \frac{R_2}{K^2}$$

$$X_{01} = X_1 + X_2' = X_1 + \frac{X_2}{K^2}$$

Load resistance, $$R_L' = R_L/K^2 = \frac{R_2}{K^2}\left(\frac{1}{s} - 1\right)$$

Gross mechanical power output,

$$P_g = 3I_2'^2 R_L'$$

But, $$I_2' = \frac{V_1}{\sqrt{(R_{01} + R_L')^2 + (X_{01})^2}}$$

$$\therefore \quad P_g = \frac{3V_1^2 R_L'}{(R_{01} + R_L')^2 + (X_{01})^2} \quad \text{...(5.28)}$$

The condition for maximum power output can be obtained by differentiating the above equation and by equating, the first derivative to zero. If it is done, it will be found that

$$R_L' = \sqrt{(R_{01})^2 + (X_{01})^2} = Z_{01}$$

Hence the gross mechanical power output of the motor is maximum when the equivalent *load resistance R_L' is equal to the standstill leakage impedance of the motor Z_{01}.*

Putting $R_L' = Z_{01}$ in eqn. (5.28), we get

$$P_{g\,(max)} = \frac{3V_1^2 Z_{01}}{(R_{01} + Z_{01})^2 + (X_{01})^2} = \frac{3V_1^2 Z_{01}}{R_{01}^2 + X_{01}^2 + Z_{01}^2 + 2R_{01}Z_{01}}$$

$$= \frac{3V_1^2}{2(R_{01} + Z_{01})} \quad \text{...(5.29)}$$

The slip corresponding to maximum gross mechanical power output will be given as

$$Z_{01} = R_L' = \frac{R_2}{K^2}\left(\frac{1}{s} - 1\right)$$

or

$$s = \frac{R_2/K^2}{R_2/K^2 + Z_{01}} \quad \text{...(5.30)}$$

Example 5.77. *In a 3-phase induction motor maximum torque occurs at a slip of 10 per cent. The equivalent secondary resistance of the motor is 0.09 Ω/phase. If the gross power output is 8.5 kW. Calculate :*

(i) Equivalent load resistance, *(ii) Equivalent load voltage, and*

(iii) Current at this slip.

Solution. Slip at maximum torque, s = 10 per cent

The equivalent secondary resistance of the motor,

$$R_2 = 0.09\ \Omega/\text{phase}$$

Gross power output, P_g = 8.5 kW or 8500 W

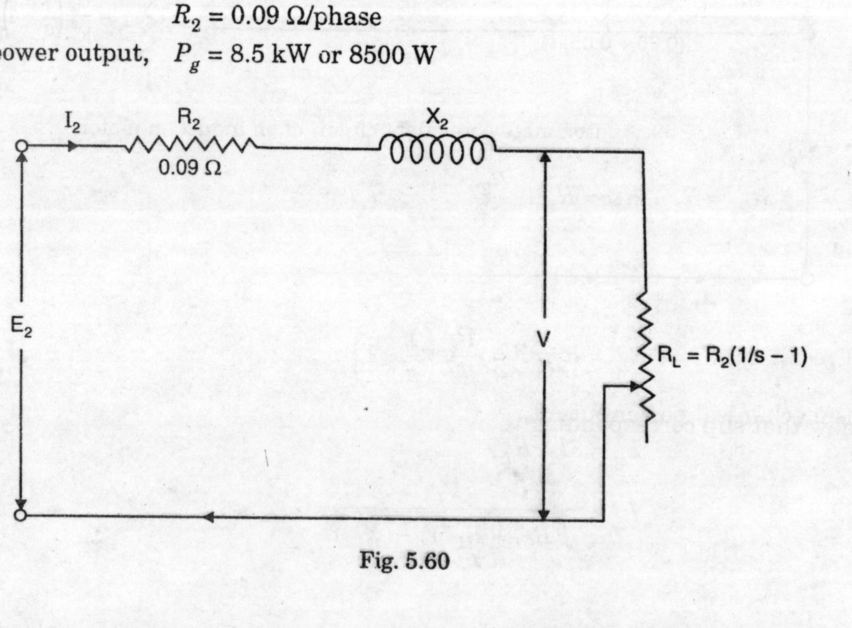

Fig. 5.60

(*i*) **Equivalent load resistance, R_L :**

We know that $$R_L = R_2\left[\left(\frac{1}{s}\right) - 1\right] = 0.09\left(\frac{1}{0.1} - 1\right) = \mathbf{0.81\ \Omega.\ (Ans.)}$$

(*ii*) **Equivalent load voltage, V :**

As shown in the equivalent circuit of the rotor in Fig. 5.60, V is a fictitious voltage drop equivalent to that consumed in the load connected to the secondary *i.e.*, rotor, $V = I_2R_L$

But, gross power, $$P_g = \frac{3V^2}{R_L}$$

$$\therefore \quad V = \sqrt{\frac{P_g \times R_L}{3}} = \sqrt{\frac{8500 \times 0.81}{3}} = 47.9 \text{ V}$$

Hence, *equivalent load voltage* = **47.9 V. (Ans.)**

(*iii*) **Equivalent load current :**

Equivalent load current $$= \frac{V}{R_L} = \frac{47.9}{0.81} = \mathbf{59.1\ A.\ (Ans.)}$$

Example 5.78. *The stator impedance and equivalent rotor impedance of a 400 V, 3-phase star-connected induction motor are (0.05 + j0.2) Ω and (0.05 + j0.24) Ω respectively. Neglecting exciting current determine the maximum gross power and the slip at which it occurs.*

Solution. Supply voltage $V_1 = 400$ V

Stator impedance $= (0.05 + j0.2)\ \Omega$

Equivalent rotor impedance $= (0.05 + j0.24)\ \Omega$

The equivalent circuit is shown in Fig. 5.61.

$$R_{01} = R_1 + R_2' = 0.05 + 0.05 = 0.1\ \Omega$$
$$X_{01} = X_1 + X_2' = 0.2 + 0.24 = 0.44\ \Omega$$
$$\therefore \quad Z_{01} = \sqrt{R_{01}^2 + X_{01}^2} = \sqrt{(0.1)^2 + (0.44)^2} = 0.45\ \Omega$$

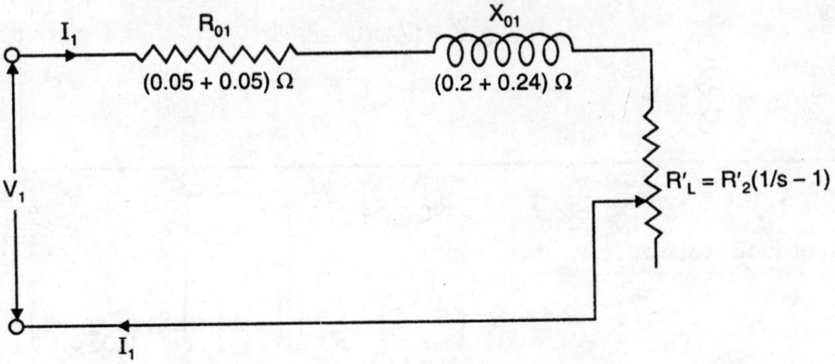

Fig. 5.61

Slip :

We know that slip corresponding to maximum gross power output (Eqn. 5.30) is given by

$$s = \frac{\frac{R_2}{K^2}}{\frac{R_2}{K^2} + Z_{01}} = \frac{R_2'}{R_2' + Z_{01}} = \frac{0.05}{0.05 + 0.45} = 0.1 \text{ or } 10\%$$

Hence, *slip at which maximum gross power output occurs*

$$= 10\%. \text{ (Ans.)}$$

Maximum gross power output

$$= \frac{3V_1^2}{2(R_{01} + Z_{01})} = \frac{3\left(\frac{400}{\sqrt{3}}\right)^2}{2(0.1 + 0.45)}$$

$$= \mathbf{145454\ W \text{ or } 145.454\ kW.} \quad \textbf{(Ans.)}$$

Example 5.79. *The following data pertain to an induction motor : stator impedance = (1.2 + j3.0) Ω ; rotor standstill impedance = (1.2 + j2.0) Ω ; no-load shunt impedance = (12 + j55) Ω ; volts/phase = 240. Determine :*

(i) Stator current, *(ii) Equivalent rotor current,*
(iii) Input power factor, *(iv) Mechanical power developed, and*
(v) Efficiency of the motor at a slip of 5%.

Solution. Stator impedance $= (1.2 + j3.0)\ \Omega$

Rotor standstill impedance $= (1.2 + j2.0)\ \Omega$

No-load shunt impedance $= (12 + j55)\ \Omega$

Volts/phase $= 240$ V

The equivalent circuit per phase of motor referred to stator is shown in Fig. 5.62. It is *assumed* that the given impedance figures are *phase values* and there is *unity turn-ratio* between stator and rotor.

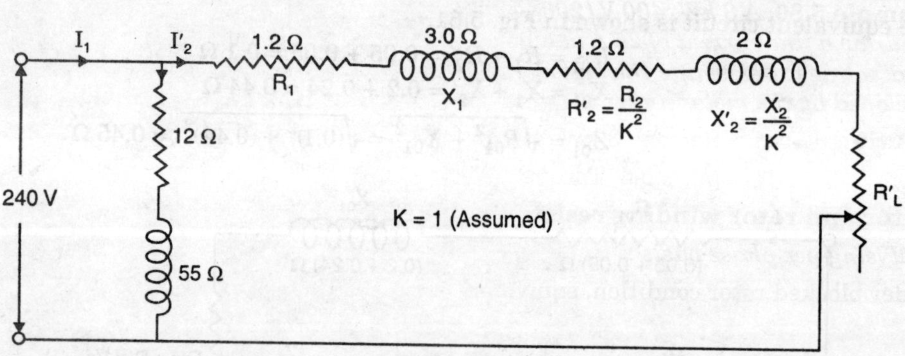

Fig. 5.62

Equivalent load resistance,

$$R_L' = R_2'\left(\frac{1}{s} - 1\right) = \frac{R_2}{K^2}\left(\frac{1}{s} - 1\right) = 1.2\left(\frac{1}{0.05} - 1\right) = 22.8\ \Omega$$

Effective impedance/phase $= (R_1 + R_2' + R_L') + j(X_1 + X_2') = (1.2 + 1.2 + 22.8) + j(3 + 2)$

$= (25.2 + j5)\ \Omega$

(i) Stator current, I_1 :

Stator load current,

$$I_2' = \frac{240}{25.2 + j5} = \frac{240(25.2 - j5)}{(25.2 + j5)(25.2 - j5)}$$

$$= \frac{240(25.2 - j5)}{660} = (9.16 - j1.82)\ \text{A}$$

Stator no-load current, $I_0 = \dfrac{240}{(12 + j55)} = \dfrac{240}{12 + j55} \times \dfrac{(12 - j55)}{(12 - j55)}$

$$= \frac{240(12 - j55)}{3169} = (0.91 - j4.16) \text{ A}$$

Total stator current, $I_1 = I_2' + I_0 = (9.16 - j1.82) + (0.91 - j4.16)$

$$= (10.07 - j5.98) \text{ A}$$

$\therefore$ $I_1 = \sqrt{(10.07)^2 + (5.98)^2} = \mathbf{11.71\ A.\ (Ans.)}$

(*ii*) **Equivalent rotor current,**

$$I_2' = \sqrt{(9.16)^2 + (1.82)^2} = \mathbf{9.34\ A.\ (Ans.)}$$

(*iii*) **Input power factor,**

$$\cos\phi = \frac{10.07}{11.71} = \mathbf{0.86.\ (Ans.)}$$

(*iv*) **Mechanical power developed,**

$$= 3(I_2')^2 R_L' = 3 \times (9.34)^2 \times 22.8$$

$$= \mathbf{5967\ W} \text{ or } \mathbf{5.967\ kW.\ (Ans.)}$$

(*v*) **Efficiency of the motor,**

$$\eta = \frac{\text{Output}}{\text{Input}} \times 100 = \frac{5967}{3 \times 240 \times 11.71 \times 0.86} \times 100$$

$$= \mathbf{82.29\%.\ (Ans.)}$$

Example 5.80. *A 3 kW, 400 V/200 V, Delta/Star, 50 Hz, three-phase, 6 pole induction motor is found to draw a line current of 25 A at a power factor of 0.4, when a blocked rotor test is conducted at the rated voltage. Determine the stator and rotor winding resistances in ohms per phase, if the torque developed by the motor under the above conditions is 25 Nm.* **(GATE, 1999)**

Solution. *Given* : Rating : 3 kW, 400 V/200 V ; Delta/star, 50 Hz ; $p = 6$; $I_L = 25$ A ; p.f. = 0.4, Torque developed, $T = 25$ N-m.

Stator and rotor winding resistances, $\mathbf{R_1}$, $\mathbf{R_2}$:

Analysing per phase basis :

Under blocked rotor condition, equivalent circuit is shown in Fig. 5.63.

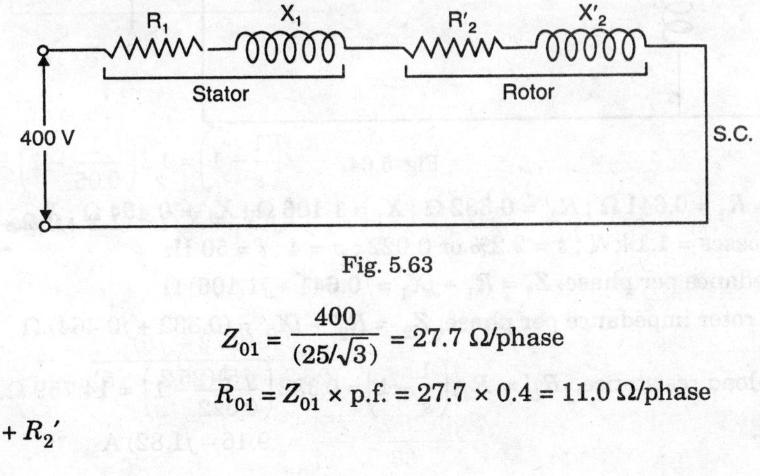

Fig. 5.63

$$Z_{01} = \frac{400}{(25/\sqrt{3})} = 27.7\ \Omega/\text{phase}$$

$$R_{01} = Z_{01} \times \text{p.f.} = 27.7 \times 0.4 = 11.0\ \Omega/\text{phase}$$

where, $R_{01} = R_1 + R_2'$

Turn ratio, $K = \frac{(200/\sqrt{3})}{400} = 0.2887$

Rotor current, $I_2 = \frac{(I_1/\sqrt{3})}{K} = \frac{(I_L/\sqrt{3})}{K} = \frac{25/\sqrt{3}}{0.2887} = 50$ A

Synchronous speed, $N_s = \frac{120f}{p} = \frac{120 \times 50}{6} = 1000$ r.p.m.

We know that, $\left(\frac{2\pi N_s}{60}\right) \times T = 3I_2^2R_2$

or,
$$\mathbf{R_2} = \frac{2\pi N_s T}{60 \times 3 \times I_2^2} = \frac{2\pi \times 1000 \times 25}{60 \times 3 \times 50^2}$$
$$= \mathbf{0.349\ \Omega/phase.\ (Ans.)}$$
$$R_2' = \frac{R_2}{K^2} = \frac{0.349}{(0.2887)^2} = 4.187\ \Omega$$

Also, $R_{01} = R_1 + R_2'$

$\therefore$ $\mathbf{R_1} = R_{01} - R_2' = 11.08 - 4.187 = \mathbf{6.893\ \Omega/phase.}$ **(Ans.)**

Example 5.81. *A 25 H.P., 400 V, 50 Hz, 4-pole, star-connected induction motor has the following impedances per phase in ohms referred to the stator side :*

$$R_s = 0.641,\ R_r = 0.332\ ;\ X_s = 1.106,\ X_r = 0.464 \text{ and } X_{mag} = 26.30.$$

Rotational losses are assumed constant and are 1.1 kW and core losses are assumed negligible. If the slip is 2.2% at rated voltage and frequency, find :

(i) Speed ; (ii) Stator current ; (iii) Power factor ; (iv) Output and input-power ; (v) Efficiency of motor. **[U.P.S.C., 1997]**

Solution. Refer Fig. 5.64.

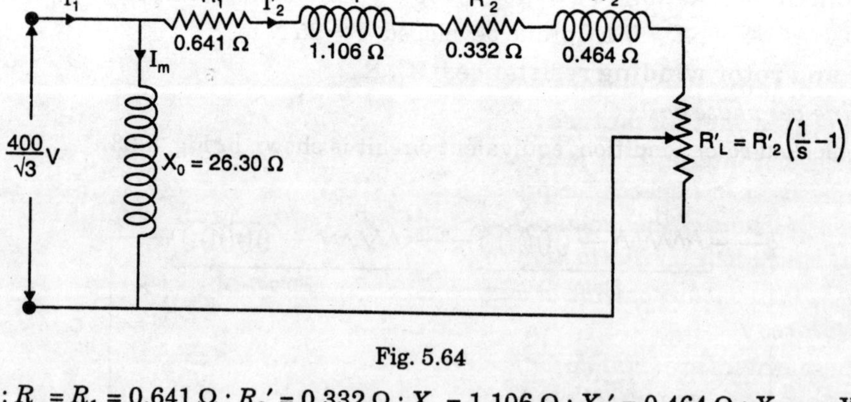

Fig. 5.64

Given : $R_s = R_1 = 0.641\ \Omega$; $R_2' = 0.332\ \Omega$; $X_1 = 1.106\ \Omega$; $X_2' = 0.464\ \Omega$; $X_{mag} = X_0 = 26.30\ \Omega$; rotational losses = 1.1 kW ; s = 2.2% or 0.022 ; p = 4 ; f = 50 Hz.

Stator impedance per phase, $Z_1 = R_1 + jX_1 = (0.641 + j1.106)\ \Omega$

Equivalent rotor impedance per phase, $Z_2' = R_2' + jX_2' = (0.332 + j0.464)\ \Omega$

Equivalent load resistance, $R_L' = R_2'\left(\frac{1}{s} - 1\right) = 0.332\left(\frac{1}{0.022} - 1\right) = 14.759\ \Omega.$

Voltage applied per phase $= \frac{400}{\sqrt{3}} = 230.9$ V

Effective impedance per phase, $Z = Z_1 + Z_2' + R_L'$

$= (0.641 + j1.106) + (0.332 + j0.464) + (14.759 + j0)$

$= (15.732 + j1.57)$ or $15.81 \angle 5.7° \ \Omega$

Counter-balancing rotor current per phase,

$$I_2' = \frac{230.9}{15.81 \angle 5.7°} = 14.6 \angle -5.7° = (14.53 - j1.45) \text{ A.}$$

Magnetising component of no-load current,

$$I_m = \frac{230.9}{X_0} = \frac{230.9}{26.30} = 8.78 \text{ A.}$$

(Energy component of no-load current, $I_w = 0$, because core losses are negligible.)

Hence no-load current, $I_0 = (0 - j8.78)$ A.

Stator current per phase, $I_1 = I_2' + I_0 = (14.53 - j1.45) + (0 - j8.78)$

$= (14.53 - j10.23)$ or $17.77 \angle -35.15°$ A.

(*i*) **Speed,** $N = N_s(1 - s) = \frac{120f}{p}(1 - s)$

$= \frac{120 \times 50}{4}(1 - 0.022) =$ **1467 r.p.m. (Ans.)**

(*ii*) **Stator current** = **17.77 A. (Ans.)**

(*iii*) **Power factor,** $\cos \phi_1 = \cos(-35.15°) =$ **0.8176 lag. (Ans.)**

(*iv*) **Output power,** $P_{out} = 3I_2'^2 R_L'$ – rotational losses

$= [3 \times (14.6)^2 \times 14.759] - (1.1 \times 1000) =$ **8338 W. (Ans.)**

Power input, $P_{in} = 3 \times 230.9 \times 17.77 \times 0.8176 =$ **10064 W. (Ans.)**

(*v*) **Efficiency of motor,** $\eta = \frac{P_{out}}{P_{in}} = \frac{8338}{10064} = 0.8285$ or **82.85%. (Ans.)**

5.13.1.14. Induction motors with special designs

High slip induction motors :

- In case of intermittent load applications which involve frequent start and stop and/or running at low speeds for prolonged periods, induction motors are *designed with high rotor resistance.* Such motors have *low starting current and high starting torque, but low full load efficiency due to high rotor copper loss.*

Because these motors operate at a large slip (between 10 and 40% of full load) *they are called* ***high slip motors.***

- These motors are *suitable for fan drives where speed is controlled by stator voltage control* and are found among both the squirrel-cage and wound rotors. The nature of speed-torque characteristic of such motors is shown in Fig. 5.65 (c)

— Good starting performance (low starting current and high starting torque), in squirrel-cage induction motors, is realised without appreciably affecting full load performance by the *use of deep-bar rotor or double-case rotor motors.*

— With the change of speed from standstill to full load, the motor frequency changes from 50 Hz to 1-3 Hz. *Variation of motor frequency is utilised in these motors to vary rotor resistance from a large value at standstill to a very small value at full speed.* Thus, while starting and low speed performance is improved, full load performance is not appreciably effected.

Deep-bar squirrel-cage rotor induction motor :

- In this type of motor, stator of the machine is identical to a general purpose induction motor, but the rotor has deep and narrow conductor bars. A rotor bar and the slot leakage fluxes produced by the current in bar are shown in Fig. 5.65 (*a*).

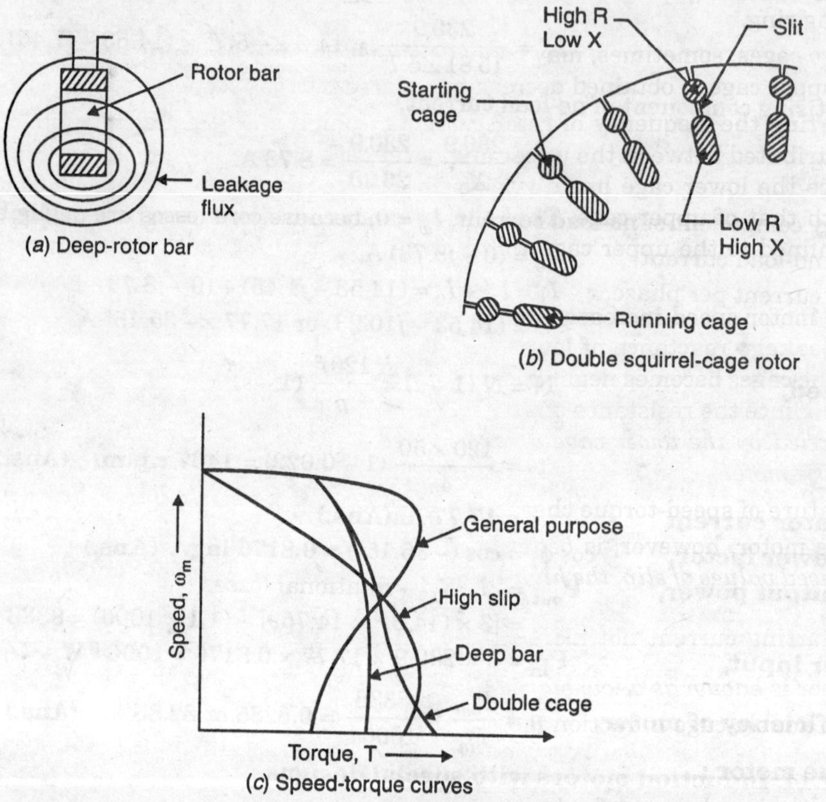

Fig. 5.65. Induction motors of special design.

— It can be imagined that the bar is made of a number of narrow layers connected in parallel. Since more leakage flux links with the bottom layer than top layer, therefore, bottom layer has a much higher leakage inductance than the top layer. As rotor frequency is high at low speeds, the reactance and impedance of bottom layer are much higher than the top layer. Therefore, at *low speeds highest amount of current is carried by topmost layer and lowest by the bottom most.*

— *Due to unequal distribution of current across cross-section of the bar, effective resistance of rotor is high and starting and low speed performance is improved.*

— At near full load speed the frequency of rotor current as well as leakage reactance are low. Therefore, currents gets equally distributed across cross-section of the bar and effective resistance has a low value. Thus, full load performance is not affected too much. Fig. 5.65 (*c*) shows the motor speed-torque curve.

Double squirrel-cage motor induction motor

Fig. 5.65 (*b*) shows a double squirrel-cage rotor. There are two sets of squirrel-cage windings. The upper cage (starting cage) arranged nearer to the air-gap is made of *high resistivity material*—brass, aluminium, bronze etc. while the lower cage (running cage) is made of copper, the two cages are separated by narrow slits. As a result the lower cage has high permeance for leakage flux due to which the leakage reactance of lower cage is considerably higher than that of upper cage. The upper and lower cages may have either common short circuiting end rings or each of the cages has its own short circuiting ring.

The two cages, sometimes, may be made of cast aluminium. In this case a high active resistance for the upper cage is obtained decreasing its area of cross-section.

At starting the frequency of rotor currents is high and is equal to supply frequency. The current is distributed between the upper and lower cages in the inverse proportion of their impedance and since the lower cage has a very high leakage reactance, its impedance is very large as compared with that of upper cage. Thus the current in the lower cage is small and therefore the current is confined to the upper cage which has a high resistance, this gives a *very good starting torque*.

As the motor speed increases, the frequency of rotor currents decreases and this results in reduction of leakage reactance of lower cage when the motor runs near about its full speed, the reactance of the cages becomes negligible and therefore the current divides itself in the inverse ratio of resistances. Since the resistance of upper cage is about 5 to 6 times that of lower cage, *most of the current is carried by the lower cage and the motor has the running characteristic of the low-slip general-purpose motor*.

The nature of speed-torque characteristics is shown in Fig. 5.65 (*c*).

- This motor, however, is *better adapted to adjust to suddenly applied loads, since at increased values of slip, the high-resistance winding again develops significant torque*. With a starting current not more than $5\frac{1}{2}$ times full-load current, the double-squirrel-cage motor is *known as a low-starting current, high-starting-torque, low-slip motor*.
- The double-cage induction motor *is ideal for compressors*.

Torque motor :

Motors designed to run for long periods in a stalled or low speed condition are known as ***torque motors.***

- They are designed to develop desired torque with low current at low speeds.
- Their speed-torque characteristics are *shaped to have negative slope so that they provide stable operation with most loads at low speeds*.
- They can be squirrel-cage or wound-rotor type. Both single-phase and three-phase motors are available.

5.13.1.15. Starting of induction motors

Small induction motors (up to 2 kW) capacity may directly be switched on to the supply mains, but those of higher capacity must use some type of starting device, or starters as they are commonly called. *The function of these starters is to restrict the initial rush of current*, which, in the case of induction motors, is about 5 times the full-load current. This excessive current has two major upsetting effects, namely, a *large voltage drop in the distributing network* and *causing stoppage of machines which are already running on the supply mains*. Hence the Electrical Undertaking Authorities *forbid the users of large capacity induction motors to directly switch on their machines*.

The principle of all starting devices is to impress lower voltage on stator phases at the time of starting, or if the motor is slip ring or wound rotor, then to include external resistance in each rotor phase to keep the initial rotor current to a low value, this consequently means less current in the stator phases and therefore in the supply mains.

Direct-on-line starting of induction motors. This method means switching the motor directly on to the supply without using any device for reducing the starting current. The method is restricted to small motors up to about 2 kW. For these small motors, the ***starting torque is about twice the full-load torque.*** Hence the starting period lasts only a few seconds.

When the motor is direct-switched on to normal voltage, then starting current is the short-circuit current I_{sc} and can be proved that,

$$\frac{T_{st}}{T_f} = \left(\frac{I_{sc}}{I_f}\right)^2 s_f \qquad \text{...[5.28 (b)]}$$

where, T_{st} = Starting torque,

I_f = Normal full-load current, and

s_f = Full load slip.

When needed, methods employed for starting induction motors are :

Squirrel-cage motors :

(*i*) Star-delta starter
(*ii*) Auto-transformer starter
(*iii*) Reactor starter
(*iv*) Saturable reactor starter
(*v*) Part winding starter
(*vi*) A.C. voltage controller starter.

Slip ring motors. Rotor resistance starter.

1. ***Star-Delta starter :***

Star-delta switching method is based upon the principle that with three windings connected in star, voltage across each winding is $\frac{1}{\sqrt{3}}$ *i.e.*, 57.7% of the line to line voltage whereas the same windings connected in delta will have full-line-to-line voltage across each.

The star-delta starter *connects the three stator windings in star* across the rated supply voltage *at the starting instant. After the motor attains speed the same windings through a change over switch are re-connected in delta across the same supply voltage.*

The basic connection diagram of a star-delta starter is shown in Fig. 5.66. An actual starter incorporates *under-voltage* and *over-load coils.* The starter is also provided with a mechanical inter-locking device to prevent the handle from being put in the 'Run' position first. Such starters are employed for starting 3-phase squirrel-cage induction motors of rating between 4 kW and 15 kW.

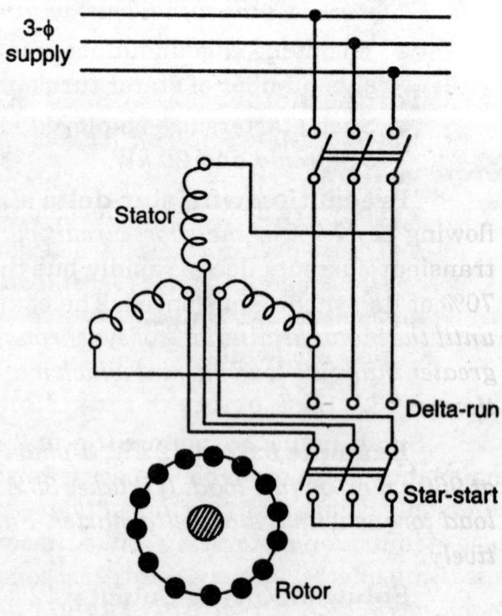

Fig. 5.66. Star-delta starter.

When *star-connected,* the applied voltage over each motor phase is reduced by a factor of $\frac{1}{\sqrt{3}}$ and hence the torque becomes $\frac{1}{3}$ of that which would have been developed if motor were directly connected in delta. The line current is reduced to $\frac{1}{3}$. Hence, during starting period when motor is star-connected, it takes $\frac{1}{3}$rd as much starting current and develops $\frac{1}{3}$rd as much torque as would have been developed were it directly connected in delta.

Relation between T_{st} and T_f :

$$I_{st} \text{ per phase} = \frac{1}{\sqrt{3}} I_{sc} \text{ per phase}$$

where I_{sc} is the current per phase which delta-connected motor would have taken if switched on to the supply directly ; however, line current at start is equal to $\frac{1}{3}$ of line I_{sc}.

Now, starting torque, $T_{st} \propto I_{st}^2 \ (s = 1)$

Full-load torque, $T_f \propto I_f^2/s_f$

$$\therefore \quad \frac{T_s}{T_f} = \left(\frac{I_{st}}{I_f}\right)^2 s_f = \left(\frac{I_{sc}}{\sqrt{3} I_f}\right)^2 s_f = \frac{1}{3}\left(\frac{I_{sc}}{I_f}\right)^2 s_f$$

i.e.,

$$\frac{T_{st}}{T_f} = \frac{1}{3}\left(\frac{I_{sc}}{I_f}\right)^2 s_f \quad \text{...(5.29)}$$

Here, I_{st} and I_{sc} represent phase values.

- This method reduces the starting line current to one-third but the starting torque is also reduced by the same amount.
- *This method is cheap but limited to applications where high starting torque is not necessary e.g., machine tools, pumps, motor-generator sets etc.*
- The method is unsuitable for motors for voltage exceeding 3000 V because of the excessive number of stator turns needed for delta connection.
- Such starters are employed for starting *3-phase squirrel cage induction motors of rating between 4 and 20 kW.*

Precaution with star-delta starting. When the motor is started in star the initial current flowing is *57.7% of the short-circuit current in delta together with a transient in each phase.* The transient currents decay rapidly but the steady state is not reached until the motor has attained 70% of its synchronous speed. The change-over from star to delta connection *should not be made until the motor attains 90% of synchronous speed, otherwise there will be a current surge considerably greater than full-load current which may even be greater than the standstill current with star-connection.*

Example 5.82. *A 12 kW, 3-phase, 6-pole, 50 Hz, 400 V, delta-connected induction motor runs at 960 r.p.m. on full-load. If it takes 85 A on direct starting, find the ratio of the starting torque to full-load torque with a star-delta starter. Full-load efficiency and power factor are 88% and 0.85 respectively.* **(Nagpur Univ. 2000)**

Solution. *Given* : Output = 12 kW ; $p = 6, f = 50$ Hz ; $V_L = 400$ V, $N_f = 960$ r.p.m.,

$\eta = 88\%$, full-load power factor = 0.85

$\frac{\mathbf{T_{st}}}{\mathbf{T_f}}$:

Full-load line current drawn by a 3-phase delta-connected induction motor

$$= \frac{\text{Output}}{\sqrt{3}\, V_L \times \text{power factor} \times \eta} = \frac{12 \times 1000}{\sqrt{3} \times 400 \times 0.85 \times 0.88} = 23.16 \text{ A}$$

Full-load current per phase, $I_f = \frac{23.16}{\sqrt{3}} = 13.37$ A

Short-circuit current per phase, $I_{sc} = \frac{85}{\sqrt{3}} = 49.07$ A

Synchronous speed, $N_s = \frac{120f}{p} = \frac{120 \times 50}{6} = 1000$ r.p.m.

Full-load slip, $s_f = \frac{N_s - N_f}{N_s} = \frac{1000 - 960}{1000} = 0.04$

Using the relation, $\frac{T_{st}}{T_f} = \frac{1}{3}\left(\frac{I_{sc}}{I_f}\right)^2 \times s_f = \frac{1}{3}\left(\frac{49.07}{13.37}\right)^2 \times 0.04 = \mathbf{0.18}$. **(Ans.)**

Example 5.83. *A squirrel cage type induction motor, when started by means of a star-delta starter takes 200% full-load current (line) and develops 44% of full-load torque at starting. Calculate the starting torque and current, if an auto-transformer with 75% tapping were employed.*

(A.M.I.E. Summer, 2000)

Solution. *Given* : $I_{st} = 200\%\ I_f$ or $2I_f$, $T_{st} = 44\%$ of $T_f = 0.44\ T_f$, $K = 75\%$ or 0.75

Supply line current at start, $I_{sc} = 3$ times, the current drawn by motor with star-delta starter

$$= 3 \times 2I_f = 6I_f$$

Starting torque with *star-delta starter*,

$$T_{st} = \frac{1}{3}\left(\frac{I_{sc}}{I_f}\right)^2 s_f = 0.44T_f$$

or, $$\frac{T_{st}}{T_f} = \frac{1}{3}\left(\frac{I_{sc}}{I_f}\right)^2 s_f = 0.44$$

or, $$\frac{1}{3} \times 6^2 \times s_f = 0.44$$

or, Full-load slip, $s_f = 0.0366$

Starting current (line value) with auto-transformer,

$$I_{st} = K^2 I_{sc} = (0.75)^2 \times 6I_f = 3.375I_f$$

or, ***337.5% of full-load current.*** **(Ans.)**

Starting torque, $$T_{st} = K^2\left(\frac{I_{sc}}{I_f}\right)^2 s_f T_f$$

$$= (0.75)^2 \times 6^2 \times 0.0366T_f = 0.7411T_f \text{ or } \mathbf{74.11\%\ of\ T_f}. \quad \textbf{(Ans.)}$$

Example 5.84. *A 3-phase induction motor has a ratio of maximum torque to full-load torque as 2.5 : 1. Determine the ratio of actual starting torque to full-load torque for star-delta starting. Given $R_2 = 0.4\ \Omega$ and $X_2 = 4\ \Omega$.*

Solution. *Given* : $\frac{T_m}{T_f} = 2.5$; $R_2 = 0.4\ \Omega$, $X_2 = 4\ \Omega$

$\frac{\mathbf{T_{st}}}{\mathbf{T_f}}$ **:**

We know that $$\frac{T_f}{T_m} = \frac{2s_f s_{mT}}{s^2{}_f + s^2{}_{mT}} \quad \text{[Eqn. (5.26)]}$$

where, s_f = full-load slip and s_{mT} (slip at maximum torque) $= \frac{R_2}{X_2} = \frac{0.4}{4} = 0.1$

or, $$\frac{1}{2.5} = \frac{2s_f \times 0.1}{s_f{}^2 + 0.1^2} = \frac{0.2s_f}{s_f{}^2 + 0.01} \quad \text{or} \quad s_f^2 - 0.5s_f + 0.01 = 0$$

or, $$s_f = \frac{0.5 \pm \sqrt{0.5^2 - 4 \times 0.01}}{2}$$

$$= \frac{0.5 \pm 0.458}{2} = 0.021, \text{ rejecting the higher value}$$

Full-load rotor current per phase, $$I_f = \frac{s_f E_2}{\sqrt{R^2 + s_f{}^2 X_2{}^2}}$$

Short-circuit rotor current per phase, $I_{sc} = \frac{E_2}{\sqrt{R_2{}^2 + X_2{}^2}}$ $(\because$ At standstill $s = 1)$

$$\therefore \quad \frac{I_{sc}}{I_f} = \frac{\sqrt{R_2{}^2 + s_f{}^2 X_2{}^2}}{s_f \sqrt{R_2{}^2 + X_2{}^2}} = \frac{\sqrt{0.4^2 + 0.021^2 \times 4^2}}{0.021\sqrt{0.4^2 + 4^2}} = \frac{0.4087}{0.0844} = 4.84$$

Starting torque with star-delta starter, $$T_{st} = \frac{1}{3} T_f \left(\frac{I_{sc}}{I_f}\right)^2 s_f \quad \text{[Eqn. (5.29)]}$$

$$= \frac{1}{3} T_f \times (4.84)^2 \times 0.021 = 0.164T_f$$

$$\therefore \quad \frac{T_{st}}{T_f} = \mathbf{0.164.} \quad \textbf{(Ans.)}$$

2. *Auto-transformers starter :*

Fig. 5.67 shows the connection diagram for auto-transformer starting of squirrel-cage induction motors.

In this method the reduced voltage is obtained by taking tappings at suitable points from a three-phase auto-transformer (Fig. 5.67). The auto-transformers are generally tapped at the 50, 60 and 80 per cent points, so that adjustment at these voltages may be made for proper starting torque requirements. Since the contacts frequently break large values of current arcing is sometimes quenched effectively by having them assembled to operate in an oil bath.

Auto-transformers may be either normally, or magnetically operated.

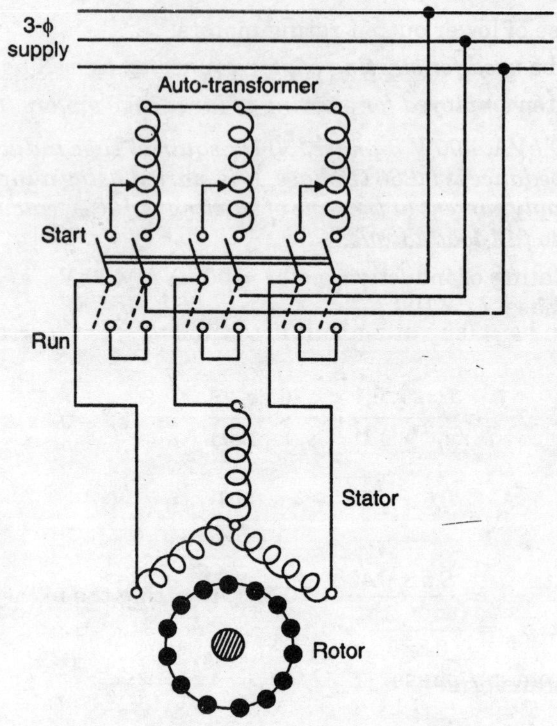

Fig. 5.67. Auto-transformer starter.

Relation between starting (T_{st}) and full-load (T_f) torques. Let the motor be started by an auto-transformer having transformation ratio K. If I_{sc} is the starting current when normal voltage is applied and applied voltage to stator winding at starting is KV then,

Motor input current, $I_{st} = KI_{sc}$

Supply current = Primary current of auto-transformer

$= K^2 I_{sc}$

and,
$$\frac{T_{st}}{T_f} = \left(\frac{I_{st}}{I_f}\right)^2 \times s_f = K^2 \left(\frac{I_{sc}}{I_f}\right)^2 \times s_f$$

Advantages :

1. Voltage is reduced by transformation and not by dropping the voltage in resistors, and therefore, the current and power drawn from the supply mains are also reduced in comparison to resistor starting.
2. Availability of highest torque per ampere of supply current.
3. Adjustment of starting voltage by selection of proper tap on the auto-transformer.
4. The method is suitable for long starting periods.
5. Motor current larger than supply current.
6. Closed transition starting.

Disadvantages :

1. Low power factor.
2. Higher cost in case of lower output rating motors.

- This method can be *used for starting of star-connected as well as delta-connected motors.*
- This method is often *employed for starting of large cage motors (rating exceeding 20 kW).*

Example 5.85. *A 50 kVA, 400 V 3-phase, 50 Hz squirrel cage induction motor has full-load slip of 5%. Its standstill impedance is 0.866 Ω/phase. It is started using a tapped auto-transformer. If the maximum allowable supply current at the time of starting is 100 A, calculate the tap position and the ratio of starting torque to full-load torque.* **(GATE, 1997)**

Solution. *Given :* Rating of induction motor = 50 kVA, 400 V ; s_f = 5% or 0.05 ; standstill impedance = 0.866 Ω per phase, I_{st} = 100 A.

Tap position ; $\frac{\mathbf{T_{st}}}{\mathbf{T_f}}$:

$$\text{Full-load current} = \frac{\text{Output in kVA} \times 1000}{\sqrt{3} \times \text{line voltage}}, \text{ neglecting losses}$$

$$= \frac{50 \times 1000}{\sqrt{3} \times 400} = 72.2 \text{ A}$$

$$\text{Short-circuit current, } I_{sc} = \frac{(400/\sqrt{3})}{0.866} = 266.7 \text{ A}$$

Tap position of the transformer,

$$K = \sqrt{\frac{I_{st}}{I_{sc}}} = \sqrt{\frac{100}{266.7}} = 0.6123 \text{ or } \mathbf{61.23\%. \text{ (Ans.)}}$$

Now,
$$\frac{T_{st}}{T_f} = K^2 \left(\frac{I_{sc}}{I_f}\right)^2 s_f = (0.6123)^2 \times \left(\frac{266.7}{72.2}\right)^2 \times 0.05 = \mathbf{0.256. \text{ (Ans.)}}$$

Example 5.86. *A 3-phase induction motor takes a starting current which is 5 times full-load current at normal voltage. Its full-load slip is 4 percent. What auto-transformer ratio would enable the motor to be started with not more than twice the full-load current drawn from the supply ? What would be the starting torque under this condition ?* **(AMIE Summer, 2000)**

Solution. *Given* : $I_{st} = 5I_f$; s_f = 4% or 0.04

Let the transformation ratio of auto-transformer be K.

Now the starting current, $I_{st} = K^2 I_{sc}$

or,
$$\frac{I_{st}}{I_f} = K^2 \frac{I_{sc}}{I_f} \quad \text{or} \quad 2 = K^2 \times 5$$

or,
$$\mathbf{K} = \sqrt{\frac{2}{5}} = 0.6324 \quad \text{or} \quad \mathbf{63.24\%.} \quad \mathbf{(Ans.)}$$

Starting torque,
$$\mathbf{T_{st}} = K^2 \left(\frac{I_{sc}}{I_f}\right)^2 s_f T_f = 0.6324^2 \times 5^2 \times 0.04 \times T_f = 0.4T_f$$

or, **40% of full-load torque. (Ans.)**

Example 5.87. *Estimate approximately the starting torque of a 3-phase induction motor in terms of full-load torque when started by (i) star-delta starter (ii) an auto-transformer starter with 50 percent tapping. The short-circuit current of motor is 5 times the full-load current and the full-load slip is 5 percent.* **(Nagpur Univ., 2001)**

Solution. *Given* : $I_{sc} = 5I_f$; $s_f = 5\%$ or 0.05.

(*i*) Starting torque with *star-delta starter*,

$$T_{st} = \frac{1}{3}\left(\frac{I_{sc}}{I_f}\right)^2 s_f \times T_f = \frac{1}{3} \times 5^2 \times 0.05 \times T_f = 0.4167T_f$$

i.e. **41.67% of full-load torque. (Ans.)**

(*ii*) Starting torque with an *auto-transformer starter* with 50% tapping (*i.e.* $K = 0.5$)

$$T_{st} = K^2 \left(\frac{I_{sc}}{I_f}\right)^2 s_f T_f = 0.5^2 \times 5^2 \times 0.05T_f = 0.3125T_f$$

i.e. **31.25% of full-load torque. (Ans.)**

Example 5.88. *A 20 H.P., 400 V, 950 r.p.m., 3-phase, 50 Hz, 6-pole cage induction motor draws 6 times the full-load current at standstill from a 400 V mains and develops 1.8 times full-load running torque. The full-load current is 30 A.*

(*i*) *What voltage must be applied to produce full-load torque at starting ?*

(*ii*) *What current will this voltage produce ?*

(*iii*) *If this voltage is obtained by an auto-transformer, what will be the line current ?*

(*iv*) *If the starting current in the line is limited to full-load current by means of an auto-transformer, what will be the starting torque as a percentage full-load torque ?*

Assume full-load slip as 5% and neglect magnetising current and the stator impedance drops.

(AMIE Summer, 1998)

Solution. *Given* : H.P. = 20 ; $V = 400$ volts ; $f = 50$ Hz ; $p = 6$; $I_{sc} = 6I_f$; $I_f = 30$ A ;

$T_{st} = 1.8T_f$; $s_f = 5\%$ or 0.05.

(*i*) Let the voltage applied to develop starting torque T_{st} equal to full-load torque be V' volts.

$\because \quad T_{st} \propto (\text{applied voltage})^2$

$$\therefore \quad \left(\frac{V'}{V}\right)^2 = \frac{T'_{st}}{T_{st}} = \frac{T_f}{1.8\,T_f} = \frac{1}{1.8}$$

or,
$$V' = \sqrt{\frac{1}{1.8}} \times V = \frac{400}{\sqrt{1.8}} = \textbf{298.14 V.} \quad \textbf{(Ans.)}$$

(*ii*) **Current produced** *at start by* $V' = I_{sc} \times \frac{V'}{V} = 6I_f \times \frac{V'}{V}$

$$= (6 \times 30) \times \frac{298.14}{400} = \textbf{134.16 A.} \quad \textbf{(Ans.)}$$

Transformation ratio,
$$K = \frac{V'}{V} = \frac{298.14}{400} = 0.745$$

(*iii*) **Line current with auto-transformer** *of ratio of 0.745*

$$= K^2 I_{sc} = 0.745^2 \times 180 = \textbf{99.9 A.} \quad \textbf{(Ans.)}$$

(iv) Auto-transformer transformation ratio to produce starting line current equal to full-load current ;

$$K = \sqrt{\frac{I_{st}}{I_{sc}}} = \sqrt{\frac{I_f}{6I_f}} = 0.408$$

$$\therefore \qquad \mathbf{T_{st}} = K^2 \left(\frac{I_{sc}}{I_f}\right)^2 sT_f = (0.408)^2 \times (6)^2 \times 0.05T_f = 0.3T_f$$

i.e. **30% of full-load torque. (Ans.)**

Example 5.89. *In a 3-phase induction motor, leakage reactance is four times the resistance for both stator and rotor circuits. But stator leakage reactance is equal to rotor standstill leakage reactance referred to stator. The motor develops a torque of 220 Nm at a slip of 0.05. Calculate the starting and maximum torques.*

The exciting current may be neglected. **(Panjab University)**

Solution. *Given* : $X_1 = X_2 = 4R_1 = 4R_2$; $T = 220$ Nm ; $s = 0.05$.

Starting and maximum torques, T_{st}, T_m :

We know that, $$T = \frac{3}{\omega_s} . I_2^2 . \frac{R_2}{s} \quad \text{(for 3-phases)}$$ ...[Refer to eqn. 5.25]

or, $$T = \frac{3}{\omega_s}\left[\frac{V^2}{\left(R_2 + \frac{R_2}{s}\right)^2 + (X_1 + X_2)^2}\right] . \frac{R_2}{s}$$

Substituting the values, we get

$$220 = \frac{3V^2}{\omega_s} \frac{1}{\left[\left(1 + \frac{1}{0.05}\right)^2 + (4+4)^2\right] R_2} . \frac{R_2}{0.05}$$

or, $$\frac{3V^2}{\omega_s} = 200\left[\left(1 + \frac{1}{0.05}\right)^2 + (4+4)^2\right] \times 0.05 = 5555$$

$$\therefore \qquad \mathbf{T_{st}} = \frac{3V^2}{\omega_s} \frac{1}{(1+1)^2 + (4+4)^2} = 5555 \times \frac{1}{(4+64)} = \mathbf{81.69\ N\text{-}m.\ (Ans.)}$$

$$s_{mT} = \frac{R_2}{\sqrt{R_1^2 + (X_1 + X_2)^2}} = \frac{1}{\sqrt{1 + (4+4)^2}} = 0.124$$

$$\therefore \qquad \mathbf{T_m} = \frac{3V^2}{\omega_s} \times \frac{1}{\left(1 + \frac{1}{0.124}\right)^2 + 8^2} \times \frac{1}{0.124}$$

$$= 5555 \times \frac{1}{146.16} \times \frac{1}{0.124} = \mathbf{306.5\ N\text{-}m.} \quad \textbf{(Ans.)}$$

3. **Reactor starter.** The starting current can also be reduced by connecting a 3-*phase reactor in series with stator.* As soon as motor attains full speed, the reactor is by passed. It is advantageous to connect reactor at the neutral end of stator winding. This minimises its voltage rating and also maintains its voltage and the voltage of the circuit breaker at neutral potential during normal motor operation.

4. **Saturable reactor starter.** In certain applications (*e.g.* textile machines) it is essential to control the starting torque steplessly. Such a starting arrangement is termed *soft start.*

- Saturable reactor has D.C. control winding. The reactance of saturable reactor can be varied steplessly by changing the control winding current. For starting, reactance is initially set at the highest value. Starting torque is close to zero. Reactance is now reduced smoothly by increasing the control winding current. This gives stepless variation of starting torque. Consequently, motor starts with jerk, accelerates smoothly.
- For soft start, thyristor voltage controller scheme is now widely used. Several existing drives also employ saturable reactor starter in which a 3-phase saturable reactor is connected in series with the stator.

5. **Part winding starter.** Some squirrel-cage motors have two or more stator windings which are connected in parallel during normal operation. During starting, only one winding is connected. This increases stator impedance and reduces stator current.

6. **A.C. voltage controller starter.** The starting speed of an induction can be reduced by reducing stator voltage through A.C. voltage controllers.

Slip-ring Induction Motors—Starting of

(*i*) **Rotor rheostat.** The slip-ring induction motors are practically always started with full line voltage applied across the stator terminals. The value of starting current is adjusted by introducing a variable resistance in the rotor circuit. The controlling resistance is in the form of rheostat connected in star (Fig. 5.68), the resistance being gradually cut out of the rotor circuit as the motor gathers speed. *By increasing the rotor resistance, not only is the rotor* (and hence stator) *current reduced at starting* but *at the same time torque is also increased due to improvement in power factor.*

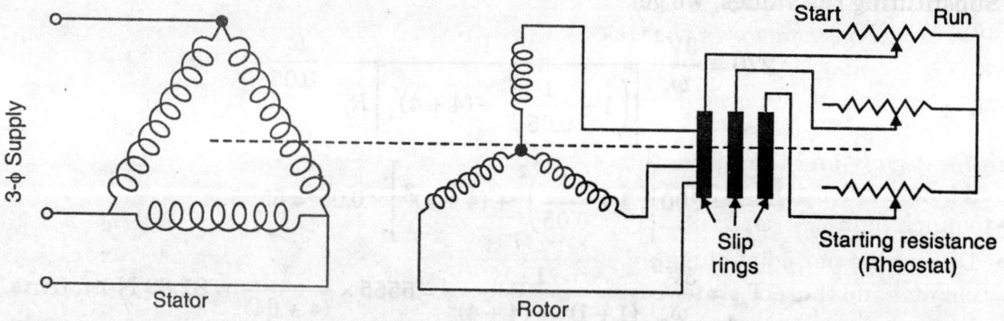

Fig. 5.68. Starting of slip-ring induction motor.

The rheostat is either of stud or contractor type and may be hand operated or automatic.

As discussed earlier, the introduction of additional external resistance in the rotor circuit enables slip-ring motor to develop a high starting torque with reasonably moderate starting current. Hence *such motors can be started under load.* When the motor runs under normal conditions the rings are short-circuited and brushes lifted from them.

5.13.1.16. Electrical braking of polyphase induction motors

The most simple way to stop an induction motor (or any other type of motor) is to *disconnect the terminals from the supply.* Torque is no longer developed, and the combined effect of the rotor and external load brings the motor to rest.

When *rapid and more positive action is required*, mechanical or electrical braking may be employed, but the *latter* has many advantages, particularly where *precise control of the stopping moment and smoothness of operation are required.*

Following are the *three* main methods of electrical braking of induction motors :

1. Plugging (or counter-current braking) ;
2. Dynamic (or rheostatic) braking ;
3. Regenerative braking.

1. Plugging :

- It is known that the rotor of a polyphase induction motor develops torque in the same direction as the rotating magnetic field set up by the stator winding. Also if any two stator leads are reversed, the rotating magnetic field is also reversed. *If, therefore, a pair of stator leads are reversed while a motor is rotating, torque is suddenly produced opposite to the original direction of rotation.* This reverse torque causes rotation in the opposite direction as soon as the motor stops, therefore provision must be made *to disconnect stator completely from the supply lines when the motor stops. A plugging controller must therefore be used in conjunction with a switch.* The plugging switch is coupled to the motor and opens when the motor stops. It is connected in the control circuit so that it permits operation of controller's reversing current when the motor rotates in a given direction only. *When the motor stops, the reversing circuit is opened by the plugging switch, and the motor is completely disconnected from the supply lines.*
- The moment the stator field reverses, the slip suddenly increases from a *small per cent to* **two hundred per cent,** since the rotor and field are now rotating in opposite directions. The rotor induced voltage also increases by a great amount, and hence there is a large in rush of stator current. Moreover, *the high slip causes a high rotor reactance,* and *hence a very low power factor in the stator.*

After plugging has started, *the best possible conditions occur when the motor stops,* slip is unity, and the current and power factor are the same as when the motor is started with full voltage. Although the above conditions occur during a very brief period of time, they may *cause disturbances in the supply* and therefore, this, as well as how frequently plugging is required, must be considered before a decision can be made to as whether or not plugging is the best way of bringing a motor to quick halt.

- During the period of plugging, since the induced e.m.f. in the rotor is very high, the rotor current and thereby stator current are very high. However, braking current can be reduced by inserting external resistance in the rotor circuit and as such wound rotor motors are beneficial as compared to squirrel cage rotor motors.
  - — Squirrel cage motors of about 20 kW output are plugged direct, using the star connection if a star-delta switch is provided. *Larger motors need stator resistors.*
  - — Wound rotor motors employ rotor resistors for current limitation as well as developing higher braking torques.
- Plugging is *advantageously employed* in the case of *reversing devices* where braking and starting up of induction motor in reverse direction comprises stages of the same continuous process.

2. Dynamic (or rheostatic) braking :

- The dynamic or rheostatic braking in case of a polyphase induction motor can be realised by disconnecting the stator winding from the A.C. supply and exciting it from (*a*) D.C. source *to produce a stationary D.C. field.*

- **In dynamic or rheostatic braking, the stator winding is employed as a D.C. field winding and the rotor winding as an armature winding.**
  - — **With a wound rotor machine, external resistors can be inserted into the rotor circuit to provide a load.**
  - — **With squirrel cage machines, however, the rotor winding itself has to form the load.**

The source of excitation may be provided by :

(*i*) an independent D.C. source, or

(*ii*) from the A.C. mains through a transformer-rectifier set.

- **The magnitude of the braking torque developed by the motor depends upon**

(*i*) The excitation (strength of the field developed by the stator winding) ;

(*ii*) The rotor circuit resistance ;

(*iii*) The speed of the motor.

Braking torque can be controlled by any or both of the following methods :

1. By controlling the D.C. excitation.
2. By varying the rotor resistance.

- A.C. dynamic braking is *not popular* owing to relatively high cost of capacitor banks.
- D.C. dynamic braking *takes little power from the supply and provides smooth braking torque, useful for mine winders and high inertia loads.*

It entails the following *advantages over plugging* :

(*i*) The absence of the reverse-rotation air-gap field (and, therefore, no tendency for the machine to run backwards).

(*ii*) Lower rotor I^2R loss.

- The advent of automatic control of dynamic braking of induction motors employing closed loop systems has made induction motors more popular than D.C. motors, especially for the drives employed in mine hoists.

3. Regenerative braking :

- When an induction motor runs at speed above synchronous speed, it operates as an induction generator and feeds power back to the supply line ; thus *regenerative braking is an inherent characteristic of an induction motor*.

The following processes may be employed to operate 3-phase induction motor at speed above synchronous speed :

(*i*) Downward motion of a loaded hoisting mechanism such as crane hoists, excavators etc.

(*ii*) Switching over to a low frequency supply in frequency controlled induction motors in order to reduce the speed of operation of the drive.

(*iii*) Switching over to a large number of pole operation from a smaller one in multi-speed squirrel cage motors.

In all the processes mentioned above the slip and torque developed become *negative* and thus the *machine acts as a generator*, receiving mechanical energy and giving it back to the supply system in the form of electrical energy.

- In case of a squirrel cage induction motor, stable speed is obtained at a speed considerably in excess of the synchronous speed and the regenerative braking cannot be applied unless the motor is specially designed to withstand the excessive speed.
- The main *disadvantage* of this method of braking is the possibility of braking only at super-synchronous speeds and, therefore, *seldom used for braking*.

- This method *can be used only in hoisting type of mechanism or with a multi-speed squirrel cage motor*. It returns about 20% of the total energy on certain railway runs and saves a great deal of brake shoe wear.

Example 5.90. *The following parameters relate to a 3-phase, 400 V, 50 Hz, 8-pole star-connected induction motor :*

$R_1 = 0.12\ \Omega$, $R_2 = 0.6\ \Omega$, $X_1 = 0.72\ \Omega$, $X_2 = 2.88\ \Omega$, $s_f = 0.05$ and the ratio of effective stator to rotor turns = 1 : 2.

The motor is to be braked at rated speed and an external resistance of 2.4 Ω per phase (referred to stator) has been inserted into the rotor circuit. Determine the initial braking torque in (i) plugging and (ii) D.C. dynamic braking.

Solution. *Given* : Supply voltage = 400 V ; $f = 50$ Hz ; $p = 8$; $R_1 = 0.12\ \Omega$; $R_2 = 0.6\ \Omega$; $X_1 = 0.72\ \Omega$; $X_2 = 2.88\ \Omega$; $s_f = 0.05$, $K = 2$; external resistance, $R' = 2.4\ \Omega$.

Rotor resistance referred to stator, $$R_2' = \frac{R_2}{K^2} = \frac{0.6}{2^2} = 0.15\ \Omega$$

Rotor leakage reactance referred to stator, $$X_2' = \frac{X_2}{K^2} = \frac{2.88}{2^2} = 0.72\ \Omega$$

(*i*) **During plugging :**

Slip, $$s = 2 - s_f = 2 - 0.05 = 1.95$$

Synchronous speed, $$N_s = \frac{120f}{p} = \frac{120 \times 50}{8} = 750 \text{ r.p.m.}$$

Initial rotor current during plugging,

$$I_2' = \frac{V}{\sqrt{\left[R_1 + \frac{R_2' + R'}{s}\right]^2 + (X_1 + X_2')^2}}$$

$$= \frac{(400/\sqrt{3})}{\sqrt{\left[0.12 + \frac{(0.15 + 2.4)}{1.95}\right]^2 + (0.72 + 0.72)^2}} = 113.9 \text{ A}$$

Initial braking torque, $$\mathbf{T_b} = \frac{3}{(2\pi N_s/60)} I_2'^2 \times \frac{R_2' + R'}{s}$$

$$= \frac{3 \times 60}{2\pi \times 750} \times (113.9)^2 \times \frac{(0.15 + 2.4)}{1.95} = \mathbf{648\ Nm.} \quad \textbf{(Ans.)}$$

(*ii*) **During D.C. dynamic braking :**

Figs. 5.69 and 5.70 show respectively the equivalent circuit of the motor corresponding to slip of 0.05 and equivalent circuit of the machine during dynamic braking at rated speed.

$$E_2' = \frac{400}{\sqrt{3}} \sqrt{\frac{\left(\frac{0.15}{0.05}\right)^2 + (0.72)^2}{\left(\frac{0.15}{0.05} + 0.12\right)^2 + (0.72 + 0.72)^2}}$$

$$= 230.9 \sqrt{\frac{9.518}{9.734 + 2.074}} = 207.3 \text{ V}$$

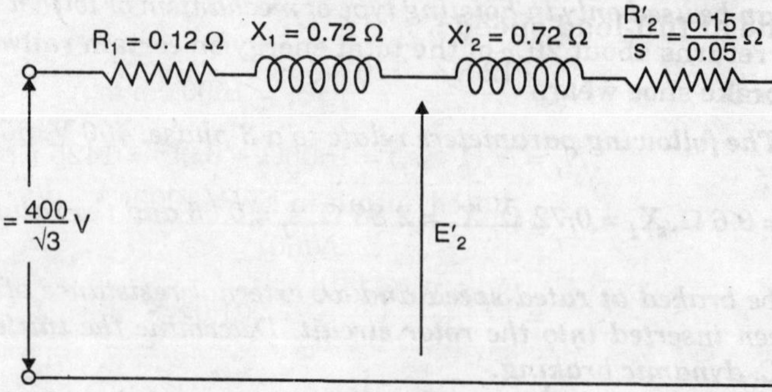

Fig. 5.69

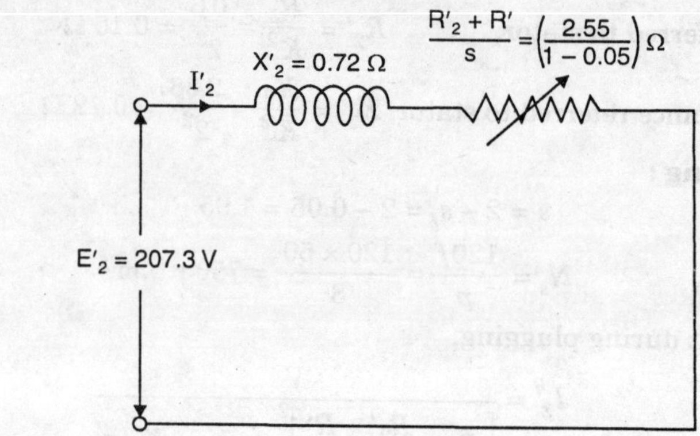

Fig. 5.70

During D.C. dynamic braking,

$$s = 1 - 0.05 = 0.95$$

$$I_2' = \frac{207.3}{\sqrt{\left(\frac{2.55}{0.95}\right)^2 + (0.72)^2}} = 74.6 \text{ A}$$

Initial braking torque, $$\mathbf{T_b} = \frac{3}{(2\pi N_s/60)} \times I_2'^2 \times \frac{R_2' + R'}{s}$$

$$= \frac{3 \times 60}{2\pi \times 750} \times (74.6)^2 \times \frac{2.55}{0.95} = \mathbf{570.6 \text{ N-m.}} \quad \textbf{(Ans.)}$$

Example 5.91. *A 30 kW rated output, 400 V, 3-phase delta-connected, 4-pole, 50 Hz induction motor has full-load slip of 5%. If the ratio of standstill reactance to resistance per rotor phase is 4, estimate the plugging torque at full-load speed. Ignore stator leakage impedance and magnetising reactance.* **(AMIE Elect. Drives, Summer, 2001)**

Solution. *Given* : Rated output = 30 kW, supply voltage = 400 V ;

$$p = 4 \; ; f = 50 \text{ Hz} \; ; s_f = 5\% \text{ or } 0.05 \; ; \frac{X_2}{R_2} = 4$$

Plugging torque at full-load speed :

Synchronous speed, $N_s = \dfrac{120f}{p} = \dfrac{120 \times 50}{4} = 1500$ r.p.m.

Full-load speed, $N_f = N_s(1 - s_f) = 1500(1 - 0.05) = 1425$ r.p.m.

Full-load torque, $T_f = \dfrac{\text{Rated output in kW} \times 1000}{(2\pi N/60)}$

$$= \frac{30 \times 1000 \times 60}{2\pi \times 1425} = 201.04 \text{ N-m}$$

Now, $$T \propto \frac{sR_2E_2^2}{R_2^2 + (sX_2)^2}$$

$$\therefore \quad \frac{T_p}{T_f} = \frac{(s_pR_2E_2^2)/(R_2^2 + s_p^2X_2^2)}{(s_fR_2E_2^2)/(R_2^2 + s_f^2X_2^2)} = \frac{s_p\,(R_2^2 + s_f^2X_2^2)}{s_f\,(R_2^2 + s_p^2X_2^2)}$$

or, $$\frac{T_p}{T_1} = \frac{s_p}{s_f}\left[\frac{1 + s_f^2\left(\dfrac{X_2}{R_2}\right)^2}{1 + s_f^2\left(\dfrac{X_2}{R_2}\right)^2}\right] = \frac{s_p}{s_f}\left[\frac{1 + 16s_f^2}{1 + 16s_p^2}\right] \qquad \left(\because \frac{X_2}{R_2} = 4\right)$$

Putting $s_f = 0.05$ and s_p = slip at plugging $= 2 - s_1 = 2 - 0.05 = 1.95$, we get

Plugging torque, $$\mathbf{T_p} = \frac{1.95}{0.05}\left[\frac{1 + 16 \times (0.05)^2}{1 + 16 \times (1.95)^2}\right]T_f$$

$$= 39 \times \frac{1.04}{61.84} \times 201.04 = \mathbf{131.86 \text{ N-m. (Ans.)}}$$

5.13.1.17. Speed control of induction motors

The rotor speed of an induction motor may be stated by the equation

$$N = \frac{120f}{p}(1 - s) \qquad \text{...(5.30)}$$

where, N = Rotor speed in r.p.m.,
f = Frequency of supply,
p = Number of poles, and
s = Fractional slip.

This shows that a change in any one of these quantities will affect its speed, namely :

(*i*) Change of frequency ;

(*ii*) Change of number of poles ; and

(*iii*) Change of slip.

1. **Control by Changing Frequency.** This method is *impractical for most applications because frequency of the supply system must remain fixed.* In special cases where the motor load is the only load connected to generators, the speed of the prime-movers may be varied to change the supply frequency and thus change the motor speed. The range over which the speed may be varied in this way, however, is limited by the range of economical speeds of the drives.

This method of control has been applied to *a limited extent in ship propulsion.*

2. **Control by changing number of poles.** Change in the number of poles is effected by making change in the stator winding connections with the help of suitable switching arrangement.

When the speed ratio is 2 : 1, the *'Consequent Pole'* method is adopted and is shown diagrammatically in Fig. 5.71.

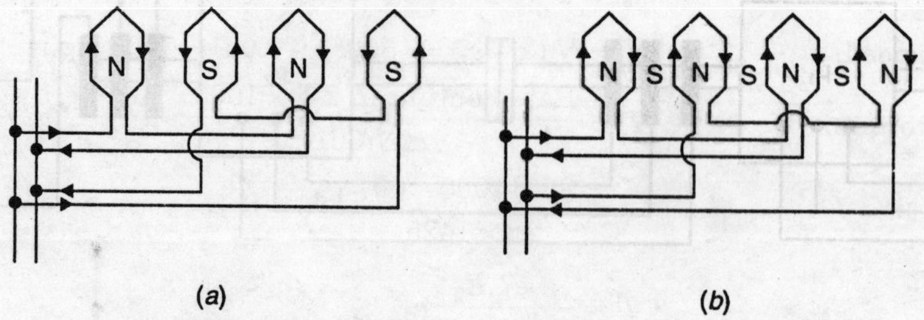

Fig. 5.71. Speed control by changing number of poles.

Two distinct windings are on the stator producing the same number of poles, one winding creates two *N-poles,* and the other two S-poles as shown in Fig. 5.71 (*a*). This gives *four* poles, and the synchronous speed at 50-Hz is 1500 r.p.m.

If the connection of one of the windings is reversed [Fig. 5.71 (*b*)], there will be in all four *N*-poles or four *S*-poles, depending upon which of the winding connection is reversed. But in between these poles other four poles of opposite polarity will be created. Thus the stator now has *eight* poles and its synchronous speed is 750 r.p.m.

The 2 : 1 change in speed is rather drastic. Hence motors are manufactured with *two windings* one is so wound that is created *8-poles* and the other *10-poles*, the synchronous speeds being 750 r.p.m. and 600 r.p.m. respectively. But the *disadvantage* is that at either speed only 50 per cent of stator copper is utilised, since only one winding is in use at a time.

In the case of squirrel-cage motors, change in the number of poles on the stator does not affect the existing arrangement on the rotor. But in case of wound rotors, the number of poles on the stator must equal to the number of poles on the rotor. Otherwise there will be a greater reduction in torque due to some rotor conductors developing a negative torque.

3. **Control by changing slip.** *Change in slip is effected by introducing an external resistance in rotor circuit of wound rotors.* But this is done at the *sacrifice of efficiency* and besides the *speed regulation is poor.*

These *disadvantages are more or less overcome by :*

- ***Supplying counter e.m.fs. to the rotor at slip frequency.*** This method requires *auxiliary commutating machine* which injects e.m.fs. into the rotor at rotor frequency through the slip rings. This method has the *advantage of providing a wide adjustment in speed without serious reduction in efficiency*. Its use in limited, however, to very special applications *that use large amounts of power in a single unit.*
- ***Supplying counter e.m.fs. to the rotor at supply frequency.*** This method requires a commutator for the rotor.

4. **Control by cascade or tandem connection.** Where multiple speeds are desired, motors are sometimes operated in tandem or cascade. When so used, two motors are rigidly coupled to the same shaft or are otherwise mechanically linked, as by means of gears. The stator winding of the first is connected to the mains in the usual way, while that of the second stator is fed from the rotor winding of the first, as shown in Fig. 5.72.

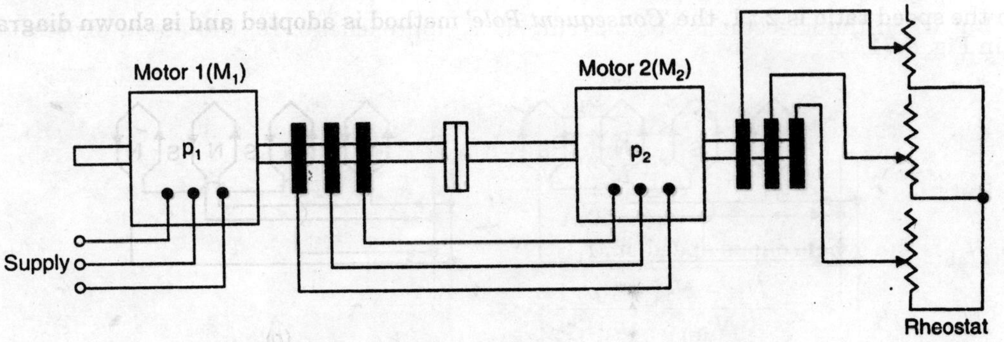

Fig. 5.72. Wound rotor motors connected in tandem.

If the two machines are designed for the same voltage, as is usually the case, the turn ratio of stator to rotor of the first machine should be *unity*. The second rotor may have a cage winding or a polyphase winding like its stator. In the latter case the rotor circuit of the second motor is connected to slip rings in the usual way, in order that resistance may be introduced while starting and for securing additional speed control when running.

The motors may be so connected that both tend to run in the same direction, or the phase rotation of one motor may be reversed, thus tending to make it rotate in the reverse direction. In either case the set will run after it is started, but in the latter case no starting torque is developed, and for this reason this connection is *little used*. If the first machine has p_1 poles and the second has p_2 poles, the synchronous speed of the set is that of a motor with *$p_1 + p_2$ poles for the first case* and *$p - p_2$ poles for the second*. If p_1 and p_2 are not equal, four synchronous speeds are possible, two with tandem operation and one for each motor separately. Some applications of this method of control are found in European railways.

The *expression for the speed of set* is derived as follows :

Let the frequency of the supply voltage be f_1 and let machines M_1 and M_2 have p_1 and p_2 number of poles respectively. Let the two machines, M_1 and M_2 run with slip of s_1 and s_2 respectively.

Speed in r.p.m. of $M_1 = N_1 = \dfrac{120 f_1}{p_1}(1 - s_1)$

Speed in r.p.m. of $M_2 = N_2 = \dfrac{120 f_2}{p_2}(1 - s_2)$

But the shafts are mechanically coupled, therefore

$$N_1 = N_2$$

Also, $f_2 = s_1 f_1$

Hence substituting the value of f_2, we get

$$N_2 = \frac{120\, s_1 f_1}{p_2}(1 - s_2)$$

Equating the expression for N_1 and N_2 and solving for s_1, we have

$$\frac{120 f_1}{N_1}(1 - s_1) = \frac{120 s_1 f_1}{p_2}(1 - s_2)$$

$$\therefore \quad s_1 = \frac{p_2}{p_1 - p_1 s_2 + p_2} = \frac{p_2}{p_1(1 - s_2) + p_2} \quad \text{...(5.31)}$$

But when the rheostat is short-circuited, s_2 approaches zero, the above expression then reduces to

$$s_1 = \frac{p_2}{p_1 + p_2} \qquad ...(5.32)$$

But,
$$s_1 = \frac{N_{s1} - N_1}{N_{s1}}$$

where, N_{s1} is the synchronous speed of M_1

$$\therefore \quad \frac{N_{s1} - N_1}{N_{s1}} = \frac{p_2}{p_1 + p_2}$$

Solving for N_1, we have $N_1 = N_{s1} \frac{p_1}{p_1 + p_2}$

However, $N_{s1} p_1 = 120 f_1$, so that substituting the value of $N_{s1} p_1$ in the above expression it becomes

$$N_1 = \frac{120 f_1}{p_1 + p_2} \text{ r.p.m.} \qquad ...(5.32\ a)$$

Equation (5.32 *a*) shows that the speed of the set is that of a single machine having the number of poles equal to the sum of the numbers of poles of the two machines. Hence the set can give four different speeds.

Example 5.92. *The full-load speed of an 8-pole, 50-Hz slip ring motor is 730 r.p.m. The rotor resistance per phase is 0.2 Ω. Calculate the external resistance per phase that must be added to lower the speed to 620 r.p.m. given that the torque is same in the two cases.*

Solution. Number of poles, $p = 8$

Supply frequency, $f = 50$ Hz

Full-load speed, $N = 730$ r.p.m.

Rotor resistance/phase, $R_2 = 0.2\ \Omega$

Final speed $= 620$ r.p.m.

External resistance/phase to be added per phase, r :

As a first approximation $T \propto \frac{s}{R_2}$

$$\therefore \quad T_1 \propto \frac{s_1}{R_2} \quad \text{and} \quad T_2 \propto \frac{s}{(R_2 + r)}$$

or,
$$\frac{T_1}{T_2} = \frac{s_1}{s_2} \times \frac{(R_2 + r)}{R_2}$$

Now,
$$N_s = \frac{120 f}{p} = \frac{120 \times 50}{8} = 750 \text{ r.p.m.}$$

$$s_1 = \frac{750 - 730}{750} \times 100 = 2.66\%$$

$$s_2 = \frac{750 - 620}{750} \times 100 = 17.33\%$$

Also, $R_2 = 0.2\ \Omega$ and $T_1 = T_2$ (*Given*)

$$\therefore \quad \frac{s_1}{R_2} = \frac{s_2}{(R_2 + r)}$$

or, $\dfrac{2.66}{0.2} = \dfrac{17.33}{(0.2 + r)}$ or $2.66(0.2 + r) = 0.2 \times 17.33$

$r = 1.1\ \Omega$

Hence, *external resistance to be added per phase* = **1.1 Ω. (Ans.)**

Example 5.93. *A 4-pole, 50-Hz, 3-phase slip ring induction motor when fully loaded runs with a slip of 4 per cent. Determine the value of the resistance to be inserted in series per phase in the rotor circuit to reduce the speed by 12 per cent and the new slip.*

The rotor resistance per phase is 0.25 Ω.

Solution. Number of poles, $p = 4$

Supply frequency, $f = 50$

Slip at full-load $s_1 = 4\%$

Speed to be reduced $= 12$ per cent

Rotor resistance/phase, $R_2 = 0.25\ \Omega$

New slip :

Resistance to be inserted/phase, r :

Synchronous speed, $N_s = \dfrac{120f}{p} = \dfrac{120 \times 50}{4} = 1500$ r.p.m.

Full-load speed, $N_1 = N_s(1 - s_1) = 1500(1 - 0.04) = 1440$ r.p.m.

New speed, $N_2 = (1 - 0.12)N_1 = 0.88 \times 1440 = 1267$ r.p.m.

∴ **New slip,** $s_2 = \dfrac{N_s - N_2}{N_s} = \dfrac{1500 - 1267}{1500} = $ **0.155 or 15.5%. (Ans.)**

Since slip, $s = \dfrac{\text{Rotor copper loss}}{\text{Input power to rotor}} = \dfrac{3I_2^2R_2}{\text{Input power to rotor}}$

For constant input power to rotor and rotor current,

slip $s \propto R_2$

∴ If a resistance of r ohms be connected in series per phase, then

$$\frac{R_2 + r}{R_2} = \frac{s_2}{s_1}$$

∴ $\dfrac{0.25 + r}{0.25} = \dfrac{15.5}{4}$

$4(0.25 + r) = 0.25 \times 15.5$

∴ $r = 0.719\ \Omega$

Hence, *external resistance/phase to be added* = **0.719 Ω. (Ans.)**

☞ **Example 5.94.** *An 8-pole, 50-Hz, 3-phase induction motor is running at 4 per cent slip when delivering full-load torque. It has standstill rotor resistance of 0.1 Ω and reactance of 0.6 Ω per phase. Calculate the speed of the motor if an additional resistance of 0.5 Ω per phase is inserted in the rotor circuit. The full-load torque remains constant.*

Solution. Number of poles, $p = 8$

Supply frequency, $f = 50$ Hz

Slip at full-load, $s_1 = 4\%$

Standstill rotor resistance/phase, $R_2 = 0.1\ \Omega$

Standstill rotor reactance/phase, $X_2 = 0.6\ \Omega$

Additional resistance added/phase, $r = 0.5\ \Omega$

Speed of the motor, N_2 :

Synchronous speed, $N_s = \frac{120f}{p} = \frac{120 \times 50}{8} = 750$ r.p.m.

The motor torque is given by the expression,

$$T = \frac{KsR_2E_2^2}{R_2^2 + s^2X_2^2}$$

Let the new slip be s_2.

Now, $$T_1 = \frac{Ks_1R_2E_2^2}{R_2^2 + s^2X_2^2} = \frac{K + 0.04 \times 0.1 \times E_2^2}{(0.1)^2 + (0.04 \times 0.6)^2} = \frac{KE_2^2}{2.644}$$

and, $$T_2 = \frac{Ks_2(R_2 + r)E_2^2}{(R_2 + r)^2 + (s_2X_2)^2}$$

$$= \frac{Ks_2 \times (0.1 + 0.5)E_2^2}{(0.1 + 0.5)^2 + (0.6s_2)^2} = \frac{0.6s_2KE_2^2}{0.36 + 0.36s_2^2}$$

Also, $T_1 = T_2$ *(Given)*

$\therefore$ $$\frac{KE_2^2}{2.644} = \frac{0.6s_2KE_2^2}{0.36 + 0.36s_2^2}$$

or, $0.6(1 + s_2^2) = 2.644s_2$ or $s_2^2 - 4.407s_2 + 1 = 0$

or, $s_2 = 0.24$ neglecting higher value, being impractical

$\therefore$ Speed of the motor, $N_2 = (1 - s_2)N_s = (1 - 0.24) \times 750 = 570$ r.p.m.

Hence, *speed of the motor* = **570 r.p.m. (Ans.)**

Example 5.95. *The rotor of a 4-pole, 50 Hz, slip ring induction motor has a resistance of 0.25 Ω per phase and runs at 1440 r.p.m. at full-load. Calculate the external resistance per phase which must be added to lower the speed to 1200 r.p.m., the torque being the same as before.*

(AMIE Advanced Elec. M/Cs, 2000)

Solution. *Given* : $p = 4$; $f = 50$ Hz ; $R_2 = 0.25\ \Omega$; $N = 1440$ r.p.m. ; $N' = 1200$ r.p.m.

External resistance/phase to be added, r :

Synchronous speed, $N_s = \frac{120f}{p} = \frac{120 \times 50}{4} = 1500$ r.p.m.

Full-load slip, $s = \frac{N_s - N}{N_s} = \frac{1500 - 1440}{1500} = 0.04$

After inserting resistance R ohms per phase in the rotor circuit,

$N' = 1200$ r.p.m.

$\therefore$ Slip, $s' = \frac{1500 - 1200}{1500} = 0.2$

Since slip, $$s = \frac{\text{Rotor copper loss}}{\text{Input power to rotor}} = \frac{3I_2^2R_2}{\text{Input power to rotor}}$$

For constant power input to rotor and rotor current, $s \propto R_2$.

Since for constant load torque power input to rotor and rotor current remains the same, therefore,

$$\frac{s'}{s} = \frac{R_2 + r}{R_2} \quad \text{or} \quad \frac{0.2}{0.04} = \frac{0.25 + r}{0.25}$$

or, $\mathbf{r = 1\ \Omega}$. **(Ans.)**

Example 5.96. *A 75 kW, 415 V, 4-pole, 3-phase, 50 Hz induction motor is delivering rated torque at 5% slip driving a pump load. The motor delivers maximum torque at 10% slip. A 3-phase A.C. regulator controls the r.m.s. supply voltage to effect limited speed control. Assume the torque-speed characteristics of the motor in the stable region to be linear. Calculate the r.m.s. voltage at which motor runs at 1350 r.p.m.* **(AMIE Electric Drives, Summer, 2002)**

Solution. *Given* : $V = 415$ volts ; $p = 4$; $f = 50$ Hz ; $s_f = 5\% = 0.05$; $s_{mT} = 10\% = 0.1$, $N' = 1350$ r.p.m.

R.M.S. voltage at which motor runs at 1350 r.p.m. :

Full-load speed, $N = N_s(1 - s_f) = \frac{120f}{p}(1 - s_f) = \frac{120 \times 50}{4}(1 - 0.05) = 1425$ r.p.m.

Also, $\frac{R_2}{X_2} = s_{mT} = 0.1$ or $X_2 = 10R_2$

When speed is $N' = 1350$ r.p.m.

$$s_f' = \frac{N_s - N'}{N_s} = \frac{(120f/p) - N'}{(120f/p)} = \frac{(120 \times 50/4) - 1350}{(120 \times 50/4)} = 0.1$$

For pump load *assuming that the torque varies as the speed so torque* corresponding to 1350 r.p.m.,

$$T' = T \times \frac{1350}{1425} \quad \text{or} \quad \frac{T'}{T} = \frac{1350}{1425}$$

Now, torque developed $= \frac{ksR_2E_2}{R_2^2 + s^2X_2^2}$ [Eqn. (5.8)]

$$\therefore \quad \frac{T'}{T} = \frac{(s_f' E_2'^2)/(R_2^2 + s_f'^2 X_2^2)}{(s_f E_2^2)/(R_2^2 + s_f^2 X_2^2)} = \frac{s_f'(V')^2(R_2^2 + s_f^2 X_2)}{s_f(V)^2(R_2^2 + s_f'^2 X_2^2)}$$

[$\because$ **$E_2 \propto V$, the supply voltage**]

or,

$$V' = V \times \sqrt{\frac{T'}{T} \times \frac{s_f}{s_f'} \times \frac{(R_2^2 + s_f'^2 X_2^2)}{(R_2^2 + s_f^2 X_2^2)}}$$

$$= 415 \times \sqrt{\frac{1350}{1425} \times \frac{0.05}{0.1} \times \frac{(R_2^2 + 0.01 \times 100R_2^2)}{(R_2^2 + 0.0025 \times 100R_2^2)}}$$

$$= 415 \times \sqrt{\frac{1350}{1425} \times 0.5 \times 1.6} = \mathbf{361.3\ volts.\ (Ans.)}$$

Example 5.97. *A 3-phase, 440 V, 1000 r.p.m. slip-ring induction motor is operating with 2% slip and taking a stator current of 50 A. Speed of the motor is reduced at constant torque to 500 r.p.m. using stator voltage control. Calculate the new value of stator current.*

(AMIE, M/Cs for Special Applications, Winter, 2000)

Solution. *Given* : Supply voltage, $V = 440$ V ; $N_s = 1000$ r.p.m., $s = 2\% = 0.02$; $I_1 = 50$ A ; $N' = 500$ r.p.m.

New stator current, I_1' :

Slip at reduced speed N', $s' = \frac{N_s - N'}{N_s} = \frac{1000 - 500}{1000} = 0.5$

Since $T \propto sV^2$, therefore, for constant torque

$$V \propto \sqrt{\frac{1}{s}} \quad \text{or} \quad V' = V \times \sqrt{\frac{s}{s'}} = 440\sqrt{\frac{0.02}{0.5}} = 88 \text{ V}$$

Also, stator current $I_1 \propto sV$

$\therefore$ New stator current, $\mathbf{I_1'} = I_1 \times \frac{s'V'}{sV} = 50 \times \frac{0.5 \times 88}{0.02 \times 440} =$ **250 A. (Ans.)**

Example 5.98. *A star-connected cage induction motor has the following data :*

440 V, 50 Hz, 4-pole, 1460 r.p.m., $R_s = 2\ \Omega$, $R_r' = 2\ \Omega$, $X_s = X_r' = 3\ \Omega$.

Calculate the starting torque and starting current of this motor at 50 Hz and 10 Hz for V/f control. **(AMIE Electric Drives Winter, 1999)**

Solution. *Given* : Supply voltage for phase, $V = \frac{440}{\sqrt{3}} = 254$ V ; $f = 250$ Hz ; $p = 4$,

$N = 1460$ r.p.m., $R_s = 2\ \Omega$, $R_r' = 2\ \Omega$, $X_s = X_r' = 3.0\ \Omega$.

At 50 Hz supply :

Equivalent resistance of motor referred to stator, $R_{01} = R_s + R_r' = 2 + 2 = 4\ \Omega$

Equivalent reactance of motor referred to stator, $X_{01} = X_s + X_r' = 3 + 3 = 6\ \Omega$

Starting current, $\mathbf{I_{st}} = \frac{V}{\sqrt{R_{01}^2 + X_{01}^2}} = \frac{254}{\sqrt{4^2 + 6^2}} =$ **35.22 A. (Ans.)**

Total copper losses in rotor $= 3I_{st}^2 \times R_r' = 3 \times 35.22^2 \times 2 = 7442.7$ W

Rotor power input at standstill, P_2 = Total copper losses = 7442.7 W

Synchronous speed, $N_s = \frac{120f}{p} = \frac{120 \times 50}{4} = 1500$ r.p.m.

Starting torque, $\mathbf{T_{st}} = \frac{P_2}{(2\pi N_s/60)} = \frac{7442.7 \times 60}{2\pi \times 1500} =$ **47.38 Nm. (Ans.)**

At 10 Hz supply :

Voltage supply per phase, $V' = V \times \frac{f'}{f} = 254 \times \frac{10}{50} = 50.8$ V

Equivalent resistance of motor referred to stator,

$R_{01}' = 4\ \Omega$, the same as before

Equivalent reactance of motor referred to stator,

$$X_{01}' = X_{01} \times \frac{f'}{f} = 6 \times \frac{10}{50} = 1.2\ \Omega$$

Starting current, $\mathbf{I_{st}'} = \frac{V'}{\sqrt{R'^2_{01} + X'^2_{01}}} = \frac{50.8}{\sqrt{4^2 + 1.2^2}} =$ **12.16 A. (Ans.)**

Power input to rotor at standstill,

P_2' = Total copper losses $= 3(I'_{st})^2 R_r'$

$= 3 \times 12.16^2 \times 2 = 887.2$ W

Synchronous speed, $N_s' = \frac{120f'}{p} = \frac{120 \times 10}{4} = 300$ r.p.m.

Starting torque, $\mathbf{T'_{st}} = \frac{p_2'}{(2\pi N_s'/60)} = \frac{887.2 \times 60}{2\pi \times 300} = \mathbf{28.24\ Nm.\ (Ans.)}$

Example 5.99. *Two induction motors with 4-poles and 6-poles respectively are connected in cumulative cascade. The frequency in the secondary circuit of the 6-pole motor is observed to be 2.0 Hz. Determine the combined speed of the set and the slip in each machine.*

Assume supply frequency as 50 Hz. **(AMIE Advanced Elec. M/Cs, Summer, 1999)**

Solution. Number of poles, $p_1 = 4$

Number of poles, $p_2 = 6$

Frequency in the secondary circuit of 6-poles motor,

$$f_2 = 2.0 \text{ Hz}$$

Supply frequency, $f_1 = 50$ Hz

Concatenated speed of the set, $N_{sc} = \frac{120 f_1}{p_1 + p_2} = \frac{120 \times 50}{4 + 6} = 600$ r.p.m.

Also, $\text{slip} = \frac{2}{f_1} = \frac{2}{50} = 0.04$ or 4%

Actual speed of the set,

$$N = (1 - 0.04) \times 600 = \mathbf{576\ r.p.m.\ (Ans.)}$$

Synchronous speed of 4-pole induction motor,

$$N_{s1} = \frac{120 f_1}{p_1} = \frac{120 \times 50}{4} = 1500 \text{ r.p.m.}$$

Slip for 4-pole induction motor,

$$s_1 = \frac{N_{s1} - N}{N_{s1}} = \frac{1500 - 576}{1500} = \mathbf{0.616\ or\ 61.6\%.\ \ (Ans.)}$$

The impressed frequency on the second motor,

$$f_2 = s_1 f_1 = 0.616 \times 50 = 30.8 \text{ Hz}$$

Synchronous speed of 6-pole induction motor with frequency f_2,

$$N_{s2} = \frac{120 f_2}{p_2} = \frac{120 \times 30.8}{6} = 616 \text{ r.p.m.}$$

Slip for 6-pole induction motor,

$$s_2 = \frac{N_{s2} - N}{N_{s2}} = \frac{616 - 576}{616} = \mathbf{0.065\ or\ 6.5\%.\ (Ans.)}$$

Example 5.100. *Two 50 Hz, 3-phase induction motors having six and four poles respectively are cumulatively cascaded, the 6-pole motor being connected to the main supply. Determine the frequencies of the rotor currents and the slips referred to each stator field if the set has a slip of 3 per cent.*

Solution. Supply frequency, $f_1 = 50$ Hz

$$p_1 = 6, p_2 = 4$$

Slip of the set, $s = 3\%$

Synchronous speed of the set,

$$N_{sc} = \frac{120 \times 50}{6 + 4} = 600 \text{ r.p.m.}$$

Actual rotor speed $= (1 - s)N_{sc} = (1 - 0.03) \times 600 = 582$ r.p.m.

Synchronous speed of the stator field of 6-pole motor,

$$N_{s1} = \frac{120 f_1}{p_1} = \frac{120 \times 50}{6} = 1000 \text{ r.p.m.}$$

Slip referred to this stator field,

$$s_1 = \frac{N_{s1} - N}{N_{s1}} = \frac{1000 - 582}{1000} = 0.418 \text{ or } \mathbf{41.8\%}. \quad \textbf{(Ans.)}$$

Frequency of the rotor currents of 6-pole motor,

$$f_2 = s_1 f_1 = 0.418 \times 50 = \mathbf{20.9\ Hz.} \quad \textbf{(Ans.)}$$

This is also the frequency of the stator currents of the 4-pole motor.

The synchronous speed of the stator of 4-pole motor

$$N_{s2} = \frac{120 f_2}{p_2} = \frac{120 \times 20.9}{4} = 627 \text{ r.p.m.}$$

The slip as referred to the 4-pole motor is

$$s_2 = \frac{N_{s2} - N}{N_{s2}} = \frac{627 - 582}{627} = 0.0717 \text{ or } \mathbf{7.17\%}. \textbf{ (Ans.)}$$

The frequency of rotor current of 4-pole motor,

$$f_2' = s_2 f_2 = 0.0717 \times 20.9 = \mathbf{1.5\ Hz\ (app.)}. \textbf{ (Ans.)}$$

[*Check* : $f_2' = s f_1 = 0.03 \times 50 = 1.5$ Hz.]

Example 5.101. *Two 50 Hz, 3-phase induction motors having 6 and 4 poles respectively are cumulatively cascaded, the 6-pole motor being connected to the main supply. Determine the frequency of the rotor currents and slips referred to each stator field if the set has a slip of 2%.*

(Madras Univ., 2001)

Solution. *Given* : $f = 50$ Hz ; $p_1 = 6$; $p_2 = 4$; $s = 2\% = 0.02$.

$\mathbf{s_1 ; f_1 ; s_2 ; f_2}$:

$$\text{Synchronous speed of the set } N_{sc} = \frac{120 f}{p_1 + p_2} = \frac{120 \times 50}{6 + 4} = 600 \text{ r.p.m.}$$

Actual speed of the set, $N = N_{sc}(1 - s) = 600(1 - 0.02) = 588$ r.p.m.

6-pole motor :

Synchronous speed of 6-pole motor stator field,

$$N_{s1} = \frac{120 f}{p_1} = \frac{120 \times 50}{6} = 1000 \text{ r.p.m.}$$

Slip referred to the stator field,

$$\mathbf{s_1} = \frac{N_{s1} - N}{N_{s1}} = \frac{1000 - 588}{1000} = 0.412 \text{ or } \mathbf{41.2\%}. \textbf{ (Ans.)}$$

The 6-pole motor current frequency (or the impressed frequency on 4-pole motor, stator field)

$$\mathbf{f_1} = s_1 f = 0.412 \times 50 = \mathbf{20.6\ Hz.} \textbf{ (Ans.)}$$

4-pole motor :

The synchronous speed of 4-pole induction motor with frequency f_1,

$$N_{s2} = \frac{120 \times f_1}{p_2} = \frac{120 \times 20.6}{4} = 618 \text{ r.p.m.}$$

The slip as referred to the 4-pole induction motor,

$$\mathbf{s_2} = \frac{N_{s2} - N}{N_{s2}} = \frac{618 - 588}{618} = 0.0485 \text{ or } \mathbf{4.85\%}. \quad \textbf{(Ans.)}$$

Frequency of rotor current of 4-pole induction motor,

$$\mathbf{f_2} = s_2 f_1 = 0.0485 \times 20.6 = \mathbf{1.0\ Hz}. \quad \textbf{(Ans.)}$$

Example 5.102. *A cascade set comprises two motors 1 and 2 with four poles and six poles respectively. The set is connected to a 50 Hz supply. Find :*

(*i*) *The synchronous speed of the set ;*

(*ii*) *The electric power transferred to motor* **2** *when the input to motor* **1** *is 30 kW. Neglect losses.* **(Utkal Univ., 2002)**

Solution. *Given* : $p_1 = 4$; $p_2 = 6$, Input to motor **1** = 30 kW.

(*i*) **The speed of the set, N_{sc} :**

$$\text{Synchronous speed of the set, } N_{sc} = \frac{120f}{p_1 + p_2} = \frac{120 \times 50}{4+6} = \mathbf{600\ r.p.m.}\ \textbf{(Ans.)}$$

(*ii*) **The electric power transferred to motor 2 :**

Since the outputs of the two motors are proportional to the number of their poles, therefore, power transferred to motor **2** with 6-poles

= power output of motor **2**, neglecting losses

$$= 30 \times \frac{p_2}{p_1 + p_2} = 30 \times \frac{6}{4+6} = \mathbf{18\ kW}. \quad \textbf{(Ans.)}$$

Eddy current drives :

- An eddy current drive consists of an *eddy current clutch placed between an induction motor running at a fixed speed and the variable speed load.* Speed is controlled by controlling D.C. excitation to magnetic circuit of the clutch. Since motor itself runs at a fixed speed it can be fed directly from A.C. mains.
- The principle of an eddy-current clutch is identical to an induction motor in which both stator and rotor are allowed to rotate. Stator, which is coupled to driving induction motor, has D.C. winding which produces magnetic field rotating at the speed of rotor. Rotor has a metal drum coupled to the load. Eddy currents are induced in rotor drum by stator magnetic field. Interaction between the stator field and eddy currents produces a torque which causes rotor to move with stator with a slip. Slip, and therefore, the load speed, can be controlled by controlling D.C. current through rotor winding. Speed torque characteristics are identical to an induction motor.
  - — Speed reduction is obtained by wasting a power equal to sP_{in} in the rotor drum. Minimum speed is usually restricted to 30 percent below the synchronous speed, because efficiency becomes too low and cooling of the rotor drum becomes difficult below this speed.
  - — Load can be developed from induction motor by setting D.C. winding current to zero. Motor can now be started on no load. Load can be smoothly started by slowly increasing D.C. winding excitation.
- Eddy current clutch is available in different constructions and sizes ranging from fraction of a kW to MW.

Advantages. The *advantages of eddy current drives are* :

(*i*) Rugged in construction.

(*ii*) Easy to maintain.

(*iii*) Reliable in operation.

(*iv*) Stepless speed control with good speed regulation.

(*v*) Controlled acceleration and soft start.

(*vi*) High starting torque.

(*vii*) High overload capacity.

(*viii*) Ability to handle impact loads.

Applications. They are widely used in :

- Blowers ;
- Conveyors ;
- Dredges ;
- Winders ;
- Compressors ;
- Cranes ;
- Elevators ;
- Line shafts and paper machines.

However, due to *poor efficiency and cooling* they are *rarely* used in new installations.

5.13.1.18. Single phase motors

General aspects :

- The number of machines operating from single-phase supplies is greater than all other types taken in total. For the most part, however, they are only used in the smaller sizes, less than 5 kW and mostly in the fractional H.P. range. They operate at *lower power-factors* and are relatively *inefficient when compared with polyphase motors*. Though simplicity might be expected in view of the two-line supply, the *analysis is quite complicated*.
- *Single phase motors* perform a great variety of useful services in the home, the office, the factory, in business establishments, on the farm, and many other places where electricity is available. Since the requirements of the numerous applications differ so widely, the motor-manufacturing industry has developed several types of such machines, each type having operating characteristics that meet definite demands. For example, one type operates satisfactorily on direct current or any frequency up to 60 cycles ; another rotates at absolutely constant speed, regardless of load ; another develops considerable starting torque and still another, although not capable of developing much starting torque, is nevertheless extremely cheap to make and very rugged.

Types of single-phase motor :

The single-phase motors may be of the following types :

1. *Single-phase Induction Motors :*

A. Split-phase motors

(*i*) Resistance-start motor (*ii*) Capacitor-start motor

(*iii*) Permanent-split (single-value) capacitor motor

(*iv*) Two-value capacitor motor.

B. Shaded-pole induction motor.

C. Reluctance-start induction motor.

D. Repulsion-start induction motor.

2. *Commutator-type, single-phase motors :*

A. Repulsion motor.
B. Repulsion-induction motor.
C. A.C. series motor.
D. Universal motor.

3. *Single-phase synchronous motors :*

A. Reluctance motor.
B. Hysteresis motor.
C. Sub-synchronous motor.

Single-phase induction motors

Applications :

- Single phase induction motors are in very wide use in industry, especially in *fractional horse-power field.*

  They are extensively used for electric drive for *low power constant speed apparatus* such as *machine tools, domestic apparatus* and *agricultural machinery* in circumstances where a three-phase supply is *not* readily available.
- There is a large demand for single-phase induction motors in sizes ranging from a *fraction of horse-power upto about 5 H.P.*

Disadvantages :

Though these machines are useful for small outputs, they are not used for large powers as they suffer from many disadvantages and are never used in cases where three-phase machines can be adopted.

The main *disadvantages* of single-phase induction motors are :

1. Their output is only 50% of the three-phase motor, for a given frame size and temperature rise.
2. They have lower power factor.
3. Lower-efficiency.
4. These motors do not have inherent starting torque.
5. More expensive than three-phase motors of the same output.

Construction and working :

Construction :

- A single phase induction motor is similar to a 3-ϕ squirrel-cage induction motor in physical appearance. Its rotor is essentially the same as that used in 3-ϕ induction motors. Except for shaded pole motors, the stator is also very similar. There is a uniform air-gap between the stator and rotor but no electrical connection between them. It can be wound for any even number of poles, two, four and six being most common. Adjacent poles have opposite magnetic property and synchronous speed equation, $N_s = \frac{120f}{p}$ also applies.
- The stator windings differ in the following *two aspects :*
  - **Firstly** single phase motors are usually provided with *concentric coils.*
  - **Secondly,** these motors normally have two stator windings. In motors that operate with both windings energised, the winding with the *heaviest wire* is known as the **main winding** and the other is called the **auxiliary winding.** If the motor run with auxiliary winding open, these windings are usually referred as *running and starting.*
  - In *most of motors the main winding is placed at the bottom of the slots and the starting winding on top but shifted 90° from the running winding.*

Working :

When the stator winding of a single phase induction motor is connected to single phase A.C. supply, a *magnetic field* is developed, *whose axis is always along the axis of stator coils.* The magnetic field produced by the stator coils is pulsating, varying sinusoidally with time. Currents are induced in the rotor conductors by transformer action, these currents being in such a direction as to oppose the stator m.m.f. Then the axis of the rotor m.m.f. wave coincides with that of the stator field, the *torque angle* is, therefore, *zero* and *no torque is developed on starting.* However, if the rotor is given a push by hand or by other means in either direction, it will pick-up the speed and continue to rotate in the same direction developing operating torque. Thus a single phase induction motor is not inherently self starting and requires some special means for starting.

The above mentioned behaviour of this type of motor can be explained by any one of the following theories :

1. Double revolving field theory. 2. Cross-field theory.

The results given by both the theories are approximately same.

Double revolving field theory is described below :

The magnetic field produced by the stator coils is *pulsating*, varying sinusoidally with time. *Ferrari* pointed out that *such a field can be resolved into two equal fields but rotating in opposite directions with equal angular velocities. The maximum value of each component is equal to half the maximum of the pulsating field.*

If the initial time is such that the rotating vectors of the two component fields are along the *Y*-axis in the positive direction, the two component waves ϕ_1 and ϕ_2 coincide. The resultant of these two is ϕ_{max}. After a short interval of time the two vectors rotate, through an angle θ in their respective directions and the waves are shown to occupy the positions in Fig. 5.73. These waves intersect at *A* on the *Y*-axis and as the waves travel *A* moves along the *Y*-axis. Hence the resultant of these two component waves at any instant is equal to 2*OA*.

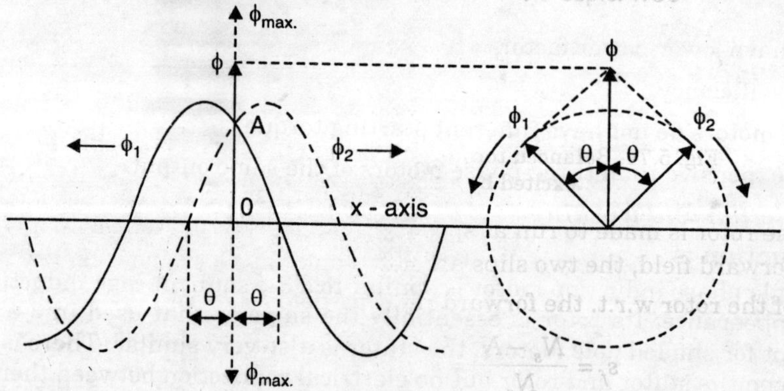

Fig. 5.73

$$\phi_1 = OA = \phi_{1\,(max)} \cos(\omega t - \theta) \qquad ...(i)$$

$$\phi_2 = OA = \phi_{2(max)} \cos(\omega t + \theta) \qquad ...(ii)$$

and, $$\phi_{1(max)} = \phi_{2(max)}$$

By expanding and adding (*i*) and (*ii*),

$$\phi_1 + \phi_2 = 2\phi_{1(max)} \cos\theta \cos\omega t$$

$$2OA = \phi_{(max)} \cos\theta \cos\omega t \qquad ...(iii)$$

which is the equation of the pulsating field and **proves Ferrari's statement.** Thus a single-phase induction motor is *not inherently self-starting*.

The existence of these two fluxes (*forward and backward*) rotating in opposite directions can be verified by supplying a fractional horsepower single-phase induction motor with rated voltage. The motor does not start, but if the shaft is turned by hand, say in clockwise direction, the rotor picks up speed. This means that the rotor conductors are rotating in the direction of that field which rotates in clockwise direction. When the motor is braked and stopped without switching off the supply, the rotor remains at rest. If now the shaft is turned by hand in anti-clockwise direction, the motor picks up speed in that direction. This means that the rotor conductors are now rotating in the direction of the other field.

This behaviour of the motor is *due to the presence of two opposing torques due to the two field.* When the rotor is at rest, (*i.e.*, slip = 1) the two torques are equal but opposite in direction. Hence the net torque is zero and therefore the rotor remains at rest. Fig. 5.74 shows the torque variations due to the two fields. If the rotor is made to speed up in one direction, say in that in which T_1 increases, T_1 exceeds the opposing torque T_2 and the motor begins to accelerate. T_2 goes on diminishing until at the working speed it is negligibly small. Hence the single-phase induction motor rotates in the direction in which it is made to run.

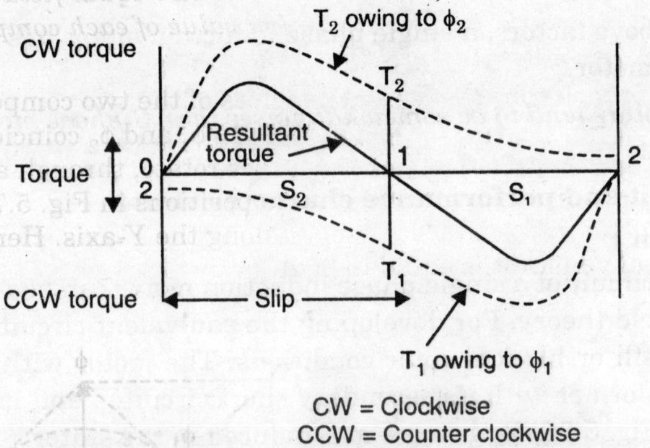

Fig. 5.74. Balanced torque at standstill in squirrel cage rotor excited by a single-phase winding.

Thus, if the rotor is made to run at speed N by some external means in any direction, say in the direction of forward field, the two slips are now s and $(2 - s)$, as shown below :

The slip of the rotor w.r.t. the forward rotating field F_f,

$$s_f = \frac{N_s - N}{N_s} = s \qquad ...(5.33)$$

The slip of the rotor w.r.t. to the backward rotating field F_b,

$$s_b = \frac{N_s - (-N)}{N_s} = \frac{2N_s - (N_s - N)}{N_s} = (2 - s) \qquad ...(5.34)$$

- Under normal running condition $(2 - s) >> s$ and as a consequence the backward field rotor currents are much larger than at standstill and have a low power factor. The corresponding opposing rotor m.m.f., owing to stator impedance, causes the backward field to be greatly reduced in strength. On the other hand, the low-slip forward rotating field induces smaller currents of a high power factor in the rotor than at standstill. This leads to greatly strengthening of forward field. Thus weakening of backward field and strengthening of forward field depends upon the slip or speed of rotor and the difference increases with the decrease in slip w.r.t. the forward field or with increase in rotor speed in forward direction.

- In a single phase induction motor, *the increase in rotor resistance increases the effectiveness of the backward field, which reduces the breakdown torque, lowers the efficiency and increases the slip corresponding to maximum torque.*
- The **performance characteristics** of a single phase induction motor are somewhat *inferior* to that of a 3-phase induction motor due to the presence of backward rotating field.
  - — A single phase induction motor has a *lower breakdown torque at larger slip and increased power losses.*
  - — Greater power input.
  - — The *speed regulation* tends to be *poorer* than that for a polyphase motor.
  - — The *power factor tends to be lower* (since the normal slip of a single phase induction motor under load conditions is rather greater than that of the corresponding 3-phase motor).
- In view of the above factors, a single phase induction motor has a *larger frame size* than that of 3-phase motor.
- Single phase motors *tend to be somewhat noiser than 3-phase motors which have no such pulsating torque.*

Equivalent circuit and performance characteristics

Equivalent circuit :

- The equivalent circuit of a single phase induction motor can be developed on the basis of two revolving field theory. For developing the equivalent circuit it is first imperative to consider standstill or blocked rotor conditions. The motor with a blocked rotor merely acts like a transformer with its secondary short circuited and its equivalent circuit will be as shown in Fig. 5.75, E_m being e.m.f. induced in the stator.

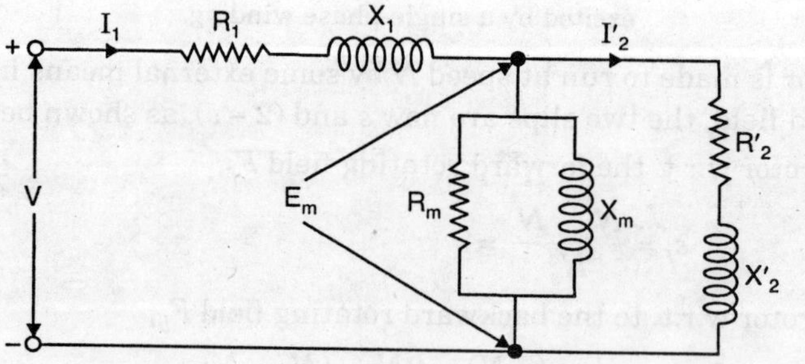

Fig. 5.75. Equivalent circuit of a single phase induction motor at standstill.

- Fig. 5.76 shows the equivalent circuit of a single phase induction motor at standstill on the basis of two revolving field theory. The phasor sum of E_{mf} and E_{mb} equals the applied voltage V (less the voltage drops in stator resistance R_1 and leakage reactance X_1).

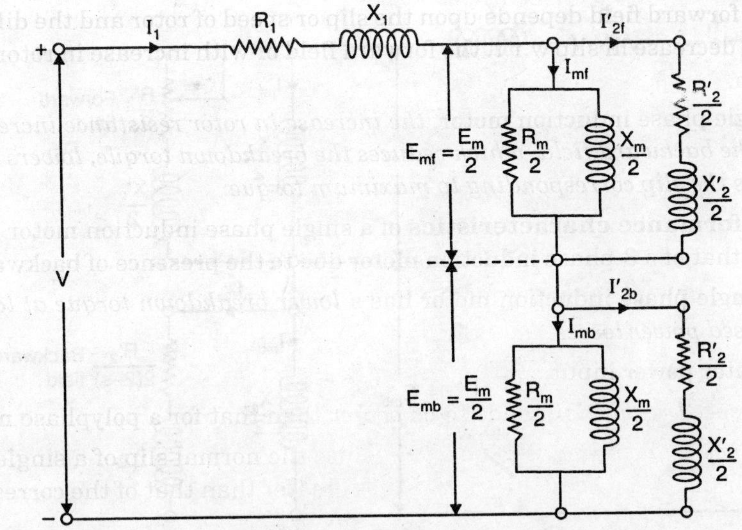

Fig. 5.76. Equivalent circuit of a single phase induction motor at standstill on the basis of two revolving field theory.

- **When the rotor runs at speed N with respect to forward field, the slip is s w.r.t. forward field and $(2 - s)$ w.r.t. backward field and the equivalent circuit becomes as shown in Fig. 5.77.**

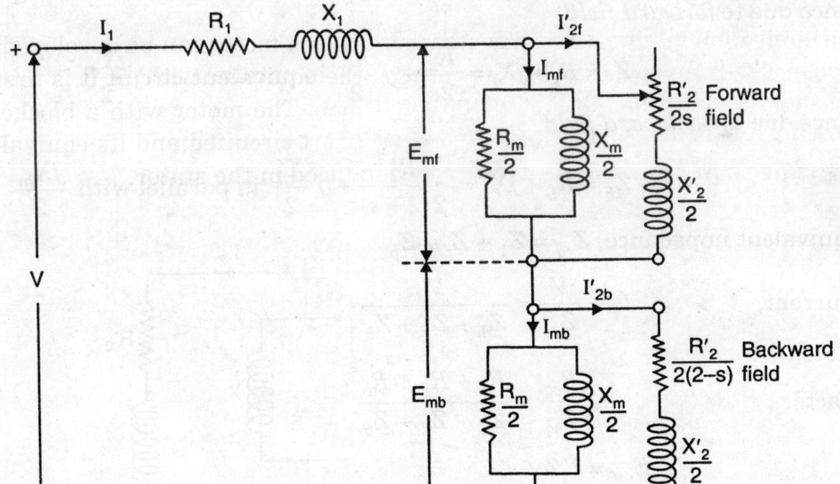

Fig. 5.77. Equivalent circuit of a single phase induction motor under normal operating conditions.

- **If the core losses are neglected the equivalent circuit is modified as shown in Fig. 5.78. The core losses, here, are handled as rotational losses and subtracted from the power converted into mechanical power ; the amount of error thus introduced is relatively small.**

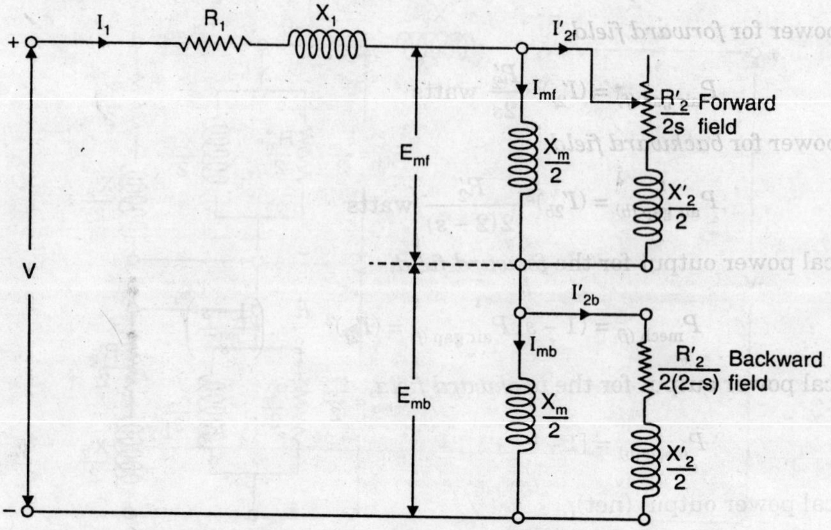

Fig. 5.78. Approximate equivalent circuit of a single phase induction motor under normal operating conditions.

Performance characteristics :

When a single phase induction motor is running with only its main winding energised, its performance characteristics can be determined from the equivalent circuit diagram *for different values of slip* ; while making calculations, the *core losses will be considered as rotational losses.*

Impedance due to *forward field,*

$$Z_f = R_f + jX_f = \frac{R_2'}{2s} + j\,\frac{X_2'}{2} \text{ in parallel with } \frac{jX_m}{2} \quad ...(5.35)$$

Impedance due to *backward field,*

$$Z_b = R_b + jX_b = \frac{R_2'}{2(2-s)} + j\,\frac{X_2'}{2} \text{ in parallel with } \frac{jX_m}{2} \quad ...(5.36)$$

Total equivalent impedance, $Z_{eq} = Z_1 + Z_f + Z_b$ ...(5.37)

Motor current, $$I_1 = \frac{V}{Z_{eq}} = \frac{V}{Z_1 + Z_f + Z_b} \quad ...(5.38)$$

Power factor $$= \frac{R_{eq}}{Z_{eq}} = \frac{R_1 + R_f + R_b}{Z_1 + Z_f + Z_b} \quad ...(5.39)$$

$$E_{mf} = I_1 Z_f \quad ...(5.40)$$

$$E_{mb} = I_1 Z_b \quad ...(5.41)$$

$$I_{2f}' = \frac{E_{mf}}{\dfrac{R_2'}{2s} + j\,\dfrac{X_2'}{2}} = \frac{I_1 Z_f}{\sqrt{\left(\dfrac{R_2'}{2s}\right)^2 + \left(\dfrac{X_2'}{2}\right)^2}} \quad ...(5.42)$$

$$I_{2b}' = \frac{E_{mb}}{\dfrac{R_2'}{2(2-s)} + j\,\dfrac{X_2'}{2}} = \frac{E_{mb}}{\sqrt{\left(\dfrac{R_2'}{2(2-s)}\right)^2 + \left(\dfrac{X_2'}{2}\right)^2}} \quad ...(5.43)$$

Air-gap power for *forward field,*

$$P_{\text{air gap }(f)} = (I'_{2f})^2 \frac{R'_2}{2s} \text{ watts} \quad ...(5.44)$$

Air-gap power for *backward field,*

$$P_{\text{air gap }(b)} = (I'_{2b})^2 \frac{R'_2}{2(2-s)} \text{ watts} \quad ...(5.45)$$

Mechanical power output for the *forward field,*

$$P_{\text{mech }(f)} = (1-s)\, P_{\text{air gap }(f)} = (I'_{2f})^2 \frac{R'_2}{2} \cdot \left(\frac{1-s}{s}\right) \quad ...(5.46)$$

Mechanical power output for the *backward field,*

$$P_{\text{mech }(b)} = [1-(2-s)]\, P_{\text{air gap }(b)} = -(I'_{2b})^2 \frac{R'_2}{2} \cdot \frac{1-s}{2-s} \quad ...(5.47)$$

Mechanical power output (net),

$$\begin{aligned} P_{\text{mech (net)}} &= P_{\text{mech }(f)} + P_{\text{mech }(b)} \\ &= (1-s)\, P_{\text{air gap }(f)} + [1-(2-s)]\, P_{\text{air gap }(b)} \\ &= (1-s)\, [P_{\text{air gap }(f)} - P_{\text{air gap }(b)}] \end{aligned} \quad ...(5.48)$$

Torque developed for the *forward field,*

$$T_f = \frac{P_{\text{air gap }(f)}}{(2\pi N_s/60)} \quad ...(5.49)$$

Torque developed for the *backward field,*

$$T_b = \frac{P_{\text{air gap }(b)}}{(2\pi N_s/60)} \quad ...(5.50)$$

Net torque developed, $T_{\text{net}} = \dfrac{[P_{\text{air gap }(f)} - P_{\text{air gap }(b)}]}{(2\pi N_s/60)}$ ...(5.51)

Net power output = $P_{\text{mech (net)}}$ – friction and windage losses – core losses ...(5.52)

Motor efficiency, $\eta = \dfrac{\text{Output power}}{\text{Input power}} \times 100$ ...(5.53)

Power delievered to the backward field

$$\begin{aligned} &= (2-s)\, P_{\text{air gap }(b)} = P_{\text{air gap }(b)} + (1-s)\, P_{\text{air gap }(b)} \\ &= P_{\text{air gap }(b)} - P_{\text{mech }(b)} \end{aligned} \quad ...(5.54)$$

All of the power delivered to the backward field is converted into heat through the rotor copper losses caused by the backward field.

Tests for determining equivalent circuit parameters :

The most common tests performed on single phase induction motors to determine various constants are as follows :

(*i*) *Measurement of stator winding resistance.* The resistance of each stator winding is measured separately ; in case of capacitor start motor the condenser is omitted.

(*ii*) *Blocked rotor test.* This test is conducted with main winding and starting winding separately and input voltage, amperes, and watts are measured in each case.

This test is sometimes performed at reduced voltage (about 40% of rated voltage) to avoid excessive short-circuit current and heating.

(*iii*) *No-load test.* No-load test is performed at rated voltage, exciting the main winding only and input current, voltage and power, as well as speed, are measured.

(*iv*) *Determination of friction and windage losses.* In this test the voltage applied to the motor is reduced until the motor will just run and power input is measured. The power input, just sufficient to keep the motor running on no-load, gives approximately the friction and windage losses.

5.13.1.19. Starting methods and types of single phase motors

Description of a few single-phase motors is given below.

Split-phase motors :

Since the single-phase induction is not self-starting, means must be provided to create an initial torque. But the initial torque is only possible if a rotating flux is created in the stator. It is known that *a rotating flux is produced when there is a difference of 90° between the currents of two stationary coils. Or if the stator possesses two fluxes having a large phase difference the result is a rotating flux.*

1. ***Split-phase resistance-start induction motor :***

- In a split-phase induction motor the stator is provided with two parallel windings displaced 90 electrical degrees in space and somewhat less than 90° in time. Fig. 5.79 (*a*) shows the winding diagram of the *two windings* of a split-phase induction motor.

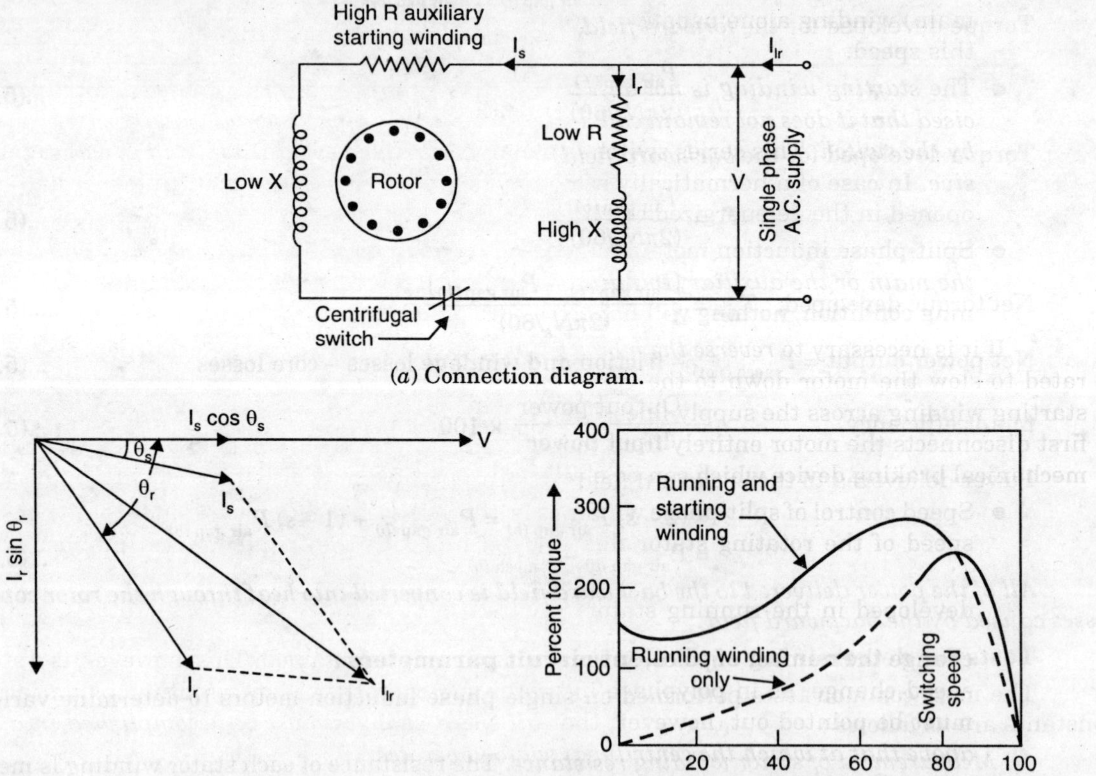

(*a*) Connection diagram.

(*b*) Phase relations. (*c*) Typical torque speed characteristic.

Fig. 5.79. Split phase resistance-start induction motor.

The *starting winding* has fewer turns and is wound of smaller diameter copper than the running winding. The starting winding, therefore, *has a high resistance and low reactance*.

The *running or main, winding* (heavier wire of more turns) has a *low resistance* and *high reactance*. Because of its lower impedance, the current in the running winding, I_r is *higher* than the current in the starting winding, I_s.

The phase relations of the lock-motor currents at the instant of starting are shown in Fig. 5.79 (*b*). The starting winding I_s lags the supply voltage by about 15°, while the greater running winding current lags the single-phase voltage by about 40°. Despite the fact that the current in the two space-quadrature windings are *not* equal, the quadrature components are practically equal.

If the windings are displaced by 90° in space, and if their quadrature current components, which are displaced by 90° in time, are practically equal, an equivalent two-phase rotating field is produced at starting which develops sufficient starting torque to accelerate the rotor, in the direction of the rotating field produced by the currents.

- As the *motor speeds up*, the *torque developed increases*. Above 85 per cent of synchronous speed, the torque developed by the running winding (main winding) alone is actually greater than that developed by both windings, and it might be advantageous to *open the auxiliary circuit at this cross over point*. To allow for individual variations among motors and switches, however, the contacts are usually designed to open at 75 per cent of synchronous speed. This does not seriously affect the operation, because the running (or main) winding alone usually develops approximately 200 per cent of full-load torque at this speed.
- The *starting winding is not designed for continuous operation, and care should be exercised that it does not remain connected to the supply after it should have been disconnected by the switch*. This series switch is usually *centrifugally operated, and is rather inexpensive*. In case of a hermatically sealed motor, the switch is magnetically operated, and is opened in the de-energized condition.
- Split-phase induction motors may be *reversed by reversing the line connections of either the main or the auxiliary winding*. If however, reversal is attempted under normal running condition, nothing will happen.

If it is necessary to *reverse the motor while it is rotating,* then some means must be incorporated to slow the motor down to the speed where the starting-switch contacts close, placing the starting winding across the supply lines. This may be done by *incorporating a timing device* which first disconnects the motor entirely from the line and then reverses one field at the proper time. A mechanical braking device which can be electrically operated may also be used.

- Speed control of split-phase windings is a *relatively difficult* matter since the synchronous speed of the rotating stator flux is determined by the frequency and number of poles developed in the running stator winding $\left(N_s = \dfrac{120f}{p}\right)$. By adding stator windings to change the number of poles, speed variation may be obtained. This, however, is a stepped speed change, as in polyphase induction motor, rather than a continuous variation. It must be pointed out, however, that *all speed changes must be accomplished in a range above that at which the centrifugal switch operates.*

  Fig. 5.79 (*c*) shows the typical torque speed characteristics.
- The starting torque is 1.5 times to twice the full-load starting torque and starting current to 6 to 8 times full-load current.
- The speed regulation is very good.

- The percent slip is about 4-6 percent.
- Such a motor may operate with a power factor of 0.55—0.65 and efficiency of 60—65 percent.
- These motors are made in fractional kW $\left(\frac{1}{20} \text{ to } \frac{1}{4} \text{ kW}\right)$ ratings with speed ranging from 2875 to 700 r.p.m.

Shortcomings and uses. The major objections to the motor are (1) its *low starting torque* ; and (2) that, when heavily loaded, the slip exceeds 5 percent, reducing the e.m.f. and producing an ellipitical or *pulsating torque which makes* the motor somewhat *annoyingly noisy*. For this reason, the split-phase motor is used in appliances to drive loads which are themselves noisy : *oil burners, machine tools, grinders, dish washers, washing machines, air blowers and air compressors.*

- *Because of low starting torque these motors are seldom employed in sizes larger than $\frac{1}{4}$ kW.*

Split-phase capacitor-start induction motor :

Another method of splitting the single-phase supply into two phases to be applied to the stator windings is *placing a capacitor in series with the starting auxiliary winding.* In this manner, the current in the starting winding may be made to lead the line voltage. Since the running winding current lags the line voltage, the phase displacement between the two currents be made to approximately 90° on starting. The circuit of capacitor-start motor is shown in Fig. 5.80 (*a*), while the

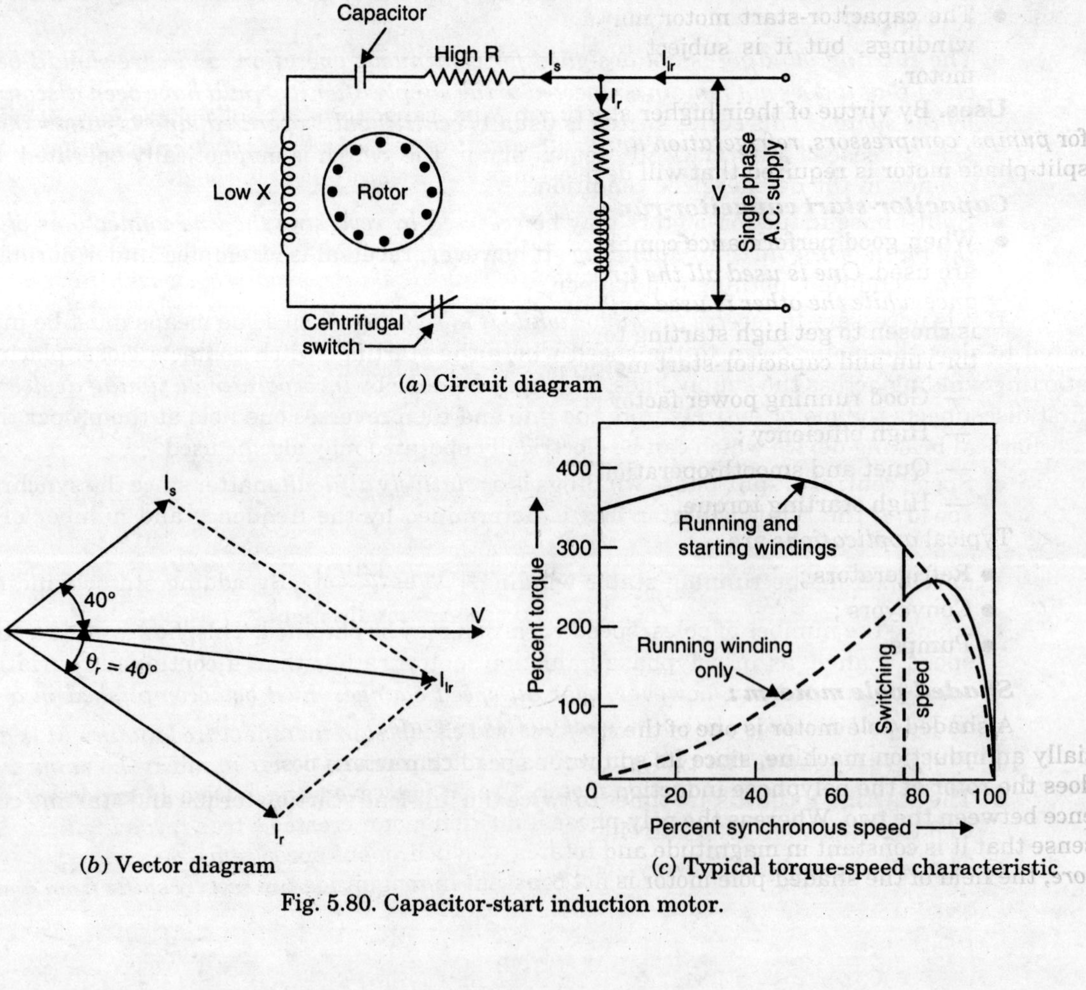

(*a*) Circuit diagram

(*b*) Vector diagram

(*c*) Typical torque-speed characteristic

Fig. 5.80. Capacitor-start induction motor.

vector diagram of the currents and voltage is shown in Fig. 5.80 (*b*). The values of the angles shown are fairly representative, and are rounded off for convenience. *One of the factors upon which the starting torque depends is the sine of the angle between the currents in the two windings.* The value of series capacitor may therefore be reduced, while maintaining a phase-shift angle of about 90°.

- The increase in phase angle between starting and running winding currents is not the only difference between the split-phase and capacitor start-motors. *The split phase motor must keep the number of starting-winding turns low, so that the current may be nearly in phase with the line voltage.* This, *however, is unnecessary in a capacitor-start motor, since the capacitor can overcome the inductance of the winding while still providing the proper phase shift.* There are thus *more auxiliary starting turns* in the capacitor-start motor than in the comparable split-phase motor. This provides a greater number of ampere-turns, hence a *larger rotating flux,* and therefore a *further increase in the starting torque.*

Also it is seen that for the same magnitudes of field currents, the *current* I_{lr} *is less in capacitor-start motor,* because of the greater angle between the two field currents. In addition, the *starting power factor is also better.* For a given line current, the starting torque is thus much higher for a capacitor-start motor than for a split-phase induction motor. *The starting torque of capacitor-start motor is from 3 to 4.5 times the full-load torque, while that of split-phase resistance-start induction motor rarely exceeds twice the full-load torque.*

- These motors are manufactured in *ratings ranging from* $\frac{1}{10}$ *kW to* $\frac{3}{4}$ *kW, but larger sizes are also available.*
- The capacitor-start motor may be reversed by changing the connections of one of the windings, but it is subject to the same limitations as the resistance-start induction motor.

Uses. By virtue of their higher starting torque, capacitor-start split-phase motors are used for *pumps, compressors, refrigeration units, air-conditioners,* and *large washing machines,* where a split-phase motor is required that will develop high starting torque under load.

Capacitor-start capacitor-run motors :

- When good performance combined with high starting torque is required, *two capacitors* are used. *One is used all the time and its value is chosen to obtain good running performance while the other is used only during start and run up.* The combined value of the two is chosen to get high starting torque. Thus the motor combines the *advantages* of capacitor-run and capacitor-start motors which are as follows :
  — Good running power factor ;
  — High efficiency ;
  — Quiet and smooth operation ;
  — High starting torque.

Typical *applications* are :

- Refrigerators ;
- Conveyors ;
- Pumps.
- Compressors ;
- Air-conditioners ;

Shaded-pole motorm :

A shaded-pole motor is one of the *simplest* and *cheapest* of manufactured motors. It is essentially an induction machine, since its squirrel-cage rotor receives power in much the same way as does the rotor of the polyphase induction motor. There is however, one extremely important difference between the two. Whereas the poly-phase induction motor creates a true revolving field, in the sense that it is constant in magnitude and rotates at synchronous speed *completely round the entire core,* the field of the shaded-pole motor is not constant in magnitude but *merely shifts from one side of the pole to the other.* Because the shaded-pole motor does not create a true revolving field, the *torque is not uniform but varies from instant to instant.*

Fig. 5.81 shows the *general construction and principle of shaded pole motor.*

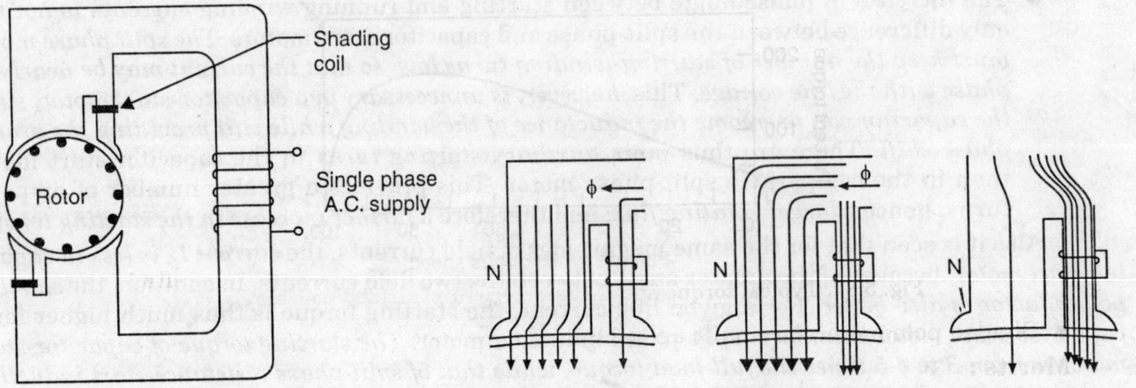

(*a*) General construction of 2 pole shaded pole motors (*b*) Increasing φ (*c*) Constant φ (*d*) Decreasing flux

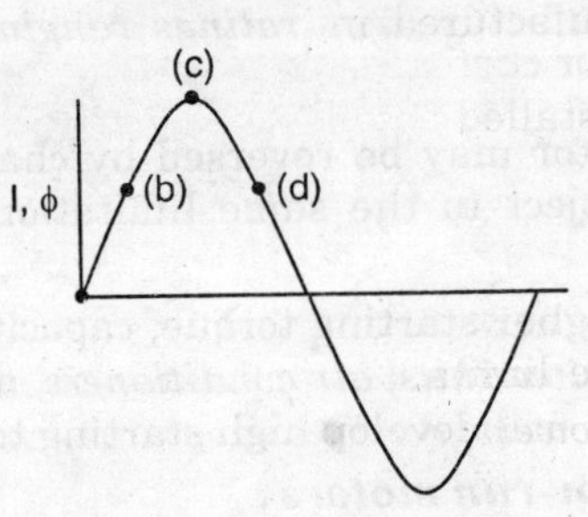

(*e*) Rate of change of current and flux in poles

Fig. 5.81. General construction and principle of shaded pole motor.

Construction. Each of the laminated poles of the stator has a slot cut *across* the laminations about one-third the distance from one edge. Around the smaller of the two areas formed by this slot is placed a heavy *copper short-circuited coil,* called a *"shading coil"* ; the iron around which the shading coil is placed is called the *shaded* part of the pole, while the free portion of the pole is the *unshaded* part. The exciting coil surrounds the entire part.

Principle of operation. When the exciting winding is connected to an A.C. source of supply, the *magnetic axis will shift from the unshaded part of the pole to the shaded part of the pole.* This shift in the magnetic axis is, in effect *equivalent to an actual physical motion of the pole* ; the result is that the squirrel cage rotor will rotate in a direction from the unshaded part to the shaded part. The shifting of flux is explained below.

- Refer to Fig. 5.81 (*b*). When the flux in the field poles tends to increase, a short-circuit current is induced in the shading coil, which by Lenz's law opposes the force and the flux producing it. Thus, as the flux increases in each field pole, there is a *concentration of flux in the main segment of each pole,* while the shaded segment opposes the main field flux.
- At point (*c*) shown in Fig. 5.81 (*e*), the rate of change of flux and of current is zero, and there is no voltage induced in the shaded coil. Consequently, the *flux is uniformly distributed across the poles* [Fig. 5.81 (*c*)].

- When the flux decreases, the current reverses in the shaded coil to maintain the flux in the same direction. The result is that the flux crowds in the *shaded segment of the pole* [Fig. 5.81 (*d*)].

A typical torque-speed characteristic is shown in Fig. 5.82.

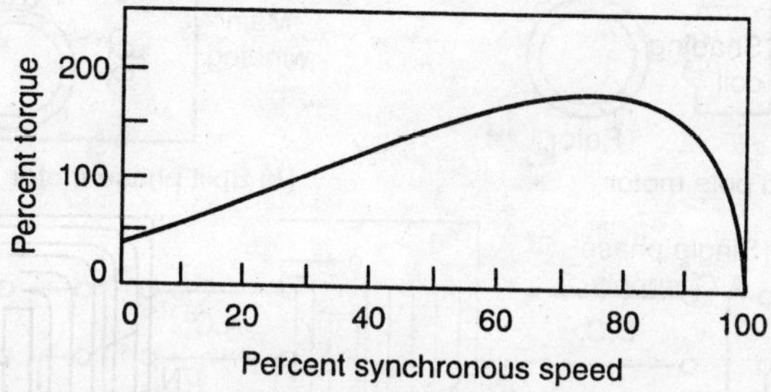

Fig. 5.82. Typical torque-speed characteristic of shaded pole motor.

- Shaded pole motors are built up to about 40 W.

Merits :

(*i*) Rugged construction (*ii*) Cheaper in cost

(*iii*) Small in size (*iv*) Requires little maintenance.

(*v*) Its stalling locked-rotor current is only slightly higher than its normal rated current, so that it can remain stalled for short periods without harm.

Demerits :

(*i*) Very low starting torque. (*ii*) Low efficiency.

(*iii*) Low power factor.

Uses. Its low starting torque limits its application to *phono-motors* or *turntables, motion picture projectors, small fans and blowers, bending machines, rotating store-window display tables, and relatively light loads.*

5.13.1.20. Braking of single-phase motors

These motor can be braked by :

(*i*) D.C. dynamic braking.

(*ii*) Plugging.

(*i*) ***D.C. dynamic braking :***

This method is commonly used for braking of single-phase induction motors. With the help of a double pole double throw switch or triple pole double throw switch, motor connection is shifted from A.C. (motoring) to D.C. source for braking. These connections for various single-phase induction motors are shown in Fig. 5.83.

- In case of split-phase, capacitor-run, and capacitor-start and capacitor-run motors, either main winding along can be connected across the D.C. source [Fig. 5.83 (*b*)] or main and auxiliary windings connected in series or parallel [Fig. 5.83 (*c*) and (*d*)].
- When in braking connection, D.C. current through the stator winding (or windings) produces a stationary field through which squirrel cage rotor moves. Current induced in rotor bars *interact with D.C. field to produce braking torque,* as in 3-phase induction motor. Motor *decelerates* and *stops.* As induced currents are zero at zero speed, the braking torque is also zero. *For braking, supply is obtained by a diode rectifier connected to A.C. mains.* Motor winding can be connected directly across diode rectifier to obtain fast braking. Winding is disconnected from D.C. supply after the motor stops.

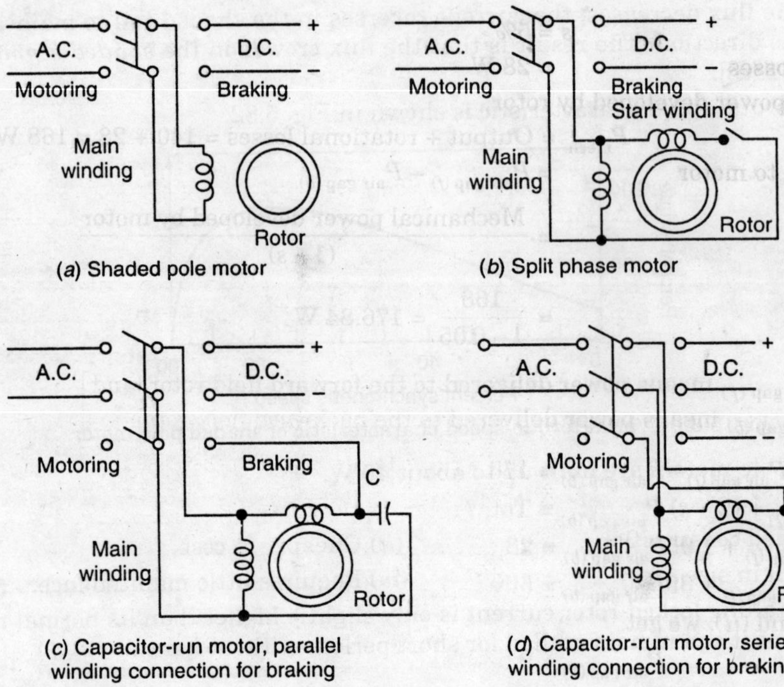

Fig. 5.83. D.C dynamic braking of single-phase induction motors.

(*ii*) ***Plugging and reversal.*** Except in case of shaded pole, motor, *plugging and reversal* is *obtained by changing phase sequence by reversing polarity of one of the windings.*

5.13.1.21. Speed control of single-phase induction motors

Speed of a single-phase induction motor can be controlled by the following methods :

(*i*) ***Stator voltage control :***

(*a*) Stator voltage can be controlled by connecting a variable resistance in series with the stator. Because of poor *efficiency* the *resistance control is now rarely used.*

(*b*) Stator voltage can also be controlled by the use of *A.C. voltage controllers.*

(*ii*) ***Variable frequency control.*** The speed of the motor can also be controlled by variable frequency control. However, it is *rarely used* for most of the variable speed applications of single-phase motors, the *stator voltage control is good enough.*

Example 5.103. *The following data pertains to a single-phase induction motor :*

Number of poles	*= 4*
Supply voltage	*= 110 V*
Rated output	*= 140 W*
Slip at rated output	*= 5 per cent*
Total copper loss at full-load	*= 28 W*
Rotational losses	*= 28 W*

Calculate the full-load efficiency and the rotor copper loss caused by the backward field. Neglect stator copper loss.

Solution. Number of poles, $p = 4$

Motor output $= 140$ W

Slip, $s = 5\%$

Rotational losses $= 28$ W

Mechanical power developed by rotor,

$$P_{mech.} = \text{Output} + \text{rotational losses} = 140 + 28 = 168 \text{ W}$$

Power input to motor $= P_{air\ gap\ (f)} - P_{air\ gap\ (b)}$

$$= \frac{\text{Mechanical power developed by motor}}{(1-s)}$$

$$= \frac{168}{1-0.05} = 176.84 \text{ W}$$

[where, $P_{air\ gap\ (f)}$ means power delivered to the forward field rotor, and $P_{air\ gap\ (b)}$ means power delivered to the backward field rotor]

i.e., $P_{air\ gap\ (f)} - P_{air\ gap\ (b)} = 176.84$ W ...(*i*)

and $sP_{air\ gap\ (f)} + (2-s)\, P_{air\ gap\ (b)} = \text{Total rotor copper loss}$

$0.05P_{air\ gap\ (f)} + 1.95P_{air\ gap\ (b)} = 28$

or $P_{air\ gap\ (f)} + 39P_{air\ gap\ (b)} = 560$ ...(*ii*)

Solving (*i*) and (*ii*), we get

$$P_{air\ gap\ (b)} = 9.58 \text{ W}$$

Rotor copper losses caused by backward field

$= (2-s)P_{air\ gap\ (b)}$

$= (2 - 0.05) \times 9.58 =$ **18.68 W. (Ans.)**

Full-load efficiency :

Input power to motor = Power input to rotor + stator copper loss

= 176.84 W, neglecting stator copper loss

$$\therefore \quad \eta_{full\text{-}load} = \frac{\text{Power output}}{\text{Power input}} = \frac{140}{176.84} = 0.79 \text{ or } \textbf{79 per cent.} \quad \textbf{(Ans.)}$$

Example 5.104. *The constants of a one-quarter H.P., 230 V, 4-pole, 60 Hz, single phase induction motor are as follows :*

Stator resistance, $R_1 = 10.0\ \Omega$; stator reactance, $X_1 = 12.8\ \Omega$; rotor resistance referred to stator, $R_2' = 11.65\ \Omega$; rotor reactance referred to stator, $X_2' = 12.8\ \Omega$; magnetising reactance, $X_m = 258.0\ \Omega$.

The total load is such that the machine runs at 3% slip, when the voltage is at 210 V. The iron losses are 35.5 W at 210 V. Calculate :

(*i*) *Input current ;* (*ii*) *Power developed ;*

(*iii*) *Shaft power (if mechanical losses are 7 W) ;*

(*iv*) *Efficiency.* **(AMIE Advance Elec. M/Cs. Summer, 2000)**

Solution. *Given* : $R_1 = 10.0\ \Omega$; $X_1 = 12.8\ \Omega$; $R_2' = 11.65\ \Omega$; $X_2' = 12.8\ \Omega$; $X_m = 258.0\ \Omega$; $s = 3\% = 0.03$; $V = 210$ volts ; $P_i = 35.5$ W ; mechanical losses = 7 W.

$$\text{Core loss current, } I_w = \frac{35.5}{210} = 0.169 \text{ A.}$$

Since under running condition V_f is very high (almost 90% of the applied voltage) and V_s is very low, so most of the core loss takes place in the 'forward motor' consisting of the common stator and forward running motor. So half value of core loss equivalent resistance, $\frac{R_m}{2} = \frac{210}{0.169} = 1242\ \Omega$.

For forward field : Refer to Figs. 5.77 and 5.78.

Conductance of core-loss branch $= \dfrac{1}{(R_m/2)} = \dfrac{1}{1242} = 0.0008$ S

Susceptance of magnetising branch $= -\dfrac{j}{(X_m/2)} = -\dfrac{j}{(285/2)} = -j0.00775$ S

Rotor admittance
$$= \frac{1}{\dfrac{R_2'}{2s} + \dfrac{jX_2'}{2}} = \frac{1}{\dfrac{11.65}{2 \times 0.03} + \dfrac{j\,12.8}{2}} = \frac{1}{194.17 + j\,6.4}$$
$$= \frac{1}{194.17 + j\,6.4} \times \frac{194.17 - j\,6.4}{194.17 - j\,6.4}$$
$$= \frac{194.17 - j\,6.4}{37743} = (0.005145 - j0.00017) \text{ S}$$

Admittance
$$Y_f = (0.0008 - j0.00775 \text{ S} + 0.005145 - j0.00017) \text{ S}$$
$$= (0.005945 - j0.00792) \text{ S}$$

Impedance,
$$Z_f = \frac{1}{Y_f} = \frac{1}{0.005945 - j\,0.00792}$$
$$= 60.62 + j80.759 = 100.979 \angle 53.1° \,\Omega$$

For backward field :

Rotor admittance
$$= \frac{1}{\dfrac{R_2'}{2(2-s)} + \dfrac{jX_2'}{2}} = \frac{1}{\dfrac{11.65}{2(2-0.03)} + \dfrac{j\,12.8}{2}}$$
$$= \frac{1}{2.957 + j6.4} = (0.0595 - j0.12876) \text{ S}$$

Admittance,
$$Y_b = (0.0008 - j0.00775 + 0.0595 - j0.12876) \text{ S}$$
$$= (0.0603 - j0.1365) \text{ S}$$

Impedance,
$$Z_b = \frac{1}{Y_b} = \frac{1}{0.0603 - j0.1365} = 2.708 + j6.13 = 6.70 \angle 66.17° \,\Omega$$

Equivalent impedance,
$$Z_{eq} = Z_1 + Z_f + Z_b$$
$$= (10 + j12.8) + (60.62 + j80.759) + j(2.708 + j6.13)$$
$$= 73.323 + j99.689 = 123.75 \angle 53.66° \,\Omega$$

(*i*) **Input current,**
$$\mathbf{I_1} = \frac{V}{Z_{eq}} = \frac{230}{123.75} = \mathbf{1.8586 \text{ A. (Ans.)}}$$

$$\text{Power input} = VI_1 \cos\phi = 230 \times 1.8586 \times \cos 53.66° = 253.3 \text{ W}$$
$$E_{mf} = I_1 \times Z_f = 1.8586 \times 100.979 = 187.68 \text{ V}$$
$$E_{mb} = I_1 \times Z_b = 1.8586 \times 6.70 = 12.45 \text{ V}$$
$$I'_{2f} = \frac{E_{mf}}{\sqrt{\left(\dfrac{R_2'}{2s}\right)^2 + \left(\dfrac{X_2'}{2}\right)^2}} = \frac{187.68}{\sqrt{(194.17)^2 + (6.4)^2}} = 0.966 \text{ A}$$

$$I'_{2b} = \frac{E_{mb}}{\sqrt{\left(\frac{R'_2}{2(2-s)}\right)^2 + \left(\frac{X'_2}{2}\right)^2}} = \frac{12.45}{\sqrt{(2.2957)^2 + (6.4)^2}} = 1.831 \text{ A}$$

Air-gap power for forward field, $P_{\text{air gap}(f)} = (I'_{2f})^2 \times \frac{R'_2}{2s} = (0.966)^2 \times \frac{11.65}{2 \times 0.03} = 181.2 \text{ W}$

Air-gap power for backward field, $P_{\text{air gap}(b)} = (I'_{2b})^2 \times \frac{R'_2}{2(2-s)} = (1.831)^2 \times \frac{11.65}{2 \times 1.97} = 9.9 \text{ W}$

(*ii*) **Power developed** $\mathbf{P_{mech\ (net)}} = (1-s)\,P_{\text{air gap}(f)} - P_{\text{air gap}(b)}$

$= (1 - 0.03)(181.2 - 9.9) = $ **166.2 W. (Ans.)**

(*iii*) **Shaft power** $= P_{\text{mech (net)}}$ – mechanical losses = 166.2 – 7 = **159.2 W. (Ans.)**

(*iv*) **Efficiency,** $\eta = \frac{\text{Power output}}{\text{Power input}} \times 100 = \frac{159.2}{253.3} \times 100 = $ **62.85%. (Ans.)**

Example 5.105. *A $\frac{1}{4}$ H.P., split-phase motor draws its starting winding current of 4 A lagging the supply voltage by 15° elec. and its running winding current is 6A lagging by 40° elec. Calculate :*

(*i*) *The total current and power factor (at steady state) ;*

(*ii*) *The component of starting winding current in phase with voltage ;*

(*iii*) *The phase angle between the starting and running winding currents ;*

(*iv*) *The component of running winding current that lags the supply voltage by 90°.*

(AMIE Advance Elect. M/Cs. Summer, 2000)

Solution. Starting winding current, $I_s = 4 \angle -15°\text{A} = (3.8637 - j1.0353)$ A

Running winding current, $I_r = 6 \angle -40°\text{A} = (4.5963 - j3.8567)$ A

Total current, $I_{lr} = I_s + I_r = (3.8637 - j1.0353) + (4.5963 - j3.8567)$

$= (8.46 - j4.892)$ A or $9.77 \angle -30°$

(*i*) **Total current** in magnitude, I_{lr} = **9.77 A. (Ans.)**

Power factor, $\cos\phi = \cos(-30°) = $ **0.866 (lag). (Ans.)**

(*ii*) **The component of starting winding current in phase with supply voltage**

= Active component of starting current

= **3.8637 A. (Ans.)**

(*iii*) **The phase angle between the starting and running winding currents,**

$\theta = \phi_r - \phi_s = 40° - 15° = $ **25°. (Ans.)**

(*iv*) **The component of running winding current that lags behind the supply voltage by 90°**

= Reactance or *j*-component of running winding current

= **3.8567 A. (Ans.)**

Example 5.106. *The following data pertain to a 230 V, 50 Hz, capacitor-start single-phase induction motor at standstill :*

Main winding excited alone = 100 V, 2 A, 40 W

Auxiliary winding excited alone = 80 V, 1 A, 50 W

Determine the value of capacitance for determining the maximum starting torque.

(AMIE Advance Elect. M/Cs. Winter, 2001)

Solution. *Given* : $V_m = 100$ V, $I_m = 2$ A ; $P_m = 40$ W ; $V_a = 80$ V ; $I_a = 1$ A ; $P_a = 50$ W

Value of capacitance for maximum torque :

Main winding :

Equivalent impedance with main winding, $Z_{01m} = \dfrac{V_m}{I_m} = \dfrac{100}{2} = 50\ \Omega$

Equivalent resistance with main winding, $R_{01m} = \dfrac{P_m}{I_m^2} = \dfrac{40}{2^2} = 10\ \Omega$

Assuming rotor resistance and reactance referred to stator, R_2' and X_2' equal to stator resistance R_1 and reactance X_1 respectively

$$R_{1m} = R_2' = \frac{R_{01m}}{2} = \frac{10}{2} = 5\ \Omega$$

and
$$X_{1m} = X_2' = \frac{X_{01m}}{2} = \frac{\sqrt{50^2 - 10^2}}{2} = 24.5\ \Omega$$

Angle of lag of main winding current, $\phi_m = \tan^{-1} \dfrac{X_{1m}}{R_{1m}} = \tan^{-1} \dfrac{24.5}{5} = 78.46°$

Auxiliary winding :

Equivalent impedance with auxiliary winding, $Z_{01a} = \dfrac{V_a}{I_a} = \dfrac{80}{1} = 80\ \Omega$

Equivalent resistance with auxiliary winding, $R_{01a} = \dfrac{P_a}{I_a^2} = \dfrac{50}{1^2} = 50\ \Omega$

Equivalent reactance with auxiliary winding, $X_{01a} = \sqrt{80^2 - 50^2} = 62.45\ \Omega$

Auxiliary winding resistance $R_a = R_{01a} - R_2' = 50 - 5 = 45\ \Omega$

Auxiliary winding reactance, $X_a = X_{01a} - X_2' = 62.45 - 24.5 = 37.95\ \Omega$

Torque developed at start will be maximum, if the phase angle between main winding current I_m and auxiliary winding current I_a is 90° or the auxiliary winding current leads the applied voltage by (90° – 78.46°) or 11.54°.

If X_c is the capacitive reactance of the capacitor connected in series with the auxiliary winding, then impedance of the auxiliary winding will be given as

$$Z_a = R_a + X_a - X_c = 45 + j37.95 - jX_c = [45 + j(37.95 - X_c)] \text{ ohms}$$

For auxiliary winding, $\tan \phi_a = \dfrac{37.95 - X_c}{45}$

or,
$$X_c = 37.95 - 45 \tan \phi_a = 37.95 - 45 \tan (-11.54°)$$
$$= 37.95 + 9.19 = 47.14\ \Omega$$

Capacitance of capacitor, $C = \dfrac{1}{2\pi f X_c} = \dfrac{1}{2\pi \times 50 \times 47.14} \times 10^6 = \mathbf{67.52\ \mu F.}$ **(Ans.)**

5.13.1.22. Linear induction motor

Introduction. A linear induction motor is a special type of induction motor which *provides linear motion* and works on the following principle (same as that of conventional induction motor).

"Whenever a relative motion occurs between the field and short-circuited conductors, currents are induced in them which results in electro-magnetic forces and under the influence of these forces,

according to Lenz's law, the conductors try to move in such a way as to eliminate the induced currents". In this case the field's movement is linear and so is the conductors' movement.

Construction and working :

Construction. A linear induction motor, in its simplest form, consists of field system having a 3-ϕ distributed winding placed in slots as shown in Fig. 5.84 (short-single primary).

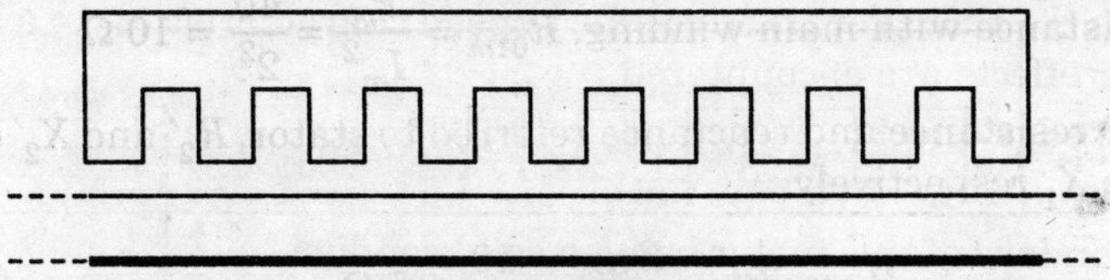

Fig. 5.84. Short-single primary.

The field system may be single or double primary system. The *secondary* of this type of induction motor is normally a *conducting plate made of either copper or aluminium in which interaction currents are induced*.

Depending upon the particular requirements either member can be the *stator*, the other being the *rotor*.

— The ferromagnetic plate, in a single primary system, is usually placed on the other side of the conducting plate to provide a path of low reluctance to the main flux. The ferromagnetic plate, however, gets attracted towards the primary when the field is energised ; consequently *unequal gap length results on the two sides of the plate. Double primary system* can be used to overcome this problem.

— The use of the motor decides which of the two primary and secondary will be shorter in length compared to other. The *primary is made shorter than secondary when the operating distance is large* (since winding a very long 3-phase primary is costly proposition) and the short secondary is used when the operating distance is limited.

Working :

When the 3-ϕ primary winding of the motor is energised from a balanced three phase source, a magnetic field moving in a straight line from one end to other at a *linear synchronous* speed V_s is produced. The linear synchronous speed V_s is given as :

$$V_s = 2\,\tau f \text{ m/s} \qquad \text{...(5.55)}$$

where, τ = Pole pitch in metres, and

f = Supply frequency in Hz.

As the flux moves linearly, it drags the rotor plate alongwith it in the same direction with speed V. Consequently the relative speed of travel of the flux w.r.t. rotor plate decreases. (In case rotor plate speed equals that of magnetic field, the latter would be stationary when viewed from rotor plate. When the rotor plate moves faster than synchronous speed of magnetic field, the direction of force is reversed and a form of regenerative braking based on the principle of induction generator, will come into existence).

Slip of the motor(s) is given as :

$$s = \frac{V_s - V}{V_s} \qquad \text{...(5.56)}$$

where, V_s = linear synchronous speed, and

V = Actual speed of the rotor plate.

Thrust or force or tractive effort (F) is given as :

$$F = \frac{P_2}{V_s} \qquad ...(5.57)$$

where, P_2 = Actual power supplied to the rotor.

Also, copper losses in rotor = sP_2, ...(5.58)

and mechanical power developed, $P_{mech.} = (1 - s)P_2$ ...(5.59)

Fig. 5.85 shows the thrust or tractive effort-speed characteristics. In a linear induction motor the following peculiar effects are encountered :

(*i*) Transverse edge effect ;

(*ii*) End effect.

Due to the secondary of this motor being a solid conducting state, the paths of the induced currents in the secondary are not well defined. The portion of the current paths parallel to the direction of motion of the secondary does not contribute anything towards the production of useful thrust but only contributes towards losses. This effect causes reduction in thrust and increases the losses and is known as **transverse edge effect** (since the current paths parallel to the direction of motion are more towards the edges of the conducting plate).

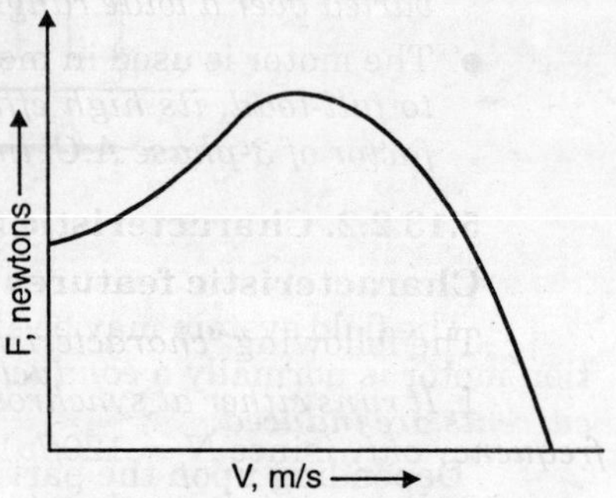

Fig. 5.85. Thrust-speed characteristics.

Advantages :

1. Simpler in construction.
2. Better power to weight ratio.
3. Low initial cost.
4. No overheating of rotor (since the motor moves continuously over cool rotor plate leaving behind heated rotor portion).
5. Owing to absence of rotating parts the maintenance cost is low.
6. No limitation of tractive effort due to adhesion between the wheel and the rail.
7. No limitation of maximum speed due to centrifugal forces.

Disadvantages :

1. Owing to transverse edge and end effects utilisation of motor is poor.
2. Capital cost of reaction rail fixed along the centre line of the track is very high.
3. Provision of three phase collector system along the track involves complications and high cost.
4. Maintaining adequate clearance at points and crossings entails a lot of difficulties.
5. Requirement of larger air-gap and non-magnetic reaction rail (rotor plate) necessitates more magnetising current resulting in poor efficiency and low-power factor.

Applications :

Following are the fields of application of linear induction motor :

- Electromagnetic pumps.
- Conveyors.
- High-speed rail traction.
- Trolley cars (for internal transport in workshops).
- As booster accelerator for moving heavy trains from rest or up the planes or on curves.
- Metallic belt conveyors etc.

Owing to design difficulties and economic considerations the use of linear induction motor is limited to only a few applications.

5.13.2. Synchronous Motor

5.13.2.1. Introduction

- The synchronous motor is the one type of 3-phase A.C. motors *which operates at a constant speed from no-load to full load.* It is similar in construction to 3-phase A.C. generator in that *it has a revolving field which must be separately excited from a D.C. source. By changing the D.C. field excitation current, the power factor of this type of motor can be varied over a wide range of lagging and leading values.*
- The motor is used in many individual applications because of its *fixed speed from no-load to full-load, its high efficiency* and *low initial cost.* It is also used *to improve the power factor of 3-phase A.C. industrial circuits.*

5.13.2.2. Characteristic features, advantages and disadvantages

Characteristic features :

The following *"characteristic features"* of a synchronous motor are worth noting :

1. *It runs either at synchronous speed or not at all.* The speed can be changed by *changing the frequency only* (since $N_s = 120f/p$).

2. *It is not inherently self-starting.* It has to be *run up to synchronous or near synchronous speed* by some means before it can be synchronized to the supply.

3. *It can operate under a wide range of power factors both lagging and leading.*

4. On no-load the motor draws very little current from the supply to meet the internal losses. With fixed excitation the input current increases with the increase in load. After the input current reaches maximum no further increase in load is possible. If the motor is further loaded, the motor will stop.

Advantages. Synchronous motors entail the following *advantages* :

1. These motors can be used for *power factor correction* in addition to supplying torque to drive loads.

2. They are more efficient (when operated at unity power factor) than induction motors of corresponding output (kW) and voltages rating.

3. The field pole rotors of synchronous motors *can permit the use of wider air-gaps* than the squirrel-cage designs used on induction motors, requiring *less bearing tolerance and permitting greater bearing wear.*

4. They may be less costly for the same output, speed, and voltage ratings as compared to induction motors.

5. They give constant speed from no-load to full-load.

6. Electromagnetic power varies linearly with the voltage.

Disadvantages. The *disadvantages* of synchronous motors are :

1. They require D.C. excitation which must be supplied from external source.

2. They have a tendency to *hunt*.

3. They cannot be used for variable speed jobs as speed adjustment cannot be done.

4. They require collector rings and brushes.

5. They cannot be started under load. Their starting torque is zero.

6. They may fall out of synchronism and stop when overloaded.

5.13.2.3. Applications

The synchronous motors have the following *fields of application :*

1. **Power houses and sub-stations.** Used in power houses and sub-stations in parallel to the bus-bars to improve the power factor.

2. **Factories.** Used in factories having large number of induction motors or other power apparatus, operating at *lagging power factor, to improve the power factor.*

3. **Mills-Industries etc.** Used in textile mills, rubber mills, mining and other big industries, cement factories for power applications.

4. **Constant speed equipments.** Used to drive continuously operating and constant speed equipment such as :

- Fans
- Blowers
- Centrifugal pumps
- Motor generator sets
- Ammonia and air compressors etc.

5.13.2.4. Types of synchronous motors

The following types of synchronous motors are commonly used :

1. Wound field synchronous motors.
2. Permanent magnet (PM) synchronous motors.
3. Synchronous reluctance motors.
4. Hysteresis synchronous motors.

— All these motors have a stator with a 3-phase winding, which is connected to an A.C. source.

— Fractional horse power synchronous reluctance and hysteresis motors employ a 1-phase stator.

1. **Wound field motors.** Wound field synchronous motor rotor has a D.C. field winding, which is supplied from a D.C. source through slip-rings and brushes. The rotor can have cylindrical or salient pole construction.

— Cylindrical rotors have *higher mechanical strength* and are *employed in high power and high speed applications.*

— Salient pole motors, due to low cost, are preferred for other applications.

2. **Permanent magnet (PM) synchronous motors.** In medium and small size motors, D.C. field can be produced by permanent magnets ; thus dispensing with D.C. source, slip rings, brushes and field winding losses. Such motors are known as *permanent magnet (PM) synchronous motors.* Usually ferrite magnets are used. Rare earth (cobalt-samarium) magnets, although very expensive, are sometimes used to *reduce* the volume and weight of the motor.

PM synchronous motors are classified as follows :

(i) ***Surface mounted :***

(a) Projecting type. In such motors magnets projects from the surface of the rotor [Fig. 5.86 *(a)*].

(b) Inset type. In this case, magnets are inserted into the rotor, providing a smooth rotor surface [Fig. 5.86 *(b)*]

- Epoxy glue is used to fix the magnets to the rotor surface in both.
- While these motors are *easy to construct and are less expensive, they are less robust compared to interior type rotors and are not suitable for high speed applications.*

(*ii*) ***Interior (or buried).*** In these motors, magnets are imbedded in the interior of the rotor (Fig. 5.86 (*ii*))

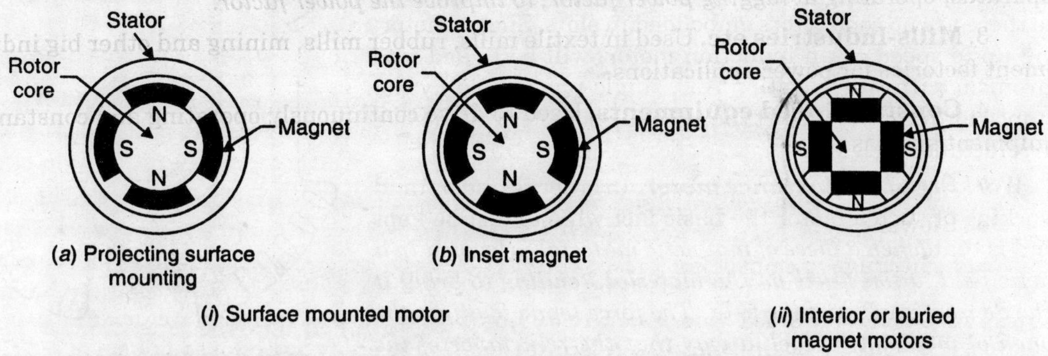

Fig. 5.86. Types of PM synchronous motors.

Features of wound field and permanent magnet synchronous motors :

The wound field and PM synchronous motors have a *higher full load efficiency and power factor than an induction motor.*

Wound field motors can be designed for a higher power rating than induction motors. Since the air-gap flux is not produced solely by the magnetising current drawn from the armature, a larger air-gap suiting the mechanical design can be chosen. The ability to control power factor is an important advantage at higher power levels. *Operating at unity power factor minimizes the inverter rating.*

PM synchronous motor, apart from the *robus construction, has low losses and high efficiency*. Because of low losses, it is possible to makes motors with very high power density and torque to inertia ratios. These make them *suitable for servo drives requiring fastest possible dynamic response.*

- One *significant difference* between the wound field and *PM* motors which are designed to operate with a source of fixed frequency is discussed below :

When a wound field motor is started as induction motor, D.C. field is kept off. In case of a *PM* motor, the field *cannot* be 'turned off'. When at a speed below synchronous speed, the rotor field induces a voltage in the stator, which has a frequency different than the frequency of stator supply. The current produced by induced voltage interacts with the rotor field to produce a braking torque, which opposes the induction motor torque due to damper winding. The permanent magnetic synchronous motor (PMSM) is designed so that the braking torque is very small compared to induction motor torque. Owing to the capability of starting direct on line these motors are called *line start PMSM*.

— PMSM are available in 3-phase and 1-phase construction.

— Although expensive to induction motor motors, they have advantages of *high efficiency, high power factor* and *low sensitivity to supply voltage variations.*

— These motors are preferred for industrial applications with *large duty cycles such as pumps, fans and compressors.*

3. **Synchronous reluctance motor :**

Single-phase salient-pole synchronous-induction motors, are generally called *reluctance motors.* If the rotor of any uniformly distributed single-phase induction motor is altered so that the laminations tend to produce *salient rotor poles*, as shown in Fig. 5.87, the reluctance of the air-gap flux path will be greater where there are no conductors embedded in slots. Such a motor, coming up to speed as an induction motor, will be pulled into synchronism with the pulsating A.C. single-phase field by the reluctance torque developed at the salient iron poles which have lower-reluctance air gaps.

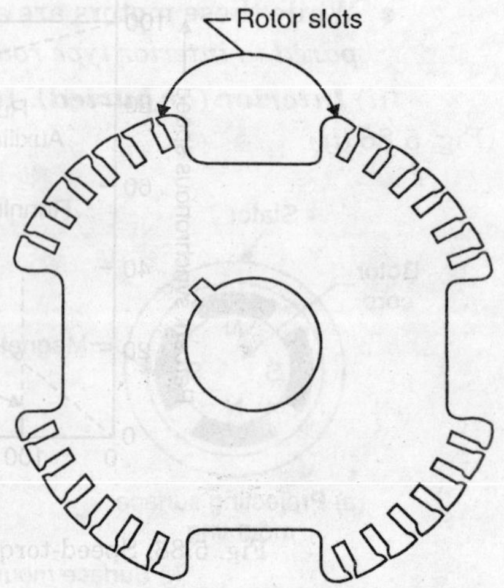

Fig. 5.87. Reluctance-motor lamination.

Working of a reluctance motor. In order to understand the working of such a motor the basic fact which must be kept in mind is that *when a piece of magnetic material is located in a magnetic field, a force acts on the material, tending to bring it into the densest portion of the field. The force tends to align the specimen of material in such a way that the reluctance of the magnetic path that passes through the material will be minimum.*

When supply is given to the stator winding, the revolving magnetic field will exert reluctance torque on the unsymmetrical rotor tending to align the salient pole axis of the rotor with the axis of the revolving magnetic field (because in this position, the reluctance of the magnetic path would be minimum). If the reluctance torque is sufficient to start the motor and its load, the rotor will pull into step with the revolving field and continue to run at the speed of the revolving field. (Actually the motor starts as an induction motor and after it has reached its maximum speed as an induction motor, the reluctance torque pulls its rotor into step with the revolving field so that the motor now runs as synchronous motor by virtue of its saliency).

Reluctance motors have approximately *one-third* the horsepower rating they would have as induction motors with cylindrical rotors, although the ratio may be increased to one-half by proper design of the field windings. *Power factor and efficiency are poorer than for the equivalent induction motor.* Reluctance motors are subject to '*cogging*', since, the locked-rotor torque varies with the rotor position, but the effect may be *minimized* by *skewing the rotor bars* and by *not* having the number of rotor slots exactly equal to an exact multiple of the number of poles.

Uses. Despite its short-comings, the reluctance motor is widely used for many *constant speed* applications such as *recording instruments, time devices, control apparatus, regulators,* and *phonograph turntables.*

- *Reversing* is obtained as in any single-phase induction motor.

Speed-torque characteristics. Fig. 5.88 shows speed-torque characteristics of a typical single-phase reluctance motor.

- The motor starts at anywhere from 300 to 400 per cent of its full-load torque (depending on the rotor position of the unsymmetrical rotor with respect to the field windings) as a two-phase motor as a result of the magnetic rotating field created by a starting and running winding (displaced) 90° in both space and time.
- At about 3/4th of the synchronous speed, a centrifugal switch opens the starting winding, and the motor continues to develop a single-phase torque produced by its running winding only.

As it approaches synchronous speed, the reluctance torque (developed as a synchronous motor) is sufficient to pull the rotor into synchronism with the pulsating single-phase field.

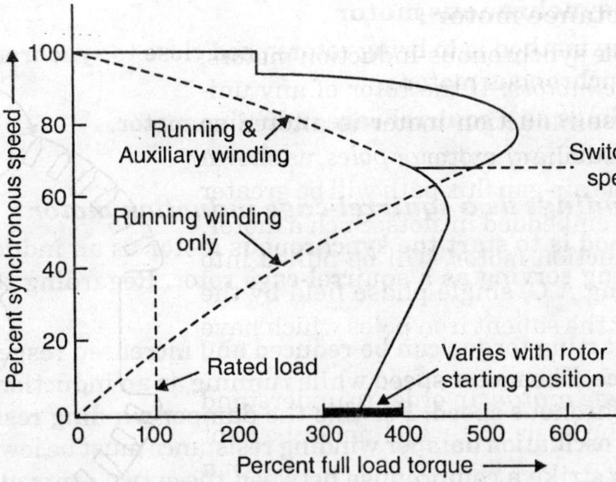

Fig. 5.88. Speed-torque characteristics of a single-phase reluctance motor.

- *The motor operates at a constant speed up to a little over 200% of its full-load torque.* If it is loaded beyond the value of pull-out torque, it will continue to operate as a single-phase induction motor up to 500% of its rated output.

Note. Some synchronous motors described above have additional winding, known as **damper or amortisseur winding**, on the rotor. It is similar to a squirrel-cage winding of an induction motor. It is provided to damp hunting oscillations which occur during transient operation of the motor. It is also used to start the motor as an induction motor. When the motor runs at synchronous speed, no voltages are induced in the damper winding. Therefore, damper winding has no effect on motor operation.

4. **Hysteresis synchronous motor.** Single-phase cylindrical (Non-salient-pole) synchronous-induction or shaded-pole motors are classed as *hysteresis motors*. A hysteresis motor has neither a salient-pole rotor nor direct excitation, but nevertheless it rotates at synchronous speed. This type of motor *runs into synchronism and runs on hysteresis torque.*

Hysteresis-type laminations, shown in Fig. 5.89, are usually made of *hardened, high retentivity steel* rather than commercial, low retentivity dynamo steel.

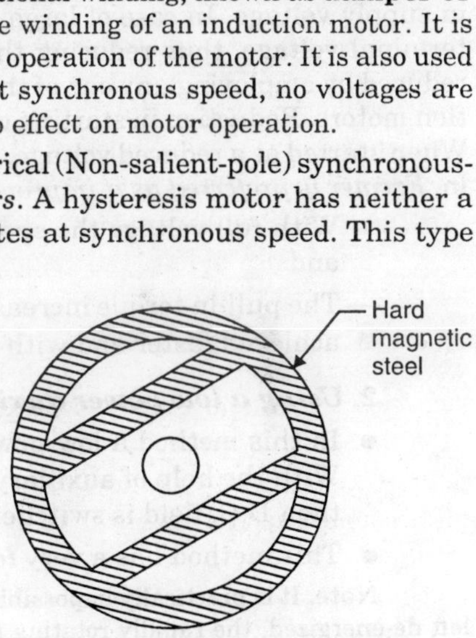

Fig. 5.89. Hysteresis rotor.

Working. As a result of a rotating magnetic field produced by phase splitting or a shaded-pole stator, *eddy currents are induced in the steel of the rotor which travel across the two bar paths of the rotor* as shown in Fig. 5.89. A high-retentivity steel produces a high hysteresis loss, and an appreciable amount of energy is consumed from the rotating field in reversing the current direction of the rotor. At the same time the *rotor magnetic field set up by the eddy currents causes the rotor to rotate. A high starting torque is produced as a result of the high resistance* (proportional to hysteresis). As the rotor approaches synchronous speed, the frequency of current reversal in the cross-bars decreases, and the rotor becomes *permanently magnetized* in one direction as a result of the high retentivity of the steel rotor. Consequently the motor continues to rotate at synchronous speed.

- An extremely important *use* of this type of motor is for the rotation of *gyroscope rotors* in inertial navigation and control systems. Here the requirement is for as near absolute accuracy as can be achieved. One major component of the instrument accuracy that contains the gyroscope is that the gyroscopic moment be *absolutely constant.* This constancy requires a synchronous motor that is driven by a regulated constant-frequency source.

5.13.2.5. Starting of synchronous motor

The purpose of starting method is to bring rotor speed close to synchronous speed. Following methods are used to start synchronous motor :

1. Using damper windings as a squirrel-cage induction motor.
2. Using a low power auxiliary motor.

1. ***Using damper windings as a squirrel-cage induction motor :***

One widely used method is to start the synchronous motor as an induction motor with field unexcited and damper winding serving as a squirrel-cage rotor. Regarding this method, following points are worth noting :

(*i*) The current and starting torque can be reduced and increased respectively, by increasing the damper winding resistance. The motor speed while running as an induction motor, for successful pull-in, must be close to synchronous speed. For this the damper winding resistance must be high. Further, for damping hunting oscillation damper winding resistance must be low. The damper winding resistance is so selected as to strike a compromise between these two contradictory requirements.

(*ii*) D.C. field should be applied only after the motor has reached close to full speed.

(*iii*) When the rotor has salient pole construction, the damper winding can have conductors only over the pole arc. This leads to a dip in the speed-torque curve at half of synchronous speed.

(*iv*) On the application of full supply voltage, the starting current in the motor can be 7 to 10 times of full load value. Except in small size motors, such a high starting current causes fluctuations in supply voltage. In case of large size motors, such a high current may cause a large drop in the terminal voltage, thus reducing the already low starting torque further. Starting current can be reduced by employing any one of the reduced voltage starting methods employed for starting induction motors. Reduction in starting current is obtained at the expense of reduction in starting torque. When started at a reduced voltage, the transition to full voltage can be made before or after the pull-in. *Former is preferred as it improves pull-in performance due to following two reasons :*

— With full voltage the speed attained as induction motor is closer to synchronous speed, and

— The pull-in torque increases in proportion to voltage squared, consequently pull-in can be achieved faster and with large motor loads.

2. ***Using a low power auxiliary motor :***

- In this method a low power auxiliary motor is coupled to the synchronous motor shaft. With the help of auxiliary motor, the rotor speed is brought near synchronous speed and then D.C. field is switched-in.
- This method has a very *low starting torque.*

Note. It is practically impossible to start a synchronous motor with its D.C. field energized. Even when left de-energized, the rapidly rotating magnetic field of the stator will induce extremely high voltages in the many turns of the field winding. It is customary, therefore, *to short-circuit the D.C. field winding during the starting period* ; whatever voltage and current are induced in it may then aid in producing induction motor action. In very large synchronous motors, field sectionalising or field splitting switches are used which short-circuit individual field windings to prevent cumulative addition of induced voltages from pole to pole.

5.13.2.6. Braking of synchronous motors

- The motor can work in regenerative braking only at synchronous speed. Therefore, *"regenerative braking" cannot be used for stepping or decelerating a load.*
- *"Dynamic braking"* is obtained by *disconnecting stator from the source and connecting it to 3-phase resistor.* Machine works as a synchronous generator and dissipates generated energy in the braking resistor.

5.13.2.7. Brushless D.C. (or Trapezoidal PMAC) motors :

Fig. 5.90 shows the cross-section of a 3-phase 2-pole trapezoidal PMAC motor.

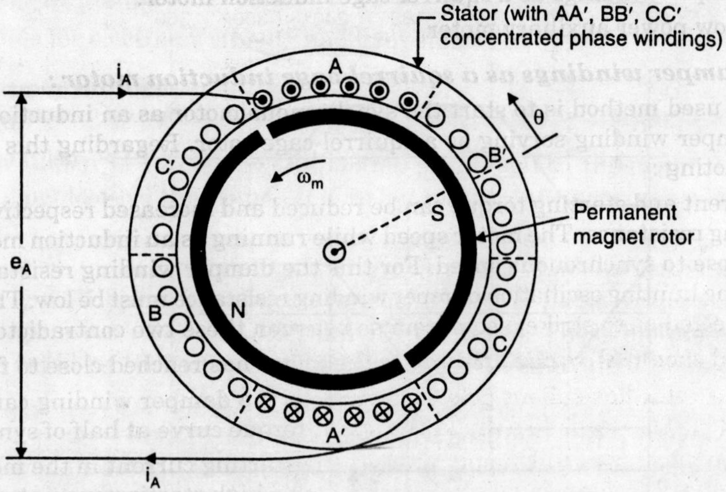

Fig. 5.90. Cross-section of a trapezoidal PMAC motor.

The stator has three concentrated phase windings (*AA′, BB′* and *CC′*) which are displaced by 120° and each phase winding spans 60° on each side. The voltages induced in three phases are shown in Fig. 5.91. The reason for getting trapezoidal waveforms is explain below :

When revolving in the counter-clockwise direction, upto 120° rotation from the position shown in Fig. 5.90, all top conductors of phase *A* will be linking the *S*-pole and all bottom conductors of phase *A* will be linking the *N*-pole. Hence the voltage induced in phase *A* will be the same during 120° rotation (Fig. 5.91). Beyond 120°, some conductors are in the top link *N*-pole and others the *S*-pole. Same happens with bottom conductors. Hence, the voltage induced in phase *A* linearly reverses in next 60° rotation. Rest of the waveform of phase *A* and waveforms of *B* and *C* can be explained on the same lines.

An inverter fed trapezoidal PMAC motor drive operating in self-controlled mode is called a **brushless D.C. motor.** This motor is also conceived as *electronically commutated D.C. motor*, because inverter here performs the same function as the brushes and commutator in a D.C. motor, *i.e.*, to shift currents between armature conductors to keep the stator and rotor fields stationary, and in quadrature, with respect to each other.

Advantages. Owing to the absence of brushes and commutator, brushless D.C. motors claim the following *advantages over the conventional D.C. motors :*

(*i*) Long life.

(*ii*) Require practically no maintenance.

(*iii*) High reliability.

(*iv*) Low inertia and friction.

(*v*) Low radio frequency interference and noise.

(*vi*) Because armature windings are on the stator, cooling is much better, *i.e. higher specific outputs* can be obtained.

(*vii*) They have a faster acceleration (due to low inertia and friction) and can be run at much higher speeds–upto 100000 r.p.m. and higher are common.

(*viii*) High efficiency, exceeding 75 percent (whereas wound field motors of low power ratings have much lower efficiency).

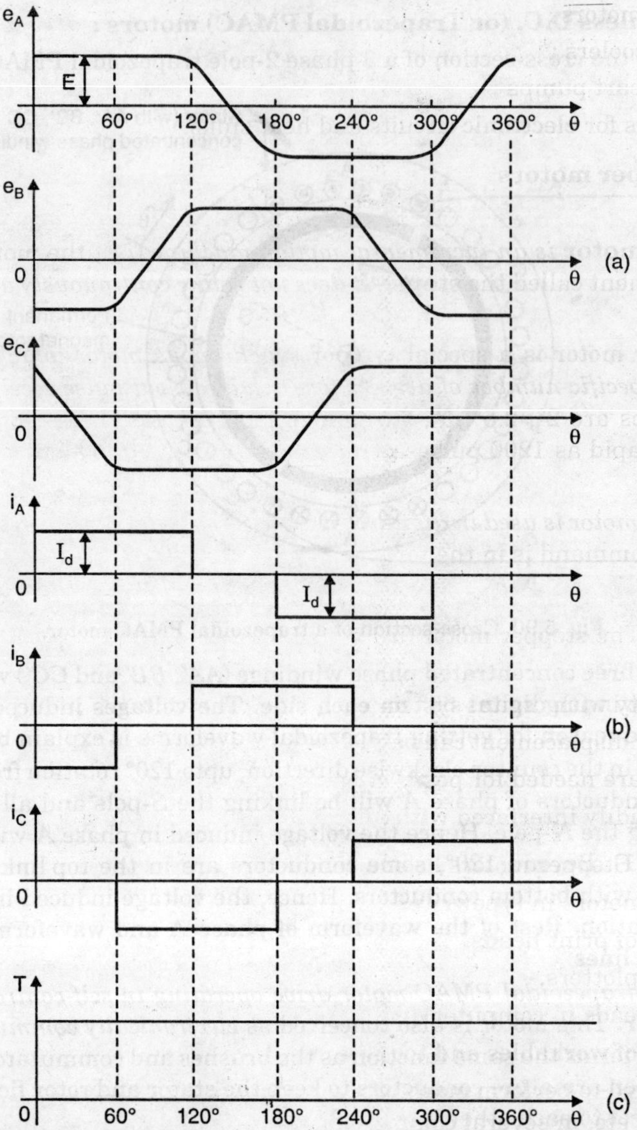

Fig. 5.91. Induced voltage, phase current and torque waveforms of a brushless D.C. motor.

Disadvantages :

(*i*) High cost

(*ii*) Low starting torque.

The size of a brushless D.C. motor is nearly the same as that of the conventional D.C. motor.

Applications. The brushless D.C. motor finds applications in :

(*i*) Tape drive for video recorders ;

(*ii*) Turn table drives in record players ;

(*iii*) Spindle drives in hard disk drives for computers ;

(*iv*) Low cost and low power drives in computer peripherals, instruments and control systems.

(*v*) Gyroscope motors ;
(*vi*) Cryogenic coolers ;
(*vii*) Artificial heart pumps ;
(*viii*) Cooling fans for electronic circuits and heat sinks.

5.13.2.8. Stepper motors

Introduction

- A **stepper motor** *is an incremental motion machine* (*i.e.,* the motor which turns in discrete movement called the **steps**). *It does not rotate continuously as a conventional motor does.*
- The stepper motor is a special type of *synchronous motor which is designed to rotate through a specific number of degrees for each electrical pulse received by its control unit.* Typical steps are 2°, 2.5°, 5°, 7.5° and 15° per pulse. These motors are built to follow signals as rapid as 1200 pulses per second and with equivalent power ratings upto several kW.
- The *stepper motor is used in digitally controlled position control system in open loop mode.* The input command is in the form of a train of pulses to turn a shaft through a specified angle.

Advantages. The stepper motor (***a position control device***) entails the following *advantages :*

1. Compatibility with digital systems.
2. The angular displacement can be precisely controlled without any feedback arrangement.
3. No sensors are needed for position and speed sensing.
4. It can be readily interfaced with microprocessor (or computer based controller).

Applications. Stepper motors have a wide range of applications, mentioned below :

1. Paper feed motors in typewriters and printers.
2. Positioning of print heads.
3. Pens in *XY*-plotters.
4. Recording heads in computer disc drives.
5. Positioning of workables and tools in numerically controlled machining equipment.
6. Also employed to perform many other functions such as *metering, mixing, cutting, blending, stirring* etc. in several commercial, military and medical applications.

Construction and working :

- A stepper motor consists of a *slotted stator having multi-pole, multi-phase winding and a rotor structure carrying no winding.* They typically use three and four phase windings, the number of poles depends upon the required angular change per input pulse.
- The rotors may be of the *permanent magnet* or *variable reluctance type.*
- Stepper motors operate with an *external drive logic circuit.* When a train of pulse is applied to the input of the drive circuit, the circuit supplies currents to the stator windings of the motor to make the axis of the air-gap field around in coincidence with the input pulses. The rotor follows the axis of air-gap magnetic field by virtue of the permanent magnet torque and/or the reluctance torque, depending upon the pulse rate and load torque (including inertia effects).

1. ***Permanent-magnet stepper motor :***

- Fig 5.92 shows the phases or stacks of a 2-phase, 4-pole permanent-magnet stepper motor.

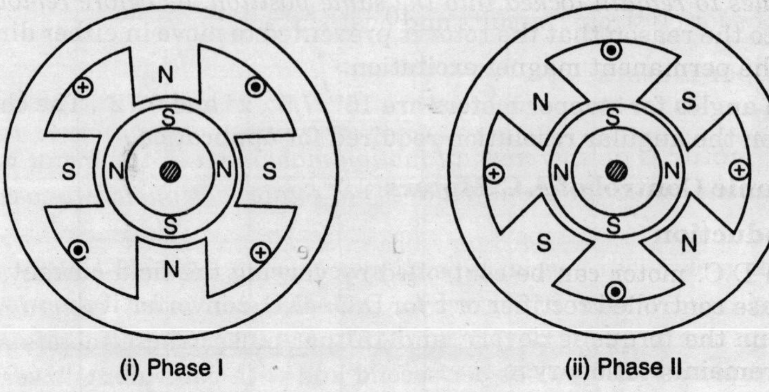

Fig. 5.92. Permanent-magnet stepper motor.

- The rotor is made of *ferrite or rare-earth material* which is *permanently magnetised.*
- The stator stack of phase II is staggered from that of phase '*I*' by an angle of 90°.
- When the phase '*I*' is excited, the rotor is aligned as shown in Fig. 5.92 (*i*). If now the phase '*II*' is also excited, the effective stator poles shift anti-clockwise by 22.5° [Fig. 5.92 (*ii*)] causing the rotor is move accordingly. Now, keeping the phase '*II*' still energised, if the phase '*I*' is now de-energised, the rotor will move another step of 22.5°. The reversal of phase '*I*' winding current will produce a further forward movement of 22.5°, and so on. It can be a easily observed/visualised as to how the direction of movement can be reversed.
- Each phase is provided with double coils to simplify the switching arrangement (which is electronically accomplished).

2. ***Variable-reluctance stepper motor :***

- A variable-reluctance stepper motor has no permanent magnet on the rotor and the rotor employed is a ferro-magnetic multi-toothed one.
- *The large differences in magnetic reluctances that exist between the direct and quadrature axes develop the torque.* The stationary field developed by the direct current in some stator coils tends to develop a torque which causes the rotor to move to the position where the reluctance of the flux path is minimum.

Stepping angle, irrespective of the type of stepper motor is given as :

$$\alpha = \frac{360°}{\text{Number of phases} \times \text{number of poles}} = \frac{360}{np}.$$

3. ***Hybrid stepper motor :***

- This is *infact a permanent-magnet stepper motor with constructional features of toothed and stacked rotor adopted from the variable-reluctance motor.*
- The stator has only one set of winding-excited poles which interact with the two rotor stacks.
- The permanent magnet is placed *axially along the rotor in the form of an annular cylinder over the motor shaft.*
- The stacks at each end of the rotor are toothed. So all the teeth on the stack at one end of the rotor acquire the *same polarity* while the teeth of the stack at the other end of the

rotor acquire the *opposite polarity*. The two sets of the teeth are displaced from each other by *one half of the tooth pitch* (also called pole pitch).

- The primary **advantage** of the hybrid motor is that if *stator excitation is removed, the rotor continues to remain locked into the same position, as before removal of excitation.* This is due to the reason that the rotor is prevented to move in either direction by torque because of the permanent magnet excitation.
- Typical step angles for stepper motors are 15°, 7.5°, 2° and 0.72°. The choice of the angle depends upon the angular resolution required for application.

5.13.3. Electronic Control of A.C. Motors

5.13.3.1. Introduction

The speed of a D.C. motor can be controlled by varying the field current or the armature voltage through a phase controlled rectifier or by a D.C.–D.C. converter if the input supply is D.C. Also, in a D.C machine the torque is developed due to the interaction of field flux and the D.C. armature flux which remains stationary in space. Whereas in A.C. machine, a 3-phase supply to the stator winding *produces rotating magnetic field of constant magnitude and which reacts with the rotor m.m.f.* to develop the torque. The rotor m.m.f. in case of an induction motor is created by the stator induction effect, whereas in case of synchronous motor the rotor m.m.f. is created by a separate field winding which carries D.C. current.

The speed of an A.C. machine depends upon the stator supply frequency which produces the synchronously rotating magnetic field. If the frequency of the stator supply is increased to increase the speed of the motor, the magnitude of air gap flux is reduced due to increased magnetising reactance and correspondingly the developed torque is reduced. This shows that the *speed and torque of an A.C. motor cannot be controlled independently by the conventional methods of speed control.* For this reason, an *A.C. motor requires 'variable voltage variable frequency', power supply for its speed control.* A '*D.C. link converter system*' consisting of a rectifier and an inverter or a '*cycloconverter*' can be used as a variable voltage variable frequency source.

However, it may be noted that the *voltage and current waves obtained by solid state devices are rich in harmonics and cause problems of harmonic heating torque pulsation.*

5.13.3.2. Speed control of a single-phase induction motor

The most common method for speed control of a single-phase induction motor is the *stator voltage control method.*

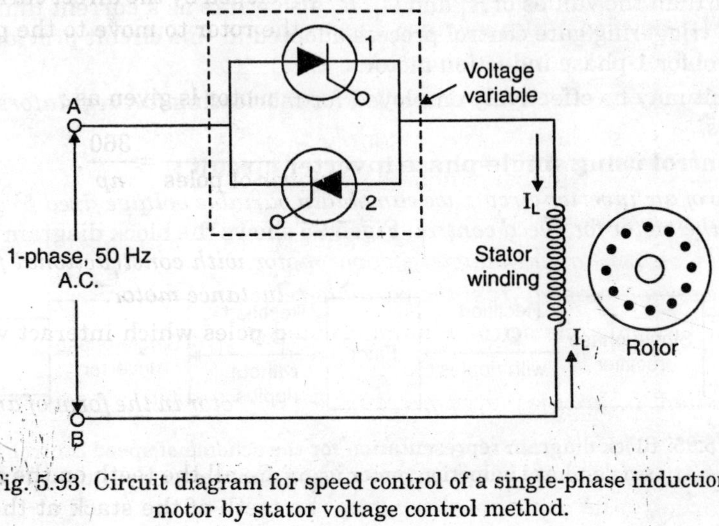

Fig. 5.93. Circuit diagram for speed control of a single-phase induction motor by stator voltage control method.

Fig. 5.93 shows the circuit diagram for speed control of a single-phase induction motor by stator voltage control method. The circuit uses two SCRs connected in antiparallel. In the +ve half cycle when point *A* is positive and point *B* is negative SCR_1 is triggered. The direction of flow of current in the stator winding is from the top to bottom. In the –ve half-cycle point *A* becomes negative and *B* becomes positive. SCR_1 is turned OFF and SCR_2 is triggered. The direction of flow of current in the stator winding is reversed. In other words, the alternating current supply becomes available across the stator winding of the motor. By varying the firing angles of SCRs 1 and 2 the magnitude of the A.C. voltage across the stator winding of the motor can be *varied* ; this in turn will *vary the speed of the motor.*

Different schemes under this method are discussed below.

1. Speed control by using triac :

By using a triac, *very smooth speed control* of a single-phase induction motor can be obtained. A diac is used as a triggering agent for the triac in the circuit. Fig. 5.94 shows the circuit diagram for this arrangement. A dia-triac pair can provide the widest range of control.

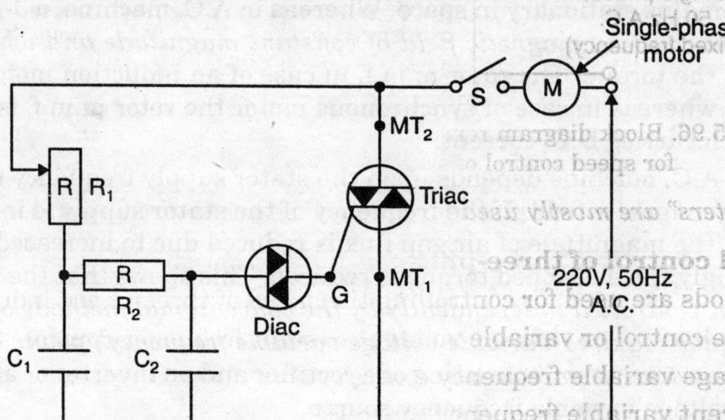

Fig. 5.94. Circuit diagram for the speed control of a single-phase induction motor using a triac.

There are two $R - C$ networks. $R_1 - C_1$ form the triggering circuit whereas $R_2 - C_2$ along with C_1 form the *Π*-network (filter) which would bypass any spike in the A.C. mains supply. The values of R_2 and C_2 are lower than the values of R_1 and C_1. R_2 also works as a current limiting resistance for the diac. The $R - C$ triggering gate control process adopted in this circuit provides a very wide and smooth speed control for 1-phase induction motor.

- This circuit may be effectively employed for fabricating *fan regulators and illumination controllers.*

2. Speed control using single-phase inverter circuit :

With the help of an 'inverter circuit' we can obtain variable voltage fixed frequency A.C. supply which can be fed to the motor for speed control. Fig. 5.95 shows the block diagram representation for this scheme.

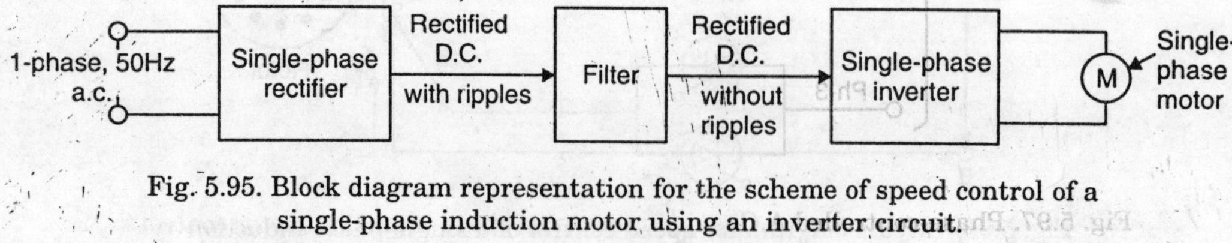

Fig. 5.95. Block diagram representation for the scheme of speed control of a single-phase induction motor using an inverter circuit.

- Single-phase A.C. is rectified with the help of a single-phase full-wave rectifier and then filtered to minimise the ripple content.
- The inverter output, which is a *fixed frequency variable A.C. voltage*, is fed to the motor whose speed is to be controlled. Inverter output (A.C. in nature) is *made variable by changing the firing time* (angle) of SCRs.

This process is known as *fixed frequency variable voltage control.*

3. Speed control by using cycloconverter circuits :

Basically, this is a *variable frequency method for speed control.* By controlling the firing sequence of the SCRs connected in a cycloconverter the frequency of the A.C. input voltage can be changed. The variable frequency A.C. supply available at the point of the cycloconverter circuit may be fed to the motor for speed control. A block diagram representation of such a scheme is shown in Fig. 5.96.

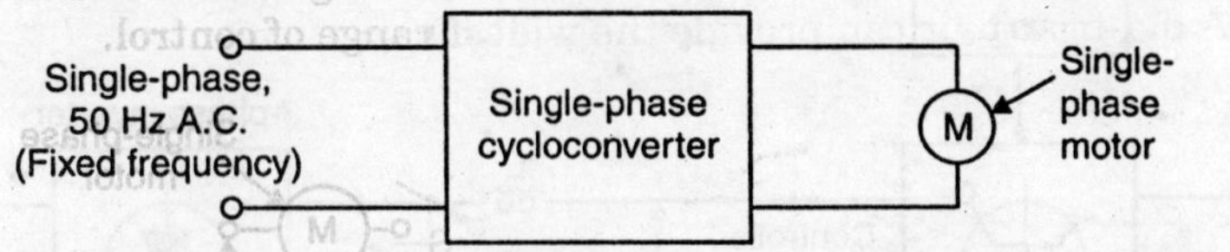

Fig. 5.96. Block diagram representation of the cycloconverter scheme for speed control of a single-phase induction motor.

- *"Cycloconverters" are mostly used for speed control of gearless drives.*

5.13.3.3. Speed control of three-phase induction motors

Following methods are used for controlling the speed of three-phase induction motors :

1. Stator voltage control or variable voltage constant frequency control.
2. Variable voltage variable frequency control.
3. Variable current variable frequency control.
4. Regulation of slip power.

The basic principles of operation of these methods are given as follows :

1. Stator voltage control :

By using a thyristor A.C. controller circuit ; constant frequency variable voltage supply can be generated. Connection diagram for such scheme is shown in Fig. 5.97.

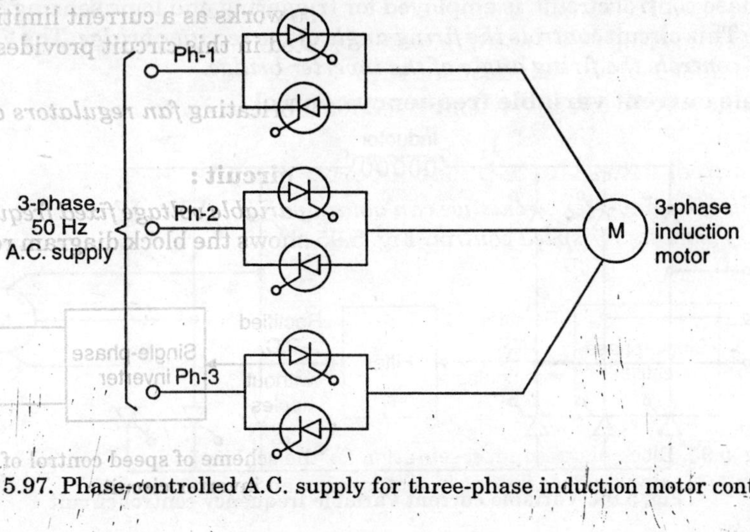

Fig. 5.97. Phase-controlled A.C. supply for three-phase induction motor control.

By changing the applied voltage, air gap flux can be changed so also the slip, and motor speed can be altered. To obtain a reasonable control a full thyristor controller is used. Two SCRs connected in antiparallel per phase are used to form three such bridges. SCRs 1, 2 form the bridge for phase-1, similarly 3, 4 form the bridge for phase-2 and 5, 6 for phase-3. The controlled (variable) three-phase voltage, when fed to the 3-phase induction motor, will result in the desired speed control of the motor.

- This arrangement is *quite costly* and *its firing circuit will also be quite complicated.*

2. Variable voltage variable frequency control :

Fig. 5.98 shows the basic block diagram for a speed control scheme of a 3-phase induction motor. This is basically a *variable voltage variable frequency supply scheme* for the speed control.

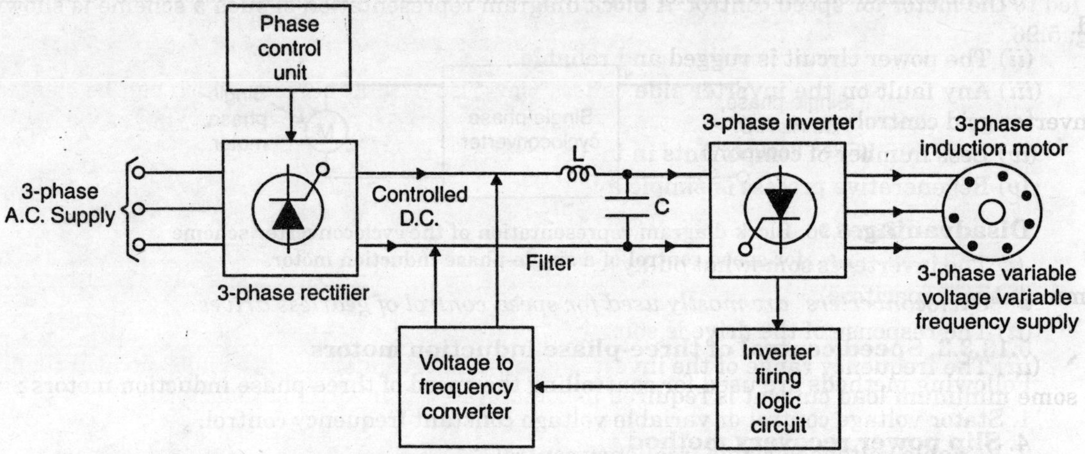

Fig. 5.98. Block diagram for basic scheme (variable voltage variable frequency control) for speed control of a 3-phase induction motor.

- 3-phase A.C. is rectified into D.C. and then filtered to minimise the ripple content. *L-C* filter is generally used for this purpose. This controlled D.C is converted into controlled pulses by means of a voltage to frequency converter. These controlled pulses are fed to the inverter bridge for producing the variable voltage variable frequency output. This output is fed to the 3-phase induction motor for controlling its speed.
- The 'phase control circuit' is employed for triggering and logic sensing of 3-phase rectifier circuit. This circuit *controls the firing angle of the rectifier bridge.* The '*inverter firing logic circuit' controls the firing angle of the inverter bridge.*

3. Variable current variable frequency control :

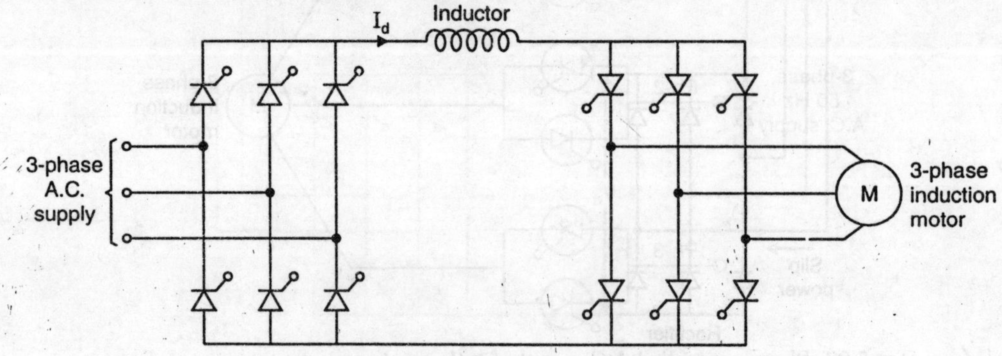

Fig. 5.99. Variable current variable frequency control circuit.

Fig. 5.99 shows a variable current variable frequency control circuit for an induction motor. A phase controlled rectifier produces variable D.C. voltage which is converted to a current source by connecting a large inductor in series.

A diode rectifier followed by a D.C. chopper can also be used as a variable voltage D.C. source. It can be shown that the voltage at the terminals of 3-phase induction motor is almost sinusoidal with superimposed voltage spikes due to commutation.

The *'converter'* used is a *line commutated* whereas the *'inverter'* is *forced commutated* as the induction motor is a lagging p.f. load.

The following are the *advantages* and *disadvantages* of this circuit :

Advantages :

(*i*) The control circuit is simpler and more reliable since only six thyristors are to be controlled.

(*ii*) The power circuit is rugged and reliable.

(*iii*) Any fault on the inverter side causes slow rise of fault current which can be cleared by converter grid control.

(*iv*) Less number of components in inverter circuit and less commutation loss.

(*v*) Regenerative process is simple and no additional component is required.

Disadvantages :

(*i*) The inverter is somewhat bulky and expensive (due to the large size of the inductance and commutation capacitors).

(*ii*) The response of the drive is somewhat sluggish.

(*iii*) The frequency range of the inverter is low and it cannot operate under no load condition as some minimum load current is required to commutate the inverter satisfactorily.

4. Slip power recovery method :

When the supply frequency and the voltage are fixed, the speed of an induction motor can be *varied by injecting a counter e.m.f. into the rotor circuit of the motor*. This method is, therefore, *used for a **wound rotor induction motor***. The inefficiency of the drive system because of large slip power dissipation can be overcome by this method.

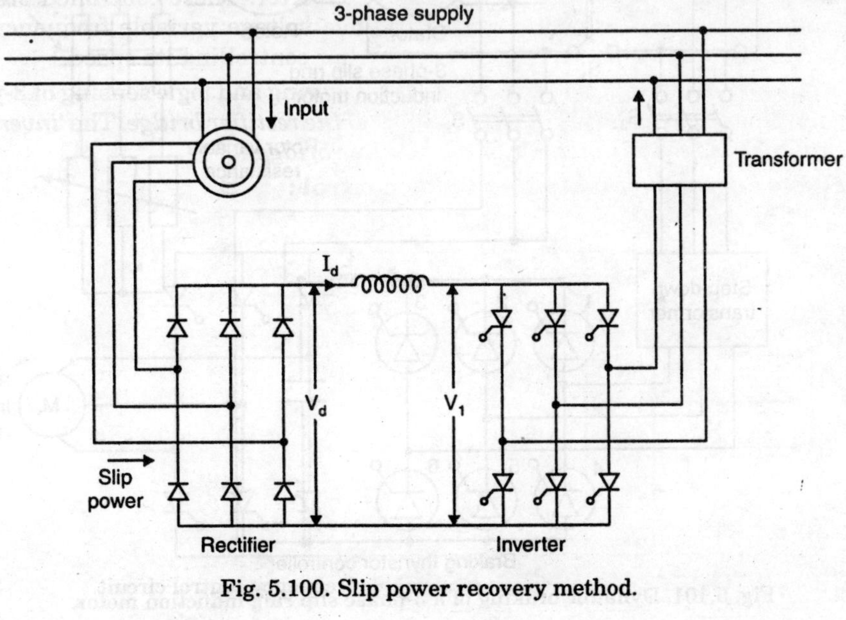

Fig. 5.100. Slip power recovery method.

In this method the slip power of the motor is rectified by a diode rectifier and is then pumped back to the A.C. line through a line commutated inverter. Fig. 5.100 shows the schematic diagram for this method.

The required power handling capability of the converter corresponds to the slip power and hence if the range of speed control is small the rating of converter is also small. However, if speed control upto standstill is the required, the converters should be rated for the full machine rating. The initial in-rush current in the converters can be avoided by connecting separate starting resistance in the slip ring circuit.

The torque pulsation and additional heating must be considered while designing the drive system.

The **disadvantage** of this system is that *regeneration and speed reversal are not possible in the drive system.*

The draw back of this method that it results in very poor p.f. can be overcome to some extent by using '*chopper control*'.

- This drive system is used in *large H.P. pump and blower type applications where limited range of speed control is required.*

Dynamic braking of a 3-phase induction motor :

The speed of an induction motor can be controlled by injecting D.C. voltage in its stator winding. A variable resistance may be used in the rotor (in case of a slip ring induction motor) for dissipating the required amount of power. Now-a-days *thyristor bridges* are used for supplying D.C. which is controllable in nature. With the help of controlled D.C. from a thyristor bridge the dynamic braking can be achieved in a more effective manner. The connection diagram for scheme is shown in Fig. 5.101.

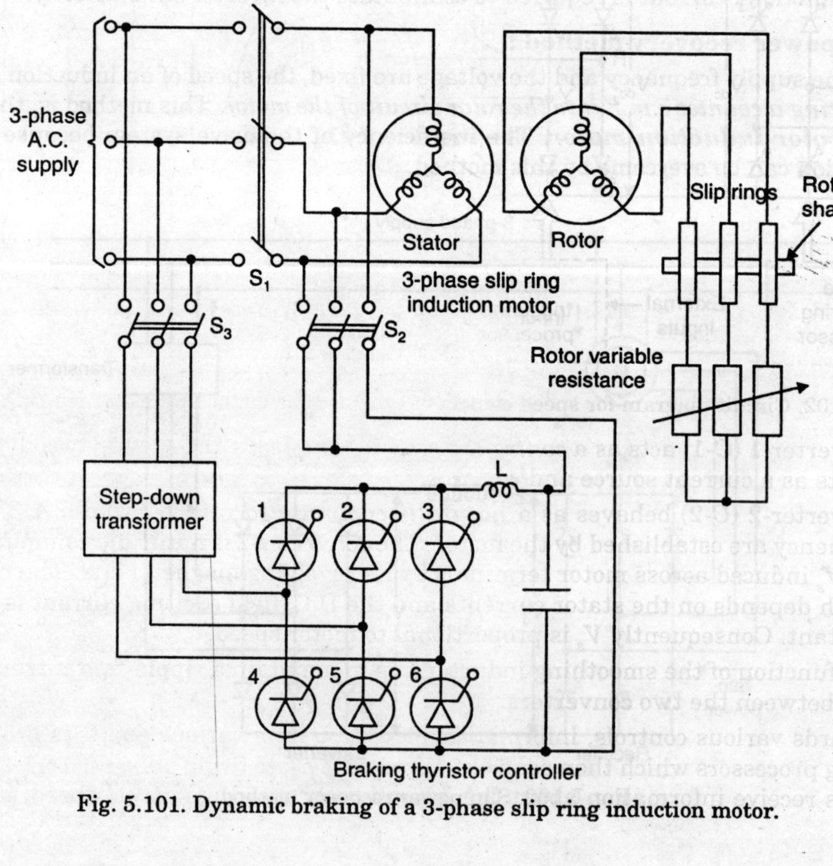

Fig. 5.101. Dynamic braking of a 3-phase slip ring induction motor.

- 3-phase A.C. is stepped down to lower voltage and fed to a 3-phase thyristor bridge which serves as the rectifier.
- This D.C. is filtered by an L.C. filter for minimising the ripples.
- Ripple free D.C. is then fed to the stator winding of the induction motor as shown in Fig. 5.101.

It is to be noted that while feeding D.C. to the stator the 3-phase A.C. input must be disconnected A.C. is disconnected with the help of S_1 and D.C. is disconnected with the help of S_2. Since, the input A.C. voltage is stepped down to a lower value, the thyristor converter may be of lower voltage rating.

5.13.3.4. Speed control of Synchronous Motors

The speed of synchronous motors can be controlled as follows :

(*i*) By using current-fed delink.

(*ii*) By using cycloconverter.

1. Speed control by current-fed D.C. link

Fig 5.102, shows the circuit diagram for speed control of synchronous motor by current-fed D.C. link.

The typical circuit consists of three converters two of which are connected between the 3-phase source and synchronous motor and the third converter (bridge rectifier) supplies D.C. field excitation for the rotor.

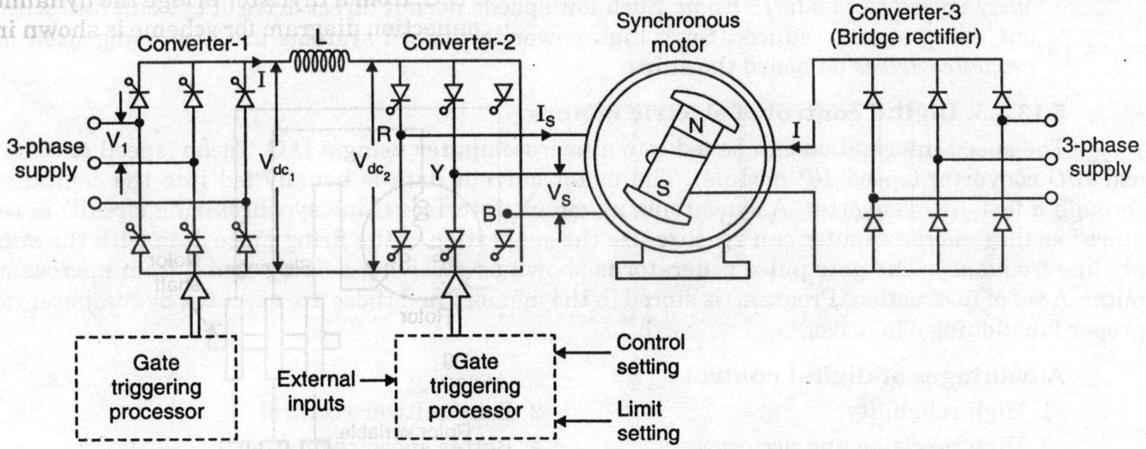

Fig. 5.102. Circuit diagram for speed control of synchronous motor by current-fed D.C. link.

— Converter-1 (*C*-1) acts as a *controlled rectifier* and feeds D.C. power to converter-2 (*C*-2). It acts as a current source and controls 1.

— Converter-2 (*C*-2) behaves as a *naturally commutated inverter* whose *A.C.* voltage and frequency are established by the motor. The converter-2 is naturally commutated by voltage V_s induced across motor terminals by its revolving magnetic flux. The revolving flux which depends on the stator currents and the D.C. field exciting current is usually kept constant. Consequently V_s is proportional to motor speed.

— The function of the smoothing inductor L is to maintain a ripple-free current in the D.C. link between the two converters.

As regards various controls, information picked up from various points is processed in the gate-triggering processors which then send out appropriate gate firing pulses to converters 1 and 2. The processors receive information about the desired rotor speed, its actual speed, instantaneous

rotor position, field current, stator voltage and current etc. The processors check whether these inputs represent normal or abnormal conditions and send appropriate gate firing pulses either to correct the situation or meet a specific demand.

Gate triggering of *C*-1 and *C*-2 is done at line frequency (50 Hz) and at motor frequency respectively. In fact, gate pulses of *C*-2 are controlled by rotor position which is plused by position transducers mounted at the end of the shaft.

The speed of the motor can be increased by increasing either D.C. link I or exciting current I_f.

- **This method of speed control is applied to *motors ranging from 1 kW to several MW.***
- ***Permanent-magnet synchronous motors used in textile industry and brushless synchronous motors for nuclear reactor pumps are controlled by this method.***

2. Speed control by cycloconverter :

This arrangement consists of *three cycloconverters* connected to the three phases of the synchronous motor and *one controlled rectifier* for supplying field exciting current I_f to the rotor. Each cycloconverter is composed of two 3-phase bridges and supplies a single-phase output. With a line frequency of 50 Hz, the cycloconverter output frequency can be varied from 0 to 10 Hz (It is well known that a cycloconverter can convert A.C. power at higher frequency to one at a lower frequency).

The cycloconverters and controlled rectifier function as *current sources.* The air-gap flux is kept constant by controlling the magnitude of the stator currents and exciting current I_f. The motor can be made to operate at unity power factor by proper timing of gate pulses.

- **The *speed of cycloconverter-driven large slow-speed synchronous motors can be continuously varied from 0 to 15 r.p.m.* Such low speeds permit direct drive of t , ball mill without using a gear reducer. Such high-power low-speed systems are also being used as *propeller drives* on board the ships.**

5.13.3.5. Digital control of electric motors

The speed information can be fed into a microcomputer using a D.C. Tacho (speed encoder) and *A/D* converter (speed *I/P* module). The motor current data is usually fed into the computer through a fast *A/D* converter. A synchronizing circuit interface (Line synchronizing circuit) is required so that microcomputer can synchronize the generation of the firing pulse data with the supply line frequency. The gate pulse generator is shown as receiving a firing signal from microcomputer. A set of instruction (Program) is stored in the memory and those are executed by computer for proper functioning of a drive.

Advantages of digital control :

1. **High reliability**
2. **Easy software control**
3. **High precision and accuracy**
4. **Better speed regulation**
5. **Faster response**
6. **Improved performance**
7. **Economical**
8. **Flexibility**
9. **Better time response**
10. **The *major advantage of the digital control is that by changing the program, desired control technique can be implemented without any change ir the hardware.***

5.14. TRANSIENT CONDITIONS IN ELECTRICAL DRIVES

5.14.1. Introduction

When an electrical drive is in steady state operation and while attaining another steady state operation due to the nature of the drive *transient conditions occur due to the following "reasons/conditions"* :

(*i*) Change of speed, torque and current ;

(*ii*) Change in the load suddenly or during the control operations on the motor during starting, braking or reversal of direction of rotation etc. ;

(*iii*) Changes in supply voltage, frequency or unbalance in supply voltage.

The exhaustive study of the transient working conditions of electrical drives helps in the following ways :

(*i*) To select the proper motor power ;

(*ii*) To reduce losses ;

(*iii*) To design control circuit ;

(*iv*) To select starting and braking equipment ;

(*v*) To attain required economy in the driving systems etc.

5.14.2. Acceleration Time

For any electric motor undergoing rotary motion, the equation of motion is given by,

$$T_{motor} - T_{load} = \frac{J\,d\omega}{dt} \text{ Nm} \qquad ...(5.59)$$

where, T = Torque (Nm),

J = Moment of inertia (kg m^2),

$d\omega$ = Change in speed (rad/sec.), and

dt = Acceleration time (sec.)

∴ The acceleration time required for changing the speed of an electric motor from ω_1 to ω_2 is given by,

$$dt = \frac{J\,d\omega}{(T_{motor} - T_{load})} \qquad ...[5.59\,(a)]$$

Thus, $T_{motor} > T_{load}$ during acceleration time and $T_{load} > T_{motor}$ during deceleration time.

The acceleration time, when the motor accelerates from ω_1 to ω_2, is given by,

$$t_{acc.} = J\int_{\omega_1}^{\omega_2} \frac{d\omega}{T_{motor} - T_{load}} \qquad ...(5.60)$$

- When the motor starts from rest, $\omega_1 = 0$ and ω_2 is the steady state speed ω_n, the acceleration time is given by,

$$t_{acc.} = J\int_{0}^{\omega_n} \frac{d\omega}{T_{motor} - T_{load}} \qquad ...(5.61)$$

As $T_{motor} - T_{load} \rightarrow$ zero we have $\dfrac{1}{T_{motor} - T_{load}} \rightarrow \infty$ (Infinity)

and therefore for practical calculations the *acceleration time* ($t_{acc.}$) is considered when $\omega_1 = 0.95\ \omega_2$ and thereby avoiding the complication in calculating the finite time when we take the value as ω_2.

- Now when the dynamic torque (T_{motor}) becomes *negative* the drive will *decelerate.* The torque equation then becomes

$$-T_{motor} - T_{load} = \frac{Jd\omega}{dt}$$

Whenever the motor develops a torque which is less than the load torque of the driven machine, the *deceleration time* ($t_{dec.}$) can be written as,

$$t_{dec.} = \int_{\omega_1}^{\omega_2} \frac{Jd\omega}{-(T_{motor} + T_{load})} = \int_{\omega_2}^{\omega_1} \frac{Jd\omega}{T_{motor} + T_{load}} \qquad ...(5.62)$$

Let us now compute the acceleration time (stated above) under different conditions of motor and load torque.

Case 1. ***When the motor and load torques are equal :***

When the motor and load torques are constant, then $(T_{motor} - T_{load})$ will also be constant.

Since $T_{motor} - T_{load} = \frac{Jd\omega}{dt}$ and as J is a constant we have $\frac{d\omega}{dt}$ as constant. This means the *speed variation will be linear w.r.t. time* and hence

$$\omega = \frac{(T_{motor} - T_{load})\, t}{J} \quad ...(5.63)$$

Hence the time t_1 taken by the motor to reach a speed of ω_1 rad/sec. is given by

$$t_1 = \frac{J\omega_1}{(T_{motor} - T_{load})} \quad ...(5.64)$$

When the motor starts without load $T_{load} = 0$, then T_{motor} = rated torque (T_{rated}). The moment of inertia J will be due to the rotor of the motor. The motor then attains a speed ω_0 which is the no-load speed of the motor. Hence $t_1 = \frac{J\omega_0}{T_{rated}}$. This is known as the *mechanical constant of the motor* ; for a rated speed of 1000 r.p.m. The time constant varies from 0.4 to 0.6 second.

Case 2. ***Load torque remains constant with linearly varying motor torque :***

Let us consider the condition when $\omega = 0$, the motor torque being the starting torque T_{start} and when $\omega = \omega_s$ the synchronous speed of induction motor with $T_{motor} = 0$. Then the relation between motor torque and slip(s) can be written as,

$$\frac{T_{start}}{1} = \frac{T_{motor}}{s} \quad \text{or} \quad s = \frac{T_{motor}}{T_{start}} \quad ...(5.65)$$

Hence, *acceleration torque*, $\frac{Jd\omega}{dt} = T_{motor} - T_{load}$ ...(5.66)

Also, $\omega = \omega_s(1 - s)$

Differentiating with respect to t, we have,

$$\frac{d\omega}{dt} = -\,\omega_s \frac{ds}{dt} \quad ...(5.67)$$

From eqns. (5.65), (5.66) and (5.67), we get

$$-\,\omega_s \frac{ds}{dt} = \frac{sT_{start} - T_{load}}{J}$$

or, $$\frac{ds}{dt} + \frac{sT_{start}}{J\omega_s} - \frac{T_{load}}{J\omega_s} = 0 \quad ...(5.68)$$

The value $\frac{J\omega_s}{T_{start}}$ is having the dimensions of seconds and is known as *starting time constant* (t_{st}) of the drive. Hence J represents the total moment of inertia of the drive system including inertia of any external rotating masses excluding the rotor of the driving motor.

$$\therefore \quad \frac{ds}{dt} + \frac{s}{t_{st}} - \frac{T_{load}}{J\omega_s} = 0 \quad ...(5.69)$$

Optimum value of $s_{max\,T}$ when accelerating time is minimum :

When the load torque and friction torque are negligibly small the equation of motion can be written as,

$$T_{motor} = \frac{Jd\omega}{dt}$$

$$= \frac{J}{dt} d\,[\omega_s(1-s)]$$

or,

$$T_{motor} = -J\omega_s \frac{ds}{dt} \qquad ...(i)$$

The electromagnetic torque developed by the induction motor neglecting stator resistance will be

$$T_{motor} = T_{max} \frac{2}{\frac{s}{s_{max\,T}} + \frac{s_{max.T}}{s}} \qquad ...(ii) \quad ...[\text{Eqn. } (5.26\,(a))]$$

In order to obtain the time taken by the motor to have slip changing from s_1 to s_2, equating (i) and (ii), we get

$$-J\omega_s \frac{ds}{dt} = \frac{2}{\frac{s}{s_{max\,T}} + \frac{s_{max\,T}}{s}}$$

$$\therefore \quad \int_0^t dt = \frac{-J\omega_s}{2T_{max}} \left[\int_{s_1}^{s_2} \frac{s}{s_{max\,T}} ds + \int_{s_1}^{s_2} \frac{s_{max\,T}}{s} ds \right]$$

or,

$$t = \frac{-J\omega_s}{2T_{max}} \left[\left| \frac{s^2}{2s_{max\,T}} \right|_{s_1}^{s_2} + \left| s_{max\,T} \cdot \log_e s \right|_{s_1}^{s_2} \right]$$

or,

$$t = \frac{J\omega_s}{2T_{max}} \left[\frac{s_1^2 - s_2^2}{2s_{max\,T}} + s_{max\,T} \log_e \left(\frac{s_1}{s_2} \right) \right] \qquad ...(5.70)$$

This equation indicates that the *time required* (t) *is a function of slip corresponding to the maximum torque.* To get the value of $s_{max\,T}$ that gives *minimum time* for the slip to change from s_1 to s_2, by differentiating t w.r.t. $s_{max\,T}$ and equating the result to zero, we have

$$(s_{max\,T})_{opt} = \left[\frac{s_1^2 - s_2^2}{2 \log_e (s_1/s_2)} \right]^{1/2} \qquad ...(5.71)$$

Substituting this value of $s_{max\,T}$ in eqn. (5.70), we get the *minimum time* required.

The optimum value of $s_{max\,T}$ to accelerate the motor from $s_1 = 1$ *i.e.* motor at standstill to any slip s in minimum time is given by,

$$(s_{max\,T})_{opt} = \left[\frac{1 - s^2}{2 \log_e (1/s)} \right]^{1/2} \qquad ...(5.72)$$

The optimum rotor resistance when the slip changes from s_1 to s_2 in minimum time is obtained by using the relation,

$$(R_2)_{opt} = (s_{max\,T})_{opt} \, (X_1 + X_2)$$

or,

$$(R_2)_{opt} = (X_1 + X_2) \left[\frac{(s_1^2 - s_2^2)}{2 \log_e (s_1/s_2)} \right]^{1/2} \qquad ...(5.73)$$

5.14.3. Energy Losses During Starting

1. D.C. shunt motor :

In a D.C. motor,

$$V = E_b + I_a R_a$$

or, $$I_a R_a = V - E_b = V - k\omega \quad \text{...(5.74)}$$

This expression is *also suitable for separately excited motor.*

The equation of motion at no load is given by,

$$T_{\text{motor}} = kI_a = \frac{Jd\omega}{dt} \quad \text{...(5.75)}$$

$$\left(\text{or } I_a = \frac{J}{k}\frac{d\omega}{dt}\right)$$

Hence, $$I_a^2 R_a = I_a(I_a R_a) = \frac{J}{k}\frac{d\omega}{dt}(V - k\omega)$$

or, $$I_a^2 R_a = \frac{JV}{k}\frac{d\omega}{dt} - \frac{J\omega\, d\omega}{dt} \quad \text{...(5.76)}$$

As $I_a R_a$ is negligible at no load, $V = k\omega_0$ at no load (ω_0 is the no-load speed of the motor)

$\therefore$ $$I_a^2 R_a dt = J\omega_0 d\omega - J\omega d\omega$$

The energy absorbed (W) by the armature for a change in speed from ω_1 to ω_2 from time t_1 to t_2 is given by,

$$W = \int_{t_1}^{t_2} I_a^2 R_a \, dt$$

$$= J\omega_0 \int_{\omega_1}^{\omega_2} d\omega - J\int_{\omega_1}^{\omega_2} \omega \, d\omega$$

or, $$W = J\omega_0(\omega_2 - \omega_1) - \frac{J}{2}(\omega_2^2 - \omega_1^2) \quad \text{...(5.77)}$$

Hence the energy change when the motor changes speed from rest to no load speed ω_0 will be

$$W_{\text{start}} = J\omega_0^2 - \frac{J}{2}\omega_0^2$$

or, $$W_{\text{start}} = \frac{J\omega_0^2}{2} \text{ joules} \quad \text{...(5.78)}$$

= K.E. absorbed by the armature in accelerating from standstill to no-load speed.

The *energy loss is independent of armature resistance.*

- The *energy loss at starting*, when the *motor is started with a constant load torque* T_{load}, is calculated as follows :

$$T_{\text{motor}} = kI_a = T_{\text{load}} + \frac{Jd\omega}{dt}$$

$\therefore$ $$I_a^2 R_a = \frac{JV}{k}\frac{d\omega}{dt} - \frac{J\omega d\omega}{dt} + \frac{V}{k}(T_{\text{load}}) - T_{\text{load}} \cdot \omega$$

The energy lost, when the motor speed changes from ω_1 to ω_2, is given by,

$$W = \int_{t_1}^{t_2} I_a^2 R_a dt$$

$$= \frac{JV}{k}\int_{\omega_1}^{\omega_2} d\omega - J\int_{\omega_1}^{\omega_2} \omega\, d\omega + \frac{V}{k} T_{load}\int_{t_1}^{t_2} dt - T_{load}\int_{t_1}^{t_2} \omega(t)\, dt$$

$$= \frac{JV}{k}(\omega_2 - \omega_1) - \frac{J}{2}(\omega_2^2 - \omega_1^2) + \frac{V}{k} T_{load}(t_2 - t_1) - T_{load}\int_{t_1}^{t_2} \omega(t)\, dt$$

Now substituting ω_0 (the no-load speed) for $\frac{V}{k}$, the *energy lost at starting on no-load* when the speed changes from 0 to ω_r is given by,

$$W_{start} = J\omega_0\omega_r - \frac{J}{2}\omega_r^2 + T_{load}\omega_0 t_{st} - T_{load}\int_0^t \omega(t)\, dt$$

or, $$W_{start} = J\left(\omega_0\omega_r - \frac{\omega_r^2}{2}\right) + T_{load}\left[\left(\omega_0\, t_{st} - \int_0^t \omega(t)\, dt\right)\right] \quad ...(5.79)$$

... where t_{st} is the accelerating time.

This equation shows that the *energy lost during starting* depends on the accelerating time and on the speed variation with time during acceleration.

2. D.C. series motor :

Since a series motor cannot be started at no load, it will be assumed that the motor starts under a constant load torque (T_{load}) to be present while finding the energy lost during starting.

The equation of motor is given by,

$$T_{motor} = kI_a^2 = \frac{Jd\omega}{dt} + T_{load}$$

The total energy dissipated in the armature circuit resistance is given by,

$$W = \int_{t_1}^{t_2} I_a^2 R_a\, dt$$

$$= \frac{JR_a}{k}\int_{\omega_1}^{\omega_2} d\omega + \frac{T_{load}}{k} R_a \int_{t_1}^{t_2} dt$$

or, $$W = \frac{JR_a}{k}(\omega_2 - \omega_1) + \frac{T_{load}\, R_a}{k}(t_2 - t_1) \quad ...(5.80)$$

Thus the *energy dissipated* when the motor speed varies from ω_1 to ω_2 will be *proportional to* R_a. (This is in contrast to the case of a D.C. shunt motor where the energy dissipated is independent of R_a).

3. Three-phase induction motor :

In a 3-phase induction motor, torque developed is given by,

$$T = \frac{3}{\omega_s}\left(I_2^2 . \frac{R_2}{s}\right) \quad ...(i) \quad ...[\text{Eqn. } (5.25)]$$

Neglecting load and friction torques, we have

$$T = J\frac{d\omega}{dt} = -J\omega_s \frac{ds}{dt} \quad ...(ii)$$

... as $\omega = \omega_s(1 - s)$.

The energy loss in the *rotor* of the induction motor when the slip changes from s_1 to s_2 is given by,

$$W = 3\int_{t_1}^{t_2} I_2^2 R_2\, dt \quad ...(iii)$$

where, t_1 and t_2 are the corresponding times when the slips are s_1 and s_2.

From (*i*), (*ii*), and (*iii*), we have

$$W = 3I_2^2R_2 = -J\omega_s^2 s \frac{ds}{dt}$$

i.e., $$W = -J\omega_s^2 \int_{s_1}^{s_2} s.ds$$

or, $$W = \frac{1}{2} J\omega_s^2 (s_1^2 - s_2^2) \text{ joules} \quad ...(5.81)$$

At the time of starting the slip changes from 1 to 0 and so the energy lost in the rotor circuit is given by,

$$W_{start} = \frac{1}{2} J\omega_s^2 \text{ joules} \quad ...(5.82)$$

The eqn. (5.81) indicates that the total energy lost in the rotor circuit depends on the moment of inertia of the rotating masses and initial and final speeds and does not depend on the rotor circuit resistance. This does not mean that the total energy lost is independent of rotor circuit resistance. The total energy lost depends on the stator and rotor resistances.

We know that, $$\frac{\text{Stator Cu less}}{\text{Rotor Cu loss}} = \frac{R_1}{R_2}$$

The total energy lost in the motor when the speed changes is given by,

$$W_{motor} = 3 \int_{t_1}^{t_2} I_2^2 (R_1 + R_2)\, dt \quad \text{... since } I_1 \simeq I_2$$

Following the same procedure of deriving W, we can write the energy lost in the motor,

$$W_{motor} = \frac{1}{2} J\omega_s^2 (s_1^2 - s_2^2) \left(1 + \frac{R_1}{R_2}\right) \quad ...(5.83)$$

At starting ($s_1 = 1$; $s_2 = 0$),

$$W_{motor} = \frac{1}{2} J\omega_s^2 \left(1 + \frac{R_1}{R_2}\right) \quad ...(5.84)$$

- When the motor starts under a load torque of T_{load}, we can write

$$T_{motor} - T_{load} = J \frac{d\omega}{dt}$$

or, $$dt = \frac{J d\omega}{T_{motor} - T_{load}}$$

Again, $$d\omega = -\omega_s ds$$

$\therefore$ $$W = 3 \int_{t_1}^{t_2} I_2^2 R_2 dt$$

$$= \omega_s \int s\, T_{motor}\, dt$$

$$= -J\omega^2 \int_{t_1}^{t_2} \frac{T_{motor}}{T_{motor} - T_{load}} s.ds$$

or, $$W = -J\omega_r^2 \int_{t_1}^{t_2} \left[1 + \frac{T_{load}}{T_{motor} - T_{load}}\right] s.ds \quad ...(5.85)$$

— Eqn. (5.85) shows that the energy lost in starting an induction motor on load in much more than that on no load.

— Again when the motor is started using direct on line starter T_{max} will be high at rated voltage and so the energy lost will be minimum. In the reduced voltage starting $V < V_{rated}$ and so the energy losses will be more. The energy loss during starting is wasted as heat in stator and rotor windings of squirrel-cage motor and in external circuit resistance added to the rotor of a wound rotor induction motor. *Whereas the external resistance added to the motor is more compared to stator and rotor resistance to have a low operating temperature, a higher resistance added to the rotor circuit will cause the starting time to increase.*

- *Reduction of energy loss during starting* is achieved by any of the following methods :

(*i*) By reducing the moment of inertia of the rotor.

(*ii*) By smooth variation of supply frequency to induction motors (*V/f* control).

(*iii*) By gradual adjustment of applied voltage to D.C. shunt motors.

(*iv*) Starting of multi-speed induction motors in discrete steps of speed.

5.14.4. Energy Losses during Braking

In an electric motor energy losses occur during braking.

1. D.C. motors with shunt or separate excitation :

Let the braking be applied where there is no load on the motor. The energy wasted in the armature circuit during rheostatic braking is calculated using the relation,

$$I_a^2 R_a dt = J\omega_0.d\omega - J\omega.d\omega$$

When rheostatic braking is applied the speed changes from ω_0 to 0 at standstill. Thus energy wasted in braking

$$= \int_{\omega_0}^{0} -J\omega.d\omega$$

$$= \int_{0}^{\omega_0} J\omega.d\omega$$

$$= \frac{J\omega_0^2}{2} = \text{K.E. of the armature at the start of braking}$$

If the motor *reverses its direction of rotation* and the speed changes from $+\ \omega_0$ to $-\ \omega_0$ the energy lost by the armature will be $4\left(\frac{J\omega_0^2}{2}\right)$.

2. Induction motor :

The braking is performed at no load and during reverse current braking (*plugging*) the slip changes from 2 to 1. Substituting in the equation for energy lost, we have

$$W_{braking} \text{(reverse current)} = \frac{1}{2} J\omega_0^2 (s_1^2 - s_2^2)$$

$$= \frac{1}{2} J\omega_0^2 (2^2 - 1^2)$$

$$= \frac{3}{2} J\omega_0^2 \quad ...(5.86)$$

Eqn. (5.86) indicates that the *power wasted in the rotor circuit is 3 times the K.E. at no load.*

Now, if the motor is *reversed* to run at ω_s the slip *varies from 2 to 0*, then the *power loss* will be

$$= \frac{1}{2} J\omega_0^2 (2^2 - 0^2)$$

$$= 4\left(\frac{J\omega_0^2}{2}\right) \quad ...(5.87)$$

These losses are in the *rotor alone*.

- The *total energy wasted* by the motor at braking $= \frac{3}{2} J\omega_0^2 \left(1 + \frac{R_1}{R_2}\right)$ joules ...(5.88)
- The *total energy wasted during the reversal* of speed $= 4\left(\frac{J\omega_0^2}{2}\right)\left(1 + \frac{R_1}{R_2}\right)$ joules ...(5.89)

The *braking energy is given by the stored energy available at the commencement of the braking process.*

5.14.5. Braking of Motors During Lowering of Loads

When load are lowered, as in the case of D.C. shunt (or separately excited) motor, the *torque due to the motor and the load act in the same direction strengthening the total torque available and thereby aiding each other.*

— With the increase in speed the induced e.m.f. in the armature also increases and reaches a value *equal to applied voltage.* At this moment the armature current and therefore the electromagnetic torque becomes *zero.* The downward motion of the hoist will be only due to the downward-moving load.

— When the speed *exceeds* the no-load speed resulting in $E_b > V$ the armature current becomes *negative* and the *machine works as an electrical generator* and *provides the braking torque.*

— When the braking torque equals the load torque the *steady speed* is obtained.

This type of braking is suitable for "regenerative braking" as power is returned to the supply.

— When the *speed is less than the no load speed braking* can be provided by adopting "*rheostatic braking*". Then the motor acts as a generator with its armature terminals short-circuited by an external resistance R_x

Hence, $$E_b = -I(R_a + R_x)$$

and, $$\omega = -T\frac{(R_a + R_x)}{k_1 k_2 \phi^2}, \text{ with } V = 0.$$

D.C. series motor :

'*Regenerative braking*' is *not* practicable with series motors since in a series motor it is not possible to make the field current exceed the armature current and hence to have E_b more than V.

'*Rheostatic braking*', however, is possible if care is taken to interconnect the armature and field windings in such a way as to *make the direction of current in the field remain unaltered although there is a change in the direction of the armature current.* Further, it should be ascertained that the external resistance added *should not* make the total resistance of the motor circuit to *exceed the critical value.* Thus the e.m.f. equation can be written as,

$$E_b = I(R_a + R_f + R_x) \quad ...(5.90)$$

where, R_a = Armature resistance,

R_f = Field resistance, and

R_x = External resistance.

5.14.6. Dynamics of Braking

When a motor is subjected to braking the following torques need to be considered :

(*i*) Torque developed in the motor [– $T(\omega)$] ;

(the –ve sign shows it is the braking torque)

(*ii*) Friction torque (T_f) due to brake friction, which is constant ;

(*iii*) Load torque (T_{load}) ;

(*iv*) Inertia torque $\left(\dfrac{Jd\omega}{dt}\right)$.

If the motor is arrested in its motion using the electromagnetic torque proportional to the speed and a form of friction torque, we can write the equation (assuming the load torque on the motor to be independent of speed) as :

$$-T(\omega) = T_{load} + T_f + \frac{Jd\omega}{dt}$$

$$-K\omega = T_{load} + T_f + \frac{Jd\omega}{dt}, \text{ where } K \text{ is constant}$$

or,
$$dt = -\frac{Jd\omega}{K\omega + T_{load} + T_f}.$$

Now, the time required to brake the motor from speed ω_1 to ω_2 is given by,

$$t = -J\int_{\omega_1}^{\omega_2} \frac{d\omega}{K\omega + T_{load} + T_f}$$

or,
$$t = \frac{J}{K}\log_e\left(\frac{K\omega_1 + T_{load} + T_f}{K\omega_2 + T_{load} + T_f}\right) \quad ...(5.91)$$

The time (t_1) required to bring the motor to rest (*i.e.*, $\omega_2 = 0$) is given by,

$$t_1 = \frac{J}{K}\log_e\left(\frac{K\omega_1 + T_{load} + T_f}{T_{load} + T_f}\right) \quad ...(5.92)$$

The number of revolutions made by the motor before coming to rest is given by,

$$N = \frac{1}{2\pi}\int_0^t \omega_2\, dt$$

where,
$$\omega_2 = \frac{1}{K}\left[(K\omega_1 + T_{load} + T_f)e^{-Kt/J} - (T_{load} + T_f)\right]$$

Hence,
$$N = \frac{1}{2\pi K}\left[\frac{J}{K}(K\omega_1 + T_{load} + T_f)(1 - e^{-Kt/J}) - (T_{load} + T_f)\,t\right] \quad ...(5.93)$$

The number of revolutions (N_r) made by the motor can be obtained by substituting t by t_r in eqn. (5.93), as given below,

$$N_r = \frac{1}{2\pi K}\left[\frac{J}{K}(K\omega_1 + T_{load} + T_f)(1 - e^{-Kt/J}) - (T_{load} + T_f)\,t_r\right] \quad ...(5.94)$$

Example 5.107. *A 6-pole, 50 Hz synchronous motor coupled to a load has a moment of inertia of 540 kg-m². If the load torque is independent of speed and the frictional torque is 300 Nm, calculate the time taken by the motor to come to stop and the number of revolutions made during : (i) Rheostatic braking, the field current and braking resistance being kept constant at values which give initial electrical braking torque of 6000 N.m ; (ii) Plugging which produces a constant braking torque of 3300 Nm.* **(Roorkee University)**

Solution. *Given* : $p = 6$; $f = 50$ Hz ; $J = 540$ kg-m² ; $T_f = 300$ Nm.

(*i*) **Rheostatic braking :**

Initial braking torque, $K\omega_1 = 6000$ Nm

$$\omega_1 = \frac{2\pi N}{60} = \frac{2\pi\left(\frac{120f}{p}\right)}{60} = \frac{2\pi \times \left(\frac{120 \times 50}{6}\right)}{60} = 104.72 \text{ rad/s}$$

Hence, $$K = \frac{T}{\omega} = \frac{6000}{104.72} = 57.3$$

Time taken by the motor to stop,

$$t_r = \frac{J}{K} \log_e\left(\frac{K\omega_1 + T_f}{T_f}\right)$$

or, $$t_r = \frac{540}{57.3} \log_e\left(\frac{6000 + 300}{300}\right) = \textbf{28.69 s.} \quad \textbf{(Ans.)}$$

Number of revolutions made before coming to rest

$$N_r = \frac{1}{2\pi K}\left[\frac{J}{K}(K\omega_1 + T_f)(1 - e^{-Kt_r/J}) - T_f.t_r\right]$$

$$= \frac{1}{2\pi \times 57.3}\left[\frac{540}{57.3}(6000 + 300)(1 - e^{-(57.3 \times 28.69/540)}) - 300 \times 28.69\right]$$

$$= 0.002777\,[59371.7(1 - e^{-3.044}) - 8607] = 133.12$$

i.e., $\mathbf{N_r = 133.12.}$ **(Ans.)**

(*ii*) **Plugging :**

Total braking torque, $T_B = 3300 + 300 = 3600$ Nm

Since the braking torque is constant, it will produce a uniform retardation of β rad/sec².

$$\beta = \frac{T_B}{J} = \frac{3600}{540} = 6.67 \text{ rad/sec}^2$$

$$\omega_1 = 104.72 \text{ rad/s} \qquad \text{(as earlier calculated)}$$

Hence time taken to come to rest,

$$\mathbf{t_r} = \frac{\omega_1}{\beta} = \frac{104.72}{6.67} = 15.7 \text{ s}$$

∴ Number of revolutions made before coming to rest,

N_r = (Average speed in rev./sec) × (Time taken to stop in sec.)

$$= \left[\frac{1}{2} \times \left(\frac{120 \times f}{60 \times p}\right)\right] \times 15.7 = 130.83$$

$$= \frac{1}{2} \times \frac{120 \times 50}{60 \times 6} \times 15.7 = 130.83$$

i.e., $N_r = 130.83$. **(Ans.)**

5.15. HEATING AND COOLING OF ELECTRICAL MACHINES

5.15.1. Heating and Cooling Curves

In an electrical machine the various energy losses which occur are finally converted into heat, this causes an increase in temperature which depends upon :

(*i*) The heat-absorbing capacity of the various parts of the machine, and

(*ii*) The facility with which heat is conducted away or radiated or otherwise dissipated from the surface of the machine.

In the beginning the temperature rise will be steep but after some time steady conditions would prevail. At that time there would be no further increase in temperature because the *rate of heat production would become equal to rate of heat dissipation.*

Let, Q = Power loss or heat developed, J/s or W,

m = Mass of active parts of machine, kg,

s = Specific heat, J/kg-°C,

A = Area of the cooling surface, m^2,

λ = Specific heat dissipation or *emissivity,* W/m^2-°C difference between the surface and the ambient medium,

$\alpha = \frac{1}{\lambda}$ = Cooling co-efficient,

θ = Temperature rise at any time t, °C,

θ_m = Final steady temperature rise while heating, °C,

θ_n = Final steady temperature rise while cooling, °C,

θ_i = Initial temperature rise over ambient medium, °C,

T_h = Heating time constant, s,

T_c = Cooling time constant, s, and

t = Time, s.

Heating of the machine. Let us consider the conditions at any time t from start.

Heat produced in the body during an infinitely small time dt

= Heat produced per second × dt

$$= Qdt \qquad ...(5.94)$$

If during this period dt the temperature of the body rises by $d\theta$, then

Heat stored in the body = Mass of the body × specific heat × temperature difference

$$= m \times s \times d\theta \qquad ...(5.95)$$

If in the process of heating, the temperature of the surface rises by $\theta°$ over the ambient medium, at the instant considered, the heat dissipated by the body into ambient medium due to radiation, connection and conduction,

= Surface in m^2 × watt dissipated per m^2 per °C × temperature rise × time

$$= A \times \lambda \times \theta \times dt \qquad ...(5.96)$$

Since, heat produced in machine

= Heat stored in parts + heat dissipated

$$\therefore \quad Qdt = msd\theta + A\lambda\theta dt \qquad ...(5.97)$$

$$dt = \frac{d\theta}{\frac{Q}{ms} - \frac{A\lambda}{ms}\theta} \quad ...(5.98)$$

Solving the differential equation (5.98), we get

$$t = -\frac{ms}{A\lambda}\log_e\left(\frac{Q}{ms} - \frac{A\lambda}{ms}\theta\right) + C \quad ...(5.99)$$

where, C is the constant of integration.

The value of C is found by using the boundary condition,

at $t = 0$, $\theta = \theta_i$

Putting this in eqn. (5.99), we get

$$0 = -\frac{ms}{A\lambda}\log_e\left(\frac{Q}{ms} - \frac{A\lambda}{ms}\theta_i\right) + C$$

or,

$$C = \frac{ms}{A\lambda}\log_e\left(\frac{Q}{ms} - \frac{A\lambda}{ms}\theta_i\right)$$

Substituting this value of C in eqn. (5.99), we get

$$t = \frac{ms}{A\lambda}\log_e\left(\frac{Q}{ms} - \frac{A\lambda}{ms}\theta\right) + \frac{ms}{A\lambda}\log_e\left(\frac{Q}{ms} - \frac{A\lambda}{ms}\theta_i\right)$$

$$= -\frac{ms}{A\lambda}\log_e\left(\frac{\frac{Q}{A\lambda} - \theta}{\frac{Q}{A\lambda} - \theta_i}\right) \quad ...(5.100)$$

When $t = \infty$ the machine reaches a final steady temperature rise θ_m. Under this condition there is no further temperature rise and the rates of heat production and dissipation are equal. This means

$$d\theta = 0$$

or,

$$msd\,\theta = 0$$

when the machine attains final steady temperature rise.

When $\theta = \theta_m$ in eqn. 5.97

$$Qdt = A\lambda\theta_m\,dt$$

or,

$$Q = A\lambda\theta_m$$

or,

$$\theta_m = \frac{Q}{A\lambda} \quad ...(5.101)$$

$$= \frac{Q\alpha}{A} \quad ...(5.102)$$

Putting $\theta_m = \frac{Q}{A\lambda}$ in eqn. (5.100), we get

$$t = -\frac{ms}{A\lambda}\log_e\left(\frac{\theta_m - \theta}{\theta_m - \theta_i}\right) \quad ...(5.103)$$

The term $\frac{ms}{A\lambda}$ is called the **heating time constant** T_h

$$\therefore \qquad T_h = \frac{ms}{A\lambda} \qquad ...(5.104)$$

Putting this value of $T_h = \dfrac{ms}{A\lambda}$ in eqn. (5.103), we get

$$t = -T_h \log_e \left(\frac{\theta_m - \theta}{\theta_m - \theta_i} \right)$$

or,
$$\frac{\theta_m - \theta}{\theta_m - \theta_i} = e^{-t/T_h}$$

or,
$$\theta_m - \theta = (\theta_m - \theta_i)\, e^{-t/T_h}$$

$$\theta = \theta_m - (\theta_m - \theta_i) e^{-t/T_h}$$

$$= \theta_m - \theta_m\, e^{-t/T_h} + \theta_i \,.\, e^{-t/T_h}$$

i.e.,
$$\theta = \theta_m (1 - e^{-t/T_h}) + \theta_i \,.\, e^{-t/T_h} \qquad ...(5.105)$$

If the machine starts from cold conditions, $\theta_i = 0$ *i.e.*, no temperature rise over the ambient medium, then,

$$\theta = \theta_m (1 - e^{-t/T_h}) \qquad ...(5.106)$$

The above relation is the equation of temperature rise with time. The temperature rise—time curve is exponential in nature as shown in Fig. 5.103.

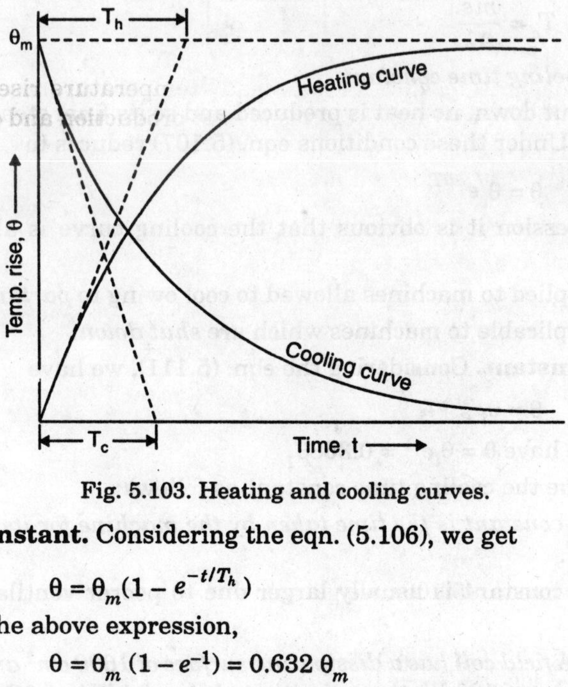

Fig. 5.103. Heating and cooling curves.

Heating time constant. **Considering the eqn. (5.106), we get**

$$\theta = \theta_m (1 - e^{-t/T_h})$$

Putting $t = T_h$ is the above expression,

$$\theta = \theta_m (1 - e^{-1}) = 0.632\, \theta_m \qquad [\because\ e^{-1} = 0.368]$$

Thus the heating time constant can be defined as follows :

***The heating time constant** is the time taken by the machine to attain 0.632 of its final steady temperature rise".*

The heating time constant of conventional electrical machines is usually within the range of $\frac{1}{2}$ to 3 to 4 hours.

Cooling of the machine. Considering the eqn. (5.97), we have

$$Qdt = msd\theta + A\lambda\theta dt$$

After solving the above equation, we get

$$t = -\frac{ms}{A\lambda}\log_e\left(\frac{Q}{ms} - \frac{A\lambda}{ms}\theta_i\right) + C$$

The value of C is obtained by using the boundary condition, when $t = 0$, $\theta = \theta_i$, we get

$$0 = -\frac{ms}{A\lambda}\log_e\left[\frac{Q}{ms} - \frac{A\lambda}{ms}\theta_i\right] + C$$

or,
$$C = \frac{ms}{A\lambda}.\log_e\left[\frac{Q}{ms} - \frac{A\lambda}{ms}\theta_i\right]$$

Substituting this value of C and proceeding as in the case of heating of the machine, we get

$$\theta = \theta_n(1 - e^{-t/T_c}) + \theta_i . e^{-t/T_c} \quad ...(5.107)$$

where,
$$\theta_n = \frac{Q}{A\lambda} \quad ...(5.108)$$

$$= \frac{Q\alpha}{A} \quad ...(5.109)$$

and,
$$T_c = \frac{ms}{A\lambda} \quad ...(5.110)$$

where, T_c is called the *cooling time constant.*

If a machine is shut down, no heat is produced and so its final steady temperature rise when cooling is zero or $\theta_n = 0$. Under these conditions eqn. (5.107) reduces to

$$\theta = \theta_i e^{-t/T_c} \quad ...(5.111)$$

From above expression it is obvious that the cooling curve is also exponential in nature (Fig. 5.103).

Eqn. (5.107) is applied to machines allowed to cool owing to *partial removal of load.*

Eqn. (5.111) is applicable to machines which are *shut down.*

Cooling time constant. Considering the eqn. (5.111), we have

$$\theta = \theta_i\, e^{-t/T_c}$$

Putting $t = T_c$, we have $\theta = \theta_i e^{-1} = 0.368\theta_1$

Thus we can define the cooling time constant as follows :

"The cooling time constant is the time taken by the machine for its temperature rise to fall to 0.368 of its initial value".

The cooling time constant is usually larger due to poorer ventilation conditions when the machine cools.

Example 5.108. *A field coil has a dissipating surface of 1600 cm² and a length of mean turn of 110 cm. It dissipates a loss of 160 W, the emissivity being 34 W/m²-°C. Estimate the final steady temperature rise of the coil and its time constant if the cross-section of the coil is 10 × 5 cm.*

Take : Space factor = 0.56, density of copper = 8900 kg/m³ and specific heat of copper = 397 J/kg-°C.

Solution. Dissipating surface of field coil,

$$A = 1600 \text{ cm}^2 = 1600 \times 10^{-4} \text{ m}^2$$

Length of mean turn, $l = 110$ cm $= 1.1$ m

Loss dissipated, $Q = 160$ W

Emissivity, $\lambda = 34$ W/m²-°C

Cross-section of the coil $= 10 \times 5$ cm $= 10 \times 5 \times 10^{-4}$ m² $= 0.005$ m²

Space factor $= 0.56$

Density of copper $= 8900$ kg/m³

Specific heat of copper, $s = 397$ J/kg-°C

Final steady temperature rise, θ_m :

Heating time constant, T_h :

Mass of copper, m = Volume of copper × density of copper

= Cross-section of coil × length of mean turn × space factor × density of copper

$= 0.005 \times 1.1 \times 0.56 \times 8900 = 27.412$ kg.

Final steady temperature rise,

$$\theta_m = \frac{Q}{A\lambda} = \frac{160}{1600 \times 10^{-4} \times 34} = \mathbf{29.41°C.\ (Ans.)}$$

Heating time constant,

$$T_h = \frac{ms}{A\lambda} = \frac{27.412 \times 397}{1600 \times 10^{-4} \times 34} = \mathbf{2000.47\ s.\ (Ans.)}$$

Example 5.109. *The initial temperature of a machine is 45°C. Calculate the temperature of the machine after 1.2 hour if its final steady temperature rise is 85°C and the heating time constant is 2.4 hours. The ambient temperature is 25°C.*

Solution. Initial temperature of a machine = 45°C

Ambient temperature = 25°C

Final steady temperature rise $\theta_m = 85$°C

Heating time constant, $T_h = 2.4$ hours

Temperature of the machine after 1.2 hour, θ :

Initial temperature rise, θ_i = Initial temperature of the machine − ambient temperature

$= 45 - 25 = 20$°C

Using the equation,

$$\theta = \theta_m(1 - e^{-t/T_h}) + \theta_i e^{-t/T_h}$$

$$= 85(1 - e^{-1.2/2.4}) + 20e^{-1.2/2.4} \qquad (\because t = 1.2 \text{ hour})$$

$$= 85(1 - e^{-1/2}) + 20 \times e^{-1/2}$$

$$= 85(1 - 0.607) + 20 \times 0.607 = 33.4 + 12.14 = 45.54°C$$

Hence, *the temperature of the machine after 1.2 hour*

$= 45.54 + 25 =$ **70.54°C. (Ans.)**

Example 5.110. *In a certain machine the temperature rise is 25°C after one hour and 37.5°C after two hours starting from cold conditions. Calculate its final steady temperature rise and the heating time constant.*

If the machine temperature falls from the final steady value to 40°C in 1.5 hour when disconnected, calculate its cooling time constant.

The ambient temperature is 30°C.

Solution. Temperature rise after one hour $= 25°C$

Temperature rise after two hours $= 37.5°C$

Ambient temperature $= 30°C.$

Final steady temperature rise, θ_m :

Cooling time constant, T_c :

When heating :

Since the machine starts from cold conditions therefore its temperature rise is given by,

$$\theta = \theta_m (1 - e^{-t/T_h}) \quad ...(i)$$

We have, $\theta = 25°C$ at $t = 1$ hour

and, $\theta = 37.5°C$ at $t = 2$ hours

Putting these values in eqn. (*i*), we get

$$25 = \theta_m(1 - e^{-1/T_h}) \quad ...(ii)$$

$$37.5 = \theta_m(1 - e^{-2/T_h}) \quad ...(iii)$$

From (*i*) and (*ii*), we get

$$\left(\frac{1 - e^{-2/T_h}}{1 - e^{-1/T_h}}\right) = \frac{37.5}{25} = 1.5$$

Let, $e^{-1/T_h} = x$

then, $\dfrac{1 - x^2}{1 - x} = 1.5$

or, $(1 + x) = 1.5$ or $x = 0.5$

or, $e^{-1/T_h} = 0.5$

Taking log on both the sides, we get

$$\log_e (e^{-1/T_h}) = \log_e 0.5$$

$$-\frac{1}{T_h} = -0.693$$

$$T_h = 1.44 \text{ hour}$$

From (*ii*), $25 = \theta_m (1 - e^{-1/T_h}) = \theta_m (1 - 0.5)$

$\therefore$ $\theta_m = 50°C$

Hence, *final steady temperature rise* **= 50°C. (Ans.)**

When cooling :

Temperature rise after 1.5 hour, $\theta = 40 - 30 = 10°C$.

Since the machine is disconnected its final steady temperature rise when cooling is $\theta_i = 0$.

Initial temperature rise, $\theta_i = 50°C$.

Using the relation, $\theta = \theta_i e^{-t/T_c}$

$$10 = 50e^{-1.5/T_c}$$

$$e^{-1.5/T_c} = 0.2$$

$$e^{1.5/T_c} = \frac{1}{0.2} = 5$$

Taking log on both the sides

$$\log_e (e^{1.5/T_c}) = \log_e 5$$

$$\frac{1.5}{T_c} = \log_e 5 = 1.609 \text{ or } T_c = 0.932 \text{ hour}$$

Hence, *cooling time constant* = **0.932 hour. (Ans.)**

Example 9.111. *The outside of a 10 kW totally enclosed motor is equivalent to a cylinder of 0.7 m diameter and 1.2 m length. The motor weighs 450 kg and may be considered to be of material having specific heat of 700 J / kg° C. The outer surface is capable of dissipating heat at the rate of 1.25 W / m²-°C. When the motor is operating at full-load with an efficiency of 90 per cent, determine :*

(i) *Final temperature rise* (ii) *Heating time constant.*

Solution. Power output of the motor = 10 kW

Efficiency = 90%

Diameter of cylinder, $d = 1.2$ m

Length of cylinder, $l = 0.7$ m

Specific heat of material, $s = 700$ J/kg°-C

Mass of active parts, $m = 450$ kg

Final temperature rise, θ_m :

Heating time constant, T_h :

$$\text{Power input} = \frac{\text{Power output}}{\text{Efficiency}} = \frac{10 \times 1000}{0.90} = 11111.11$$

Power loss, Q = Power input—power output $= 11111.11 - 10000 = 1111.11$

(i) Final temperature rise, $\theta_m = \dfrac{Q}{A\lambda}$

where $A = \pi dl = \pi \times 0.7 \times 1.2 = 2.64 \text{ m}^2$

$$\therefore \quad \theta_m = \frac{1111.11}{2.64 \times 12.5} = 33.67°\text{C}$$

Hence, *final temperature rise*, θ_m = **33.67°C. (Ans.)**

(iii) *Heating time constant,* $T_h = \dfrac{ms}{A\lambda} = \dfrac{450 \times 700}{2.64 \times 1.25} = 9545.45 \text{ s} = 2.6515 \text{ hours}$

Hence, *heating time constant*, T_h = **2.6515 hours. (Ans.)**

Example 5.112. *The following rises were observed in a temperature-rise test on a D.C. machine at full-load :*

After one hour 15°C

After two hours 25°C.

Find : (i) Final steady temperature rise and the time constant.

(ii) The steady temperature rise after one hour at 50 per cent overload, from cold. Assume that the final temperature rise on 50 per cent overload is 90°C.

Solution. Temperature rise after one hour = 15°C

Temperature rise after two hours = 25°C

(i) From the given data :

When $t = 1$ hour, $\theta = 15°$C

and, when $t = 2$ hours, $\theta = 25°$C

Using the relation, $\theta = \theta_m (1 - e^{-t/T_h})$ ...(i)

$$15 = \theta_m(1 - e^{-1/T_h}) \quad ...(ii)$$

$$25 = \theta_m(1 - e^{-2/T_h}) \quad ...(iii)$$

From (ii) and (iii), we get $\frac{25}{15} = \frac{(1 - e^{-2/T_h})}{(1 - e^{-1/T_h})} = (1 + e^{-1/T_h})$

$$\therefore \quad e^{-1/T_h} = \frac{25}{15} - 1 = \frac{2}{3}$$

i.e., $$e^{-1/T_h} = \frac{2}{3} \quad ...(iv)$$

Putting $e^{-1/T_h} = \frac{2}{3}$ in eqn. (ii), we get

$$15 = \theta_m \left(1 - \frac{2}{3}\right)$$

$$\therefore \quad \theta_m = 45°C$$

Hence, *final steady temperature rise* = **45°C. (Ans.)**

From eqn. (iv), we get $e^{-1/T_h} = \frac{2}{3}$

or $$e^{1/T_h} = \frac{3}{2} = 1.5$$

or $$\frac{1}{T_h} = \log_e 1.5 = 0.405 \text{ or } T_h = 2.466 \text{ hours}$$

Hence, *heating time constant* = **2.466 hours. (Ans.)**

(ii) On 50% overload θ_m = 90°C and *heating time constant remaining the same, as it is independent of rate of production, i.e.,* T_h = 2.466 hours.

Therefore, final steady temperature rise after one hour (*i.e.*, when t = 1 hour) is given by the relation,

$$\theta = \theta_m (1 - e^{-t/T_h}) = 90(1 - e^{-1/2.466}) = 30°C$$

Hence, **final steady temperature rise after one hour at 50% overload = 30°C. (Ans.)**

Example 5.113. *A D.C. machine has a final temperature of 80°C on continuous rating and a thermal time constant of 45 minutes.*

(i) What is the temperature rise after one hour at this load ?

(ii) If the temperature rise on one hour rating is 80°C, find the maximum steady temperature at this rating ?

(iii) When working at its one-hour rating, how long does it take the temperature to increase from 70°C to 80°C ?

Solution. Final temperature rise on continuous rating,

$$\theta_m = 80°C$$

Thermal time constant, T_h = 45 minutes = $\frac{45}{60}$ = 0.75 hour

(i) Temperature rise after one hour (*i.e.*, when $t = 1$ hour) can be found by using the following relation

$$\theta = \theta_m (1 - e^{-t/T_h})$$
$$= 80(1 - e^{-1/0.75}) = 58.9°C$$

Hence, *temperature rise after one hour* = **58.9°C. (Ans.)**

(ii) Let θ'_m = maximum steady temperature rise at one hour rating.

From given data : $T_h = 0.75$ hour, $\theta = 80°C$, $t = 1$ hour

Now, $$\theta = \theta'_m (1 - e^{-t/T_h})$$

i.e., $$80 = \theta'_m (1 - e^{-1/0.75})$$

or, $$\theta'_m = \frac{80}{(1 - e^{-1/0.75})} = 108.7°C$$

Hence, *the maximum steady temperature rise at one hour rating* = **108.7°C. (Ans.)**

(iii) Assuming initial temperature as 0°C, time taken to attain temperature of 80°C is one hour or 60 minutes.

Maximum temperature rise $\theta_m = 108.7°C$.

Now let the time taken to attain temperature of 70°C from 0°C be t hours, then

$$70 = 108.7 (1 - e^{-t/0.75})$$

or, $$1 - e^{-t/0.75} = \frac{70}{108.7} \quad \text{or} \quad e^{-t/0.75} = 1 - \frac{70}{108.7}$$

or, $$e^{-t/0.75} = 2.81 \quad \text{or} \quad -\frac{t}{0.75} = \log_e 0.356 = -1.0328$$

[Taking log on both sides]

or, $$t = 0.775 \text{ hour} = 46.5 \text{ minutes}$$

Hence, *time taken to increase the temperature from 70°C to 80°C* = **46.5 minutes. (Ans.)**

Example 5.114. *A 40 kW motor when run continuously on full-load attains a temperature of 35°C above the surrounding air. Its heating time constant is 90 minutes. What would be the half hour rating of the motor for this temperature rise ?*

Assume that the machine cools down completely between each load period and the losses are proportional to the square of the load.

Solution. Rating of the motor, $P = 40$ kW

Final temperature rise, $\theta_m = 35°C$ in case of continuously rating

Heating time constant, $T_h = 90$ minutes = 1.5 hour

Let P' be the hour rating of the motor.

Then final temperature rise, $$\theta'_m = \frac{\theta}{1 - e^{-t/T_h}} = \frac{35}{1 - e^{-0.5/1.5}} = 123.47°C$$

Since losses are proportional to the square of the load, therefore

$$\frac{\theta'_m}{\theta_m} = \frac{\text{losses at } P' \text{ kW}}{\text{losses at } P \text{ kW}}$$

$$\frac{123.47}{35} = \left(\frac{P'}{40}\right)^2 \quad \text{or} \quad P' = 75.13 \text{ kW}$$

Hence, *half hour rating* = **75.13 kW. (Ans.)**

Example 5.115. *A D.C. machine which is totally enclosed has a temperature rise of 25°C after half an hour and 40°C after one hour on full-load. Determine the temperature rise after two hours on load. If a fan is fitted driving cooling air over the external surface of the machine, which reduces the final steady state temperature rise of the machine to 82 per cent of the original value, determine the temperature rise from cold after one hour on full-load.*

Solution. Temperature rise after $\frac{1}{2}$ hour = 25°C

Temperature rise after 1 hour = 40°C

i.e., when $t = \frac{1}{2}$ hour, $\theta = 25°C$

and $t = 1$ hour, $\theta = 40°C$

Using the relation, $\theta = \theta_m(1 - e^{-t/T_h})$ ...(*i*)

where, θ = Temperature rise after time '*t*',

θ_m = Final steady temperature rise, and

T_h = Heating time constant of the motor.

Now, $25 = \theta_m(1 - e^{-0.5/T_h})$ ...(*ii*)

and, $40 = \theta_m(1 - e^{-1/T_h})$ ...(*iii*)

From (*ii*) and (*iii*), we have $\frac{40}{25} = \frac{1 - e^{-1/T_h}}{1 - e^{-0.5/T_h}}$

Let, $e^{-0.5/T_h} = x$

$\therefore$ $1.6 = \frac{1 - x^2}{1 - x} = 1 + x = 1 + e^{-0.5/T_h}$

i.e., $e^{-0.5/T_h} = 0.6$ ...(*iv*)

Substituting $e^{-0.5/T_h} = 0.6$ in (*ii*), we get

$25 = \theta_m(1 - 0.6)$ or $\theta_m = 62.5°C$

$\therefore$ After two hours of full-load, $\theta = 62.5(1 - e^{-2/T_h})$

$= 62.5[1 - (e^{-0.5/T_h})^4]$

$= 62.5 - [1 - (0.6)^4]$ [$\because$ $e^{-0.5/T_h} = 0.6$] [Calculated above]

$= 54.4°C$

Hence, *temperature rise after two hours on load* = **54.4°C. (Ans.)**

Now, in both cases, the rate of production is same, hence initial rate of temperature rise is same.

Now, $\frac{d\theta}{dt} = \frac{d}{dt}[\theta_m(1 - e^{-t/T_h})] = \frac{\theta_m}{T_h} \cdot e^{-t/T_h}$

Initial rate of temperature rise can be found by putting $t = 0$

$$\therefore \left(\frac{d\theta}{dt}\right)_{t=0} = \frac{\theta_m}{T_h}$$

$$\therefore \quad \frac{\theta_m}{T_h} = \frac{\theta_{m2}}{T_{h2}} \quad \text{or} \quad T_{h2} = \frac{\theta_{m2}}{\theta_m} . T_h$$

Now, $$\frac{\theta_{m2}}{\theta_m} = 0.82$$

Since, $$e^{-0.5/T_h} = 0.6$$

$$\therefore \quad T_h = 0.979 \text{ hour}$$

and, $$T_{h2} = 0.82 \times 0.979 = 0.803 \text{ hours.}$$

This is the value of thermal time constant under new cooling conditions

$$\theta_{m2} = 0.82\theta_m = 0.82 \times 62.5 = 51.25°\text{C}$$

Hence, *the temperature rise from cold after one hour is,*

$$\theta = 51.25\,(1 - e^{-1/.803}) = \mathbf{36.5°C.} \quad \textbf{(Ans.)}$$

Example 5.116. *A 50 kW motor with a heating time constant of 100 minutes has a final temperature rise of 50°C on continuous rating. Find the half-hour rating of the motor for this temperature rise assuming that it cools down completely between each load period. The motor has maximum efficiency of 80% at its full load.* **(A.M.I.E. Elec. Drives & their control, Winter, 2000)**

Solution. *Given :* $P = 50$ kW, $T_h = 100$ minutes ; $\theta_m = 50$°C.

Half-hour rating of the motor :

Let P kW be the half-hour (30 minutes) rating, then final temperature rise,

$$\theta'_m = \frac{\theta}{1 - e^{-t/T_h}} = \frac{50}{1 - e^{(-30/100)}} = \frac{50}{1 - e^{-0.3}} = 193°\text{C}$$

Since maximum efficiency occurs at 80% of full load *i.e.*, a load of 40 kW, therefore, at 40 kW load *copper loss is equal to constant losses*. Let each be equal to W watts.

Losses at 50 kW $$= W + \left(\frac{50}{40}\right)^2 W = 2.5625\text{ W}$$

Losses at P kW $$= W + \left(\frac{P}{40}\right)^2 W = W + \frac{P^2}{1600} W = W\left(1 + \frac{P^2}{1600}\right)$$

[Since copper loss varies as square of the load.]

$$\frac{\text{Total losses at P kW}}{\text{Total losses at 50 kW}} = \frac{\theta'_m}{\theta_m} \qquad [\because \ \theta_m \propto \text{losses}]$$

or, $$\frac{W\left(1 + \frac{P^2}{1600}\right)}{2.5625\,W} = \frac{193}{50} \quad \text{or} \quad 1 + \frac{P^2}{1600} = 2.5625 \times \frac{193}{50}$$

or, $$\mathbf{P} = \left[\left(2.5625 \times \frac{193}{50} - 1\right) \times 1600\right]^{1/2} = \mathbf{119.3\ kW.} \quad \textbf{(Ans.)}$$

Example 5.117. *The 15 minute rating of a motor is 400 watts. The heating time constant is 60 minutes. Determine the continuous rating of the motor if the maximum efficiency occurs at 80 percent of full load.* **(A.M.I.E. Electric Drives & their control, Winter, 2001)**

Solution. *Given :* 15 minutes rating of motor = 400 W, T_h = 60 min.

Continuous rating of the motor :

Since maximum efficiency occur at 80% of full load so at 80% of full load iron loss is equal to copper loss, let each be equal to P_c watts.

$$\text{Losses on full load} = P_c + \left(\frac{1}{0.8}\right)^2 P_c = 2.5625P_c$$

$$\text{Loss at load of 400 W} = P_c + \left[\frac{400}{0.8 \times \text{full load}}\right]^2 P_c$$

Also,
$$\frac{\theta'_m}{\theta_m} = \frac{1}{1 - e^{t/T_h}} = \frac{\text{Total loss on 15 min. rating}}{\text{Total loss on full load}}$$

or,
$$\frac{1}{1 - e^{-15/60}} = \frac{P_c + \left[\frac{400}{0.8 \times \text{full load}}\right]^2 P_c}{2.5625P_c}$$

$$4.52 = \frac{1 + \left(\frac{500}{\text{full load}}\right)^2}{2.5625}$$

or,
$$\frac{500}{\text{full load}} = [(4.52 \times 2.5625) - 1]^{1/2} = 3.253$$

$\therefore$
$$\text{Full load} = \frac{500}{3.253} = 153.7 \text{ W}$$

Hence, *continuous rating of motor* = **153.7 W. (Ans.)**

Example 5.118. *A motor has a thermal time constant of 45 minutes. When the motor runs continuously on full load, its final temperature rise is 80°C. (i) What is the temperature rise after 1 hour if the motor runs continuously on full load ? (ii) If the temperature on one-hour rating is 80°C, find the maximum steady state temperature at this rating. (iii) How much time does the motor take in its temperature to rise from 50°C to 80°C if it is working at its 1 hour rating ?*

(A.M.I.E. Electric Drives & Their Control, Winter, 1999)

Solution. *Given :* T_h = 45 min. = 0.75 hour ; θ_m = 80°C

(*i*) **Temperature rise after 1 hour, θ :**

We know that,
$$\theta = \theta_m (1 - e^{-t/T_h})$$

or,
$$\theta = 80(1 - e^{-1/0.75}) = \mathbf{58.9°C.} \quad \textbf{(Ans.)}$$

(*ii*) **Maximum steady state temperature, θ′$_m$:**

$$\theta'_m = \frac{\theta_m}{1 - e^{-t/T_h}} = \frac{80}{1 - e^{-1/0.75}} = \mathbf{108.64°C.} \quad \textbf{(Ans.)}$$

(*iii*) **Time taken to increase temperature from 50°C to 80°C, t :**

Time taken to attain temperature of 80°C (assuming initial temperature as 0°C) is one hour or 60 minutes.

Maximum temperature rise, θ'_m = 108.64°C.

Let the time taken to attain temperature of 50°C from 0°C be *t* hours, then

$$50 = 108.64(1 - e^{-t/0.75})$$

or, $$e^{-t/0.75} = 1 - \frac{50}{108.64} = 0.5397$$

or, $$e^{t/0.75} = 1.853$$

or, $$t = 0.463 \text{ hour or } 27.78 \text{ min.}$$

Hence, time taken to increase temperature from 50°C to 80°C

$$= 60 - 27.78 = \mathbf{32.22\ minutes.}\quad \textbf{(Ans.)}$$

Example 5.119. *The outside of a 20 H.P. totally enclosed motor is equivalent to a cylinder of 0.75 m diameter and 1.2 m in length. The motor weighs 450 kg and its material has a specific heat of 700 J/kg-°C. Estimate final temperature of the machine and its time constant when operating on full load with an efficiency of 90 per cent. Assume that the outer cylindrical surface dissipates heat at 15 W/m²°C.*

If this machine were semienclosed with a resultant improvement in cooling to 20 W/m²-°C, determine the load which could be delivered for the same temperature rise. Assume the efficiency to remain unchanged. **(Nagpur University)**

Solution. *Given* : Power of the motor = 20 H.P. ; $d = 0.75$ m ; $l = 1.2$ m ; $m = 450$ kg

$s = 700$ J/kg-°C, $\eta = 0.9$; $\lambda = 15$ W/m²-°C (= 20 W/m²°C in the second case)

***Case 1.* Heating time constant, T_h :**

$$T_h = \frac{ms}{A\lambda} \qquad \text{...[Eqn. (5.104)]}$$

$$= \frac{450 \times 700}{(\pi dl \times \lambda)} = \frac{450 \times 700}{(\pi \times 0.75 \times 1.2) \times 15} = 7427.2 \text{ s} = \mathbf{2.06\ h.}\quad \textbf{(Ans.)}$$

Final temperature, θ_m :

The loss in motor, $$Q = \frac{20 \times 735.5}{0.9} - 20 \times 735.5 = 1634 \text{ W}$$

$$\theta_m = \frac{Q}{A\lambda} = \frac{1634}{(\pi \times 0.75 \times 1.2) \times 15} = \mathbf{38.53°C.}\quad \textbf{(Ans.)}$$

***Case 2.* Load which could be delivered for the same temperature rise :**

If temperature rise is to remain same, the loss in the second case (semi-enclosed)

$$= 38.53 \times (\pi \times 0.75 \times 1.2 \times 25) = 2723.5 \text{ W}$$

Let P be the rating in watts.

Then, $$\frac{P}{P + 2723.5} = 0.9$$

or, $$P = 0.9(P + 2723.5) = 0.9P + 2451.15$$

∴ $$P = 24511.5 \text{ W or } 33.3 \text{ H.P. say } 33 \text{ H.P.}$$

or, *Rating is* **33 H.P. (Ans.)**

5.15.2. Insulating Materials

The temperature rise which electric motors safely withstand is determined by the termal stability (limiting temperature) of the insulating materials used in them. With the use of materials of higher thermal stability, motors of the same physical size may be rated for large power output.

The insulating materials can be classified according to temperature as follows :

Class	*Insulating materials included*	*Assigned limiting insulating temperature*
Y (Formerly *O*)	Cotton, silk, paper, cellulose, wood, etc., neither impregnated nor immersed in oil. Materials of *Y* class are *unsuitable for electrical machines and apparatus as they deteriorate rapidly and are extremely hygroscopic*	90°C
A	Materials of class *Y* impregnated with natural resins, cellulose esters, insulating oils etc. Also included in this list are laminated wool, varnished paper.	105°C
E	Synthetic resin enamels, cotton and paper laminates with formaldehyde bonding etc.	120°C
B	Mica, glass fibres, asbestos with suitable bonding substances, built up mica, glass fibre, and asbestos laminates.	130°C
F	Materials of class *B* with bonding materials of higher thermal stability.	155°C
H	Glass fibre and asbestos materials, and built up mica, with silicon resins.	180°C
C	Mica, ceramics, glass, quartz without binders or with silicon resins of higher thermal stability.	above 180°C

5.16. SIZE AND RATING OF MOTORS

For any particular service the following *factors govern the size and rating of motor* :

(*i*) The maximum temperature rise of the motor under given load conditions ;

(*ii*) The maximum torque required.

It has been observed that a motor which is satisfactory from the point of view of maximum temperature rise usually satisfies the requirement of maximum torque as well.

The maximum permissible temperature rise with insulations *A* and *B* are 40°C and 50°C respectively.

5.16.1. Standard rating of motors :

The different ratings for electrical motors are as under ;

1. **Continuous rating.** *It is an output which a motor can deliver continuously without exceeding the permissible temperature.* It can deliver *25% overload for two hours.*

2. **Continuous maximum rating.** It is similar to continuous rating but *not allowing any overload.*

- It is used for motors of capacity larger than 2 kW per r.p.m.
- These motors are bit inferior to the continuous-rated motors.

3. **Short time rating.** This is an output, which a motor can deliver for a specified period (say 1 hour, $\frac{1}{2}$ hour, $\frac{1}{4}$ hour etc.) without exceeding the specified temperature rise.

4. **Intermittent rating.** This rating indicates the maximum load of the motor for the specified time *followed by a no-load period during which the machine cools down to its original temperature.*

- For intermittent loads short time rated motors can be used if the loading condition corresponds to the short time rating of the motor or if it can be assumed that the heating of the motor is proportional to the square of current, the equivalent continuous rating can be found by taking the r.m.s value of the load curve and choosing a motor having a continuous rating equal to this. This method is *very frequently used*.

5.16.2. Classes of Duty

The rating of a motor selected from the viewpoint of heating depends on the load conditions or duty to which it is subjected. According to *ISS* : 4722, these operating conditions are classified into the following *eight classes of duty* :

(*i*) ***Continuous duty.*** It denotes operation at constant load of sufficient duration for thermal equilibrium to be reached.

Examples. Centrifugal pumps, fans, compressors and conveyors are some examples of equipment which run continuously with a constant load.

(*ii*) ***Short time duty.*** It denotes the operation at constant load during a given time, *less* than that required to reach thermal equilibrium, followed by a rest of sufficient duration to re-establish equality of temperature with the cooling medium.

The recommended values for short time duty are 10, 30, 60 and 90 minutes.

Examples. Motors used for opening and closing weirs, lockgates and bridges, motors employed in battery-charging units etc. are rated for such a duty.

(*iii*) ***Intermittent periodic duty.*** It denotes a sequence of identical duty cycles, each consisting of a period of operation of constant load and a rest period, these periods being too short to attain thermal equilibrium during one duty cycle. The starting current does not affect the temperature rise for this type of duty. Unless otherwise specified, the duration of duty cycle is 10 minutes.

The recommended values for the cycle duration factors are 15, 25, 40 and 60 percent.

Examples. Motors that are used in different kinds of hoisting mechanisms and those used in trams, trolley buses etc. are subjected to intermittent periodic duty.

(*iv*) ***Intermittent periodic duty with starting.*** It indicates a sequence of identical duty cycles each consisting of a period of starting, a period of operation at constant load and a rest period, the operating and rest periods being too short to attain thermal equilibrium during one duty cycle.

— The stopping of motor, in this type of duty, is obtained by natural deceleration after disconnecting the electrical supply by means of mechanical brake which does not cause additional heating of the windings.

— This type of duty is characterised by the cyclic duration factor, the number of duty cycles per hour and *factor of inertia* which is the *ratio of the combined inertia of the motor and load to the motor inertia*.

Examples. Motors that drive metal cutting lathes and certain auxiliary equipment of rolling mills are subjected to such operating conditions.

(*v*) ***Intermittent periodic duty with starting and braking.*** It denotes a sequence of identical duty cycles each consisting of a period of starting, a period of operation at a constant load, a period of braking and a rest period. The operating and rest periods are too short to obtain thermal equilibrium during one duty cycle.

In this duty *braking is rapid and is carried out electrically*.

Examples. Certain auxiliary equipment used in rolling mills and metal cutting lathes offer such operating conditions to their driving motors.

(*vi*) ***Continuous duty with intermittent periodic loading***. It denotes a sequence of identical duty cycles each consisting of a period of operation at constant load and a period of operation at no load, machines with excited windings having normal no-load rated voltage excitation. The *operation and no load periods are too short to attain thermal equilibrium during one duty cycle.*

Unless otherwise specified the duration of the duty cycle is 10 minutes.

The recommended values of cyclic duration factor are 15, 25, 40 and 60 per cent.

- This type of duty is distinguished from intermittent periodic duty by the fact that *after a period of operation at constant load follows a period of no-load operation instead of rest.*

(*vii*) ***Continuous duty with starting and braking.*** It denotes a sequence of identical duty cycles each consisting of a period of starting, a period of operation at constant load and a period of electrical braking. There is *no period of rest.*

This type of duty is also indicated by the *number of cycles per hour and the factor of inertia.*

(*viii*) ***Continuous duty with periodic speed changes.*** It indicates a sequence of identical duty cycles each consisting of a period of operation at constant load corresponding to a specified speed of rotation, followed immediately by a period of operation at another load corresponding to a different speed of rotation, the operating periods being too short to attain thermal equilibrium during one duty cycle. There is *no rest period.*

For this duty the number of duty cycles per hour and a factor of inertia together with the load have to be indicated. In addition, the cyclic duration factor should be indicated for each speed.

5.16.3. Ambient Temperature and Ratings

- The operating temperature of a motor should *never* exceed the maximum permissible temperature because it will result in *deterioration and breakdown of insulation and will shorten the service life of motors.*
- For simplifying the heating calculations, it is general practice to base the motor power ratings on a standard value of temperature say 35°C. Accordingly, the power given on the name plate of a motor corresponds to the power which the motor is capable of delivering without overheating at an ambient temperature of 35°C. At an ambient temperature considerably *below 35°C,* the motor can safely deliver a *somewhat higher output* than its ratings because there is a greater difference between the ambient temperature and the maximum allowable temperature of insulation used in the motor. Similarly at ambient temperatures greater than 35°C, the load on the motor should be smaller than that indicated on its name plate unless of course special measures are taken to improve the cooling to bring down the temperature rise.
- *The duty cycle is closely related to temperature and is generally taken to include the environmental factors also.*

5.16.4. Overload Capacity of Motors

The rating of a machine can be determined from heating considerations. However the motor so selected should be *checked* for its overload capacity and starting torque. This is because the motor selected purely on the basis of heating may not be able to meet the mechanical requirements of the load to be driven by it. Table 5.3 lists the coefficients used to determine the permissible instantaneous overload torque of different electric motors.

Table 5.3. Motor instantaneous torque overload co-efficients

S.No.	*Type of motor*	*Torque overload co-efficient*
1.	D.C. motors	2 (for special types upto 3 or 4)
2.	Squirrel cage induction motors	2—2.5
3.	Double cage and deep bar squirrel cage motors	1.8—3
4.	Synchronous motors	2—2.5 (For special types upto 3—4)
5.	A.C. commutator motors	2—3

- In case of *D.C. machines* the maximum overload capacity is determined on the basis of *commutation,* while in case of synchronous and induction motors it is determined by maximum electromagnetic torque.

5.16.5. Determination of Power Rating of Electric Motors for Different Applications

1. Continuous duty and constant load :

- For *most of the applications,* the rating can be determined from the equation given as under :

$$P = \frac{TN}{975\,\eta} \text{ kW} \quad ...(5.112)$$

where, T = Load torque, kg-m,
N = Speed, r.p.m., and
η = Product of the efficiency of the driven equipment and that of transmitting device.

- In case of *linear motion*, the rating of the motor required is given by,

$$P = \frac{F \times v}{102\,\eta} \text{ kW} \quad ...(5.113)$$

where, F = Force caused by the load, kg, and
v = Velocity of motion of the load, m/s.

Eqn. (5.113) is directly applicable in case of *hoisting mechanisms.* It is also suitable for *lifts or elevators* ; but since a counterweight which balances the weight of the cage or car as well as one-half of the useful load will be always present, it should be modified as follows :

$$P = \frac{F \times v}{2 \times 102\,\eta} \text{ kW} \quad ...(5.114)$$

The velocity of normal passenger lift cabins vary from 0.5 to 1.5 m/s.

- In case of *pumps*, the rating can be determined from the following relation :

$$P = \frac{\rho Q H}{102\,\eta} \text{ kW} \quad ...(5.115)$$

where, ρ = Density of liquid pumped, kg/m^3,
Q = Delivery of pump, m^3/s, and
H = Gross head (static head + friction head), m.

η varies from 0.8 to 0.9 for *reciprocating pumps* and from 0.4 to 0.8 for *centrifugal pumps.*

- The rating of a *fan motor* is given by, $P = \frac{Qh}{102\,\eta}\ kW$ ...(5.116)

where, Q = Volume of air or any other gas, m^3/s, and
h = Pressure in mm of water or kg/m^2.

For small power fans, the efficiency η may be taken as 0.6 and for large power ones it may reach a value up to 0.8.

- The rating of a motor used in *metal shearing lathes* can be found from the relation.

$$P = \frac{F \times v}{102 \times 60 \times \eta} \text{ kW} \quad ...(5.117)$$

where, F = shearing force, kg,
v = Velocity of shearing, m/min, and
η = Mechanical efficiency of the lathe.

2. Motor rating for variable load :

The following are the commonly used methods for determination of motor rating for variable load drives :

(*i*) Method of average losses ; (*ii*) Equivalent current method ;

(*iii*) Equivalent torque method ;

(*iv*) Equivalent power method.

(*i*) **Method of averages losses (Q_{av}).** The method consists of finding average losses $Q_{av.}$ in the motor when it operates according to the given load diagram. These losses are then compared with Q_{nom}, the losses corresponding to the continuous duty of the machine when operated at its nominal rating. This method presupposes that when $Q_{av} = Q_{nom}$, the motor will operate without temperature rise going above the maximum permissible for the particular class of insulation.

In this case,
$$Q_m = \frac{Q_{av}}{A\lambda} = \frac{Q_{nom}}{A\lambda} \quad ...(5.118)$$

Fig. 5.104 shows a simplified load diagram for a certain drive.

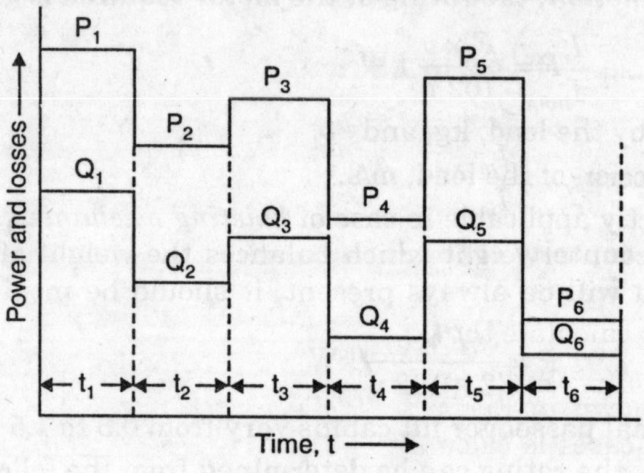

Fig. 5.104. A simplified power load diagram and loss diagram for variable load conditions.

The loss diagram (loss Q plotted versus time) of the electric motor is shown. The rating of the electric motor can be found from method of successive approximations. To start with a motor is selected in accordance with the required capacity calculated by multiplying average power load and a factor of safety of 1.1 to 1.3. The *losses of the motor are calculated for each portion of the load diagram by referring to the efficiency curve of the motor.* The average losses are given by

$$Q_{av} = \frac{Q_1 t_1 + Q_2 t_2 + Q_3 t_3 + ... + Q_n t_n}{t_1 + t_2 + t_3 + ... + t_n} \quad ...(5.119)$$

The average losses as found from eqn. (5.119) are compared with losses of selected motor at rated frequency. In case the two losses are equal or differ by a small amount the motor is selected. However, in case the losses differ considerably, another motor is selected and the calculations repeated till the motor having almost the same losses or the average losses is found. It should be checked that the motor selected has a sufficient overload capacity and starting torque.

- This method does *not* take into account the maximum temperature rise under variable load conditions. However, this method is *accurate and reliable for determining the average temperature rise of the motor during one work cycle.*

The *disadvantage* of the method of equal losses is that *it is tedious to work with and also many a times the efficiency curve is not readily available and the efficiency has to be calculated by means of empirical formulae which may not be accurate to work with.*

(*ii*) **Equivalent current method.** This method is based on the *assumption that the actual variable current may be replaced by an equivalent current I_{eq} which produces the same losses in the motor as the actual current.*

$$\therefore \quad I_{eq} = \sqrt{\frac{I_1^2 t_1 + I_2^2 t_2 + I_2^2 t_3 + ... + I_n^2 t_n}{t_1 + t_2 + t_3 + ... + t_n}} \quad ...(5.120)$$

The heating and cooling conditions in self ventilated machines depend upon its speed. At low speeds the cooling conditions are poorer than at normal speeds. Therefore, if the work cycle involves slow speed operation, it must be taken into consideration when using eqn. (5.120).

The equivalent current as found from eqn. (5.120) should be compared with the rated current of the motor selected and the conditions $I_{eq} \le I_{nom}$ should be met (I_{nom} is the rated current of the machine).

The machine selected should also be checked for its overload capacity.

— For D.C. motors ...... $\frac{I_{max}}{I_{nom}} \le 2$ to 2.5

— For induction motors ...... $\frac{I_{max}}{I_{nom}} \le 1.65$ to 2.75

In case the overload capacity of the motor selected is not sufficient it becomes necessary to select a motor of high power rating.

It may not be easy to calculate the equivalent current especially in cases where the current load diagram is irregular as shown in Fig. 5.105. The equivalent current in such cases is calculated from the following expression :

$$I_{eq} = \sqrt{\frac{1}{\sum_1^n t} \int^{\Sigma t} i^2 \, dt} \quad ...(5.121)$$

The value of the integral may be found with the help of integral.

An easier way is to break up the current load diagram into a series of straight line geometrical figures.

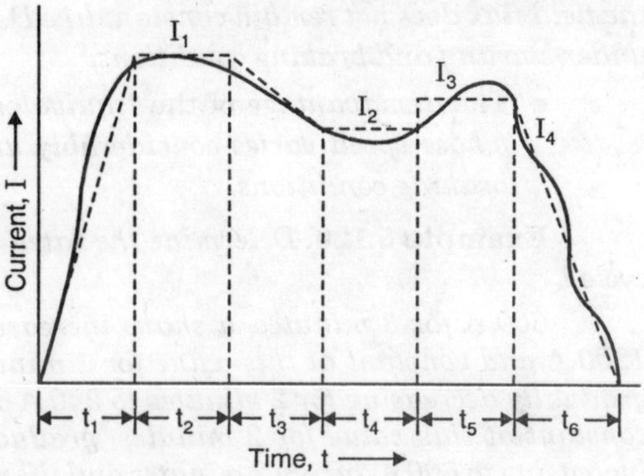

Fig. 5.105. Simplification of a regular current load diagram by short segments of straight lines.

For a *triangular shaped diagram :*

$$I_{eq} = \sqrt{\frac{I_1}{3}} \quad ...(15.122)$$

For a trapezoidal diagram :

$$I_{eq} = \sqrt{\frac{I_1^2 + I_1 I_2 + I_2^2}{3}} \quad ...(5.123)$$

The current values obtained by this method are sufficiently accurate for practical purposes.

(iii) and *(iv)* **Equivalent torque and equivalent power methods.** For the selection of suitable capacity of the motor it often becomes necessary to use torque or power load diagrams. The equivalent torque or power is found in the same manner as the equivalent current.

Assuming constant flux and constant power factor, the *torque is directly proportional to current* and, therefore, the equivalent torque is :

$$T_{eq} = \sqrt{\frac{T_1^2 t_1 + T_2^2 t_2 + T_3^2 t_3 + \ldots\ldots + T_n^2 t_n}{t_1 + t_2 + t_3 + \ldots\ldots + t_n}} \quad \ldots(5.124)$$

The equation for equivalent power follows directly from eqn. (5.124) as power is directly proportional to the torque. At constant speed or where the changes in speed are small, the equivalent power is given by :

$$P_{eq} = \sqrt{\frac{P_1^2 t_1 + P_2^2 t_2 + P_3^2 t_3 + \ldots\ldots + P_n^2 t_n}{t_1 + t_2 + t_3 + \ldots\ldots + t_n}} \quad \ldots(5.125)$$

- The "*equivalent current method*" is the *most accurate* out of all the above methods discussed above.

This method may be used to determine the motor capacity for all uses except where it is necessary to take into account the *changes in so called 'constant losses' i.e.* the iron and mechanical losses.

- The "*equivalent torque method*" *cannot be used for cases where equivalent current method can not be applied.*

The "*equivalent torque method*" *can not be used for selection of motor rating for cases in which the field flux does not remain constant like D.C. series motors and for squirrel-cage induction motors under starting and braking conditions.*

- The *disadvantage* of the "*equivalent power method*" is that it *cannot be used for motors whose speed varies considerably under load, especially when dealing with starting and braking conditions.*

Example 5.120. *Determine the rated current of an electrical machine for the following duty cycle :*

600 A for 3 minutes, a sharp increase to 1200 A and constant at this value for 1 minute, gradually decreasing for 2 minutes to 240 A and constant at this value for 2 minutes, gradually increasing to 600 A during 2 minutes and the repetition of the cycle.

Solution. The load diagram is plotted in Fig. 5.106.

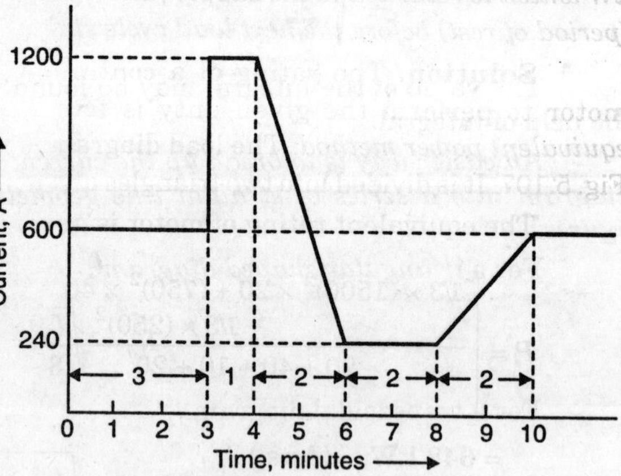

Fig. 5.106

The load diagram can be divided into rectangles and trapeziums (in this case) with the help of line segments. The equivalent current with sides I_1 and I_2 is :

$$I_{eq} = \left(\frac{I_1^2 + I_1 I_2 + I_2^2}{3} \right)^{1/2}$$

$\therefore$ The equivalent current for the entire cycle is :

$$I_{eq} = \left[\frac{(600)^2 \times 3 + 1200^2 \times 1 + (1/3)(1200^2 + 1200 \times 240 + 240^2) \times 2 + 240^2 \times 2 + (1/3)(240^2 + 240 \times 600 + 600)^2 \times 2}{(3 + 1 + 2 + 2 + 2)} \right]^{1/2}$$

$$= \left(\frac{1080000 + 1440000 + 1190400 + 115200 + 374400}{10} \right)^{1/2} = \textbf{648 A.} \quad \textbf{(Ans.)}$$

Example 5.121. *An induction motor has to perform the following duty, cycle, which is repeated indefinitely :*

75 kW for 10 minutes ; No load for 5 minutes ;

45 kW for 8 minutes ; No load for 4 minutes

Determine a suitable capacity of a continuously rated motor. Motors of standard (continuous) ratings of 45, 55, 75 kW are available. The ratio of maximum torque to nominal torque should be less than 1.8.

Solution. *Given :* $P_1 = 75$ kW, $t_1 = 10$ min. ; $P_2 = 0$, $t_2 = 5$ min ; $P_3 = 45$ kW,

$t_3 = 8$ min., $P_4 = 0$, $t_4 = 4$ min. ; $\frac{T_{max}}{T_{nom}} < 1.8$

The capacity of motor is given by :

$$P = \left[\frac{P_1^2 t_1 + P_2^2 t_2 + P_3^2 t_3 + P_4^2 t_4}{t_1 + t_2 + t_3 + t_4} \right]^{1/2}$$

$$= \left(\frac{75^2 \times 10 + 0^2 \times 5 + 45^2 \times 8 + 0^2 \times 4}{10 + 5 + 8 + 4} \right)^{1/2} = 51.8 \text{ kW.}$$

Example 5.122. *A motor driving a colliery winder has to deliver a load rising uniformly from zero to maximum of 1500 kW in 20 s during the accelerating period, 750 kW for 40 s during the full speed period and during the decelerating period of 10 s, when the regenerative braking takes place the kW which is returned to the supply falls from an initial value of 250 to zero. The interval for decking (period of rest) before the next load cycle starts is 20 s. Estimate a suitable rating for the motor.*

Solution. The rating of a continuously rated motor to perform the given duty is found by using *equivalent power method.* The load diagram is shown in Fig. 5.107. It is divided into two triangles and a rectangle.

The equivalent rating of motor is given by :

$$P = \left[\frac{1/3 \times (1500)^2 \times 20 + (750)^2 \times 40 + 1/3 \times (250)^2 \times 10}{20 + 40 + 10 + 20} \right]^{1/2}$$

$= \textbf{648 kW.} \quad \textbf{(Ans.)}$

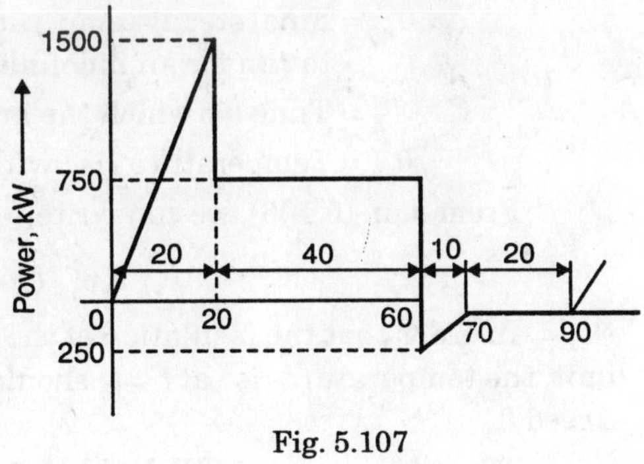

Fig. 5.107

Examples 5.123. *A motor works on 2 minutes load cycle constituted as follows :*

0 to 15 sec	......	*Load rising from 0 to 1260 H.P. ;*
15 to 85 sec	......	*Constant load of 720 H.P. ;*
85 to 95 sec	......	*Regenerative braking with the H.P. returned falling uniformly from 240 to 0 H.P. ;*
95 to 120 sec	......	*Motor is at rest.*

Determine the continuous rating of the motor that would be suitable for the load cycle. Assume the rating to depend upon :

(*i*) *The r.m.s. value of loading ;*

(*ii*) *The average value of loading.*

Solution. The loading is shown in Fig. 5.108.

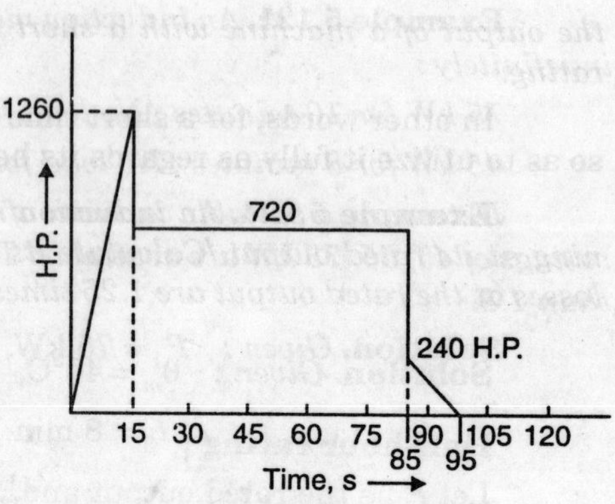

Fig. 5.108

(*i*) **The r.m.s. value of loading :**

$$(\text{H.P. rating})^2 \times 120$$

$$= \int_0^{15} \left(\frac{1260}{15} t\right)^2 dt + (720^2 \times 70) + \int_0^{10} \left(\frac{240}{10} t\right)^2 dt$$

$$= \left[7056 \frac{t^3}{3}\right]_0^{15} + 36288000 + \left[576 \frac{t^3}{3}\right]_0^{10}$$

$$= 7938000 + 36288000 + 192000 = 44418000$$

$\therefore$ H.P. = **608.4. (Ans.)**

(*ii*) **The average value of loading :**

$$\text{H.P. rating} \times 120 = \frac{1}{2} \times 1260 \times 15 + 720 \times 70 + \left(\frac{1}{2} \times 240 \times 10\right)$$

$$= 61050$$

$\therefore$ H.P. $\simeq$ **509. (Ans.)**

5.16.6. Temperature Rise with Short Time Ratings

Let, θ_m = **Final steady temperature rise with continuous rating. This may be assumed as maximum allowable temperature rise, θ_{per},**

T_h = **Heating time constant,**

θ'_m = **Final steady temperature rise that would be attained if the machine were allowed to run for an indefinite time at its short time rating,**

T_h = **Time for which the machine is run at its short time rating, and**

θ_{sh} = **Temperature rise with short time rating $t = t_{sh}$.**

From eqn. (5.106), we can write,

$$\theta_{sh} = \theta'_m \left(1 - e^{-t_h/T_h}\right) \quad \text{...(5.126)}$$

In order that the insulation of the machine is not operated beyond the specified temperature limit, the temperature rise at $t = t_h$ should not exceed the final steady temperature or θ_{sh} should not exceed θ_m.

Thus the maximum allowable temperature rise during short time rating is, $\theta_{sh} = \theta_m$

Inserting this value in eqn. (5.126), we get :

$$\theta_m = \theta'_m \left(1 - e^{-t_h/T_h}\right) \quad \text{...(5.127)}$$

or,
$$\frac{\theta'_m}{\theta_m} = \frac{1}{1 - e^{-t_h/T_h}}$$

Obviously θ'_m is $\dfrac{1}{1 - e^{-t_h/T_h}}$ times θ_m

Thus the permissible losses in the machine during short time rating can also be as many times greater. Since $\frac{1}{1-e^{-t_h/T_h}}$ has a value greater than unity therefore *with the given dimensions, the output of a machine with a short time rating is larger than that of machine with continuous rating.*

In other words, for a short time load t_h, a motor with a lower power rating should be selected so as to utilize it fully as regards its heating.

Example 5.124. *An induction motor has a final steady temperature rise of 40°C when running at its rated output. Calculate its half-hour rating for the same temperature rise if the copper losses at the rated output are 1.25 times its constant losses. The heating time constant is 90 minutes.*

Solution. *Given* : $\theta_m = 40°\text{C}$, $P_c = 1.25\,P_i$; $T_h = 90$ min. = 1.5 hour ; $t_h = \frac{1}{2}$ hour.

Half-hour rating :

Let P be the rated output and xP be the half hour rating. Also let P_c be the copper losses at full load and P_i be the other constant losses.

$\therefore$ Total losses at full load $= P_c + P_i$

and, total losses with *short time rating* $= x^2P_c + P_i$

Specific heat dissipation and surface remaining the same the temperature rise is proportional to the losses. Suppose θ'_m is the final steady temperature rise with short time rating, then,

$$\frac{\theta'_m}{\theta_m} = \frac{x^2P_c + P_i}{P_c + P_i} = \frac{1.25x^2 + 1}{2.25} \quad \text{as } P_c = 1.25P_i$$

The temperature rise after $\frac{1}{2}$ hour of operation at short time rating should not exceed $\theta_m = 40°\text{C}$.

Now, using the relation :

$$\theta_{sh} = \theta'_m\,(1 - e^{-t_h/T_h}) \qquad \text{...[Eqn. (5.126)]}$$

$$40 = \theta'_m\,[1 - e^{-1/(2\times 3/2)}]$$

$$= \theta'_m\,(1 - 0.716) = 0.284\,\theta'_m$$

or, $$\theta'_m = 141.1°\text{C}$$

$\therefore$ $$\frac{1.25\,x^2 + 1}{2.25} = \frac{141.4}{40}$$

or, $$x = 2.36$$

Thus half-hour rating of machine is *2.36 times its continuous rating.* **(Ans.)**

Example 5.125. *If one hour rating of a machine is $\sqrt{3}$ times the continuous rating and if steady temperature rise for one-hour rating is twice that on normal load, find the ratio of iron to copper loss at full load.*

Solution. Let one-hour rating of transformer be x times its continuous rating.

Losses with continuous rating $= P_c + P_i$

Loss with short time rating $= x^2P_c + P_i$

Let θ_m and θ'_m be final steady temperature rise with continuous and one-hour rating respectively.

$$\therefore \quad \frac{\theta'_m}{\theta_m} = \frac{x^2 P_c + P_i}{P_c + P_i}$$

Now, $\frac{\theta'_m}{\theta_m} = 2$ and $x = \sqrt{3}$ ...[*Given*]

$$\therefore \quad 2 = \frac{(\sqrt{3})^2 P_c + P_i}{P_c + P_i} = \frac{3P_c + P_i}{P_c + P_i}$$

or, $2P_c + 2P_i = 3P_c + P_i$

$\therefore \quad P_c = P_i$

Thus *iron losses are equal to full-load copper losses for continuous rating.* **(Ans.)**

5.16.7. Load Equalisation

In several industrial drives (*e.g.* electric hammers, presses, rolling mills, reciprocating pumps, planing machines etc.) the load fluctuates between wide limits over a span of few seconds. It is very essential to smooth out the fluctuating load, otherwise during intervals of peak load it will draw a heavy current from the supply *either producing large voltage drop in the system or requiring cables and wires of heavy section. The process of smoothing out the fluctuating load is known as* **load equalisation.**

The process of load equalisation involves the storage of energy during light load periods which can be given out during the peak load period, so that the demand for the system is more less constant. Load equalisation is often achieved by means of a ***Flywheel***, which is mounted on the motor shaft, if the speed of the motor is not to be reversed. In case of reversing rolling mills, the flywheel is mounted on the motor-generator set feeding the driving motor. In order that the flywheel operates effectively, the *driving motor should have a drooping speed characteristic.* During intervals of heavy load, speed of the motor decreases. This enables the flywheel to release a portion of the stored kinetic energy, which together with the energy drawn from supply, will meet the demand of load. During off peak load period, the motor draws more energy than required by the load. The surplus energy taken is again stored as kinetic energy in the flywheel, whose speed, then increases. The load torque required and motor torque developed as well as speed variations with time are shown in Fig. 5.109.

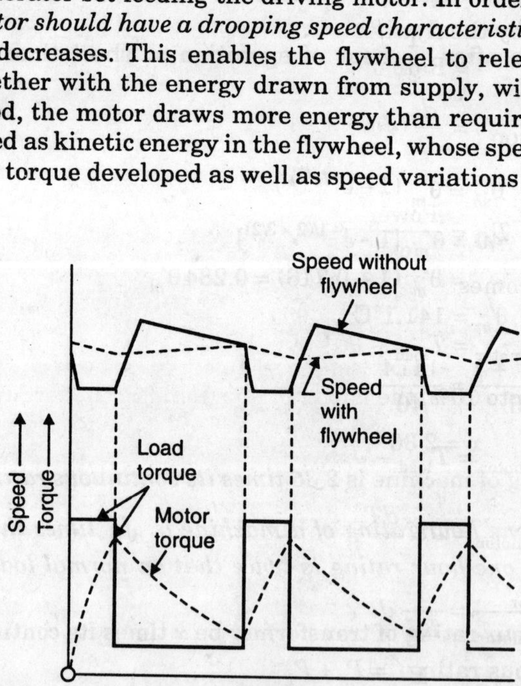

Fig. 5.109. Variations of speed, load torque and motor torque against time.

Flywheel calculations :

Let, T_{load} = Load torque (assumed constant during the period), Nm,

T_{fw} = Flywheel torque, Nm,

T_0 = No-load torque, Nm,

T_{motor} = Motor torque at any instant, Nm,

ω_0 = Motor speed on no-load, rad/s,

ω = Motor speed at any instant, rad/s,

$s = (\omega_0 - \omega)$ = Motor slip,

J = Moment of inertia of flywheel, kg-m², and

t = Time, *s*.

Case 1. *Load increasing (flywheel decelerating) :*

During the period the *flywheel declerates and gives up a part of stored energy in it.*

The torque required to be supplied by the motor,

$$T_{motor} = T_{load} - T_{fw} \qquad ...(5.128)$$

Energy given out by the flywheel when its speed is reduced from

$$\omega_0 \text{ to } \omega = \frac{1}{2} J(\omega_0{}^2 - \omega^2) = J\left(\frac{\omega_0 + \omega}{2}\right)(\omega_0 - \omega)$$

But, $\frac{\omega_0 + \omega}{2} = \omega(\text{mean speed})$ and $(\omega_0 - \omega_1) = s$ (slip)

$\therefore$ Energy given out $= J\omega s$

The power given out by the flywheel

= Rate of energy given up

$$= \frac{d}{dt}(J\omega s) = J\omega \frac{ds}{dt}$$

Flywheel torque,
$$T_{fw} = \frac{\text{Power}}{\omega} = \frac{J\omega \frac{ds}{dt}}{\omega} = J\frac{ds}{dt} \qquad ...(5.129)$$

Eqn. (5.128) thus becomes

$$T_{motor} = T_{load} - J\frac{ds}{dt}$$

As for value of slip upto 10%, the slip is proportional to the torque, *i.e.*, $s = KT_{motor}$

$$\therefore \qquad T_{motor} = T_{load} - JK\frac{dT_{motor}}{dt}$$

or,
$$T_{load} - T_{motor} = JK\frac{dT_{motor}}{dt}$$

or,
$$\frac{dT_{motor}}{T_{load} - T_{motor}} = \frac{1}{KJ}dt$$

or,
$$-\log_e(T_{load} - T_{motor}) = \frac{1}{KJ} + C_1$$

From initial conditions, when $t = 0$; $T_{motor} = T_0$

$$C_1 = -\log_e(T_{load} - T_0)$$

$$\therefore \quad -\log_e (T_{load} - T_{motor}) = \frac{1}{KJ} t - \log_e (T_{load} - T_0)$$

or, $$-\frac{1}{KJ} t = \log_e \left(\frac{T_{load} - T_{motor}}{T_{load} - T_0} \right)$$

or, $$e^{-t/KJ} = \frac{T_{load} - T_{motor}}{T_{load} - T_0}$$

or, $$T_{motor} = T_{load} - (T_{load} - T_0)\, e^{-t/KJ} \qquad \text{...(5.130)}$$

Case 2. *Load decreasing (Flywheel accelerating)* :

When *load decreases*, the *motor accelerates the flywheel to its normal speed, slip is therefore decreased* and $\frac{ds}{dt}$ is –ve.

$$T_{motor} = T_0 + T_{fw}$$

or, $$T_{motor} = T_0 - J \frac{ds}{dt} \qquad \text{... since } \frac{ds}{dt} \text{ is –ve}$$

By solving the above expression, we get

$$T_{motor} = T_0 + (T'_{motor} - T_0)\, e^{-t/KJ} \qquad \text{...(5.131)}$$

Example 5.126. *A 440 V, 50 H.P., 600 r.p.m. D.C. shunt motor takes 80 A at full load. The current during the starting period varies between the limits 1.1 to 1.5 times the full load current. The moment of inertia of the rotating system is 220 kg-m².*

Calculate the time taken to accelerate the motor to the rated speed against full load torque.

(Gorakhpur University)

Solution. *Given* : Supply voltage = 440 V ; *Power* = 50 H.P. ; N = 600 r.p.m. ;

I_{fl} = 80 A ; I = 220 kg-m².

Time taken to accelerate the motor, t :

Full load torque, $$T_{fl} = \frac{\text{H.P.} \times 735.5}{(2\pi N/60)}$$

$$= \frac{50 \times 735.5}{(2\pi \times 600/60)} = 585.3 \text{ Nm}$$

The motor is required to accelerate against full-load. The starting current varies from 1.1 to 1.5 times full load current, therefore starting torque fluctuates between 1.1 and 1.5 times the full-load torque.

$$\therefore \quad \text{Average starting torque} = \frac{1.1 + 1.5}{2} \times \text{full-load torque}$$

$$= 1.3 \times 585.3 = 760.9 \text{ Nm}$$

Torque available for acceleration,

$$T_{motor} = \text{Average starting torque} - \text{full load torque}$$
$$= 760.9 - 585.3 = 175.6 \text{ Nm}$$

Angular acceleration, $$\alpha = \frac{T_{motor}}{J} = \frac{175.6}{220} = 0.798 \text{ rad/s}^2$$

Time taken to accelerate the motor to the rated speed,

$$t = \frac{\omega}{\alpha} = \frac{(2\pi \times 600/60)}{0.798} = \mathbf{78.7\ s.}\quad \textbf{(Ans.)}$$

Example 5.127. *A 60 H.P., 500 r.p.m shunt motor is used for shop. The stored energy of the machinery averages 4090 Nm / H.P. Assuming the load torque during starting to be equal to the full-load torque, and starting current to be twice the full-load current, estimate the time taken to start the motor.* **(Panjab University)**

Solution. *Given* : Power of the motor = 60 H.P. ; $N = 500$ r.p.m. ;

Average stored energy of the machinery = 4090 Nm/H.P. ; $T_{st} = 2 \times T_{fl}$

Time taken to start the motor, t :

Full-load torque, $$T_{fl} = \frac{60 \times 735.5}{(2\pi \times 500/60)} = 842.8 \text{ Nm}$$

Torque developed during starting $$= \frac{\text{Starting current}}{\text{Full - load current}} \times \text{Full-load torque}$$
$$= 2 \times 842.8 = 1685.6 \text{ Nm}$$

Torque available for acceleration,

$$T_{motor} = 1685.6 - 842.8 = 842.8 \text{ Nm}$$

Stored energy of the machinery = 4090 × 60 = 245400 Nm

Since stored energy $$= \frac{1}{2} J\omega^2 = 245400 = \frac{1}{2} \times J \times \left(\frac{2\pi \times 500}{60}\right)^2 = 245400 \text{ (calculated above)}$$

$$\therefore \quad J = 179 \text{ kg-m}^2$$

Angular acceleration, $$\alpha = \frac{T_{motor}}{J} = \frac{842.8}{179} = 4.708 \text{ rad/s}^2$$

Starting period, $$t = \frac{\omega}{\alpha} = \frac{(2\pi \times 500/60)}{4.708} = \mathbf{11.12\ s.}\quad \textbf{(Ans.)}$$

Example 5.128. *A motor fitted with a flywheel supplies a load of torque 900 Nm for 2 seconds. During no-load period the flywheel regains its original speed. The motor torque is required to be limited to 450 Nm. The no-load speed of the motor is 500 r.p.m. and its full-load slip is 10 percent.*

Determine the moment of inertia of the flywheel. **(Panjab University)**

Solution. *Given* : $T_{load} = 900$ Nm (for 2 seconds) ; $T_0 = 0$; $T_{motor} = 450$ Nm,

$N = 500$ r.p.m. ; $s_f = 10\%$.

Moment of inertia of the flywheel, J :

Slip speed, $$s = 0.1 \times 500 = 50 \text{ r.p.m.} = \frac{50 \times 2\pi}{60} = 5.236 \text{ rad/s}$$

$$K = \frac{s}{T_{motor}} = \frac{5.236}{450} = 0.0116$$

Now, $$T_{motor} = T_{load} - (T_{load} - T_0)\, e^{-t/KJ}$$
$$= T_{load}(1 - e^{-t/KJ}) \qquad \ldots(\because\ T_0 = 0)$$

or, $$450 = 900\,[1 - e^{-2/(0.0116 \times J)}]$$

or, $$0.5 = 1 - e^{(-172.4/J)}$$

or, $$e^{-172.4/J} = 0.5$$

or, $$e^{172.4/J} = 2$$

$$\therefore \quad J = \mathbf{248.7\ kg\text{-}m^2.}\quad \textbf{(Ans.)}$$

Example 5.129. *A 25 H.P., 3-phase, 10-pole, 50 Hz induction motor provided with a flywheel has to supply a load torque of 800 Nm for 10 seconds followed by a no-load period during which the flywheel regains the full speed. The full-load slip of the motor is 4 per cent and the torque-speed curve may be assumed linear over the working range. Find the moment of inertia of the flywheel if the motor torque is not to exceed twice the full-load torque.* **(A.M.I.E.)**

Solution. *Given* : Power of the motor = 25 H.P. ; $p = 10$; $f = 50$ Hz ;

$T_{load} = 800$ Nm for 10 seconds ; slip, $s_f = 4\%$; $T_{motor} = 2T_{fl}$.

Moment of inertia of flywheel, J :

Synchronous speed, $N_s = \dfrac{120f}{p} = \dfrac{120 \times 50}{10} = 600$ r.p.m.

Full load speed, $N_f = N_s(1 - s_f) = 600(1 - 0.04) = 576$ r.p.m.

Full load torque, $T_{fl} = \dfrac{25 \times 735.5}{(2\pi \times 576/60)} = 304.8$ Nm

$$T_{motor} = 2T_{fl} = 2 \times 304.8 = 609.6 \text{ Nm}$$

Slip speed, $s = 0.04 \times 600 = 24 \text{ r.p.m} = \dfrac{24 \times 2\pi}{60} = 2.5133$ rad/s

As the slip at full load is proportional to full-load torque,

$$\therefore \quad K = \frac{s}{T_{ft}} = \frac{2.5133}{304.8} = 0.008246.$$

Also, $T_{motor} = T_{load} - (T_{load} - T_0)\, e^{-t/KJ}$ ...[Eqn. (5.130)]

$= T_{load}\,[1 - e^{-t/KJ}]$ ...($\because$ $T_0 = 0$)

or, $609.6 = 800\,[1 - e^{-10/(0.008246 \times J)}]$

or, $0.762 = 1 - e^{(-1212.7/J)}$

or, $e^{-(1212.7/J)} = 0.238$

or, $e^{(1212.7/J)} = 4.2$

$\therefore$ $J = 845$ **kg-m². (Ans.)**

Example 5.130. *The following data refer to a rolling mill, induction motor equipped with a flywheel :*

Power of the motor	*...500 H.P.*
No-load speed	*...40 r.p.m.*
Slip at full-load (torque)	*...12 per cent*
Load torque during actual rolling	*...407115 Nm*
Duration of each rolling period	*...10 sec.*

Determine inertia of the flywheel required in the above case to limit motor torque to twice its full-load torque. Neglect no-load losses and assume that the rolling mill torque falls to zero between each rolling period. Assume motor slip proportional to full-load torque. **(Nagpur University)**

Solution. *Given* : Power of the motor, $P = 500 \times 735.5 = 367750$ W ; $N_0 = 40$ r.p.m.

i.e., $\omega_0 = \dfrac{2\pi \times 40}{60} = 4.189$ rad./s ; $s = 12\%$, $T_{load} = 407115$ Nm ; Duration of each rolling period,

$t = 10$ s, $T_{motor} = 2 \times T_{fl}$

Inertia of flywheel J :

Now, full load torque, $T_{fl} = \dfrac{P}{\omega} = \dfrac{P}{(1-s)\,\omega_0} = \dfrac{367750}{(1-0.12)\times 4.189} = 99761$ Nm ;

$\therefore \quad T_{motor} = 2 \times T_{fl} = 2 \times 99761 = 199522$ Nm

$$s = (\omega_0 - \omega) = \frac{2\pi}{60}(N_0 - N) = \frac{2\pi}{60}(40 - 0.88 \times 40) = 0.503 \text{ rad/s}$$

Also, $\quad s = KT_{fl}$

$$0.503 = K \times 99761 \quad \therefore \quad K = 5.04 \times 10^{-6}$$

We know that, $\quad T_{motor} = T_{load} - (T_{load} - T_0)\, e^{-t/KJ}$ ...(Eqn. 5.130)

Since $\quad T_0 = 0,$

$\therefore \quad T_{motor} = T_{load}\,(1 - e^{-t/KJ})$

$$199522 = 407115\,[1 - e^{-10/(5.04 \times 10^{-6} \times J)}]$$

or, $\quad 0.49 = 1 - e^{-10^7/(5.04J)}$

or, $\quad e^{-10^7/(5.04J)} = 0.51$

or, $\quad e^{10^7/(5.04J)} = \dfrac{1}{0.51} = 1.96$

$\therefore \quad \dfrac{10^7}{5.04\,J} = \log_e 1.96 = 0.673$

or, $\quad \mathbf{J} = \dfrac{10^7}{5.04 \times 0.673} = \mathbf{2.948 \times 10^6}$ **kg-m². (Ans.)**

Example 5.131. *A 6-pole, 50 Hz, 3-phase wound rotor induction motor has a flywheel coupled to its shaft. The total moment of inertia is 900 kg-m². Load torque is 900 Nm for 10 seconds followed by a no-load period which is long enough for the motor to reach its no-load speed. Motor has a slip of 5 percent at a torque of 450 Nm. Calculate :*

(*i*) *Maximum torque developed by the motor.*

(*ii*) *Speed at the end of deceleration period.* **(Panjab University)**

Solution. *Given :* $p = 6$; $f = 50$ Hz ; $J = 900$ kg-m² ; $T_{load} = 900$ Nm

Slip = 5% at torque of 450 Nm (T_{motor}).

(*i*) **Maximum torque developed by the motor :**

$$N_s = \frac{120f}{p} = \frac{120 \times 50}{6} = 1000 \text{ r.p.m. ;}$$

$$s = \frac{2\pi}{60}(1000 \times 0.05) = 5.236 \text{ rad/s}$$

Now, $\quad s = KT_{motor}$

or, $\quad 5.236 = K \times 450 \quad \therefore \quad K = 0.0116$

$(T_{motor})_{max} = T_{load}\,(1 - e^{-t/KJ})$ ...($T_0 = 0$)

$= 900\,[1 - e^{-10/(0.0116 \times 900)}] = \mathbf{554.6}$ **Nm. (Ans.)**

(*ii*) **Speed at the end of deceleration period, N**

$$s = KT_{fl}$$

$$\frac{2\pi}{60}(1000 - N) = 0.0116 \times 554.6$$

$$\therefore \qquad \mathbf{N = 938.6\ r.p.m.} \quad \textbf{(Ans.)}$$

Example 5.132. *A 6-pole, 50 Hz induction motor has a flywheel of 1200 kg-m² as moment of inertia. Load torque is 980 Nm for 10 seconds. No-load period is long enough for the flywheel to regain its full speed. Motor has a slip of 6 percent at a torque of 490 Nm. Calculate :*

(i) Maximum torque exerted by the motor.

(ii) Speed at the end of deceleration period. **(Nagpur University)**

Solution. *Given :* $p = 6, f = 50$ Hz, $J = 1200$ kg-m², $T_{load} = 980$ Nm, $t = 10$ s ; $s = 6\%$; $T = 490$ Nm.

(*i*) **Maximum torque exerted by the motor, T_m :**

$$T_{motor} = T_{load} - (T_{load} - T_0)\, e^{-t/KJ0} \qquad \text{...[Eqn. (5.130)]}$$

Assuming, $T_0 = 0$, we have

$$T_{motor} = T_{load}\,(1 - e^{-t/KJ})$$

$$N_s = \frac{120f}{p} = \frac{120 \times 50}{6} = 1000 \text{ r.p.m.} = N_0$$

$$s = (\omega_0 - \omega) = \frac{2\pi}{60}(N_0 - N) = \frac{2\pi}{60}(1000 - 940) = 6.283 \text{ rad/s.}$$

$$[\because \; N = N_s(1 - s) = 1000(1 - 0.06) = 940 \text{ r.p.m.}]$$

Now,

$$K = \frac{s}{T} = \frac{6.283}{490} = 0.0128$$

$$\therefore \qquad T_{motor} = 980\,[1 - e^{-10/(0.0128 J)}]$$

$$= 980\,[1 - e^{-10/(0.0128 \times 1200)}]$$

$$= 980(1 - e^{-0.651}) = 468.9 \text{ Nm}$$

i.e., $$T_{motor} = \mathbf{468.9\ Nm.} \quad \textbf{(Ans.)}$$

(*ii*) **Speed at the end of deceleration period :**

Slip speed,

$$s' = KT_{motor}$$

$$= 0.0128 \times 468.9 = 6.0 \text{ rad/s}$$

$$= \frac{6 \times 60}{2\pi} = 57.3 \text{ r.p.m.}$$

$\therefore$ *Actual speed* $= 1000 - 57.3 = \mathbf{942.7\ r.p.m.}$ **(Ans.)**

Example 5.133. *A 3-phase, 45 kW, 6-pole, 960 r.p.m. induction motor has a constant load torque of 270 Nm and at wide intervals additional torque of 1350 Nm for 10 seconds. Calculate :*

(i) The moment of inertia of the flywheel used for load equalization, if motor torque is not to exceed twice the rated torque.

(ii) Time taken after removal of additional load, before the motor torque becomes 630 Nm.

Solution. *Given :* $P = 45$ kW ; $p = 6$; $N = 960$ r.p.m. ; $T_{load} = 270 + 1350 = 1620$ Nm ; $T_{motor} = 2T_{fl}$.

(*i*) **Moment of inertia of the flywheel, J :**

$$P = T \times \omega \qquad \therefore \quad T = \frac{P}{\omega}$$

$$\therefore \quad T_{fl} = \frac{45 \times 10^3}{\left(\frac{2\pi \times 960}{60}\right)} = 447.6 \text{ Nm}$$

and, $$T_{motor} = 2T_{fl} = 2 \times 447.6 = 895.2 \text{ Nm}$$

$$N_s = \frac{120f}{p} = \frac{120 \times 50}{6} = 1000 \text{ r.p.m.}$$

$\therefore$ Full load slip = 1000 – 960 = 40 r.p.m. = 4%

Also, $$s = KT_{fl}$$

$$\frac{2\pi \times 40}{60} = K \times 447.6 \quad \therefore \quad K = 0.00936$$

$$T_{motor} = T_{load} - (T_{load} - T_0)\, e^{-t/KJ}$$

$$895.2 = 1620 - (1620 - 270)\, e^{-10/(0.00936J)}$$

$$= 1620 - 1350 \times e^{(-1068.4/J)}$$

or, $$e^{-(1068.4/J)} = 0.537 \quad \text{or} \quad \frac{1068.4}{J} = \log_e\left(\frac{1}{0.537}\right) = 0.622$$

$$\therefore \quad J = \mathbf{1717.7\ kg\text{-}m^2}. \quad \textbf{(Ans.)}$$

(*ii*) **Time taken after removal of additional load :**

$$T_{motor} = T_0 + (T'_{motor} - T_0)\, e^{-t/KJ}$$

$$630 = 270 + (895.2 - 270)\, e^{-t/(0.00936 \times 1717.7)}$$

$$(630 - 270) = 625.2 \times e^{-t/16.08}$$

or, $$e^{t/16.08} = 1.737$$

$$\therefore \quad t = 16.08 \log_e (1.737) = \mathbf{8.879\ s.} \quad \textbf{(Ans.)}$$

Example 5.134. *An induction motor equipped with a flywheel is driving a rolling mill which requires a load torque of 1900 Nm for 10 sec. followed by 280 Nm for 30 sec. This cycle being repeated indefinitely. The synchronous speed of motor is 750 r.p.m. and it has a slip of 10 percent when delivering 1400 Nm torque. The total moment of inertia of the flywheel and other rotating parts is 2100 kg-m². Draw the curve showing the torque exerted by the motor and the speed for five complete cycles, assuming that initial torque is zero.* **(Nagpur University, Summer, 2000)**

Solution. *Given* : T_{load} = 1900 Nm for 10 s ; T_{load} = 280 Nm for 30 s ; N_s = 750 *r.p.m* ; *slip* = 10% at 1400 Nm torque (T_{motor}) ; J = 2100 kg-m².

$$s = 0.1 \times 750 = 75 \text{ r.p.m.} = \frac{75 \times 2\pi}{60} = 7.85 \text{ rad/s}$$

Also, $$s = KT_{motor}$$

$$\therefore \quad K = \frac{s}{T_{motor}} = \frac{7.85}{1400} = 0.0056$$

(*i*) **During Ist cycle :**

(*a*) *Flywheel decelerating* (On-load period)

$$T_{motor} = T_{load} - (T_{load} - T_0)\, e^{-t/KJ}$$

$$= 1900 - (1900 - 0)\, e^{-10/(0.0056 \times 2100)}$$

$$= 1088 \text{ Nm} \qquad \text{...assuming } T_0 = 0$$

$$\text{Slip} = 0.0056 \times 1088 = 6.09 \text{ rad/s} = \frac{6.09 \times 60}{2\pi} = 58 \text{ r.p.m}$$

$$\therefore \quad \text{Speed} = 750 - 58 = 692 \text{ r.p.m.}$$

(*b*) *Flywheel accelerating* (*Off-load period*) :

$$T_{motor} = T_0 + (T'_{motor} - T_0)\, e^{-t/KJ}$$

T_0 = No load torque = 280 Nm

T'_{motor} = 1088 Nm (T_{motor} at the beginning of the period *i.e.*, the motor torque at the instant when the load is removed)

After 30 sec., $T_{motor} = 280 + (1088 - 280)\, e^{-30/(0.0056 \times 2100)}$

$= 280 + (1088 - 280) \times 0.078 = 343$ Nm

∴ Slip at this $T_{motor} = 0.0056 \times 343 = 1.92$ rad/s = 18.33 r.p.m.

∴ Speed $= 750 - 18.33 \simeq 731$ r.p.m.

(*ii*) **During 2nd cycle :**

(*a*) *Flywheel decelerating :*

T_0 = 343 Nm

$$T_{motor} = 1900 - (1900 - 343)\, e^{-10/(.0056 \times 2100)} = 1235 \text{ Nm}$$

Slip at this $T_{motor} = 0.0056 \times 1235 = 6.92 \text{ rad/s} = \dfrac{6.92 \times 60}{2\pi} = 66$ r.p.m.

∴ Speed $= 750 - 66 = 684$ r.p.m.

(*b*) *Flywheel accelerating* (*Off-load period*) :

$$T_{motor} = 280 + (1235 - 280) \times e^{-30/(0.0056 \times 2100)} = 354.5 \text{ Nm}$$

∴ Slip at this $T_{motor} = 0.0056 \times 354.5 = 1.99 \text{ rad/s} = \dfrac{1.99 \times 60}{2\pi} = 19$ r.p.m.

Speed $= 750 - 19 = 731$ r.p.m.

(*iii*) **During 3rd cycle**

(*a*) *Flywheel decelerating* :

T_0 = 354.6 Nm

$$T_{motor} = 1900 - (1900 - 354.6)\, e^{-10/(0.0056 \times 2100)} = 1239 \text{ Nm}$$

∴ Slip at this $T_{motor} = 0.0056 \times 1239 = 6.94 \text{ rad/s} = \dfrac{6.94 \times 60}{2\pi} = 66$ r.p.m.

Speed $= 750 - 66 = 684$ r.p.m.

(*b*) *Flywheel accelerating* (*Off-load period*) :

$$T_{motor} = 280 + (1239 - 280)\, e^{-30/(0.0056 \times 2100)} = 354.8 \text{ Nm}$$

Slip at this $T_{motor} = 0.0056 \times 354.8 = 1.99 \text{ rad/s} = \dfrac{1.99 \times 64}{2\pi} = 19$ r.p.m.

Speed $= 750 - 19 = 731$ r.p.m.

Initial condition at the beginning of the 3rd peak load are thus practically the same as that at the beginning of the 2nd. Therefore motor torque in this and all succeeding load cycles will follow a similar curve to that in second period (Fig. 5.110).

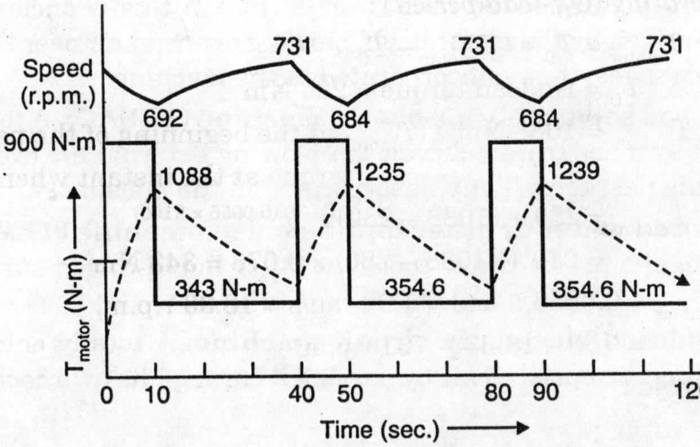

Fig. 5.110

5.17. ENCLOSURES FOR ROTATING ELECTRICAL MACHINES

The problem of ventilation in rotating electrical machines is closely linked with the types of enclosures used. The various types of enclosures used are :

1. **Open machine.** One in which there is no restriction to ventilation other than that necessitated by good mechanical construction.

2. **Open pedestal machine (OP).** An open machine which has pedestal bearing supported independently of the machine frame.

3. **Open end-bracket machine (OEB).** An open machine having end brackets of which the bearings form an integral part :

4. **Protected machine (P).** A machine in which the internal rotating parts and live parts are protected mechanically from accidental or inadvertent contact, while ventilation is not materially impeded.

5. **Screen-protected machine (SP).** A protected machine in which the ventilating openings are not less than 64.5 mm^2 in area. Such protection may be provided by screens of wire mesh, expanded metal, perforated metal or other suitable covers. The use of openings smaller than 64.5 mm^2 is not recognized, as such openings are liable to become closed in service.

6. **Drip proof machine (DP).** A protected machine in which the opening for ventilation are so protected as to exclude vertically falling water or dirt.

7. **Splash-proof machine (SPLP).** A protected machine in which the ventilating openings are so constructed that drops of liquid of solid particles falling on or reaching any part of the machine at any angle between the vertically downward direction and 100° from that direction cannot enter the machine, whether the machine is running or at rest, by splashing, or otherwise, either directly or by striking and running along a surface.

8. **Hose-proof machine (HSP).** A protected machine so enclosed as to exclude water whether the machine is running or at rest, when washed by a hose having 9.5 mm diameter nozzle with a maximum pressure of 3.5 bar for a period not exceeding 30 seconds, from a minimum distance of 1.8 metres.

9. **Pipe-ventilated or Duct-ventilated machine.** A machine in which there is a continuous supply of fresh ventilating air, the frame being so arranged that the ventilating air may be conveyed to and/or from the machine through pipes or ducts attached to the enclosing case.

10. **Totally enclosed machine (TEM).** A machine so constructed that the enclosed air has no connection with the external air but is not necessarily 'air tight'.

11. **Totally enclosed fan-cooled machine (TEFC).** A totally enclosed machine with cooling augmented by a fan, driven by the motor itself, blowing external air over the cooling surface and/or through the cooling passages, if any, incorporated in the machine.

12. **Totally enclosed separately air-cooled machine (TESAC).** A totally enclosed machine with cooling augmented by a separately-driven fan blowing external air over the cooling surfaces and/or through the cooling passages, if any, incorporated in the machine.

13. **Totally enclosed water or other liquid-cooled machine (TEWC).** A totally enclosed machine with cooling augmented by water-cooled at other liquid-cooled surfaces embodied in the machine itself.

14. **A totally enclosed closed air circuit machine.** A totally enclosed machine having special provision for cooling the enclosed air by passing it through its own cooler, usually external to machine.

15. **Totally enclosed closed gas circuit machine.** A totally enclosed machine cooled by gas other than air, the cooling gas being circulated through associated water-cooled gas coolers.

16. **Weather-proof machine (WP).** A machine so constructed that it can work without further protection from weather conditions specified by the purchaser.

17. **Watertight machine (WT).** A machine so constructed that it will withstand without damage sign of leakage, complete immersion in water to a depth of not less than 1 m, or subjection to an external water pressure of 0.1 bar for a period of one hour. The test for watertightness shall be made with the machine stationary and the temperature of the machine shall not exceed the temperature of the water in which it is immersed.

18. **Submersible machine.** A machine capable of working for an indefinitely long period when submerged under a specified head of water.

19. **Flame proof machine (FPM).** A machine which complies with the requirements of IS : 2148-1962 "Specifications for flame-proof enclosures of electrical apparatus".

5.18. BEARINGS

A **bearing** *is a device which supports, guides and restrains moving elements.*

The *material used* for bearing is commonly *'cast-iron' for slow speeds, bronze and brass lining being fitted for higher speeds.* White metal or antifriction metal is used as a lining for the bronze, or it may be held directly in the cast-iron or the steel of a connecting rod. The value of soft metals such as these is that they *do not roughen the journal, and they are able to flow slightly under pressure if insufficient clearance has been allowed or if the shaft is very slightly out of line.*

The bearings way be *classified* as follows :

1. *Plain bearings :*

(*i*) Journal bearing

(*ii*) Pivot bearing

(*iii*) Collar or thrust bearing.

2. *Ball and roller bearings*

1. Plain bearings :

- These are in the form of self-aligning porous bronze bushes for fractional kW motors and in the form of journal bearings for large motors.
- Since they run very silently, they are fitted on super-silent motors used for driving fans and lifts in offices or other applications where noise must be reduced to the absolute minimum.

2. Ball and roller bearings :

Upto about 75 kW motors, the ball bearings are preferred to other because of the following *advantages :*

Advantages :

1. They have low frictional resistance, particularly starting from rest.
2. They are usually much shorter than plain bearings for the same load.
3. They hold the shaft more accurately in position, since with plain sliding bearings there must always be a finite difference in size between the journal and the shaft to permit lubrication.
4. There is less power loss due to friction.
5. They can withstand heavy loads for a long time without undergoing failure.
6. Invite less attention for lubrication since once they are lubricated they need no lubrication for a long period.
7. Their maintenance is easy.
8. Easy replacement.
9. Their selection from manufacturer's catalogue for a particular load is easy.
10. Extremely reliable.

Disadvantages : Ball and roller bearings entail the following *disadvantages :*

1. They are more expensive in their initial cost.
2. They produce more noise at high speeds.
3. They cannot sustain shock loads effectively.
4. The lubrication fails without any warning.
5. Sometimes serious troubles come up due to minor defects.
6. Easy ingress of dirt and grit.
7. High friction at very high speeds.

Uses of bearings in electrical equipment :

The uses of bearings in electrical equipment are given in tabular form below :

S.No.	*Equipment*	*Bearings*
1.	High H.P. motors, generators or alternators whose shafts are horizontal and have *no thrust* (end pressure)	*Journal bearing*
2.	High H.P. electrical machines with horizontal shafts and having end *thrust.*	*Thrust or roller bearings*
3.	Turbogenerator sets with vertical shafts.	*Foot step or pivot bearings*
4.	Table fans.	*Ball bearings* (*Radial type*)
5.	Ceiling fans.	*Ball bearings* (*Thrust type*)
6.	Medium H.P. motors or generators (shafts with horizontal axis without end thrust)	*Roller bearings* (*Radial type*)
7.	Medium H.P. motors or generators (shafts with vertical axis).	*Roller bearing (Thrust type)*
8.	Medium H.P. motors and generators (shafts with horizontal axis and having end thrust or shafts placed in an inclined position).	*Roller bearing* (*Tapered type*)

5.19. TRANSMISSION OF DRIVE

The mechanical power available at the motor shaft has to transmitted to the driven machine(s) ; some of the methods to do so are as follows :

1. Direct drive
2. Belt drives
   (*i*) Flat belt drive
   (*ii*) V-belt drive
3. Rope drive
4. Chain drive.
5. Gear drive.
6. Vertical drive.

1. **Direct drive.** In a direct drive the driving member is connected direct by means of solid or flexible coupling (without any interposed gearing).

Advantages :

(*i*) Simplest method.

(*ii*) Space required is less than belt drive.

(*iii*) Efficiency is 100 percent.

Disadvantages :

(*i*) The driving and driven members must run at the same speed, which is usually not desirable.

(*ii*) End thrust is exerted on the motor by load if it happens to be a centrifugal pump.

(*iii*) Solid coupling requires very accurate aligning otherwise there is a possibility of damage to shafts. Another shortcoming of solid coupling is that sudden jerks of load are transmitted to the motor. By employing flexible coupling both of these shortcomings can be overcome to some extent.

Applications. This type of drive is employed where there is a possibility of arranging the driving member in line with the driven member where speed of driven member is same as that of driving member.

2. **Belt drives :**

A **belt** *is a continuous band of flexible material passing over pulleys to transmit motion from one shaft to another.*

Belts are available :

(*i*) with a narrow *rectangular cross-section*—***Flat belts*** [Fig. 5.111 (*i*)].

(*ii*) with a *trapezoidal cross-section*—***V-belts*** [Fig. 5.111 (*ii*)] and multiple ***V-belts*** [Fig. 5.111 (*iv*)].

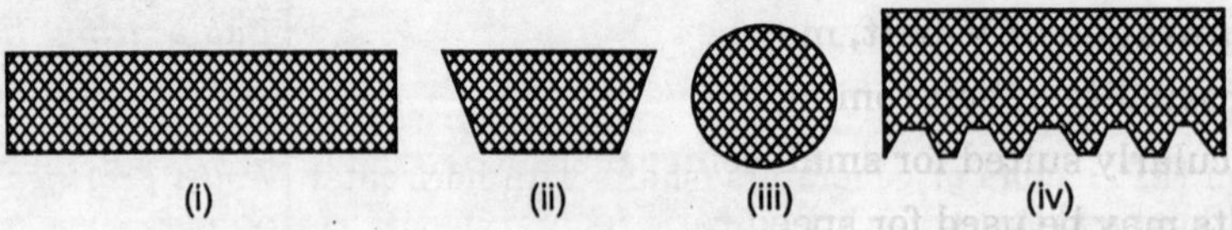

Fig. 5.111

(*iii*) *round cross-section*—***Round belts*** [Fig. 5.111 (*iii*)]

Chiefly used in machinery are *flat and V-belts.*

Flat belts :

- Flat belts are used for their simplicity and because they are subjected to minimum bending stress on the pulleys. The load capacity of flat belts is varied by varying their width, and only one is used in each drive. They are made of *leather, rubber, textile, balata and steel.*

- *Leather belts* have the best *pulling capacity.* Because of high cost of leather they are used very rarely.
- *Rubber belts* made of rubber on a cotton-duck base are used where the belt is exposed to the weather or steam, as they do not absorb moisture so readily as leather. They get destroyed if kept in contact with oil or grease.
- *Textile belts* are made of cotton and are used for rough end short service.
- *Balata belts* are acid and water proof and cannot withstand temperature higher than 100°C.
- *Steel belts* are claimed to transmit more horse power per cm width, and to remain unaffected by dampness or heat and be immune from stretching and slipping. The pulleys on which they are mounted do not have camber. Steel belts are sometimes used, the belt being subjected to considerable initial tension, to maintain the pressure on the pulley, on which the friction depends.

Note. The pulley of the flat belts is made *convex* at centre. This feature of the pulley is called **camber** or *crown* and due to it the lateral displacement of belt is prevented.

V-belts :

A V-belt is a belt of trapezoidal section running on pulleys with grooves cut to match the belt. The normal angle between the sides of the groove is 40 deg. Fig. 5.112 (*a, b*).

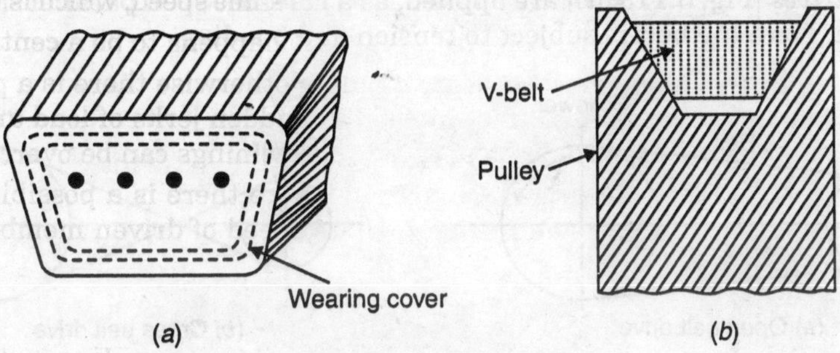

Fig. 5.112. V-belt.

- V-belts are usually made of fabric coated with rubber. They are silent and resilient. They are used when the distance between the shafts is too short for flat-belt drives. Owing to the *wedge action* between the belt and the sides of the groove in the pulley, the V-belt is *less likely to slip*, hence *more power can be transmitted for the same belt tension.*

The **advantages of V-belt,** may be summed up as follows :

1. No possibility of belt coming out of grooves.
2. Particularly suited for small centre distances requiring no idler.
3. V-belts may be used for speed ratio as high as 10 : 1 and belt speeds upto 2100 m/min.
4. Wedging action permits a smaller arc of contact.
5. The gripping action results in lower belt tension.
6. Power out can be increased by use of multiple belts.
7. In case of multiple-belt drive, if one belt fails, the machine does not come to a stop.
8. As V-belts are made endless the splicing problem is eliminated.
9. V-belts offer a *more positive drive because of reduced slippage.*

10. As these can be used over small pulleys, large reductions in speed are possible in a single drive.
11. V-belt drive may be inclined at any operating angle, slack side, top or bottom.
12. Drives are quiet at high speed.
13. The drive is capable of absorbing high shock.
14. Standardisation of V-belts results in better initial installation and replacement.

Round belts :

Round belts are employed to *transmit low power, mainly in instruments, table-type machine tools, machinery of the clothing industry and household appliances*. Round belts are used singly, as a rule. They may be made of *leather, canvas and rubber*. The diameter range is from 3 to 12 mm, usually from 4 to 8 mm. The minimum allowable ratio of the diameter of smaller pulley to the belt diameter is about 20, the recommended ratio is 30.

Belt drive :

A belt drive consists of the driving and driven pulleys and the belt which is mounted on the pulleys with a certain amount of tension and transmits peripheral force the friction.

- Belt drives may be :

(*i*) Open belt drive

(*ii*) Crossed belt drive.

Open belt drives [Fig. 5.113 (*a*)] are applied, as a rule, between parallel shafts which *rotate in the same direction.* Here the belt is subject to tension and bending.

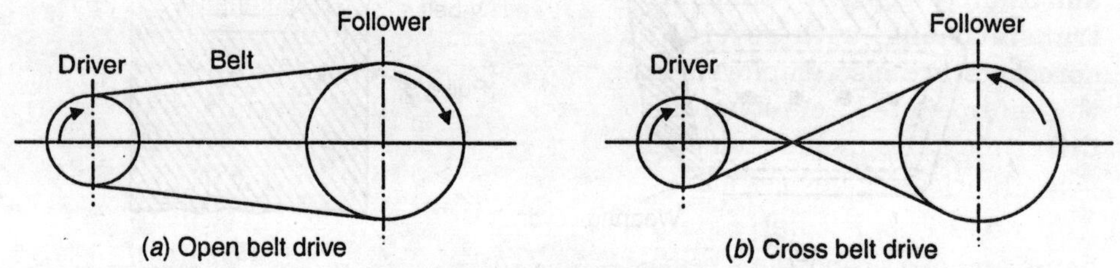

Fig. 5.113. Belt drives.

In *crossed belt drives* [Fig. 5.113 (*b*)] the power is transmitted between small shafts *rotating in opposite direction.* Since *the angle of contact in this type of drive is more,* it can transmit *more power than open belt drive.* However there is *more wear and tear of the belt in this drive.*

Applications of belt drives :

The main applications of belt drives are :

(*i*) To transmit power from low or medium capacity electric motors to operative machines.

(*ii*) To transmit power from small prime movers (internal combustion engines) to electric generators, agricultural and other machinery.

Advantages and disadvantages of belt drives :

Advantages :

(*i*) They can transmit motion over medium distances.

(*ii*) They give smooth operation (they cushion shocks and are silent).

(*iii*) They can operate at high speeds of rotation.

(*iv*) Their cost is relatively low.

Disadvantages :

(*i*) Their considerable overall size, usually several times larger than toothed gearing.

(*ii*) The inevitability of some elastic slipping (creep) of the belt.

(*iii*) The necessity for belt tensioning devices.

(*iv*) The necessity to keep oil from getting on the belt.

(*v*) The relatively short service life in high speed drives.

A primary condition in the operation of belt drives is proper belt tension which should be maintained in service.

3. **Rope drive :**

- Rope drive is the drive by means of ropes running over pulleys having a number of ropes, and is used for power ranges beyond the limit of V-belt drive.
- It is a long centre drive.
- Where the speed ratio is very high rope drive is indispensable.

Advantages : (*i*) Negligible slip ;

(*ii*) Ability to take sudden loads.

Application. Used on small and medium sized machines.

4. **Chain drive :**

- A *chain consists of links connected by joints which provide for articulation or flexibility of the chain.*
- A chain drive consists of two sprockets and chain (Fig. 5.114). Chain drives, or transmissions, with several driven sprockets are also employed. Besides the enumerated components, chain drives may also include tensioning devices, lubricating devices and guards.

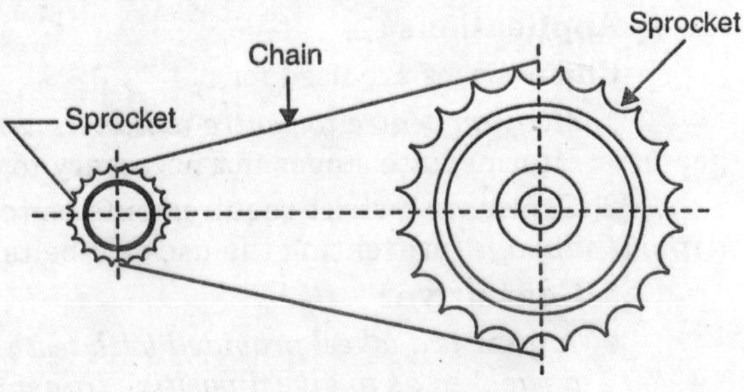

Fig. 5.114. Chain drive.

- This drive is particularly suitable for damp and dirty conditions because the chain case protects the drive from outside influences.
- This method is suitable for speeds upto 900 m/min. needing positive drives without any slipping.

There are two principal types of chain drives :

(*i*) Roller chain drive, and

(*ii*) Inverted tooth or silent chain drive.

A chain drive entails the following *advantages* and *disadvantages* :

Advantages :

1. Efficiency of transmission is very high (upto 98%).
2. More durable.
3. Can carry heavy loads.
4. Can be used where exact timing in movement is desired.
5. No slip takes place.
6. No initial tension required for operation.

7. A single chain can be used for transmitting motion to several shafts.
8. Smaller overall size than a belt drive.
9. Small forces acting on the shafts since no initial tension is required.
10. Easy replacement of the chain.
11. These permit high speed ratio upto 8 in one step.
12. These can be operated under adverse temperature and atmospheric conditions.

Disadvantages :

1. Chain drive requires housing.
2. It needs careful maintenance and accurate mounting.
3. The velocity of the chain, especially with small numbers of teeth on the sprockets, is not constant leading to non-uniform rotation of sprockets.
4. The chain gets stretched and requires a tensioning device.
5. Cost is high comparatively.

Uses of chain drives :

Chain drives are extensively employed in :

(*i*) Motor cycles (*ii*) Bicycles
(*iii*) Automobiles
(*iv*) Conveyers (*v*) Agricultural machinery
(*vi*) Oil-well drilling rigs (*vii*) Machine tools etc.

Applications :

Chain drives are used for :

1. *Medium centre to centre distances* which, in the case of a gear drive, would require idle gears, or intermediate stages not necessary to obtain the required speed ratio.

2. Drives with strict requirements as to overall size or ones *requiring positive transmission without slippage* (preventing the use of V-belts drives).

5. **Gear drive :**

- A **gear** *is a wheel provided with teeth which mesh with the teeth on another wheel, or on to a rack, so as to give a positive transmission of motion from one component to another.*
- Gears constitute the most commonly used device for power transmission or for changing power-speed ratios in a power system. They are used for transmitting motion and power from one shaft to another *when they are not too far apart and when a constant velocity ratio is desired.* Gears also afford a convenient way of changing the direction of motion.
- A number of devices such as *differentials, transmission gear boxes, planetary drives* etc., used in many construction machines employ gears as basic component.

The following are the *advantages* and *disadvantages* of toothed gearing/gear drive :

Advantages :

1. High efficiency
2. Long service life.
3. High reliability.
4. More compact.
5. Can operate at high speeds.
6. Can be used where precise timing is required.
7. Large power can be transmitted.

8. Constant speed ratio owing to absence of slipping.

9. Possibility of being applied for a wide range of torques, speeds and speed ratios.

10. The force required to hold the gears in position is much less than in an equivalent friction drive. This results in lower bearing pressure, less wear on the bearing surface and efficiency.

Disadvantages :

1. Special equipment and tools are required to manufacture the gears.
2. When one wheel gets damaged the whole set up is affected.
3. Noisy in operation at considerable speeds.

6. **Vertical drive.** In a vertical drive motor is arranged with its axis vertical. This arrangement frequently proves to be convenient.

5.20. CHOICE OF DRIVE

Choice of drive is governed by the following *factors :*

(*i*) Speed of driving and driven machines :

(*ii*) Convenience ;

(*iii*) Space available ;

(*iv*) Clutching arrangement required ;

(*v*) Cost.

- The choice of motor speed is the most important factor as it not only affects the performance of motor but also overall cost. The dimensions and, therefore, the first cost of a motor for a given output are approximately inversely proportional to the speed, so for the some output kW the cost of a high speed motor is *less* than that of a slow speed motor. In case of induction motor, the efficiency and power factor decrease with decrease in speed. Thus for a low-speed drive high speed motor using a reduction gear is usually found *cheaper* than a low-speed direct-coupled motor.

5.21. NOISE

In the workshops, the noise should be kept as low as possible to avoid fatigue to workers. Also, it is essential to use noiseless motors for domestic appliances and the appliances to be used in hospitals and theatres.

The noise of a motor may be due to :

(*i*) Bearings ; (*ii*) Vibrations ;

(*iii*) Magnetic pulsations ; (*iv*) Bad foundation.

- The noise due to bearings, vibrations and magnetic pulsation is a matter for the manufacturer, but noise transmitted from one part of a building to other part can be minimised by mounting the motor on a heavy concrete or cast iron block or alternatively on a flexible support provided by rubber pads or springs.

— To avoid transmission of vibrations, electrical connections to the motor should be made by means of flexible pipe rather than by conduit pipe.

5.22. SELECTION OF ELECTRIC MOTOR FOR ANY APPLICATION

The selection of an electric motor for any application depends on the following *factors :*

I. Electrical characteristics :

(*i*) Running characteristics (*ii*) Starting characteristics

(*iii*) Speed control (*iv*) Braking characteristics.

II. Mechanical characteristics :

(*i*) Power transmission (*ii*) Cooling

(*iii*) Noise (*iv*) Type of enclosure.

III. Size and rating of motors :

(*i*) Load cycle :

(*a*) Continuous

(*b*) Intermittent

(*c*) Variable.

(*ii*) Overload capacity

IV. Cost

(*i*) Initial cost (*ii*) Cost of control gear

(*iii*) Running cost.

Characteristics of electric motors is given in Table 5.4.

Table 5.4. Characteristics of Electric Motors

Motors	*Methods of starting*	*Methods of speed control*	*Voltage limit*	*H.P. limit*	I_{st}/I_{fl}	T_{st}/T_{fl}
D.C. Motors						
1. *Shunt*	Series resistance in armature	(*i*) Variable resistance in field circuit (*ii*) Variable resistance in armature circuit	3 kV	25000	2	2
2. *Series*	(*i*) Series resistance (*ii*) Series-parallel method	(*i*) By tapping the field (*ii*) By field diverter (*iii*) Variable resistance in series (*iv*) Series-parallel control	1.5 kV	3000	2	3
3. *Compound*	Series resistance in armature	(*i*) Variable resistance in shunt field (*ii*) Series field diverter	1.5 kV	3000	2	2-3
A.C. Motors						
1. *Single phase induction motor*	(*i*) Repulsion start (*ii*) Pole shading (*iii*) Phase splitting by *C*, *L* or *C* and *R*	Voltage drop in series impedance	250 V	1	2	1.5-3
2. *Single phase series motor*	Variable voltage supply	Voltage variation	500 V	3000 2 3		

3. *Squirrel cage induction motor*	(*i*) Star-delta starter (*ii*) Auto-transformer (*iii*) Resistance in stator circuit	(*i*) Pole changing (*ii*) Variation of frequency	11 kV	300	1	1
4. *Slip ring induction motor*	Resistance in rotor circuit	(*i*) Variation in applied voltage (*ii*) Cascading (*iii*) Pole changing (*iv*) Variation of frequency (*v*) motor rheostat control (*vi*) Injecting e.m.f. in rotor	11 kV	11000	2	2
5. *Pole changing induction motors*	(*i*) Direct switching (*ii*) Start-delta starter (*iii*) Auto-transformer	(*i*) Pole changing (*ii*) Variation of frequency	11 kV	300	6	1
6. *Schrage motor*	(*i*) Low voltage application (*ii*) Direct switching	(*i*) Shifting the brushes (*ii*) Inserting impedance in secondary winding	1 kV	1000	2	2.6
7. *Synchronous motor*	(*i*) Damper winding (*ii*) Prony brake	Fixed speed	15 kV	10000	2	3

5.23. MOTORS FOR PARTICULAR APPLICATIONS

For a particular application, the type of electric drive and control gear are determined by the following considerations :

1. Starting torque
2. Duty (whether light, medium, heavy)
3. Limitations on starting current
4. Speed control range and its nature
5. Need for automatic control
6. Conditions of environment.

S.No.	Applications	Type of motor(s) used
1.	***Domestic uses*** (*e.g.* Vacuum cleaners, sewing machines, mixy, electric shavers, fans, cloth-washing machines, refrigerators etc.)	• *Small universal motors*, usually series type.
2.	***Cranes***	• *D.C. series or compound motors* (on account of their high starting torque and smooth speed control)... for cranes — The motors used for lifting, travelling and reversing, and for conveying and hoisting are D.C. compound wound, D.C. series wound and A.C. slip ring induction motors respectively.
3.	***Lifts***	• *D.C. compound wound and A.C. slip-ring induction motors.* • Induction repulsion and variable speed A.C. commutator motors are *also* used.
4.	***Lathes, milling and grinding machines***	• *D.C. shunt or squirrel cage induction motors*
5.	***Planers***	• *Shunt or compound wound motors* (for D.C. supply), • *Slip-ring induction motor* (for A.C. supply)
6.	***Punches, presses and shears***	• *High slip squirrel-cage motors, slip ring induction motors or D.C. cumulative compound motors.*
7.	***Shapers and slotters***	• *Constant speed squirrel-cage induction motors*
8.	***Drilling machines***	• *Constant speed squirrel-cage or D.C. shunt or multispeed A.C. motors.*
9.	***Blowers and fans***	• *Squirrel cage induction motors.* • *Synchronous motors.*
10.	***Air compressors***	• *Squirrel cage induction motors*-for small compressors only • *Slip-ring induction motors and synchronous motors*-for large units
11.	***Electric traction***	• *D.C. series motors*-for all types of services but more particularly for suburban services where high rate of acceleration is essential. • *Single-phase A.C. compensated series motors* extensively used for main-line work. Due to poor starting these motors are *not* suitable for suburban services where stops are frequent.
12.	***Hoist work***	• *3-phase slip-ring induction motors.*
13.	***Pumps***	• *Polyphase induction motors* (Drip proof or totally enclosed surface cooled. — *Squirrel-cage motors*-for centrifugal pumps where T_{st} = 40 to 50% T_{fl} is required. — *Slip-ring motors*-for reciprocating pumps where T_{st} = 100 to 200% T_{fl} is required.

S.No.	Applications	Type of motor (s) used
14.	***Refrigeration and air-conditioning***	• *1-phase, 230 V, capacitor type induction motor* with D.O.L. starters-for small units. • *High torque squirrel-cage induction motors or slip-rings induction motors*-for large installations. • *Synchronous motor*, driving turbo-compressor may suit especially if power factor correction is also required.
15.	***Machine tools***	• *Squirrel-cage motors*
16.	***Belt conveyors***	• *Double-cage induction motor* with D.O.L. or *wound-rotor induction motor* — These motors are totally enclosed, surface cooled type as grit and dust remain present in the atmosphere.
17.	***Flour mills***	• *Pipe ventilated or totally enclosed type motors.*
18.	***Mines***	• *Flame proof squirrel-cage induction motor* • *Slip ring induction motor* • *D.C. motor* • *D.C. cumulative motor*
19.	***Rolling mill drives***	• *D.C. motors* • *A.C. motors with speed control*
20.	***Kiln drives***	• *Slip-ring induction motor* • *Three-phased shunt wound commutator motor* • *Cascade-controlled A.C. motor* • *Ward Leonard controlled D.C. motor* • *D.C. motor with transformer step switch control*
21.	***Paper industry***	• *Squirrel-cage motors with clutches, slip ring*
22.	***Iron and steel industry***	• *D.C. shunt wound motor* with stabilizing, compensating and interpole windings.
23.	***Textile industry***	• *Three-phase A.C. motor*
24.	***Drives for numerical control of machine tools***	• Numerical control is used to keep a part or tool under continuous specific varying control, producing a family of cuttings. Numerical control enables a cutter through a programmed path in the work to give a specific shape of accurate dimensions. • Programmed controlled drives are used to get a command signal from punched card, punched tape or magnetic tape and feed the information to a digital computer which controls the feed drives. Accurate positioning and position measurement is carried out by using *transducers* which provide digital position data by a current of pulses which are compared with the instructions punched on a tape. If there is any change in the two a feed back signal corrects the position and adjusts the drive. D.C. motors are used for numerical control (NC) positioning and they are controlled by SCR in electronic amplifier circuits. These days stepper motors are used to move a motor by pulses through incremental angular changes to drive a tool or slide through gearing. The use of numerical controlled machines has greatly helped in getting great precision in machine tool works.

Motors for Different Industrial Drives

S.No.	*Type of motor*	*Characteristics*	*Fields of application*
1.	***D.C. Shunt motor***	— Medium starting torque — Nearly constant speed	● Constant-speed line shafts ● Lathes ● Vacuum cleaners ● Wood-working machines ● Laundry washing machines ● Elevators ● Conveyors ● Small printing presses ● Grinders etc.
2.	***D.C. series motor***	— High starting torque — Variable speed	● Electric locomotives ● Steel rolling mills ● Lifts and cranes ; hoists
3.	***D.C. cumulative compound motor***	— High starting torque — Varying speed motor	● Compressors ● Rotary presses ● Circular saws ● Shearing machines ● Elevators ● Continuous conveyors ● Variable-head centrifugal pumps etc.
4.	***Three-phase synchronous motor***	— Speed remains constant under varying loads	● Ammonia and air compressors ● Motor-generator sets ● Continuous rolling mills ● Paper and cement industries
5.	***Squirrel-cage induction motor***	— Simple and rugged in construction — High overload capacity — Used for low and medium power drives where speed control is not required.	● Water pumps ● Tube wells ● Lathes ● Drills ● Griders ● Polishers ● Fans and blowers ● Laundry washing machines ● Compressors ● Wood planers etc.
6.	***Double squirrel-cage motor***	— High starting torque — Large overload capacity — Nearly constant speed — Used for driving loads which require high starting torque	● Compressor pumps ● Reciprocating pumps ● Large refrigerators ● Crushers ● Textile machinery ● Lathes ● Boring mills ● Punches etc.

S.No.	Type of motor	Characteristics	Fields of application
7.	***Slip-ring induction motor***	— High starting torque — Large overload capacity — Its speed can be changed upto 50% of its normal speed — Used for those drives which require high starting torque and speed control	• Lifts • Pumps • Winding machines • Line shafts • Printing presses • Compressors • Elevators etc.
8.	***Single-phase synchronous motor***	— Constant speed	• Teleprinters • Recording instruments • All kinds of timing devices • Clocks etc.
9.	***Single-phase series motor***	— High starting torque — Speed can be controlled over a wide range	• Refrigerators • Vacuum cleaners etc.
10.	***Repulsion motor***	— High starting torque — Wide speed control — High speed at high loads	• Coil winding machines
11.	***Capacitor-start induction run motor***	— Fairly constant speed — Moderately high starting torque — Speed control not possible	• Compressors • Refrigerators • Small portable hoists
12.	***Capacitor-start-and-run motor***	— Operating characteristics similar to above motor except that it has better power factor and higher efficiency	• Drives requiring quiet operations

5.24. ENERGY CONSERVATION IN ELECTRICAL DRIVES

- For keeping the manufacturing cost of a product minimum, and make its price competitive in the market, it is necessary to *minimise energy conscrvation* at all stages of manufacturing process *including electrical drives.*
- Conservation of energy is also necessary because with the ever increasing demand, need for electrical power can only be met by conserving electrical power in addition to installation of new generating units. A major portion of electrical power in a plant is consumed by electrical drives.

A large amount of electrical energy can be saved by the use of efficient and right type of electrical drives.

Energy conservation in electrical drive is achieved by *reduction of following losses* in its various parts :

(*i*) *Electrical transmission losses* ... These losses depend on the drive power factor and harmonics in the line current.

(*ii*) *Electrical motor losses to convert electrical power into mechanical power* ... These are determined by choice of motor and quality of supply.

(*iii*) *Conversion losses in the power modulator or converter.*

(*iv*) *Mechanical losses in the parts of transmission system such as bearings, gears, clutches and belts.*

(*v*) *Losses in the load.*

(*vi*) *Losses caused by throttling* or *by other means that control material flow by absorbing or bypassing excess output.*

(*vii*) *Mechanical transmission losses.*

Energy conservation in electrical drives can be materialised by the following *measures* :

- Use of efficient motors.
- Use of efficient semiconductor converters.
- Use of a motor of right rating.
- Improvement of power factor.
- Use of single-to-three-phase semiconductor converters in rural applications.
- Regular and preventive maintenance of motors, transformers and coupled equipment.
- Improvement of quality of supply.

HIGHLIGHTS

1. A form of machine equipment designed to convert electrical energy into mechanical energy and to provide electrical control of this process is known as *electrical drive.*
2. Same of the important factors on which the selection of electrical drives depend are :
   (*i*) Steady state operation requirements
   (*ii*) Transient operation requirements
   (*iii*) Requirement related to the source
   (*iv*) Capital and running cost, maintenance needs, life
   (*v*) Reliability
   (*vi*) Environment and location
   (*vii*) Space and weight restrictions, if any.
3. The loads which are due to forces of gravity, tension or compression in a spring or any elastic body are called *active loads.*
   Passive torques are due to friction, cutting and deformation of inelastic bodies.
4. *Commonly used D.C. motors are :*
   (*i*) Shunt and separately excited motor (*ii*) Series motor
   (*iii*) Compound motor (*iv*) Universal motor
   (*v*) Permanent magnet motors (*vi*) D.C. servo motors
   (*vii*) Moving coil motors.
5. *A.C. motors* used in electrical drives are :
   (*i*) 3-phase induction motors (*ii*) Single-phase induction motors
   (*iii*) Synchronous motors (*iv*) Brushless D.C. motors
   (*v*) Stepper motors (*vi*) Switched reluctance motors.
6. *Heating time constant* (T_h) is the time taken by the machine to attain 0.632 of its final steady temperature rise.
7. *Cooling time constant* (T_c) is the time taken by the machine for its temperature rise to fall to 0.368 of its initial value.

8. The different ratings for electrical motors are :

(i) Continuous rating (ii) Continuous maximum rating

(iii) Short time rating (iv) Intermittent rating.

9. The process of smoothing out the fluctuating load is known as *load* **equalisation.**

10. A *bearing* is a device which supports, guides and restrains moving elements.

11. The various types of drives (transmission) are :

(i) Direct drive (ii) Belt drive

(iii) Rope drive (iv) Chain drive

(v) Gear drive (vi) Vertical drive.

OBJECTIVE TYPE QUESTIONS

A. Choose the Correct Answer :

1. The selection of an electric motor for any application depends on which of the following factors ?
   (a) Electrical characteristics (b) Mechanical characteristics
   (c) Size and rating of motors (d) Cost
   (e) All of the above.
2. For a particular application the type of electric and control gear are determined by which of the following considerations ?
   (a) Starting torque (b) Conditions of environment
   (c) Limitation on starting current (d) Speed control range and its nature
   (e) All of the above.
3. Which of the following motors is preferred for traction work ?
   (a) Universal motor (b) D.C. series motor
   (c) Synchronous motor (d) Three-phase induction motor.
4. Which of the following motors always starts on load ?
   (a) Conveyor motor (b) Floor mill motor (c) Fan motor (d) All of the above.
5. ...... is preferred for automatic drives.
   (a) Squirrel cage induction motor (b) Synchronous motors
   (c) Ward-Leonard controlled D.C. motors (d) Any of the above.
6. When the load is above ...... a synchronous motor is found to be more economical ?
   (a) 2 kW (b) 20 kW (c) 50 kW (d) 100 kW.
7. The load cycle for a motor driving a power press will be ......
   (a) variable load (b) continuous
   (c) continuous but periodical (d) intermittent and variable load.
8. Light duty cranes are used in which of the following ?
   (a) Power houses (b) Pumping stations
   (c) Automobile workshops (d) All of the above.
9. While selecting an electric motor for a floor mill, which electrical characteristics will be of *least* significance ?
   (a) Running characteristics (b) Starting characteristics
   (c) Efficiency (d) Braking.
10. Which of the following motors are pre-ferred for overhead travelling cranes ?
    (a) Slow speed motors (b) Continuous duty motors
    (c) Short time rated motors (d) None of the above.
11. ...... is preferred for synthetic fibre mills.
    (a) Synchronous motor (b) Reluctance motor
    (c) Series motor (d) Shunt motor.

12. Ward-Leonard controlled D.C. drives are generally used for ...... excavators.
(*a*) Light duty (*b*) Medium duty (*c*) Heavy duty (*d*) All of the above.

13. Which of the following motors is used for elevators ?
(*a*) Induction motor (*b*) Synchronous motor
(*c*) Capacitor start single phase motor (*d*) Any of the above.

14. Which part of a motor needs maximum attention for maintenance ?
(*a*) Frame (*b*) Bearing (*c*) Stator winding (*d*) Rotor winding.

15. ...... need frequent starting and stopping of electric motors.
(*a*) Paper mills (*b*) Grinding mills (*c*) Air-conditioners (*d*) Lifts and hoists.

16. Which feature, while selecting a motor for centrifugal pump, will be of *least* significance ?
(*a*) Starting characteristics (*b*) Operating speed
(*c*) Horse power (*d*) Speed control.

17. ...... motor is a constant speed motor.
(*a*) Synchronous motor (*b*) Schrage motor (*c*) Induction motor (*d*) Universal motor.

18. The starting torque in case of centrifugal pumps is generally
(*a*) less than running torque (*b*) same as running torque
(*c*) slightly more than running torque (*d*) double the running torque.

19. Which of the following motors are best for the rolling mills ?
(*a*) Single phase motors (*b*) Squirrel cage induction motors
(*c*) Slip ring induction motors (*d*) D.C. motors.

20. ...... is not a part of ball bearing ?
(*a*) Inner race (*b*) Outer race (*c*) Cage (*d*) Bush.

21. The starting torque of a D.C. motor is independent of which of the following ?
(*a*) Flux (*b*) Armature current
(*c*) Flux and armature current (*d*) Speed.

22. Rotor of a motor is usually supported on ...... bearings.
(*a*) ball or roller (*b*) needle (*c*) bush (*d*) thrust.

23. For which of the following applications D.C. motors are still preferred ?
(*a*) High efficiency operation (*b*) Reversibility
(*c*) Variable speed drive (*d*) High starting torque.

24. In a paper mill where constant speed is required
(*a*) synchronous motors are preferred (*b*) A.C. motors are preferred
(*c*) individual drive is preferred (*d*) group drive is preferred.

25. A reluctance motor ......
(*a*) is provided with slip rings (*b*) requires starting gear
(*c*) has high cost (*d*) is compact.

26. The size of an excavator is usually expressed in terms of
(*a*) 'crowd' motion (*b*) angle of swing (*c*) cubic metres (*d*) travel in metres.

27. For blowers which of the following motors is preferred ?
(*a*) D.C. series motor (*b*) D.C. shunt motor
(*c*) Squirrel cage induction motor (*d*) Wound rotor induction motor.

28. Belted slip ring induction motor is almost invariably used for
(*a*) water pumps (*b*) jaw crushers (*c*) centrifugal blowers (*d*) none of the above.

29. Which of the following is essentially needed while selecting a motor ?
(*a*) Pulley (*b*) Starter (*c*) Foundation pedal (*d*) Bearings.

30. Reluctance motor is a ......
(*a*) variable torque motor (*b*) low torque variable speed motor
(*c*) self starting type synchronous motor (*d*) low noise, slow speed motor.

31. ...... method of starting a three phase induction motor needs six terminals.
(*a*) Star-delta (*b*) Resistance starting (*c*) Auto-transformer (*d*) None of the above.

32. In ...... method of starting three phase induction motors the starting voltage is not reduced.
(*a*) auto-transformer (*b*) star-delta (*c*) slip ring (*d*) any of the above.

33. In jaw crushers a motor has to often start against ...... load.
(*a*) heavy (*b*) medium (*c*) normal (*d*) low.

34. For a motor-generator set which of the following motors will be preferred ?
(*a*) Synchronous motor (*b*) Slip ring induction motor
(*c*) Pole changing induction motor (*d*) Squirrel cage induction motor.

35. Which of the following motors is usually preferred for kiln drives ?
(*a*) Cascade controlled A.C. motor (*b*) Slip ring induction motor
(*c*) Three phase shunt wound commutator motor (*d*) Any of the above.

36. Heat control switches are used in ......
(*a*) transformers (*b*) cooling ranges
(*c*) three phase induction motors (*d*) single phase.

37. ...... has relatively wider range of speed control.
(*a*) Synchronous motor (*b*) Ship ring induction motor
(*c*) Squirrel cage induction motor (*d*) D.C. shunt motor.

38. In squirrel cage induction motors which of the following methods of starting cannot be used ?
(*a*) Resistance in rotor circuit (*b*) Resistance in stator circuit
(*c*) Auto-transformer starting (*d*) Star-delta starting.

39. In which of the following applications the load on motor changes in cyclic order ?
(*a*) Electric shovels (*b*) Cranes (*c*) Rolling mills (*d*) All of the above.

40. Flame proof motors are used in
(*a*) paper mills (*b*) steel mills
(*c*) moist atmospheres (*d*) explosive atmospheres.

41. Which of the following machines has heavy fluctuation of load ?
(*a*) Printing machine (*b*) Punching machine
(*c*) Planer (*d*) Lathe.

42. For derries and winches which of the following drives can be used ?
(*a*) Pole changing squirrel cage motors (*b*) D.C. motors with Ward-leonard control
(*c*) A.C. slip ring motors with variable resistance (*d*) Any of the above.

43. Battery operated scooter for braking uses
(*a*) plugging (*b*) mechanical braking
(*c*) regenerative braking (*d*) rheostatic braking.

44. ...... has least range of speed control.
(*a*) Slip ring induction motor (*b*) Synchronous motor
(*c*) D.C. shunt motor (*d*) Schrage motor.

45. ...... has the least value of starting torque to full load torque ratio.
(*a*) D.C. shunt motor (*b*) D.C. series motor
(*c*) Squirrel cage induction motor (*d*) Slip ring induction motor.

46. In case of ...... speed control by injecting e.m.f. in the rotor circuit is possible.
(*a*) d.c. shunt motor (*b*) schrage motor
(*c*) synchronous motor (*d*) slip ring induction motor.

47. A pony motor is used for the starting which of the following motors ?
(*a*) Squirel cage induction motor (*b*) Schrage motor
(*c*) Synchronous motor (*d*) None of the above.

48. In ...... the speed can be varied by changing the position of brushes.
(*a*) slip ring motor (*b*) schrage motor (*c*) induction motor (*d*) repulsion motor.

49. In which of the following applications variable speed operation is preferred ?
(*a*) Exhaust fan (*b*) Ceiling fan (*c*) Refrigerator (*d*) Water pump.

50. Heavy duty cranes are used in
(*a*) ore handling plants (*b*) steel plants
(*c*) heavy engineering workshops (*d*) all of the above.

51. The travelling speed of cranes varies from
(*a*) 20 to 30 m/s (*b*) 10 to 15 m/s (*c*) 5 to 10 m/s (*d*) 1 to 2.5 m/s.

52. Besides a constant speed a synchronous rotor possesses which of the following advantages ?
(*a*) Lower cost (*b*) Better efficiency (*c*) High power factor (*d*) All of the above.

53. By the use of which of the following D.C. can be obtained from A.C. ?
(*a*) Silicon diodes (*b*) Mercury arc rectifier
(*c*) Motor generator set (*d*) Any of the above.

54. Which of the following motors is preferred when quick speed reversal is the main consideration ?
(*a*) Squirrel cage induction motor (*b*) Wound rotor induction motor
(*c*) Synchronous motor (*d*) D.C. motor.

55. Which of the following motors is preferred when smooth and precise speed control over a wide range is desired ?
(*a*) D.C. motor (*b*) Squirrel cage induction motor
(*c*) Wound rotor induction motor (*d*) Synchronous motor.

56. For crane travel which of the following motors is normally used ?
(*a*) Synchronous motor (*b*) D.C. differentially compound motor
(*c*) Ward-Leonard controlled D.C. shunt motor (*d*) A.C. slip ring motor.

57. The capacity of a crane is expressed in terms of
(*a*) type of drive (*b*) span (*c*) tonnes (*d*) any of the above.

58. The characteristics of drive for crane hoisting and lowering are which of the following ?
(*a*) Precise control (*b*) Smooth movement (*c*) Fast speed control (*d*) All of the above.

59. Which of the following motors is preferred for boom hoist of a travelling crane ?
(*a*) Single phase motor (*b*) Synchronous motor
(*c*) A.C. slip ring motor (*d*) Ward-Leonard controlled D.C. shunt motor.

60. A wound rotor induction motor is preferred, as compared to squirrel cage induction motor, when major consideration is
(*a*) slop speed operation (*b*) high starting torque (*c*) low windage losses (*d*) all of the above.

61. Which of the following motors has series characteristics ?
(*a*) Shaded pole motor (*b*) Repulsion motor
(*c*) Capacitor start motor (*d*) None of the above.

62. Which of the following happens when star-delta starter is used ?
(*a*) Starting voltage is reduced (*b*) Starting current is reduced
(*c*) Both (*a*) and (*b*) (*d*) None of the above.

63. For a D.C. shunt motor which of the following is *incorrect* ?
(*a*) Unsuitable for heavy duty starting (*b*) Torque varies as armature current
(*c*) Torque-armature current is a straight line (*d*) Torque is zero for zero armature current.

64. For which of the following applications motor has to start with high acceleration ?
(*a*) Oil expeller (*b*) Floor mill
(*c*) Lifts and hoists (*d*) Centrifugal pump.

65. Which of the following types of motor enclosure is *safest* ?
(*a*) Totally enclosed (*b*) Totally enclosed fan cooled
(*c*) Open type (*d*) Semi closed.

66. While selecting motor for an air conditioner which of the following characteristics is of great importance ?
(a) Type of bearings
(b) Type of enclosure
(c) Noise
(d) Arrangement for power transmission.
(e) None of the above

67. The diameter of the rotor shaft for an electric motor depends on which of the following ?
(a) r.p.m. only
(b) Horse power only
(c) Horse power and r.p.m.
(d) Horse power, r.p.m. and power factor.

68. Which of the following alternatives will be cheaper ?
(a) A 100 H.P. A.C. three phase motor
(b) Four motors of 25 H.P. each
(c) Five motors of 20 H.P. each
(d) Ten motors of 10 H.P. each.

69. The cost of an induction motor will increase as
(a) horsepower rating increases but r.p.m. decreases
(b) horsepower rating decreases but r.p.m. increases
(c) horsepower rating and operating speed increases
(d) horsepower rating and operating speed decreases

70. In series motor which of the following methods can be used for changing the flux per pole ?
(a) Tapped field control
(b) Diverter field control
(c) Series-parallel control
(d) Any of the above.

B. Fill in the Blanks/Say 'Yes' or 'No' :

71. Electroplating essentially reeds ...... current.
72. For traction work D.C. shunt motor is preferred.
73. Totally enclosed fan cooled type of motor enclosure is the safest.
74. The capacity of a crane is expressed in terms of type of drive.
75. When quick speed reversal is the consideration ...... motor is preferred.
76. In a ball bearing a bush is used.
77. The starting torque of a D.C. motor is independent of ......
78. D.C. series motors are preferred for traction work.
79. In overhead travelling cranes ...... rated motors are preferred.
80. Bolted slip ring induction motor is almost invariably used for water pumps.
81. In jaw crushers, a motor has to often start against ...... load.
82. For synthetic fibre mills shunt motor is preferred.
83. The size of an excavator is usually expressed in ......
84. Star-delta method of starting a three phase induction motor needs six terminals.
85. A pony motor is used for the starting of a ...... motor.
86. Belted wound rotor induction motors are preferred for gyratory crushers.
87. The ratio of starting current to full load current can be highest in case of pole changing induction motor.
88. Flame proof motors are used in ...... atmospheres.
89. A punching machine has heavy fluctuation of load.
90. The resistance of earth wire should be ......
91. In automobiles the sound is produced by vibrating diaphragm.
92. The earth wire should not be than a ...... wire.
93. Non-metallic conduits for wiring are generally made of ......
94. Premature blowing of a fuse may occur due to heating at ferrule contacts.
95. Inside the earth or pit, the earthing electrode should be placed ......
96. Continuous operation of automobile horn will damage the operating coil.
97. Power factor in case of reluctance motor is nearly unity.
98. Reluctance motor is a self starting type synchronous motor.

99. Ward-Leonard controlled D.C. drives are generally used for heavy duty excavators.

100. ...... operated electromagnet is preferred for noiseless operation.

Answers

A. Choose the Correct Answer :

1. (*e*)	**2.** (*e*)	**3.** (*b*)	**4.** (*d*)	**5.** (*c*)
6. (*d*)	**7.** (*d*)	**8.** (*d*)	**9.** (*d*)	**10.** (*c*)
11. (*b*)	**12.** (*c*)	**13.** (*a*)	**14.** (*b*)	**15.** (*d*)
16. (*d*)	**17.** (*a*)	**18.** (*a*)	**19.** (*d*)	**20.** (*d*)
21. (*d*)	**22.** (*a*)	**23.** (*c*)	**24.** (*c*)	**25.** (*d*)
26. (*c*)	**27.** (*c*)	**28.** (*b*)	**29.** (*b*)	**30.** (*c*)
31. (*a*)	**32.** (*c*)	**33.** (*a*)	**34.** (*a*)	**35.** (*d*)
36. (*b*)	**37.** (*d*)	**38.** (*a*)	**39.** (*d*)	**40.** (*d*)
41. (*b*)	**42.** (*d*)	**43.** (*b*)	**44.** (*b*)	**45.** (*c*)
46. (*d*)	**47.** (*c*)	**48.** (*b*)	**49.** (*b*)	**50.** (*d*)
51. (*d*)	**52.** (*c*)	**53.** (*d*)	**54.** (*d*)	**55.** (*a*)
56. (*d*)	**57.** (*c*)	**58.** (*d*)	**59.** (*c*)	**60.** (*b*)
61. (*b*)	**62.** (*c*)	**63.** (*a*)	**64.** (*c*)	**65.** (*b*)
66. (*c*)	**67.** (*c*)	**68.** (*a*)	**69.** (*a*)	**70.** (*d*)

B. Fill in the Blanks/Say 'Yes' or 'No' :

71. direct	**72.** No	**73.** Yes	**74.** No	**75.** D.C.
76. No	**77.** speed	**78.** Yes	**79.** short-time	**80.** No
81. heavy	**82.** No	**83.** cubic meters	**84.** Yes	**85.** synchronous
86. Yes	**87.** Yes	**88.** explosive	**89.** Yes	**90.** very low
91. Yes	**92.** 8 SWG	**93.** PVC	**94.** Yes	**95.** vertical
96. Yes	**97.** No	**98.** Yes	**99.** Yes	**100.** D.C.

THEORETICAL QUESTIONS

1. What is an electional drive ?

2. List the major parts of an electrial drive.

3. Sate the advantages and disadvantages of electrical drives.

4. Name the motors which are generally used for a electrical drive.

5. State the important factors on which the selection of electrical drives depends.

6. Explain briefly the status of D.C. and A.C. drives.

7. How are electrical drives classified ?

8. Explain briefly the following types of electrical drives :
(*i*) Group drive ; (*ii*) Individual drive ;
(*iii*) Multi motor drive.

9. Discuss briefly the following :
(*i*) Active or potential loads (*ii*) Passive loads.

10. What are the advantages and disadvantages of D.C. motors ?

11. List the commonly used D.C. motor for electrical drives.

12. Discuss the merits and demerits of individual and group drives.

13. Draw and explain different electrical and mechanical characteristics of various D.C. motors.

14. Give the characteristics of D.C. shunt motors. Why are such motors not suitable for traction purpose ?

15. Compare the speed-torque characteristics of cumulatively-compounded D.C motor and a 3-phase slip ring induction motor.

16. Using the basic e.m.f. and torque equations of a D.C. machine deduce of the shape of the following characteristics of both shunt and series motors :
(*i*) Speed-versus armature current (*ii*) Torque versus armature current.
(*iii*) Speed versus torque.
17. Discuss with the help of curves, the characteristics and applications of the following electric motors :
(*i*) 3-phase, cage type induction motor (*ii*) 3-phase, wound rotor induction motor
(*iii*) Single-phase induction motor (*iv*) Synchronous motor
(*v*) D.C. shunt motor.
18. Explain the characterics and applications of
(*i*) Universal (*ii*) Squirrel-cage induction, and
(*iii*) Scharge motors.
19. With relevant diagrams, explain how the speed of D.C. shunt motor is controlled using Ward-Leonard scheme.
20. Describe with the help of connection diagram the Ward-Leonard method of speed control and explain. its working. What are the merits and demerits of this method ? State its fields of application.
21. Explain briefly the following motors :
(*i*) Permanent magnet D.C. motors (*ii*) D.C. servo-motors
(*iii*) Moving coil motors.
22. Discuss briefly the following :
(*i*) Starting of D.C. motors (*ii*) Reversing of D.C. motors.
23. Explain briefly electric braking of D.C. motors.
24. Discuss briefly the following methods of speed control of D.C. motors :
(*i*) Field control (*ii*) Rheostatic control
(*iii*) Voltage control (*iv*) Thyristor control.
25. What do you understand by "Self governing propoerties of D.C. properties" ?
26. Explain briefly the following motors :
(*i*) Brushless D.C. motors (*ii*) Stepper motors.
27. Discuss briefly reluctance motors.
28. Explain briefly "Linear induction motor".
29. Define the following :
(*i*) Heating time constant. (*ii*) Cooling time constant.
30. State the factors which govern the size and rating of motor for any particular service.
31. Explain briefly the following ratings for electrical motors :
(*i*) Continuous rating (*ii*) Continuous maximum rating
(*iii*) Short time rating (*iv*) Intermittent rating.
32. What do you mean by load equilisation ?
33. Name the different methods for the speed control of a 3-phase induction motor and discuss briefly the principles underlying each. Describe in brief speed control of induction motor by ,pole changing.
34. Explain with suitable diagrams two methods each for controlling the speed of (*i*) a cage type induction motor, and (*ii*) a slip-ring type induction motor.
35. Write short notes on the following methods of speed control of induction motors :
(*i*) By pole changing (*ii*) By injecting e.m.f. into rotor circuit
(*iii*) By cascade connections (*iv*) By change of supply frequency.
36. Describe the main features of (*i*) plugging and (*ii*) rheostatic braking. What are their limitations ?
37. Describe various methods of electric braking used for braking of induction motors. Compare their advantages and disadvantages.
38. Explain with necessary diagrams the difference between regenerative braking and plugging with D.C. shunt motors.

39. What is regenerative braking ? Why it cannot be accomplished with a series wound D.C motor without modifications in the circuit, explain. Give its merits and demerits.

40. Specify the different types of enclosures of motors, highlighting the special feature of each type. Mention the area of application of each.

41. Discuss the different types of drives required for transmission of power from the driving machines to the loads.

42. Discuss the following features pertaining to the selection of a motor :

(*i*) Type of enclosures (*ii*) Bearings
(*iii*) Transmission of drive (*iv*) Noise.

43. State the factors that contribute to mechanical vibrations and noise.

44. Describe the following :

(*i*) Flat belt drive (*ii*) V-belt drive
(*iii*) Rope drive (*iv*) Chain drive
(*v*) Gear drive (*vi*) Direct drive.

45. Draw the heating and cooling curves of a motor and give the meaning of heating and cooling time constants.

46. What is heating time constant ? Explain how the rating of a motor is affected by the temperature rise ?

47. What is meant by rating of motors ? Discuss how the type and size of motors for intermittent loads is determined ?

48. Define (*i*) continuous rating and (*ii*) short time rating of a machine.
Explain how temperature rise is one of the chief features in fixing the size of a motor for a given purpose.

49. What do you understand by load equilisation ? Where do you find its applications ? Derive the expression for instantaneous motor torque, M.O.I. of the flywheel, and the motor slip. State any assumptions made.

50. What is the function of a flywheel in rolling mill drive ? Deduce an expression for the motor torque driving a rolling mill when equipped with a flywheel.

51. Give the applications of various types of motors.

52. Explain the factors that must be taken into account while selecting a motor for a particular work. On the basis of those factors suggest and justify motors for the following :

(*i*) Paper mill drive (*ii*) Cranes
(*iii*) Cement works (*iv*) Printing machinery.

53. Suggest suitable type of motors with reasons for any of the following three applications :

(*i*) Air compressor (*ii*) Crane
(*iii*) Electric lifts (*iv*) Ceiling fans
(*v*) Rolling mills.

54. Suggest the possible choice of motor (*b*) for the following services :

(*i*) Domestic uses (*ii*) Punches, presses and schears
(*iii*) Textile mills (*iv*) Belt conveyors.

UNSOLVED EXAMPLES

1. An unsaturated 220 V shunt motor, having an armature resistance of 0.3 ohm and a shunt field resistance of 50 ohms is driving a load of constant torque. The armature current taken by the motor at this torque and at a speed of 1000 r.p.m. is 60 A. Determine the additional resistance to be included in the shunt field circuit to raise the speed to 1200 r.p.m. **[Ans.** 11.24 ohms]

2. An unsaturated 200 V shunt motor has a no-load speed of 750 r.p.m. At full-load it takes an armature current of 90 A, running at 720 r.p.m. Armature and shunt field resistances are 0.1 ohm and 50 ohms respectively. If the load torque remains unaltered find the resistance of the field regulator to be included in the field winding to increase its speed on full-load to no-load speed. **[Ans.** 2.19 ohm]

3. A D.C. shunt motor runs at 900 r.p.m. from a 460 V supply when taking an armature current of 25 A. Calculate the speed at which it will run from a 230 V supply when taking an armature current of 15 A. The resistance of the armature circuit is 0.8 ohm. Assume the flux per pole at 230 V to have decreased to 75% of its value at 460 V. **[Ans.** 595 r.p.m.]

4. A 250 V shunt motor has an armature resistance of 0.5 ohm and runs at 1200 r.p.m. when the armature current is 80 A. If the torque remains unchanged, find the speed and armature current when the field is strengthened by 25 per cent. **[Ans.** 998 r.p.m., 64 A]

5. A 250 V D.C. shunt motor has an armature resistance of 0.2 ohm and shunt field resistance of 100 ohms. It takes 60 A when running at a speed of 600 r.p.m. If the speed is to be increased to 800 r.p.m., find the resistance of the field regulator to be inserted. Assume the magnetisation characteristic to be a straight line. **[Ans.** 33.4 ohms]

6. A 250 V shunt motor has a total armature resistance of 0.12 ohm and a shunt field resistance of 100 ohms. It runs at 750 r.p.m when taking a current of 150 A. If the machine is to be run as a generator delivering 150 A at 250 V, find :
   (*i*) The field current required at 750 r.p.m
   (*ii*) The speed at a field current of 2 A.
   Assume the magnetisation characteristic to be a straight line. **[Ans.** (*i*) 2.885 A ; (*ii*) 1081 r.p.m.]

7. A 500 V shunt motor has an armature resistance of 0.5 ohm and has a full-load intake current of 200 A. Determine the percentage change in flux required if the speed is to be increased to 1.5 times the full-load speed.
   Assume torque of the motor to be constant. **[Ans.** 50% reduction]

8. A 6-pole 220 V series motor takes 15 A when running at 800 r.p.m. with field coils connected in series. If the field coils are connected in two parallel groups of three in series find the speed and current taken by the motor. Armature and series field resistances are 0.2 ohm and 0.1 ohm respectively. The load torque remains unaltered. Assume that the flux is proportional to the current when the field coils are connected in series. **[Ans.** 1130 r.p.m., 21.21 A]

9. A 4-pole series wound fan motor runs normally at 600 r.p.m. on a 250 V D.C. supply taking 20 A. The field coils are all connected in series. Estimate the speed and current taken by the motor if the coils are reconnected in two parallel groups of two in series. The load torque increases as the square of the speed. Assume that the flux is directly proportional to the current and ignore losses. **[Ans.** 714 r.p.m., 33.64 A]

10. A 4-pole, 230 V series motor runs at 1000 r.p.m. when the load current is 12 A. The series field resistance is 0.8 ohm and armature resistance is 1.0 ohm. The series field coils are now regrouped from all in series to two in series with two parallel paths. The line current is now 20 A. If the corresponding weakening of the field is 15 per cent, calculate the speed of the motor. **[Ans.** 1163 r.p.m.]

11. A 500 V series motor has an armature resistance of 0.3 ohm and series field resistance of 0.2 ohm. It takes a current of 120 A at a speed of 600 r.p.m. Find the speed of motor if a divertor of resistance 0.6 ohm is connected across the series field, the load torque being kept constant.
    Neglect armature reaction and assume that the flux is proportional to the current. **[Ans.** 689 r.p.m.]

12. A D.C. series motor drives a load the torque of which varies as the square of speed. The motor takes a current of 15 A when the speed is 600 r.p.m. Calculate the speed and current when the motor field winding is shunted by a divertor of the same resistance as that of the field winding.
    Neglect all motor losses and assume that the magnetic circuit is unsaturated. **[Ans.** 25.23 A, 713.6 r.p.m.]

13. A series motor runs at 500 r.p.m. when taking a current of 60 A at 460 V. The resistance of the armature circuit is 0.2 ohm and that of the field winding is 0.1 ohm. Calculate the speed when a 0.15 ohm divertor is connected in parallel with the field winding.
    Assume the torque to remain unaltered and flux to be proportional to the field current. **[Ans.** 386 r.p.m.]

14. A 220 V D.C. series motor takes 40 A when running at 700 r.p.m. Calculate the speed at which the motor will run and the current taken from the supply, if the field is shunted by a resistance equal to the field resistance and the load torque is increased by 50 per cent. Armature resistance = 0.15 ohm, field resistance = 0.1 ohm.
    Assume that the flux per pole is proportional to the field current. **[Ans.** 794 r.p.m.]

15. A 250 V D.C. series motor takes a current of 50 A at 1000 r.p.m. Determine the speed and current when a divertor of resistance equal to double that of series field winding is connected across it. The load torque varies as cube of the speed. Neglect all losses.
Assume that flux is proportional to current. [**Ans.** 1085 r.p.m. 69.15 A]

16. A 250 V shunt motor has an armature resistance of 0.2 ohm and shunt field resistance of 250 ohms. It takes a current of 66 A at 1000 r.p.m. If the current taken remains unaltered find the resistance to be included in series with the armature circuit to reduce the speed to 750 r.p.m. [**Ans.** 0.911 ohm]

17. What value of resistance should be inserted in the armature circuit of a 200 V, 1000 r.p.m. shunt motor to reduce its speed to 800 r.p.m. when drawing 20 A. Armature resistance is 0.5 ohm. [**Ans.** 1.9 ohm]

18. A 200 V D.C. shunt motor takes 20 A at rated voltage and runs at 1000 r.p.m. Its armature resistance is 0.1 ohm. Calculate the value of additional resistance required in the armature circuit to reduce the speed to 800 r.p.m. when the load torque is independent of the speed.
Ignore the field current. [**Ans.** 1.98 ohm]

19. The speed of a 230 V shunt motor is to be reduced from 900 r.p.m. to 600 r.p.m. The resistance of the armature circuit is 0.3 ohm. When the motor current is 30 A, speed is 900 r.p.m. If the current reduces to 20 A at lower speed, what arrangement should be made ?
Neglect any effect not included in the data given. [**Ans.** 3.833 ohms]

20. The armature and shunt field resistances of a 600 V shunt motor are 1.5 ohm and 600 ohms respectively. When running unloaded the current taken is 3 A and the speed is 1000 r.p.m. Calculate the speed when motor is full-loaded and the total current drawn from the supply is 30 A. Estimate the speed at this load if :

(*i*) a resistance of 3 ohms is connected in series with the armature ; and

(*ii*) the shunt current is reduced by 10 per cent. [**Ans.** 932 r.p.m., 786.5 r.p.m., 873 r.p.m.]

21. A 220 V D.C. shunt motor with armature resistance of 0.5 ohm is excited to produce a constant main field. At full-load, the motor runs at 500 r.p.m. and consumes an armature current of 30 A. If a resistance of 1 ohm is placed in series with the armature current, find the speed at :

(*i*) full-load torque

(*ii*) thrice full-load torque. [**Ans.** 427 r.p.m., 207.3 r.p.m.]

22. A 200 V, 7.46 kW shunt motor has a full-load efficiency of 90 per cent and has an armature resistance of 0.25 ohm. It is running against the full-load torque at 600 r.p.m. Determine :

(*i*) the resistance to be connected in series with the armature to reduce the speed to 400 r.p.m if the torque was constant.

(*ii*) The speed if the armature current is reduced to 20 A and if the resistance found in (*i*) is in circuit.

[**Ans.** 1.525 ohm, 520 r.p.m.]

23. A 220 V, 75 kW D.C. shunt motor has a full-load speed of 1000 r.p.m. The resistance of the armature and field circuit are 0.25 ohm and 146 ohms respectively. The full-load efficiency of the motor is 85.2%. Neglecting brush drop and effects of armature reaction. Calculate :

(*i*) The speed at which the motor will run on load, if under these conditions its total input is 550 W ;

(*ii*) The value of the resistance to be added to the armature circuit to reduce the speed to 800 r.p.m. when giving full-load torque. [**Ans.** 1044 r.p.m., 1.093 ohm]

24. A 250 V, 11.19 kW shunt motor takes a current of 50 A on full-load, the full-load speed being 1000 r.p.m. The armature and shunt field resistances are 0.1 ohm and 250 ohms respectively. If a resistor is to be connected in series with the armature to reduce the speed to 0.75 times the full-load speed, find the resistance of the resistor and the efficiency at this speed.
Assume that the field and armature currents remain constant and constant losses vary as $(\text{speed})^{1.2}$.

[**Ans.** 1.251 ohm ; 67.41%]

25. A 250 V, 18.65 kW shunt motor has armature and shunt field resistances of 0.2 ohm and 125 ohms respectively. At no-load it runs at 800 r.p.m. taking a current of 5 A. At full-load it has an efficiency of 93.2 per cent. If the armature reaction weakens the field by 5 per cent on full-load, determine :

(*i*) Full-load speed ;

(*ii*) Full-load speed when a resistance of 2 ohms is connected in series with the armature.

[**Ans.** 790 r.p.m., 250.5 r.p.m.]

26. The armature and shunt field resistances of a 500 V shunt motor are 0.1 and 100 ohms respectively. A resistance of 1.4 ohm is connected in series with the armature and it runs at 1000 r.p.m., taking a current of 75 A. What must be the resistance of a divertor connected across the armature to reduce the speed to 600 r.p.m.

Assume armature current to be constant. **[Ans.** 2.161 ohms]

27. A 200 V D.C. series motor runs at 500 r.p.m. when taking a current of 25 A. The resistance of the armature is 0.5 ohm and that of field is 0.3 ohm. If the current remains constant, calculate the resistance necessary to reduce the speed to 250 r.p.m. **[Ans.** 3.6 ohms]

28. A 230 V series motor has a total resistance of 0.6 ohm. It runs at 1000 r.p.m. taking an input current of 15 A. Find the series resistance required to reduce the speed to 750 r.p.m., the input current being kept constant.

Assume the magnetisation characteristic to be a straight line. **[Ans.** 3.683 ohms]

29. A 240 V series lift motor has resistance of 0.2 ohm. At at speed of 1800 r.p.m. it takes 40 A. Find the resistance to be added in series with the motor to limit the speed to 3600 r.p.m. when the current is 10 A. Assume the flux to be proportional to the current. **[Ans.** 12.2 ohms]

30. A D.C. series motor when loaded to 50 A runs at 1500 r.p.m. on a 220 V supply. The motor resistance is 0.1 ohm and the flux may be taken as proportional to the current. Calculate the value of resistance to be inserted in series to bring down the speed to 900 r.p.m. at which the load torque is half the previous value. **[Ans.** 3.544 ohms]

31. A 200 V D.C. series motor runs at 1000 r.p.m. and takes 20 A. Combined resistance of armature and field is 0.4 ohm. Calculate the resistance to be inserted in series so as to reduce the speed to 800 r.p.m, assuming torque to vary as square of the speed and linear magnetisation curve. **[Ans.** 3.62 ohms]

32. A 440 V series motor has an armature resistance of 0.6 ohm and series field resistance of 0.4 ohm. It takes 60 A, when running at a speed of 1000 r.p.m. If a resistance of 20 ohms is connected across the armature and the torque remains constant determine the speed of the motor.

Assume straight line magnetisation. **[Ans.** 846. r.p.m.]

33. A series motor having a resistance of 1 ohm between terminals drives a fan for which the torque varies as square of the speed. On 200 V, it runs at 300 r.p.m and takes 25 A. The speed is to be raised to 400 r.p.m. by increasing the voltage. Find the voltage assuming flux is directly proportional to current.

[Ans. 380 V]

34. A D.C. series motor with unsaturated magnetic circuit has a resistance of 1 ohm between terminals. It drives a load whose torque is proportional to the square of the speed. At 400 V the motor runs at 600 r.p.m. and takes 36 A. It is required to reduce the speed of the motor to 500 r.p.m. by reducing the applied voltage. Calculate the voltage to be applied and the current drawn by the motor.

[Ans. 282.77 V, 30 A]

35. A series motor has an armature resistance of 0.3 ohm and series field resistance of 0.2 ohm. It runs at 600 r.p.m. when taking a current of 40 A at 400 V. If the speed is to be increased to 900 r.p.m. by increasing the supply voltage, determine the new supply voltage and current if the field is

(*i*) saturated (*ii*) unsaturated.

Assume that the torque varies as the cube of the speed. **[Ans.** 637.5 V, 135 A, 1088.8 V, 73.52 A]

36. A squirrel cage induction motor when started by means of a star-delta starter takes 180% of full load line current and develops 35% of full load torque at starting. Calculate the starting torque and current in terms of full load values, if an autotransformer with 70 percent tapping were employed.

[Ans. 0.5145 T_f, 2.646 I_f]

37. A 3 phase, 400 V, 4 pole, star-connected induction motor takes a full-load current of 10 A and a short circuit current of 60 A at 0.25 power factor. The full-load slip is 5%. Find the minimum starting current taken from the supply if the starting torque is not to be less than 60% of full-load torque, the motor being started by

(*a*) an auto transformer

(*b*) a series resistance in the stator circuit.

Find also the percentage tapping in case (*a*) and the magnitude of the resistance to be added per phase in case (*b*). **[Ans.** (*a*) Ist = 20 A ; *K* = 57.8% (*b*) Ist = 34.64 A ; 2.81 Ω]

38. A 220 V d.c. shunt motor has an armature resistance of 0.5 Ω and takes a current of 40 A on full load. By how much must the main flux be reduced to raise the speed by 50% if the developed torque is constant. **[Ans. 37.5%]**

39. A 500 V, d.c. series motor with a terminal resistance of 0.5 Ω drives a load, the torque of which varies as the square of the speed. When the speed is 600 r.p.m., the current drawn by the motor is 30 A. If the speed is to be reduce to 400 r.p.m., find the value of the resistance to be inserted in the motor circuit. Assume the magnetic circuit to be unsaturated. **[Ans. 13.72 Ω]**

40. A series motor working on 500 volts d.c. supply mains runs at a speed of 800 r.p.m. when the load is such that the current is 120 A. The resistance of the motor is 0.15 Ω of which 0.04 Ω is the resistance of field circuit. Assuming an unsaturated field calculate the speed of the motor when the torque is one half of the full-load value and the field winding is connected in parallel with a diverter resistance of 0.08 Ω.

[Ans. 1,395 r.p.m.]

41. A 240 V, d.c. series motor being used in a lift has a total resistance of 0.2 Ω. At a speed of 1800 r.p.m. it takes 40 A. Find the resistance to be added to limit the speed to 3,600 r.p.m. When the current is 10 A, assuming the flux to be proportional to the field current between 10 A and 40 A. **[Ans. 12.2 Ω]**

42. A 400 V, d.c. shunt motor takes full-load armature current of 150 A and has an armature resistance of 0.1 Ω. It is braked by plugging while running at 500 r.p.m. Calculate :

(*i*) The value of the resistance to be placed in series with the armature to limit the initial current to 200 A.

(*ii*) Initial value of the braking torque.

(*iii*) Value of the braking torque when the speed has fallen to 200 r.p.m.

[Ans. (*i*) 3.825 Ω (*ii*) 1,528 N-m (*iii*) 1,078 N-m]

43. A 50 h.p., 440 V, d.c. shunt motor is braked by plugging. Calculate the value of the resistance to be placed in series with the armature circuit to limit the initial braking current to 150 A. Calculate the braking torque so obtained and also the torque when the speed of the motor has fallen by 40%. Assume the following data :

Armature resistance = 0.1 Ω; Full load armature current = 100 A Full load speed = 600 r.p.m.

[Ans. 5.7 Ω; 878, N-m ; 1704 N-m]

44. The armature and field resistances of a 250 V, 25 h.p., 750 r.p.m. d.c. shunt motor are 0.2 Ω and 100 Ω respectively. The motor has a full-load efficiency of 83%. The motor is to be braked by plugging. If the initial current is to be limited to 1½ times the full load current, what resistance must be inserted in series with the armature ? Calculate the initial braking torque and also the torque when the speed of the motor has fallen by 40%. **[Ans. 3.54 Ω; 351.2 N-m ; 283.5 N-m]**

45. A 400 V, 50 Hz, 4-pole cage type induction motor is braked by plugging, while working at full load delivering 50 h.p. ; the full load slip being 4%. The ratio of rotor resistance to stand still reactance = 1.5. Calculate the initial braking torque in terms of full load torque. **[Ans. 128 N-m]**

46. A 40 h.p., 400 V. 3-ϕ, 4-pole, 50 Hz induction motor has full load slip of 5%. If the ratio of standstill reactance to resistance per motor phase is 4, estimate the plugging torque at full speed.

[Ans. 129.3 N-m]

47. Rheostatic braking is used to limit the speed of a series type d.c. crane motor to 500 r.p.m. when the supply is cut off. The motor has a resistance of 1.35 Ω and the descending load exerts a torque of 442 N-m. Determine the resistance to be connected across the motor terminals.

The magnetisation characteristic of the machine at 500 r.p.m. is as follows :

Current (A)	20	30	40	50
E.M.F. (V)	360	440	490	520

State assumptions made if any. **[Ans. 10.005 Ω]**

48. Magnetisation curve of a d.c. series motor at 500 r.p.m. is given as follows :

Field current in Amps	20	40	60	80	100
Induced voltage	50	100	125	140	150

Total resistance of armature circuit including the field is 0.03 Ω. Motor is driving a hoist-carrying a load of 30 kg at a radius of 50 cm. Estimate the resistance to be connected in the motor circuit for applying rheostatic braking to limit the speed to 400 r.p.m. **[Ans.** 1.63 Ω]

49. A machine on a short test run gave the following mean temperature rise :

Time hours	0	0.25	0.5	0.75	1.0	1.25	1.5
Temp. rise °C	42.5	45.0	47.4	49.5	51.4	53.0	54.4

Estimate graphically the final temperature rise and the time constant. Justify the procedure adopted by you. **[Ans.** 62°C, 1.6 hours]

50. A motor driving some haulage equipment for a mine has to deliver a load which follows the following duty cycle :

100 h.p. for 10 minutes, no-load for 4 minutes.

50 h.p., for 10 minutes, no-load for 6 minutes,

The cycle being repeated indefinitely. Estimate the suitable size of continuously rated motor for the purpose. **[Ans.** 65 h.p.]

51. The load cycle of a haulage system is given below :

(*a*) uniform acceleration for 30 s with load increasing from 0 to a maximum of 2,500 h.p.

(*b*) Steady full load of 1,500 h.p. for 50 s.

(*c*) Regenerative braking for 15 s with the power returned to supply reducing uniformly from 500 h.p. to zero.

(*d*) Re-loading for 25 s.

Estimate the suitable rating for the motor. **[Ans.** 1,200 h.p.]

52. A 10 h.p. motor has a final temperature rise of 20°C at normal rated load. Find the half hour rating and the corresponding temperature rise if heating time constant of the motor is 30 minutes. The losses are proportional to the square of the load ? **[Ans.** 12.6 h.p. 31.64°C]

53. A 25 h.p. motor has a heating time constant of 90 minutes and when run continuously on full load attains temperature of 40°C above the surrounding air. What would be the ½ hour rating of the motor for this a temperature rise assuming that it cools down completely between each load period and that the losses are proportional to the square of the load ? **[Ans.** 47 h.p.]

54. A transformer has a heating time constant of 4 hours. Compare the permissible loadings of the transformer, when rated continuously and on one-hour-basis if iron losses are negligible in comparison with copper loss. **[Ans.** One-hour-basis rating will be 2.12 times of continuous rating]

55. Assuming an exponential law of temperature rise, determine the final steady temperature on full load and the time constant for an electric motor whose temperature rise after one hour is 25°C and after two hours, it is 45°C. **[Ans.** 125°C ; 4½ hours]

56. A 500 r.p.m. synchronous motor and its load have a total moment of inertia of 420 kg-m^2 and a frictional torque of 140 N-m. The motor is braked by rheostatic method of field current and the braking resistance being kept constant at value which gives an initial electric braking torque of 5400 N-m. Find the time taken to bring the motor to rest. **[Ans.** 15 seconds]

57. A motor fitted with a flywheel supplies a load torque of 150 kg-m for 15 seconds. During no load period the flywheel regains the original speed. The motor torque is required to be limited to 85 kg-m. Determine the moment of inertia of the flywheel. The no-load speed of the motor is 500 r.p.m and it has a slip of 10% on full load,. Draw time versus speed, load torque and motor torque curves. Derive the formula used. **[Ans.** 2,860 kg-m^2]

58. A motor fitted with a flywheel supplies a load torque of 100 N-m for 20 seconds. During the no-load period, the flywheel regains its original speed. The motor torque is required to be limited to 500 N-m. Determine the moment of inertia of the flywheel. The no-load speed of the motor is 500 r.p.m. and it has a slip of 10% on full load. **[Ans.** 2,755 kg-m^2]

59. A d.c. shunt motor 30 h.p., 400 V, 500 r.p.m. draws a current of 40 A when running at full load. The moment of inertia of the rotating system is 12.4 kg-m^2. If the current during the starting period varies from 1.05 to 1.25 times of full load current, calculate (*i*) the full load torque, and (*ii*) the time required for the motor to attain the rated speed against full load. **[Ans.** (*i*) 421.4 N-m (*ii*) 10.27 seconds]

60. A 6-pole, 50 Hz induction motor has a flywheel of 1200 kg-m^2 moment of inertia. Load torque is 100 kg-m for 10 seconds. No load period is long enough for the flywheel to regain its full speed. Motor has a slip of 6% at a torque of 50 kg-m. Calculate (*i*) maximum torque exerted by motor (*ii*) speed at the end of deceleration period. **[Ans.** (*i*) 469 N-m (*ii*) 925 r.p.m.]

61. A motor driving a colliery winding equipment has to deliver a load rising uniformly from zero to a maximum of 1500 kW in 20 sec. during the accelerating period. 750 kW for 40 sec. during the full speed period and during the deceleration period of 10 sec. when regenerative braking is taking place from an initial value of 250 kW to zero and then a no load period of 20 sec. Estimate remittable kW rating of the motor. **[Ans.** 648 kW]

62. A constant speed drive has the following duty cycle :

Load rising from 0 to 400 kW – 5 minutes

Uniform load of 400 kW – 5 minutes

Regenerative power of 400 kW returned to supply – 4 minutes

Remains idle for – 2 minutes

Estimate power rating of motor. **[Ans.** 380 H.P.] **[Nagpur University Winter 1996]**

63. Determine the rated current of a transformer for the following duty cycle :

- 500 A for 3 minutes
- Sharp increase to 1000 A and constant at this value for 1 minute
- Gradually decreasing to 200 A, for 2 minutes
- Constant at this value for 2 minutes
- Gradually increasing to 500 A during 2 minutes repeated indefinitely. **[Ans.** 540 A]

64. An induction motor has to perform the following duty cycle :

75 KW for 10 minutes. No load for 5 minutes

45 KW for 8 minutes. No load for 4 minutes

which is repeated indefinitely.

Determine suitable capacity cf a continuously rated motor. **[Ans.** 570 H.P.]

65. A 25 H.P. motor has heating time constant of 90 min. and when run continuously on full load attains a temperature of 45°C, above the surrounding air. What would be the half hour rating of the motor for this temperature rise, assuming that it cools down completely between each load period and that the losses are proportional to square of the load. **[Ans.** 47 H.P.]

66. At full load of 10 H.P., temperature rise of a motor is 25°C, after 1 hr. and 40°C after 2 hrs. Find (*a*). Heating time constant of motor, (*b*) Final temperature rise on full load. **[Ans.** T = 1,96 hrs, θf = 62.50]

67. A totally enclosed motor has a temperature rise of 20°C after half an hour and 35°C after one hour on full load. Determine temperature rise after 2 hours on full load. **[Ans.** 54.680 C]

68. A 25 H.P., 3-ϕ, 10 pole, 50 c.p.s. induction motor provided with a flywheel has to supply a load torque of 800 N-m for 10 sec, followed by a no load period, during which the flywheel regains its full speed. Full load slip of motor is 4% and torque-speed curve may be assumed linear over the working range. Find moment of inertia of flywheel, if the motor torque is not to exceed twice the full load torque. Assume efficiency = 90%. **[Ans.** 718 kg-m^2]

69. A motor fitted with a flywheel has to supply a load torque of 200 kg-m for 10 sec. followed by a no-load period. During the no load period, the motor regains its speed. It is desired to limit the motor torque to 100 kg-m. What should be the moment of inertia of flywheel. No load speed of motor is 500 r.p.m. and has a slip of 10% at a torque of 100 kg-m. **[Ans.** I = 2703 kg-m^2]

Heating and Cooling Curves

1. A field coil has a dissipating surface of 1500 cm^2 and a length of mean turn of 100 cm. It dissipates a loss of 150 W, the emissivity being 34 W/m^2-°C. Estimate the final steady temperature rise of the coil and its time constant if the cross-section of the coil is 10 × 5 cm. Specific heat of copper is 397 J/kg-°C, space factor = 0.56, density of copper = 8.9 gm/cm³. **[Ans. 29.4°C, 1940 s]**

2. The initial temperature of a machine is 40°C. Calculate the temperature of the machine after 1 hour if its final steady temperature rise is 80°C and the heating time constant is 2 hours. The ambient temperature is 30°C. **[Ans. 67.51°C]**
3. A D.C. machine is heated to a temperature of 60°C and is then shut down. Calculate its temperature at a time 20 minutes after the shut down if the cooling time constant is 60 minutes. The ambient temperature is 30°C. **[Ans. 51.5°C]**
4. During a test on a D.C. machine the load was kept constant and the rates of temperature rise obtained were 0.075°C per minute when temperature rise was 19°C and 0.055°C when the temperature rise was 27°C. Estimate the thermal constant and the final steady temperature rise. **[Ans. 400 minutes, 49°C]**
5. The outside of a 8.8 kW totally enclosed motor is equivalent to a cylinder of 65 cm diameter and 1 metre length. The motor weighs 400 kg and may be considered to be of a material having specific heat of 700 J/kg-°C. The outer surface is capable of dissipating heat at the rate of 12.5 W/m²-°C. Determine the final temperature rise and heating time constant of the motor when operating at full-load with an efficiency of 90%. **[Ans. 38.3°C, 3.0486 hours]**
6. Assuming an exponential law of temperature rise, calculate the final steady temperature on full-load and the time constant for an electric motor whose temperature rise after one hour is 25°C and after 2 hours, it is 45°C. **[Ans. 125°C, 4.5 hours]**
7. In a temperature-rise test on a D.C. machine at full-load, the following rises were observed : 16°C after one hour, 28°C after 2 hours. Find :
   (*i*) Final steady temperature rise and the time constant.
   (*ii*) The steady temperature rise after one hour at 50% over-load, from cold.
   Assume that the final temperature rise on 50% overload is 92°C. **[Ans. 64°C, 3.481 hours ; 23°C]**
8. A D.C. machine with a thermal constant of 45 minutes has a final temperature rise of 75°C on continuous rating. Find :
   (*i*) What is temperature rise after one hour at this load ?
   (*ii*) If the temperature rise on one hour rating is 75°C, find the maximum steady temperature at this rating.
   (*iii*) When working at its one-hour rating, how long does it take the temperature to increase from 60°C to 75°C ? **[Ans. 55°C, 102°C, 20 min.]**
9. A 50 kW motor has a heating time constant of 100 minutes and when run continuously on full-load attains a temperature of 40°C above the surrounding air. What would be the $\frac{1}{2}$ hour rating of the motor for this temperature rise, assuming that it cools down completely between each load period and the losses are proportional to the square of the load ? **[Ans. 98.3 kW]**
10. A totally-enclosed motor has a temperature rise of 20°C after half an hour and 35°C after one hour on full-load. Determine the temperature rise after two hours on load. If a fan is fitted driving cooling air over the external surface of the machine, which reduces the final steady state temperature rise of the machine to 80 per cent of the original value, determine the temperature rise from cold after one hour on full-load. **[Ans. 54.7°C, 32.8°C]**

6

Electric Traction

6.1. Introduction. 6.2. Traction systems. 6.3. Requirements of an ideal traction system. 6.4. Different systems of traction. 6.5. Systems of railway electrification. 6.6. Comparison between pure A.C. and D.C. systems. 6.7. Comparison between D.C. and A.C systems of railway electrification from the point of view of main line and suburban line railway service. 6.8. Electric traction systems-power supply—Transmission lines to sub-stations—Sub-stations—Feeding and distribution system on A.C. traction—Feeding and distribution system for D.C. tramways—Electrolysis by currents through earth—Negative boosters—Problems associated with single-phase traction system. 6.9. **A.C. locomotive. 6.10. Tramways.** 6.11. **Trolley-bus.** 6.12. **Diesel electric traction. 6.13. Overhead equipment**—Collector gear for overhead equipment—The conductor-rail equipment/system. 6.14. **Power factor and harmonics**—Highlight—Objective Type Questions—Theoretical Questions.

6.1. INTRODUCTION

The locomotion in which the driving or tractive force is obtained from electric motors is called **Electric traction.**

Electric traction has many *advantages* as compared to other non-electrical systems of traction including steam traction.

Electric traction is *used* in :

(i) *Electric trains*
(ii) *Tram cars*
(iii) *Trolley buses*
(iv) *Diesel-electric vehicles etc.*

A very *brief history of railways in India* (leading to present status of Indian railways among electrified railway in the world) is as follows :

- In India, railways first appeared in April, 1853, when a section from Bombay to Kalyan (53 km) was opened to traffic. The management of railways was left to private companies (on lease basis).
- In 1920 the Government of India set up a committee to go into the justification of nationalisation of the whole railway system. Subsequently, as per the recommendations of the committee, railways were gradually taken over by the Central Government and eventually (within about 25 years) the railways were gradually reorganised by the government and renamed on zonal basis.
- In the year 1890, the electric traction was *first introduced on British Railways at 600 V D.C. using third rail.* This system was not considered desirable specially for open lines, since although the cross-section of rail was enough to carry heavy traction currents because of low voltage, the use of third rail even at 600 V was fraught with danger to life ; furthermore large currents lead to heavy losses. Then efforts were made *towards increasing the voltage.* Raising of the line voltage to 1500 V was possible only due to the advances made in the design and manufacturing techniques of rotary converters.
- During the period 1925 to 1932 the *first electrification was undertaken in India.*

Ample progress took place in the following years through a lot of experimentation on the various types of motors to be used for electric traction.

- Railway board in 1957 after making sure that single-phase traction load will not produce adverse effects of unbalance and harmonics in three-phase power supply, decided to adopt 25 kV, 50 cycle single-phase system of electrification for all future schemes. (The various systems of railway electrification are discussed in detail in art. 6.5).

6.2. TRACTION SYSTEMS

All traction systems, broadly speaking, can be *classified* as follows :

1. Non-electric traction systems. These systems *do not use* electrical energy at any stage.

Examples : (i) Steam engine drive used in railways.

(ii) Internal combustion-engine-drive used for road transport.

2. Electric traction systems. These systems *involve the use* of electric energy at some stage or the other.

These are further sub-divided into the following two groups :

(*a*) ***Self contained vehicles or locomotives***

Examples : (*i*) Battery-electric drive

(*ii*) Diesel-electric drive etc.

(*b*) ***Vehicles which receive electric power from a distribution network or suitably placed sub-stations.***

Examples : (*i*) Railway electric locomotive fed from overhead A.C. supply ;

(*ii*) Tramways and trolley buses supplied with D.C. supply.

6.3. REQUIREMENTS OF AN IDEAL TRACTION SYSTEM

The requirements of an ideal traction system are :

1. *High adhesion coefficient,* so that high tractive effort at the start is possible to have rapid acceleration.
2. The locomotive or train unit should be *self contained* so that it can run on any route.
3. *Minimum wear* on the track.
4. It should be *possible to overload the equipment for short periods.*
5. The equipment required should be *minimum*, of high efficiency and low initial and maintenance cost.
6. It should be *pollution free.*
7. Speed control should be *easy.*
8. Braking should be such that *minimum wear* is caused on the brake shoes, and if possible the energy should be regenerated and returned to the supply during braking period.
9. There should be no interference to the communication lines (Telephone and telegraph lines) running the track.

6.4. DIFFERENT SYSTEMS OF TRACTION

The description of commonly used *traction systems* is given below.

1. Steam engine drive. Steam engine drive, though losing ground gradually due to various reasons, it is still the amply adopted means of propulsion of railway work in underdeveloped countries. In this type of drive, the reciprocating engine is invariably used for getting the necessary motive power.

Advantages. Following are the *advantages* of steam engine drive :

1. Simplicity in design.
2. Simplified maintenance.
3. Easy speed control.
4. Simplicity of connections between the cylinders and the driving wheels.
5. No interference with communication network.
6. Low capital cost as track electrification is not required.
7. The locomotive and train unit is self contained, therefore, it is not tied to a route.
8. It is cheap for low density traffic areas and in initial stages of communication by rail.
9. Operational dependability.

Disadvantages. This system, because of the following *disadvantages*, is being replaced even by the underdeveloped countries by either diesel electric or straight electric systems :

1. Low thermal efficiency.
2. The steam engine system is available for haulage for about 60 percent of its working time, the remainder of the time is spent in preparing for service, maintenance and overhaul.
3. Owing to unbalanced reciprocating parts there is a considerable wear on the track, and also riding qualities are not good.
4. Due to the reason of low adhesion coefficient, power-weight ratio of steam locomotive is low.
5. It has strictly limited overload capacity.
6. Steam locomotive cannot be put into service at any moment as time is required for raising of steam.
7. Owing to high centre of gravity of steam locomotive, speed is limited.
8. Steam locomotive requires more repair and maintenance.
9. Steam locomotive has to carry sufficient quantity of coal and water which correspondingly reduces the pay load that can be hauled.
10. Extensive and costly auxiliary equipment.
11. Since driving wheels are very close, hence more concentrated adhesive weight is required.
12. Steam locomotive service is not clean owing to coal dust, cinder ash etc.
13. Bigger sizes of running sheds and workshop are required.

2. Internal combustion (I.C.) engine drive. This drive is widely used for road transport (buses, trucks, cars etc.). It has an efficiency of about 25 percent when operating at normal speed.

Advantages :

1. Low initial investment.
2. It is self-contained unit and, therefore, it is not tied to any route.
3. Easy speed control.
4. Very simple braking system.
5. It is cheap drive for the outer suburbs and country districts.

Disadvantages :

1. Limited overload capacity.
2. A gear box is essential for speed control.
3. Higher running and maintenance costs.
4. Operation at any but the normal speed is uneconomical.
5. The life of propulsive equipment is much shorter than that of electrical equipment of a [illegible] car or a trolley bus.

6. Unsuitable for heavy railway work as the arrangement possible for speed control and utilizing the engine torque to full capacity when running to full speed is not altogether satisfactory for it.

3. I.C. engine electric drive. In an I.C. engine electric drive the reduction gear and gear box are eliminated as the diesel engine is to drive the D.C. generator coupled to it at a constant speed. This type of drive is finding considerable favour for railway work and *locomotives* of this type are being widely used.

Advantages :

1. Low initial investment (since no overhead structure distribution system and equipment are required).

2. No modification of existing tracks is required while converting from steam to diesel electric traction.

3. As the locomotive and train is a self contained unit, therefore, it is not lied to any route.

4. Can be put into service at any moment.

5. Loss of power in speed control is very low (since it can be carried out by the field control of generator).

6. It is available for hauling for about 90% of its working days.

7. Overall efficiency (about 25%) is greater than that of steam locomotives (about 8% or so).

Disadvantages :

1. Limited overload capacity (since diesel engine cannot be overloaded).

2. High running and maintenance cost.

3. Higher dead weight of locomotives ; more axles required comparatively (six for diesel electric locomotives in comparison with four for electric locomotive).

4. Comparatively costlier than steam or electric locomotives.

5. In such drives, regenerative braking cannot be used (but there is no bar in making use of rheostatic braking).

6. The life of the diesel engine is comparatively shorter.

7. There is a necessity to provide special cooling system for the diesel engine in addition to motor-generator set.

4. Petrol-electric traction :

- This system, due to electric conversion, *provides a very fine and continuous control* which makes the vehicle capable of moving slowly at an imperceptible speed and creeping up the steepest slope without throttling the engine.
- Petrol-electric traction is employed in heavy lorries and buses.

5. Battery electric drive :

- In this system the locomotive carries the secondary batteries which supply power to D.C. motors employed for driving the vehicle.
- This type of drive is well suited for *frequently operated service* such as for local delivery of goods in large towns with maximum daily run of 50 to 60 km, shunting and traction in industrial works and mines.
- Battery vehicles are started by series-parallel grouping of batteries in parallel for starting and running at the speed upto half maximum speed and in series for running at full maximum speed.

Advantages :

1. Battery driven vehicle is easy to control and very convenient to use.

2. Low maintenance cost.

3. Absence of fumes.

Disadvantages :

1. The major disadvantage of this type of drive is the *small capacity of batteries and the necessity for frequent charging.*

2. Limited speed range.

6. Electric drive. Here the drive is by means of electric motors which are fed from overhead distribution system. The drive of this type is most widely used.

Advantages :

1. As it has *no smoke*, electric traction is most suited for the underground and tube railways.

2. The motors used in electric traction have a very high starting torque. Hence, it is possible to achieve higher accelerations of 1.5 to 2.5 km/h/s as against 0.6 to 0.8 km/h/s in steam traction. Consequently, we have the following *advantages* :

(*i*) *High schedule speed ;*

(*ii*) *Increased traffic handling capacity ;*

(*iii*) Due to high schedule speed and high traffic handling capacity, *less terminal space is required*, this is an important factor in urban areas.

3. An electric locomotive is *ready to start at moment's notice* against about two hours required for steam locomotive to heat up.

4. An electric locomotive can *negotiate curves at higher speeds quite safely*, since its centre of gravity is low in comparison to a steam locomotive.

5. The *maintenance cost* of an electric locomotive is *50 per cent* of that of steam locomotive, its maintenance time is also much less comparatively.

6. In electric traction the *adhesion coefficient is more,* due to the absence of unbalanced forces produced by reciprocating masses. This not only reduces the weight to H.P. ratio of the locomotive but also improves the riding qualities and reduces the wear and tear of the track.

7. By the use of electric traction *high grade coal can be saved*, since electric locomotives can be fed either from hydroelectric stations or pit-head thermal power stations which use cheap low-grade coal.

8. In electric traction system it is *possible to use regenerative braking*, which leads to the following *advantages* : (*i*) About 80 percent of the energy taken from the supply during ascent is returned to it during descent ; (*ii*) Goods traffic on gradients becomes safer and speedier ; (*iii*) Since mechanical brakes are used to a very small extent, maintenance of brake shoes, wheels tyres and track rails is considerably reduced due to less wear and tear.

9. Owing to complete absence of smoke and fumes, this system is *healthier from the hygienic point of view.*

10. The *vibrations* in electrically operated vehicles are *less* as the torque exerted by the electric motor is continuous.

11. Electric equipment *can withstand large temporary overloads* and *can draw relatively large power* from the distribution system.

12. There is *no need to provide Rosenberg generators*, since the electrical energy required for lights and fans can be drawn from the lines directly.

13. Railway electrification *encourages rural electrification* as no special transmission lines have to be run for this purpose.

14. In electric section it is *possible to instal power units* in two or more power cars in the train and to control them from one point whereas in steam traction each locomotive is to be manned by its own crew.

Disadvantages :

1. High initial cost of laying out overhead electric supply system. Unless the traffic to be handled is heavy, electric traction becomes uneconomical.

2. Power failure for a few minutes can cause traffic dislocation for hours.

3. The electric traction system is tied up to only electrified routes.

4. Communication lines which usually run parallel to the power supply lines suffer from electrical interference. Hence, these communication lines have either to be removed away from the rail track or else underground cables have to be used for the purpose which makes the system still more expensive.

5. Additional equipment is required for regeneration. In case of D.C. series motors regeneration is not a simple process.

6. In case of electric traction provision of a negative booster is essential. By avoiding the flow of return currents through earth, it curtails corrosion of underground pipe work and interference with telegraph and telephone circuits.

7. Whereas steam locomotives can use their steam for heating the compartments in cold weather very cheaply, the electric locomotives have to do it at an extra cost.

8. In cold countries a service locomotive is required to run up and down the line in order to prevent the formation of layer of ice on the conductor rails.

6.5. SYSTEMS OF RAILWAY ELECTRIFICATION

Following are the *four types of track electrification systems presently available* :

1. D.C. system ...... 600 V, 750 V, 1500 V, 3000 V
2. Single-phase A.C. system ...... 15 to 25 kV, $16\frac{2}{3}$, 25 and 50 Hz
3. Three-phase A.C. system ...... 3.3 to 3.6 kV at $16\frac{2}{3}$ Hz
4. Composite system ...... involving conversion of single-phase A.C. into 3-phase A.C. or D.C.

1. D.C. system :

- In D.C. system the electric motors employed for getting the necessary motive power are usually D.C. series motors, although compound motors are coming into favour for tramways and trolley buses where regenerative braking is desired.
- Direct current at 600 to 750 V is universally employed for tramways in urban areas and for many suburban railways while 1500 to 3000 V D.C. is used for main line railways. The current collection is from third rail or conductor rail upto 750 V, where large currents are involved and from overhead wire for 1500 V and 3000 V, where small currents are involved. Both of these systems are fed from substations which are spaced 3 to 5 km for heavy suburban traffic and 40 to 50 km for main lines operating at higher voltages of 1500 to 3000 V. These sub-stations themselves receive power from 110/132 kV, 3-phase network or grid. This high-voltage 3-phase supply, at these substations, is converted into low-voltage 1-phase supply with the help of Scott-connected or V-connected 3-phase transformers. Then, this low A.C. voltage is converted into the required D.C. voltage by using suitable rectifiers or converters (*e.g.,* rotary converter, mercury-arc, metal or semiconductor rectifiers). These sub-stations are usually *automatic* and are *remote-controlled.* The D.C. supply so obtained is fed via suitable contact system to the traction motors which are either D.C. series motors for electric locomotive or compound motors for tramway and trolley buses where regenerative braking is desired.

For *heavy suburban service, low voltage D.C. system claims the following* ***advantages*** *over single-phase A.C. system mentioned below :*

(*i*) When operating under similar conditions, D.C. train consumes less energy than a single-phase A.C. train.

(*ii*) No interference with overhead communication lines.

(*iii*) Equipment of D.C. train is lighter, less costly and more efficient than similar A.C. equipment.

(*iv*) For frequent and rapid acceleration of heavy trains, D.C. motors are better suited than A.C. motors.

(*v*) For distribution system, the conductor rail is less costly, both initially and in maintenance than the high-voltage overhead A.C. distribution system.

The main ***disadvantage*** of D.C. system is the *necessity of locating A.C./D.C. conversion substations at relatively short distances apart.*

2. Single-phase A.C. system :

In this system *A.C. series motors are used* for getting the necessary motive power.

The voltage used for distribution network is 15000 to 25000 V at $16\frac{2}{3}$ or 25 Hz, which is stepped down on the locomotive to a low voltage of 300 to 400 V suitable for supplying to single-phase A.C. series motors by means of a step-down transformer carried in the locomotive.

When the extent of electrification is within a radius of about 30 km from the generating station, the distribution network may be fed directly, at high voltage, from a *generating station.* For longer distances the economic voltage for the power transmission system is higher than that desirable for traction system, and therefore transformer sub-stations are employed for feeding the distribution network. Owing to low current requirement at high voltage the spacing of sub-stations is 50 to 80 km. The sub-stations receive power at high voltage either from the existing extra high voltage transmission lines or from a separate system employing single-phase generating stations and step down the voltage to the required value and/or change the supply frequency through frequency converters. The change of supply frequency becomes necessary because of better performance of the A.C. series motors at low frequency.

Low frequency A.C. supply entails the following ***advantages :***

(*i*) Low frequency A.C. supply, apart from *improving the commutation properties of A.C. motors, increases their efficiency* and *power factor.*

(*ii*) At low frequency, line reactance is less so that line impedance drop and hence line voltage drop is reduced. Because of this *reduced line drop,* it is feasible to space the sub-stations 50 to 80 km apart.

(*iii*) Telephonic interference is reduced.

- *Single-phase A.C. system is invariably adopted for the main line service where the cost of overhead structure is of more importance and rapid acceleration and retardation is not so important as for suburban railways :*

3. Three-phase A.C. system. In this system the required motive power is obtained by the use of 3-phase induction motors operating at 3.3 kV to 3.6 kV, $16\frac{2}{3}$ Hz supply. Sub-stations receive power at a very high voltage from 3-phase transmission lines at the usual industrial frequency of 50 Hz. The high voltage is stepped down to 3.3 kV or so by transformers whereas frequency is reduced from 50 Hz to $16\frac{2}{3}$ Hz by frequency converters installed at the sub-stations.

The distribution system consists of *two overhead wires and track rail for the third phase* and receives the power either directly from the generating station or through the transformer sub-stations. The sub-stations receive power from high voltage lines at power frequency and step down the voltage and change the frequency.

Induction motors employed in this system are *quite simple* and *robust* and give *trouble-free operation*. They possess merit of high efficiency and of operating as a generator when driven at speeds above the synchronous speed. Hence, they have the property of *automatic regenerative braking during the descent on gradients*. However, besides these advantages, the system is no longer likely to be adopted because of the following inherent ***disadvantages*** :

(*i*) The constant-speed characteristics of induction motors are not suitable for traction work.

(*ii*) Since the speed/torque characteristics of induction motors are similar to D.C. shunt motors, therefore, they are not suitable for parallel operation (because motors will become overloaded very unevenly, even with little difference in rotational speeds caused by unequal diameters of the wheels).

(*iii*) At crossings and junctions the overhead contact wire system becomes complicated.

- This system is *used* in *some hilly areas*, where *output power required is high and regeneration on large scale is possible.*

4. Composite system. Two composite systems, presently, in use are :

(*a*) *Single-phase to three-phase system* (*or Kando system*).

(*b*) *Single-phase to D.C. system.*

(a) Single-phase to three-phase system or Kando system. This system (developed in Hungary in 1932) consists of 16 kV, 50 Hz single-phase overhead supply being converted to three-phase supply by means of phase converter equipment carried on the locomotive. This three-phase supply is fed to three-phase induction motors.

The *advantage* of this system is that complicated overhead trolley wire equipment of ordinary three-phase system is replaced by a single-wire system. By using S.C.R. (Silicon Controlled Rectifier) as inverter, it is possible to get variable-frequency 3-phase supply at $\frac{1}{2}$ to 9 Hz frequency. At this low frequency, 3-phase motors develop high starting torque without taking excessive current. There exists good possibility of this traction system to develop for further adoption.

(b) Single-phase A.C. to D.C. system. This system employs overhead single-phase 25 kV, 50 Hz supply which is stepped down by the transformer in the locomotive. This single-phase supply is then converted to D.C. and used in driving D.C. series traction motor.

This system has been adopted for all future track electrification in India.

Advantages :

(*i*) Higher starting efficiency.
(*ii*) Higher adhesion coefficient.
(*iii*) Light overhead catenary.
(*iv*) Simplicity of sub-station design.
(*v*) Flexibility in location of sub-stations.
(*vi*) Less number of sub-stations.
(*vii*) Lower cost of fixed installations.
(*viii*) A.C. locomotive has less kW demand at starting than that of a D.C. locomotive owing to elimination of starting resistances in A.C. locomotives (This is of particular importance at peak traffic hours specially when a number of locomotives are to be started simultaneously after an interruption of supply).

Disadvantages :

(*i*) Single-phase A.C. system produces both current and voltage unbalancing effect on the supply.

(*ii*) It produces interference in telecommunication circuits.

Both of the *shortcomings* mentioned *can be minimised* by taking suitable *precautions* mentioned below :

- Drawing supply for the traction sub-stations at very high voltage (say 110 kV or 132 kV) from a supply system having high capacity.
- Balancing traction loads equally on all the three phases, by connecting different traction sub-stations across different phases in rotation.
- Using the Scott or T-connected three phase/two phase transformers in the traction sub-stations.
- Providing isolating transformers which reduce the induced voltages and keep them below 60 V under operating conditions.
- Using booster transformer and return conductor for suppressing interference at the source in case of thickly populated areas.
- Replacing open aerial communication lines by lead or aluminium sheathed underground cables and earthing the sheathing of the cables at regular intervals.

6.6. COMPARISON BETWEEN PURE A.C. AND D.C. SYSTEMS

The comparison between pure A.C. and D.C. systems is given below :

S. No.	*Aspects*	*D.C. series motors*	*A.C. series motors*
1.	*Development of starting as well as running torque for the same size*	More	Less
2.	*Number of speeds*	Limited number of speeds (except by chopper method)	Many speeds (By tap changing method)
3.	*Cost, weight, efficiency for the same H.P.*	Less costly, lighter, more efficient	More costly, heavier, less efficient comparatively
4.	*Maintenance required*	Less	More
5.	*Regenerative braking*	More efficient and less complications	Less efficient and more complications comparatively
6.	*Negative boosting required*	Less elaborate	More elaborate (due to high impedance of track)
7.	*Number of sub-stations required for a given track kilometerage*	More in number	Less in number
8.	*Overhead distribution system*	Heavier and more costly comparatively	Lighter and less costly
9.	*Interference with communication lines*	Low	High
10.	*Rail conductor system of track electrification*	Possible with D.C. system	Not possible

- *The composite system* (as stated earlier) *of employing A.C. for track electrification and rectifying the same in locomotive for feeding D.C. motors is the best arrangement which combines good points of both A.C. and D.C. systems.*

6.7. COMPARISON BETWEEN D.C. AND A.C. SYSTEMS OF RAILWAY ELECTRIFICATION FROM THE POINT OF VIEW OF MAIN LINE AND SUBURBAN LINE RAILWAY SERVICE

A. **Main line railway service.** Following are the most important *requirements* of main line railway service :

1. *Minimum cost* of overhead structure.
2. *Higher maximum speed.*
3. *Acceleration and retardation are not so important* for main line service as in suburban service.

For main line railway service *single-phase A.C. system is preferred* due to following *advantages :*

(*i*) Due to the use of high voltage, the *cost of distribution system* employing conductor of smaller section to carry the low current required is *reduced.*

(*ii*) Less number of sub-stations required.

(*iii*) The initial, maintenance and operating costs of A.C. sub-stations are less comparatively.

B. **Suburban line railway service.** Following are the most important *requirements* of suburban line railway service :

1. *Rapid acceleration and retardation* (since frequent starting and stopping is required)
2. The working of motors should *not be affected* by the *fluctuations in voltage.*
3. There should be *no interference to the communication network* running along the track on account of power system adopted.

All the above requirements are fulfilled by the D.C. system and is invariably used for suburban service.

The *advantages* and *disadvantages* of D.C. system are listed below :

Advantages :

(*i*) For exerting same torque, current required in D.C. system is less than that in A.C. system.

(*ii*) Less energy consumption, in comparison to A.C. system.

(*iii*) The D.C. locomotive and motor coach equipment are lighter in weight, cheaper in initial as well as in maintenance cost and more efficient.

Disadvantages :

(*i*) Less efficiency of the sub-stations.

(*ii*) In order to keep the voltage of return rail within limits, additional equipment such as negative booster etc. is required.

(*iii*) Number of sub-stations required is more.

(*iv*) Sub-stations are more costly as converting machinery is required.

The above mentioned disadvantages are partly eliminated by using composite system-single-phase A.C. to D.C. system.

6.8. ELECTRIC TRACTION SYSTEMS—POWER SUPPLY

6.8.1. Transmission Lines to Sub-stations

The transmission lines to sub-stations depend on the location of the source of power supply. When the power station is situated at one extreme end, it is then probably the *worst* location. At the receiving station, the correct voltage must be made available.

In *duplicate transmission lines* arrangement, two separate transmission lines are run, one on each side of the railway track. Under normal operating conditions these lines operate together but when some fault develops in one line, the power can be transmitted by the other. For a double line, the reactance voltage drop is jIX, since each line carries a current I. In the event of fault occurring in one line, the other has to carry a current equal to $2I$ and therefore the reactance voltage drop becomes $j(2IX)$ and resistance drop equals $2IR$ (where R is the resistance of one line), then total drop becomes $[2IR + j(2IX)]$. If both the lines operate in parallel then the voltage drop is equal to $[IR + jIX]$.

- A duplicate transmission line feeding three sub-stations is shown in Fig. 6.1.

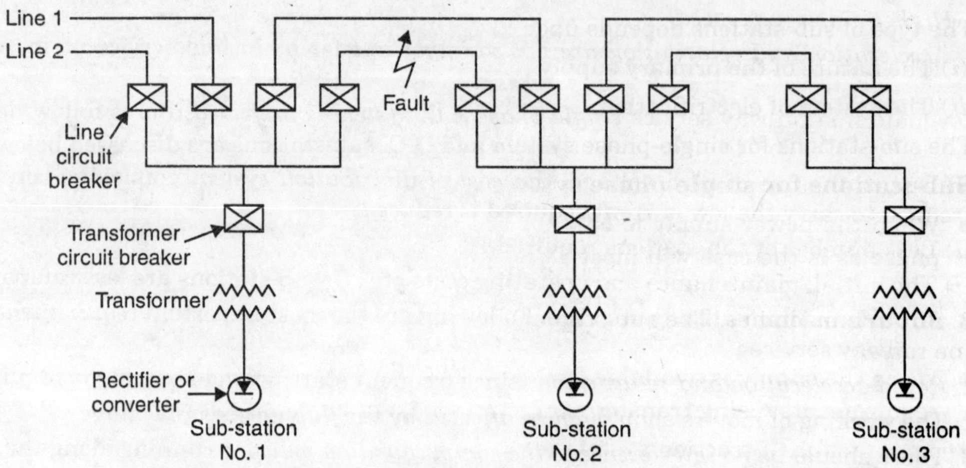

Fig. 6.1 Duplicate transmission lines.

— On the occurrence of a fault the line circuit breakers open the faulty section and the supply is continued through other line. In the event of a fault occurring over a particular section, the voltage drop in the other line of this section is doubled since it has to carry double the current.

— For every single-unit sub-station there are four line circuit breakers and one transformer circuit breaker ; since the cost of high voltage equipment increases very rapidly with higher voltages, it is evident that a transmission line of voltage 66 to 132 kV *must be tapped as seldom as possible.*

Where supply for traction work is available from national grid, the arrangement shown in Fig. 6.2 may be adopted.

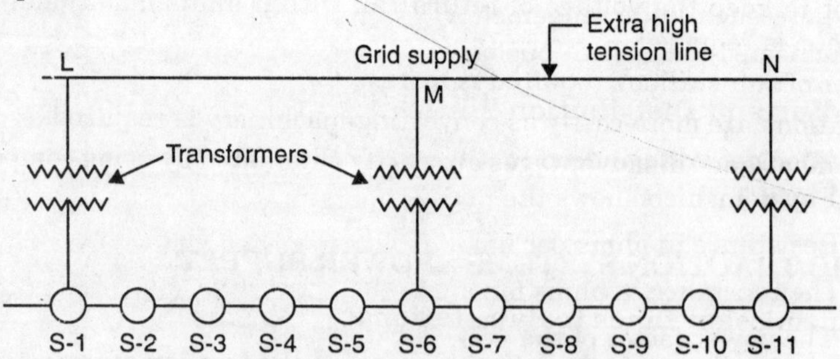

Fig. 6.2. Alternative method of power supply.

— The extra high tension supply is tapped at suitable points L, M and N and the power is converted to a lesser voltage at the transformer stations L, M and N. A single high tension line may be taken to the sub-stations from the transformer stations.

— An alternative proposal will be to have no transmission line between *sub-stations* S-3 and S-4 and S-8 and S-9. Duplicate transmissions lines can then be run between (S-1, S-2, S-3), (S-4, S-5, S-6) and (S-7, S-8) and (S-9, S-10, S-11). The transformer station at L will feed *sub-stations* : S-1, S-2, S-3 ; that at M : S-4, S-5, S-6, S-7, S-8 and that at N : S-9, S-10, S-11.

6.8.2. Sub-stations

The type of sub-stations depends upon :

(*i*) The nature of the primary supply ;

(*ii*) The system of electrification.

The sub-stations for single-phase system and D.C. sub-stations are discussed below :

Sub-stations for single-phase system :

- When the power supply is only for the purpose of feeding a traction system on single-phase as is the case will most European railways, the generators may be single-phase, $16\frac{2}{3}$ Hz machines. The sub-stations may be supplied at 66–132 kV.
- When the supply is available from a national grid, it has to be converted to suit traction requirements, using transformers and frequency changers.
- In case of A.C. traction system at standard 50 Hz frequency, the grid sub-stations are tapped from the 66–132 kV busbars to obtain 25 kV single-phase supply either through Scott-connected transformer or straight transformers.
- The spacing of the sub-stations is based on the permissible voltage drop expressed as percentage of the normal voltage. It should not exceed 20% and 40% under normal and abnormal conditions respectively.

— The sub-station spacing on 25 kV is about 1.5 times the spacing on 3000 V D.C. system.

D.C. sub-stations :

The D.C. sub-stations equipped with mercury arc rectifiers, as compared to rotary converter, have much higher all-day efficiency, reliable operation acceptability to remote control. Since the rectifier is non-reversible, suitable arrangements need to be made to receive the energy return during regenerative braking. The rectifier working as inverter feeds back power into the A.C. system ; however, where such an arrangement is not employed, the energy may be dissipated in a loading resistor connected to the D.C. busbars.

6.8.3. Feeding and Distribution System on A.C. Traction

The calculations of voltage drop in a distributor for single-phase A.C. system, are made as follows : Refer to Fig. 6.3 which shows the phasor diagram for current and voltage.

Let, R = Resistance in ohms per km of trolley wire and rail return,

X = The reactance in ohms for the same,

$\cos \phi$ = The power factor of the system,

V = Voltage at the receiving end,

E = Voltage at the sending end, and

I = The current flowing in the feeder/distributor.

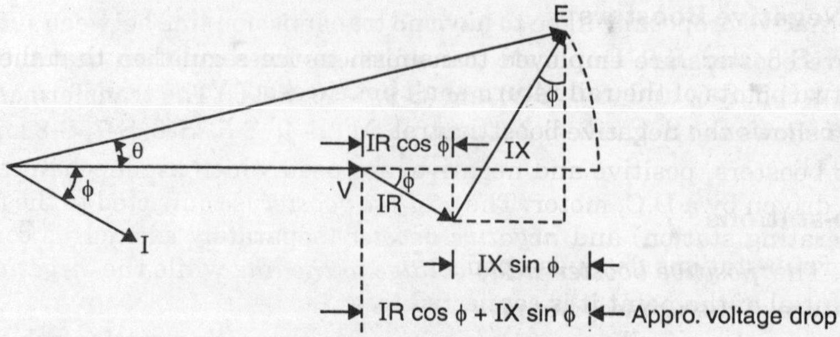

Fig. 6.3. Voltage drop in a distributor.

Then, since θ is small, the voltage drop in the feeder/distributor is *approximately* given as : $IR \cos \phi + IX \sin \phi = I(R \cos \phi + X \sin \phi)$.

- For conductor rails the value of resistance at $16\frac{2}{3}$ Hz is 3.3 times the actual D.C. value on account of *"skin effect"* and 7.8 times the D.C. value at 50 Hz. The inductance is about 1.55 mH per km of the route.

6.8.4. Feeding and Distribution System for D.C. Tramways

The tramways need to conform the following Indian *regulations* :

1. The voltage at the trolley is not to exceed 550 V and at the sending-end stations 650 V.
2. The trolley wire shall be divided into sections not exceeding 1.61 km in length.
3. The potential difference between any two points of the rail return shall not exceed 7 volts.

6.8.5. Electrolysis by Currents through Earth

When the track is used as return conductor, the currents flow through the rail as well as through the earth, as shown in Fig. 6.4. As these currents spread out into the earth, they follow low-resistance paths provided by gas mains, water pipes, cable sheaths etc. ; these currents enter such conductors and leave them causing corrosion due to electrolytic action. Such effects can be minimised by using the following methods : (*i*) By providing a *return path of very low resistance* (by providing good bonding and using insulated negative feeder) ; (*ii*) By discouraging the entry of currents into the pipes by incorporating insulating joints.

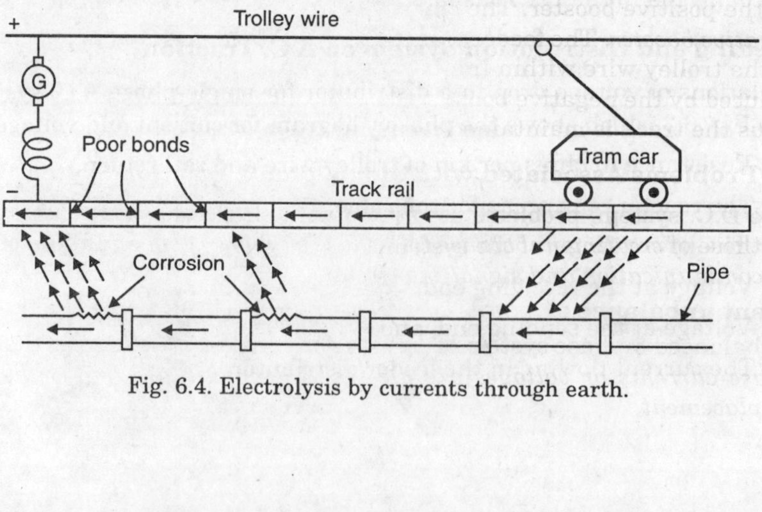

Fig. 6.4. Electrolysis by currents through earth.

6.8.6. Negative Boosters

Negative boosters are employed to conform to the regulation that the potential difference between any two points of the rail return shall not exceed 7 V.

Fig. 6.5 shows the negative boost control.

- Two boosters, positive and negative, are used which are mechanically coupled together and driven by a D.C. motor. The *positive booster* is connected to the trolley wire (near the generating station) and *negative booster* (separately excited) is connected to the track rail. The *'positive booster' adds voltage to the line* while the *'negative booster' lowers the* potential of the point it is connected to.
- As we go along trolley wire away from the generating station/sub-station, the potential drop increases and the voltage of the trolley wire falls. Since the current returns via the track rail, points away from the generating station acquire high potentials.

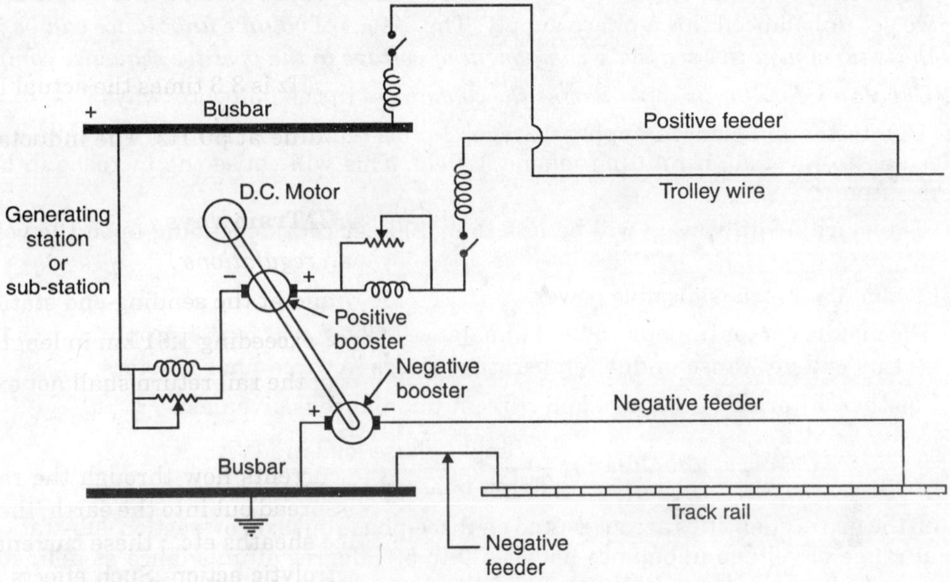

Fig. 6.5. Negative boost control.

This potential is brought down by the negative boost provided by the negative booster.

- When the load is sufficiently far away from the generating station, the trolley wire is fed by the positive booster. The *current in the positive booster provides the excitation for the negative booster.* The feeder current as it flows through the booster maintains the voltage of the trolley wire within limits. The potential at the corresponding point on the tack is reduced by the negative booster since its voltage is regulated by the same feeder current. Thus the track is maintained nearly at earth potential.

6.8.7. Problems Associated with Single-phase Traction System

Unlike D.C. system, problems associated with single-phase traction system can be broadly classified as those of *creating unbalance of 3-phase supply system* and those of *producing induction effects in telecommunication and signalling circuits.*

Current unbalance :

- Unbalanced 3-phase system of currents or voltages is defined as that *in which individual phase currents or voltages are unequal in magnitudes and or have no regular phase displacement.*

The extent of unbalance can be indicated by the *ratio* of *negative phase sequence component to positive phase sequence component of the current.*

- When considering generators which supply unbalanced load, current unbalance is defined as the *ratio of effective value of negative sequence component to the effective value of nominal rated current of machine.*
  - — It is observed that it would be safe for the generators to take *continuous unbalance of current upto 8 per cent* and this would have negligible effect on them.
  - — The permissible limit of allowable current unbalance is higher for salient pole machines than cylindrical rotor machines because of improved rotor cooling in the former.

Voltage unbalance :

- Owing to flow of unbalanced currents, unequal voltage drop takes place in line conductors. This gives rise to angle of distortion to three-phase voltage star and consequently we get unbalanced three-phase supply. The extent of *voltage unbalance can be judged by the ratio of negative sequence component of voltage to the positive sequence component.*
- *Unbalanced voltage supply harms the consumers* in the following ways :

(*i*) Due to the unbalanced supply voltage, 3-phase induction motors will take unbalanced current leading to reversal of rotating magnetic field. This will cause the increase in losses and decrease in output.

(*ii*) The machine utilisation will be less than 100 per cent depending upon the voltage balance.

(*iii*) Reduction in transmissible power.

- The maximum continuous voltage unbalance of *2 per cent* in the supply system is allowed as this will not cause undue temperature stress in three-phase motors.
- Another criterion for voltage and current unbalance is given as :

$$\textit{Degree of unbalance} = \frac{\text{Single-phase load in MVA}}{\text{Three-phase short circuit MVA}}$$

Both the above quantities are measured on three-phase busbars of traction sub-station. Within the normal range of voltage unbalance between 0 to 5%, degree of unbalance as defined above is equal in magnitude to voltage unbalance. The *following points are worth noting* :

(*i*) The degree of unbalance will decrease as short circuit capacity increases. This will occur : (*a*) If unbalance occurs near generating station, (*b*) If unbalanced load is supplied from high voltage transmission system, and (*c*) If unbalanced load is supplied from system having high MVA capacity.

(*ii*) For a given constant load, degree of unbalance decreases if power factor of load increases.

It may be noted that *voltage unbalance is maximum at the point of unbalance load and attenuates as it penetrates into the system towards the generation.*

Production of harmonics :

- The effect of harmonics on the system is *to increase slightly both current and voltage unbalance thereby giving rise to increased disservice to other consumers connected to the system beyond that associated with fundamental unbalance.*
- The presence of harmonics in supply system also *causes interference on telephone, radio and television systems.* The aerial telegraph and telephone lines running in proximity to A.C. single phase 50 Hz overhead contact lines *would require replacement by cables.*

Induction effects. In *communication circuits* we are very much concerned with the interference from harmonics or high frequency currents, whereas in *signalling circuits* we are concerned with the final magnitude of all induced-voltages. In signalling cables, the voltage is induced by the following *two techniques* :

(*i*) By electrostatic induction ;

(*ii*) By electromagnetic induction.

(*i*) *Electrostatic induction :*

- The statically induced e.m.f. in signalling cable is produced by virtue of its location in the voltage gradient between high voltage contact wire and earth. Magnitude of this voltage depends upon the relative value of capacitances between contact wire and signalling cable and signalling cable and earth and the voltage of the power line.

In view of several reasons, the use of cables in overhead except in unavoidable circumstances is not a desirable practice and therefore, discouraged.

(*ii*) *Electromagnetic induction :*

When current flows through contact wire, it produces magnetic field around itself. This magnetic field will induce e.m.f. in the nearby circuits with which it links. This is how longitudinal voltages appear in telecommunication and signalling lines.

The magnitude of induced voltage in communication line can be reduced by the use of screen which may be a cable sheath, additional earthed conductor or rail.

Measures adapted to limit the voltage rise in communication circuits :

The international Consultative Committee on Telephones and Telecommunication, for safety of equipment and personnel, has specified voltage rise of not more than 430 V (r.m.s.) in telecommunication cable under conditions of short-circuits in power circuit and not more than 60 V (r.m.s.) under normal conditions of maximum current. In order to limit the magnitude of the induced voltages stipulated above, the following *steps are taken* :

1. Long telecommunication circuits are broken by using insulating transformers.
2. All telecommunication circuits are taken in cables instead of aerial circuits. This arrangement reduces the chances of pilferage of copper conductor thus ensuring uninterrupted service in addition to saving in losses due to theft.
3. Use of booster transformer.
4. Telecommunication cables are provided with aluminium sheathing instead of lead.
5. Telecommunication circuits may be removed from the vicinity of railway track, if practicable.
6. Use of P.V.C. or rubber covering to prevent sheath corrosion.
7. Use of additional conductors or steel tapes underneath the cable for augmenting the screening effect. Running rails also provide additional screening.

6.9. A.C. LOCOMOTIVE

The block diagram of an A.C. locomotive is shown in Fig. 6.6. Its working/operation is given below :

— Power at 25 kV is taken from the overhead contact wire, via a pantograph, and fed to the step-down transformer in the locomotive.

— The low A.C. voltage so obtained is then converted into pulsating D.C. voltage by means of rectifier.

— The *pulsations* in the D.C. voltage are then *removed by the smoothing choke* before it is fed to D.C. series motors which are mounted between the wheels.

— The *circuit breakers* are provided to immediately disconnect the locomotive from the overhead supply in case of any faults in the electrical system.

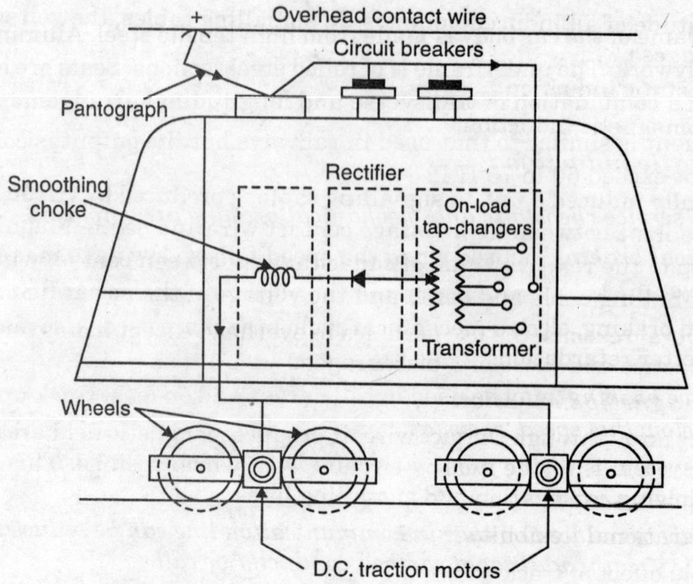

Fig. 6.6. Block diagram of an A.C. locomotive.

— The no-load tap changers are employed to change the voltage across the motors and *regulate their speed*.

- A very significant feature of the A.C. locomotive in comparison with the D.C. locmotive is the *greater amount of adhesion obtained with A.C. locomotive.* The adhesion factor is of the order of 30 to 33% for A.C. locomotive as compared with 23 to 25% for D.C. locomotive. The *greater adhesion is on account of connecting all D.C. motors in parallel permanently.* Their speed is controlled by changing the voltage across each motor by tap-changing auto-transformer on the A.C. side. *If a particular motor slips, the voltages across other motors is not affected and adhesion is maintained.* In D.C. locomotive resistance and series-parallel control is employed and if one motor slips, its load is thrown on the other motor which reduces the voltage across other motor and may cause slipping of this motor also. The adhesion with the track is seriously affected.

However, there are some "*drawbacks*" with *A.C. traction like induced voltage in aerial signalling equipment and interference with neighbouring communication lines.* Some remedial devices have to be provided to overcome these drawbacks.

6.10. TRAMWAYS

- The tramway is perhaps the *cheapest* type of transport available in very dense traffic.
- It receives power through a bow collector or a grooved wheel from an overhead conductor at about 600 V D.C., the running rail forming the return conductor.
- It is provided with *atleast two driving axles* in order to *secure necessary adhesion, start it from either end and use two motors with series-parallel control. Two drum-type controllers,* one at each end, are used for controlling the tramcar. Though these controllers are connected in parallel, they have suitable interlocking arrangement to prevent their being used simultaneously.

- The main frame of the car body is made from high tensile steel. Aluminium is extensively used for bodywork. The underframe is of rolled steel sections. Seats are either in transverse direction or a combination of transverse and longitudinal arrangement is used.
- The equipment is similar to that used in railways but the output is considerably smaller and does not exceed 60 to 75 H.P.
- For *normal service rheostatic and mechanical braking are employed.*
  — '*Mechanical braking*' is used when the speed falls so low as to make electrical braking ineffective.

For mechanical braking, electro-mechanical drum brakes are used. Also magnetic track brakes are used for giving better retardation.

— '*Regenerative braking*' may be used in hilly areas and on level track upto a speed of about 20 km/h. *Below this speed 'rheostatic braking' is used.*

This traction system is *losing ground to trolley bus or internal combustion engined ominus system* due to the following *reasons* :

(*i*) Lack of operational flexibility in areas of heavy congestion.

(*ii*) Undesirable effect of track on other road vehicles.

(*iii*) It is an economical method of transportation *only for large cities having dense traffic.*

6.11. TROLLEY-BUS

Serious drawback of tramway is the lack of manoeuvrability in congested areas and noise ; this is overcome by the trolley-bus drive.

- It is an electrically-operated pneumatic-tyred vehicle which needs *no track in the roadway.* It receives its power at 600 V D.C. from two overhead contact wires. Since adhesion between a rubber-tyred wheel and ground is sufficiently high, only a single driving axle and, hence, a single motor is used. The trolley bus can manoeuvre through a metre or two on each side of the centre line of the trolley wires.
  — A D.C. compound motor of output of 50 to 100 kW is normally used.
  — Speed control is obtained by field weakening (by providing a resistance in the shunt field or the tappings on series field or divertor with the series field) method. Foot-operated master controllers are used so that drive may have his hands free to steer the vehicle and apply hand brake. One pedal controls the starting, speed control and regenerative braking, if any and second pedal control rheostatic and compressed air brakes. Regenerative braking is usually not employed in trolley-bus drive because of difficulty of ensuring that supply system is always in a position to absorb the energy regenerated.
- The lighting system in the car is low-voltage D.C. supplied from a motor-generator set connected in parallel with a battery. The vehicles are usually provided with secondary batteries so that the vehicles can be manoeuvred in case of emergency.
- Since the body of the car is insulated from earth on account of the rubber-tyred wheels, it must be properly checked for adequate insulation resistance lest it leaks and causes electric shocks to the passengers while boarding and alighting from the bus. The insulation resistance is checked at the end of the day.
- Trolley-buses have *more passanger carrying capacity*, higher *acceleration* and *braking retardation than oil-engined buses.* These are, therefore, used for medium traffic density

as obtained in inner suburbs. Oil engined buses, on the other hand, are used for outer suburbs and country side where there is low traffic density.

6.12. DIESEL ELECTRIC TRACTION

The diesel electric traction can be economically adopted in areas where the cost of diesel oil is low (*e.g.*, U.S.A.). This system is also adopted in areas where coal for steam traction is not available in plenty and water is sometimes scarce as in Rajasthan Countries rich in coal may like to have electric traction for urban areas and steam traction for suburban areas.

Diesel locomotives were introduced for the first time in our country in 1945 for shunting purposes on B.G. sections and in 1956 for main line services on M.G. sections. Indian Railways started using diesel locomotive extensively for main line service since 1958.

- Diesel electric traction system employs a self-contained motive power unit using diesel engine which drives D.C. generator. This supplies current to the traction motors which are geared to driving axles. The efficiency of diesel electric locomotive (26 to 30%) is more than that of steam locomotive (5 to 7%) and less than that of electric locomotive (34 to 38%).

Following four types of the diesel-electric locomotives are employed in practice :

(*i*) The main-line diesel electric locomotive having engines of output not exceeding 1500 kW and speeds 160 km/h.

(*ii*) The shunting diesel electric locomotive having an engine of 225 to 375 kW output and speed between 25 to 50 km/h.

(*iii*) The diesel-electric multiple unit stock of which each motor has an engine of 135 to 150 kW output and train is capable of having speeds between 80 to 110 km/h.

(*iv*) The diesel electric rail car having an engine of 75-450 kW output which may operate as a single car or with one or more trailer coaches.

- The diesel engine employed for diesel electric locomotives is usually high speed type with three or four different running speeds with different output power. The generator may be self excited type or separately excited. In case of separately excited generators excitation is obtained from an auxiliary generator connected in parallel with the battery, so that standby supply is availabe. Similarly diesel engine is made self starting employing the supply from the battery. The voltage control of main generator becomes easier.
- Diesel electric locomotive with all its advantages of incorporating most versatile power unit is *very expensive in the initial cost.* Life of diesel engine is low. Locomotive requires cooling of engine in addition to motors and generators. Regenerative braking cannot be employed with diesel electric traction but there is no bar on employing rheostatic braking.

Characteristics of diesel engine. The internal combustion engine employed for rail transport is of compression ignition (C.I.) type, using heavy diesel oil. While 4-stroke engines have been used, there is a scope for employment of 2-stroke engines which will result in the reduction in weight per H.P. developed. For main line application, supercharging of diesel engine is employed. This increases the engine H.P. with less increase in weight thereby giving overall low engine weight to H.P. ratio.

- The characteristic of a diesel engine is that its torque is fairly constant for various speeds and H.P. increases fairly uniformly with speed. Also it has a small overload capacity of only 10 per cent. For *traction work*, the torque is *inversely proportional to speed* and therefore constant torque of the diesel engine has to be modified to suit traction requirement by the use of a suitable torque converter. This can be achieved either by *mechancial or electrical transmission.*

— "*Mechanical transmission system*" employes a variable gear between the engine shaft ands the driving wheels. This method is almost universally employed for road and light railway vehicles but it is not suitable for railway work where output required exceeds 300 kW, due to mechanical difficulties with the gears and gear changing equipment.

This system of transmission is *light and more efficient* than electrical transmission system but it is not as flexible as the latter one.

— "*Electrical transmission system*" employs a diesel engine coupled to a D.C. generator which supplies power to the D.C. traction motors. This system makes use of the simple speed control properties and desirable speed-torque characteristic of the D.C. series motor.

- Normally, a diesel engine is adjusted to operate at a constant speed. Now, if it is coupled to a D.C. generator producing constant voltage, an increase in the *tractive effort* (F_t) of the train reduces the speed of the D.C. motors on account of the shape of their speed-torque curve. The horsepower ($\propto F_t \times N$) increases since a decrease in speed does not cause an increase in tractive effort to a value to keep the product ($F_t \times N$) *i.e.*, H.P. constant, and consequently the engine will be overloaded.

In order that *power* delivered by the engine remains the *same*, the product ($F_t \times N$) must remain constant, where N is *speed* in r.p.m.

Hence, speed $N \propto \dfrac{1}{F_t}$

In series motors, the flux $\phi \propto I$,

$$\therefore \quad N \propto \frac{V}{I} \quad \text{and } F_t \propto I^2$$

$$\therefore \quad \text{For constant power,} \quad \frac{V}{I} \propto N \propto \frac{1}{F_t} \quad \text{or} \quad V \propto \frac{1}{I}$$

Thus, for keeping the power constant, the generator voltage should be varied *inversely* as the current. This is accomplished by using a series winding on the generator or exciter and arranging it in such a way manner that the current of traction motors produces demagnetising ampere turns so-as to weaken the generator field and consequently vary V. Various manufacturers have their own refinements of this basic principle to have a close agreement with the constant power output curve.

- *In diesel electric locomotives, regenerative braking is not possible but, if required, the rheostatic braking can be employed.*

6.13. OVERHEAD EQUIPMENT

The overhead system is used with D.C. tramways, trolley-buses and locomotives and also with A.C. locomotives. The track is used as the return conductor.

Broadly speaking, following are the two *systems of current collection* by a traction unit :

(*i*) Overhead wire system.

(*ii*) Third rail system.

It has been found that current collection from an overhead wire is far *superior* to that from the third rail. Moreover, insulation of third rail at high voltage becomes an impracticable proposition and endangers the safety of the working personnel.

- The *simplest type of overhead equipment* consists of a *single contact wire* of hard drawn copper or silicon-bronze supported either by bracket or an overhead span. The wire is grooved (Fig. 6.7) to facilitate connection to the support. Owing to high sag of the wire betweeen supports, the speed of the traction unit is limited to about 30 km/h. Hence single contact wire system is suitable for tramways and in complicated yards and terminal stations where speeds are low and simplicity of layout is desirable.

In case of high-speed trains, for collection of current the contact or trolley-wire has to be kept level without any abrupt changes in its height between the supporting structures. It can be accomplished by using the single catenary system which consists of one catenary or messanger wire of steel with high sag and the contact or trolley-wire by means of droppers clipped to both wires (Fig. 6.8). In the single catenary construction the span may be from 45 to 99 m with sags from 1 to 2 m, and droppers spaced 3 to 5 m.

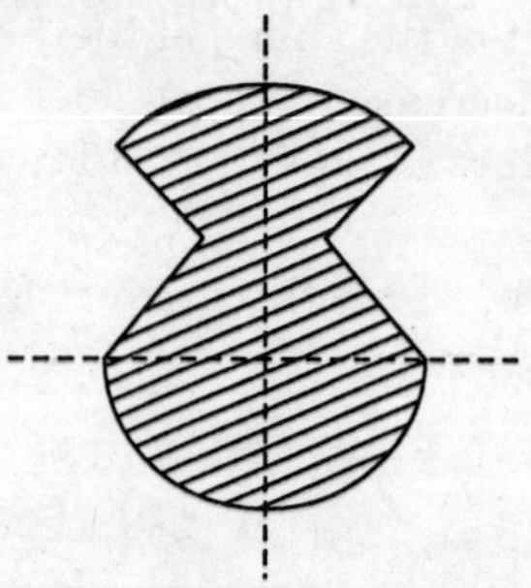

Fig. 6.7. Trolley-wire section.

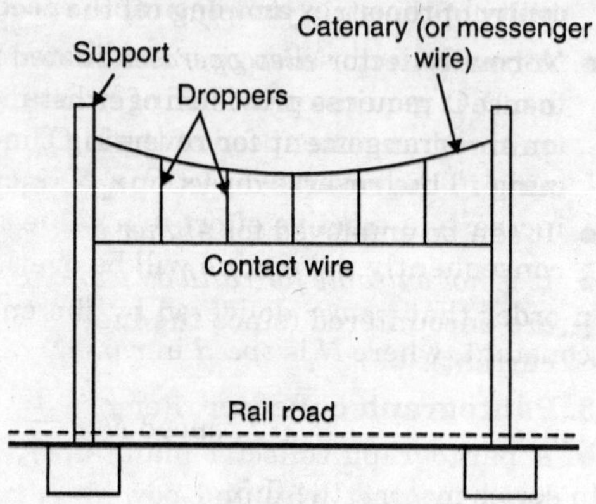

Fig. 6.8

6.13.1. Collector Gear for Overhead Equipment

The primary requirement of a collector is that it should maintain a continuous contact with trolley-wire at all speeds. *Three types of commonly used gear are :*

1. Trolley collector ;
2. Bow collector ;
3. Pantogrph collector.

1. Trolley collector. Refer to Fig. 6.9.

- A trolley collector is used on tramways and trolley-buses and is mounted on the roof of the vehicle.
- A contact with the overhead wire is made by means of either a grooved wheel or a sliding shoe carried at the end of a light trolley pole attached to the top of the vehicle and held in contact with overhead wire by means of a spring. The pole is hinged to a swivelling base so that it may be reversed for reverse running thereby making it unnecessary for the trolley wire to be accurately maintained above the centre of the track.
- Trolley collectors *always operate in the trailing position.*
- The trolley collector is employed upto speeds of about 32 km/h as beyond this speed there is every possibility of the wheel jumping off the trolley wire.

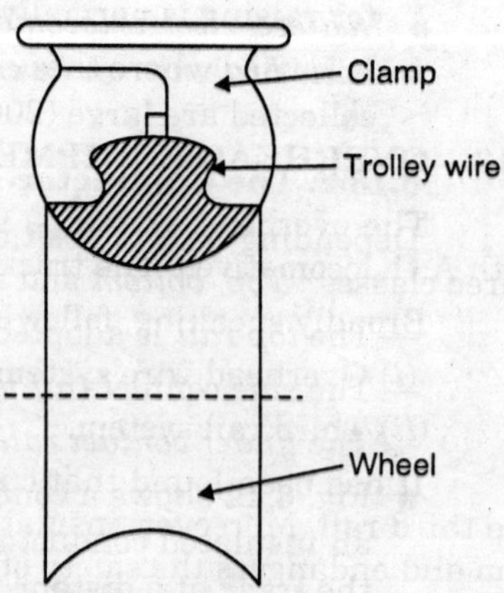

Fig. 6.9. Collector wheel and trolley-wire.

2. Bow collector. Refer to Fig. 6.10.

- It consists of a light metal strip or bow (about 1 m long) pressing against the trolley wire (for collection of current) and the framework is mounted on the roof of the car. The strip is purposely made of soft material (*e.g.*, copper, aluminium or carbon) in order that most of the wear may occur on it rather than on the trolley wire. There is no possibility of the strip jumping off the trolley wire.
- A bow collector *also operates in trailing position*, hence it requires provision of either duplicate bows or an arrangement for reversing the bow for running in the reverse direction.

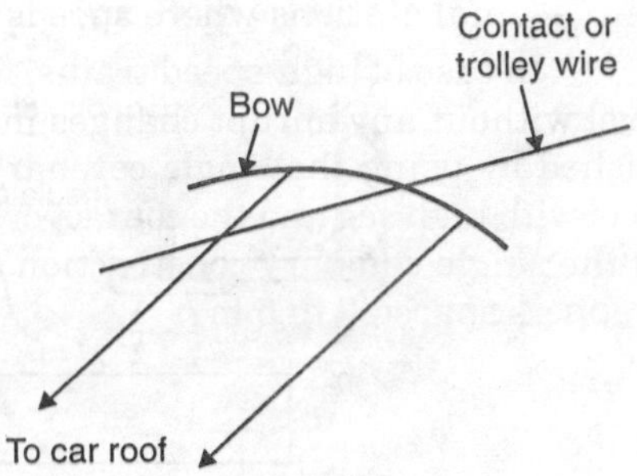

Fig. 6.10. The bow collector.

- It can be employed for *higher speeds*.
- It is *not suitable for railway work* where speeds upto 120 km/h and currents upto 3000 A are encountered (since the inertia of bow is too high to ensure satisfactory collection of current).

3. Pantograph collector. Refer to Fig. 6.11.

- A pantograph collector maintain link between overhead contact wire and power circuit of the electric locomotive at different speeds under all wind conditions and stiffness of overhead equipment. This necessitates that positive pressure must be maintained at all times to avoid loss of contact and sparking but the pressure must be as low as possible so that wear of overhead contact wire is minimum.
- It is mounted on a pentagonal framework (of high-tensile alloy-steel tubing) which can be raised or lowered by compressed air or springs. Compressed air for raising is normally used.

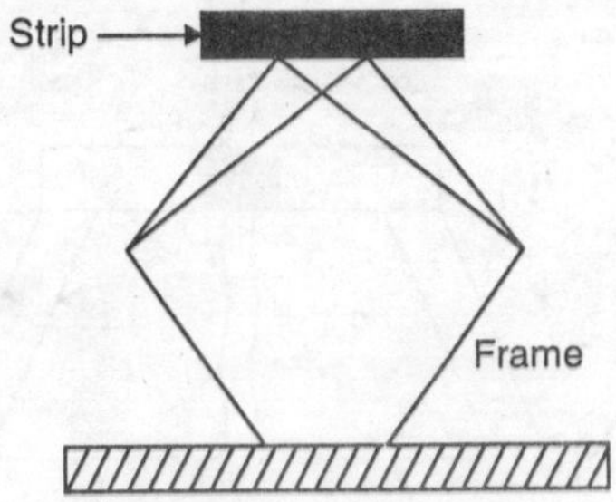

Fig. 6.11. Pantograph collector.

- It is used where the *vehicles run at high speeds ; i.e.*, in railways and where currents to be collected are large (2000 to 3000 A).

6.13.2. The Conductor-Rail Equipment/System

Depending on the position of the contact surface, the conductor rails may be divided into three classes : *Top, bottom* and *side.*

— The *top rail* is adopted *universally for 600 V D.C. electrification.*

— The *side contact rail* is used for *1200 V D.C. supply.*

— The *under contact rail* has the advantage of being *protected from snow, sleet and ice.*

- Fig. 6.12 shows a conductor-rail system when electric supply is collected from the top of an insulated conductor rail (made of special high-conductivity steel) running parallel to the track at a distance of 0.3 to 0.4 m from the running rail which forms the return path.

For collecting the current a shoe, as shown in Fig. 6.13 is used. The shoe presses on to the rail with a force of about 150 N with a current collection ranging from 300 to 500 A per shoe. Since the conductor rail is not always laid on the same side of the track shoes are attached to the motor-coach or locomotive on both sides.

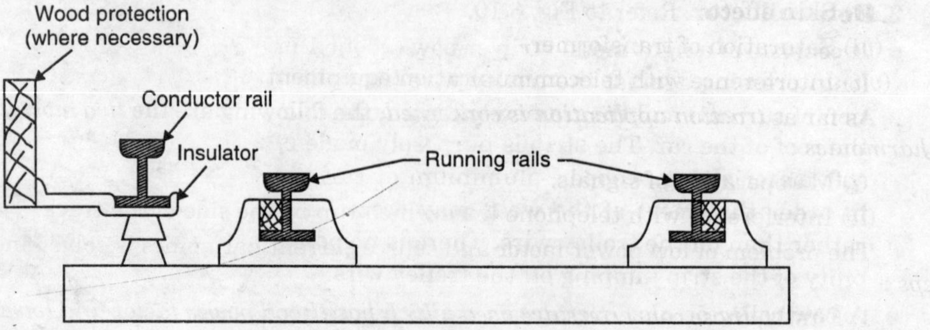

Fig. 6.12. Conductor-rail system.

Fig. 6.14 shows the *side contact rail* and method of mounting.

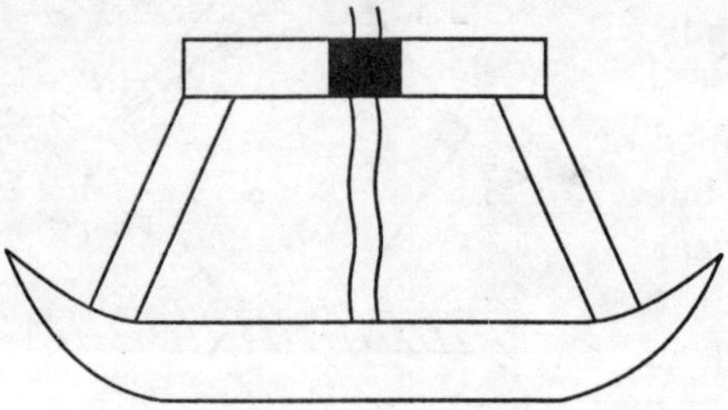

Fig. 6.13. Current collecting shoe.

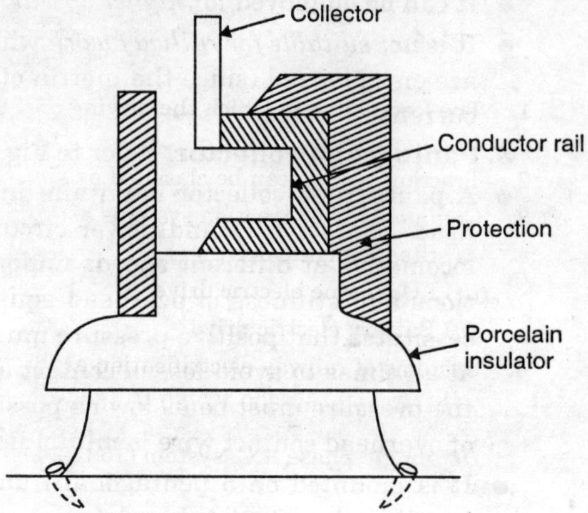

Fig. 6.14. Conductor rail with side running contact.

6.14. POWER FACTOR AND HARMONICS

In several traction systems A.C. to D.C. thyristor converters are employed which have the following *drawbacks* :

(*i*) Low power factor at low output voltage.

(*ii*) Generation of harmonics in source current and voltage. The locomotives with capacities 5 MW and higher are powered from a single-phase supply, resulting into large voltage drop and high transmission line losses during acceleration. If the power factor is allowed to fall low, the voltage drop and high transmission line losses will be too high and transmission line and substation equipment may be overloaded. Thus, measures are taken to ensure that the power factor does not fall below 0.8.

Harmonics in source current and voltage have the following *undesirable effects* :

(*i*) Malfunction of electronic equipment connected to the line.

(*ii*) Excitation of system resonances.

(*iii*) Over loading of capacitors.

(*iv*) Decrease in efficiency owing to increase in losses due to harmonic currents.

(*v*) Skin effect.

(*vi*) Saturation of transformer.

(*vii*) Interference with telecommunication equipment.

As far as *traction application is concerned*, the following are the *two most undesirable effects of harmonics* :

(*i*) Maloperation of signals.

(*ii*) Interference with telephone lines which run by the side of the track.

The problem of low power factor and source current harmonics is solved in the following *two steps* :

1. Firstly, *those converters are used which have good power factor and lower harmonics.*

2. Secondly, static VAR compensator is used to maintain power factor above 0.8 and filters to reduce harmonics to the extent that the possibility of maloperation of signals is completely eliminated and the noise is reduced to tolerable level in the transmission lines.

HIGHLIGHTS

1. The locomotion in which the driving or tractive force is obtained from electric motors is called *"Electric traction"*.
2. Traction system can be classifed as *non-electric* and *electric traction.*
3. Commonly used traction systems are :
   (*i*) Steam engine drive
   (*ii*) I.C. engine drive
   (*iii*) I.C. engine electric drive
   (*iv*) Petrol-electric traction
   (*v*) Battery electric drive
   (*vi*) Electric drive.
4. Systems of railway electrification are :
   (*i*) D.C. system (600 V, 750 V, 1500 V, 3000 V)
   (*ii*) Single-phase A.C. system (15 to 25 kV, $16\frac{2}{3}$, 25 and 50 Hz)
   (*iii*) Three-phase A.C. system (3.3 kV to 3.6 kV) at $16\frac{2}{3}$ Hz.
   (*iv*) Composits system (involving conversion of single-phase A.C. into 3-phase A.C. or D.C.)
5. *Negative boosters* are employed to conform to the regulation that the potential difference between any two points of the rail return shall not exceed 7 V.
6. Problems associated with single-phase traction system are :
   (*i*) Current unbalance
   (*ii*) Voltage unbalance
   (*iii*) Production of harmonics
   (*iv*) Induction effects.
7. Commonly used collector gear for overhead equipment are :
   (*i*) Trolley collector ;
   (*ii*) Bow collector;
   (*iii*) Pantograph collector.

OBJECTIVE TYPE QUESTIONS

Fill in the Blanks or Say "Yes" or "No" :

1. The locomotion in which the driving tractive force is obtained from electric motors is called ...... traction.
2. In India, Railways first appeared in April, 1753.
3. During the period 1925 to 1932 the first electrification was undertatken in India.
4. Steam engine drive is simple in design.

5. Steam locomotive requires ...... repair and maintenance.
6. I.C. engine drive has ...... initial investment.
7. I.C. engine drive in unsuitable for heavy railway work.
8. In I.C. engine electric drive loss of power in speed control is very low.
9. Petrol-electric traction is employed in ...... lorries and buses.
10. Battery driven vehicle is easy to control and very convenient to use.
11. The maintenance cost of an electric locomotive is 80 per cent of that of steam locomotive.
12. In electric traction system it is possible to use regenerate braking.
13. 3-phase A.C. system is used in some hilly areas, where output power required is high and regeneration on large scale is possible.
14. Single-phase A.C. to D.C. system has low starting efficiency.
15. Negative boosters are employed to conform to the regulation that the potential difference between any two poitns of the rail return shall not exceed ....
16. The magnitude of induced voltage in communication line can be reduced by the use of ...... which may be a cable sheath, additional earthed conductor or rail.
17. The ...... is the cheapest type of transport available in very dense traffic.
18. Trolley-bus in an electrically-operated pneumatic-tyred vehicle which needs no track in the roadway.
19. The primary requirement of a ...... is that it should maintain a continuous contact with the trolley-wire at all speeds.
20. Trolley collectors never operate in trailing position.

Answers

1. electric	2. No	3. Yes	4. Yes	5. more
6. low	7. Yes	8. Yes	9. heavy	10. Yes
11. No	12. Yes	13. Yes	14. No	15. 7 V
16. screen	17. tramway	18. Yes	19. collector	20. No.

THEORETICAL QUESTIONS

1. What do you mean by "Electric traction" ?
2. State the advantages of electric traction over other non-electrical systems of traction including steam traction.
3. Give a brief history of railways in India.
4. How are traction systems classified ?
5. What are the requirements of an ideal traction system ? How are they met in an electric traction system ?
6. What are the various traction systems in practice in our country ? Give the advantages of electric drives with its limitations and discuss briefly the factors governing the final choice of traction system.
7. State the main requirements for an ideal traction system. Name the different traction systems. Give merits and demerits of electric traction over steam engine traction.
8. What are the different systems of track electrification ?
9. What are the merits and demerits of D.C. system of track electrification ?
10. What are the disadvantages of third rail system of the track electrification ?
11. State the system of track electrification generally employed for the following services. Justify your answer and mention the voltages adopted in each case.
   (*i*) Electric suburban railway service.
   (*ii*) Main line services.
   (*iii*) Tram-way service in a busy town area.

12. What are the advantages of single-phase low-frequency system of track electrification ? What are the factors due to which its wide spread application remains limited ?
13. What are the advantages of composite system of traction employing 25 kV A.C. supply and D.C. traction motors ?
14. What are the disadvantages of 25 kV A.C. traction system ?
15. Explain briefly electrostatic and electromagnetic induction in signalling and telecommunication lines ?
16. Give the comparison between pure A.C. and D.C. systems.
17. Compare the D.C. and A.C. systems of railway electrification from the point of main line and suburban line railway service.
18. Write a short note on negative boosters.
19. Discuss various factors on which final choice of traction depends ?
20. What is the scope of application of battery drive ?
21. Why are tramways losing ground to other systems of traction ?
22. Discuss the problems associated with diesel-electric traction and indicate how these are overcome in practice.
23. Write short notes on the following :
    (*i*) Trolley collector
    (*ii*) Conductor-rail system
    (*iii*) Third rail system of current collection.

7

Train Movement and Energy Consumption

7.1. **Types of railway services.** 7.2. **Speed-time curves for train movement.** 7.3. **Crest speed, average speed and schedule speed.** 7.4. **Simplified speed-time curves.** 7.5. **Mechanism of train movement**—Adhesive weight and coefficient of adhesive—Driving axle code for locomotive—Nature of traction load. 7.6. **Tractive effort for propulsion of train.** 7.7. **Power output from driving axles.** 7.8. **Energy output from driving axles**—Specific energy output—Energy consumption—Highlights—Objective Type Questions—Theoretical Questions—Unsolved Examples.

7.1. TYPES OF RAILWAY SERVICES

Railways offer the following *three* types of passenger services :

1. City or urban service.
2. Suburban service.
3. Main line service.

1. **City or urban service.** In this type of service there are *frequent stops*, the distance between stops being nearly 1 km or less. Thus, in order to achieve moderately high schedule speed between the stations, it is *essential to have high acceleration and retardation.*

2. **Suburban service.** Here, the distance between stops averages from 3 to 5 km over a distance of 25 to 30 km from the city terminus. In this case *also, high rates of acceleration and retardation are necessary.*

3. **Main line service.** In this case, operation is *over long routes* and *stops are infrequent.* Here, operating speed is high and acceleration and braking periods are relatively less important.

The three types of services available on *goods traffic side are :*

(*i*) Main-live freight service.
(*ii*) Local or pick-up freight service.
(*iii*) Shunting service.

7.2. SPEED-TIME CURVES FOR TRAIN MOVEMENT

The movement of trains and their energy consumption can be conveniently studied by means of *speed-time* and *speed-distance* curves. As their names suggest, former gives speed of the train at various *times* after the start of the run and later gives speed at various *distances* from the starting point. Out of the two *speed-time curve is more important* because of the *reasons* : (*i*) Its *slope gives acceleration or retardation* as the case may be, (*ii*) *Area* between it and horizontal (*i.e.*, time) axis *represents the distance travelled*, and (*iii*) *Energy required for propulsion can be calculated* if resistance to the motion of the train is known.

Fig. 7.1 shows a typical *speed-time curve* for electric trains operating on passenger services. It mainly consists of the following five parts :

(*i*) Constant acceleration period.
(*ii*) Acceleration on speed curve.
(*iii*) Free-running period.
(*iv*) Coasting.
(*v*) Braking.

1. **Constant acceleration period (O to t_1) :**

- During this period the traction motors accelerate from rest, the *current taken by the motors and the tractive effort are practically constant.*

- This is also known as *notching* up period.
- It is represented by portion *OL*.

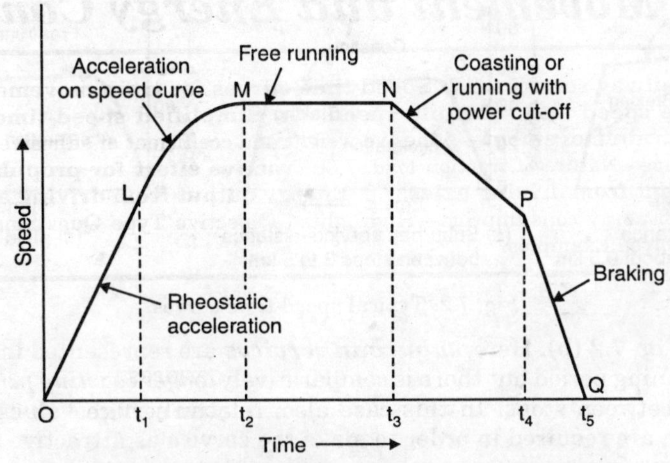

Fig. 7.1. Typical speed-time curve for main line service.

2. **Acceleration on speed curve (t_1 to t_2) :**

- After the starting operation of the motors is over, the train still continues to accelerate along the curve LM. During this period, the motor current and torque decrease as train speed increases. Hence, acceleration gradually *decreases* till torque developed by the motors exactly balances that due to resistance to the train motion.
- The shape LM of the speed-time curve depends primarily on the torque-speed characteristics of the traction motors.

3. **Free-running period (t_2 to t_3) :**

At the end of speed curve running *i.e.*, at t_2 the train attains the *maximum speed.* During this period the train runs at constant speed attained at t_2 and *constant power is drawn.*

- It is represented by the portion MN.

4. **Coasting (t_3 to t_4) :**

- At the end of free running period (*i.e.*, at t_3) the *power supply is cut off* and the train is allowed to run under its own momentum. The speed of the train starts decreasing on account of resistance to the motion of train. The rate of decrease of speed during coasting period is known as *coasting retardation* (which practically remains constant).
- Coasting is *desirable since it utilizes some of the kinetic energy* of the train which would, otherwise, be wasted during braking. Hence, it helps to reduce the energy consumption of the train.
- It is represented by portion *NP*.

5. **Braking (t_4 to t_5)**

- During this period brakes are applied and the train is brought to a stop.
- It is represented by the portion *PQ*.

Typical speed-time curves for different services

Fig. 7.2 (*a*, *b*, *c*) shows the typical speed-time curves for different services.

Refer to Fig. 7.2 (a). It represents ***city service*** where relative values of acceleration and retardation are high in order to achieve moderately high average speed between stops. Due to short distances between the steps, there is no possibility of free-running period though a short coasting period is included to save on energy consumption.

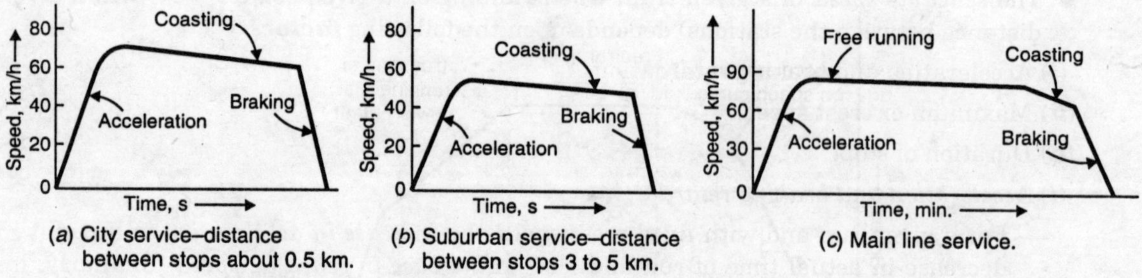

Fig. 7.2. Typical speed-time curves.

- **Refer to Fig. 7.2 (b).** Here, ***suburban services*** are represented in which again there is no free-running period but there is comparatively *longer coasting period* because of longer distances between stops. In this case also, relatively high values of acceleration and retardation are required in order to make the service as attractive as possible.
- **Refer to Fig. 7.2 (c).** It represents ***main line service.*** Here, there are long periods of free-running at high speeds. The acceleration and retardation periods are relatively less important.

Approximate values of acceleration and retardation. The values of acceleration and retardation are governed by the fact they should not cause discomfort to the passengers. Some typical values are given below :

— The *acceleration* on suburban and urban services is between 1.6 to 4 kmphps. With trolley-buses it varies from 0.4 to 6.5 kmphps, the higher values being employed in U.S.A.

— The *coasting retardation* is about 0.16 kmphps.

— The braking retardation is between 3 and 5 kmphps.

7.3. CREST SPEED, AVERAGE SPEED AND SCHEDULE SPEED

Crest speed. *It is the maximum speed (V_m) attained by the vehicle during the run.*

Average speed. *The distance covered between two stops divided by the actual time of run is known as* ***average speed.***

$$\therefore \qquad \text{Average speed} = \frac{\text{Distance between two stops}}{\text{Actual time of run}}$$

Schedule speed. *It is defined as the ratio of distance covered between two stops and total time of run including time of stop.*

i.e., $$\text{Schedule speed, } V_{sch} = \frac{\text{Distance between stops}}{\text{Actual time of run + Stop time}}$$

— This shows that schedule speed is always *smaller* than average speed. The difference is large in case of urban and suburban services and is negligibly small in case of main line service.

— Further, this suggests that *in order to have a fairly good schedule speed* for urban or suburban services the *stops must be reduced.* It is recommended to have stop time of 15 to 20 seconds for small services. The effect of stop time on the schedule speed for main line service is negligibly small if tne stop time is in seconds or a few minutes.

— Again, for high schedule speed, especially for suburban or urban services, it is desirable to have high acceleration and high value of maximum velocity V_m.

- The *schedule speed* of a given train when running on a given service (*i.e.*, with a given distance between the stations) depends upon the following ***factors.***

(*i*) Acceleration and braking retardation.

(*ii*) Maximum or crest speed.

(*iii*) Duration of stop.

(*i*) *Acceleration and braking retardation :*

— For a given run and with fixed crest speed the *increase in acceleration* will result in decrease in actual time of run and lead to *increase in schedule speed.* Similarly, increase in braking retardation will affect speed.

— The Variation in acceleration and retardation will have more effect on schedule speed in case of shorter distance run in comparison to longer distance run.

(*ii*) *Maximum speed :*

— With fixed acceleration and retardation, for a constant distance run, the actual time of run will decrease and, therefore, schedule speed will *increase* with increase in crest speed.

— In case of long distance run, the effect of variation in crest speed on schedule speed is considerable.

(*iii*) *Duration of stop :*

— The schedule speed, for a given average speed, will increase by reducing the duration of stop.

— The variation in duration of stop will affect the schedule speed more in case of shorter distance run in comparison to longer distance run.

7.4. SIMPLIFIED SPEED-TIME CURVES

— The speed-time curve of an *urban service* can be replaced by an equivalent speed-time of *simple quadrilateral shape.*

— The speed-time curve of a *main line* service is best and most easily replaced by a *trapezoid.*

Since the area of speed-time curve represents the total distance travelled hence the *areas of the two curves should be same.*

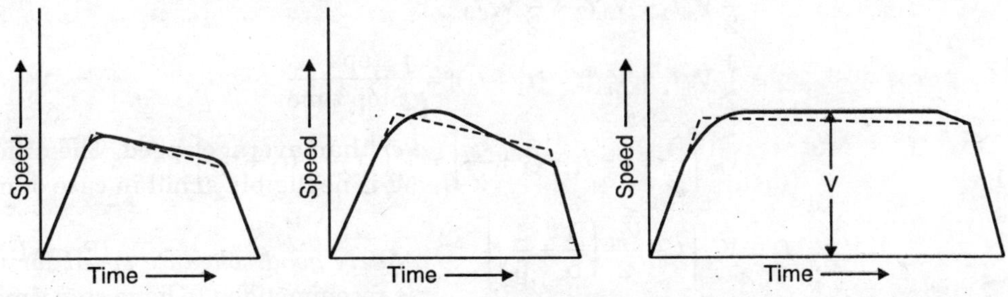

Fig. 7.3. Approximate speed-time curves.

1. Trapezoidal speed-time curves. Fig. 7.4 shows a trapezoidal speed-time curve *OABC*.

Let, D = Distance between stops (*metres*),

t = Actual time of run between stops (*seconds*),

α = Acceleration during starting period (m/s^2),

β = Retardation during braking (m/s^2),

V_m = Maximum (or crest) speed (m/s),

V_a = Average speed (= D/t), *m/s,

t_1 = Time of acceleration (*seconds*),

t_3 = Time of braking (*seconds*), and

t_2 = Time of free running = $t - (t_1 + t_3)$, in *seconds*.

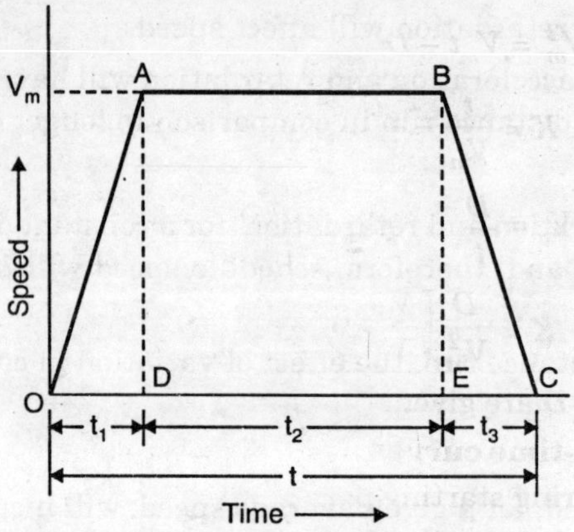

Fig. 7.4. Trapezoidal speed-time curve.

From Fig. 7.4, we have

$$\alpha = \frac{V_m}{t_1} \quad \text{or} \quad t_1 = \frac{V_m}{\alpha} \qquad \ldots(i)$$

$$\beta = \frac{V_m}{t_3} \quad \text{or} \quad t_3 = \frac{V_m}{\beta} \qquad \ldots(ii)$$

Since the total distance D between the stops is given by the area of trapezium *OABC*, therefore,

$$D = \text{Area } OABC$$

$$= \text{Area } OAD + \text{Area } ABED + \text{Area } BCE$$

$$= \frac{1}{2} V_m t_1 + V_m t_2 + \frac{1}{2} V_m t_3$$

$$= \frac{1}{2} V_m t_1 + V_m [t - (t_1 + t_3)] + \frac{1}{2} V_m t_3$$

$$= V_m \left[\frac{t_1}{2} + t - t_1 - t_3 + \frac{t_3}{2}\right] = V_m \left[t - \frac{1}{2}(t_1 + t_3)\right]$$

or,
$$D = V_m \left[t - \frac{V_m}{2}\left(\frac{1}{\alpha} + \frac{1}{\beta}\right)\right] \qquad \text{[From } (i) \text{, and } (ii)\text{]}$$

Let,
$$K = \frac{1}{2}\left(\frac{1}{\alpha} + \frac{1}{\beta}\right) \quad \text{or} = \frac{\alpha + \beta}{2\alpha\beta} \qquad \ldots(7.1)$$

Substituting the value of K in the above eqn., we get

$$D = V_m(t - KV_m)$$

or,
$$KV_m^2 - V_m t + D = 0 \quad \text{...(iii)}$$

$$\therefore \quad V_m = \frac{t \pm \sqrt{t^2 - 4KD}}{2K}$$

The +ve sign will not be used, as it will give much higher value of V_m which will not be met with in practice. Therefore, we have

$$V_m = \frac{t - \sqrt{t^2 - 4KD}}{2K} \quad \text{...(7.2)}$$

From eqn. (*iii*), we get

$$KV_m^2 = V_m t - D$$

or,
$$K = \frac{t}{V_m} - \frac{D}{{V_m}^2} = \frac{D}{{V_m}^2}\left(V_m \cdot \frac{t}{D} - 1\right)$$

Now,
$$V_a = \frac{D}{t}$$

$$\therefore \quad K = \frac{D}{V_m^2}\left(\frac{V_m}{V_a} - 1\right) \quad \text{...(7.3)}$$

Obviously, if V_m, V_a and D are given, then value of K and hence of α and β can be found.

2. **Quadrilateral speed-time curves.** Fig. 7.5 shows a quadrilateral speed-time curve OABC.

Let, α = Acceleration during starting period,

β_c = Retardation during coasting period,

β = Retardation during braking,

V_1 = Maximum speed at the end of acceleration,

V_2 = Speed at the end of coasting, and

t = Total time of run.

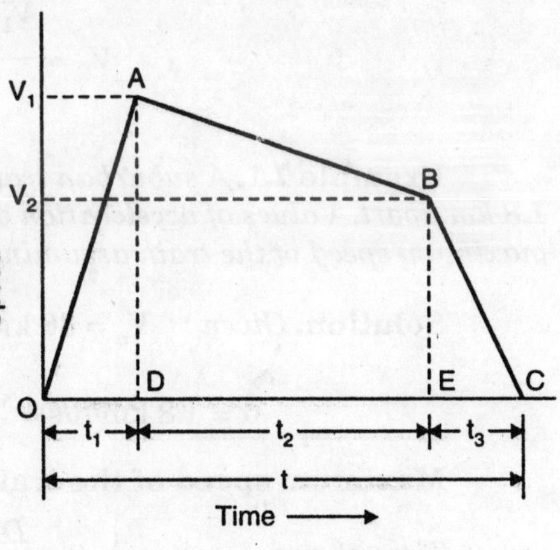

Fig. 7.5. Quadrilateral speed-time curve.

Time of acceleration, $t_1 = \dfrac{V_1}{\alpha}$

Time of coasting, $t_2 = \dfrac{V_1 - V_2}{\beta_c}$

Time of braking, $t_3 = \dfrac{V_2}{\beta}$

Total distance travelled, D = Area $OABC$

= Area OAD + Area $ABED$ + Area BCE

$$= \frac{1}{2} V_1 t_1 + \left(\frac{V_1 + V_2}{2}\right) t_2 + \frac{1}{2} V_2 t_3$$

---

* Velocity and acceleration given in km/h and km/h/s units may be converted to m/s and m/s² respectively by multiplying the factor $\frac{1000}{3600}$ or 0.2778. **Example.** 36 km/h = $36 \times \frac{1000}{3600}$ = 10 m/s² ; 1.8 km/h/s = $1.8 \times \frac{1000}{3600}$ = 0.5 m/s².

$$= \frac{1}{2}V_1(t_1 + t_2) + \frac{1}{2}V_2(t_2 + t_3)$$

$$= \frac{1}{2}V_1(t - t_3) + \frac{1}{2}V_2(t - t_1)$$

$$= \frac{1}{2}t(V_1 + V_2) - \frac{V_2 t_1}{2} - \frac{V_1 t_3}{2}$$

$$= \frac{1}{2}t(V_1 + V_2) - \frac{1}{2}V_1 V_2\left(\frac{1}{\alpha} + \frac{1}{\beta}\right)$$

or,
$$D = \frac{1}{2}t(V_1 + V_2) - KV_1V_2 \qquad ...(7.4)$$

where $K = \frac{1}{2}\left(\frac{1}{\alpha} + \frac{1}{\beta}\right) = \frac{\alpha + \beta}{2\alpha\beta}$, Also $\beta_c = \frac{V_1 - V_2}{t_2}$

$\therefore$
$$V_2 = V_1 - \beta_c t_2 = V_1 - \beta_c(t - t_1 - t_3)$$

$$= V_1 - \beta_c\left(t - \frac{V_1}{\alpha} - \frac{V_2}{\beta}\right)$$

or,
$$V_2 = V_1 - \beta_c\left(t - \frac{V_1}{\alpha}\right) + \beta_c\frac{V_2}{\beta}$$

or,
$$V_2\left(1 - \frac{\beta_c}{\beta}\right) = V_1 - \beta_c\left(t - \frac{V_1}{\alpha}\right)$$

$\therefore$
$$V_2 = \frac{V_1 - \beta_c\left(t - \frac{V_1}{\alpha}\right)}{\left(1 - \frac{\beta_c}{\beta}\right)} \qquad ...(7.5)$$

Example 7.1. *A suburban trains run with an average speed of 36 kmph between two stations 1.8 km apart. Values of acceleration and retardation are 1.8 km/h/s and 3.6 km/ph/ps. Calculate the maximum speed of the train assuming trapezoidal speed-time curve.* **(Panjab University)**

Solution. *Given :* $V_a = 36 \text{ km/h} = \frac{36 \times 1000}{3600} = 10 \text{ m/s}$; $D = 1.8 \text{ km} = 1800 \text{ m}$;

$$\alpha = 1.8 \text{ km/h/s} = \frac{1.8 \times 1000}{3600} = 0.5 \text{ m/s}^2 ; \beta = 3.6 \text{ km/h/s} = \frac{3.6 \times 1000}{3600} = 1.0 \text{ m/s}^2.$$

Maximum speed of the train, V_m :

Time of run, $t = \frac{D}{V_a} = \frac{1800}{10} = 180 \text{ s}$

Also, $K = \frac{\alpha + \beta}{2\alpha\beta} = \frac{0.5 + 1.0}{2 \times 0.5 \times 1} = 1.5$

Using the relation : $V_m = \frac{t - \sqrt{t^2 - 4KD}}{2K}$ ...[Eqn. (7.2)]

$$= \frac{180 - \sqrt{(180)^2 - 4 \times 1.5 \times 1800}}{2 \times 1.5}$$

$$= 11.01 \text{ m/s} = \frac{11.01 \times 3600}{10000} = \mathbf{39.64 \text{ km/h.}} \quad \textbf{(Ans.)}$$

Example 7.2. *A train is required to run between two stations 2 km apart at a schedule speed of 36 km/h, the duration of stops being 20 seconds. The braking retardation is 2.7 km/h/s. Assuming a trapezoidal speed-time curve, calculate the acceleration if the ratio of maximum speed to average speed is 1.2.* **(Bombay University)**

Solution. *Given* : $D = 2$ km $= 2000$ m ; Schedule speed $= 36$ km/h $= \dfrac{3600 \times 1000}{3600} = 10$ m/s ;

Duration of stop $= 20$ s ; $\beta = 2.7$ km/h/s $= \dfrac{2.7 \times 1000}{3600} = 0.75$ m/s^2 ; $\dfrac{V_m}{V_a} = 1.2$.

Acceleration, α :

Schedule time of run $= \dfrac{\text{Distance between two stops}}{\text{Schedule speed}} = \dfrac{2000}{10} = 200$ s

Actual time of run, $t = 200 - 20 = 180$ s

$\therefore$ Average speed, $V_a = \dfrac{D}{t} = \dfrac{2000}{180} = 11.11$ m/s

$\therefore$ $V_m = 1.2\, V_a = 1.2 \times 11.11 = 13.33$ m/s

Now, $$K = \frac{D}{V_m^2}\left(\frac{V_m}{V_a} - 1\right) \quad \text{...[Eqn. 7.3]}$$

$$= \frac{2000}{(13.33)^2}\left(\frac{13.33}{11.11} - 1\right) = 2.25$$

Also, $$K = \frac{1}{2}\left(\frac{1}{\alpha} + \frac{1}{\beta}\right) \quad \text{...[Eqn. (7.1)]}$$

$$2.25 = \frac{1}{2}\left(\frac{1}{\alpha} + \frac{1}{0.75}\right)$$

or, $$\frac{1}{\alpha} + \frac{1}{0.75} = 4.5$$

or, $$\frac{1}{\alpha} = 4.5 - \frac{1}{0.75} = 3.167$$

$\therefore$ $$\alpha = 0.316 \text{ m/s}^2 = \frac{0.316 \times 3600}{1000} = \mathbf{1.138\ km/h/s.}$$ **(Ans.)**

Example 7.3. *A schedule speed of 45 km/h is required between two stops 1.5 km apart. Find the maximum speed over the run if the stop is of 20 seconds duration. The values of acceleration and retardation are 2.4 km / h / s and 3.2 km / h / s respectively. Assume a simplified trapezoidal speed-time curve.*

Solution. *Given :* Schedule speed $= 45$ km/h $= \dfrac{4.5 \times 1000}{3600} = 12.5$ m/s ; $D = 15$ km $= 1500$ m ;

Duration of stop $= 20$ s ; $\alpha = 2.4$ km/h/s $= \dfrac{2.4 \times 1000}{3600} = 0.667$ m/s^2 ;

$\beta = 3.2$ km/h/s $= \dfrac{3.2 \times 1000}{3600} = 0.888$ m/s^2.

Maximum speed, V_m :

Schedule time of run, $t = \dfrac{\text{Distance between two stops}}{\text{Schedule speed}} = \dfrac{1500}{12.5} = 120$ s

Actual time of run, $t = 120 - 20 = 100$ s

Now, $$K = \frac{1}{2}\left(\frac{1}{\alpha} + \frac{1}{\beta}\right) \quad \text{...[Eqn. (7.1)]}$$

$$= \frac{1}{2}\left(\frac{1}{0.667} + \frac{1}{0.888}\right) = 1.313$$

$\therefore$ $$V_m = \frac{t - \sqrt{t^2 - 4KD}}{2K} \quad \text{...[Eqn. (7.2)]}$$

$$= \frac{100 - \sqrt{(100)^2 - 4 \times 1313 \times 1500}}{2 \times 1313} = 20.54 \text{ m/s}$$

$$= 20.54 \text{ m/s} = \frac{20.54 \times 3600}{1000} = \mathbf{74 \text{ km/h.}} \quad \textbf{(Ans.)}$$

Example 7.4. *An electric train is to have acceleration and braking retardation of 0.8 km/h/s and 3.2 km/h/s respectively. If the ratio of maximum to average speed is 1.3 and time for stop 26 seconds, find the schedule speed from a run of 1.5 km. Assume simplified trapezoidal speed-time curve.* **(Gorakhpur University)**

Solution. *Given* : $\alpha = 0.8 \text{ km/h/s} = \frac{0.8 \times 1000}{3600} = 0.222 \text{ m/s}^2$;

$\beta = 3.2 \text{ km/h/s} = \frac{3.2 \times 1000}{3600} = 0.888 \text{ m/s}^2, \frac{V_m}{V_a} = 1.3$; Time for stops = 26 s ;

Distance of run, D = 1.5 km = 1500 m.

Schedule speed, V_{sch} :

Now, $$K = \frac{1}{2}\left(\frac{1}{\alpha} + \frac{1}{\beta}\right) = \frac{\alpha + \beta}{2\alpha\beta} = \frac{0.222 + 0.888}{2 \times 0.222 \times 0.888} = 2.815$$

We know that, $KV_m{}^2 - V_m t + D = 0$

or, $$2.815 \times \left(\frac{1950}{t}\right)^2 - \frac{1950}{t} \times t + 1500 = 0$$

or, $$2.815 \times \left(\frac{1950}{t}\right)^2 - 1950 + 1500 = 0 \quad \text{or} \quad 2.815 \times \left(\frac{1950}{t}\right)^2 = 450$$

$\therefore$ $$t = 154 \text{ s}$$

$\therefore$ Schedule time, $t_{sch} = t + 26 = 154 + 26 = 180$ s

Hence, *Schedule speed*, $V_{sch} = \frac{D}{t_{sch}} = \frac{1500}{180} = 8.333$ m/s

$$= \frac{8.333 \times 3600}{1000} = \mathbf{30 \text{ km/h.}} \quad \textbf{(Ans.)}$$

Example 7.5. *A suburban electric train has a maximum speed of 72 km/h. The schedule speed including a stop of 25 seconds is 45 km/h. If the acceleration is 1.8 km/h/s, find the value of retardation when average distance between stops is 4.5 km.*

Solution. *Given* : $V_m = 70 \text{ km/h} = \frac{72 \times 1000}{3600} = 20 \text{ m/s}$; Duration of stop = 25 s ;

Schedule speed, $V_{sch} = 45 \text{ km/h} = \frac{45 \times 1000}{3600} = 12.5 \text{ m/s}$;

$\alpha = 1.8 \text{ km/h/s} = \frac{1.8 \times 1000}{3600} = 0.5 \text{ m/s}^2$; D = 4.5 km = 4500 m.

Retardation, β :

Schedule time of run, $t_{sc} = \dfrac{\text{Distance } (D)}{\text{Schedule speed } (V_{ch})} = \dfrac{4500}{12.5} = 360$ s

Actual time of run, $t = t_{sch} - 25 = 360 - 25 = 335$ s

We know that, $KV_m^2 - V_m t + D = 0$

or, $K \times (20)^2 - 20 \times 335 + 4500 = 0$

or, $K = 5.5$

or, $K = \dfrac{1}{2}\left(\dfrac{1}{\alpha} + \dfrac{1}{\beta}\right) = 5.5$ or $\dfrac{1}{2}\left(\dfrac{1}{0.5} + \dfrac{1}{\beta}\right) = 5.5$

∴ $\beta = 0.1111 \text{ m/s}^2 = 0.1111 \times \dfrac{3600}{1000} = 0.4$ **km/h/s. (Ans.)**

Example 7.6. *An electric train has an average speed of 42 km/h on a level track between stops 1400 apart. It is accelerated at 1.7 km/h/s and is braked at 3.3 km/h/s. Draw the speed-time curve for the run.* **(A.M.I.E.)**

Solution. *Given :* $V_a = 42 \text{ km/h} = \dfrac{42 \times 1000}{3600} = 11.67$ m/s ; $D = 1400$ m ;

$\alpha = 1.7 \text{ km/h/s} = \dfrac{1.7 \times 1000}{3600} = 0.472 \text{ m/s}^2$; $\beta = 3.3 \text{ km/h/s} = \dfrac{3.3 \times 1000}{3600} = 0.917 \text{ m/s}^2$.

Speed-time curve for the run :

Actual time of run, $t = \dfrac{D}{V_a} = \dfrac{1400}{11.67} = 120$ s

Now, $K = \dfrac{1}{2}\left(\dfrac{1}{\alpha} + \dfrac{1}{\beta}\right) = \dfrac{1}{2}\left(\dfrac{1}{0.472} + \dfrac{1}{0.917}\right) = 1.6$

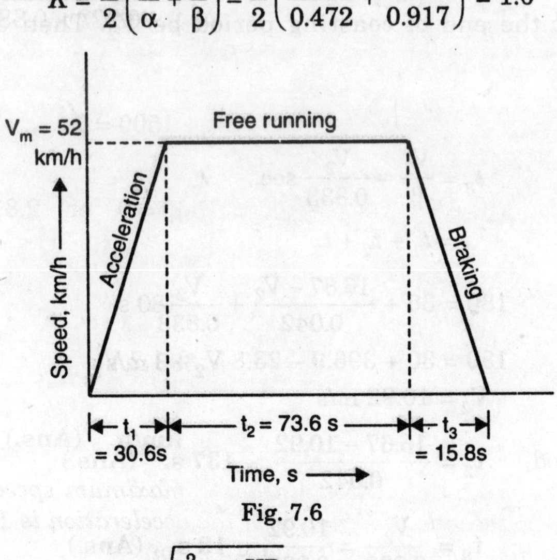

Fig. 7.6

Also, maximum speed, $V_m = \dfrac{t - \sqrt{t^2 - 4KD}}{2K}$ ...[Eqn. (7.2)]

∴ $V_m = \dfrac{120 - \sqrt{(120)^2 - 4 \times 1.6 \times 1400}}{2 \times 1.6} = 14.45$ m/s

$= 14.45 \times \dfrac{3600}{1000} = 52$ km/h

Acceleration period, $t_1 = \frac{V_m}{\alpha} = \frac{14.45}{0.472} = 30.6$ s

Braking period, $t_3 = \frac{V_m}{\beta} = \frac{14.45}{0.917} = 15.8$ s

Free running period, $t_2 = t - (t_1 + t_3)$

$= 120 - (30.6 + 15.8) = 73.6$ s

The speed-time curve for the run is shown in Fig. 7.6.

Example 7.7. *A train runs between two stations 2 km apart at an average speed of 40 km / h. The run is to be made according to simplified quadrilateral speed-time curve. If the maximum speed to be limited to 60 km / h, acceleration to 2km / h / s, coasting retardation to 0.15 km / h / s and braking retardation 3 km / h / s, determine the duration of acceleration, coasting and braking periods.*

Solution. *Given :* $D = 2$ km $= 2000$ m ; $V_a = 40$ km/h $= \frac{40 \times 1000}{3600} = 11.11$ m/s ;

$V_1 = 60$ km/h $= \frac{60 \times 1000}{3600} = 16.67$ m/s ; $\alpha = 2$ km/h/s $= \frac{2 \times 1000}{3600} = 0.555$ m/s^2 ;

$\beta_c = 0.15$ m/h/s $= \frac{0.15 \times 1000}{3600} = 0.042$ m/s^2 ; $\beta = 3$ km/h/s $= \frac{3 \times 1000}{3600} = 0.833$ m/s^2

Duration of acceleration, coasting and braking periods, t_1, t_2, t_3 :

Refer to Fig. 7.5.

Duration of acceleration, $t_1 = \frac{V_1}{\alpha} = \frac{16.67}{0.555} = \mathbf{30\ s.}$ **(Ans.)**

Actual time of run, $t = \frac{D}{V_a} = \frac{2000}{11.11} = 180$ s

Let the speed at the end of coasting period be V_2. Then duration of coasting, $t_2 = \frac{V_1 - V_2}{\beta_c} = \frac{16.67 - V_2}{0.042}$ sec.

Braking period, $t_3 = \frac{V_2}{\beta} = \frac{V_2}{0.833}$ sec.

Now, $t = t_1 + t_2 + t_3$

or $180 = 30 + \frac{16.67 - V_2}{0.042} + \frac{V_2}{0.833}$

or $180 = 30 + 396.9 - 23.8\ V_2 + 1.2\ V_2$

or $V_2 = 10.92$ m/s

$\therefore$ Coasting period, $\mathbf{t_2} = \frac{16.67 - 10.92}{0.042} \simeq \mathbf{137\ s.}$ **(Ans.)**

and, Braking period, $\mathbf{t_3} = \frac{V_2}{0.833} = \frac{10.92}{0.833} \simeq \mathbf{13\ s.}$ **(Ans.)**

Example 7.8. *A train is required to run between two stations 1.6 km apart at the average speed of 40 km / h. The acceleration, retardation during coasting and braking are 2 km / h / s, 0.16 km / h / s and 3.2 km / h / s respectively. Assuming quadrilateral approximation of speed-time curve, determine :*

(i) The duration of acceleration, coasting and braking periods, and

(ii) The distance covered during these periods.

Solution. *Given :* $D = 1.6\text{ km} = 1600\text{ m}$; $V_a = 40\text{ km/h} = \dfrac{40 \times 1000}{3600} = 11.11\text{ m/s}$;

$\alpha = 2\text{ km/h/s} = \dfrac{2 \times 1000}{3600} = 0.555\text{ m/s}^2$; $\beta_c = 0.16\text{ km/h/s} = \dfrac{0.16 \times 1000}{3600} = 0.0444\text{ m/s}^2$;

$\beta = 3.2\text{ km/h/s} = \dfrac{3.2 \times 1000}{3600} = 0.888\text{ m/s}^2$.

(*i*) **Duration of acceleration, coasting and braking periods, t_1, t_2, t_3 :**

Refer to Fig. 7.5.

Actual time of run, $t = \dfrac{D}{V_a} = \dfrac{1600}{11.11} = 144\text{ s}$

We know that, $$V_2 = \frac{V_1 - \beta_c\left(t - \dfrac{V_1}{\alpha}\right)}{\left(1 - \dfrac{\beta_c}{\beta}\right)} \quad \text{...[Eqn. (7.5)]}$$

or, $$V_2 = \frac{V_1 - 0.0444\left(144 - \dfrac{V_1}{0.555}\right)}{\left(1 - \dfrac{0.0444}{0.888}\right)} = \frac{V_1 - 6.394 + 0.08\,V_1}{0.95}$$

or, $$V_2 = 1.1368\,V_1 - 6.73$$

Using the relation : $$D = \frac{1}{2}t(V_1 + V_2) - KV_1V_2 \quad \text{...[Eqn. (7.4)]}$$

where $K = \dfrac{1}{2}\left(\dfrac{1}{\alpha} + \dfrac{1}{\beta}\right) = \dfrac{1}{2}\left(\dfrac{1}{0.555} + \dfrac{1}{0.888}\right) = 1.464$

Substituting the various values in the above eqn., we get

$$1600 = \frac{1}{2} \times 144\,[V_1 + (1.1368V_1 - 6.73)] - 1.464V_1\,(1.1368V_1 - 6.73)$$

or, $$1600 = 72(2.1368V_1 - 6.73) - 1.664V_1^2 + 9.853V_1$$

or, $$1600 = 153.85V_1 - 484.56 - 1.664V_1^2 + 9.853V_1$$

or, $1.664V_1^2 - 163.7V_1 + 2084.6 = 0$

or, $V_1^2 - 98.4\,V_1 + 1252.8 = 0$

$$\therefore \quad V_1 = \frac{98.4 \pm \sqrt{(98.4)^2 - 4 \times 1 \times 1252.8}}{2} = \frac{98.4 \pm 68.35}{2} = 15\text{ m/s}$$

(considering –ve sign)

and, $$V_2 = 1.1368 \times 15 - 6.73 = 10.322\text{ m/s}$$

$\therefore$ *Duration of acceleration,* $\mathbf{t_1} = \dfrac{V_1}{\alpha} = \dfrac{15}{0.555} = \mathbf{27.03\ s.}$ **(Ans.)**

Duration of coasting, $\mathbf{t_2} = \dfrac{V_1 - V_2}{\beta} = \dfrac{15 - 10.322}{0.0444} = \mathbf{105.4\ s.}$ **(Ans.)**

Duration of retardation, $\mathbf{t_3} = \dfrac{V_2}{\beta} = \dfrac{10.322}{0.888} = \mathbf{11.62\ s.}$ **(Ans.)**

(*ii*) **Distance covered during the periods t_1, t_2 and t_3 :**

$$(S)_{t_1} = \frac{1}{2} \times 15 \times 27.03 = \mathbf{202.7\ m} \quad \textbf{(Ans.)}$$

$$(S)_{t_2} = \frac{1}{2} \times (15 + 10.322) \times 105.4 = \mathbf{1334..} \quad \textbf{(Ans.)}$$

$$(S)_{t_3} = \frac{1}{2} \times 10.322 \times 11.62 = \mathbf{60\ m} \quad \textbf{(Ans.)}$$

Example 7.9. *A train runs between two stations 1.92 km apart. The schedule speed and the duration of stops respectively are 40 km/h and 20 sec. Assume the quadrilateral approximation of speed-time curve and the coasting and tracking retardation are 0.16 km/h/s and 3.2 km/h/s respectively. Determine :*

(i) *The acceleration if the speed at the end of acceleration period is 60.8 km/h.*

(ii) *The duration of coasting period.*

Solution. *Given :* $D = 1.92$ km = 1920 m ; Schedule speed = 40 km/h = $\frac{40 \times 1000}{3600} = 11.11$ m/s ;

Duration of stops = 20 s ; $\beta_c = 0.16$ km/h/s = $\frac{0.16 \times 1000}{3600} = 0.0444$ m/s² ;

$\beta = 3.2$ km/h/s = $\frac{3.2 \times 1000}{3600} = 0.888$ m/s² ; $V_1 = 60.8$ km/h = $\frac{60.8 \times 1000}{3600} = 16.89$ m/s.

(i) **The acceleration, α :**

The schedule time, $t_{sch} = \frac{\text{Distance}}{\text{Schedule speed}} = \frac{1920}{1111} = 172.8$ s

∴ Actual time of run, $t = t_{sch} - 20 = 172.8 - 20 = 152.8$ s.

We know that,
$$V_2 = \frac{V_1 - \beta_c\left(t - \frac{V_1}{\alpha}\right)}{\left(1 - \frac{\beta_c}{\beta}\right)} \qquad \text{...[Eqn. 7.5)]}$$

or,
$$V_2 = \frac{16.89 - 0.0444\left(152.8 - \frac{16.89}{\alpha}\right)}{\left(1 - \frac{0.0444}{0.888}\right)} = \frac{16.89 - 6.78 + \frac{0.75}{\alpha}}{0.95}$$

or,
$$V_2 = 10.64 + \frac{0.79}{\alpha} \qquad \text{...}(i)$$

Using the relation :
$$D = \frac{1}{2}t(V_1 + V_2) - KV_1V_2 \qquad \text{...[Eqn. (7.4)]}$$

where, $K = \frac{1}{2}\left(\frac{1}{\alpha} + \frac{1}{\beta}\right) = \frac{1}{2}\left(\frac{1}{\alpha} + \frac{1}{0.888}\right) = \left(\frac{1}{2\alpha} + 0.563\right)$

Substituting the various values in the above eqn., we get

$$1920 = \frac{1}{2} \times 152.8\left(16.89 + 10.64 + \frac{0.79}{\alpha}\right) - \left(\frac{1}{2\alpha} + 0.563\right) \times 16.89 \times \left(10.64 + \frac{0.79}{\alpha}\right)$$

or,
$$1920 = 76.4\left(27.53 + \frac{0.79}{\alpha}\right) - \left(\frac{1}{2\alpha} + 0.563\right) \times \left(179.7 + \frac{13.34}{\alpha}\right)$$

or,
$$1920 = 2103.3 + \frac{60.35}{\alpha} - \left[\frac{89.85}{\alpha} + \frac{6.67}{\alpha^2} + 101.17 + \frac{7.51}{\alpha}\right]$$

or,
$$1920 = 2103.3 + \frac{60.35}{\alpha} - \frac{89.95}{\alpha} - \frac{6.67}{\alpha^2} - 101.17 - \frac{7.51}{\alpha}$$

or, $$\frac{6.67}{\alpha^2} + \frac{37.11}{\alpha} - 82.13 = 0$$

or, $$82.13\,\alpha^2 - 37.11\,\alpha - 6.67 = 0$$

or, $$\alpha = \frac{37.11 \pm \sqrt{(37.11)^2 + 4 \times 82.13 \times 6.67}}{2 \times 82.13}$$

or, $$= \frac{37.11 \pm 59.73}{164.26} = 0.589 \text{ m/s}^2 \quad \text{or} \quad \frac{0.589 \times 3600}{1000} = 2.12 \text{ km/h/s}$$

i.e., *Acceleration,* α = **2.12 km/h/s (Ans.)**

(*ii*) **Duration of coasting period, t_2 :**

From eqn. (*i*), we have

$$V_2 = 10.64 + \frac{0.79}{0.589} = 11.98 \text{ m/s} \quad \text{(or 43.12 km/h)}$$

$$\therefore \quad t_2 = \frac{V_1 - V_2}{\beta_c} = \frac{16.89 - 11.98}{0.0444} = 110.6 \text{ s} \quad \textbf{(Ans.)}$$

7.5. MECHANISM OF TRAIN MOVEMENT

Fig. 7.7 shows the essentials of driving mechanism in an electric vehicle. The armature of the driving motor has a pinion which meshes with the gear wheel keyed to the axle of the driving wheel. The gear wheel transfers the tractive effort at the edge of the pinion to the driving wheel.

Let, T = Torque exerted by the motor,

F_1 = Tractive effort at the pinion,

F_t = Tractive effort at the wheel,

γ = Gear ratio,

d_1, d_2 = Diameters of the pinion and gear wheel respectively,

D = Diameter of the driving wheel, and

η = Efficiency of power transmission from the motor to driving axle.

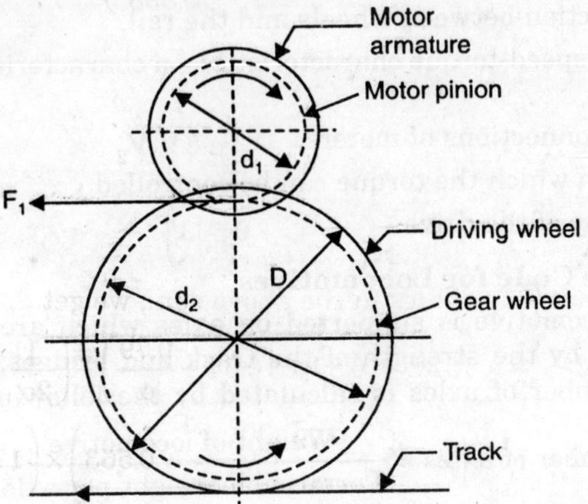

Fig. 7.7. Essentials of driving mechanism in an electric vehicle.

Now, torque, $$T = F_1 \times \frac{d_1}{2} \quad \text{or} \quad F_1 = \frac{2T}{d_1} \qquad ...(7.6)$$

Tractive effort transferred to the driving wheel,

$$F_t = \eta F_1\left(\frac{d_2}{D}\right) = \eta.\frac{2T}{d_1}\left(\frac{d_2}{D}\right) = \eta T\left(\frac{2}{D}\right)\left(\frac{d_2}{d_1}\right) = 2\gamma\eta\frac{T}{D} \quad ...(7.7) \quad \left(\because \gamma = \frac{d_2}{d_1}\right)$$

7.5.1. Adhesive Weight and Coefficient of Adhesion

The **"adhesive weight"** of *a train in equal to the total weight to be carried on the driving wheels.* It is less than the dead weight by about 20 to 40 per cent.

If $x = \dfrac{\text{Adhesive weight } (W_a)}{\text{Dead weight } (W)}$, then $W_a = xW$

The **"coefficient of adhesion"** (μ_a) is defined as the *ratio of tractive effort to slip the wheels and adhesive weight.*

i.e., $\mu_a = \dfrac{\text{Tractive effiort to slip wheels}}{\text{Adhesive weight}} = \dfrac{F_t}{W_a}$

$\therefore$ $F_t = \mu_a W_a = \mu_a xW = \mu_a xMg$

If M is in tonnes, then

$$F_t = 1000 \times 9.8\, x\mu_a M = 9800\, \mu_a xM \text{ newton} \quad ...(7.7\ a)$$

- The normal value of coefficient of adhesion with clean ***dry rails*** is *0.25* and with wet or ***greasy rails*** the value may be as low as 0.08.
- Electric traction has a significant advantage over steam traction and that is *due to greater adhesive weight*—in motor coach 100% of the weight is on the driving wheels, in electric locomotive 70% whereas in a steam locomotive less than 50%.

The coefficient of adhesion in electric traction is also *greater* than that in steam traction due to follow *reasons* :

(*i*) The torque exerted in electric traction is *continuous*, whereas in steam traction the torque is *pulsating in nature which cause jolting and skidding.*

(*ii*) The driving wheels in electric traction are distributed over a much greater length whereas in steam traction they are close to each other.

Co-efficient of adhesion, which is a measure of the tendency for free wheel slip, depends upon the following *factors* :

(*i*) Coefficient of friction between wheels and the rail.

(*ii*) Nature of motor speed-torque characteristics—a characteristic with low speed regulation is preferred.

(*iii*) Series-parallel connections of motors.

(*iv*) Smoothness with which the torque can be controlled.

(*v*) Speed of response of the drive.

7.5.2. Driving Axle Code for Locomotives

The weight of a locomotive is supported on axles which are coupled to the wheels. The weight per axle is limited by the strength of the track and bridges, and usually varies between 15 and 30 tons. Total number of axles is calculated by the following equation :

$$\text{Number of axles} = \frac{\text{Weight of locomotive}}{\text{Permissible weight per axle}}$$

- *The number of driving axles and coupled motors are described using a code as follows :*

2 driving axles ...... category B

3 driving axles ...... Category C

4 driving axles ...... Category BB
6 driving axles ...... Category CC

— When each axle is driven by an individual motor ; a subscript 'O' is used along with these symbols.

— When axles are divided into groups and each group is driven by a single motor, only letters B and C are appropriately used.

— The number of dummy (non-driving) axles is denoted by numerals.

7.5.3. Nature of Traction Load

When a train runs on level track the following types of frictional forces oppose its motion :

(*i*) *Internal friction.* It consists of friction at bearings, guides etc.

(*ii*) *External friction.* It is the friction between wheel-flanges and rails.

(*iii*) *Air friction.* It is independent of weight of train but depends upon its size and shape, and velocity and relative direction of wind.

All these frictional forces collectively are termed as ***"train resistance".***

Fig. 7.8 shows the variation of train resistance (F_r) with speed (V) ; load torque *v/s* speed curve will have similar nature.

— The train resistance or load torque can also be identified in terms of common classification of friction such as windage, viscous friction, coulomb friction and stiction. *Stiction* has a large value and the influence of air friction, which *varies as the square of speed*, is quite prominent at high speeds.

- While deciding torque requirements of driving motors, the *torque components required to provide acceleration and to overcome gravity* must also be taken into consideration.

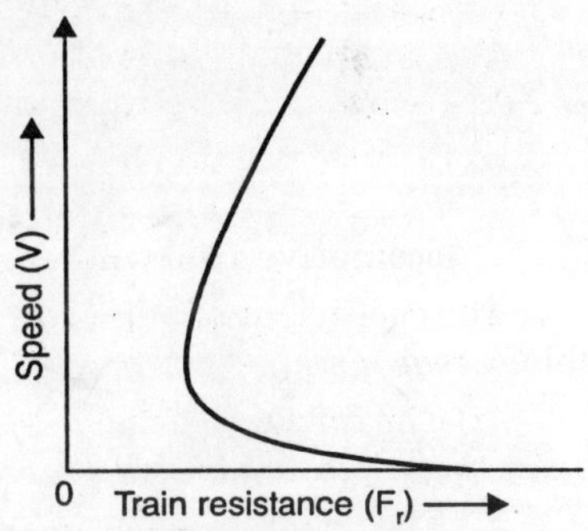

Fig. 7.8. Relation between V and train resistance F_r.

Owing to the large values of stiction and accelerating torque, the torque requirement at start and during acceleration is much higher than the torque needed for running at the highest speed. Therefore, *only those drives are suitable for traction application which develop large torque from zero to the base speed.*

7.6. TRACTIVE EFFORT FOR PROPULSION OF TRAIN

The **tractive effort** is *defined as the effective force necessary to propel the train at the wheels of locomotive.*

The tractive effort (F_t) required for train propulsion is given by :

$$F_t = F_a + F_r \quad \text{...For Level track ;} \quad ...(7.8)$$

$$F_t = F_a \pm F_g + F_r \quad \text{... When gradients are involved} \quad ...(7.9)$$

{+ *sign* for *ascending gradient*, and – *sign* for *descending gradient*]

where, F_a = Force required for giving linear acceleration to the train,

F_g = Force required to overcome the gravitational effect, and

F_r = Force required to overcome resistance to the motion of train.

1. **Value of F_a.** If M is the dead or stationary mass of the train and α is the linear acceleration, then

$$F_a = M\,\alpha \quad ...(7.10)$$

Owing to the fact that a train has rotating parts likes wheels, axles, motor armatures and gearing etc., its *effective* or *accelerating* mass M_e is more (about 8–15 percent) than its stationary mass. These parts need to be given angular acceleration at the same time as the whole train is accelerated in linear direction. Hence,

$$F_a = M_e \,.\, \alpha \qquad ...(7.11)$$

- If M_e is in kg and α in m/s², then $F_a = m.\alpha$ newton ...(7.11 *a*)
- If M_e is in tonne and α in km/h/s, then converting them in absolute units, we have

$$F_a = (1000\ M_e) \times \left(\frac{1000}{3600}\right) \alpha = 277.8\ \alpha\ M_e \text{ newton} \qquad ...(7.11\ b)$$

2. **Value of F_g** : Refer to Fig. 7.9.

$$F_g = W \sin\theta = Mg \sin\theta \qquad ...(7.12)$$

$$\text{Now gradient} = \sin\theta = \frac{\text{Elevation } (BC)}{\text{Distance along truck } (AC)}$$

$$\%\ \text{gradient } G = 100 \sin\theta$$

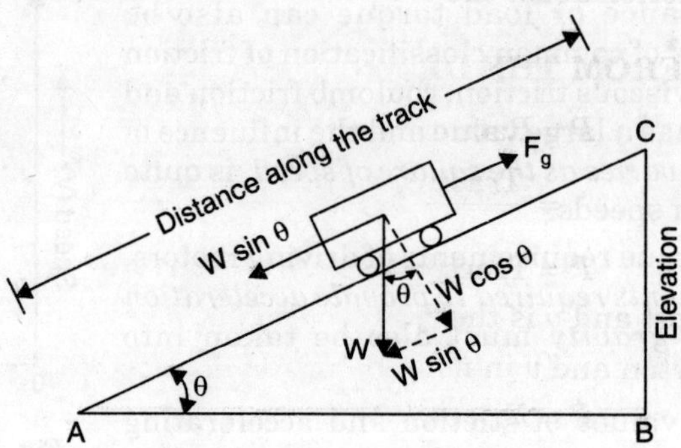

Fig. 7.9

Substituting the value of sin θ in eqn. (7.12), we get

$$F_g = \frac{Mg.G}{100} \qquad ...(7.13)$$

- When M is in kg, $F_g = 0.098\ MG$ ...(7.13 *a*)
- When M is in tonne, then

$$F_g = 0.098\ (1000\ M)G = 98\ MG \text{ newton} \qquad ...(7.13\ b)$$

3. **Value of F_r** :

Train resistance consists of all the forces which resist the motion of a train. It consists of the following :

(*i*) ***Mechanical resistance :***

(*a*) *Internal resistance.* It comprises friction at journals, axles, guides and buffers etc.

(*b*) *External resistance.* It consists of friction between wheels and rails and flange friction etc.

— Mechanical resistance is ***almost independent of train speed but depends on its weight.***

(*ii*) ***Wind resistance.*** The wind resistance varies ***directly as the square of the train speed.***

If r is '*specific resistance*' of the train *i.e.*, ***resistance offered per unit mass of the train, then***

$$F_r = M.r \qquad ...(7.14)$$

- When r is in newton per kg of train mass and M is the train mass in kg, then

$$F_r = M.r \text{ newton} \quad ...(7.14\ a)$$

- When r is in newton per tonne train mass (N/t) and M is in tonne (t), then

$$F_r = M \text{ tonne} \times r = Mr \text{ newton} \quad ....(7.14\ b)$$

[If r is in kg (wt.) per tonne train mass and M is in the tonne, then $F_r = M$ tonne × (r × 9.8) newton/tonne = 9.8 Mr newton]

Hence, expression for total tractive effort becomes,

$$F_t = F_a \pm F_g + F_r$$

or,
$$F_t = (277.8\alpha M_e \pm 98\, MG + Mr) \text{ newton} \quad ...(7.15)$$

It may noted that in this eqn. (7.15),

- M is in *tonne*
- α is in *km/h/s*
- G is in *metres per 100 m of track length* (*i.e.*, % G)
- r is in *newton/tonne (N/t) of train mass.*
- *+ve* sign for the motion *up* the gradient, and
  –ve sign for the motion *down* the gradient.

7.7. POWER OUTPUT FROM THE DRIVING AXLES

Power output, P = Rate of doing work

$$= \frac{\text{Tractive effort} \times \text{distance}}{\text{time}}$$

or,
$$P = \text{Tractive effort} \times \text{velocity} = F_t \times v \quad ...(7.16)$$

where, F_t is the tractive effort and v is the train velocity

- When F_t is in newton and v in m/s, then

$$P = F_t \times v \text{ watt} \quad ...(7.16\ a)$$

- When F_t is in newton and v is in km/h, then converting v into m/s, we have

$$P = F_t \times \left(\frac{v \times 1000}{3600}\right) \text{watt} = \frac{F_t \times v}{3600} \text{ kW} \quad ...(7.16\ b)$$

If η is the efficiency of transmission gear, then, power output of motors,

$$P = \frac{F_t \times v}{\eta} \text{ watt} \quad ... v \text{ in m/s} \quad ...(7.17\ a)$$

and,
$$P = \frac{F_t \times v}{3600\ \eta} \text{ kW} \quad ... v \text{ in km/h} \quad ...(7.17\ b)$$

7.8. ENERGY OUTPUT FROM DRIVING AXLES

Energy (*E*) *defined as capacity to do work, is given by the product of power and time.*

i.e.,
$$E = (F_t \times v) \times t = F_t \times (v \times t) = F_t \times D \quad ...(7.18)$$

where, D is the distance travelled in the direction of tractive effort.

Total energy output from driving axles for the run,

$$E = \text{Energy during acceleration} + \text{Energy during free run}$$

Referring to Fig. 7.4, we have :

$$E = F_t \times \text{area } OAD + F_t' \times \text{area } ABED$$

or, $$E = F_t \times \frac{1}{2} V_m t_1 + F'_t \times V_m t_2 \quad ...(7.18a)$$

where, F_t = The tractive effort during acceleration periods (It consists of all the three components given in Art. 7.6), and

$F'_t = (98\,MG + Mr)$ provided there is an ascending gradient.

7.8.1. Specific Energy Output

- The ***specific energy output*** *is the energy output of the driving wheel expressed in watt-hour (Wh) per tonne-km (t-km) of the train.*
- It can be found by first converting the energy output into *Wh* and then dividing it by the mass of the train in tonne and route distance in km. Hence, unit of specific energy output generally used in railway is : *Wh/tonne-km (Wh/t-km).*
- The specific energy output is *used for comparing the dynamical performances of trains operating to different schedules.*

Evaluation of specific energy output :

In order to find out the specific energy output let us first calculate the total energy output of the driving wheels and then divide it by train mass in *tonne* and route length in *km* ; it is *presumed that there is a gradient of G throughout the run and power remains 'ON' upto the end of free run in case of trapezoidal curve (Fig. 7.4) and upto the accelerating period in case of quadrilateral curve (Fig. 7.5).*

The output of the driving axles is used for *accelerating the train, overcoming the gradient and overcoming the train resistance.*

(i) ***Energy required to accelerate the train*** **(E_a) :**

From trapezoidal diagram of Fig. 7.4, we have

$$E_a = F_a \times \text{Distance } OAD$$

$$= 277.8\,\alpha\, M_e \times \frac{1}{2} V_m t_1 \text{ joules}$$

$$= 277.8\,\alpha\, M_e \times \frac{1}{2} V_m \times \frac{V_m}{\alpha} \text{ joules} \qquad \left(\because\ t_1 = \frac{V_m}{\alpha}\right)$$

$$= 277.8\,\alpha\, M_e \times \left[\frac{1}{2} \times \frac{V_m \times 1000}{3600} \times \frac{V_m}{\alpha}\right] \text{ joules}$$

It may be noted that since V_m in km/h, it has been converted into m/s by multiplying it with the conversion factor of $\left(\frac{1000}{3600}\right)$. In the case of $\frac{V_m}{\alpha}$, conversion factor for V_m and α being the same, they cancel out.

Since $$1 \text{ Wh} = 3600 \text{ J},$$

$$\therefore \quad E_a = 277.8\,\alpha\, M_e \left[\frac{1}{2} \times \frac{V_m \times 1000}{3600} \times \frac{V_m}{\alpha}\right] \times \frac{1}{3600} \text{ Wh}$$

or, $$E_a = 0.01072\, V_m^2 M_e\, Wh \quad ...(7.19)$$

(ii) ***Energy required to overcome gradient*** **(E_g) :**

$$E_g = F_g \times D' \quad ...(7.20)$$

where D' is the *total distance over which the power remains 'ON'*. Its maximum value equals the distance represented by the area *DABE* in Fig. 7.4 *i.e.,* from the *start to the end of free-running period in the case of trapezoidal curve.*

Inserting the value of F_g [Eqn. 7.13 *b*] in eqn. (7.20), we get

$$E_g = 98\,MG\,(1000\,D')\text{ joules} = 98000\,MGD'\text{ joules} \quad ...(7.20\,a)$$

...where D' is in *km*.

or, $$E_g = 98000\,MGD' \times \frac{1}{3600}\text{ Wh} = 27.25\,MGD'\,Wh \quad ...(7.20\,b)$$

(*iii*) ***Energy required to overcome resistance*** **(E_r) :**

$$E_r = F_r \times D' = M.r \times (1000\,D')\text{ joules} \quad ...(7.21\,a)$$

or, $$E_r = \frac{M.r \times 1000\,D'}{3600}\,Wh = 0.2778\,MrD'Wh \quad ...(7.21\,b)$$

∴ Total energy output of the driving axles, ...where D' is in *km*

$$E = E_a + E_g + E_r$$

or, $$E = (0.01072\,V_m^{\,2}\,M_e + 27.25\,MGD' + 0.2778\,MrD')\,Wh \quad ...(7.22)$$

Specific energy output, $E_{spo} = \dfrac{E}{M \times D}$ ...where D is the *total* run length

or, $$E_{spo} = \left[0.01072\,\frac{V_m^{\,2}}{D} \cdot \frac{M_e}{M} + 27.25\,G\,\frac{D'}{D} + 0.2778r\,\frac{D'}{D}\right] Wh/t-km \quad ...(7.22\,a)$$

If there is no gradient, then,

$$E_{spo} = \left[0.01072\,\frac{V_m^{\,2}}{D} \cdot \frac{M_e}{M} + 0.2778\,r\,\frac{D'}{D}\right] Wh/t-km \quad ...(7.22\,b)$$

Alternative method :

Considering again the trapezoidal speed-time curve, let us calculate energy output *period wise* (not force-wise).

(*i*) ***Energy output during accelerating period*** **(E_a) :**

$$E_a = F_t \times \text{Area } OAD\text{ (i.e., distance travelled during accelerating period)}$$

...(Fig. 7.4)

$$= F_t \times \frac{1}{2} V_m t_1 = \frac{1}{2} F_t \cdot V_m \cdot \frac{V_m}{\alpha}$$

$$= \frac{1}{2} F_t \left(V_m \times \frac{1000}{3600}\right) \frac{V_m}{\alpha}\text{ joules}$$

or, $$E_a = \frac{1}{2} F_t \left(V_m \times \frac{1000}{3600}\right) \frac{V_m}{\alpha} \times \frac{1}{3600}\,Wh$$

Inserting the value of F_t, we get

$$E_a = \frac{1000}{(3600)^2} \cdot \frac{V_m^{\,2}}{2\alpha}\,(277.8\,\alpha\,M_e + 98\,MG + Mr)\,Wh \quad ...(7.23)$$

It must be remembered that during this period, all the three forces (F_a, F_g, F_r) are at work.

(*ii*) ***Energy output during free-running period,*** **E_{fr} :**

$$E_{fr} = F_t' \times \text{Area } ABED \quad ...(\text{Fig. } 7.4)$$

$$= F_t' \times (V_m \times t_2)$$

$$= F_t' \times \left(V_m \times \frac{1000}{3600}\right) t_2\text{ joules}$$

$$= \left(\frac{1000}{3600}\right) F_t' \times V_m \times t_2 \times \frac{1}{3600}\,Wh$$

$$= \left(\frac{1000}{3600}\right) F_t' \times D_{fr} \; Wh$$

where D_{fr} is the distance in km travelled during the free-running period

$$\left[\begin{array}{l} D_{fr} = \text{Velocity in km/h} \times \text{time in hours} = V_m \times \left(\dfrac{t_2}{360}\right) \\ \text{because times are always taken in seconds} \end{array}\right]$$

or, $$E_{fr} = \frac{1000}{3600} (98MG + Mr) D_{fr} \; Wh \qquad ...(7.24)$$

Total energy required,

$$E = E_a + E_{fr}$$

$$= \left[\frac{1000}{(3600)^2} \cdot \frac{V_m^{\ 2}}{2\alpha} (277.8\,\alpha\, M_e + 98MG + Mr) + \frac{1000}{3600} (98MG + Mr)\, D_{fr}\right] Wh$$

$$= \left[\frac{1000}{(3600)^2} \cdot \frac{V_m^{\ 2}}{2\alpha} \times 277.8\,\alpha\, M_e + \frac{1000}{(3600)^2} \cdot \frac{V_m^{\ 2}}{2\alpha} (98MG + Mr) + \frac{1000}{3600} (98MG + Mr) D_{fr}\right] Wh$$

$$= \left[0.01072\, V_m^{\ 2} . M_e + \frac{1000}{3600} (98MG + Mr) \left\{\frac{V_m^{\ 2}}{2\,\alpha \times 3600} + D_{fr}\right\}\right] Wh$$

Now, $$\frac{V_m^{\ 2}}{2\,\alpha \times 3600} = \frac{1}{2}\left(\frac{V_m}{3600}\right) . \frac{V_m}{\alpha} = \frac{1}{2}\left(\frac{V_m}{3600}\right) t_1$$

$$= \text{Distance travelled during accelerating period } i.e., D_a$$

$\therefore$ $$E = \left[0.01072\, V_m^{\ 2} . M_e + \frac{1000}{3600} (98MG + Mr)\, (D_a + D_{fr})\right] Wh$$

or, $$E = \left[0.01072\, V_m^{\ 2} . M_e + (27.25\, MG + 0.2778\, Mr)\, D'\right] Wh$$

It is the same expression found above.

7.8.2. Energy Consumption

"Energy consumption" *is defined as* the *total energy input to the traction motors from the supply.* It can be found by dividing the energy output of the driving wheels by the combined efficiency of transmission gear and motor.

i.e., $$\textit{Energy consumption} = \frac{\text{Output of driving axles}}{\eta_{motor} \times \eta_{gear}}$$

Specific energy consumption, E_{spc} :

"Specific energy consumption" *is defined as the energy consumed* (*Wh*) *per tonne mass of the train per km length of the run.*

i.e., $$E_{spc} = \frac{\text{Total energy consumed } (Wh)}{\text{Train mass (tonne)} \times \text{run length (km)}}$$

$$= \frac{\text{Specific energy output}}{\eta}$$

where, η = Overall efficiency of transmission motor and gear

$$= \eta_{motor} \times \eta_{gear}$$

From eqn. (7.22*a*), we have *specific energy consumption* as :

$$E_{spc} = \left[0.01072 . \frac{V_m^2}{\eta D} . \frac{M_e}{M} + 27.25 \frac{G}{\eta} . \frac{D'}{D} + 0.2778 \frac{r}{\eta} . \frac{D'}{D}\right] Wh/t\text{-}km$$

The value of E_{spe} when gradient is *not* involved is given as :

$$E_{spe} = \left[0.01072 .. \frac{V_m^2}{\eta D} . \frac{M_e}{M} + 0.2778 \frac{r}{\eta} . \frac{D'}{D}\right] Wh/t\text{-}km \quad ...(7.25\ a)$$

- The *specific energy consumption* of a train running at a given schedule speed is *influenced by the following factors :*

(*i*) Acceleration. (*ii*) Retardation.
(*iii*) Maximum speed. (*iv*) Distance between stops.
(*v*) Track configuration. (*vi*) Type of train and equipment.

- Figs. 7.10 and 7.11 show the fall of specific energy consumption with the increase in value of acceleration (or retardation) and distance of run.

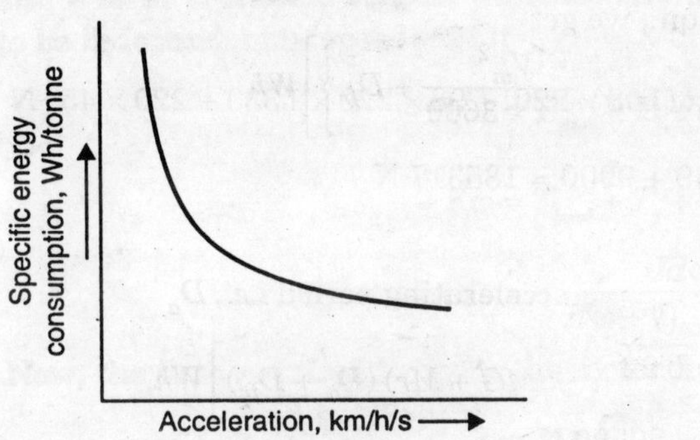

Fig. 7.10

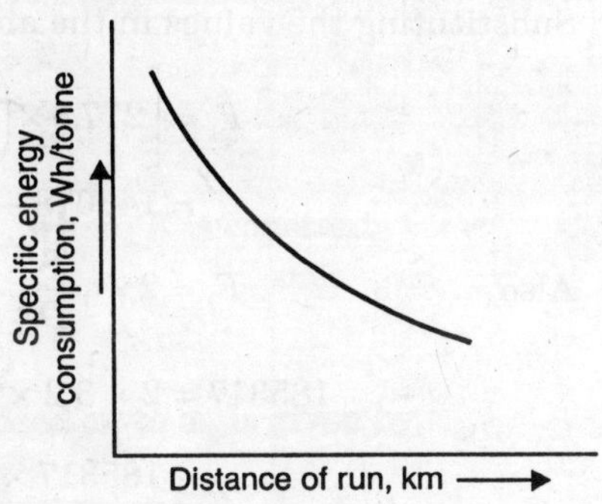

Fig. 7.11

- Typical values of specific energy consumption are :

For suburban service ... 50-75 Wh/tonne-km
For main line service ... 20-30 Wh/tonne-km.

Example 7.10. *The peripheral speed of a railway traction motor cannot be allowed to exceed 42 m/s. If gear ratio is 16/64, motor armature diameter 40 cm and wheel diameter 88 cm, calculate the limiting value of the train speed.*

Solution. Peripheral speed = 42 m/s ; Gear ratio, γ = 16/64, Motor armature diameter, d_m = 40 cm = 0.4 m and wheel diameter, D = 88 cm = 0.88 m.

Limiting value of the train speed :

$$\pi d_m n_m = 42$$

where, n_m is the number of resolutions made by the motor per second.

$$\therefore \quad n_m = \frac{42}{\pi d_m} = \frac{42}{\pi \times 0.4} = 33.42 \text{ r.p.s.}$$

Maximum number of revolution per second made by the driving wheel, $n_w = n \times \frac{16}{64}$

$$= 33.42 \times \frac{16}{64} = 8.36 \text{ r.p.s.}$$

Maximum distance travelled by the driving wheel in one second

$$= 8.36 \times (\pi D) = 8.36 \times \pi \times 0.88 = 23.11 \text{ m/s}$$

Hence, limiting value of train speed = $23.11 \times \frac{3600}{1000}$ = **83.2 km/h. (Ans.)**

Example 7.11. *A 220-tonne motor coach driven by four motors takes 18 seconds to attain a speed of 40 km / h, starting from rest on an ascending gradient of 1 in 75. The gear ratio is 3.2, gear efficiency 90%, wheel diameter 92 cm, train resistance 45 N / t and rotational inertia 8 percent of the dead weight. Find the torque developed by each motor.*

Solution. *Given :* $M = 220$ t ; $t_1 = 18$ s ; $V_m = 40$ km/h ; % $G = \frac{1}{75} \times 100 = 1.333$, $\gamma = 3.2$;

$\eta = 90\%$; $D = 92$ cm $= 0.92$ m ; $r = 45$ N/t ; $M_e = 1.08$ M

Torque developed by each motor, T :

Total tractive effort,

$$F_t = (277.8\, \alpha\, M_e + 98\, MG + Mr) \text{ newton} \quad \text{...(Eqn. 7.15)}$$

In this eqn. M is in tonne, $\alpha \left(= \frac{V_m}{t_1}\right)$ in km/h/s, G in percent and r in N/t of train mass.

Substituting the values in the above eqn., we get

$$F_t = \left[277.8 \times \left(\frac{40}{18}\right) \times (1.08 \times 220) + 98 \times 220 \times 1.333 + 220 \times 45\right] \text{N}$$

$$= 146678 + 28739 + 9900 = 185317 \text{ N}$$

Also,
$$F_t = 2\gamma\eta \frac{T}{D} \quad \text{...(Eqn.7.7)}$$

or,
$$185317 = 2 \times 3.2 \times 0.9 \times \frac{T}{0.92}$$

or,
$$T = \frac{185317 \times 0.92}{2 \times 3.2 \times 0.9} = 29599 \text{ N-m}$$

$\therefore$ Torque developed by each motor = $\frac{29599}{4}$ = **7399.8 Nm. (Ans.)**

Example 7.12. *A 220-tonne motor coach having four motors, each developing a torque of 7500 Nm during acceleration, starts from rest. If up-gradient is 25 in 1000, gear ratio 3.2, gear transmission efficiency 90 percent, wheel diameter 92 cm, train resistance 45 N / t, rotational inertia effect 8 percent,*

(*i*) Calculate the time taken by the coach to attain a speed of 72 km/h.

(*ii*) If the supply voltage is 3000 V and motor efficiency 87 percent, estimate the current taken by each motor during the acceleration period.

Solution. *Given* $M = 220$ *t* ; T (each motor) = 7500 Nm ; $G = \frac{25}{1000} \times 100 = 2.5$; $\gamma = 3.2$;

$\eta_t = 90\%$, $D = 92$ cm $= 0.92$ m ; $r = 45$ N/t ; $M_e = 1.08$ M ; $V_m = 72$ km/h ;

Supply voltage = 3000 V ; $\eta_{motor} = 87\%$.

(*i*) **Time taken by the coach to attain a speed of 72 km/h, t_1 :**

Total tractive effort at the wheel,

$$F_t = 2\gamma\eta_t \frac{T}{D} \text{ newton} \quad \text{(Eqn. 7.7)}$$

or,
$$F_t = 2 \times 3.2 \times 0.9 \times \frac{(7500 \times 4)}{0.92} = 187826 \text{ N} \quad \text{...}(i)$$

Also, $F_t = [277.8\ \alpha\ M_e + 98\ MG + Mr]$ newton ...(7.15)

$= [277.8\ \alpha \times (1.08 \times 220) + 98 \times 220 \times 2.5 + 220 \times 45]$ N

or, $F_t = (66005\ \alpha + 63800)$ N ...(*ii*)

Equating (*i*) and (*ii*), we get

$$187826 = 66005\ \alpha + 63800$$

or, $\alpha = 1.88$ km/h/s

$$\therefore \quad \mathbf{t_1 = \frac{V_m}{\alpha} = \frac{72}{1.88} = 38.3\ s.} \quad \textbf{(Ans.)}$$

(*ii*) **The current taken during the acceleration period, I :**

Power taken by the motors

$$= \frac{F_t \times v}{\eta_{motor}} = \frac{F_t \times V_m}{\eta_{motor}} = F_t \times \left(\frac{V_m \times 1000}{3600}\right) \times \frac{1}{\eta_{motor}} \text{ watts}$$

$$= 187826 \times \left(\frac{72 \times 1000}{3600}\right) \times \frac{1}{0.87} = 43.178 \times 10^5 \text{ W}$$

Total current drawn $= \dfrac{43.178 \times 10^5}{3000} = 1439.3$ A

Current drawn by each motor $= \dfrac{1439.3}{4} =$ **359.8 A. (Ans.)**

Example 7.13. *A goods train weighing 450 tonnes is to be hauled by a locomotive up an ascending gradient of 2.2 percent with an acceleration of 0.8 km/h/s. If the coefficient of adhesion is 0.24, train resistance 42 N/t and effect of rotational inertia 10 percent, find the weight of locomotive and number of axles if the load is not to increase beyond 18 tonnes/axle.*

Solution. *Given* : $M = 450$ t ; $G = 2.2\%$; $\alpha = 0.8$ km/h/s ; $\mu_a = 0.24$; $r = 42$ N/t ; $M_e = 1.1$ M ; load limit = 18 t/axle.

Mass of locomotive, M_L :

Tractive effort required,

$$F_t = (277.8\ \alpha\ M_e + 98\ MG + Mr) \text{ newton} \quad \text{(Eqn. 7.15)}$$

$$= M\left(277.8\ \alpha \times \frac{M_e}{M} + 98\ G + r\right) \text{ newton}$$

$$= M(277.8 \times 0.8 \times 1.1 + 98 \times 2.2 + 42) = 502.1 \text{ M newton}$$

If M_L is the mass of the locomotive, then

$$F_t = 502.1(M + M_L) = 502.1(450 + M_L) \quad ...(i)$$

Maximum effort is given by,

$$F_t = 9800\ \mu_a\ xM \text{ newton} \quad ...(\text{Eqn. } 7.7\ a)$$

or

$$F_t = 9800 \times 0.24 \times M_L = 2352\ M_L \quad ...(ii)$$

From (*i*) and (*ii*), we get (Here $x = 1$)

$$502.1(450 + M_L) = 2352\ M_L$$

$\therefore$ M_L (weight) = **122 tonnes. (Ans.)**

Number of axles required, N_{axles} :

$$N_{\mathbf{axles}} = \frac{122}{18} = \mathbf{7.} \quad \textbf{(Ans.)}$$

Example 7.14. *A 500-tonne goods train is to be hauled by a locomotive up a gradient of 1 in 40 with an acceleration of 1.5 km/h/s. Determine the mass of the locomotive and number of axles, if* *axle load is not to exceed 23 tonnes. Coefficient of adhesion is 0.28, track resistance 44 N/t and effective rotating masses 10 percent of dead weight.* **(Bombay University)**

Solution. *Given :* $M = 500$ tonnes ; $G = \frac{1}{40} \times 100 = 2.5\%$; $\alpha = 1.5$ km/h/s. Limiting load of an axle = 23 tonnes ; $\mu_a = 0.28$, $r = 44$ N/t ; $\frac{M_e}{M} = 1.1$.

(*i*) **Mass of the locomotive :**

Let the mass of the locomotive be M_L tonnes.

Dead mass of train and locomotive, $M = (500 + M_L)$ tonnes.

Tractive effort, $F_t = (277.8\ \alpha\ M_e + 98\ MG + Mr)$ newton ...(Eqn. (7.15)

$$= M\left(277.8 \times \frac{M_e}{M} + 98G + r\right) \text{ newton}$$

$$= M(277.8 \times 1.1 + 98 \times 2.5 + 44) = 594.6 \text{ M newton}$$

or, $F_t = 594.6(500 + M_L)$ newton ...(*i*)

Maximum value of tractive effort that can be possible with M_L adhesive mass of locomotive,

$F_t = 9800\ \mu_a\ x\ M_L$ newton ...(Eqn. 7.7 *a*)

or, $F_t = 9800 \times 0.28 \times M_L = 2744\ M_L$ ...(*ii*)

Comparing (*i*) and (*ii*), we get ...($x = 1$)

$594.6(500 + M_L) = 2744\ M_L$

∴ or $\mathbf{M_L = 138.32}$ **tonnes. (Ans.)**

(*ii*) **Number of axles required, $\mathbf{N_{axles}}$:**

$$N_{axles} = \frac{138.32}{23} = 6. \quad \textbf{(Ans.)}$$

Example 7.15. *An electric locomotive is required to haul a train of 10 coaches each weighing 36 tonnes on the main line service requiring an initial acceleration of 0.8 km/h/s up a gradient of 1 in 100. Estimate the adhesive weight and hence the number of driving axles the locomotive must have if the permissible axle loading is 22 tonnes per axle assuming rotational inertia to be 4 percent for coaches and 15 percent for locomotive. Maximum coefficient of adhesion is 0.2 and the tractive resistance is 50 N/t.* **(B.H.U.)**

Solution. Dead mass of train, $M = 10 \times 36 = 360$ tonnes ; $\alpha = 0.8$ km/h/s ; $G = \frac{1}{100} \times 100 = 1\%$;

Permissible axle loading = 22 t/axle ; Rotational inertia of coaches and locomotive = 4 and 15 percent respectively ; $\mu_a = 0.2$; $r = 50$ N/t.

Adhesive weight and no. of axles required :

Accelerating mass of train, $M_e = 1.04 \times 360 = 374.4$ tonnes

Let adhesive mass of locomotive = M_L tonnes

Accelerating mass of locomotive = $1.15\ M_L$ tonnes

Tractive effort, $F_t = [277.8\ \alpha\ (M_e + 1.15\ M_L) + 98(M + M_L)G + (M + M_L)r]$ newton

$= [277.8\ \alpha\ (M_e + 1.15\ M_L) + (M + M_L)(98\ G + r)]$ newton

$= [277.8 \times 0.8 \times (374.4 + 1.15\ M_L) + (360 + M_L)(98 \times 1 + 50)]$ newton

$= [(83207 + 255.6\ M_L) + (53280 + 148\ M_L)$ newton

or, $F_t = (136487 + 403.6\ M_L)$ newton ...(*i*)

Maximum value of tractive effort that can be possible with M_L adhesive mass of locomotive,

$F_t = 9800\ \mu_a\ x\ M_L$ newton ...(Eqn. 7.7 *a*)

or $F_t = 9800 \times 0.2 \times M_L = 1960\ M_L$ ...(*ii*)

Comparing (*i*) and (*ii*), we get ...($x = 1$)

$$1960\ M_L = 136487 + 403.6\ M_L$$

$\therefore$ $\mathbf{M_L = 87.7}$ **tonnes. (Ans.)**

Number of axles $= \dfrac{87.7}{22} = \mathbf{4.}$ **(Ans.)**

Example 7.16. *An ore-carrying train weighing 5000 tonnes is to be hauled down a gradient of 1 : 50 at a maximum speed of 30 km/h and started on a level track at an acceleration of 0.29 km/h/s. How many locomotives each weighing 75 tonnes, will have to be employed ?*

Train resistance during starting = 29.4 N/t ; Train resistance at 30 km/h = 49 N/t ; Coefficient of adhesion = 0.3 ; Rotational inertia = 10 percent. **(A.M.I.E.)**

Solution. *Given :* $M = 5000$ tonnes ; $G = \dfrac{1}{50} \times 100 = 2\%$; $V_m = 30$ km/h ; $\alpha = 0.29$ km/h/s (on level track) ; $M_L = 75$ tonnes ; $r = 29.4$ N/t (during starting) ; $r = 49$ N/t (at 30 km/h), $\mu_a = 0.3$; Rotational inertia = 10 per cent.

No. of locomotives required :

Downward force due to gravity $= Mg \sin\theta = (5000 \times 1000) \times 9.8 \times \dfrac{1}{50} = 980000$ N

Train resistance (at 30 km/h) $= 49 \times 5000 = 245000$

Braking force to be applied by brakes

$$= 980000 - 245000 = 735000 \text{ N}$$

Maximum braking force which one locomotive can provide

$$= 1000\ \mu_a M_L g \text{ newton} \qquad \text{... } M \text{ in tonnes}$$

$$= 1000 \times 0.3 \times 75 \times 9.8 = 220500 \text{ N}$$

No. of locomotives required for *braking*

$$= \frac{735000}{220500} = 3.33$$

It means (fraction being meaningless) *4 locomotives* are required. Tractive effort required to *haul the train on level track,*

$$F_t = (277.8\ \alpha M_e + Mr) \text{ newton} \qquad \text{...(Eqn. 7.15)}$$

$$= 277.8 \times 0.29 \times (5000 \times 1.1) + 5000 \times 29.4 = 590091 \text{ N}$$

$\therefore$ No. of locomotives required $= \dfrac{590091}{220500} = 2.68 \simeq 3$

It means that **4 locomotives** *are enough to look after braking as well starting.* **(Ans.)**

Example 7.17. *A 203-tonne electric train accelerates uniformly from rest to a speed of 45 km/h up a gradient of 1 in 500, the time taken being 30 seconds. The power is then cut off and the train coasts down a uniform gradient of 1 in 1000 for a period of 40 seconds when brakes are applied for a period of 15 seconds so as to bring the train uniformly to rest on this gradient. Assuming train resistance to be 44 N/t and allowing 10 percent for rotational inertia, calculate :*

(*i*) *The maximum power output from the driving axles.*

(*ii*) *The energy taken from the conductor rails in kWh, assuming an efficiency of 62 percent.* **(Agra University)**

Solution. *Given :* $M = 203$ t ; $V_m = 45$ km/h $= \frac{45 \times 1000}{3600} = 12.5$ m/s ; $G = \frac{1}{500} \times 100 = 0.2\%$;

$$t_1 = 30 \text{ s} ; r = 44 \text{ N/t}, \frac{M_e}{M} = 1.1 ; \eta = 62\%.$$

(*i*) The maximum power output from the driving axles :

Acceleration, $\alpha = \frac{V_m}{t_1} = \frac{45}{30} = 1.5$ km/h/s

Tractive effort required,

$$F_t = (277.8 \, \alpha \, M_e + 98.1 \, MG + M.r) \text{ newton} \quad \text{...(Eqn. 7.15)}$$
$$= 277.8 \times 1.5 \times (203 \times 1.1) + 98.1 \times 203 \times 0.2 + 203 \times 44$$
$$= 93049 + 3983 + 8932 = 105964 \text{ N}$$

$\therefore$ The maximum power output from driving axles,

$$P = \frac{F_t \times V_m}{3600} \text{ kW, where } V_m \text{ is in km/h} \quad \text{...(Eqn. 7.16 (b))}$$
$$= \frac{105964 \times 45}{3600} = \mathbf{1324.55 \text{ kW.}} \quad \textbf{(Ans.)}$$

(*ii*) The energy taken from the conductor rails :

The energy required for the run

= Energy required during acceleration as there is no free run

$$= \frac{1}{2} \times P \times t_1$$
$$= \frac{1}{2} \times 1324.55 \times \frac{30}{3600} = 5.52 \text{ kWh}$$

$\therefore$ The energy taken from conductor rails

$$= \frac{5.52}{0.62} = \mathbf{8.9 \text{ kWh.}} \quad \textbf{(Ans.)}$$

Example 7.18. *The average distance between stops on a level section of a railway is 1.25 km. Motor-coach train weighing 200 tonne has a schedule speed of 30 km / h, the duration of stops being 30 seconds. The acceleration is 1.9 km / h / s and the braking retardation is 3.2 km / h / s. Train resistance to traction is 45 N / t. Allowance for rotational inertia is 10 percent. Calculate the specific energy output in Wh / t-km. Assume a trapezoidal speed-time curve.* **(Bombay University)**

Solution. *Given :* $D = 1.25$ km ; $M = 200$ t ; Schedule speed, $V_{sch} = 30$ km/h ;

Duration of stops = 30 s ; $\alpha = 1.9$ km/h/s $= \frac{1.9 \times 1000}{3600} = 0.528$ m/s^2,

$\beta = 3.2$ km/h/s $= \frac{3.2 \times 1000}{3600} = 0.889$ m/s^2 ; $r = 45$ N/t ; $M_e = 1.1$ M.

Specific energy output, E_{spo} : Refer to Fig. 7.4.

$$K = \frac{1}{2}\left(\frac{1}{\alpha} + \frac{1}{\beta}\right) = \frac{(\alpha + \beta)}{2\alpha\beta} = \frac{(0.528 + 0.889)}{2 \times 0.528 \times 0.889} = 1.51$$

Schedule time, $t_{sch} = \frac{D}{V_{sc}} = \frac{1.25 \times 3600}{30} = 150$ s

$\therefore$ Running time, $t = 150 - 30 = 120$ s

Also, $$V_m = \frac{t - \sqrt{t^2 - 4KD}}{2K}$$ ...(Eqn. 7.2)

$$= \frac{120 - \sqrt{(120)^2 - 4 \times 1.51 \times (1.25 \times 1000)}}{2 \times 1.51} = 12.33 \text{ m/s}$$

or, 44.39 km/h

Braking distance $$= \frac{1}{2} V_m t_3 = \frac{1}{2} V_m \times \frac{V_m}{\beta} = \frac{V_m^2}{2\beta}$$

$$= \frac{12.33^2}{2 \times 0.889} \times \frac{1}{1000} = 0.0855 \text{ km}$$

$\therefore$ $D' = D -$ braking distance $= 1.25 - 0.0855 = 1.1645$ km

Specific energy output (E_{spo}) is given by :

$$E_{spo} = \left[0.01072 \frac{V_m^2}{D} \cdot \frac{M_e}{M} + 0.2778\, r \frac{D'}{D}\right] \text{Wh/t-km} \qquad \text{...(Eqn. 7.22 } b\text{)}$$

$$= \left[0.01072 \times \frac{(44.39)^2}{1.25} \times 1.1 + 0.2778 \times 45 \times \frac{1.1645}{1.25}\right] = \mathbf{30.23 \text{ Wh/t-km.}} \quad \textbf{(Ans.)}$$

Example 7.19. *A 250-tonne EMU is started with a uniform acceleration and reaches a speed of 42 km/h in 25 seconds, on a level track. Assuming trapezoidal speed-time curve, find specific energy consumption if rotational inertia is 10 percent, retardation is 2.8 km/h/s, distance between stops is 3.2 km, motor efficiency is 88 percent and train resistance is 45 N/t.*

Solution. *Given :* $M = 250$ t ; $V_m = 42$ km/h ; $t_1 = 25$ s ; $M_e = 1.1$ M ; $\beta = 2.8$ km/h/s ; $D = 3.2$ km, $\eta = 0.88$; $r = 45$ N/t

Specific energy consumption, E_{spc}:

Refer to Fig. 7.4. Let us first find distance D', the distance upto which energy is consumed from the supply. It is the distance travelled upto the end of the free-running period. It is equal to the *total distance minus the distance travelled during the braking.*

Braking period, $$t_3 = \frac{V_m}{\beta} = \frac{42}{2.8} = 15 \text{ s}$$

Distance travelled during braking period

$$= \frac{1}{2} V_m t_3 = \frac{1}{2} \times \left(\frac{42}{3600}\right) \times 15 = 0.0875 \text{ km}$$

$\therefore$ $D' = D -$ braking distance $= 3.2 - 0.0875 = 3.1125$ km

Specific energy consumption is given by,

$$E_{spc} = \left[0.01072 \frac{V_m^2}{\eta D} \cdot \frac{M_e}{M} + 0.2778 \frac{r}{\eta} \cdot \frac{D'}{D}\right] \text{Wh/t-km} \qquad \text{...(Eqn. 7.25 } a\text{)}$$

$$= \left[0.01072 \frac{42^2}{0.88 \times 3.2} \times 1.1 + 0.2778 \times \frac{45}{0.88} \times \frac{3.1125}{3.2}\right]$$

$= \mathbf{21.2}$ **Wh/t-km. (Ans.)**

Example 7.20. *A 320-tonne electric train runs up an ascending gradient of 1.2 per cent with the following speed-time curve : (i) Uniform acceleration of 2.0 km/h/s for 20 seconds, (ii) Constant speed for 45 seconds, (iii) Coasting for 28 seconds, and (iv) Braking at 2.82 km/h/s to rest.*

Calculate the specific energy consumption if train resistance is 45 N/t, effect of rotational inertia 8 percent, overall efficiency of transmission gear and motor 78 percent.

Solution. *Given :* $M = 320$ t ; $G = 1.2\%$; $\alpha = 2$ km/h/s ; $t_1 = 20$ s ; $t_2 = 45$ s ; $t_3 = 28$ s ;
$\beta = 2.82$ km/h/s ; $r = 45$ N/t ; $M_e = 1.08$ M, $\eta = 78\%$.

Specific energy consumption, E_{spc} :

Fig. 7.12 shows the speed-time curve.

$$V_1 = \alpha.t_1 = 2 + 20 = 40 \text{ km/h}$$

Tractive force during coasting,

$$F_t = (98\,MG + M.r) \quad \text{...(i)}$$

Also, $\quad F_t = 277.8\, M_e\, \beta_c$, during coasting ...(ii)

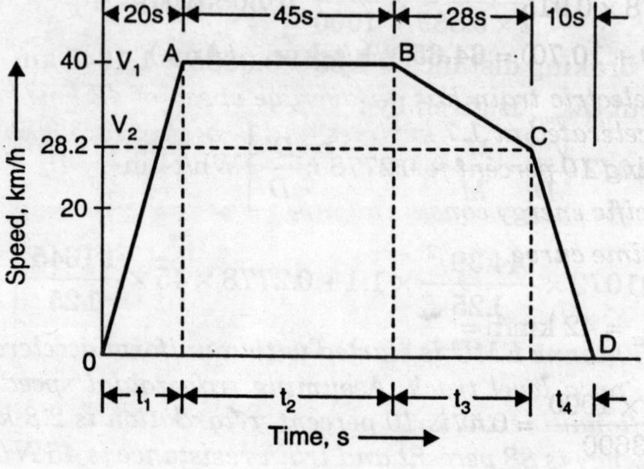

Fig. 7.12

Equating (*i*) and (*ii*), we get

$$277.8\, M_e \beta_c = 98\, MG + M.r$$
$$= M(98 \times 1.2 + 45) = 126.6 \text{ M}$$

$$\therefore \quad \beta_c = \frac{126.6}{277.8} \times \frac{M}{M_e} = \frac{126.6}{277.8} \times \frac{1}{1.08} = 0.422 \text{ km/h/s}$$

Now, $\quad V_2 = V_1 + \beta_c \,.\, t_3$

$$= 40 + (-0.422) \times 25 = 28.2 \text{ km/h}$$

$$\therefore \quad t_4 = \frac{V_2}{\beta} = \frac{28.2}{2.82} = 10 \text{ s}$$

Distance travelled during *acceleration period*

$$= \frac{1}{2} V_1 t_1 = \frac{1}{2} \times 40 \left(\frac{\text{km}}{\text{h}}\right) \times \frac{20}{3600} (h) = 0.111 \text{ km}$$

Distance travelled *during constant speed* (or *free-running*) *period*

$$= V_1 \times t_2 = 40 \times \frac{45}{3600} = 0.5 \text{ km}$$

Distance travelled *during coasting period*

$$= \left(\frac{V_1 + V_2}{2}\right) t_3 = \left(\frac{40 + 28.2}{2}\right) \times \frac{28}{3600} = 0.265 \text{ km}$$

Distance travelled *during braking period*

$$= \frac{1}{2} V_2\, t_4 = \frac{1}{2} \times 28.2 \times \frac{10}{3600} = 0.039 \text{ km}$$

Total distance between stops,

$$D = 0.111 + 0.5 + 0.265 + 0.039 = 0.915 \text{ km}$$

Distance travelled *during acceleration and free-running period,*

$$D' = 0.111 + 0.5 = 0.611 \text{ km}$$

Specific energy consumption (E_{spc}) is given by :

$$E_{spc} = \left[0.01072 \frac{V_m^2}{\eta D} \cdot \frac{M_e}{M} + 27.25 \frac{G}{\eta} \cdot \frac{D'}{D} + 0.2778 \frac{r}{\eta} \cdot \frac{D'}{D}\right] \text{Wh/t-km} \quad \text{...(Eqn. 7.25)}$$

$$= \left[0.01072 \times \frac{40^2}{0.78 \times 0.915} \times 1.08 + 27.25 \times \frac{1.2}{0.78} \times \frac{0.611}{0.915} + 0.2778 \times \frac{45}{0.78} \times \frac{0.611}{0.915}\right] \text{Wh/t-km}$$

$= (25.96 + 27.99 + 10.70) =$ **64.65 Wh/t-kw. (Ans.)**

Example 7.21. *An electric train has an average speed of 42 km/h on a level track between stops 1.4 km apart. It is accelerated at 1.7 km/h/s and is braked at 3.3 km/h/s. Assuming tractive resistance as 50 N/t, allowing 10 percent for rotational inertia, and efficiency of motors 85 per cent :*

(*i*) *Estimate the specific energy consumption.*

(*ii*) *Draw the speed-time curve.* **(Roorkee University)**

Solution. *Given :* $V_a = 42 \text{ km/h} = \frac{42 \times 1000}{3600} = 11.67 \text{ m/s}$; $D = 1.4$ km ;

$\alpha = 1.7 \text{ km/h/s} = \frac{1.7 \times 1000}{3600} = 0.472 \text{ m/s}^2$; $\beta = 3.3 \text{ km/h/s} = \frac{3.3 \times 1000}{3600} = 0.917 \text{ m/s}^2$;

$r = 50$ N/t ; $\frac{M_e}{M} = 1.1$; $\eta = 85\%$.

(*i*) **Specific energy consumption, E_{spc} :**

Time of run, $t = \frac{D}{V_a} = \frac{1.4 \times 1000}{11.67} = 120$ s

$$K = \frac{\alpha + \beta}{2\alpha\beta} = \frac{0.472 + 0.917}{2 \times 0.472 \times 0.917} = 1.605$$

$\therefore$ Maximum speed, $V_m = \frac{t - \sqrt{t^2 - 4KD}}{2K}$ ...(Eqn. 7.2)

$$= \frac{120 - \sqrt{(120)^2 - 4 \times 1.605 \times 1400}}{2 \times 1.605} = 14.47 \text{ m/s}$$

$$= \frac{14.47 \times 3600}{1000} = 52 \text{ km/h}$$

Now, Duration of acceleration, $t_1 = \frac{V_m}{\alpha} = \frac{52}{1.7} = 30.6$ s

and, Duration of braking, $t_3 = \frac{V_m}{\beta} = \frac{52}{3.3} = 15.8$ s

$\therefore$ Duration of free running, $t_2 = t - (t_1 + t_3) = 120 - (30.6 + 15.8) = 73.6$ s

Distance travelled during braking

$$= \frac{1}{2} V_m \times t_3 = \frac{1}{2} \times 52 \times \frac{15.8}{3600} = 0.114 \text{ km}$$

$\therefore$ Distance travelled during acceleration and free run,

$$D' = 1.4 - 0.114 = 1.286 \text{ km}$$

Specific energy output,

$$E_{spo} = \left[0.01072 \frac{V_m^2}{D} \cdot \frac{M_e}{M} + 0.2778\, r \frac{D'}{D}\right] \text{Wh/t-km}$$

...where V_m is in km/h

$$= \left[0.01072 \times \frac{(52)^2}{1.4} \times 1.1 + 0.2778 \times 50 \times \frac{1.286}{1.4}\right] \text{Wh/t-km}$$

$$= 22.77 + 12.76 = 35.53 \text{ Wh/t-km}$$

$\therefore$ Specific energy consumption,

$$E_{spc} = \frac{E_{spo}}{\eta} = \frac{35.53}{0.85} = \mathbf{41.8\ Wh/t\text{-}km.} \quad \textbf{(Ans.)}$$

(*ii*) **Speed-time curve :**

Fig. 7.13 show the speed-time curve for the given electric train.

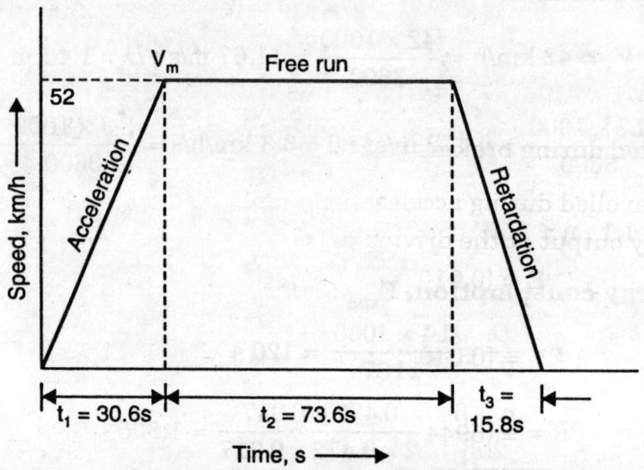

Fig. 7.13. Speed-time curve.

Example 7.22. *A 100-tonne electric train has a rotational inertia of 10 percent. This train while running between two stations which are 2.5 km apart has an average speed of 50 km/h. The acceleration and retardation during braking are respectively 1 km/h/s and 2 km/h/s. The percentage gradient between these two stations is 1 percent and the train is to move up the incline. The track resistance is 40 N/t. If the combined efficiency of the electric train is 65 percent, determine :*

(*i*) *Total energy output at the driving axles.*

(*ii*) *Total energy consumption.*

(*iii*) *Specific energy consumption.*

Assume that journey estimation is being made in simplified trapezoidal speed-time curve.

Solution. *Given :* $M = 100$ tonnes ; $\frac{M_e}{M} = 1.1$; $D = 2.5$ km ;

$$V_a = 50 \text{ km/h} = \frac{50 \times 1000}{3600} = 13.89 \text{ m/s} ; \alpha = 1 \text{ km/h/s} = \frac{1 \times 1000}{3600} = 0.278 \text{ m/s}^2$$

$$\beta = 2 \text{ km/h/s} = \frac{2 \times 1000}{3600} = 0.556 \text{ m/s}^2 ; G = 1 \text{ percent} ; r = 40 \text{ N/t} ; \eta = 65 \text{ per cent.}$$

(*i*) **Total energy at the driving axles :**

Refer to Fig. 7.4 (Trapezoidal speed-time curve).

Distance of run, $t = \dfrac{D}{V_a} = \dfrac{2.5}{50} \times 3600 = 180$ s

$$K = \frac{(\alpha + \beta)}{2\alpha\beta} = \frac{(0.278 + 0.556)}{2 \times 0.278 \times 0.556} = 2.7$$

Maximum speed, $V_m = \dfrac{t - \sqrt{t^2 - 4KD}}{2K}$ ...(Eqn. 7.2)

$$= \frac{180 - \sqrt{(180)^2 - 4 \times 2.7 \times (2.5 \times 1000)}}{2 \times 2.7}$$

$$= 19.73 \text{ m/s} = \frac{19.73 \times 3600}{1000} = 71 \text{ km/h}$$

Now, Acceleration period, $t_1 = \dfrac{V_m}{\alpha} = \dfrac{71}{1} = 71$ s

Braking period, $t_3 = \dfrac{V_m}{\beta} = \dfrac{71}{2} = 35.5$ s

Time of free-run, $t_2 = t - (t_1 + t_3) = 180 - (71 + 35.5) = 73.5$ s

Distance travelled during braking period $= \dfrac{1}{2} V_m \times t_3 = \dfrac{1}{2} \times 71 \times \dfrac{35.5}{3600} = 0.35$ km

∴ Distance travelled during acceleration and free-run, $D' = 2.5 - 0.35 = 2.15$ km

∴ Total energy output at the driving axles,

$$\mathbf{E} = [0.01072\, V_m^2\, M_e + 27.25\, MGD' + 0.2778\, MrD'] \text{ Wh}$$

...where V_m is in km/h and M in tonnes.

$$= [0.01072 \times (71)^2 \times (1.1 \times 100) + 27.25 \times 100 \times 1 \times 2.15 + 0.2778 \times 100 \times 40 \times 2.15] \text{ Wh}$$

$= (5944 + 5859 + 2389) = 14192$ Wh or **14.192 kWh.** **(Ans.)**

(*ii*) **Total energy consumption :**

Total energy consumption $= \dfrac{14.192}{\eta} = \dfrac{14.192}{0.65} = \mathbf{21.834\ kWh.}$ **(Ans.)**

(*iii*) **Specific energy consumption $\mathbf{E_{spc}}$:**

$$\mathbf{E_{spc}} = \frac{\text{Total energy consumption (Wh)}}{\text{Mass of train (tonnes)} \times \text{Distance of run (km)}}$$

$$= \frac{21.834 \times 1000}{100 \times 2.5} = \mathbf{87.34\ Wh/t\text{-}km.} \quad \textbf{(Ans.)}$$

Example 7.23. *The following data relate to a 200-tonne electric train running according to the following quadrilateral speed-time curve :*

(*i*) *Uniform acceleration from rest at 2 km / h / s for 30 seconds ;*

(*ii*) *Coasting for 50 seconds, and*

(*iii*) *Duration of baking : 15 second.*

Up-gradient ...... *1 percent*

Train resistance ...... *40 N / t*

Rotational inertia effect ...... *10 percent*

Duration of stops ...... *15 s*
Overall efficiency of gear and motor ...... *75 percent*

Find the following :

(i) Schedule speed.

(ii) Specific energy consumption.

(iii) How will the value of specific energy consumption change if there is a down-gradient of 1 percent rather than the up-gradient. **(Panjab University)**

Solution. *Given :* $M = 200$ tonnes ; $\alpha = 2$ km/h/s, $t_1 = 30$ s ; $t_2 = 50$ s ; $t_3 = 15$ s ; Up-gradient $= 1\%$; $r = 40$ N/t ; Rotational inertia $= 10\%$ *i.e.,* $\frac{M_e}{M} = 1.1$; Duration of stops $= 15$ s, $\eta = 75\%$.

(i) Schedule speed, V_{sch} :

Refer to Fig. 7.14.

$$V_1 = \alpha \, . \, t_1 = 2 \times 30 = 60 \text{ km/h}$$

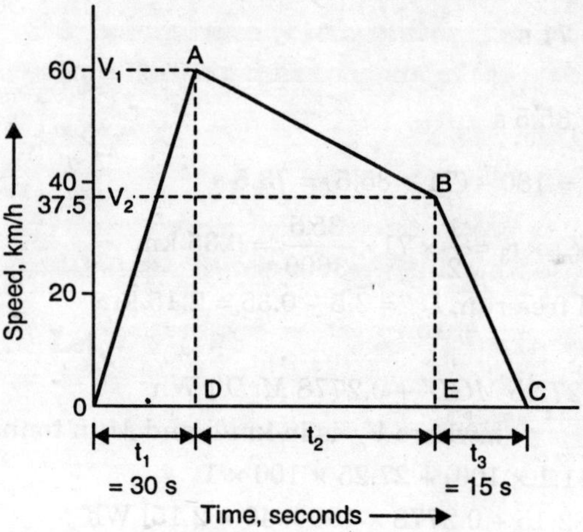

Fig. 7.14. Up-gradient (1%)

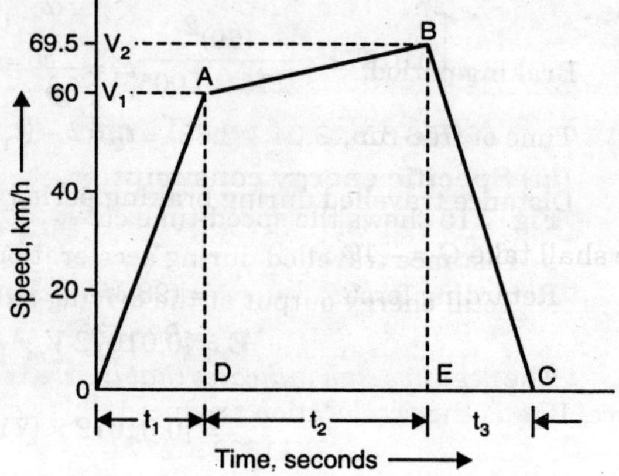

Fig. 7.15. Down-gradient (1%)

During coasting, gravity component and train resistance will cause coasting retardation β_c.

Retarding force $= (98\, MG + Mr)$ newton

$= (98 \times 200 \times 1 + 200 \times 40) = 27600$ N

Also, $277.8\, M_e . \beta_c = 27600$

or, $277.8 \times (1.1 \times 200) \times \beta_c = 27600$

$$\therefore \quad \beta_c = \frac{27600}{277.8 \times (1.1 \times 200)} = 0.45 \text{ km/h/s}$$

Now, $V_2 = V_1 + \beta_c t_2$

$= 60 + (-0.45) \times 50 = 37.5$ km/h

$$\therefore \quad \text{Braking retardation, } \beta = \frac{V_2}{t_3} = \frac{37.5}{15} = 2.5 \text{ km/h/s}$$

Distance travelled during acceleration (area *OAD*)

$$= \frac{1}{2} V_1 . t_1 = \frac{1}{2} \times 60 \times \left(\frac{30}{3600}\right) = 0.25 \text{ km}$$

Distance travelled during coasting (area *ABED*)

$$= \left(\frac{V_1 + V_2}{2}\right) \times t_2 = \left(\frac{60 + 37.5}{2}\right) \times \frac{50}{3600} = 0.677 \text{ km}$$

Distance travelled during braking (area *BCE*)

$$= \frac{1}{2} V_2 . t_3 = \frac{1}{2} \times 37.5 \times \left(\frac{15}{3600}\right) = 0.078 \text{ km}$$

Total distance travelled, $D = 0.25 + 0.677 + 0.078 = 1.005$ km

Total schedule time, $t_{sch} = 30 + 50 + 15 + 15 = 110$ s

$\therefore$ Schedule speed, $V_{\mathbf{sch}} = \frac{D}{t_{sch}} = \frac{1.005}{(110/3600)} =$ **32.9 km/h. (Ans.)**

(*ii*) **Specific energy consumption (E_{spc}) :**

$$E_{spc} = \left[0.01072 \frac{V_m^{\ 2}}{\eta D} . \frac{M_e}{M} + 27.25 \frac{G}{\eta} . \frac{D'}{D} + 0.2778 . \frac{r}{\eta} . \frac{D'}{D}\right] \text{Wh/t-km} \quad \text{...(Eqn. 7.25)}$$

$$= \left[0.01072 \times \frac{(60)^2}{0.75 \times 1.005} \times 1.1 + 27.25 \times \frac{1}{0.75} \times \frac{0.25}{1.005} + 0.2778 \times \frac{40}{0.75} \times \frac{0.25}{1.005}\right] \text{Wh/t-km}$$

$= (56.32 + 9.04 + 3.68) =$ **69 Wh/t-km. (Ans.)**

(*iii*) **Specific energy consumption change if there is a down-gradient of 1% :**

Fig. 7.15 shows the speed-time curve for the case when there is a down-gradient of 1%. Here we shall take $G = -1\%$.

Retarding force $= (98\,MG + Mr)$ newton

$= 98 \times 200(-1) + 200 \times 40 = -11600$ N

The *negative sign indicates that instead of being a retarding force, it is, in fact, an accelerating force.* If α_c is the acceleration produced, then

$$11600 = 277.8 \times (200 \times 1.1) \times \alpha_c$$

or,
$$\alpha_c = \frac{11600}{277.8 \times (200 \times 1.1)} = 0.19 \text{ km/h/s}$$

Also,
$$V_2 = V_1 + \alpha_c t_2 = 60 + 0.19 \times 50 = 69.5 \text{ km/h}$$

$$\beta = \frac{V_2}{t_3} = \frac{69.5}{15} = 4.63 \text{ km/h/s}$$

Distance travelled during acceleration = 0.25 km ... as before.

Distance travelled during coasting

$$= \left(\frac{60 + 69.5}{2}\right) \times \frac{50}{3600} = 0.9 \text{ km}$$

Distance travelled during braking $= \frac{1}{2} \times 69.5 \times \frac{15}{3600} = 0.145$ km

$\therefore$
$$D = 0.25 + 0.9 + 0.145 = 1.295 \text{ km}$$

Hence, specific energy consumption,

$$E_{spc} = \left[0.01072 \times \frac{60^2}{0.75 \times 1.295} \times 1.1 - 27.25 \times \frac{1}{0.75} \times \frac{0.25}{1.295} + 0.2778 \times \frac{40}{0.75} \times \frac{0.25}{1.295}\right] \text{Wh/t-km}$$

$= (43.71 - 7.014 + 2.86) =$ **39.56 Wh/t-km. (Ans.)**

From above calculation, we find that *specific energy consumption has decreased from 69 to 39.56 Wh/t-km.* **(Ans.)**

Example 7.24. *An electric locomotive weighing 100 tonnes can just accelerate a train of 500 tonnes (trailing weight) with an acceleration of 1 km/h/s on an up-gradient of 0.1 percent. Train resistance is 45 N/t and rotational inertia is 10 percent. If this locomotive is helped by another locomotive of weight 120 tonnes, find :*

(*i*) *The trailing weight that can now be hauled up the same gradient under the same conditions.*

(*ii*) *The maximum gradient, if the trailing hauled load remains unchanged.*

Assume adhesive weight expressed as percentage of total dead weight as 0.8 for both locomotives.

(A.M.I.E.)

Solution. *Given :* Dead weight of the train and locomotive combined = 100 + 500 = 600 t (same is the value of dead mass) ; α = ·1 km/h/s ; $G = 0.1$; $r = 45$ N/t ; $M_e = 1.1$ M ; Adhesive weight = 0.8 × total dead weight (of both locomotives)

$$F_t = (277.8\,\alpha\,M_e + 98\,MG + Mr) \text{ newton}$$

$$= [277.8 \times 1 \times (1.1 \times 600) + 98 \times 600 \times 0.1 + 600 \times 45] \text{ N} = 216228 \text{ N}$$

Maximum tractive effort of the first locomotive

$$= 9800\ x\ \mu_a\ M_L = 9800 \times 0.8 \times \mu_a \times 100 = 784000\ \mu_a$$

$$\therefore \quad 216228 = 78400\ \mu_a$$

or,
$$\mu_a = \frac{216228}{784000} = 0.276$$

With two locomotives, $M_L' = (100 + 120) = 220$ tonnes

$$\therefore \quad F_t = 9800\ x \mu_a M_L' = 9800 \times 0.8 \times 0.276 \times 220 = 476045 \text{ N} \qquad ...(i)$$

(*i*) **The trailing weight that can be hauled.** Let the trailing load which the two combined locomotives can haul be *M* tonnes. In that case, total dead mass becomes, $M' = (100 + 120 + M) = (220 + M)$ tonnes. Tractive effort required.

$$F_t = (277.8\ \alpha M_e' + 98 M'G + M'r) \text{ newton}$$

$$= M'(277.8 \times 1 \times 1.1 + 98 \times 0.1 + 45) = 360.4 \text{ M' newton}$$

$$\therefore \quad 360.4 \text{ M'} = 476045, \quad \text{or} \quad \text{M'} = 1321 \text{ tonnes}$$

$\therefore$ *Trailing load,* $M = 1321 - 220 =$ **1101 tonnes. (Ans.)**

(*ii*) **The maximum gradient, G :**

Total hauled load = 500 + 100 + 120 = 720 tonnes

Let *G* be the value of maximum percentage gradient. Then,

$$F_t = (277.8\,\alpha\,M_e + 98\,MG + Mr) \text{ newton}$$

$$= M\left(277.8\,\alpha\,\frac{M_e}{M} + 98\,G + r\right) \text{ newton}$$

$$= 720(277.8 \times 1 \times 1.1 + 98\,G + 45) \text{ newton}$$

or,
$$F_t = (252418 + 70560\,G) \text{ newton} \qquad ...(ii)$$

From (*i*) and (*ii*), we get

$$476045 = 252418 + 70560\,G$$

$$\therefore \quad G = \frac{476045 - 252418}{70560} = \textbf{3.17 percent. (Ans.)}$$

HIGHLIGHTS

1. Railways provide the following three types of passengers services. (*i*) City or urban service, (*ii*) Suburban service, and (*iii*) Main line service.

2. **The typical speed-time curve for electric trains, operating on passenger services, consists of the following *five* parts :**

(*i*) Constant acceleration period, **(*ii*) Acceleration on speed curve,**
(*iii*) Free-running period, **(*iv*) Coasting, and,**
(*v*) Braking.

3. ***Coefficient of adhesion*** **is defined as the ratio of tractive effort to slip the wheels and adhesive weight.**

4. **Tractive effort (F_t) for propulsion of train,**

$$F_t = (277.8\ \alpha\ M_e \pm 98\ MG + Mr) \text{ newton}$$

where, α = **Linear acceleration,**
M = **Dead or stationary mass,**
M_e = **Effective mass (about 8 to 15% more than M)**
G = **Percent gradient, and**
r = **Specific resistance of the train.**

5. **The "*specific energy output*" is the energy output of the driving wheel expressed in watt-hour (Wh) per tonne-km (t-km) (*i.e.,* Wh/t-km).**

6. **Specific energy output,**

$$E_{spo} = \left[0.01072\frac{V_m^{\ 2}}{D}\cdot\frac{M_e}{M} + 27.25\,G\frac{D'}{D} + 0.2778\,r\frac{D'}{D}\right] \text{Wh/t-km}$$

where V_m is in km/h, D' and D are in km and M and M_e in tonne.

7. ***Specific energy consumption,***

$$E_{spc} = \left[0.01072\frac{V_m^{\ 2}}{\eta D}\cdot\frac{M_e}{M} + 27.25\frac{G}{\eta}\cdot\frac{D'}{D} + 0.2778\frac{r}{\eta}\cdot\frac{D'}{D}\right] \text{Wh/t-km}$$

where, η = **overall efficiency** = $\eta_{motor} \times \eta_{gear}$

OBJECTIVE TYPE QUESTIONS

A. Choose the Correct Answer :

1. **Which of the following is an advantage of electric traction over other methods of traction ?**
(*a*) Faster acceleration **(*b*) No pollution problems**
(*c*) Better braking action **(*d*) All of the above.**

2. **Which of the following is the voltage for single phase A.C. system ?**
(*a*) 22 V **(*b*) 440 V** **(*c*) 5 kV** **(*d*) 15 kV**
(*e*) None of the above.

3. **Long distance railways use which of the following ?**
(*a*) 200 V D.C. **(*b*) 25 kV single phase A.C.**
(*c*) 25 kV two phase A.C. **(*d*) 25 kV three phase A.C.**

4. **The speed of a locomotive is controlled by**
(*a*) flywheel **(*b*) gear box**
(*c*) applying brakes **(*d*) regulating steam flow to engine.**

5. **Main traction systems used in India are, those using**
(*a*) electric locomotives **(*b*) diesel engine locomotives**
(*c*) steam engine locomotives **(*d*) diesel electric locomotives**
(*e*) all of the above.

6. **In India diesel locomotives are manufactured at**
(*a*) Ajmer **(*b*) Varanasi** **(*c*) Bangalore** **(*d*) Jamalpur.**

7. **For diesel locomotive the range of horsepower is**
(*a*) 50 to 200 **(*b*) 500 to 1000** **(*c*) 1500 to 2500** **(*d*) 3000 to 5000.**

8. ...... locomotive has the highest operational availability.
(*a*) Electric (*b*) Diesel (*c*) Steam.

9. The horsepower of steam locomotives is
(*a*) upto 1500 (*b*) 1500 to 2000 (*c*) 2000 to 3000 (*d*) 3000 to 4000.

10. The overall efficiency of steam locomotive is around
(*a*) 5 to 10 percent (*b*) 15 to 20 percent (*c*) 25 to 35 percent (*d*) 35 to 45 percent.

11. In tramways which of the following motors is used ?
(*a*) D.C. shunt motor (*b*) D.C. series motor
(*c*) A.C. three phase motor (*d*) A.C. single phase capacitor start motor.

12. In a steam locomotive electric power is provided through
(*a*) overhead wire (*b*) battery system
(*c*) small turbo-generator (*d*) diesel engine generator.

13. Which of the following drives is suitable for mines where explosive gas exists ?
(*a*) Steam engine (*b*) Diesel engine (*c*) Battery locomotive (*d*) Any of the above.

14. In case of locomotives the tractive power is provided by
(*a*) Single cylinder double acting steam engine (*b*) double cylinder, single acting steam engine
(*c*) double cylinder, double acting steam engine (*d*) single stage steam turbine.

15. Overload capacity of diesel engines is usually restricted to
(*a*) 2 percent (*b*) 10 percent (*c*) 20 percent (*d*) 40 percent.

16. In case of steam engines the steam pressure is
(*a*) 1 to 4 kgf/cm² (*b*) 5 to 8 kgf/cm² (*c*) 10 to 15 kgf/cm² (*d*) 25 to 35 kgf/cm².

17. The steam engine provided on steam locomotives is
(*a*) single acting condensing type (*b*) single acting non-condensing type
(*c*) double acting condensing type (*d*) double acting non-condensing type.

18. Electric locomotives in India are manufactured at
(*a*) Jamalpur (*b*) Bangalore (*c*) Chittranjan (*d*) Gorakhpur.

19. The wheels of a train, engine as well as bogies, are slightly tapered to
(*a*) reduce friction (*b*) increase friction
(*c*) facilitate braking (*d*) facilitate in taking turns.

20. Automatic signalling is used for which of the following trains ?
(*a*) Mail and express trains (*b*) Superfast trains
(*c*) Suburban and Urban electric trains (*d*) All trains.

21. The efficiency of diesel locomotives is nearly
(*a*) 20 to 25 percent (*b*) 30 to 40 percent (*c*) 45 to 55 percent (*d*) 60 to 70 percent.

22. The speed of a superfast train is
(*a*) 60 kmph (*b*) 75 kmph (*c*) 100 kmph (*d*) more than 100 kmph.

23. The number of passanger coaches that can be attached to a diesel engine locomotive on broad gauge is usually restricted to
(*a*) 5 (*b*) 10 (*c*) 14 (*d*) 17.

24. Which of the following state capitals is not on broad gauge track ?
(*a*) Lucknow (*b*) Bhopal (*c*) Jaipur (*d*) Chandigarh.

25. Which of the following is the advantage of electric braking ?
(*a*) It avoids wear of track (*b*) Motor continues to remain loaded during braking
(*c*) It is instantaneous
(*d*) More heat is generated during braking.

26. Which of the following braking systems on the locomotives is coslty ?
(*a*) Regenerative braking on electric locomotives (*b*) Vacuum braking on diesel locomotives
(*c*) Vacuum braking on steam locomotives (*d*) All braking systems are equally costly.

27. Tractive effort is required to
(*a*) overcome the gravity component of train mass (*b*) overcome friction, windage and curve resistance
(*c*) accelerate the train mass (*d*) do all of the above.

28. For given maximum axle load tractive efforts of A.C. locomotive will be
(*a*) less than that of D.C. locomotive (*b*) more than that of D.C. locomotive
(*c*) equal to that of D.C. locomotive (*d*) none of the above.

29. Co-efficient of adhesion reduces due to the presence of which of the following ?
(*a*) Sand on rails (*b*) Dew on rails (*c*) Oil on the rails (*d*) both (*b*) and (*c*).

30. Due to which of the following co-efficient of adhesion improves ?
(*a*) Rust on the rails (*b*) Dust on the rails (*c*) Sand on the rails (*d*) All of the above.

31. Quadrilateral speed-time curve pertains to which of the following services ?
(*a*) Main line service (*b*) Urban service
(*c*) Sub-urban service (*d*) Urban and sub-urban service.

32. Which of the following is the disadvantage of electric traction over other systems of traction ?
(*a*) Corrosion problems in the underground pipe work
(*b*) Short time power failure interrupts traffic for hours
(*c*) High capital outlay in fixed installations beside route limitation
(*d*) Interference with communication lines
(*e*) All of the above.

33. Co-efficient of adhesion is
(*a*) high in case of D.C. traction than in the case of A.C. traction
(*b*) low in case of D.C. traction than in the case of A.C. traction
(*c*) equal in both A.C. and D.C. traction
(*d*) any of the above.

34. Speed-time curve of main line service differs from those of urban and sub-urban services on following account
(*a*) it has longer free running period (*b*) it has longer coasting period
(*c*) accelerating and braking periods are comparatively smaller
(*d*) all of the above.

35. The rate of acceleration on sub-urban or urban services is restricted by the consideration of
(*a*) engine power (*b*) track curves
(*c*) passanger discomfort (*d*) track size.

36. The specific energy consumption of a train depends on which of the following ?
(*a*) Acceleration and retardation (*b*) Gradient
(*c*) Distance covered (*d*) All of the above.

37. The friction at the track is proportional to
(*a*) 1/speed (*b*) $1/(\text{speed})^2$ (*c*) speed (*d*) none of the above.

38. The air resistance to the movement of the train is proportional to
(*a*) speed (*b*) $(\text{speed})^2$ (*c*) $(\text{speed})^3$ (*d*) 1/speed.

39. The normal value of adhesion friction is
(*a*) 0.12 (*b*) 0.25 (*c*) 0.40 (*d*) 0.75.

40. The pulsating torque exerted by steam locomotives causes which of the following ?
(*a*) Jolting and skidding (*b*) Hammer blow (*c*) Pitching (*d*) All of the above.

41. Which of the following braking systems is used on steam locomotives ?
(*a*) Hydraulic system (*b*) Pneumatic system (*c*) Vacuum system (*d*) None of the above.

42. Vacuum is created by which of the following ?
(*a*) Vacuum pump (*b*) Ejector (*c*) Any of the above (*d*) None of the above.

43. The resistance encountered by a train in motion is on account of
(*a*) resistance offered by air (*b*) friction at the track
(*c*) friction at various parts of the rolling stock (*d*) all of the above.

44. Battery operated trucks are used in
(*a*) steel mills (*b*) power stations
(*c*) narrow gauge traction (*d*) factories for material transportation.

45. ...... method can bring the locomotive to dead stop.
(*a*) Plugging braking (*b*) Rheostatic braking
(*c*) Regenerative braking (*d*) None of the above.

46. The value of co-efficient of adhesion will be high when rails are
(*a*) greased (*b*) wet (*c*) sprayed with oil (*d*) cleaned with sand.

47. The voltage used for suburban trains in D.C. system is usually
(*a*) 12 V (*b*) 24 V (*c*) 220 V (*d*) 600 to 750 V.

48. For three-phase induction motors which of the following is the least efficient method of speed control ?
(*a*) Cascade control (*b*) Pole changing
(*c*) Rheostatic control (*d*) Combination of cascade and pole changing.

49. Specific energy consumption becomes
(*a*) more on steeper gradient (*b*) more with high train resistance
(*c*) less if distance between stops is more (*d*) all of the above.

50. In main line service as compared to urban and suburban service
(*a*) distance between the stops is more (*b*) maximum speed reached is high
(*c*) acceleration and retardation rates are low (*d*) all of the above.

51. Locomotive having monomotor bogies
(*a*) has better co-efficient of adhesion
(*b*) are suited both for passanger as well as freight service
(*c*) has better riding qualities due to the reduction of lateral forces
(*d*) has all above qualities.

52. Series motor is not suited for traction duty due to which of the following account ?
(*a*) Less current drain on the heavy load torque
(*b*) Current surges after temporary switching off supply
(*c*) Self relieving property
(*d*) Commutating property at heavy load.

53. When a bogie negotiates a curve, reduction in adhesion occurs resulting in sliding. Thus sliding is acute when
(*a*) wheel base of axles is more (*b*) degree of curvature is more
(*c*) both (*a*) and (*b*) (*d*) none of the above.

54. Energy consumption in propelling the train is required for which of the following ?
(*a*) Work against the resistance to motion
(*b*) Work against gravity while moving up the gradient
(*c*) Acceleration
(*d*) All of the above.

55. An ideal traction system should have ....
(*a*) easy speed control (*b*) high starting tractive effort
(*c*) equipment capable of with standing large temporary loads
(*d*) all cf the above.

56. ...... have maximum unbalanced forces
(*a*) Diesel shunters (*b*) Steam locomotives (*c*) Electric locomotives (*d*) Diesel locomotives.

57. Specific energy consumption is affected by which of the following factors ?
(*a*) Retardation and acceleration values (*b*) Gradient
(*c*) Distance between stops (*d*) All of the above.

58. In case of ...... free running and coasting periods are generally long.
(*a*) main-line service (*b*) urban service (*c*) sub-urban service (*d*) all of the above.

59. Overhead lines for power supply to tramcars are at a minimum height of
(*a*) 3 m (*b*) 6 m (*c*) 10 m (*d*) 20 m.

60. The return circuit for tramcars is through ......
(*a*) neutral wire (*b*) rails (*c*) cables (*d*) common earthing.

61. Specific energy consumption is least in ...... service.
(*a*) main line (*b*) urban (*c*) suburban.

62. Locomotives with manometer bogies have
(*a*) uneven distribution of tractive effect
(*b*) suitability for passanger as well as freight service
(*c*) lot of skidding
(*d*) low co-efficient of adhesion.

63. ...... was the first city in India to adopt electric traction.
(*a*) Delhi (*b*) Madras (*c*) Calcutta (*d*) Bombay.

64. ...... frequency is not common in low frequency traction system
(*a*) 40 Hz (*b*) 25 Hz (*c*) $16\frac{2}{3}$ Hz.

65. For 25 kV single phase system power supply frequency is ......
(*a*) 60 Hz (*b*) 50 Hz (*c*) 25 Hz (*d*) $16\frac{2}{3}$ Hz.

66. Power for lighting in passanger coach, in a long distance electric train, is provided
(*a*) directly through overhead electric line
(*b*) through individual generator of bogie and batteries
(*c*) through rails
(*d*) through locomotive.

67. In India, electrification of railway track was done for the first time in which of the following years ?
(*a*) 1820—1825 (*b*) 1880—1885 (*c*) 1925—1932 (*d*) 1947—1954.

68. Suri transmission is ......
(*a*) electrical-pneumatic (*b*) mechanical-electrical
(*c*) hydro-mechanical (*d*) hydro-pneumatic.

69. In case of a steam engine an average coal consumption per km is nearly
(*a*) 150 to 175 kg (*b*) 100 to 120 kg (*c*) 60 to 80 kg (*d*) 28 to 30 kg.

70. Which of the following happens in Kando system ?
(*a*) Three phase A.C. is converted into D.C.
(*b*) Single phase A.C. is converted into D.C.
(*c*) Single phase supply is converted into three phase system
(*d*) None of the above.

71. For which of the following locomotives the maintenance requirements are the least ?
(*a*) Steam locomotives (*b*) Diesel locomotives
(*c*) Electric locomotives (*d*) Equal in all of the above.

72. Which of the following methods is used to control speed of 25 kV, 50 Hz single phase traction ?
(*a*) Reduced current method (*b*) Tap changing control of transformer
(*c*) Series parallel operation of motors (*d*) All of the above.

73. If the co-efficient of adhesion on dry rails is 0.26, which of the following could be the value for wet rails ?
(*a*) 0.3 (*b*) 0.26 (*c*) 0.225 (*d*) 0.16.

74. ...... watt-hours per tonne km is usually the specific energy consumption for suburban services.
(*a*) 15—20 (*b*) 50—75 (*c*) 120—150 (*d*) 160—200.

75. The braking retardation is usually in the range
(*a*) 0.15 to 0.30 km phps (*b*) 0.30 to 0.6 km phps (*c*) 0.6 to 2.4 km phps (*d*) 3 to 5 km phps
(*e*) 10 to 15 km phps.

76. The rate of acceleration on suburban or urban service is in the range
(*a*) 0.2 to 0.5 km phps (*b*) 1.6 to 4.0 km phps (*c*) 5 to 10 km phps (*d*) 15 to 25 km phps.

77. The coasting retardation is around
(*a*) 0.16 km phps (*b*) 1.6 km phps (*c*) 16 km phps (*d*) 40 km phps.

78. Which of the following track is electrified ?
(*a*) Delhi—Bombay (*b*) Delhi—Madras (*c*) Delhi—Howrah (*d*) Delhi—Ahmedabad.

79. ...... is the method of braking in which motor armature remains connected to the supply and draws power from it producing torque opposite to the direction of motion.
(*a*) Rheostatic braking (*b*) Regerative braking (*c*) Plugging.

80. For 600 V D.C. line for tramcars, track is connected to ......
(*a*) positive of the supply (*b*) negative of the supply
(*c*) mid voltage of 300 V (*d*) none of the above.

B. Fill in the Blanks/Say 'Yes' or 'No' :

81. Direct internal combustion engine drive is widely employed for road transport.
82. In single phase A.C. system A.C. series motors are used for getting the necessary motive power.
83. The overhead distribution system in the case of traction is identical with that of power system.
84. The cost involved in substations in case of A.C. traction system is ...... than involved in D.C. system.
85. Cost of D.C. electrification is ...... than that of A.C. electrification.
86. Steam locomotive gives high rate of acceleration.
87. Weight transfer is a desirable quality.
88. Regenerative braking is more efficient and less complicated in D.C. system as compared to that with A.C. system.
89. Steam locomotive is used for under-ground railway.
90. Steam traction involves ...... capital cost.
91. ...... transformers are necessary in A.C. traction to confine the return current through the rail and return feeder than through the ground.
92. Shunt motors are better suited for running them in parallel mechanically.
93. Increase of motor torque beyond adhesive limit does not increase the tractive effort but causes wheel ......, on the other hand increasing the braking effort beyond adhesive limit produces wheel ......
94. Interference with communication system is more with A.C. than with D.C.
95. Operating voltage for D.C. traction system is ......
96. In D.C. traction ...... type pantograph is used.
97. In A.C. traction single ended diamond type pantograph is used.
98. The pressure between pantograph and contact wire, for better current collection, should be ......
99. In A.C. traction ...... transformers are necessary to confine the return current through the rails.
100. During notching up the acceleration is constant.
101. In India, steam locomotives use ...... acting reciprocating engines.
102. Steep gradient will involve less energy.
103. The maximum speed attained by the vehicle during the run is known as crest speed.
104. During free run speed remains constant.
105. The co-efficient of adhesion is better in steam traction than that in electric fraction.
106. During ...... the speed of the train decreases.

107. Coasting period precedes braking.

108. In Calcutta city tramways are still a means of public transport.

109. A diesel engine has an overload capacity of 30%.

110. Least pollution can be expected from electric locomotives.

111. Stagger in the contact wire is provided to even out the wear on the contact strip of pantograph.

112. In D.C. traction system negative boosters have to be employed to confine ground currents to rails.

113. Initial rush of current on temporary interruption of supply is more with series motor than with shunt motor.

114. Speed control of motors in A.C. traction is difficult and wasteful than that of D.C. system.

115. During coasting the power supply is zero.

116. Maintenance and repair cost of steam locomotives is ...... as compared to that of electric locomotives.

117. Power output from the driving axles is the product of tractive effort and speed.

118. Braking is followed by coasting.

119. Initial expenditure in steam system is ...... as compared to that in electric railway traction.

120. Dead weight is smaller than acceleration weight.

ANSWERS

A. Choose the Correct Answer :

1. (*d*)	**2.** (*d*)	**3.** (*b*)	**4.** (*d*)	**5.** (*e*)
6. (*b*)	**7.** (*c*)	**8.** (*a*)	**9.** (*a*)	**10.** (*a*)
11. (*b*)	**12.** (*c*)	**13.** (*c*)	**14.** (*c*)	**15.** (*b*)
16. (*c*)	**17.** (*d*)	**18.** (*c*)	**19.** (*d*)	**20.** (*c*)
21. (*a*)	**22.** (*d*)	**23.** (*d*)	**24.** (*c*)	**25.** (*a*)
26. (*a*)	**27.** (*d*)	**28.** (*b*)	**29.** (*d*)	**30.** (*d*)
31. (*d*)	**32.** (*e*)	**33.** (*b*)	**34.** (*d*)	**35.** (*c*)
36. (*d*)	**37.** (*c*)	**38.** (*b*)	**39.** (*b*)	**40.** (*a*)
41. (*c*)	**42.** (*c*)	**43.** (*d*)	**44.** (*d*)	**45.** (*a*)
46. (*d*)	**47.** (*d*)	**48.** (*c*)	**49.** (*d*)	**50.** (*d*)
51. (*d*)	**52.** (*b*)	**53.** (*c*)	**54.** (*d*)	**55.** (*d*)
56. (*b*)	**57.** (*d*)	**58.** (*a*)	**59.** (*c*)	**60.** (*b*)
61. (*a*)	**62.** (*b*)	**63.** (*d*)	**64.** (*a*)	**65.** (*b*)
66. (*b*)	**67.** (*c*)	**68.** (*c*)	**69.** (*d*)	**70.** (*c*)
71. (*c*)	**72.** (*b*)	**73.** (*d*)	**74.** (*b*)	**75.** (*d*)
76. (*b*)	**77.** (*a*)	**78.** (*c*)	**79.** (*c*)	**80.** (*b*).

B. Fill in the Blanks/Say 'Yes' or 'No' :

81. Yes	**82.** Yes	**83.** No	**84.** less	**85.** more
86. No	**87.** No	**88.** Yes	**89.** No	**90.** less
91. Booster	**92.** No	**93.** Skidding, slipping		**94.** Yes
95. 1500 V	**96.** diamond	**97.** No	**98.** positive	**99.** booster
100. Yes	**101.** double	**102.** No	**103.** Yes	**104.** Yes
105. No	**106.** coasting	**107.** Yes	**108.** Yes	**109.** No
110. Yes	**111.** Yes	**112.** Yes	**113.** Yes	**114.** No
115. Yes	**116.** more	**117.** Yes	**118.** No	**119.** less
120. Yes.				

THEORETICAL QUESTIONS

1. What do you understand by speed-time curves ? What is its use in practice ?
2. Draw the speed-time curves for urban and main line service.
3. Explain clearly 'free running', 'coasting' and 'braking' with reference to electric traction systems.
4. Explain speed-time curve of a train running on main line. Define 'crest speed', 'average speed' and 'schedule speed'.
5. Draw and explain a typical speed-time curve for an electric train and explain what do you understand by crest speed, average speed and schedule speed.
6. Define 'crest speed' and 'schedule speed' and discuss the factors which affect the schedule speed of a train.
7. Explain how an actual speed-time curve for an electric train service can be replaced by a curve having a simple geometric shape. Deduce from first principles the relation between acceleration, retardation, maximum speed, running time, and distance between stops, assuming a simplified speed-time curve.
8. Sketch the typical speed-time curves for the train movement and from these derive a simplified speed-time curve.
9. What are the different methods of approximation of speed-time curves ? Derive expressions for distance travelled using quadrilateral approximation method of $V(t)$ curve.
10. Draw the speed-time curve for an electric train accelerating uniformly to speed V_m, the power is cut off and after a time t when it attains V_m coasting is allowed to continue up to the end of the run, the braking being neglected. Derive a suitable equation to determine V_m.
11. Explain the terms : (*i*) Adhesive weight, and (*ii*) Train resistance.
12. What is tractive effort of a train and what are its functions ? Derive an expression for the tractive effort developed by a train unit.
13. How does the 'train resistance' play its part in the mechanics of train motion ?
14. Deduce the expressions for : (*i*) the tractive effort exerted by road wheel in terms of wheel diameter, motor torque, gear ratio and efficiency of transmission of power through gears, (*ii*) The tractive effort for propulsion of train up and down the gradient.
15. What is the coefficient of adhesion ? How does it affect slipping of driving wheels of traction unit ?
16. What is specific energy consumption ? Enumerate the factors which affect the specific energy consumption of trains operating at a given schedule speed.
17. Describe the procedure of calculating the specific energy consumption of an electric train.
18. Derive equation of motion of a vehicle. How does the effective mass of a vehicle differ with its physical (actual) mass. Explain significance of equation of motion in energy consumption calculations.
19. Derive an expression for specific energy output on level track using a simplified speed-time curve.
20. Differentiate between adhesive and total weight of an electric locomotive.
21. Explain the terms 'dead weight', 'effective weight' and 'adhesive weight' in a locomotive.
22. Write short notes on the following :
    (*i*) Factors affecting energy consumption in propelling a train.
    (*ii*) Mechanics of 'train movement'.
    (*iii*) Dead weight, accelerating weight and adhesive weight of a train.
    (*iv*) Speed-time curves in respect of train motion.
    (*v*) Tractive effort for propulsion of train.
23. How does the value of acceleration and retardation affect the specific energy consumption for a given run at a given schedule speed ?

UNSOLVED EXAMPLES

1. A train is required to run between two stations 1.5 km apart at a schedule speed of 36 km/h, the duration of stops being 25 seconds. The braking retardation is 3 km/h/s. Assuming a trapezoidal speed-time curve, calculate the acceleration if the ratio of maximum speed to average speed is to be 1.25. **[Ans.** 1.7 km/h/s]
2. A train has schedule speed of 60 km/h between the stops which are 6 km apart. Determine the crest speed over the run, assuming trapezoidal speed-time curve. The train accelerates at 2 km/h/s and retards at 3 km/h/s. Duration of stops is 60 seconds. **[Ans.** 81.2 km/h]
3. The distance between two stations is 1 km, and the schedule speed is 30 km/h. Station stopping time is 20 seconds. Assume braking retardation of 3 km/h/s and maximum speed 1.25 times the average speed. Determine the acceleration required to run the service if the speed-time curve is approximated by a trapezoidal curve. **[Ans.** 1.8 km/h/s]
4. A suburban electric train has a maximum speed of 70 km/h. The schedule speed including a station stop of 30 seconds is 45 km/h. If the acceleration is 1.5 km/h/s, find the value of retardation when the average distance between stops is 4 km. **[Ans.** 0.575 km/h/s]
5. An electric train has a schedule speed of 25 km/h between stations 800 m apart. The duration of station stop is 20 seconds, the maximum speed is 20 percent higher than the average running speed and the braking retardation is 3 km/h/s. Calculate the rate of acceleration required to operate this service. **[Ans.** 1.85 km/h/s]
6. A train runs between two stations 1.6 km apart at an average speed of 36 km/h. If the maximum speed is limited to be 72 km/h, acceleration to 2.7 km/h/s, coasting retardation to 0.18 km/h/s and braking retardation to 3.2 km/h/s, compute the duration of acceleration, coasting and braking periods. Assume a simplified speed-time curve. **[Ans.** t_1 = 27 s ; t_2 = 120 s ; t_3 = 14 s]
7. The peripheral speed of a railway traction motor cannot be allowed to exceed 44 m/s. If the gear ratio is 18/75, motor armature diameter 42 cm and wheel diameter 91 cm, calculate the limiting value of train speed. **[Ans.** 82 km/h]
8. A 250-tonne motor coach driven by four motors takes 20 seconds to attain a speed of 42 km/h, starting from rest on an ascending gradient of 1 in 80. The gear ratio is 3.5, gear efficiency 92%, wheel diameter 92 cm, train resistance 40 N/t and rotational inertia 10 percent of the dead weight. Find the torque developed by each motor. **[Ans.** 7186 Nm]
9. A 250-tonne motor coach having four motors, each developing a torque of 8000 Nm during acceleration, starts from rest. If up-gradient is 30 in 1000, gear ratio 3.5, gear transmission efficiency 90 percent, wheel diameter 90 cm, train resistance 50 N/t, rotational inertia effect 10 percent, calculate the time taken by each coach to attain a speed of 80 km/h. Also estimate the current taken by each motor during the acceleration period if supply voltage is 3000 V and motor efficiency 85 percent. **[Ans.** 44.4 s ; 488 A]
10. A goods train weighing 500 tonnes is to be hauled by a locomotive up an ascending gradient of 2 percent with an acceleration of 1 km/h/s. If the coefficient of adhesion is 0.25, train resistance 40 N/t and effect of rotational inertia 10 percent. Find the weight of locomotive and number of axles if load is not to increase beyond 21 tonnes/axle. **[Ans.** 142 tonnes, 7]
11. A 300-tonne EMU is started with a uniform acceleration and reaches a speed of 40 km/h in 24 seconds on a level track. Assuming trapezoidal speed-time curve, find specific energy consumption if rotational inertia is 8 percent, retardation is 3 km/h/s, distance between stops is 3 km, motor efficiency is 0.9 and train resistance is 40 N/tonne. **[Ans.** 21.6 Wh/t-km]
12. A 350-tonne electric train runs up an ascending gradient of 1 percent with the following speed-time curves : (*i*) Uniform acceleration of 1.6 km/h/s for 25 seconds, (*ii*) Constant speed for 50 seconds, (*iii*) Coasting for 30 seconds, and (*iv*) Braking at 2.56 km/h/s to rest.

    Calculate the specific energy consumption if train resistance is 50 N/t, effect of rotational inertia 10 percent, overall efficiency of transmission gear and motor, 75 percent. **[Ans.** 63.1 Wh/t-km]

8

Traction Drives

8.1. SIGNIFICANT FEATURES OF TRACTION DRIVES

One major application of electric drives is in electric traction, *i.e., to transport men and materials from one place to another.*

The ***important features of traction drives*** are :

1. During start and acceleration, large torque is required for accelerating the heavy mass.

2. During acceleration and when negotiating up gradients, the motor is subjected to torque overloads.

3. In A.C. traction single-phase supply is used due to economic reasons (25 kV, 50 Hz single-phase supply is used in Indian Railways).

4. When the locomotive crosses from one supply section to another, there is a sharp voltage fluctuations (including discontinuity) in the supply. The locomotive ratings can be 6000 H.P. and higher.

5. In view of the fact that the supply is weak in nature and the reactive power has very adverse effect, the power factor should not be allowed to be lower than 0.8 (The *power factor should never be allowed to be leading to avoid over voltages*).

6. Both in A.C. and D.C. traction, the injection of harmonics can lead to maloperation of signals and interference in the lines.

7. Dynamic braking is widely used. However, when the train is stationary mechanical brakes are also provided.

8. Regenerative braking is employed when the energy saved is large enough to justify the additional cost of drive and transmission lines and is possible only when the system is able to absorb

the energy generated. Mostly, system is not able to absorb all the energy that is generated. Therefore, *dynamic braking is combined with regenerative braking.* The energy which cannot be absorbed is dissipated by dynamic braking. Such a braking combination is termed as "*composite braking*".

— When regenerative braking is employed, the distribution system is subjected to higher voltages during regeneration. Thus, the distribution system and drive equipment must be designed to withstand such high voltages.

9. In a locomotive more than one motors are fed from a converter. The load sharing between motors is more uniform when the motors have a large speed regulation. However, to avoid wheel slip, a motor with low speed regulation is preferred.

8.2. DESIRABLE PROPERTIES OF TRACTION MOTORS

Traction motors are expected to have both electrical and mechanical characteristics, as far as possible, to suit the stringent service conditions of traction work which are as follows :

I. Electrical characteristics :

1. High starting torque.

2. Simple speed control.

3. Self relieving property (The speed-torque characteristics of the motor should be such that the speed may fall with the increase in load. The motors having such characteristics are *self protective against overloading* as power output of a motor is proportional to the product of torque and speed.

4. Better commutation.

5. Possibility of *dynamic or regenerative braking.*

6. Capability of *withstanding voltage fluctuations.*

7. Capability to *withstand temporary interruption of supply.*

8. Capability to *take heavy loads without flashover.*

9. Capability to *operate in parallel* (for which they should have suitable speed-torque and current-torque characteristics).

II. Mechanical characteristics :

1. Traction motors must be *robust and capable to withstand continuous vibrations.*

2. Must be *totally enclosed type*, particularly when mounted beneath the locomotive or motor coach, to provide protection against ingress of dirt, dust, water, mud etc.

3. Must be *small in overall dimensions* specially in its overall diameter (so that it may easily be located underneath a motor coach, when required).

4. *Minimum weight* (so that pay load capacity of the vehicle may be increased).

8.3. TRACTION MOTORS

None of the motors can have all the desirable operating characteristics mentioned above. Some of the motors which find application in traction are listed below :

1. D.C. series motors supplied with straight D.C. or rectified D.C.
2. A.C. series motors-single-phase.
3. Repulsion motors.
4. Three-phase induction motors.
5. Linear induction motors.

- *Earlier* **D.C. series motor** *was widely used in traction.*
  - — A series D.C. motor has *high starting torque and capability for high torque overloads.*
  - — With an increase in torque, the flux also increases ; therefore, for the same increase in torque, the *increase in motor current is less* compared to other motors. Thus during heavy torque overloads, power overload on the source and thermal overloading of the motor are *kept limited to reasonable values.*
  - — The speed-torque characteristic is also *suitable for better sharing of loads between motors.*
  - — Further due to a large inductance in the field, sharp fluctuations in supply voltage *do not produce sharp peaks in armature current.* Thus the *motor commutation remains satisfactory* which does not happen in separately excited motor, unless an additional inductance is connected in armature circuit.

Following are the ***limitations*** of *a D.C. series motor :*

(*i*) The field of a series motor *cannot be easily controlled by semiconductor switches.* If field control is not employed, the series motor must be designed with its speed equal to the highest desired speed of the drive. The higher base speeds are obtained by using fewer turns in the field winding. This, however, *reduces the torque per ampere at start and therefore, acceleration.*

(*ii*) There are a number of problems with regenerative and dynamic brakings of a series motor.

- With the advent of semiconductor converters, **separately excited motor** is now *preferred over series motor.*
  - — With independent control of armature and field, the speed-torque characteristic of separately excited motor can be shaped to satisfy the traction requirements in the optimum manner.
  - — Further, *because of low regulation of its speed-torque characteristics, the coefficient of adhesion has higher value.*
  - — The *regenerative and dynamic brakings* of a separately excited motor *are fairly simple and efficient, and can be carried out down to very low speeds.*
- These days **compound motor** is being preferred for traction application since *it incorporates the advantages of both series and separately excited motors.*
- Due to the *availability of reliable variable frequency semiconductor inverters,* **Squirrel-cage induction motor** and **synchronous motor** are now *finding application in traction.*

Following are the *advantages of squirrel-cage induction motor over D.C. motors :*

(*i*) Ruggedness

(*ii*) Lower maintenance

(*iii*) Better reliability

(*iv*) Lower cost, weight, volume and inertia

(*v*) Higher efficiency

(*vi*) Ability to operate satisfactorily with sharp supply voltage fluctuations and in dirty environment.

The *major disadvantage of D.C. motor* is the *presence of commutator and brushes, which require frequent maintenance, particularly when the flashovers at the commutator occur due to sharp voltage fluctuations.*

- The **synchronous motor,** in terms of advantages mentioned for squirrel-cage motor in comparison with D.C. motors, lies in-between the two and has one important advantage over squirrel-cage induction motor, that *it can be operated at leading power factor.* This *permits the use of load-commutated thyristor inverter which is cheaper and occupies less volume and weight compared to forced commutated thyristor inverter required by induction motors.*

— The weight and volume of an induction motor drive can also be kept low by using GTO (gate turn-off thyristor) inverter but is more expensive than a load commutated thyristor inverter.

8.4. D.C. SERIES MOTORS

8.4.1. Characteristics of D.C. Motors

The properties of all motors and, in particular, D.C. motors are defined as a totality of the following characteristics :

(*i*) *Starting* ; (*ii*) *Operating and mechanical* ;

(*iii*) *Braking ; and* (*iv*) *Regulation.*

Starting characteristics. The starting characteristics *determine the operation of starting from the moment the motor begins running to the moment when steady-state operation is established* and *include* :

- The starting current I_{start} generally determined by the ratio $\frac{I_{start}}{I_{run}}$;
- The starting torque T_{start}, determined by the ratio $\frac{T_{start}}{T_{run}}$;
- The duration of starting t_{start} ;
- The economy of operation determined by the amount of energy consumed in starting ; and
- The cost and reliability of the starting equipment.

Operating characteristics. The operating characteristics are those that give the relation between *speed, torque* and *efficiency* as *functions* of the *useful power* or the *armature* current for V = constant and constant resistances in the armature and field circuit.

Mechanical characteristics. Of major importance for industrial drive mechanisms are the mechanical characteristics, which are the relation $N = f(T)$ (where N and T stand for speed and torque respectively) for conditions of constant voltage and resistances in the armature and field circuits. These also include the *braking characteristics.*

Regulation characteristics. These characteristics determine the *properties of motors when their speed is controlled.* These include :

- The regulation range determined by the ratio $\frac{N_{max}}{N_{min}}$;
- The efficiency of regulation from the point of view of the initial cost of the equipment and maintenance ;
- The nature of regulation—continuous or stepped ;
- The simplicity of the control apparatus and methods.

The *D.C. motors possess versatile* and diverse regulation characteristics, and for this reason are indispensable in installations where wide-range control of speed is necessary.

The characteristic curves of a motor are those curves which show relation between the following quantities :

1. *Torque and armature current i.e.,* T_a/I_a characteristics. This is also known as **electrical characteristic.**

2. *Speed* and *armature current i.e.,* N/I_a characteristic.

3. *Speed* and *torque i.e.,* N/T_a characteristic. This is also known as **mechanical characteristic.** This can be obtained from (1) and (2) above.

Following relations are worth *keeping in mind* while discussing motor characteristics :

$$N \propto \frac{E_b}{\phi} \text{ and } T_a \propto \phi I_a.$$

8.4.2. Torque Developed by a D.C. Motors

The expression for the torque developed by the motor armature may also be deduced as follows :

Let T_a be the torque developed in Nm by the motor armature running at N r.p.m.

Power developed = Work done per second

$$= T_a \times 2\pi N \text{ watts} \quad ...(i)$$

Electrical equivalent of mechanical power developed by the armature also

$$= E_b I_a \text{ watts} \quad ...(ii)$$

Equating (*i*) and (*ii*), we get

$$T_a \times \frac{2\pi N}{60} = E_b I_a$$

or,
$$T_a = \frac{E_b I_a}{2\pi\left(\frac{N}{60}\right)}$$

Also since
$$E_b = \frac{p\phi ZN}{60a} \text{ volts}$$

$$\left[\begin{array}{ll} \text{where,} & p = \text{Number of poles,} \\ & \phi = \text{Flux per pole, Wb,} \\ & Z = \text{Number of armature conductors,} \\ & N = \text{Speed of armature, r.p.m., and} \\ & a = \text{Number of parallel path.} \end{array}\right]$$

$\therefore$
$$T_a \times 2\pi \frac{N}{60} = \frac{p\phi ZN}{60a} . I_a$$

or,
$$T_a = \frac{Z\phi p}{2\pi} . \frac{I_a}{a} \text{ Nm}$$

i.e.,
$$T_a = 0.159\, Z\phi p . \frac{I_a}{a} \quad ...(8.1)$$

Note. From the above equation for torque, we find that

$$T \propto \phi I_a$$

Then,

(*i*) In the case of shunt motors, ϕ is practically constant,

hence,
$$T \propto I_a$$

(*ii*) In the case of series motors, ϕ is proportional to I_a before saturation (because field windings carry full armature current)

$\therefore$
$$T \propto I_a^2.$$

(*i*) **Shaft torque (T_{sh}).** The torque developed by the armature is the *gross torque.* Whole of this torque is not available at the pulley, since certain percentage of torque developed by the armature is lost to overcome the iron and friction losses. *The torque which is available for useful work is* *known as* **shaft torque** T_{sh}. It is so called because it is *available at the shaft.* The horse power obtained by using shaft torque is called brake horse power (B.H.P.).

$$\text{B.H.P. (metric)} = \frac{T_{sh} \times 2\pi N}{735.5}$$

$$\therefore \qquad T_{sh} = \frac{\text{B.H.P. (metric)} \times 735.5}{\frac{2\pi N}{60}} \qquad \text{...(8.2)}$$

where, N = speed of armature in r.p.m.

The difference $T_a - T_{sh}$ is known as **lost torque** (*i.e.*, torque lost in iron and friction losses)

$$= 0.519 \times \frac{\text{Iron and friction losses}}{\frac{N}{60}} \text{ Nm},$$

(*ii*) **Relation between speed (N) and armature current (I_a) :**

Back e.m.f., $$E_b = \frac{p\phi ZN}{60a} \qquad \text{...(8.3)}$$

Also, $$E_b = V - I_a R_a$$

where, R_a is the armature resistance ...(*ii*)

From (*i*) and (*ii*), we get

$$\frac{p\phi ZN}{60a} = V - I_a R_a$$

or, $$N = \frac{(V - I_a R_a)}{\phi Z} \times \frac{60a}{p} \qquad \text{...(8.4)}$$

or, $$N \propto \frac{V - I_a R_a}{\phi}$$

since Z, p and a are constant for a particular machine,

$\therefore$ $$N \propto \frac{V}{\phi}, \text{ since } I_a R_a \text{ is very small as compared to applied voltage}$$

or, $$N \propto \frac{1}{\phi}, \text{ if applied voltage is constant}$$

(*iii*) **Relation between speed and torque :**

Torque developed, $$T = 0.159\, Z\phi p \,.\, \frac{I_a}{a} \qquad \text{[Eqn. (8.1)]}$$

or, $$\frac{a}{\phi Zp} = \frac{0.159\, I_a}{T}$$

Substituting $\frac{a}{\phi Zp} = \frac{0.159\, I_a}{T}$ in eqn. (8.4), we get

$$N = \frac{(V_1 - I_a R_a) \times 60 \times 0.159\, I_a}{T}$$

or, Speed, $$N = \frac{9.55\,(V - I_a R_a)\, I_a}{T} \qquad \text{...(8.5)}$$

The torque-current (T/I_a) and speed-torque (N/T) curves for D.C. motors are shown in Fig. 8.1 and 8.2 respectively.

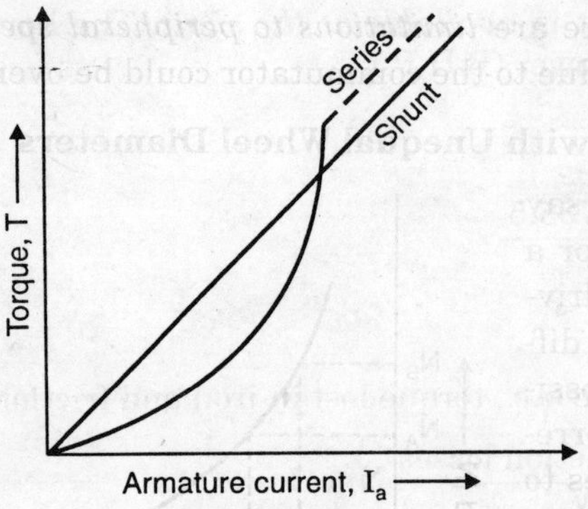

Fig. 8.1. *T/I*$_a$ curves.

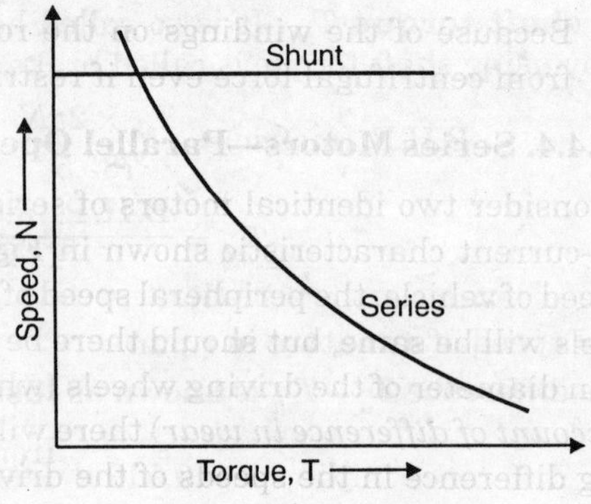

Fig. 8.2. *N/T* curves.

8.4.3. Suitability of Series Motor for Traction Duty

The suitability of series motor for traction duty is discussed point wise below :

1. D.C. series motor develops high torque at low speeds and low torque at high speeds. This is the exact requirement of the traction units. For accelerating the vehicle, development of high starting torque is essential.

2. Since the torque and speed control characteristics of series motor are *steep,* therefore, for a given difference in rotational speed of various motors running in parallel caused by unequal wear of different driving wheels, torque developed and current taken by each motor *will not differ much.*

3. In case of series motor (where field flux varies as the armature current) torque corresponding to given current will be unaffected by the variations in the line voltage. Voltage variations will, however, affect the speed and efficiency of the motor.

4. *Series speed-torque leads to stability in control.* This is so because in case the driving torque happens to be more than resisting torque, speed of the motor will increase. Increase in speed will lead to falling off in the torque so that stable situation is rapidly restored.

5. The *commutating properties of series motor are quite good.* With the increase in the armature current as a result of heavy load torque results in the decrease of armature speed, this reduces the magnitude of e.m.f. induced in the coils undergoing commutation which helps in achieving sparkless collection of current.

6. Series motor of a given rating is *capable of withstanding more heavy load torque.*

7. Series motor to *some extent has self relieving property.*

8. The speed of series motor can be controlled by various methods (as discussed in chapter 5).

9. Basically, series motor is *not* suitable for adopting it for regenerative braking as it suffers from electrical instability unless certain measures are taken.

- Owing to *low weight and high starting torque developed, D.C. series motor are capable of producing high rates of acceleration.* As such D.C. series motors are *better suited for urban and suburban services.*
- In *1500 V D.C. system* whilst motors are insulated for 1500 V, they may operate at 1500 V or at 750 V by connecting two motors permanently in series. *In 3000 V D.C. system*, motors are insulated for 3000 V but operated at 1500 V by connecting two in series.

Drawbacks :

- The drawbacks/shortcomings of series motors arise first from *commutator which restricts speed, current and voltage.* There is a *risk of flashover and the brush gear requires considerable maintenance.*

- Because of the windings on the rotor there are *limitations to peripheral speed* arising from centrifugal force even if restrictions due to the commutator could be overcome.

8.4.4. Series Motors—Parallel Operation with Unequal Wheel Diameters

Consider two identical motors of series type, say of speed-current characteristic shown in Fig. 8.3. For a given speed of vehicle, the peripheral speed of all the driving wheels will be same, but should there be a slight difference in diameter of the driving wheels (which is possible *on account of difference in wear*) there will be a corresponding difference in the speeds of the driving axles to which they are connected and also, therefore, a difference in speed of the motors. The speed of the motors will be given by the relation : $\frac{N_A}{N_B} = \frac{D_B}{D_A}$

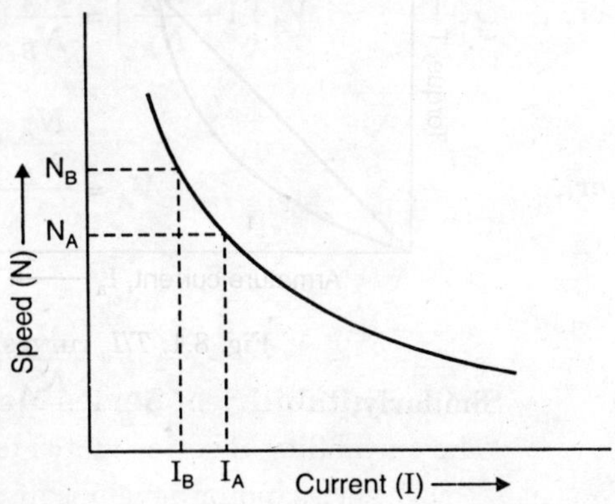

Fig. 8.3

When the motors are working in parallel the *current drawn by each motor* (determined speed-current characteristic) due to slight difference in speed *does not differ much.* As voltage across each motor is same, therefore, motors *share the load almost equally.*

8.4.5. Series Motors—Series Operation with Unequal Diameters

When the series motors are working in series, the current drawn by each motor will be same but due to slight difference in speeds (due to unequal wheel diameters) the back e.m.fs will differ slightly. This will result in slight difference in voltage acting across each motor, and also therefore slight difference in load shared by each motor.

Let the motors *A* and *B* be identical having armature resistance *R* in series, as shown in Fig. 8.4. Since they are in series, the same current *I* will flow through them. But due to unequal wheel diameters (*i.e.*, slight difference in speeds), they will deliver different loads *i.e.*, *voltage across each will be different.*

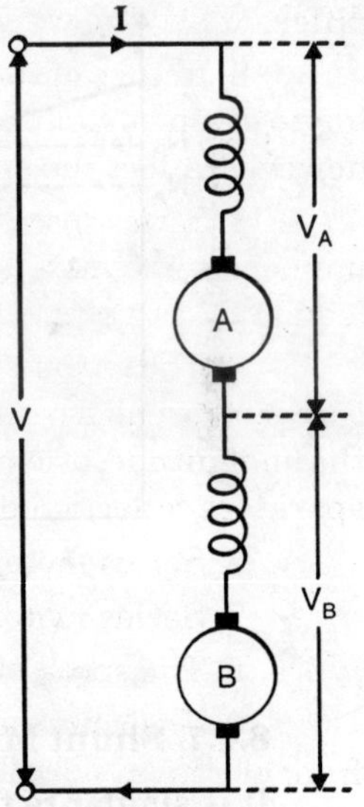

Fig. 8.4. Series operation of series motors.

$$V = V_A + V_B, \quad \text{and} \quad N \propto (V - IR)$$

$$\therefore \quad \frac{N_A}{N_B} = \frac{V_A - IR}{V_B - IR}$$

$$\therefore \quad V_A - IR = \frac{N_A}{N_B}(V_B - IR)$$

or, $$V_A = \frac{N_A}{N_B}(V_B - IR) + IR$$

or, $$V_A = \frac{N_A}{N_B}(V - V_A - IR) + IR$$

or, $$V_A = \frac{N_A}{N_B}(V - IR) + IR - \frac{N_A}{N_B} V_A$$

or, $$V_A + \frac{N_A}{N_B} V_A = \frac{N_A}{N_B}(V - IR) + IR$$

or, $$V_A\left(1 + \frac{N_A}{N_B}\right) = \frac{N_A}{N_B}(V - IR) + IR$$

or, $$V_A = \frac{\frac{N_A}{N_B}(V - IR) + IR}{1 + \frac{N_A}{N_B}} \quad \text{...(8.6)}$$

Similarly, $$V_B = \frac{\frac{N_B}{N_A}(V - IR) + IR}{1 + \frac{N_B}{N_A}} \quad \text{...(8.7)}$$

8.4.6. Shunt Motors—Parallel Operation with Unequal Wheel Diameters

Refer to Fig. 8.5. A small difference in speeds of two motors, *causes motors to be loaded very unequally due to flat speed current curve of D.C. shunt motor.*

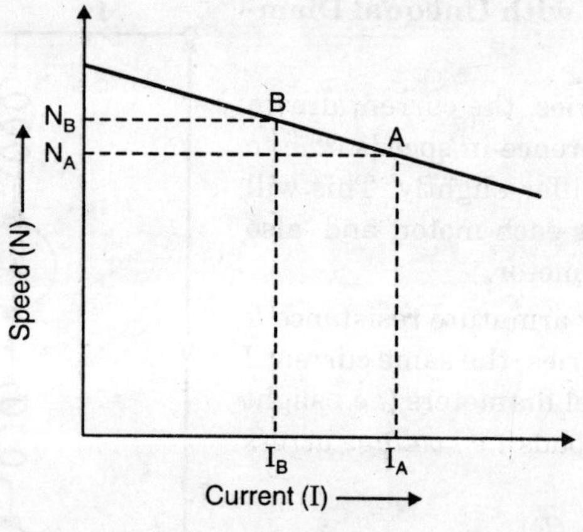

Fig. 8.5

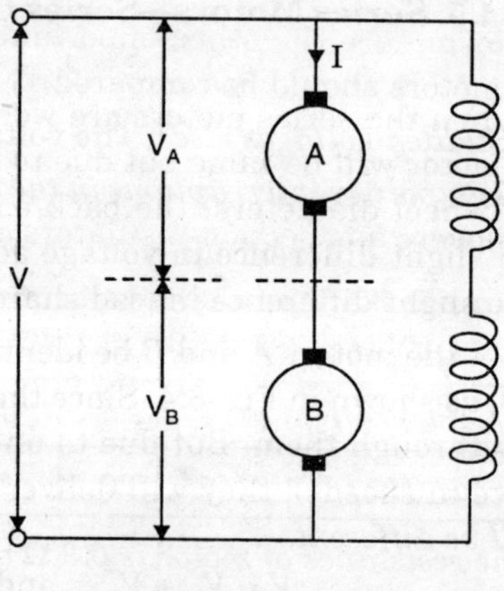

Fig. 8.6

8.4.7. Shunt Motors—Series Operation with Unequal Wheel Diameters

It is similar to the case of series operation of series motors, and hence the same equation(s) hold(s) good.

8.4.8. Parallel Operation, Equal Wheel Diameters, but Dissimilar Speed—Current Curves

Refer to Figs. 8.7 and 8.8. As the diameters of the driving wheels are equal, for the particular speed of the car, the r.p.m. of both the motors will be same. If the *characteristics of the two motors are slightly different,* it can be seen that :

(*i*) In case of *series motor* the loading of the two motors is *slightly different ;*

(*ii*) For *shunt motors* the loading of the motors is *materially different.*

For traction purposes, it can be concluded from the above analysis that, the *permissible deviation from the standard speed-current curves and the difference in the diameters of driving wheels, in case of shunt motors are much smaller as compared to series motor.* However, these restrictions are *not* applicable as rigorously for D.C. compound motors and they are *usually preferred for main line service on undulating routes where the chances of wear and tear of wheel are more.*

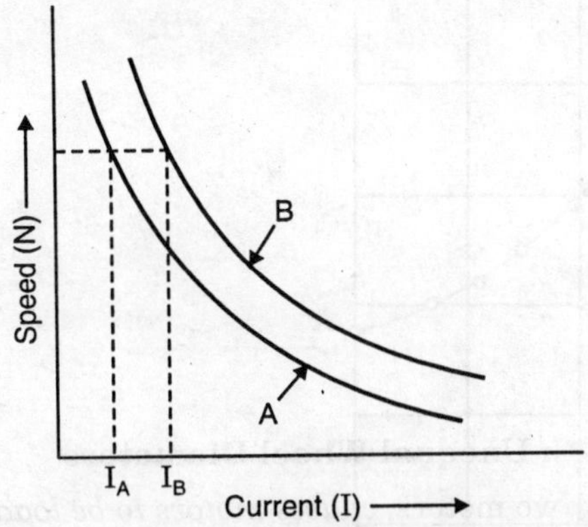

Fig. 8.7. Series motor

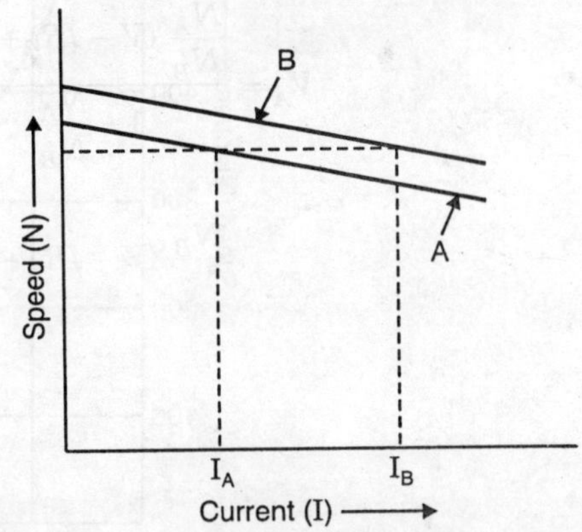

Fig. 8.8. Shunt motor.

Following are the other important considerations against which the performance of series and shunt motors should be compared :

(*i*) *Sudden voltages rises.* The voltage rise may occur due to opening of a circuit.

(*ii*) *Temporary interruptions in the supply.* Temporary interruption of supply to motors occurs when cross-overs and section-insulators are crossed with the controller 'ON'.

Example 8.1. *For a series traction motor following torque-armature current characteristics are given :*

Armature current (A)	5	10	15	20	25	30	35	40
Torque (Nm)	20	50	100	155	215	290	360	430

If the resistance of motor is 0.3 Ω and the motor is connected across 230 V, deduce the speed-armature current characteristics. **(Nagpur University)**

Solution. *Given :* Supply voltage, $V = 230$ V ; Total resistance of series motor, $R_m = R_a + R_{se} = 0.3\ \Omega$.

Armature current, I_a in amperes	5	10	15	20	25	30	35	40
Torque, T in Nm	20	50	100	155	215	290	360	430
Back e.m.f. $E_b = (V - I_a R_m)$ in volts	228.5	227.0	225.5	224.0	222.5	221.0	219.5	218.0
Speed, N $\frac{9.55\,(V - I_a R_m)}{T} I_a$ in r.p.m.	545	434	323	276	247	218	204	194

The speed-armature current (N/I_a) characteristics are shown in Fig. 8.9.

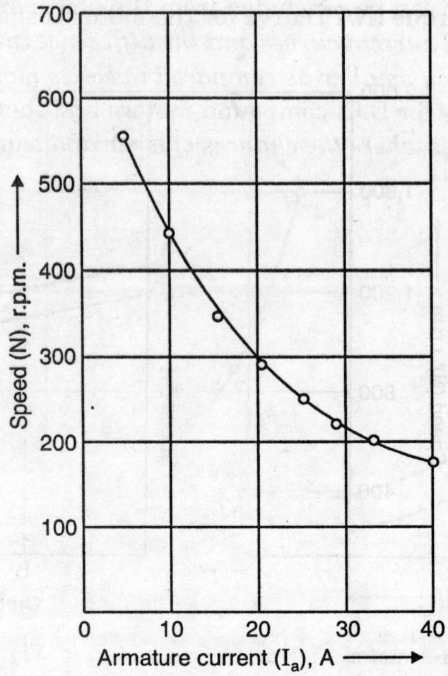

Fig. 8.9 N/I_a characteristics.

Example 8.2. *The following data relate to the magnetisation curve of D.C. series motor when working as a separately excited generator at 600 r.p.m.*

Field current (A)	*20*	*40*	*60*	*80*
E.m.f. (V)	*215*	*381*	*485*	*550*

Deduce the speed-torque curve for this motor when operating at 600 V. The total resistance of the motor is 0.8 Ω. **(Gorakhpur University)**

Solution. *Given* : Supply voltage = 600 V ; $R_m = R_a + R_{se} = 0.8\ \Omega$; $N_1 = 600$ r.p.m.

Field current (amperes)	20	40	60	80
Back e.m.f., E_{b1} (volts) at speed N_1 (600 r.p.m.) = e.m.f. generated by the armature (given)	215	381	485	550
Back e.m.f., E_b (volts) at speed N (to be determined) = $V - I_aR_m$	584	568	552	536
Speed, N (to be determined in r.p.m.) = $N_1/E_{b1} \times (V - I_aR_m)$	1630	895	683	585
Torque, $T = \dfrac{9.55\,(V - I_aR_m)I_R}{N}$ Nm	68.4	240	462	700

The deduced speed-torque (*N/T*) curve for the motor is shown in Fig. 8.10.

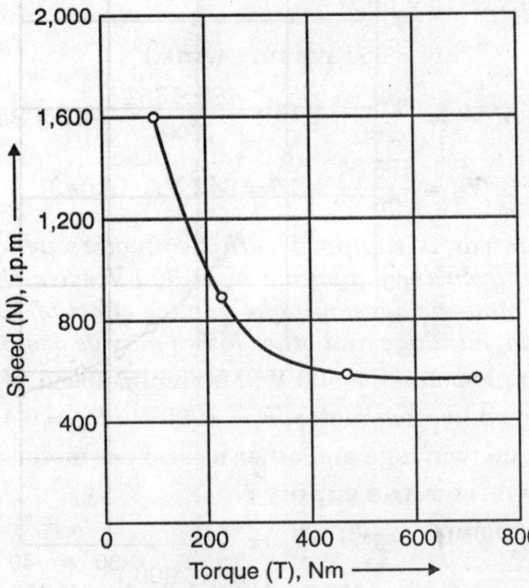

Fig. 8.10. *N/T* curve.

Example 8.3. *Two D.C. traction motors run at speeds 700 r.p.m. and 750 r.p.m. respectively when each takes a current of 50 A from 500 V mains. Each motor has an effective resistance of 0.3 Ω. Calculate the speed and voltage across each machine when mechanically coupled and electrically insulated in series and taking a current of 50 A from 500 V mains, the resistance of each motor being unchanged.* **(A.M.I.E.)**

Solution. *Given* : N_A = 700 r.p.m. ; N_B = 750 r.p.m. ; $I_A = I_B$ = 50 A ;

Supply voltage = 500 V ; R_m = 0.3 Ω (each motor)

Speed (N) and voltage across each machine (V_A, V_B) :

Back e.m.f. of motor *A* when taking a current of 50 A,

$$E_{bA} = V - IR_m = 500 - 50 \times 0.3 = 485 \text{ V}$$

Similarly, $$E_{bB} = V - IR_m = 500 - 50 \times 0.3 = 485 \text{ V}$$

When the machines are mechanically coupled and connected in series, the speed of each motor will be same, say *N*, current will be same and equal to 50 A (given) and the sum of voltage across the two motors will be equal to 500 V.

Let the voltages across motors *A* and *B* be V_A and V_B respectively.

Now, $$V_A + V_B = 500 \quad ...(i)$$

Back e.m.f. of motor *A*, $$E'_{bA} = E_{bA} \times \frac{N}{N_A} = 485 \times \frac{N}{700}$$

Voltage across motor *A*, $$V_A = E'_{bA} + IR_m = \frac{485}{700} N + 50 \times 0.3 = \frac{485}{700} N + 15$$

Back e.m.f. of motor *B*, $$E'_{bB} = E'_{bB} \times \frac{N}{N_B} = 485 \times \frac{N}{750}$$

Voltage across motor *B*, $$V_B = E'_{bB} + IR_m = \frac{485}{700} N + 50 \times 0.3 = \frac{485}{700} N + 15$$

Substituting the values of V_A and V_B in (*i*), we get

$$\frac{485}{700}N + 15 + \frac{485}{750}N + 15 = 500$$

or, **N = 351 r.p.m. (Ans.)**

Voltage across motor A, $V_A = \frac{485\,N}{700} + 15 + \frac{485 \times 351}{700} + 15 =$ **258 V. (Ans.)**

Voltage across motor B, $V_B = \frac{485\,N}{750} + 15 =$ **242 V. (Ans.)**

Example 8.4. *A tram car is equipped with two motors which are operating in parallel. Calculate the current drawn from the supply main at 500 V when the car is running at a steady speed of 45 km/h and each motor is developing a tractive effort of 1900 N. The resistance of each motor is 0.4 ohm. The friction, windage and other losses may be assumed as 3100 watts per motor.*

Solution. *Given* : Supply voltage = 500 V ; Maximum speed, V_{max} = 45 km/h.

Tractive effort developed by each motor, F_t = 1900 N ; R_m = 0.4 Ω (each motor) ;

Constant losses (friction, windage and other losses) per motor = 3100 W.

Total Current drawn from the supply :

Power output of each motor $= F_t \times V_{max}$

$$= 1900 \times \left(\frac{45 \times 1000}{3600}\right) = 23750 \text{ W}$$

Copper losses per motor $= I^2R_m = 0.4I^2$

... where I is the current drawn from the supply

∴ Input to each motor = Motor output + constant losses + copper losses

$$VI = 23750 + 3100 + 0.4I^2$$

or, $$500I = 23750 + 3100 + 0.4I^2$$

or, $$0.4I^2 - 500I + 26850 = 0$$

∴ $$I = \frac{500 \pm \sqrt{(500)^2 - 4 \times 0.4 \times 26850}}{0.8} = \frac{500 \pm 455}{0.8}$$

$$= 1193.75 \text{ A}, 56.25 \text{ A}$$

Current drawn by each motor = 56.25 A (since current 1193.75 A being unreasonably high can not be accepted)

∴ *Total current drawn from the supply* = 2 × 56.25 = **112.5 A. (Ans.)**

Example 8.5. *A motor coach is being driven by two identical series motors. First motor is geared to driving wheel having diameter of 90 cm and other motor to driving wheel having diameter of 86 cm. The speed of the first motor is 500 r.p.m. when connected in parallel with the other across 600 V supply. Find the motor speeds when connected in series across the same supply. Assume armature current to remain same and armature voltage drop of 10 percent at this current.* **(A.M.I.E.)**

Solution. D_1 = 90 cm or 0.9 m ; D_2 = 86 cm or 0.86 m ; N_1 = 500 r.p.m. ;

Supply voltage = 600 V ; Armature voltage drop = 10%.

Motor speeds when connected in series :

Back e.m.f. of the first motor, $E_{b1} = 600 - \frac{10}{100} \times 600 = 540$ V

When the motors are connected in series across 600 V supply, as shown in Fig. 8.11, let the supply voltage across motors I and II be V_1 and V_2 volts and speeds N'_1 and N'_2 respectively.

Since speed, $N \propto \frac{V - IR}{\phi}$, current through the motors remains the same, therefore flux produced by it also remains the same and as such $N \propto (V - IR)$.

$$\therefore \quad \frac{N'_1}{N'_2} = \frac{V_1 - IR}{V_2 - IR} = \frac{V_1 - \frac{10}{100} \times 600}{V_2 - \frac{10}{100} \times 600} = \frac{V_1 - 60}{V_2 - 60} \quad ...(i)$$

Also, $N'_1 D_1 = N'_2 D_2$ ... Since peripheral speeds are equal

$$\therefore \quad \frac{N_1'}{N_2'} = \frac{D_2}{D_1} = \frac{0.86}{0.9} = 0.956 \quad ...(ii)$$

From (*i*) and (*ii*), we get

$$\frac{V_1 - 60}{V_2 - 60} = 0.956$$

or, $V_1 - 60 = 0.956\,(V_2 - 60) = 0.956 V_2 - 57.36$

or, $V_1 - 0.956\, V_2 = 2.64$ ...(*iii*)

Also, $V_1 + V_2 = 600$ ...(*iv*)

Solving (*iii*) and (*iv*), we get

$$V_2 = 305.4 \text{ V} ;\ V_1 = 294.6 \text{ V}$$

The speeds of the motors can be calculated as follows :

$$\frac{N'_1}{N_1} = \frac{E'_{b1}}{E_{b1}}$$

or, $\mathbf{N'_1} = N_1 \times \frac{E'_{b1}}{E_{b1}} = 500 \times \frac{294.6 - 60}{600 - 60}$

$= \mathbf{217.2\ r.p.m.}$ **(Ans.)**

and, $\mathbf{N'_2} = N'_1 \times \frac{D_1}{D_2} = 217.2 \times \frac{0.9}{0.86}$

$= \mathbf{227.3\ r.p.m.}$ **(Ans.)**

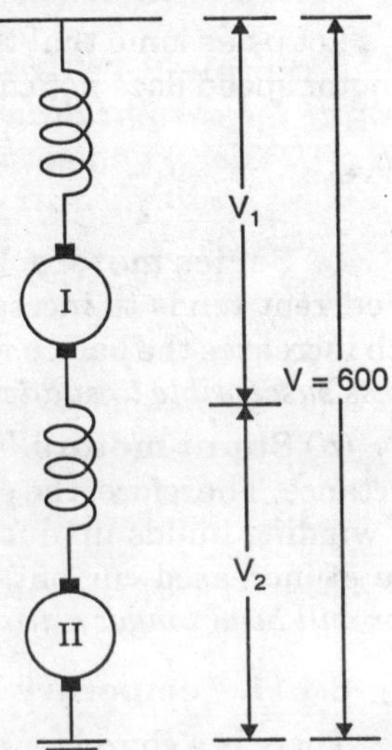

Fig. 8.11. Motors connected in series.

8.4.9. Tractive Effort and Horse Power

In the case of D.C. series motor, since flux is proportional to the current, we have,

Speed $N \propto \frac{V}{\phi} \propto \frac{V}{I} \propto \frac{V}{I} \propto \frac{1}{I}$ when V is constant at 1.00 p.u.

and, Torque, $T \propto \phi I \propto I^2$

or, $I \propto \sqrt{T}$

$$\therefore \quad N \propto \frac{1}{\sqrt{T}} \quad ...(8.8)$$

The horse power, H.P. (or power P) $\propto TN \propto T \times \frac{1}{\sqrt{T}} \propto \sqrt{T}$ ...(8.9)

Thus, the *power* taken from the supply is *proportional to the square root of torque.*

— The *same equation holds good for the A.C. series motor.*

— In case of D.C. shunt motor or 3-*phase induction motor* where the *speed is fairly constant,*

$$\text{H.P.} \propto TN \propto T \quad ...(8.10)$$

From the above analysis it can be seen that *series motor is ideally suited for traction service due to the following reasons :*

1. It can give a high starting torque.
2. It has a high free-running speed.
3. The speed automatically decreases as the torque increases.
4. For the same torque, the power and therefore the current drawn from the supply is more in the case of shunt motor than series motor.

8.4.10. Effect of Sudden Change in Supply Voltage

Let us assume that the change (say increase) in voltage is so sudden that during that period the motor speed does not change because of electro-magnetic inertia, then armature current,

$$I_a = \frac{V - E_b}{R_a} .$$

(*i*) **Series motors.** In series motors, with the sudden increase in supply voltage, the armature current tends to increase. But with the increase in armature current the flux also increases which increases the back e.m.f. E_b and thus bring current back to its initial value. Thus *series motors are less susceptible to sudden change in supply voltage.*

(*ii*) **Shunt motors.** The rate of change of current in the field winding is slow due to its high inductance. Therefore, the current increases due to increase in supply voltage. As the current in the field winding builds up due to increase in supply voltage, the back e.m.f. increases and hence the value of increased current starts decreasing. Consequently, the *current in-rush in case of shunt motor will be of longer duration and higher initial value as compared to series motor.*

8.4.11. Temporary Interruption of Supply

There is a characteristic difference in the operation of D.C. series and shunt motors when the supply is interrupted briefly and is restored at full value.

— Owing to the high time-constant of field winding of *shunt motors* and because it is connected across the armature, the back e.m.f. *during interruption* in maintained even though it is decreasing, but in case of *series motor* back e.m.f. ceases to exist.

— Now when the *supply voltage is restored*, the in-rush current in the shunt motor will be determined by the resultant voltage in the circuit and the impedance of the circuit whereas for the series motor, it will depend only on the impedance of the rotor circuit and will be relatively much higher as compared to shunt motor.

Example 8.6. *A locomotive exerts a tractive effort of 34 kN in hauling a train at 45 kmph on the level track. If the locomotive has to haul the same train on a gradient and the effort required is 54 kN, determine the H.P. delivered by the locomotive when the motors used are :*

(*i*) *D.C. series motors ;* (*ii*) *Induction motors.*

Solution. *Given* : Tractive effort, F_t = 34 kN ; V = 45 kmph ; F_t = 54 kN (second case)

H.P. required on level track

$$= F_t \times V = (34 \times 1000) \times \left(\frac{45 \times 1000}{60 \times 60}\right) = 425000 \text{ W}$$

$$= \frac{425000}{735.5} \text{ H.P.} = 577.84 \text{ H.P. (metric)}$$

(*i*) **D.C. motors :**

For D.C. series motor H.P. $\propto \sqrt{T}$

$\therefore$ H.P. required on upgradient track $= 577.84 \sqrt{\frac{54}{34}} =$ **728.22 H.P. (Ans.)**

(*ii*) **Induction motors :**

For a 3-phase induction motor H.P. $\propto T$

$\therefore$ H.P. required on upgradient track $= 577.84 \times \frac{54}{34} =$ **917.74 H.P. (Ans.)**

Example 8.7. *The following date refer to the speed-current and tractive effort-current characteristics of a D.C. series traction motor :*

Current (amps.)	*50*	*100*	*150*	*200*	*250*
Speed (kmph)	*73.5*	*48.0*	*41.0*	*37.2*	*35.0*
Tractive effort (N)	*1500*	*5250*	*9300*	*13350*	*17500*

The diameter of the car wheel is one metre and the gear ratio is 3 : 1. The car wheel is changed to 1.05 metres and the gear ratio 3.5 : 1. Give the new characteristics. **(A.M.I.E.)**

Solution. Diameter of car wheel, $D_1 = 1$ m ; Diameter of new wheels, $D_2 = 1.05$ m ;

Gear ratio, $\gamma_1 = 3 : 1$; $\gamma_2 = 3.5 : 1$.

New speed, $$V_2 = \frac{D_2}{D_1} \times \frac{\gamma_1}{\gamma_2} V_1 = \frac{1.05}{1} \times \frac{3}{3.5} V_1 = 0.9\, V_1$$

New tractive effort, $$F_{t2} = \frac{D_1}{D_2} \times \frac{\gamma_2}{\gamma_1} F_{t1} = \frac{1}{1.05} \times \frac{3.5}{3} F_{t1} = \frac{F_{t1}}{0.9}$$

New characteristics are :

Current (amps)	50	100	150	200	250
Speed (kmph) $V_2 = 0.9V_1$	66.15	43.2	36.9	33.48	31.5
Tractive effort (N), $F_{t2} = \frac{F_{t1}}{0.9}$	1666.6	5833.3	10333.3	14833.3	19444.4

Example 8.8. *Give a comparison of D.C. series and shunt motors with reference to their use in electric traction. The characteristics of a D.C. series motor when operating at 600 V and driving 700 cm wheels are as follows :*

Current (amps.)	*500*	*400*	*300*	*200*
Train speed (km/h)	*35.6*	*38.5*	*42.5*	*50.5*
Tractive effort (kg)	*2580*	*1960*	*1380*	*725*

Two of such motors are mounted on a motor coach. The wheels driven by one motor A are 107 cm in dia. and other motor B are 102 cm in dia.

Calculate the power input and tractive efforts of each motor when :

(*i*) *Motors are connected in parallel and train speed is 40 km/h ;*

(*ii*) *Motors are in series and current is 400 A.*

Assume resistance of each motor as 0.08 Ω. **(Panjab University)**

Solution. The characteristics of a D.C. series motor when operated at 600 V and driving 107 cm diameter wheels are drawn from the given data, as shown in Fig. 8.12.

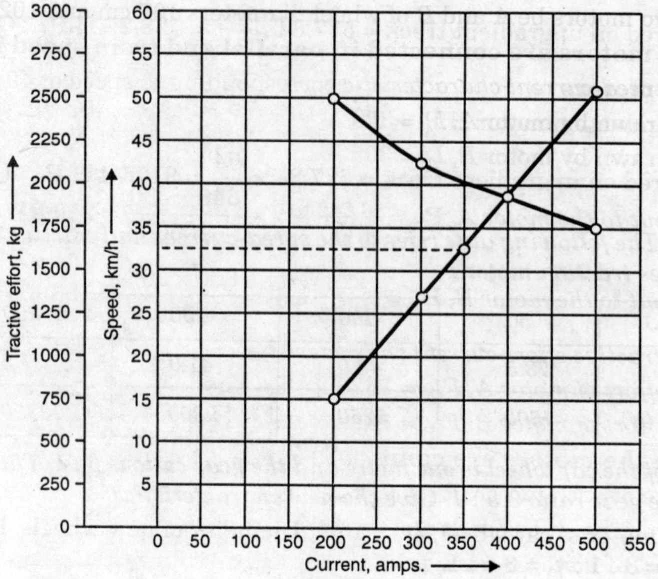

Fig. 8.12

- **The characteristics of a D.C. series motor when operated at 600 V and driving 102 cm diameter wheel are derived as follows and drawn, as shown in Fig. 8.13.**

Current (amps.)	500	400	300	200
Train speed (km/h), $V_2 = \frac{V_1 \times 1.02}{1.07}$	33.9	36.7	40.5	48.2
Tractive effort (kg), $F_{t2} = F_{t1} \times \frac{1.07}{1.02}$	2706	2056	1448	760

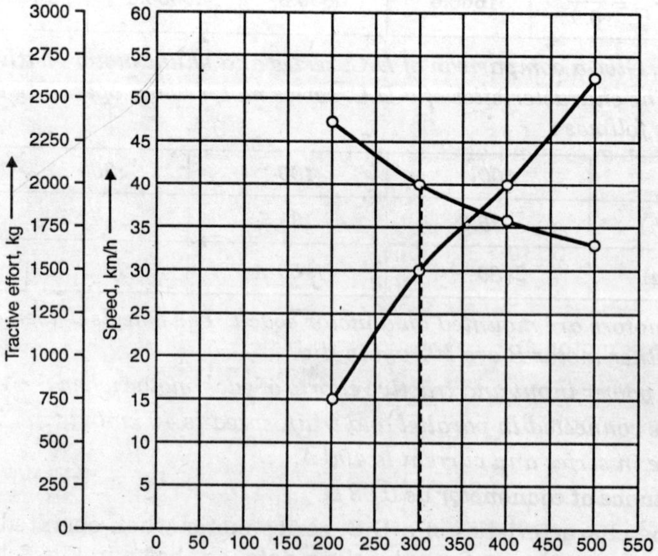

Fig. 8.13

Let the two motors be A and B of wheel diameters 107 cm and 102 cm respectively.

(*i*) **When motors are connected in parallel and train speed is 40 km/h :**

(*a*) From *speed-current characteristic* corresponding to speed of 40 km/h,

Current drawn by motor A, $I_A = 350$ A

Current drawn by motor B, $I_B = 305$ A

Power input to the motor A, $\mathbf{P_A} = \frac{VI_A}{1000}$ kW $= \frac{600 \times 350}{1000}$ = **210 kW. (Ans.)**

Power input to the motor B, $\mathbf{P_B} = \frac{VI_B}{1000}$ kW $= \frac{600 \times 305}{1000}$ = **183 kW. (Ans.)**

(*b*) From *tractive effort-current characteristics*,

Tractive effort of motor A, F_{tA} = **1628 kg. (Ans.)**

Tractive effort of motor B, F_{tB} = **1490 kg. (Ans.)**

(*ii*) **When the motors are connected in series and current is 400 A :**

Resistance of each motor, $R_m = 0.08\ \Omega$

Applied voltage across each motor, $V = 600$ volts

Back e.m.f. of motor A when taking current of 400 A and at a speed of 38.5 km/h,

$$E_{bA} = 600 - 400 \times 0.08 = 568 \text{ V}$$

Back e.m.f. of motor B when taking current of 400 A at a speed of 36.7 km/h,

$$E_{bB} = 600 - 400 \times 0.08 = 568 \text{ V}$$

After motors being connected in series, let the peripheral speed be v km/h, which will be same for both motors.

New back e.m.f. of motor A, $\quad E'_{bA} = E'_{bA} \times \frac{v}{v_A} = 568 \times \frac{v}{38.5}$

$\therefore$ Voltage across motor A, $\quad V_A = E'_{bA} + IR_m$

$$= 568 \times \frac{v}{38.5} + 400 \times 0.08 = \frac{568v}{38.5} + 32$$

Back e.m.f. of motor B, $\quad E'_{bB} = \frac{568}{36.7} \times v$

$\therefore$ Voltage across motor B, $\quad V_B = E'_{bB} + IR_m = \frac{568}{36.7} v + 32$

Also, $\quad V_A + V_B = 600$

$$\therefore \quad \frac{568v}{38.5} + 32 + \frac{568v}{36.7} + 32 = 600 \quad \text{or} \quad v = 17.7 \text{ km/h}$$

P.D. across motor A, $\quad V_A = \frac{568}{38.5} \times 17.7 + 32 = 294$ V

P.D. across motor B, $\quad V_B = \frac{568}{36.7} \times 17.7 + 32 = 306$ V

(*a*) Power input to motor A, $\quad P_A = \frac{V_A \times I}{1000}$ kW $= \frac{294 \times 400}{1000} = 117.6$ kW

Power input to motor B, $\quad P_B = \frac{V_B \times I}{1000}$ kW $= \frac{306 \times 400}{1000} = 122.4$ kW

(*b*) From *tractive effort-current characteristic* corresponding 40 A current,

Tractive effort of motor A, F_{tA} = **1965 kg. (Ans.)**

Tractive effort of motor B, F_{tB} = **2065 kg. (Ans.)**

8.5. A.C. SERIES MOTOR

The series motor due to its desirable speed-torque characteristics *is almost exclusively used in railway service.* While the D.C. motor is entirely satisfactory for this class of work service and is generally used on street railway cars and trolley coaches, the fact that it is more convenient and more economical to transmit power and to transform voltages in A.C. systems than with direct currents has lead to the development of the A.C. series motor for use on some of the important steam-road electrifications.

Working principle. The working principle of an A.C. series motor is the same as that of the D.C. series motor. The armature and field are wound and interconnected in the same manner as the D.C. series motor.

When an alternating e.m.f. is applied to the terminals, since field and armature windings are connected in series, the field flux and armature current reverse simultaneously every half cycle, but the direction of the torque remains unchanged. The torque is pulsating, but its average value is equal to that which a D.C. motor will develop if it had the same r.m.s. value of flux and current. Motor connections, direction of torque, etc. for two successive half cycles are shown in Fig. 8.14. If the field and armature core are run at low saturation, the air-gap flux is approximately proportional to the current and the torque is approximately proportional to current squared.

Although it is theoretically possible to operate a D.C. series motor from an A.C. circuit, the following *structural changes must be made in the motor to make it a practical and reasonable efficient machine :*

- The entire magnetic circuit must be *laminated*, and materials with low iron-loss coefficients should be used as in transformers.

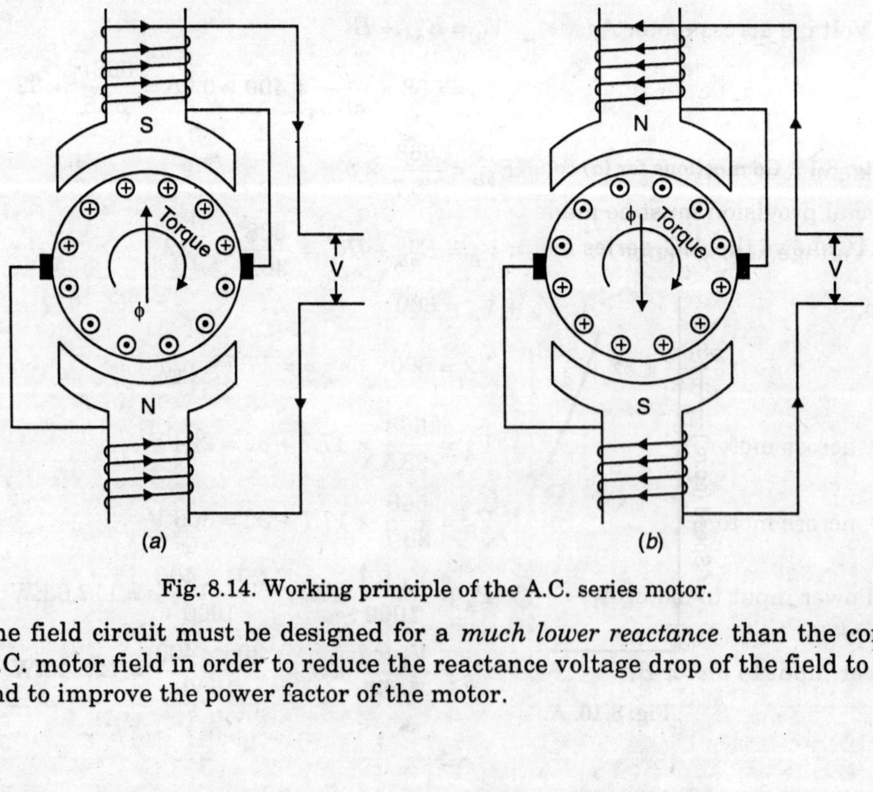

Fig. 8.14. Working principle of the A.C. series motor.

- The field circuit must be designed for a *much lower reactance* than the corresponding D.C. motor field in order to reduce the reactance voltage drop of the field to a minimum and to improve the power factor of the motor.

- A *distributed compensating winding* is required to reduce the reactance of the armature winding by reducing the leakage flux and to neutralize the cross-magnetising effect of the armature ampere turns.

The compensating winding may be connected in series with the series-field and armature windings, or it may be short-circuited upon itself and receive its excitation voltage by transformer action, since it is inductively coupled with the armature cross-field (Fig. 8.15). In the first case, the motor is said to be *conductively* compensated, while in the second it is *inductively* compensated. *Conductive compensation* is required on motors which are intended for operation in *D.C. as well as A.C. circuits.*

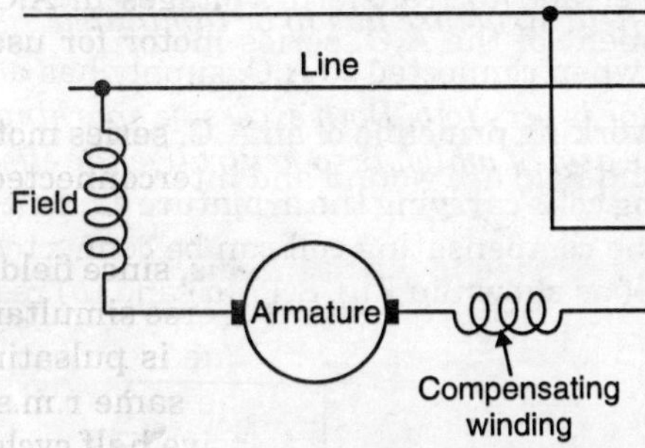

(*a*)

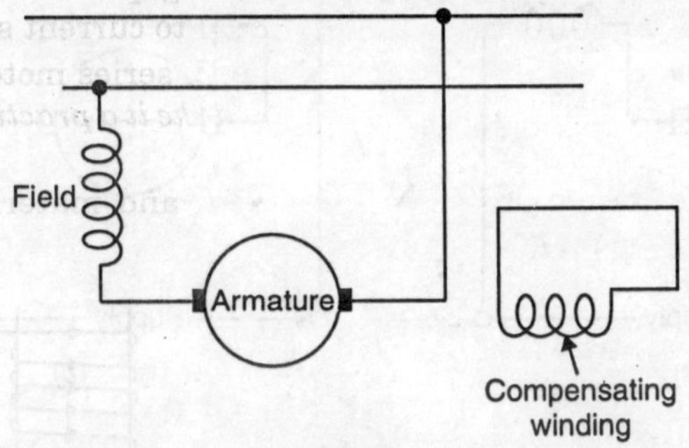

(*b*)

Fig. 8.15. Connections for (*a*) conductive compensation ; (*b*) inductive compensation.

- Special provision must be made to secure *satisfactory commutation.*

Fig. 8.16 shows the A.C. series motor characteristics.

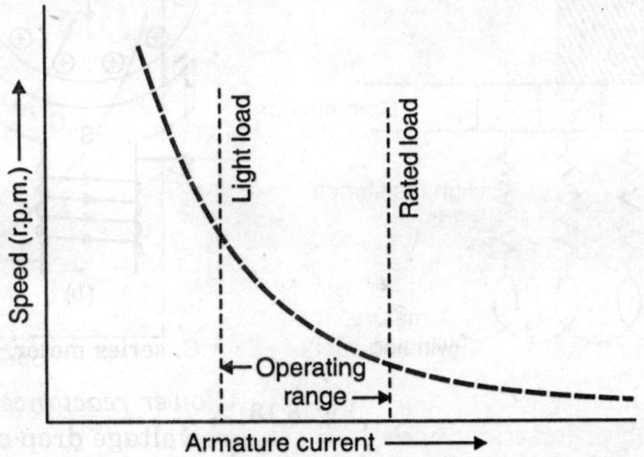

Fig. 8.16. A.C. series motor characteristics.

Example 8.9. *Mention the problems usually encountered when a D.C. series motor is operated on A.C. What design modifications are to be incorporated for its satisfactory operation on A.C. ? Compare the speed for D.C. and A.C. operations.*

Solution. If an alternating current is passed through an ordinary D.C. series motor, the field and armature currents reverse simultaneously every half cycle, so that a unidirectional torque is produced and the armature rotates. The motor has to be considerably modified for running it with alternating currents. The various operational difficulties and the required design modifications are given below :

1. With operation on alternating currents the *eddy current loss increases considerably*. Hence in *addition to the armature, the field structure has to be laminated.*

2. The D.C. series motor when connected to A.C. supply has a *much greater tendency to produce sparking.* Also it *power factor is very low. Both excessive sparking and poor factor are prevented by* **compensating** *the motor, the aim of which is to neutralize the effect of self induction.* The armature is compensated by employing coils carrying the armature current and placed in such a way as to neutralize the armature flux. The compensating coil can be connected in series with the main field or can be short circuited on itself as shown in Fig. 8.17 (*a*) and (*b*) respectively.

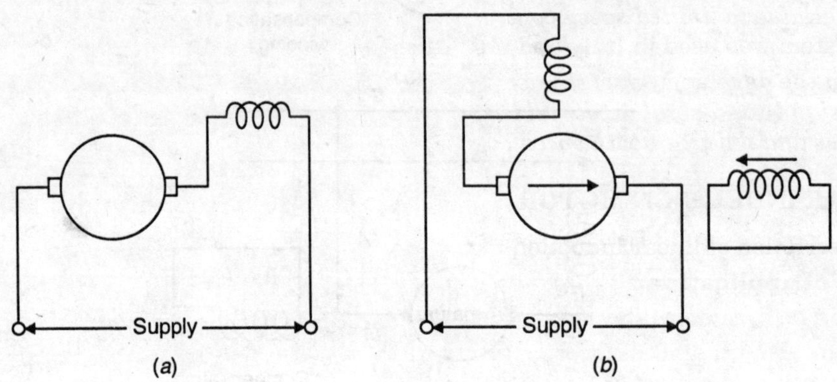

Fig. 8.17

3. Sparking at the brushes is further reduced by *using leads of high resistance between the windings and the commutator segments* as shown in Fig. 8.18.

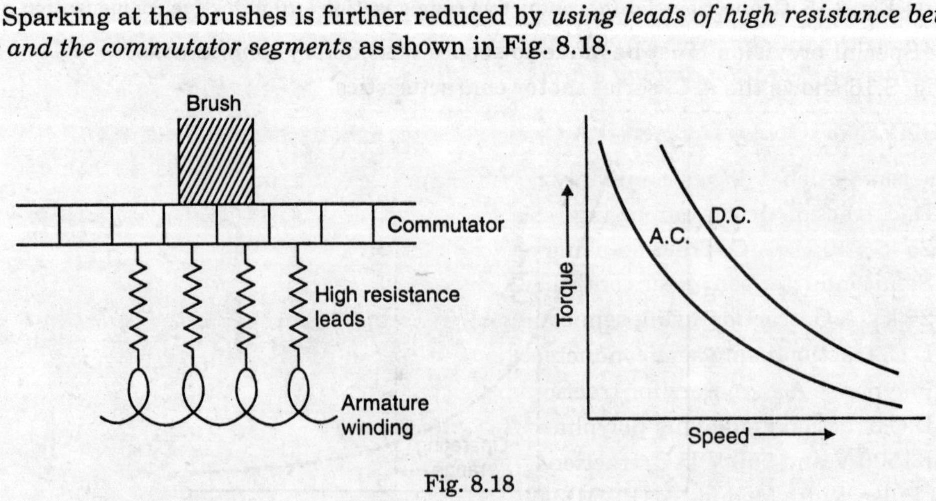

Fig. 8.18

4. *The air gap is kept small because of the weak main field.* The number of turns in the field winding has to be *kept low to reduce the reactance of the field winding.*

5. Because of commutation difficulties the operation of large A.C. motors is confined to *low frequencies* as $16\frac{2}{3}$ and 25 Hz.

Torque-speed characteristics for the series motor when run on A.C. and D.C. supply are shown in Fig. 8.18. The torque obtained with D.C. is somewhat higher than that obtained with A.C. supply.

8.6. THREE-PHASE INDUCTION MOTORS

Some of the factors which are *favourable* for the application of 3-phase induction motors *for traction purposes* are :

(*i*) Simple and robust construction.
(*ii*) Absence of commutator.
(*iii*) High efficiency.
(*iv*) Simple regenerative braking.

However, *due to limited speed control, this motor could not become popular* for *traction purposes.* This system was used in Italy and now is more or less replaced by 3000 V D.C. system.

- With the development of *thyristorised inverter circuits*, it has now been possible to *invert the supply and obtain a variable frequency which could be used for the 3-phase induction motor and a very smooth speed control can be obtained.*

8.7. LINEAR INDUCTION MOTOR

For description of linear induction motor refer Art. 5.13.1.22.

Scope of application :

- It can be *used on trolley cars for internal transport in workshops.*
- It can also be used as *booster accelerator* for moving heavy trains from rest or *up the inclines or on curves or as a propulsion unit* in *marshalling yards in place of shunting locomotives.*
- Linear induction motor has superiority over conventional motor for speeds over 200 km/h.
- This motor forms *excellent source of motive power for magnetically suspended trains when conventional motor which depends for torque conversion, to linear tractive force upon the adhesive weight on driving wheels, fails.*

8.8. CONVENTIONAL D.C. AND A.C. DRIVES USED IN INDIA

The conventional drives commonly used in India are enumerated below :

1. D.C. traction drives employing resistance control.
2. 25 kV, 50 Hz A.C. Traction using on-load transformer tap changer.
3. Semiconductor converter controlled drives.
4. 25 kV A.C. traction using semiconductor converter controlled D.C. motors.
5. D.C. traction using semiconductor chopper controlled D.C. motors.
6. Polyphase A.C. motors for traction drives.
7. D.C. traction employing polyphase A.C. motors.

For 1500 V and 750 V D.C. tractions, following *A.C. drives are widely used* :

(*i*) Pulse Width Modulated (PWM) Voltage Source Inverter (VSI) squirrel-cage induction motor drives ;
(*ii*) Load Commutated Inverter (LCI) synchronous Motor Drives.

8. A.C. traction employing polyphase A.C. motors.

Following *A.C. drives are widely used in 25 kV A.C. traction* :

(*i*) Current Source Inverter (CSI) Squirrel-Cage Induction Motor Drive ;

(*ii*) PWM Voltage Source Inverter (VSI) squirrel-cage Induction Motor (IM) Drive ;

(*iii*) Load Commutated Inverter (LCI) Synchronous Motor Drive.

9. Diesel electric traction :

(*i*) Diesel engine driven D.C. generator feeding D.C. series motors ;

(*ii*) Diesel engine driven three-phase alternator supplying D.C. motors ;

(*iii*) Diesel engine driven alternator feeding squirrel-cage induction motors.

Example 8.10. *Two three-phase locomotives are coupled to haul a heavy train. The motors of both the locomotives have a slip of 5 percent when the loco is delivering its full load tractive effort of 5750 kg. Loco-A has driving wheels of 1.27 metres in diameter whereas loco-B has 1.244 metres in diameter. Determine the tractive effort shared by the two locomotives if the total tractive effort is : (i) 12458 kg (ii) 1440 kg.*

How would the above tractive efforts have been shared, if in addition to the inequality of the driving wheel diameters, the loco-B had a slip of 4 percent instead of 5 percent ?

Solution. *Given* : $D_A = 1.27$ m ; $D_B = 1.244$ m

***Case I.* When both the locomotives have a slip of 5% :**

Let F_{tA} *be the tractive effort exerted by loco-A.*

(*i*) *Total tractive effort = 12458 kg :*

$$\%\text{age drop in speed of loco-}A = \frac{5F_{tA}}{5750}$$

$$\therefore \quad \%\text{ speed} = \left[100 - \frac{5F_{tA}}{5750}\right]$$

$$\%\text{ speed of loco-}B = \left[100 - \frac{5}{5750}(12458 - F_{tA})\right]$$

$$\text{Ratio of speed, } \frac{N_B}{N_A} = \frac{\left[100 - \frac{5}{5750}(12458 - F_{tA})\right]}{\left[100 - \frac{5F_{tA}}{5750}\right]} = 1.0209 \quad \text{...(1)}$$

$$\left(\because \frac{N_B}{N_A} = \frac{D_A}{D_B} = \frac{1.27}{1.244} = 1.0209\right)$$

or, $$\left[100 - \frac{5}{5750}(12458 - F_{tA})\right] = 1.0209\left[100 - \frac{5F_{tA}}{5750}\right]$$

or, $$100 - 10.833 + \frac{5F_{tA}}{5750} = 102.09 - \frac{5 \times 1.0209\, F_{tA}}{5750}$$

or, $$\frac{5F_{tA}}{5750} + \frac{5 \times 1.0209\, F_{tA}}{5750} = 102.09 - 100 + 10.833 = 12.923$$

or, $$F_{tA}\left(\frac{5}{5750} + \frac{5 \times 1.0209}{5750}\right) = 12.923$$

or, $$\mathbf{F_{tA} = 7354\ kg.} \quad \textbf{(Ans.)}$$

$\therefore$ Tractive effort shared by loco-*B*, $\mathbf{F_{tB}} = 12458 - F_{tA}$

$= 12458 - 7354 = \mathbf{5104\ kg.}$ **(Ans.)**

(ii) Total effort = 1440 kg :

Substituting 1440 kg in place of 12458 kg in eqn. (1), we have

$$\frac{\left[100 - \frac{5}{5750}(1440 - F_{tA})\right]}{\left[100 - \frac{5F_{tA}}{5750}\right]} = 1.0209$$

or, $$100 - 1.252 + \frac{5F_{tA}}{5750} = 102.09 - \frac{5 \times 1.0209\, F_{tA}}{5750}$$

$$\frac{2.0209 \times 5F_{tA}}{5750} = 102.09 - 100 + 1.252 = 3.342$$

or, $\mathbf{F_{tA} = 1901.8\ kg.}$ **(Ans.)**

$\therefore$ Tractive effort of loco-*B* $= 1440 - 1901.8 = \mathbf{-461.8\ kg.}$ **(Ans.)**

***Case II.* When loco-B has a slip of 4% instead of 5% :**

(i) Total tractive effort = 12458 kg :

Again let F_{tA} kg be the effort shared by loco-*A*.

$$\% \text{ speed of loco-}A = \left[100 - \frac{5F_{tA}}{5750}\right]$$

and, $$\% \text{ speed of loco-}B = \left[100 - \frac{4}{5750}(12458 - F_{tA})\right]$$

$$\therefore \quad \frac{N_B}{N_A} = \frac{\left[100 - \frac{4}{5750}(12458 - F_{tA})\right]}{\left[100 - \frac{5F_{tA}}{5750}\right]} = 1.0209 \quad ...(2)$$

or, $$100 - 8.666 + \frac{4F_{tA}}{5750} = 102.09 - \frac{5 \times 1.0209\, F_{tA}}{5750}$$

or, $$\frac{4F_{tA}}{5750} + \frac{5 \times 1.0209\, F_{tA}}{5750} = 102.09 - 100 + 8.666 = 10.756$$

or, $\mathbf{F_{tA} = 6793\ kg.}$ **(Ans.)**

(ii) Total effort = 1440 kg :

Substituting 1440 kg in place of 12458 kg in eqn. (2), we get

$$100 - \frac{4}{5750}(1440 - F_{tA}) = 1.0209\left(100 - \frac{5\,F_{tA}}{5750}\right)$$

or, $$100 - 1.002 + \frac{4\,F_{tA}}{5750} = 102.09 - \frac{5.1045\,F_{tA}}{5750}$$

or, $$\frac{4F_{tA}}{5750} + \frac{5.1045\,F_{tA}}{5750} = 102.09 - 100 + 1.002 = 3.092$$

or, $\mathbf{F_{tA} = 1952.8\ kg.}$ **(Ans.)**

$\therefore$ Tractive effort of loco-*B*, $F_{tB} = 1440 - 1952.8 = \mathbf{-512.8\ kg.}$ **(Ans.)**

- It may be noted that *while sharing a load of 12458 kg, loco-A is overload as compared to loco-B because of small difference in the diameters of the driving wheels.*

The significance of the loading of loco-*B* is explained as under :

This means that loco-*A* is exerting an effort more than the require, the balance (*–ve* loading of loco-*B*) is used to drive loco-*B* due to the motion of the train and thus loco-*B* is returning an equivalent amount of power back to the supply *i.e., loco-A is* **motoring** whereas *loco-B is* **generating.**

8.9. TRACTION MOTOR CONTROL

The traction motor control is required for starting without drawing excessive current from the supply for the following *purposes* :

(*i*) To provide smooth acceleration without sudden shock.

(*ii*) To avoid damage to couplings and inconvenience to the passengers.

(*iii*) To control speed, depending upon the type of service.

The control equipment may be operated manually by master controller or it may be fully automatic even though usually master controller is employed for the purpose.

8.9.1. Control of D.C. Motors

The starting current taken by a D.C. motor during its starting period is limited to a value approximately equal to the normal rated current by the resistance of starter. In the starting resistance there is a considerable loss of energy.

At the *instant of switching on the motor*, back e.m.f. $E_b = 0$

$$\therefore \quad \text{Supply voltage} = V = IR + V_{R_s} \qquad ...(8.1)$$

where, IR = Voltage drop in the armature, and

V_{R_s} = Voltage drop across the starting resistance.

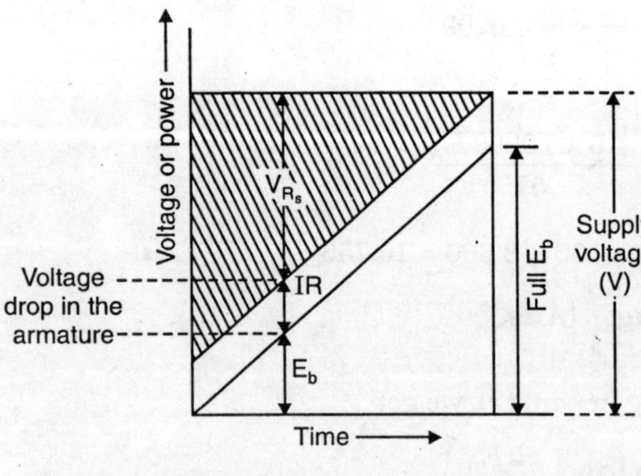

Fig. 8.19. Voltage during starting of a D.C. motor.

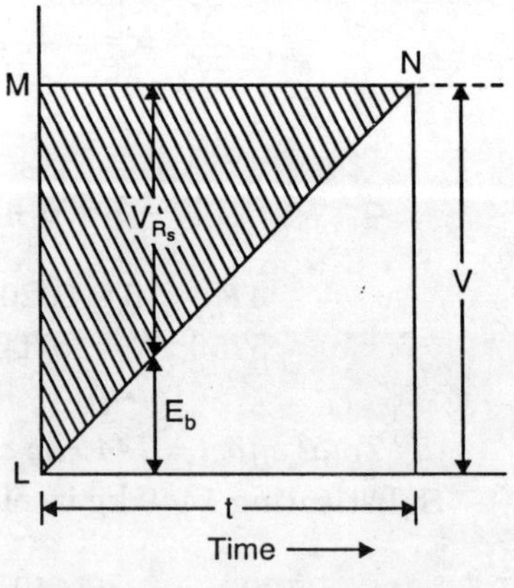

Fig. 8.20. Voltage during starting neglecting *IR* drop.

At *any other instant during starting* (Fig. 8.19),

$$V = IR + V_{R_s} + E_b \qquad ...(8.2)$$

At the end of accelerating period when total R_s (*i.e.* starting resistance) is *cut-off*,

$$V = E_b + IR \qquad ...(8.3)$$

If 't' is the time in sec. for starting and neglecting IR drop, *total energy supplied* = $V.I.t.$ watt-sec.

From Fig. 8.20, *energy wasted in* R_s

$$= \text{Area of } \Delta LMN \times I$$

$$= \frac{1}{2} \times t \times V \times I = \frac{1}{2} VIt \text{ watt-sec.}$$

This means half the energy (total energy supplied being $V.I.t.$) is wasted in starting.

$$\therefore \quad \eta_{starting} = 50 \text{ percent.}$$

8.9.1.1. Series-parallel control. This method is so named since this method of control assumes that at least two motors are being used which are *connected in series at start for low speed and in parallel for full speed running.*

In traction work, usually two or more similar motors are employed. Considerable saving in energy can be effected by employing series-parallel starting. Consider the use of two series motors. They are started in series with the help of a starting resistance till each of them develops a back e.m.f. equal to half the supply voltage minus the IR drop. The *motors give one running speed when they are in full series position.* The starting resistance is *again re-inserted* in the circuit and the *motors are switched in parallel. The starting resistance is cut out in steps and back e.m.f. of each motor develops from about half the value to the normal value.* In the full parallel position the motors give another running speed which is obviously *higher* than that when the motors are in full series.

Let us consider the case of two similar motors started by series-parallel method (Fig. 8.21).

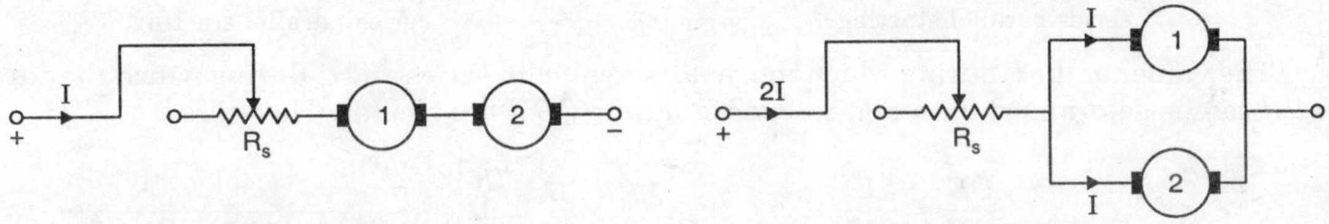

(*a*) Motors in series (*b*) Motors in parallel

Fig. 8.21. Series-parallel starting.

(*a*) **Series operation.** The two motors (1 and 2) are started in series with the help of R_s (Fig. 8.21 (*a*)). The current during starting is limited to normal rated current 'I' per motor. During series operation, current 'I' is drawn from supply. At the instant of starting $OA = AB = IR$ drop in each motor, OK = supply voltage V. The back e.m.fs of two motors jointly develop along OM as shown in Fig. 8.22. At point E, supply voltage V = Back e.m.fs of two motors + IR drops of two motors. Any point on the line BC (at any instant) represents the sum of back e.m.fs of two motors + IR drops of two motors + voltage cross resistance R_s of two motors. OE is the time taken for the series running (t_s). At point 'E' at the end of series running period, each motor has developed a back e.m.f. equal to $\frac{V - 2IR}{2} = \frac{V}{2} - IR.$

The back e.m.f. of each motor is represented by ordinate EL

$$= ED - LD = \left(\frac{V}{2} - IR\right)$$

(*b*) **Parallel operation.** At the instant E the motors are switched on in parallel, with R_s reinstated as shown in Fig. 8.21 (*b*). Current drawn is $2I$ from supply. Back e.m.f. across each motor $= EL$. So the back e.m.f. now develops along LG. At H when the motors are in full parallel, ($R_s = 0$ and both the motors are running at rated speed).

Supply voltage $= V = HF = HG + GF$

$=$ Normal back e.m.f. of each motor + IR drop in each motor.

To find t_s, t_p and η of starting :

Fig. 8.23 shows the current during the series and parallel starting periods. During series period OE, the current is I while during the parallel period EH it is $2I$.

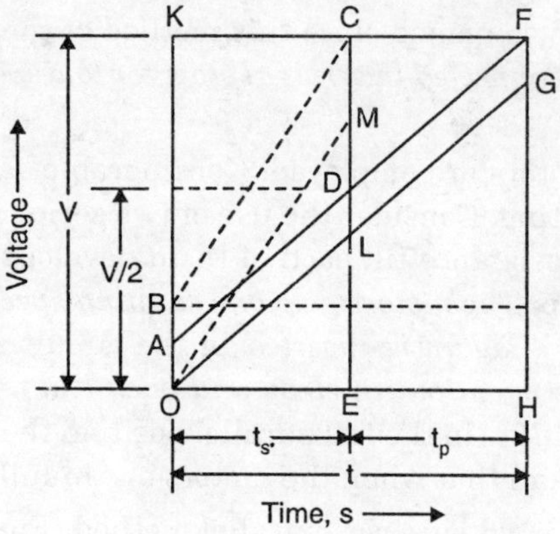

Fig. 8.22. Voltage built-up in series-parallel starting.

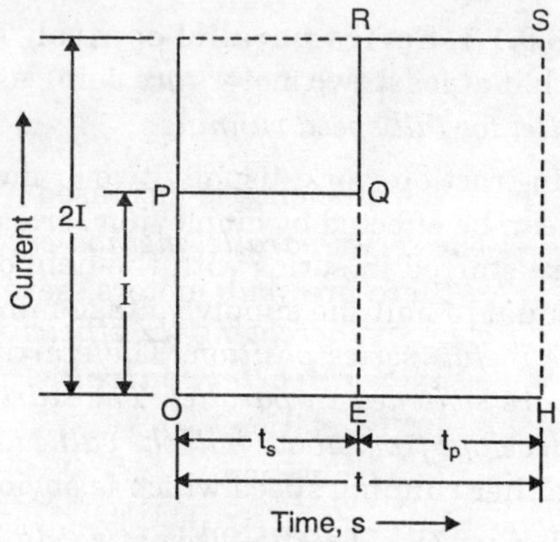

Fig. 8.23. Variation of current in series-parallel starting.

The values of the t_s during which the motors remain in series and t_p during which they are in *parallel* can be determined from Fig. 8.22. Δs OLE and OGH are similar. Therefore

$$\frac{OE}{OH} = \frac{LE}{GH} \; ; \quad \therefore \quad \frac{t_s}{t} = \frac{LE}{GH} = \frac{\frac{V}{2} - IR}{V - IR}$$

$$\therefore \qquad \mathbf{t_s} = \left[\frac{1}{2}\left(\frac{V - 2IR}{V - IR}\right)t\right] \qquad \text{...(8.4)}$$

and, $$\mathbf{t_p} = t - t_s = t - \left[\frac{1}{2}\left(\frac{V - 2IR}{V - IR}\right)t\right]$$

or, $$\mathbf{t_p} = t\left[1 - \frac{1}{2}\left(\frac{V - 2IR}{V - IR}\right)\right] \qquad \text{...(8.5)}$$

Let us now calculate the efficiency (η) of this method. For this purpose neglect IR drop in the armature circuit as E_b developed practically equals the voltage impressed across the motor. This modifies Fig. 8.23 to Fig. 8.24. Since D is the mid-point of CE and back e.m.f. motor develops along DF in the parallel combination, $KC = CF$ *i.e.*, time for series combination = time for the parallel combination.

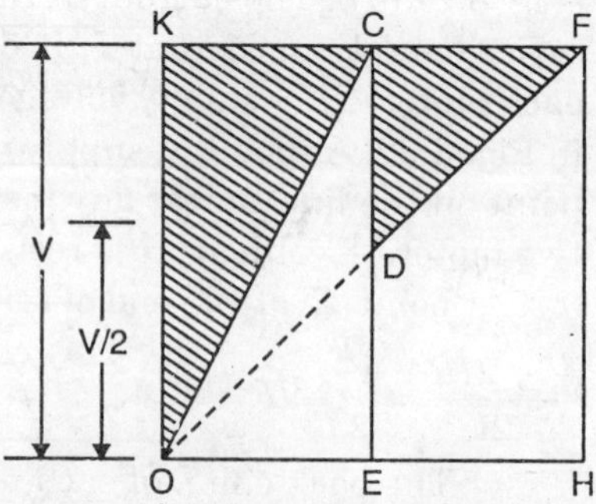

Fig. 8.24. Efficiency of starting by series-parallel method.

Let $t_s = t_p = \dfrac{t}{2}$ and the average starting current be I per motor, $t_s = OE$, $t_p = EH$.

Energy lost in R_s = Shaded area under ΔOKC + Shaded area under ΔCDF

$$= \frac{1}{2} VI \times \frac{t}{2} + \left(\frac{1}{2} \times \frac{V}{2} \times 2I\right) \times \frac{t}{2} = \frac{VIt}{2}$$

But, total energy supplied $= VI \times \dfrac{t}{2} + V \times 2I \times \dfrac{t}{2} = \dfrac{3VIt}{2}$

(series) (parallel)

$\therefore$ *Efficiency of starting,* $$\eta = \frac{\dfrac{3VIt}{2} - \dfrac{VIt}{2}}{\dfrac{3VIt}{2}} = \frac{2}{3} \quad \text{or} \quad \mathbf{66.66\%.}$$

Thus the efficiency is increased by about 17%.

- The *series-parallel method enables a saving of about 15 to 20 percent in the energy.*
- If there are *four* motors, series, series parallel and parallel starting can be employed thereby *increasing the efficiency of starting to 73 per cent.*

The ***series-parallel control*** is carried out by the following two methods :

1. Shunt transition.
2. Bridge transition.

1. Shunt transition :

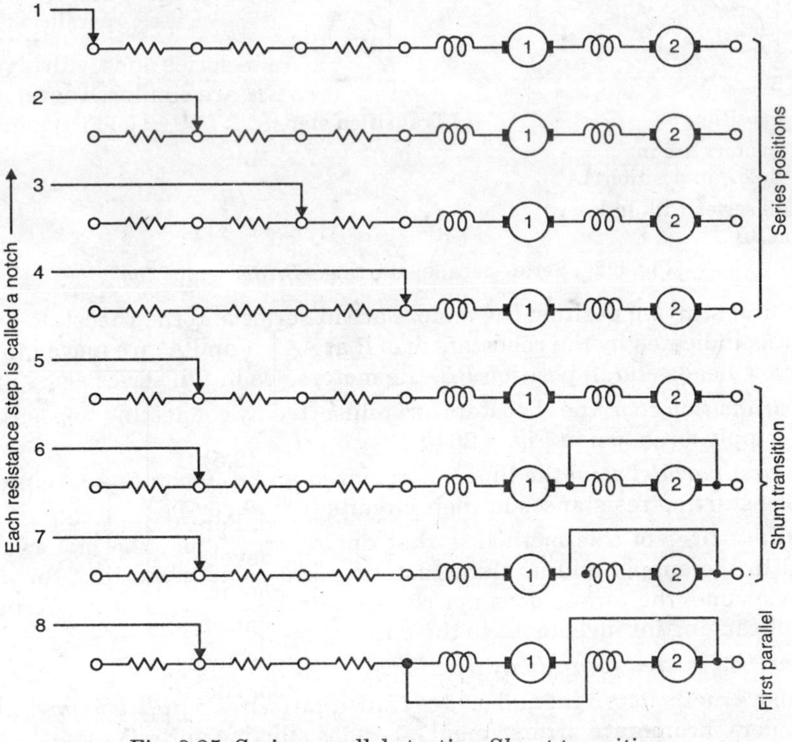

Fig. 8.25. Series-parallel starting-*Shunt transition.*

In this method of series-parallel control the following steps/stages are involved :

— In *steps 1, 2, 3, 4* the motors are in series and are accelerated by cutting out the starting resistance (R_s) in steps. In *step-4*, the motors are in *full series.*

— During *transition from series to parallel*, the resistance is reinserted in the motor circuit ... *step-5.*

— One of the motors is by-passed (*step-6*) (the total torque developed by the system is reduced by 50% during this period, thereby a noticeable jerk is experienced by the vehicle) and disconnected from the main circuit (*step-7*). It is then connected in parallel with the other motor (*step-8*) giving us the first parallel position.

- Since during transition the torque developed by the motor is reduced suddenly, the vehicle *experiences a sudden jerk and causes inconvenience to the passangers, this method is normally employed for high vehicles like trams etc.*

2. **Bridge transition :**

— The motor and the starting rheostats are connected in the form of a Wheatstone bridge as shown in Fig. 8.26 (a).

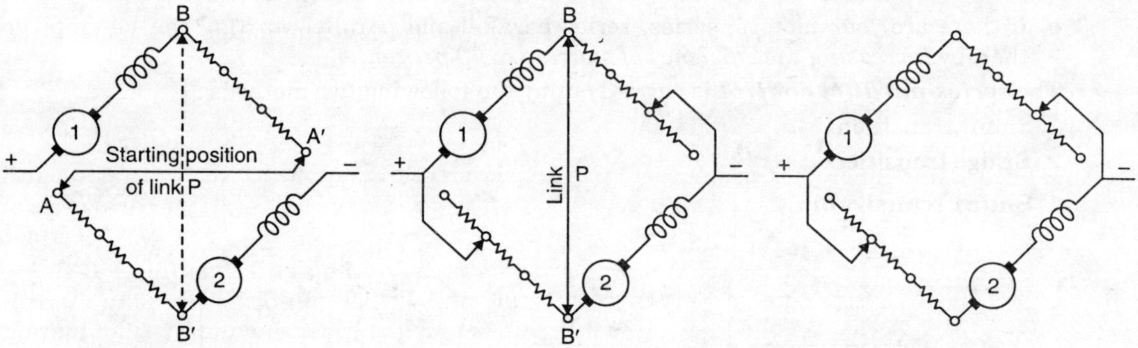

(a) Series position.

(i) At starting, motors are in series with link P in position AA'.

(ii) Motors in full series with link P in position BB'.

(b) Transition step.

(c) First parallel position.

Fig. 8.26. Series-parallel starting—*Bridge transition.*

— In the first starting position the motors are in series and the rheostats are completely in circuit as indicated by the rheostats arm P at AA'. A and A' are moved in the direction of the arrow heads and in position BB' the motors are in full series.

— In the *transition step*, the rheostats are reinserted by connecting to positive and negative of the supply as shown in Fig. 8.26 (b).

— In the *first parallel step*, the link P is removed and the motors are connected in parallel with the starting resistances in their circuits [Fig. 8.26 (c)]

- The ***advantage*** of this method is that during transition, the motors are always connected to the supply and as the resistances are so adjusted that the value of current remains same, the torque does not change and hence *uniform acceleration* is obtained without causing inconvenience to the passengers.
- *This method is used for railway traction.*

For the above methods (shunt and bridge transition) *series parallel drum-type controllers are used.* The controllers incorporate arrangements, besides starting, for reversing, and braking of the motors also.

Example 8.11. *A motor-coach bogie having two series motors is accelerated uniformly to a speed of 45 km/h in 25 sec. with the help of a series-parallel controller. If the average tractive effort per motor is 12750 N, calculate the approximate loss of energy in the starting rheostats.*

Solution. *Given :* V_{max} = 45 km/h = 12.5 m/s ; t = 25 s ; Total F_t = 2 × 12750 = 25500 N

$$\text{Useful energy for acceleration} = \frac{1}{2} \times (F_t \times V_{max}) \times \frac{1}{1000} \times \frac{t}{3600} \text{ kWh}$$

$$= \frac{1}{2} \times 25500 \times 12.5 \times \frac{1}{1000} \times \frac{25}{3600} = 1.107 \text{ kWh}$$

The loss of energy in starting rheostats is *approximately equal to half the useful energy.*

$$= \frac{1.107}{2} = 0.5535 \text{ kWh.} \quad \textbf{(Ans.)}$$

Example 8.12. *Two motors of a motor coach are started on series-parallel system, the current per motor being 350 A (considered as being constant) during the starting period which is 18 sec. If the acceleration during starting period is uniform, the line voltage is 600 V and resistance of each motor is 0.1 Ω. Calculate :*

(i) The time during which the motors are operated in series.

(ii) The energy loss in the rheostat during starting period. **(Nagpur University)**

Solution. *Given* : Current per motor, I = 350 A ; t = 18 s ;
V = 600 V, R = 0.1 Ω.

(*i*) **The time during which the motors are operated in series t_s :**

$$t_s = \left[\frac{1}{2}\left(\frac{V - 2IR}{V - IR}\right)t\right] \quad \text{...[Eqn. (8.4)]}$$

$$= \left[\frac{1}{2}\left(\frac{600 - 2 \times 350 \times 0.1}{600 - 350 \times 0.1}\right) \times 18\right] = \textbf{8.44 s.} \quad \textbf{(Ans.)}$$

Time during which motors are in parallel,

$$t_p = t - t_s = 18 - 8.44 = \textbf{9.56 s} \quad \textbf{(Ans.)}$$

(*ii*) **The energy loss in the rheostat during starting period :**

Back e.m.f. (E_b) of each motor, in *series operation,*

$$E_{bs} = \frac{V}{2} - IR \quad \text{(Refer to Fig. 8.22)}$$

$$= \frac{600}{2} - 350 \times 0.1 = 265 \text{ V}$$

When two motors are in series,

$$E_b = 265 + 265 = 530 \text{ V}$$

Back e.m.f. (E_b) of each motor in *parallel operation*,

$$E_{bp} = V - IR$$

$$= 600 - 350 \times 0.1 = 565 \text{ V}$$

Energy lost when motors are connected in *series.*

$$= \frac{1}{2} E_b I t_s = \frac{1}{2} \times 530 \times 350 \times \frac{8.44}{3600} = 217.4 \text{ W-h}$$

Energy lost when motors are connected in *parallel*.

$$= \frac{1}{2} \cdot \frac{E_b}{2} \times (2I) t_p$$

$$= \frac{1}{2} \times \frac{565}{2} \times (2 \times 350) \times \frac{9.56}{3600} = 262.6 \text{ W-h}$$

∴ *Total energy lost* = 217.4 + 262.6 = **480 W-h.** **(Ans.)**

Example 8.13. *A multiple-unit train weighing 210 tonnes is equipped with 6 motors giving a total tractive effort of 82800 N. The motors operate at a line voltage of 600 V and each motor takes an average current of 200 A during starting. Series-parallel control is used for starting. The full line voltage is applied to each motor when the speed reaches 38.6 km/h. Resistance of each motor is 0.1 ohm. Calculate :*

(*i*) *Energy supplied during the starting period.*

(*ii*) *Energy lost in the starting resistance.*

(*iii*) *Useful energy supplied to the train.* **(Roorkee University)**

Solution. *Given* : W = 210 tonnes ; No. of motors = 6 ;

Total tractive effort = 82800 N, Line voltage = 600 volts ;

Current taken by each motor, I = 200 A ; V = 38.6 km/h ;

$$R = 0.15\ \Omega.$$

Refer to Fig. 8.27.

(*i*) **Energy supplied during the starting period :**

Tractive effort, $F_t = 277.8\ \text{M}.\alpha$ ... M is in tonnes

$$\therefore \quad \alpha = \frac{F_t}{277.8\ \text{M}} = \frac{82800}{277.8 \times 210} = 1.42 \text{ km phps}$$

Time for acceleration, $t = \dfrac{38.6}{1.42} = 27.2$ s

We know that, $t_s = \dfrac{1}{2}\left(\dfrac{V - 2IR}{V - IR}\right) t$ ...[Eqn. (8.4)]

$$= \frac{1}{2}\left(\frac{600 - 2 \times 200 \times 0.15}{600 - 200 \times 0.15}\right) \times 27.2 = 12.9 \text{ s}$$

$$t_p = t - t_s = 27.2 - 12.9 = 14.3 \text{ s}$$

Total energy supplied = Energy supplied – *series position* (2 motors in series and 3 such circuits in parallel)
+ Energy supplied – *parallel position* (all motors in parallel)

$= 3 \times 600 \times 200 \times 12.9 + 6 \times 600 \times 200 \times 14.3$

$= 4.64 \times 10^6 + 10.3 \times 10^6 = 14.94 \times 10^6$ W-s

= **4.15 kWh.** **(Ans.)**

(*ii*) **Energy lost in starting resistance :**

Series position $= 3 \times \dfrac{1}{2} \times 540 \times 200 \times 12.9 = 2.09 \times 10^6$ W-s

Parallel position $= 6 \times \dfrac{300}{2} \times 200 \times 14.3 = 2.57 \times 10^6$ W-s

Total energy lost $= 2.09 \times 10^6 + 2.57 \times 10^6$

$= 4.66 \times 10^6$ W-s $\simeq$ **1.3 kWh** **(Ans.)**

Energy lost in motor resistance

$$= 6 \times (200)^2 \times 0.15 \times 27.2 = 0.98 \times 10^6 \text{ W-s} \simeq \mathbf{0.272\ kWh}$$

(*iii*) **Useful energy supplied to the train :**

Useful energy supplied to the train $= \dfrac{1}{2} \times (F_t \times V) \times t$

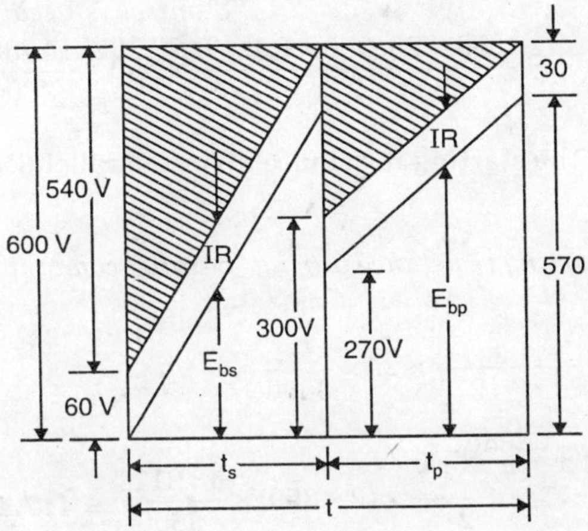

Fig. 8.27

$$= \frac{1}{2} \times 82800 \times \left(\frac{38.6 \times 1000}{3600}\right) \times \frac{1}{1000} \times \frac{27.2}{3600} \text{ kWh}$$

$$= \mathbf{3.35\ kWh.} \quad \textbf{(Ans.)}$$

Example 8.14. *Two 650 V motors each having a resistance of 0.1 ohm are started on the series-parallel system, the mean current per motor throughout the starting period being 350 A. The starting period is 25 seconds and the train speed at the end of this period is 40 km/h. Calculate :*

(*i*) *The rheostat losses during* (*a*) *the series and* (*b*) *the parallel combinations of motors.*

(*ii*) *The train speed at which transition from series to parallel must be made.*

(Banglore University)

Solution. *Given* : Rated voltage of each motor = 650 V, $R = 0.1\ \Omega$ per motor ; Mean current, $I = 350$ A ; Starting period, $t = 25$ s ; Speed at the end of starting period = 40 km/h.

(*i*) **Rheostat losses :**

Back e.m.f. of each motor in full series position,

$$E_{bs} = \frac{V}{2} - IR = \frac{650}{2} - 350 \times 0.1 = 290 \text{ V}$$

Back e.m.f. of each motor in full parallel position,

$$E_{bp} = V - IR = 650 - 350 \times 0.1 = 615 \text{ V}$$

Assuming smooth acceleration, back e.m.f. will be built up at constant rate.

Since motors take 25 seconds to build up 615 V, therefore, time taken to build up 290 V e.m.f. will be :

$$t_s = 25 \times \frac{290}{615} = 11.7886 \text{ s}$$

$$\therefore \quad t_p = 25 - 11.7886 = 13.2114 \text{ s}$$

(*a*) *Voltage drop* in the starting rheostat in *series combination at the starting instant*

$$= V - 2\,IR = 650 - 2 \times 350 \times 0.1 = 580 \text{ V},$$

which reduces to zero in full series operation.

Energy dissipated in starting resistance during "series combination".

$$= \frac{(V - 2IR) + 0}{2} \times I \times \frac{t_s}{3600}$$

$$= \frac{(650 - 2 \times 350 \times 0.1) + 0}{2} \times 350 \times \frac{11.7886}{3600} = \mathbf{332.37\ Wh.} \quad \textbf{(Ans.)}$$

(*b*) Voltage drop across the starting resistance in first parallel position is $\frac{V}{2}$ *i.e.* 325 V which gradually reduces to zero.

Energy dissipated in starting resistance during "parallel combination"

$$= \frac{\frac{V}{2} + 0}{2} \times 2I \times \frac{t_p}{3600}$$

$$= \frac{\frac{650}{2} + 0}{2} \times (2 \times 350) \times \frac{13.2114}{3600} = \mathbf{417.44\ Wh.} \quad \textbf{(Ans.)}$$

(*ii*) **Speed at the end of series period :**

$$\text{Acceleration, } \alpha = \frac{\text{Maximum speed}}{\text{Starting period}} = \frac{40}{25} = 1.6 \text{ km ph ps}$$

$\therefore$ Speed at the end of series period $= \alpha \times t_s$

$$= 1.6 \times 11.7886 = \mathbf{18.86\ km/h.} \quad \textbf{(Ans.)}$$

Example 8.15. *Two 750 V D.C. series motors each having a resistance of 0.1 Ω are started on series-parallel system. Mean current throughout the starting period is 300 A. Starting period is 15 sec. and train speed at the end of this period is 25 km/h. Calculate :*

(*i*) *Rheostatic losses during series and parallel combination of motors.*

(*ii*) *Energy lost in motors.*

(*iii*) *Motor output.*

(*iv*) *Starting efficiency.*

(*v*) *Train speed at which transition from series to parallel must be made.*

(Nagpur University, Summer 2000)

Solution. Rated voltage of each series motor = 750 V ; $R = 0.1\ \Omega$ per motor ; $I = 300$ A ; $t = 15$ s ; Speed after starting period = 25 km/h.

(*i*) **Rheostatic losses :**

$$t_s = \frac{1}{2}\left(\frac{V - 2IR}{V - IR}\right)t = \frac{1}{2}\left(\frac{750 - 2 \times 300 \times 0.1}{750 - 300 \times 0.1}\right) \times 15 = 7.1875 \text{ s}$$

$$\therefore \quad t_p = 15 - 7.1875 = 7.8125 \text{ s}$$

Energy lost in rheostat

$$= \frac{1}{2} E_{bs} \times I \times t_s + \frac{1}{2}\frac{E_{bp}}{2} \times 2I \times t_s$$

$$= \frac{1}{2}\left[2\left(\frac{V}{2} - IR\right)\right] I \times t_s + \frac{1}{2}\left[\left(\frac{V - IR}{2}\right)\right] 2I \times t_p$$

$$= \frac{1}{2}\left[2\left(\frac{750}{2} - 300 \times 0.1\right)\right] \times 300 \times 7.1875 + \frac{1}{2}\left[\left(\frac{750 - 300 \times 0.1}{2}\right)\right] \times 2 \times 300 \times 7.8125$$

$= 743906 + 843750$

$= 1587656$ W-s = **441 Wh. (Ans.)**

(*ii*) **Energy lost in motors :**

Energy lost in two motors $= (I_a^2 \times R_a) \times 2 \times 15$

$= [(300)^2 \times 0.1] \times 2 \times 15 = 270000$ W-s = **75 W-h. (Ans.)**

(*iii*) **Motor output :**

Total energy supplied $= V \times I \times t_s + 2I \times V \times t_p$

$= 750 \times 300 \times 7.1875 + 2 \times 300 \times 750 \times 7.8125$

$= 1617187 + 3515625$

$= 5132812$ W-s = 1425.78 Wh

Motor output = Total energy supplied – energy lost in rheostat – energy lost in armatures

$= 1425.78 - 441 - 75 =$ **909.78 Wh. (Ans.)**

(*iv*) **Starting efficiency, η_{start} :**

$$\eta_{start} = \frac{\text{Total energy supplied} - \text{energy lost in rheostat}}{\text{Total energy supplied}}$$

$$= \frac{1425.78 - 441}{1425.78} = 0.6907 \text{ or } \mathbf{69.07\%}. \text{ (Ans.)}$$

(*v*) **Train speed-transition**

Acceleration is uniform during starting period of 15 sec. Therefore speed after which series to parallel transition must be made

$$= \frac{\text{Speed after starting period}}{\text{Total starting period}} \times t_s$$

$$= \frac{25}{15} \times 7.1875 = \mathbf{11.98 \text{ km/h.}} \text{ (Ans.)}$$

Example 8.16. *The following data refer to two traction motors rated at 1500 V :*

Armature resistance of each motor	*= 0.15 Ω*
Current drawn by each motor during starting	*= 500 A*
Effective weight of the train	*= 140 tonnes*
Dead weight of the train	*= 120 tonnes*
Specific resistance	*= 50 N/tonne*
Tractive effort per motor	*= 38000 N*
Speed at the end of starting period	*= 40 km/h*

Calculate :

(*i*) *Duration of starting period.*

(*ii*) *Speed of train at transition.*

(*iii*) *Rheostatic loss.*

Solution. *Given* : Rated voltage of each motor = 1500 V ; $R = 0.15\ \Omega$; $I = 500$ A ;
$M_e = 140$ tonnes ; $M = 120$ tonnes ;
Specific resistance $r = 50$ N/t ; F_t per motor = 38000 N ; $V = 40$ km/h.

(*i*) **Duration of starting period, t_s :**

Tractive effort required by the train,

$$F_t = 277.8\, M_e\, \alpha + M.r = 38000 \times 2 = 76000 \text{ N}$$

(where, α is the acceleration in km ph ps)

or, $$277.8 \times 140 \times \alpha + 120 \times 50 = 76000$$

or, $$\alpha = 1.8 \text{ km ph ps}$$

∴ *Starting time,* $$t_s = \frac{V}{\alpha} = \frac{40}{1.8} = \mathbf{22.22\ s.}\quad \textbf{(Ans.)}$$

(*ii*) **Speed of train at transition, V_{trans} :**

Resistance drop per motor = $IR = 500 \times 0.15 = 75$ V

Also, $$t_s = \left[\frac{1}{2}\left(\frac{V - 2IR}{V - IR}\right)t\right] \qquad \text{...(Eqn. 8.4)}$$

$$= \left[\frac{1}{2}\left(\frac{6500 - 2\times 500\times 0.15}{1500 - 500\times 0.15}\right)\times 22.22\right] = 10.53 \text{ s}$$

∴ $$t_p = t - 10.53 = 22.22 - 10.53 = 11.69 \text{ s}$$

Since acceleration during starting is assumed to be constant, therefore, speed at transition, $V_{trans} = \alpha \times t_s = 1.8 \times 10.53 =$ **18.95 km/h. (Ans.)**

(*iii*) **Rheostat loss :**

Refer to Fig. 8.28. Loss of energy in controller per motor

$$= \frac{\left(\frac{1}{2}\times 675\times 10.53\right)\times 500 + \left(\frac{1}{2}\times 750\times 11.69\right)\times 500}{3600\times 1000} = 1.102 \text{ kWh}$$

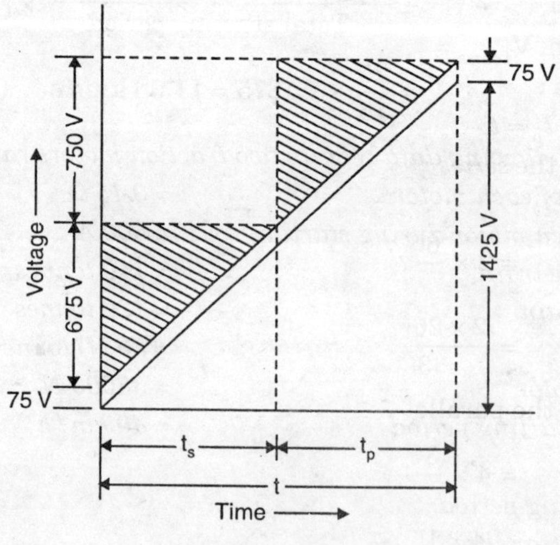

Fig. 8.28

∴ *Loss of energy in controllers for two motors* = 2 × 1.102 = **2.204 kWh. (Ans.)**

Example 8.17. *An electric train weighing 145 tonnes is equipped with four 600-V motors arranged in two pairs for series-parallel control. During series-parallel starting, the current per motor is maintained at 400 A. At 400 A and 600 V, the tractive effort per motor is 21200 N and the train speed is 40 km / h. Assume that the train is started up a gradient of 1 m 100 and resistance to traction is 45 N / tonne. Allow 10 percent for the effect of rotational inertia. If the resistance of each motor is 0.1 ohm, calculate :*

(i) *Duration of the starting period.*

(ii) *Speed of the train at transition.*

(iii) *Rheostatic losses during (a) series and (b) parallel steps of starting.*

Solution. *Given* : $M = 145$ t ; $M_e = 1.1 \times 145 = 159.5\ t$;

Rated voltage of each motor = 600 V ;

I per motor = 400 A ; Tractive effort per motor = 27200 N ;

Train speed, $V = 40$ km/h ; $G = \dfrac{1}{100}$; $r = 45$ N/t ; $R = 0.1\ \Omega$ per motor.

(i) **Duration of the starting period, t :**

We know that,
$$F_t = 277.8\ M_e \,.\, \alpha + M.r + 98\ \text{MG}$$

$$4 \times 21200 = 277.8 \times 159.5 \times \alpha + 145 \times 45 + 98 \times 145 \times \frac{1}{100}$$

or
$$\alpha = 1.76 \text{ km ph ps}$$

$$\therefore \quad \text{t} = \frac{40}{1.76} = \textbf{22.73 s.} \quad \textbf{(Ans.)}$$

(ii) **Speed of the train at transition, V_{trans} :**

$$t_s = \left[\frac{1}{2}\left(\frac{V - 2IR}{V - IR}\right) t\right]$$

$$= \frac{1}{2}\left(\frac{600 - 2 \times 400 \times 0.1}{600 - 400 \times 0.1}\right) \times 22.73 = 10.55 \text{ s}$$

$\therefore$ Speed at transition, $V_{trans} = \alpha \times t_s = 1.76 \times 10.55 =$ **18.6 km/h. (Ans.)**

(iii) **Rheostatic losses :**

$$t_p = t - t_s = 22.73 - 10.54 = 12.19 \text{ s}$$

(a) Energy lost during the *series period* (assuming two motors in series and two such circuits in parallel).

$$= 2 \times \frac{600 - 80}{2} \times 400 \times t_s \text{ watt-sec.}$$

$$= \frac{2 \times 260 \times 400 \times 10.55}{3600 \times 1000} = \textbf{0.61 kWh.} \quad \textbf{(Ans.)}$$

(b) Energy lost during the parallel period (all the motors in parallel giving 4 circuits)

$$= 4 \times \frac{300}{2} \times 400 \times t_p \text{ watt-sec.}$$

$$= \frac{4 \times 150 \times 400 \times 12.19}{3600 \times 1000} = \textbf{0.813 kWh.} \quad \textbf{(Ans.)}$$

Example 8.18. *Two motors each rated at 600 V drive a coach. One of the motors is geared to driving wheels having diameter of 88 cms and other motor to driving wheels having diameter of 86 cms. If the speed of the first motor is 100 r.p.m. when connected in parallel with other motor across 650 V supply, determine the speed of motors when connected in series.*

Assume armature drop to be 10 percent and current to remain constant during starting.

Solution. *Given* : $D_1 = 88$ cms ; $D_2 = 86$ cms ; $N = 1000$ r.p.m., Armature drop = 10 per cent

Speed of motor when connected in series :

As the linear velocity of two driving wheels in same,

$$\therefore \qquad N_1D_1 = N_2D_2 \quad \text{or} \quad \frac{N_1}{N_2} = \frac{D_2}{D_1} = \frac{86}{88}$$

We know that,

$$V_1 = \frac{\frac{N_1}{N_2}(V - IR) + IR}{1 + \frac{N_1}{N_2}} \qquad \text{...[Eqn. (8.6)]}$$

$$= \frac{\frac{86}{88}(650 - 0.1 \times 650) + (0.1 \times 650)}{1 + \frac{86}{88}} = 322 \text{ V}$$

or,

$$V_2 = 650 - V_1 = 650 - 332 = 328 \text{ V}$$

$$\therefore \qquad N_1 = N \times \left(\frac{V_1 - IR}{V - IR}\right)$$

or,

$$\mathbf{N_1} = 1000 \times \left(\frac{322 - 0.1 \times 650}{650 - 0.1 \times 650}\right) = \mathbf{439 \text{ r.p.m.}} \quad \textbf{(Ans.)}$$

$$\therefore \qquad N_2 = N_1 \times . \frac{D_1}{D_2} = 439 \times \frac{88}{86} = \mathbf{449 \text{ r.p.m.}} \quad \textbf{(Ans.)}$$

8.9.1.2. Tapped field control (or control by field weakening).

Even when the traction motors have run upto speed, an increase in speed is still possible to the extent of 15 to 30 percent by *weakening the field.* Since speed is inversely proportional to flux, by weakening the field, the speed increased. For weakening the field is *either a divertor is provided or the field is tapped.*

The '*advantage*' *of field control is that it makes the equipment very flexible.*

- Adequate *flexibility* in the operation of traction motors can be obtained with a *combination of both the field and voltage control.* For *city service*, where the *speed required is low* and frequent starting and stopping are required, the equipment can operate with the full field. However, when the same vehicle is to operate for *suburban services* where *higher speeds* are required, field weakening can be used, thus eliminating the need for change in gear ratio which otherwise would be required. Thus, one type of equipment can be used to operate under different service conditions with reasonable energy consumption.

8.9.1.3. Buck and boast method

The description of this method is as follows :

- The armatures of the two traction motors and the motor-generator set are connected in *series* and *across the supply.*

- When the generator voltage *equals* the supply voltage and is in opposition to it, the *main contactor is closed.* Under this condition, there is no voltage across the traction motors and hence their speed is zero. If now the generator voltage is reduced, voltage across traction motors starts increasing and their speed rises.
- When generator voltage is zero, full supply voltage appears across the motors *i.e.*, each motor receives one-half of the supply voltage. If the polarity of the generated e.m.f. equals supply voltage, each traction motor will receive voltage equal to supply voltage. Thus by adjusting the generator excitation the equivalent supply voltage can be reduced or boosted up.

Advantages :

This method entails the following *advantages* :

1. Any operating speed of traction motors can be obtained (In case of resistance controllers only a few speeds are possible).
2. No energy loss in the starting resistance of the traction motors (of course, loss does take place in the starting resistance of motor generator set).
3. If there is a temporary interruption in the supply, the kinetic energy of the flywheel can utilized in generating energy from the motor-generator set and fed to traction motors.

8.9.1.4. Metadyne control

In the series-parallel control of D.C. traction motors, there is a considerable loss of energy in the starting resistances. The *metadyne system of control estimates the energy loss and achieves a very smooth control during accelerating period. Machines with more than two brush sets per pair of poles are called **metadynes.** It is a device which converts power at constant voltage and variable current into one with constant current and variable voltage.*

The primary *advantage* of Metadyne control is that the *loss is much lower than in case of resistance starting method and since in this case current throughout the starting period can be maintained constant hence uniform tractive effort is developed which avoids jerky environment of the train* (which other exists in resistance starting method where the current varies between a certain maximum and minimum value whenever notch position is changed).

Description of the 'Metadyne' :

- Consider a D.C. armature with brushes and two poles. If the current is supplied to the brushes A_1A_2 the armature cross-flux will be as shown and mainly confined to the poles as shown in Fig. 8.29 (*i*).

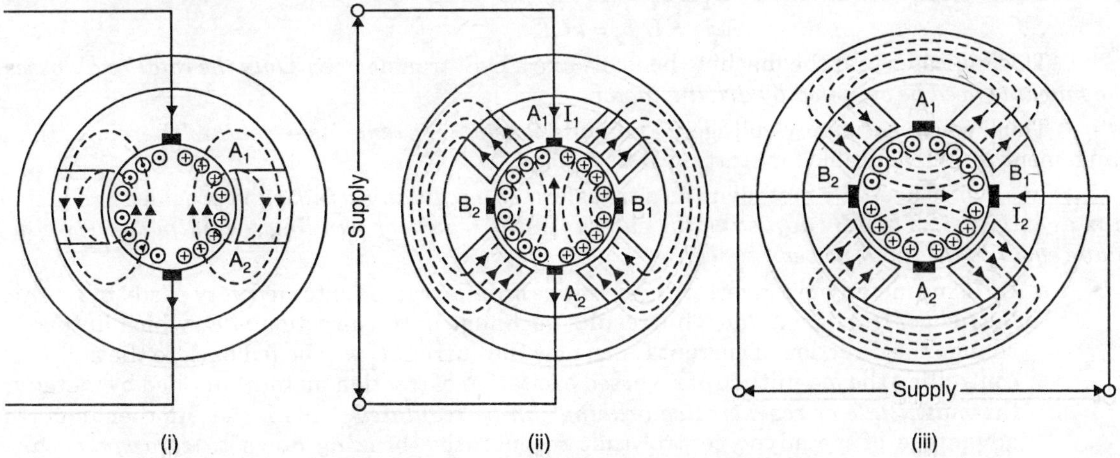

Fig. 8.29. Illustration of metadyne.

- In case there are four brushes, current is supplied to A_1A_2 and the armature cross-flux will take up the path as shown in Fig. 8.29 (*ii*)
- Now, if the current is supplied to brushes B_1B_2 as shown in Fig. 8.29 (*iii*), the armature cross-flux takes up path as indicated. If the armature is rotated at constant speed and current I_1 is fed into the brushes A_1A_2, an e.m.f. is induced in the winding between B_1B_2 due to flux produced by I_1. No e.m.f. is induced between A_1A_2 and the voltage between A_1A_2 is on account of the voltage drop due to I_1. Since an e.m.f. is induced across B_1B_2 a current I_2 will flow in a load connected between them (Fig. 8.30 (*i*)). The *resultant flux distribution* on account of I_1 and I_2 is as shown in Fig. 8.30 (*ii*).

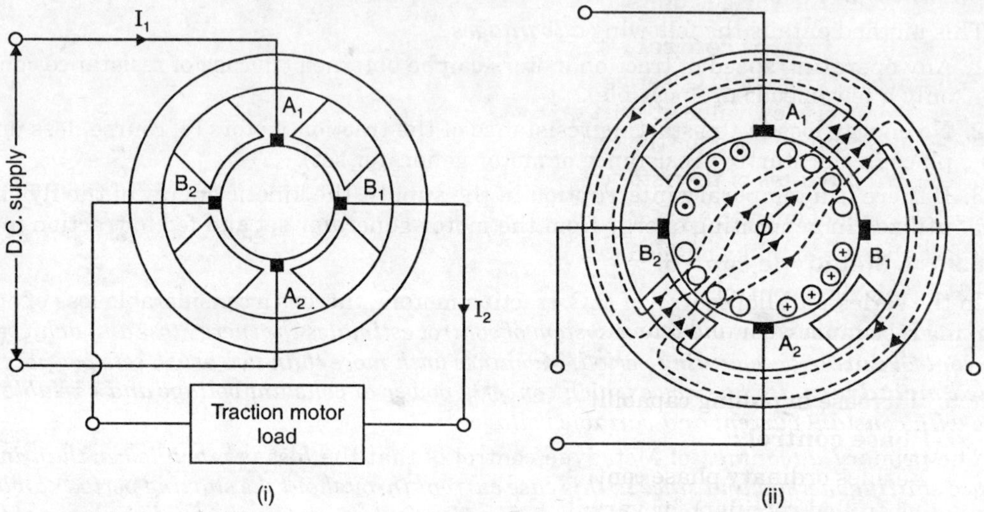

Fig. 8.30

— The total flux ϕ may be assumed to be made up of two components ϕ_1 and ϕ_2 at right angles and directed along A_2A_1 and B_2B_1. The rotation of armature in ϕ_2 induces an e.m.f. E_1 between A_1 and A_2 which opposes the supply voltage.

— Since the current is to be kept at its original value of I_1, the supply voltage must be induced to overcome E_2.

Under steady conditions : $E_1 \alpha \phi_2 = kI_2 \,;\, E_2 \alpha \phi_1 = kI_1$

$\therefore \quad E_1I_1 = E_2I_2 = kI_1I_2$

This indicates that the machine behaves like a D.C. transformer. *Only the rotational losses of the machine need be supplied by driving motor.*

Thus, when the supply voltage E_1 *remains constant*, I_2 *remains constant.* Therefore, the arrangement is quite suitable for starting D.C. motors.

In order to keep the metadyne as a transformer a *regulator winding* is provided producing a flux as in Fig. 8.29 (*ii*). *By adjusting the current in the regulator winding the output power can be made equal to the input power.*

- By using metadyne converter, *regenerative braking* can be obtained very easily by *reversing the field of traction motor.* This results in change in the direction of E_2 which in turn will change the direction of current I_1 *i.e.*, now the current I_1 will be fed back to the supply. By controlling the magnitude of reversed excitation of traction motors supplied by metadyne, the *magnitude or regenerative braking can be regulated.* This is yet another important advantage of metadyne control that regenerative braking down to *zero speed can be obtained. This saving in energy during acceleration and braking may easily counter balance the additional cost of the more complicated equipments required in metadyne control.*

8.9.1.5. Multiple unit control

For city and suburban services, motor coaches are usually employed ; each motor-coach may have one or two trailer coaches. These motor coaches *must be capable of control from a single point.* This object is achieved in the *'multiple unit control'*.

A unit of train equipment consists of a group of two or four motors in every motor-coach and provided with a *series-parallel controller, reverser, starting resistance and controller gear.* The controllers and reverser are controlled by a *'master controller'* in the cab of any motor-coach. The electric cables joining the various units are called coupling *cables.*

8.9.1.6. Thyristor control

The acceleration of D.C. series traction motor is controlled by varying the voltage applied to the motors from very low value at start to the full value at high speed. Thyristors (instead of conventional methods) can be used for the control of D.C. motors.

The ***advantages*** of thyristor control are :

(*i*) Saving of energy.

(*ii*) Absence of bulky on-load tap changer and electromagnetic switching devices.

(*iii*) Notchless control.

(*iv*) Minimum wear and tear due to absence of conventional moving parts in the motor control circuit.

(*v*) Increase in pulling capability of the motive power.

1. **Phase control :**

- Besides ordinary phase control methods *'cycle selection methods'* of control of SCR (silicon controlled rectifier) for varying the voltage applied across traction motors is employed. *This is done by accepting* or *rejecting a certain number of complete half cycles to give the required mean voltage.* In practice, at the start only one half cycle in eight can be accepted. As the locomotive picks up the speed, this is gradually increased to (2/8), (3/8) and finally (8/8) for full power operation.
- Cycle selection method of control of thyristor is *advantageous* due to :

— Lower rate of rise of current ;

— Better power factor ;

— Low frequency harmonics etc.

2. **Chopper control :**

The *conventional method* of starting and speed control of D.C. series motor employing external resistance is *most inefficient due to heat loss in the external resistor.* Besides this, the *torque transmissions are abrupt in case of contactor type controller.* In order to eliminate these power losses and provide an infinitely variable speed control, ***pulsed input voltages*** can be used.

The pulsed input voltages are obtained by using a *chopper controller.* At start, the 'ON' period of pulse is very short which lengthens during the period of controlled acceleration. Thus, the average voltage to the traction motors is gradually increased keeping the mean value of the accelerating current close to the required value.

8.9.2. Control of Single-phase Series Motors

The speed and torque of a single phase series motor can be varied by *applying variable voltage.* Usually the operating voltage of the motor is smaller than the supply voltage. Various voltages for starting and speed control be obtained from a transformer (which is the integral part of the equipment) which is provided with a number of taps on the secondary side. With this not only the starting resistances are avoided (Thus there being no loss of power during starting period) but also each control point (tapping) becomes a running point so that a number of economical speeds are available.

The *voltage tapping is effected either by a 'group of contactors'* or a *'tap changing switch'*

— While changing the tap it should be seen that the *successive short circuiting* of the sections of the transformer winding between the tappings *should be avoided.*

— For large capacity traction motors, a large number of contactors should be used simultaneously so that the current is divided between them and each contactor has to deal with not more that a prespecified design value (In general D.C. contactor is much lighter than the A.C. contactor and requires less power for its operation).

8.9.3. Control of Three-phase Motors

In polyphase induction motors the following methods of speed control are used :

1. Rheostatic control
2. Pole changing
3. Cascade control
4. Combination of cascade and pole-changing arrangement.

Refer Art. 5.13.1.17.

8.10. BRAKING

8.10.1. Introduction

- In order to meet with the heavy traffic requirements as a result of phenomenal increase of both the industrial activity and population, we have to resort to have heavy and faster trains. With the modern traction equipment, achievement of rapid acceleration does not present any problem and also failure during acceleration will not involve any danger. Whereas, resorting to heavy and faster trains involves the *problem of dissipation of large amount of K.E. in shortest possible lines as in case of emergency braking.* Any failure during emergency braking will be a catastrophe. Thus the importance of braking system cannot be underestimated.
- In ***traction work*** both *electrical and mechanical braking* are employed for bringing the vehicle to *rest.* Electrical braking can not do away with mechanical brakes since a vehicle cannot be held stationary by its use ; it nevertheless forms a very important part of a traction system.

In "***electric braking***" *The braking energy is converted into electrical energy instead of converting it into heat energy at the brake shoes and either dissipated in the resistances mounted on the vehicle or returned to the supply system.*

8.10.2. Advantages and Disadvantages of Electrical Braking over Mechanical Braking

Advantages :

1. Electrical brake is *smooth.*
2. Electrical braking is *more economical* than mechanical braking (In mechanical braking, due to excessive wear on brake drum, liner etc. it needs frequent and costly replacement).
3. Mechanical braking produces metal dust, which can damage bearings. There are no such problems in electrical braking.
4. In mechanical braking the noise produced is very high.
5. In regenerative braking electrical energy can be returned back to supply which is not possible in mechanical braking.

6. If mechanical brakes are not correctly adjusted it may result in shock loading of machine or machine parts in case of lifts, trains which may result in discomfort to the occupants.
7. Due to wear and tear of brake liner equipment adjustments are needed thereby making the maintenance costly.
8. In mechanical braking the heat is produced at brake liner or brake drum, which may be a source of failure of the brake. In electrical braking the heat is produced at convenient place which is no way is harmful to a braking system.
9. By employing electrical braking the capacity of the system can be increased by way of higher speeds and haulage of heavy loads.

Disadvantages/Limitations :

1. Since the motor has to function as a generator during braking period, therefore, it must have suitable braking characteristics *i.e.*, the choice of motor is limited.
2. In electrical braking, the driving motor operates as a generator during the period of braking, and motor ceases to operate as a generator at standstill so that although an electric brake can almost stop a machine or load, but it can not hold it stationary, therefore, a *friction brake is required in addition.*
3. Additional complications, high initial cost, special motors capable of generating electrical energy, make electric braking *costly.*

8.10.3. Requirements of a Braking System

The following are the *requirements* which a braking system should *satisfy* :

1. Brake *actuation time* should be as *small as possible.*
2. The system should *apply brakes simultaneously* over all the vehicles.
3. It should be *robust, simple and easy* for driver to control and operate.
4. It should require *less maintenance.*
5. It should be *reliable.*
6. Normal service application of brakes should be *very gradual and smooth.*
7. In order to obtain uniform deceleration, braking force applied to each axle should be *proportional to axle load.*
8. In case of emergency braking, *safety consideration is the prime most consideration.*
9. In the case of mechanical braking, there should be *automatic slack adjustment for constant piston stroke* as a result of wear on the rim and brake blocks.
10. The braking should be *inexhaustible* (*i.e.*, repeated quick application of brake should be possible without needing any relaxation, recuperation or normalising time in between consecutive operations.
11. K.E. of the train should as far as possible be *stored during braking which could subsequently be utilized for accelerating the train.*

8.10.4. Types of Braking

Broadly speaking, the braking is of the following three types :

1. *Electric braking :*
   - (*i*) Plugging
   - (*ii*) Rheostatic or dynamic braking
   - (*iii*) Regenerative braking
   - (*iv*) Electromagnetic braking
   - (*v*) Eddy current braking.

2. *Mechanical braking :*
   (*i*) Pneumatic braking
   (*ii*) Vacuum braking
   (*iii*) Electro-pneumatic braking
   (*iv*) Disc braking.
3. *Hydraulic braking.*

8.10.5. Electric Braking

1. **Plugging :**

- This method of braking involves *reconnection of motor to supply in such a way that motor now develops torque in opposite direction to the movement of the rotor.* System speed will decrease till zero speed is reached and then it will accelerate in opposite direction. Therefore, it becomes *necessary to disconnect the supply as soon as system comes to rest.*
- This method is *most inefficient* as *not only the K.E. of the moving masses is wasted but it requires additional energy from supply for the development of counter torque.*

For more details refer to Chapter 5.

2. **Rheostatic or dynamic braking :**

During rheostatic braking, *motor is made to work as generator and all the kinetic energy of the moving masses is converted to electrical energy which is dissipated in resistance connected as electrical load.* (Refer also to Chapter 5)

In ***traction work*** when two or three series motors are used, these motors are connected in parallel across a resistance. The K.E. of the vehicle is used in driving the motors as generators which dissipate this energy in the form of heat in the rheostats to which they are connected.

Equaliser connection. The two machines in parallel, amount to two series generators in parallel and in order that they may self-excite, an equiliser connection as shown in Fig. 8.31 has to be used. If this connection were not used, the machine that would build first would send a current through the other in the opposite direction with the result that the second machine would excite with reversed voltage. The two machines would be short-circuited on themselves and might even burnt out on account of large currents. The equiliser prevents such a condition.

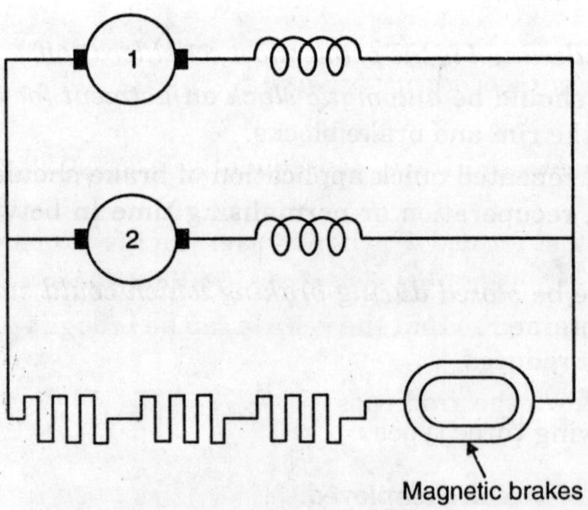

Fig. 8.31. Equiliser connection.

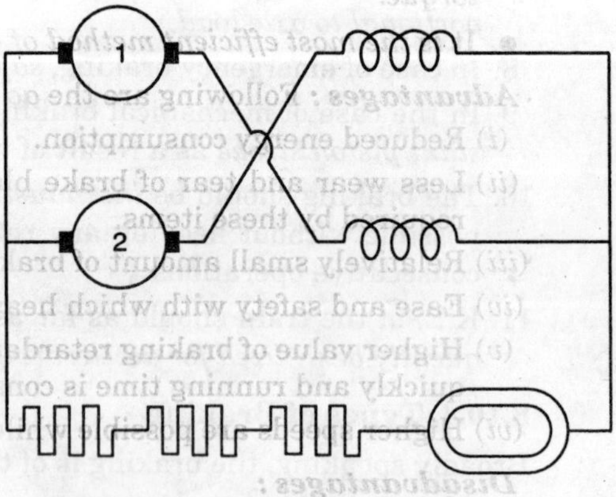

Fig. 8.32. Cross connection.

Cross-connection. Another alternative method of avoiding the above mentioned situation is cross-connecting fields of the two machines, as shown in Fig. 8.32. Suppose the voltage of machine-1 is greater than that of machine-2. It will send a greater current through field of the machine-2 causing it to excite to a higher voltage and its own excitation will be kept down because of the lesser induced e.m.f of machine-2. Thus automatic compensation is provided and the two machines undergo satisfactory operation.

The second method is advantageous to first one since if the direction of rotation of the armatures reverses, say, due to run-back, the machines fail to excite in the first method and no braking effect can be produced but with the cross-connected fields the machines build up in series and since they are short-circuited upon themselves, they provide emergency braking and would not allow the car to run-back on a gradient.

Rheostatic braking of induction motors :

- If an *induction motor* is disconnected from the supply for rheostatic braking, there would be no magnetic flux and, hence, no generated e.m.f. in the rotor and no braking torque. However, if after disconnection, *direct current is passed* through the stator, steady flux would be set up in the air-gap which will induce current, in the short-circuited rotor. This current which is proportional to the rotor speed, will produce the required braking torque whose value can be regulated by either controlling D.C. excitation or varying the rotor resistance.
- In case of *A.C. series motors* the rheostatic braking is obtained by operating the machine as generators excited from the supply or as self-excited D.C. generators supplying power to resistance load. In the former case the fields are energised at low voltage from a suitable tapping on the main transformer while in the latter case the field of the motors are excited from one of the motors acting as a series generator and in this case D.C will be generated in rotors of the motors and the kinetic energy of the rotors will be dissipated as D.C. power in the loading resistors.

3. **Regenerative braking :**

- In regenerative braking, *motor is not disconnected from the supply but is made to run as a generator by utilising the K.E. of the moving train. Electrical energy is fed back to the supply. The magnetic drag produced on account of generator action offers the braking torque.*
- It *is the most efficient method of braking.*

Advantages : Following are the *advantages of regenerative braking* :

(*i*) Reduced energy consumption.

(*ii*) Less wear and tear of brake blacks, wheels and track, consequently less maintenance required by these items.

(*iii*) Relatively small amount of brake block dust formed which increases the life of bearings.

(*iv*) Ease and safety with which heavy loads can be hauled over steep gradients.

(*v*) Higher value of braking retardation is obtained so that the vehicle can be brought to rest quickly and running time is considerably reduced.

(*vi*) Higher speeds are possible while going down the gradients.

Disadvantages :

(*i*) Increased capital cost as motors of heavier size are employed.

(*ii*) Additional cost on control equipment with consequent complications in methods of operation and control.

(*iii*) Mechanical braking is required at a speed of 6.5 km/h to bring the locomotive to rest.

(*iv*) Due to recuperated energy the operation of the sub-stations become complicated and difficult.

- *Generally regenerative braking is desirable and necessary for service lines having long gradients exceeding 0.6 percent.*

(i) Regenerative braking with D.C. motors

For regeneration to take place, the terminal voltage of the D.C. motor must exceed the supply voltage and this voltage must be kept at this value irrespective of the variation in speed or braking torque.

The D.C. series motor *cannot* be used for regenerative braking *without modification*, due to the following *reason* :

"During regeneration the current through the armature reverses and since the excitation has to be maintained, the field connection must be reversed, if a short circuit condition is to be avoided. For, if the field connections were not reversed the regenerated current in it would reverse the field which would reverse the e.m.f. of the motor and the supply voltage and back e.m.fs would aid each other setting up a short-circuit condition".

French method. One method of obtaining regenerative braking with series motors is the *French method* ; this method is discussed below :

- If there is a single series motor as in the case of a trolley-bus it is equipped with a main series field winding placed in parallel with the auxiliary field windings, as shown in Fig. 8.33.

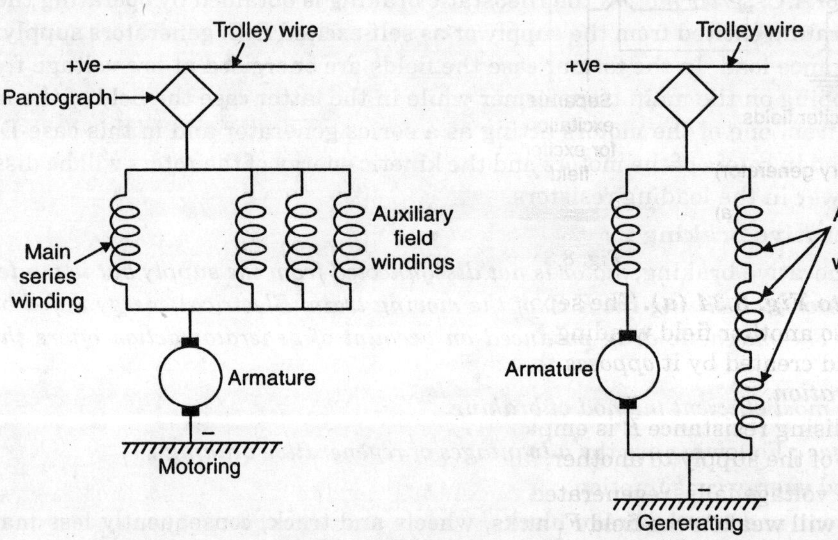

Fig. 8.33. Regenerative braking.

- During regenerative braking the auxiliary field windings are placed in series with each other and switched over in parallel across the armature and the main series field (Fig. 8.33). The machine acts as a *compound generator with slight differential* compounding. When there is a change in the line voltage, the shunt excitation being sensitive to such changes, immediately *causes the e.m.f. of the generator to increase or decrease thus providing the necessary balance.*
- In case of locomotives where four or six series motors are used, there need not be any auxiliary windings. *During normal working all the motors are in series* with their respective field windings but *during regeneration*, the *motor armatures are in parallel* with field windings of all other motors except one. This arrangement is now similar to the previous one.

Alternative method. An alternative method of regenerative braking is by *using an exciter for controlling the excitation of the field windings during regeneration.* The exciter may either be axle driven or worked from an auxiliary supply. Refer to Fig. 8.34.

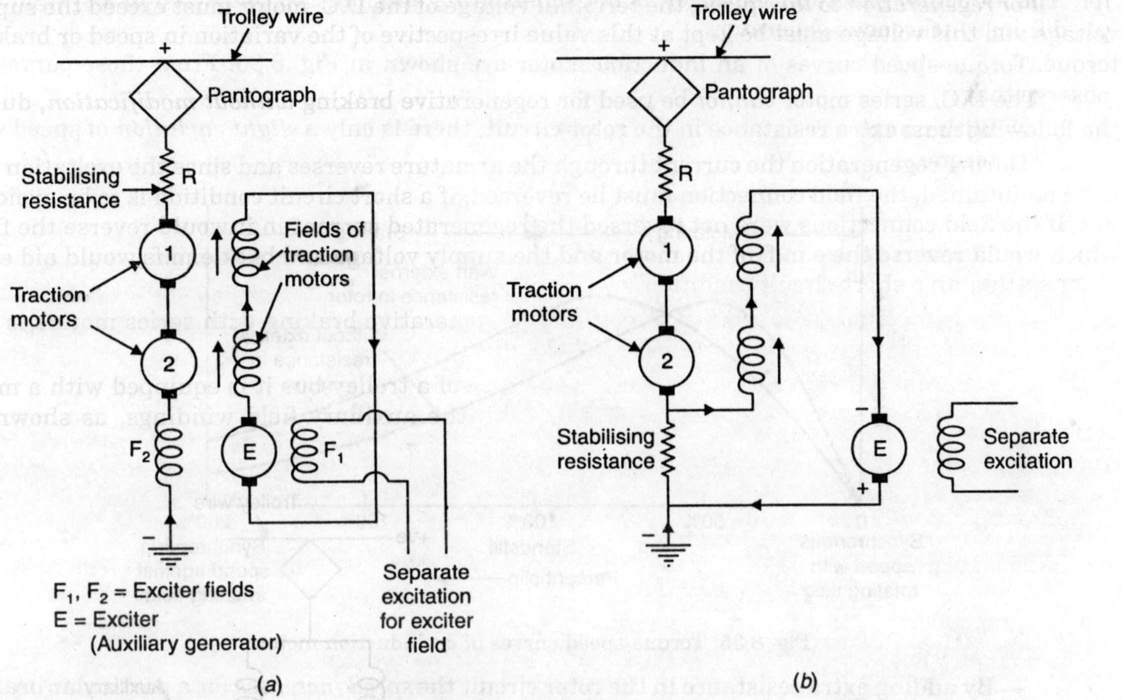

Fig. 8.34. Regenerative braking.

- ***Refer to Fig. 8.34 (a).*** The separate driven excitor has a separately excited winding F_1 and also another field winding F_2 is connected in the main circuit in such a fashion that the field created by it *opposes* the field created by separately excited winding F_1 *during regeneration.*

The stabilising resistance R is employed *to prevent current surges* when the tram crosses from one section of the supply to another, and *to compensate for variable line voltage.*

If the line voltage falls, regenerated current will tend to increase resulting in strengthening of field F_2 which will weaken the field F_1 and therefore reduction in e.m.f. generated by the exciter. Thus the field of the traction motors will be *weakened* resulting in reduction of e.m.fs generated by the traction motors operating as D.C. series generators. Thus *compensation for a decrease in line voltage is automatically provided.*

Refer to Fig. 8.34 (b). This arrangement has the exciter armature connected in the circuit of the field windings of traction motors and the stabilising resistance. The current flowing through the stabilising resistance is the *sum of exciter current and regenerated current.* The voltage of the exciter circuit can be regulated either by varying its field strength or manipulating resistances in series with armature.

When the line voltage falls the regenerated current will tend to increase resulting in increase in voltage drop across the stabilising resistance and therefore reduction in the voltage available in the exciter armature circuit *causing a reduction in the excitation current of traction motors operating as D.C. generators.* This reduces the e.m.fs generated and thus the compensation is automatically provided.

(ii) Regenerative braking with three-phase induction motors

In case of a 3-phase induction motors regenerative braking occurs automatically when the motor runs at a speed slightly above the synchronous speed ; it then works as an *induction generator.* The induction generator, however, is not self-exciting and must be connected to a system supplied from synchronous generator.

Torque-speed curves of an induction motor are shown in Fig. 8.35. From these curves we observe that :

— With no extra resistance in the rotor circuit, there is only a *slight variation* of speed with torque.

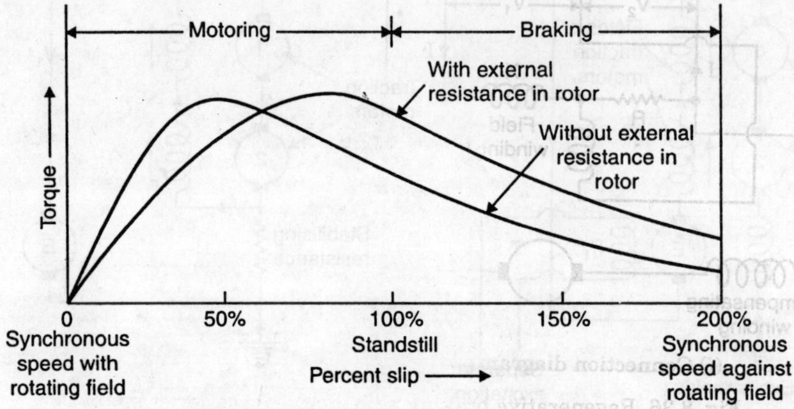

Fig. 8.35. Torque-speed curves of an induction motor.

— By adding extra resistance in the rotor circuit the *speed increases* for a particular braking torque.

Therefore, while braking without any extra resistance in the rotor circuit, the speed will be kept almost constant independent of the gradient and the load of the train. This is a *great advantage with the induction motor when used for traction.* But if increased speeds are necessary with light loads, these can be obtained by inserting external resistance in the rotor circuit. It is advantageous on *mountain* railways. It returns about 20 percent of the total energy on certain railway run and saves a great-deal of brake shoe wear.

(iii) Braking with single-phase series motors

(i) ***Rheostatic braking :***

- In this type of braking the motors are worked as *separately excited generators supplying energy to resistance load.*

  The fields are energised at low voltage from a suitable tapping on the train transformer. The K.E. of the rotor is dissipated as electrical energy in the load resistances.

- Also, the *fields of the motors may be excited from one of the motors acting as a series generator.* In this case D.C. will be generated in the rotors of the motors and K.E. of the rotors will be dissipated as D.C. power in the loading resistors.

(ii) ***Regenerative braking :***

Regenerative braking with A.C. series motors is more difficult than that with the D.C. series motor. The main difficulties relate to prevention of self excitation and attainment of a high power factor. Circuits for recuperation at high power factor usually suffer from self-excitation, and those which are inherently stable and free from self excitation usually operate at a poor p.f.

In case of "*regenerative braking*" the regenerated power should be *at the frequency of the main supply.* This necessitates the *energising of the field winding from the main supply.* Secondly, the regenerated current must be *in phase opposition* to the applied voltage and also the flux ϕ, so that power may be fed back into the supply system. The voltage supplied to the field winding must be 90° out of phase w.r.t. the supply voltage. The necessary arrangement to realise these conditions is shown in Fig. 8.36.

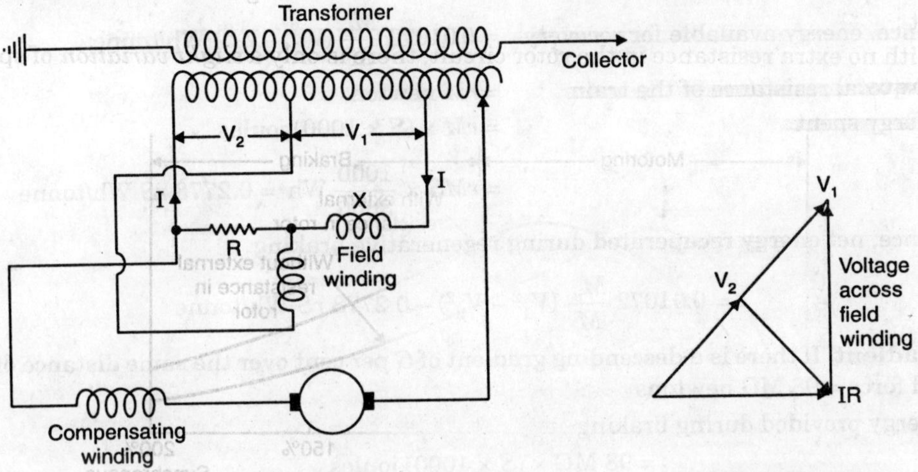

(*i*) Connection diagram (*ii*) Phasor diagram

Fig. 8.36. Regenerative braking with single-phase series motor.

8.10.6. Energy Saving in Regenerative Braking

When a train is accelerated upto a certain speed, it acquires K.E. corresponding to that speed. During the coasting period, a part of this stored energy is utilised in propelling the train against frictional and other resistances to motion and therefore the speed falls. Similarly, when the train is *going down the gradient or moving on level* track, the speed remaining the same or reduced, this *stored energy can be converted into electrical energy and returned back to the lines.*

The amount of energy returned to the supply lines depends on :

(*i*) The initial and final velocities of the train during braking.

(*ii*) The train resistance, and gradient of track also in case the train is moving down the gradient.

(*iii*) The efficiency of the system.

Let, V_1 = Initial train velocity, *km/h*,

V_2 = Final train velocity, *km/h*,

M_e = Effective mass of the train, *tonnes*,

r = Specific resistance of the train, *N/t*, and

S = Distance travelled during braking, *km*.

then, K.E. of the train at $V_1 = \frac{1}{2} M_e V_1^2$

$$= \frac{1}{2}(M_e \times 1000)\left(\frac{V_1 \times 1000}{3600}\right)^2 \text{ joules}$$

$$= \frac{1}{2}(M_e \times 1000)\left(\frac{V_1 \times 1000}{3600}\right)^2 \times \frac{1}{3600} \text{ Wh}$$

$$= 0.01072\, M_e V_1^2 \text{ Wh}$$

$$= 0.01072 \frac{M_e}{M} V_1^2 \text{ Wh/tonne}$$

Similarly, K.E. at $V_2 = 0.01072 \dfrac{M_e}{M} V_2^2$ Wh/tonne

Hence, energy available for recovery $= 0.01072 \dfrac{M_e}{M} (V_1^2 - V_2^2)$ Wh/tonne

Now total resistance of the train $= rM$ newton

Energy spent $= rM \times (S \times 1000)$ joules

$$= rMS \times \frac{1000}{3600} \text{ Wh} = 0.2778\, rS \text{ Wh/tonne}$$

Hence, net energy recuperated during regenerative braking

$$= 0.01072 \frac{M_e}{M} (V_1^2 - V_2^2) - 0.2778\, rS \text{ Wh/tonne} \quad \text{...(8.6)}$$

Gradient. If there is a descending gradient of G per cent over the same distance of S km, the downward force = 98 MG newtons

Energy provided during braking

$$= 98 \text{ MG} \times (S \times 1000) \text{ joules}$$

$$= 98 \text{ MG } S \times \left(\frac{1000}{3600}\right) \text{ Wh} = 27.25 \text{ G}S \text{ Wh/tonne}$$

Hence, *net energy recuperated* in this case

$$= \left[0.01072 \frac{M_e}{M} (V_1^2 - V_2^2) - 0.2778\, rS + 27.25\, GS\right] \text{ Wh/tonne}$$

$$= 0.01072 \frac{M_e}{M} (V_1^2 - V_2^2) + S(27.25\, G - 0.2778\, r) \text{ Wh/tonne} \quad \text{...(8.7)}$$

If η is the system efficiency, net energy returned to the line :

(*i*) ***For level track*** $= \eta \left[0.01072 \dfrac{M_e}{M} (V_1^2 - V_2^2) - 0.2778\, rS\right]$ Wh/tonne ...(8.8)

(*ii*) ***For gradient*** $= \eta \left[0.01072 \dfrac{M_e}{M} (V_1^2 - V_2^2) + S\,(27.25\, G - 0.2778\, r)\right]$ Wh/tonne ...(8.9)

Example 8.19. *A 500-tonne train travels down a gradient of 1 in 70 for 120 secs. during which period its speed is reduced from 80 km/h to 50 km/h by regenerative braking. Find the energy returned to lines if the tractive resistance is 49 N/tonne and allowance for rotational inertia is 7.5 percent. Over efficiency of the motor is 80 percent.*

Solution. *Given* : $M = 500$ tonnes ; $G = \dfrac{1}{70} \times 100 = \dfrac{100}{70}\,\%$; $t = 120$ s ; $V_1 = 80$ km/h ;

$$V_2 = 50 \text{ km/h} ;\ r = 49 \text{ N/tonne} ;\ \frac{M_e}{M} = 1.075 ;\ \eta = 80\%.$$

Energy returned to the lines :

Energy returned to lines for *gradient* is

$$= \eta \left[0.01072 \frac{M_e}{M} (V_1^2 - V_2^2) + S\,(27.25\,G - 0.2778\,r) \right] \text{Wh/tonne}$$

$$= 0.8 \left[0.01072 \times 1.075\,(80^2 - 50^2) + 2.167 \left(27.25 \times \frac{100}{70} - 0.2778 \times 49 \right) \right] \text{Wh/tonne}$$

...Eqn. (8.9)

$$= 0.8\,[44.94 + 54.86] = 79.84 \text{ Wh/tonne} \qquad \left[\because S = \left(\frac{V_1 + V_2}{2} \right) \times t = \left(\frac{80 + 50}{2} \right) \times \frac{120}{3600} = 2.167 \right]$$

$\therefore$ **Total energy returned to the line during regenerative braking**

= 500 × 79.4 = 39700 Wh or **39.7 kWh. (Ans.)**

Example 8.20. *A 400-tonne electric train has its speed reduced by regenerative braking from 60 to 40 km/h over a distance of 2 km along down gradient of 1.5 percent. Assuming specific train resistance as 50 N/t, rotational inertia effect 10% and conversion efficiency of the system 75 percent, calculate :*

(*i*) *Electrical energy returned to the line.*

(*ii*) *Average power returned to the line.* **(Bombay University)**

Solution. *Given* : M = 400 tonnes ; V_1 = 60 km/h ; V_2 = 40 km/h ; S = 2 km ; G = 1.5% ;

$$r = 50 \text{ N/tonne} ; \frac{M_e}{M} = 1.1 ; \eta = 75\%.$$

(*i*) **Electrical energy returned to the line :**

Electrical energy returned to the line

$$= \eta \left[0.01072 \frac{M_e}{M} (V_1^2 - V_2^2) + S\,(27.25\,G - 0.2778\,r) \right] \text{Wh/tonne}$$

$$= 0.75\,[0.01072 \times 1.1(60^2 - 40^2) + 2(27.25 \times 1.5 - 0.2778 \times 50)] \text{ Wh/tonne}$$

$$= 58.16 \text{ Wh/tonne} = 58.16 \times 400 \times 10^{-3} = \mathbf{23.26 \text{ kWh.}} \quad \textbf{(Ans.)}$$

(*ii*) **Average power returned to the line :**

$$\text{Average speed} = \frac{60 + 40}{2} = 50 \text{ km/h}$$

$$\therefore \quad \text{Time taken} = \frac{\text{Distance(s)}}{\text{Average speed}} = \frac{2}{50} = 0.04 \text{ h}$$

$$\text{Hence, power returned} = \frac{23.26}{0.04} = \mathbf{581.5 \text{ kW.}} \quad \textbf{(Ans.)}$$

Example 8.21. *A train weighting 400 tonnes is going down a gradient of 20 in 1000. It is desired to maintain train speed at 40 km/h by regenerative braking. If the tractive resistance is 40 N/t, rotational inertia 10% and efficiency of conversion of 75 percent, calculate the power fed into the line.* **(A.M.I.E.)**

Solution. *Given* : M = 400 tonnes ; $G = \frac{20}{1000} \times 100 = 2\%$; Speed, V = 40 km/h ;

$$r = 40 \text{ N/t} ; \frac{M_e}{M} = 1.1 ; \eta = 75\%$$

Power fed into the line :

Down-gradient tractive effort that drives the motor as generator is,

$$F_t = (98\ MG - Mr)$$
$$= (98 \times 400 \times 2 - 400 \times 40) = 62400 \text{ N}$$

Power that can be recuperated $= F_t\left(\dfrac{V \times 100}{3600}\right) = 0.2778\ F_t V$

$$= 0.2778 \times 62400 \times 40 = 693389 \text{ W}$$

Since $\eta = 0.75$, the power that can is actually fed into the lines

$$= 0.75 \times 693389 \times 10^{-3} = \mathbf{520\ kW.} \quad \textbf{(Ans.)}$$

Example 8.22. *A 2400 tonnes train (including loco) proceeds down a gradient of 1 in 80 for 5 minutes, during which period, its speed gets reduced from 60 km/h to 36 km/h by application of regenerative braking. Find the energy returned to the lines, if the tractive resistance is 49 N/tonne, rotational inertia 10 percent and overall efficiency of the motors during regeneration as 75 percent.*

(Nagpur University)

Solution. *Given* : $M = 2400$ tonnes ; $G = \dfrac{1}{80} \times 100 = 1.25\%$; $t = 5$ min. $= 300$ s ;

$V_1 = 60$ km/h ; $V_2 = 36$ km/h ; $r = 49$ N/t ; $\dfrac{M_e}{M} = 1.1$; $\eta = 75\%$

Energy returned to the lines :

Energy available due to reduction in speed

$$= 0.01072\ M_e(V_1^2 - V_2^2)$$
$$= 0.01072 \times 1.1 \times 2400\ (60^2 - 36^2) = 65205 \text{ Wh or } 65.205 \text{ kWh}$$

Tractive effort required while going down the gradient

$$= Mr - 98\ MG$$
$$= M(r - 98G) = 2400(49 - 98 \times 1.25) = -176400 \text{ N}$$

i.e., Tractive effort of 176400 N is now *available.*

Distance moved during regeneration

$$= \left[\left(\frac{V_1 + V}{2}\right) \times \frac{1000}{3800}\right] \times t = \left[\left(\frac{60 + 36}{2}\right) \times \frac{1000}{3600}\right] \times 300 = 4000 \text{ m}$$

Energy available on account of moving down the gradient over distance of 4000 m

$$= \frac{176400 \times 4000}{1000 \times 3600} = 196 \text{ kWh}$$

Total energy available $= 65.205 + 196 = 261.205$

Energy returned to the lines $= 0.75 \times 261.205 =$ **195.9 kWh. (Ans.)**

Example 8.23. *The following figures refer to the speed-current and torque-current characteristics of a 600 V D.C. series traction motor :*

Current, amperes :	*50*	*100*	*150*	*200*	*250*
Speed, km/h :	*73.6*	*48*	*41.1*	*37.3*	*35.2*
Torque, Nm :	*150*	*525*	*930*	*1335*	*1750*

Determine the braking torque at a speed of 48 km/h when operating as self-excited D.C. generator. Assume the resistance of motor and braking rheostat to be 0.6 Ω and 3.0 Ω respectively.

(Gorakhpur University)

Solution.

As motor :

Terminal voltage, $V = 600$ volts,

The motor current at a speed of 48 km/h (from speed-current characteristic curve), $I = 100$ A

Back e.m.f. developed by the motor, $E_b = V - IR_m = 600 - 100 \times 0.6 = 540$ V

As generator :

At the instant of applying rheostatic braking at a speed of 48 km/h, the terminal voltage of the machine will be equal to e.m.f. developed by the machine *i.e.* 540 V

Total resistance in the circuit $= R_m + R_{rheostat} = 0.6 + 3.0 = 3.6\ \Omega$

Current delivered by the machine, $I = \dfrac{540}{3.6} = 150$ A

The braking torque (the torque corresponding to 150 A from torque-current curve)

$= \mathbf{930\ Nm.}$ **(Ans.)**

Example 8.24. *The magnetisation characteristics of a D.C. series motor at 600 r.p.m. is given below :*

Field current, A :	*20*	*40*	*60*	*80*	*100*
E.m.f. V :	*252*	*522*	*750*	*900*	*951*

The motor is subjected to rheostatic braking against a load torque of 350 Nm. The total resistance of the armature and field circuits is 0.05 Ω. Neglecting rotational losses determine the value of the resistance to be connected in the motor circuit to limit the speed to 400 r.p.m.

Solution. The magnetisation characteristic corresponding to 400 r.p.m. is given below :

Field current, A :	20	40	60	80	100
E.m.f. V :	168	348	500	600	634

In order to determine E and I_a separately we plot besides magnetisation characteristic, the EI_a *vs* I_a characteristic

Field current, A :	20	40	60	80	100
EI_a, *watts :*	3360	13920	30000	48000	63400

Power input to the motor during braking

$$= \frac{2\pi NT}{60} = \frac{2\pi \times 400 \times 350}{60} = 14660 \text{ W}$$

Since rotational losses are neglected, power input to the motor corresponds to electromagnetic power developed by the generator during braking *i.e.* the product EI_a.

Assuming *linear variation* between successive points the current corresponding to 14660 W is

$$= 40 + \frac{14660 - 13920}{30000 - 13920} \times 20 = 40.92 \text{ A}$$

$$\therefore \qquad E = \frac{14660}{40.92} = 358.26 \text{ V}$$

Now, Total resistance in the circuit $= \dfrac{358.26}{40.92} = 8.755\ \Omega$

∴ External resistance to be connected in the motor circuit

$$= 8.755 - 0.05 = \mathbf{8.705\ \Omega}. \quad \textbf{(Ans.)}$$

Example 8.25. *The following figures give the magnetisation curve of a D.C. series motor when working as a separately excited generator of 640 r.p.m.*

Field current, amperes	*20*	*40*	*60*	*80*
E.m.f., volts	*215*	*381*	*485*	*550*

The total resistance of the motor is 0.8 Ω. When connected for rheostatic braking against a load giving a torque of 392.4 Nm it is desired to limit the speed to 600 r.p.m. What resistance must be connected across the motor during braking ?

Solution. The magnetisation characteristic corresponding to 600 r.p.m. will be a follows :

Field current, amperes	*20*	*40*	*60*	*80*
E.m.f., volts	*202*	*357*	*455*	*516*

Input of load $= \dfrac{2\pi NT}{60} = \dfrac{2\pi \times 600 \times 392.4}{60} = 24655$ W

During rheostatic braking the machine operates as a series generator, assuming no mechanical losses, the generated power will be equal to the input of the load *i.e.*, 24655 W.

From the graph (Fig. 8.37) we find that for excitation of 56 A the induced voltage is about 440 V, which gives generated power of 24655 W.

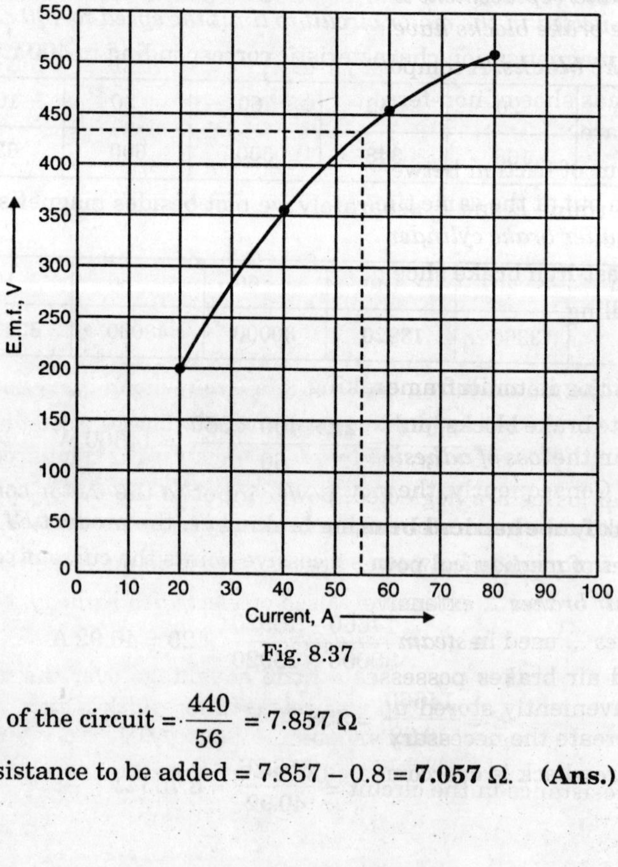

Fig. 8.37

Total resistance of the circuit $= \dfrac{440}{56} = 7.857\ \Omega$

∴ External resistance to be added $= 7.857 - 0.8 = \mathbf{7.057\ \Omega}$. **(Ans.)**

8.10.7. Mechanical Braking

8.10.7.1. Introduction

All types of electrical braking are effective only when the train is *moving above certain speed.* They are although helpful in reducing the speed of train but *cannot keep it stationary on down gradients.* As such electric braking has to supplemented with mechanical braking.

The modern tendency is to employ *regenerative braking down to a speed of 16 km/h, then rheostatic braking down to about 6.5 km/h, and finally mechanical braking to a standstill.*

Mechanical brakes are usually applied by *brake block shoes* pressed with force against the tread of wheels. Braking force is produced by the movement of piston which is transmitted to the brake blocks through the system of levers.

Brakes blocks. "*Grey cast-iron*" is the traditional material from which brake blocks have been made for all railway systems of the world. This is due to the reason that cast-iron has good *mechanical strength to withstand the rough treatment which it receives in service and is also easy to be produced in railway workshops.*

— *The treads braked by cast-iron maintain a reasonably good surface and, therefore, coefficient of adhesion between wheel and rail is not impaired.*

— Relative hardness of brake blocks and the tyre has important influence on the wear properties and coefficient of friction. If the brake material is hard it will smoothen the wheel tread with consequent reduction of coefficient of friction. This will also have wearing effect on the tread of the wheels. Therefore, from the point of view of *economic and easy maintenance and replacement it is desirable to wear brake blocks and not the wheel tyres and, therefore brake blocks have to be soften.*

Composite brake blocks. A composite brake block, which is an improvement on the integral cast-iron brake block, has shoe of non-ferrous material such as *ferrobestos*. The main *features* of a composite brake block are :

(*i*) The coefficient of friction between the shoe and the tyre is not only considerably *higher* than that with cast iron but at the same time *it remains practically constant.* This ultimately *results in economy of using smaller brake cylinder and lighter brake rigging.*

(*ii*) Absence of cast-iron brake shoe dust.

(*iii*) Ease of handling.

(*iv*) Longer life.

(*v*) No fire occurring in underframes, no thermal damage to tyres and no noise is produced.

(*vi*) The composite brake blocks, unlike cast-iron brake blocks, do not have any abrasive action on tyres which results in the *loss of adhesion by about 20 percent* as compared with treads braked by cast-iron brake blocks. Consequently, the *maximum available rate of acceleration is lowered.*

8.10.7.2. Types of mechanical brakes

Mainly two types of mechanical power brakes have been developed, these are :

1. *Compressed air brakes* ... extensively used on *electrified railway.*

2. *Vacuum brakes* ... used in *steam railways.*

The compressed air brakes possesses a little advantage over the vacuum brakes as compressed air can be conveniently stored up and released for quick action whereas in the vacuum brake, a pump has to create the necessary vacuum.

However, this drawback is overcome by the use of a vacuum reservoir.

1. The compressed air brakes system :

Fig. 8.38 shows a compressed air brake system. It consists of a reservoir of compressed air, a brake cylinder, compressed air pipe, a valve, spring, piston and a piston rod. The piston is connected to brake shoe through piston rod and levers. The brakes are kept in the "off" position by means of spring provided for the purpose.

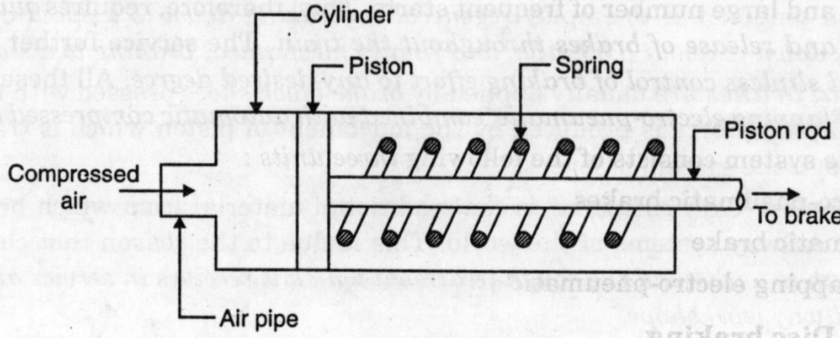

Fig. 8.38. Compressed air brake system.

- While applying brakes, compressed air is allowed to enter into brake cylinder through the air pipe and valve, which presses the piston against the force of springs. The force with which brakes are applied depends upon the quantity of compressed air allowed to enter into the cylinder. The flow of air is controlled by a valve placed at the inlet of brake cylinder.
- The *brakes are released by exhausting the air*. The piston along with the brakes return to its original position under the influence of spring force on exhausting the air. The compressed air is used at about 5.5 bar pressure.

2. The vacuum brake system :

In this system of braking, *force is applied by piston when vacuum underside piston is destroyed.*

The vacuum brake system, (Fig. 8.39) consists of a vertical cylinder having a piston and a piston rod which operates the braking arrangement through a system of levers. Vacuum is created on the top and underside of the system so that in normal condition, the piston rests at the bottom of the cylinder.

When brakes are to be applied, the vacuum is broken from the underside by admitting air at atmospheric pressure. The *piston moves up and applies the brakes.* The brakes may be *released by either recreating the vacuum or by making the pressure equal on both sides of the piston.*

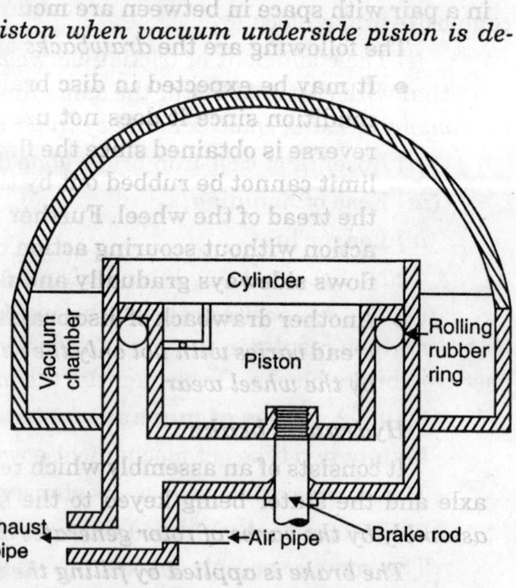

Fig. 8.39. The vacuum brake system.

Drawbacks. The vacuum brake system suffers from the following *drawbacks* ;

(*i*) The air propagation time on long goods train is very long, this makes the response to the drivers control very much sluggish (and is also very much undesirable from the point of view of safety). To reduce this time of propagation, use is made of *direct admission valves or accelerator valves.*

(*ii*) After the application of brakes, it is only after recreating the vacuum that brakes are released. This takes some time. Release of brakes on long trains can be speeded up by *quick release valves.*

8.10.7.3. Electro-pneumatic braking

Suburban service employing EMU stock is characterised by very high values of acceleration and retardation and large number of frequent starts. This, therefore, requires *quick and simultaneous application and release of brakes throughout the train.* The service further requires *accurate stops which need slipless control of braking effort to any desired degree.* All these requirements are met with in *"self lapping electro-pneumatic combined with automatic compressed air"* brake system.

The above system consists of the following *three units* :

(*i*) Electro-pneumatic brakes

(*ii*) Automatic brake

(*iii*) Self lapping electro-pneumatic drivers brake valve.

8.10.7.4. Disc braking

Braking at high speeds requires special cooling arrangement so as to conduct away the heat to avoid thermal troubles on the wheel tyres. Therefore, *for very high acceleration rates separate disc brakes with special cooling are used.*

In disc brake, frictional surfaces are the sides of disc which is either a part of the wheel or separately mounted on the wheel axle. Force is applied on pads rubbing against the disc surface.

The discs are assembled in the following ways : (*i*) Wheel web is made straight and machined on both sides to provide contact surface for brake pads. This method is *used for low energy dissipation.* (*ii*) Improvement on the method as at (*i*) is obtained by bolting cast-iron discs either in solid contact with web or with air space left to produce cooling on either side of the wheel. (*iii*) Two discs in a pair with space in between are mounted on the axle and pads being applied from outside.

The following are the *drawbacks* of disc braking :

- It may be expected in disc braking that the wheel tread will be maintained in a better condition since it does not use tread as braking surface. In actual practice, exactly the reverse is obtained since the flats produced on the wheel tread by exceeding the adhesion limit cannot be rubbed out by the friction forces as in the case of brake blocks applied on the tread of the wheel. Further the metal of the tyres is subjected to the continual rolling action without scouring action of the brake shoes to keep them in order so that the metal flows sideways gradually and flakes off and the *surface of tyre becomes uneven.*
- Another drawback of disc braking is that the equivalent tangential braking force at wheel tread *varies with not only the radius of application of the friction pad but it is affected also by the wheel wear.*

Hydraulic brake :

It consists of an assembly which resembles a *hydraulic coupling*, the rotor being keyed to the axle and the stator being keyed to the bogie frame. *The circulation of fluid (oil or water) in the assembly by the vanes of rotor generates a braking torque.*

The brake is applied by filling the coupling by means of a pump.

8.10.7.5. Eddy current brakes

Eddy current brakes are of the following two types :

(*i*) Linear type (*ii*) Rotary type.

(*i*) ***Linear type.*** In linear type eddy current brakes, primary member is the shoes which carries excitation current and the rail itself forms the secondary element of the eddy current system.

While braking, traction motors can be made to work as generators and supply power to the shoe of eddy current brakes.

(*ii*) ***Rotary type.*** In this type of brake, the secondary member *i.e.*, the rotor is attached to the motor shaft and primary member which is supplied with D.C. excitation is fixed in space. This is a case similar to D.C. dynamic braking as applied to induction motor.

The torque produced by the induced currents in the rotor will tend to reduce the relative speed between the rotor and stationary field *i.e.*, it will be a *braking torque.* This torque can be varied only by varying the D.C. excitation of the stator.

8.10.7.6. Magnetic track brake

Magnetic brake is fitted in between the wheels of the bogie and runs longitudinal along the track ; it is used in tram cars.

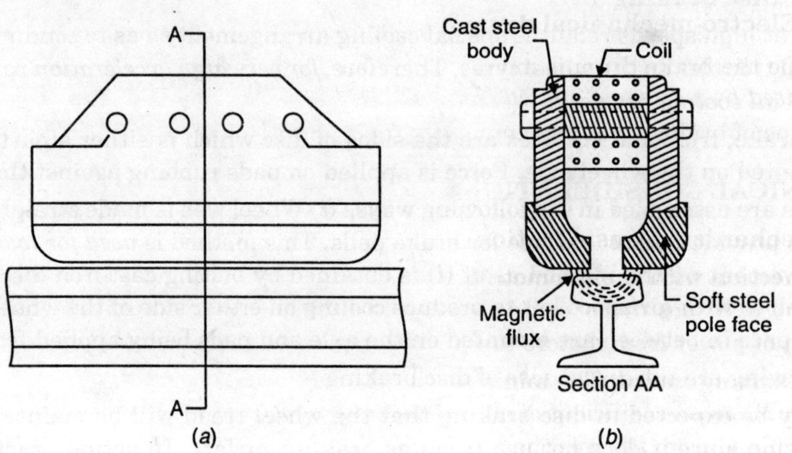

Fig. 8.40. Magnetic track brake.

Fig. 8.40 shows a magnetic track brake. It consists of a bipolar magnet with elongated pole faces a short distance apart and along with rails. The body is made of cast steel and the pole faces are made of soft steel and can be renewed. Pole faces are parallel to the rails. The exciting coil is enclosed in a water-tight case. The magnetic flux is perpendicular to the pole faces and track. The force of attraction between the magnet and track is given by

$$F = \frac{B^2 A}{2 \times \mu_0} = \frac{B^2 A}{2 \times 4\pi \times 10^{-7}} \text{ newtons} \quad (8.10)$$

where, B = Flux density, Wb/m^2, and

A = Area of pole face, sq.m.

This magnetic force in effect increases the weight on braking wheels with the result that braking force of magnitude "μF" is produced.

Example 8.26. A *tramcar is fitted with two magnetic track brakes with an area of 17.75 cm^2 per pole. Find the braking effect if the flux is 2.6 mWb and coefficient of friction 0.2. Calculate the rate of retardation produced by this braking effect in a car weighing 11 tonnes.*

Solution. *Given* : $A = 17.75\ cm^2 = 17.75 \times 10^{-4}\ m^2$; $\phi = 2.6$ mWb $= 2.6 \times 10^{-3}$ Wb ;
$\mu = 0.2$; $M = 11$ tonnes.

Braking effect F :

There are two track brakes. Each brake has two pole faces. Therefore there are four pole faces.

$$F = \frac{4B^2 A}{2 \times 4\pi \times 10^{-7}} = \frac{(\phi/A)^2 A}{2\pi \times 10^{-7}} = \frac{\phi^2}{2\pi \times 10^{-7} \times A}$$

$$= \frac{(2.6 \times 10^{-3})^2}{2\pi \times 10^{-7} \times 17.75 \times 10^{-4}} = \mathbf{6061\ N.}\ \textbf{(Ans.)}$$

Rate of retardation β :

Braking force $= \mu F = 0.2 \times 6061 = 1212.2$ N

Now, $1212.2 = 11 \times 1000 \times \beta$

$\therefore$ $\beta = \mathbf{0.11\ m/s^2}$. **(Ans.)**

8.10.7.7. Electro-mechanical drum brakes

In this type the brake drum is mounted on the motor shaft and *brake shoes are applied by springs and released by a solenoid excited from a battery.*

- This type of brake is usually employed on *trolley buses* and *cars.*

8.11. MECHANICAL CONSIDERATIONS, CONTROL AND AUXILIARY EQUIPMENT

8.11.1. Mechanical Considerations

- In connection with the locomotive, the main considerations are the *wheel arrangement* and *general disposition of equipment* in order to have good riding properties and the transmission of power from motors to the driving wheels.

1. **Wheel arrangement.** The minimum number of driving axles (N_{axles}) required, for a certain tractive offer F_t, given coefficient of adhesion μ and the maximum permissible weight per driving axle, is determined from the relation :

$$N_{axles} = \frac{F_t}{\mu \times \text{maximum permissible weight per driving axle}}.$$

- In case of locomotive since the electrical equipment is lighter in weight, so whole of the locomotive weight is placed on the driving axles so that required effort may be obtained.
- In case of A.C. locomotives, since the electrical equipment is comparatively heavy, so the required tractive effort can be obtained by placing only a portion of the weight on the driving wheels and supporting the rest on trailing axles, the riding qualities of high speed locomotives are improved by the trailing axles, which are often used on D.C. locomotives for this purpose.

2. **Drive.** The electric train may have the following two types of drives :

(*i*) Individual drive (*ii*) Collective drive.

In *individual drive* separate motor of smaller size is employed for each driving axle whereas in collective drive a single large size motor is employed for driving all the axles through connecting rods as in steam locomotives.

The *collective drive* has the following *advantages and disadvantages over individual* drive :

Advantages :

(*i*) The motor for collective drive need not be paid so much attention in designing with regard to dimensions as much that for individual drive.

(*ii*) The wheels when coupled have got less tendency to slip.

(*iii*) The motor can be mounted well up in the main body of the locomotive therefore, the centre of gravity is raised thereby reducing the wear on track.

Disadvantages. Collective drive is not suitable for high speed locomotives because of stresses and vibrations in the connecting rods of high speeds.

- *Individual drive is employed for high speed locomotives and collective drive for low speed and heavy (goods) locomotives.*

3. **Transmission of drive.** There are several methods of transmitting power from the armature of electric motors to the driving wheels, a few of these are described below.

(*i*) ***Gearless drive :***

(*a*) *Direct drive.* In this type of drive bipolar motors are employed whose armatures are mounted directly on the driving axle with the field attached to the frame of the locomotive. The pole faces of the motors are approximately flat so that the armature can move relatively to them without affecting the operation. The *size of the motor is limited by the diameter of driving wheels.* This method has *limited application* due to following *reasons :*

— Relatively large unsprung weight ;

— Poor utilisation of material ;

— Low centre of gravity of the locomotives.

(*b*) *Direct quill drive.* The *quill* is a *hollow shaft* which surrounds the driving axle and is connected to it by springs or some other flexible device. The armature of the motor is directly mounted on the quill and therefore it is entirely *spring borne.*

(*ii*) ***Geared drive.*** The motor size is reduced with the increase in speed for a given output, so the gear drive is necessary to reduce the motor size for given output by running it at a higher speed. Commonly used gear ratio is 3 to 5 : 1.

(*a*) *Nose suspension.* A nose suspended motor is the most common type of geared drive. In this case motor is *partly supported from the driving axles and partly between springs which are put on the vehicle frame.* More than 50% of the weight of the motor and gear is thus spring borne. The motor is coupled to the driving axles by means of gears, one being on the same axle and the other on the shaft of the motor armature.

This method is usually employed for *motor coaches and tramways.* This system is *not* used on large locomotives due to the relatively large unsprung weight.

(*b*) *Geared quill drive.* This type of drive is employed when the motor is *to be placed higher in the frame* than is possible with the nose suspension so that a motor of *larger* size (*output*) *can be employed, centre of gravity can be raised and when unsprung weight is to be reduced.* In this case quill surrounds the driving axle (as in direct quill drive) and gear is mounted on the quill instead of motor armature.

8.11.2. Control and Auxiliary Equipment

Control equipment. The control equipment is required for the following purposes :

(*i*) To connect the motor(s) to the supply mains without taking excessive current at start and *provide smooth acceleration without causing sudden shock so as to avoid damage to the coupling.*

(*ii*) To provide *rheostatic or regenerative braking.*

The operation may be :

(*a*) By hand ; (*b*) Semi-automatic.

(*a*) "*Operation by hand*". In this method the closing of the accelerating contactors is governed by the position of the controller handle and the rate of acceleration remains completely under the control of the driver.

The hand operation is preferred in *locomotives where there is a probability of divergences in the hauling loads and possibility of slipping of wheels.*

(*b*) *"Semi-automatic operation"*. In this type of operation the position of the controller handle governs the number of contactors which will close, but the sequence of operation and the time interval between their individual operations are *governed by time-lag devices*, as in fully automatic equipment.

Semi-automatic operation is *sometimes employed for suburban services in order to ensure uniform acceleration with a minimum energy consumption.*

- For *trams and small industrial locomotives plain drum controller is used.*
- For *railway work* where very high currents are to be dealt with, *'master controller system'* is employed. In this system the contacts instead of being made or broken by hand are closed or opened by the switches known as *contactors* operated by solenoids.

Following methods are usually employed for closing and opening of the contactors by the master controller :

(*i*) *"All electrical method"*. Here, the contactors are *closed or opened by the solenoids from the master controller.*

The main disadvantage of this method is that there may be large fluctuations in the supply voltage which may cause uncertainty of action of the contactors.

(*ii*) *"Electric-pneumatic method"*. In this case contactors are *operated by the compressed air.* The electrically operated solenoids operate the valves, which admit the air into the operating cylinders.

This system is preferred to all electric operation system because of the following *reasons* :

(*i*) The contactor groups are more compact and robust.

(*ii*) Each valve magnet requires only a few watts for operation whereas the solenoids in all electric operation system may require from 100 to 200 V.

(*iii*) A low voltage control can be employed which reduces the size of master controller and gives reliability to the system.

- All *electric trains are almost universally equipped with compressed air brakes, the supply of air for electro-pneumatic operation is therefore always available.*

(*iii*) *"Camshaft drive"*. In this method the contactors are closed or opened directly by cams, which are mounted on a camshaft driven by an electric or pneumatic motor. This method is *widely used* since electrical connections are very simple.

This method has the following *advantages* :

(*i*) Occupies less space.

(*ii*) Less wear and tear.

(*iii*) Easy maintenance.

(*iv*) Light in weight.

Auxiliary equipment :

In addition to main traction motors required to be installed on the electric locomotives, the auxiliary equipment provided include the following :

1. Motor-generator set
2. Battery
3. Air compressor
4. Heating equipment
5. Pantograph equipment
6. Traction motor blowers.

1. Motor-generator set

- It consists of a *series motor and a shunt generator*. It is of the two-bearing type.
- At light load the speed is limited by ventilating fan and the voltage of the generator is controlled by an automatic regulator.

- It is required to supply lighting, control system and other power circuits at low voltage from 30 to 100 V.
- The design of the set must be specific so that fluctuations of line voltage do not have any effect on the low voltage supply.

2. **Battery :**

- Batteries are provided to serve the following purposes :

(*i*) To operate the air blast circuit breaker.

(*ii*) To raise the pantograph.

(*iii*) To run the auxiliary compressor.

Arno converter provides cabin lighting.

- Batteries may be lead type or alkaline type. The battery is either charged by separate rectifier or from D.C. generator.
- The capacity of the battery depends on the vehicle and may be between 175 and 375 Ah at the 5 hours.
- The batteries are usually connected in parallel with the motor-generator set.

3. **Air compressor :**

- The air compressor is usually of two cylinder type and is directly connected to the motor.
- With compressors for motor coaches, a moderate speed motor is employed so that weight is reduced and the compressor is driven through double helical spur gearing.

4. **Heating equipment :**

- Heat equipment is required for heating the locomotive and train if desired.
- This is supplied directly from the line upto 3000 V.

5. **Pantograph operating equipment.** The pantograph is usually raised and lowered by compressed air. This is accomplished by the provision of suitable switches for controlling the solenoid operated valves of the air supply.

6. **Traction motor blowers :**

- The traction motor blowers are required on locomotives when *forced ventilated motors are used*. In some cases a single blower is used which delivers air at a low pressure into a central duct built into the underframe of the locomotive body from which it is distributed to the motors.
- In case of large frame mounted motors, each motor is provided with a separate blowers. The blowers are then mounted on the motor frames and two or more blowers are coupled together and are driven by a single motor.

HIGHLIGHTS

1. Some of the motors which find application is traction are :

(*i*) D.C. series motors supplied with straight D.C. or rectified D.C ;

(*ii*) A.C. series motors-single-phase ;

(*iii*) Repulsion motors ;

(*iv*) Three-phase induction motors ;

(*v*) Linear induction motors.

2. The series motors are less susceptible to sudden change in supply voltage.

3. Following factors are favourable for the application of 3-phase induction motors for traction purposes :

(*i*) Simple and robust construction ;

(*ii*) Absence of commutator ;

(*iii*) High efficiency ;

(*iv*) Simple regenerative braking.

4. *Linear induction motor* forms excellent source of motive power for magnetically suspended trains when conventional motor which depends for torque conversion to linear tractive force upon the adhesive weight or driving wheels, fails.
5. *Series-parallel control method* is so named since the this method of control assumes that at least two motors are being used which are connected in series at start for low speed and in parallel for full speed running.
6. In poly-phase induction motors the following methods of speed control are used :

   (*i*) Rheostatic control ;

   (*ii*) Pole changing ;

   (*iii*) Cascade control ;

   (*iv*) Combination of cascade and pole-changing arrangement.
7. In *electric braking* the braking energy is converted into electrical energy instead of converting it into heat-energy at the brake shoes and either dissipated in the resistances mounted on the vehicle or re-turned to the supply system.
8. The *plugging* method of braking involves reconnection of motor to supply in such a way that motor now develops torque in opposite direction to the movement of the motor.
9. During *rheostatic braking*, motor is made to work as generator and all the K.E. of moving masses is converted to electrical energy which is dissipated in resistance connected as electrical load.
10. In *regenerative braking*, motor is not disconnected from the supply but is made to run as a generator by utilising the K.E. of the moving train. Electrical energy is *fed back to the supply.* The magnetic drag produced on account of generator action affects the braking torque. It is the *most efficient method of braking.*
11. The main types of mechanical brakes are :

    (*i*) Compressed air brakes ; (*ii*) Vacuum brakes.

OBJECTIVE TYPE QUESTIONS

Fill in the Blanks or Say 'Yes' or 'No' :

1. In A.C. traction single-phase supply is used due to economic reason.
2. Both in A.C. and D.C. traction, the injection of harmonics can lead to maloperation of signals and interference in the lines.
3. D.C. series motors are rarely used in traction.
4. The major disadvantage of D.C. motor is the presence of commutator and ...... , which require frequent maintenance, particularly when the flashover at the commutator occur due to sharp voltage fluctua-tions.
5. The ...... properties of series motors are quite good.
6. Owing to low weight and high starting torque developed, D.C. series motors are capable of producing high rates of acceleration.
7. In series motor the speed automatically decreases as the torque increases.
8. The series motors are ...... susceptible to sudden change in supply voltage.
9. The series motor due to its desirable speed-torque characteristics is almost exclusively used in railway service.
10. A linear induction motor can be used on trolley cars for internal transport in workshops
11. ...... induction motor has superiority over conventional motor for speeds over 200 km/h.
12. In traction work, usually two or more similar motors are employed.
13. Machines with more than two brush sets per pair of poles are called ....
14. ...... is a device which converts power at constant voltage and variable current into one with constant current and variable voltage.

15. By using metadyne converter, regenerative braking can be obtained very easily by reversing the field of traction motor.
16. Brake actuation time should be as ...... as possible.
17. The braking should be exhaustible.
18. The ...... method of braking involves reconnection of motor to supply in such a way that motor now develops torque in opposite direction to the movement of the rotor.
19. During ...... braking, motor is made to work as generator and all the K.E. of the moving masses is converted to electrical energy which is dissipated in resistance connected as electrical load.
20. In ...... braking, motor is not disconnected from the supply but is made to run as a generator by utilising the K.E. of the moving train. Electrical energy is fed back to the supply.
21. Generally regenerative braking is desirable and necessary for service lines having long gradients exceeding 0.6 percent.

Answers

1. Yes	2. Yes	3. No	4. Brushes	5. Commutating
6. Yes	7. Yes	8. less	9. Yes	10. Yes
11. Linear	12. Yes	13. metadynes	14. Metadyne	15. Yes
16. small	17. No	18. plugging	19. rheostatic	20. regenerative
21. Yes.				

THEORETICAL QUESTIONS

1. State the significant features of traction drives.
2. Discuss briefly the desirable properties of traction motors.
3. What are the chief requirements of a traction motor with regards to electrical and mechanical features ?
4. Give the essential electrical and mechanical characteristics of traction motors.
5. State the mechanical and electrical features of electric traction motors and discuss the relative suitability of (*i*) D.C. series motor, (*ii*) A.C. series motor.
6. Enumerate the motors which commonly find application in traction.
7. State the advantages of squirrel-cage induction motor over D.C. motors.
8. What is the major disadvantage of a D.C. motor ?
9. Discuss the suitability of series motors for traction duties with the help of characteristic curves.
10. Discuss in detail why series motors are ideal for D.C. or A.C. traction.
11. What is the effect of changing wheel diameter and gear ratio on the characteristics of a motor ?
12. State the effects of wheels that are worn out when used along with new wheels to drive a train.
13. What speed-torque characteristics are desirable for traction motors operating (*i*) suburban services ; (*ii*) main line service ?
14. Explain how the difference in driving wheel diameters due to unequal wear affects the sharing of load by two similar series motors, working in parallel, driving an electric train.
15. What type of A.C. motor is usually employed for single-phase electric traction ? Discuss briefly the principal features in the construction of the motor and analytically how good commutation and high power factor are obtained. For what frequency and voltage are such motors usually built and why ?
16. Explain briefly the construction and characteristics of A.C. series motor, pointing out how they differ from the D.C. type. In what way is the good commutation and high power factor assured ?
17. Discuss with neat sketches the construction and working principle of high acceleration linear induction motor. Discuss its advantages and disadvantages.
18. State the merits and demerits of the induction motor for traction duties.
19. Discuss the advantages of series-parallel starting against the ordinary rheostatic starting for a pair of D.C. traction motors.

20. Explain with the help of suitable circuit diagrams (*i*) shunt transition, (*ii*) bridge transition as applied to a pair of D.C. traction motors.
21. Give reasons for the preference of motor coach trains to locomotive hauled trains for suburban service.
22. What type of series-parallel transition is used for suburban trains ? Give reasons.
23. State the advantages of 'Metadyne' control of D.C. traction motors. Explain the working of 'Metadyne' with the help of neat circuit diagram.
24. Discuss (*i*) Multiple unit control and (*ii*) Metadyne system of control of D.C. motors.
25. Discuss briefly the methods of speed control of single-phase A.C. series motors and 3-phase induction motors.
26. What are requirements which an ideal braking system should possess ?
27. What is a composite brake block ? How does it differ from integral brake block ?
28. Explain vacuum brake system and give the drawbacks this system of braking suffers from.
29. What is disc braking and what are its advantages and disadvantages ?
30. Explain how hydraulic and eddy current brakes work. What are their advantages ?
31. Explain how magnetic brake works.
32. Name the various methods of electric braking. Give the merits and limitations of regenerative braking.
33. What is the basic principle involved in the rheostatic braking of D.C. motors used in traction ?
34. Explain the basic principle involved in electric braking of traction motors.
35. What are the advantages and disadvantages of the regenerative braking of electric traction motor ?
36. With the help of neat diagram, briefly explain 'French method of regenerative braking' with series motors.
37. Discuss and distinguish between rheostatic and regenerative braking applied in electric traction. Give the advantages of regenerative braking.
38. Show how regenerative braking is applied to a D.C. series motor when there are at least two motors working in parallel. How to compensate the effect of variation of voltage in the overhead line ?
39. Explain briefly how regenerative braking differs from other methods of braking traction motors. With neat and explanatory notes, demonstrate the usefulness of this method of braking with (*a*) D.C series motors and (*b*) A.C. series motors in traction.
40. Explain how regenerative braking can be obtained in D.C. locomotives.
41. Explain with proper diagrams the following braking systems used in traction :
    (*i*) Vacuum brakes
    (*ii*) Compressed air brake
    (*iii*) Magnetic track brake
    (*iv*) Electro-mechanical drum brake
    (*v*) Regenerative braking with 3-phase induction motors.
42. Describe the various methods of drives which have been applied for transmitting motive power from motor shaft to driving wheels.
43. Explain the method of quill drive of transmitting power from motor to the driving wheels.
44. State with sketches, the three important methods of transmission of mechanical power from traction motor to the driving wheels.

UNSOLVED EXAMPLES

1. Two D.C. traction motors run at speeds of 900 r.p.m. and 950 r.p.m. respectively when each takes a current of 50 A from 500 V mains. Each motor has an effective resistance of 0.3 Ω. Calculate the speed and voltage across each machine when mechanically coupled and electrically connected in series and taking a current of 50 A from 500 V mains, the resistance of each motor being unchanged.

[**Ans.** 447.87 r.p.m. ; 256.35 V, 243.65 V]

2. A tram car is equipped with two motors which are operating in parallel. Calculate the current drawn from the supply main at 500 V when the car is running at a steady speed of 40 km/h and each motor is developing a tractive effort of 1800 N. The resistance of each motor is 0.4 Ω. The friction, windage and other losses may be assumed as 3200 W per motor. **[Ans.** 96.5 A]

3. The supply fed to the series connection is 650 V. If this motor is geared to driving wheel of radius 45 cms and other to 43 cms and if the speed of first motor when connected in parallel to second motor across the main supply lines is 400 r.p.m., determine the speeds of the motors when connected in series. Assume armature current to remain same and armature voltage drop of 10 per cent at this current. **[Ans.** 175 r.p.m ; 184 r.p.m.]

4. The tractive effort exerted by a locomotive while hauling a train on level track at 50 kmph is 35 kN. If the locomotive has to haul the same train at the same speed on a gradient and the tractive effort required is 55 kN, determine the H.P. delivered by the locomotive when the motors used are : (*i*) D.C.-series motors ; (*ii*) Induction motors. **[Ans.** 828.5 H.P. ; 1101.5 H.P.]

5. A motor-coach bogie having two series motor is accelerated uniformly to a speed of 48 km/h in 30 secs. with the help of a series-parallel controller. If the average tractive effort per motor is 13350 N, calculate the approximate loss of energy in the starting rheostats. **[Ans.** 0.741 kWh]

6. A 400-tonne train travels down a gradient 1 in 70 for 120 secs. during which period its speed is reduced from 80 km/h to 50 km/h by regenerative braking. Find the energy returned to lines if the tractive resistance is 49 N/tonne and allowance for rotational inertia is 7.5 percent. Overall efficiency of motors is 75 percent. **[Ans.** 29.94 kWh]

7. A train weighing 500 tonne is going down a gradient of 20 in 1000. It is desired to maintain train speed at 40 km/h by regenerative braking. Calculate the power fed into the line. Tractive resistance is 40 N/t and allow rotational inertia of 10 percent and efficiency of conversion of 75 percent. **[Ans.** 650 kW]

8. A tram car is fitted with two magnetic track brakes each with an area of 16.13 cm^2 per pole face. Find the braking effect if the flux is 2.5 mWb and the coefficient of friction is 0.2. Calculate the rate of retardation produced by this braking effect in a car weighing 12 tonnes. **[Ans.** 10.28 m/s^2]

9. The following figures gives the magnetisation curve of a D.C. series motor when working as a separately excited at 500 r.p.m.

Field current (amps.)	10	20	30	40
E.M.F. (volts)	144	254	324	366

The total resistance of the motor is 1.2 Ω. Deduce the speed-torque curve for this motor when operating at a constant voltage of 400 volts.

10. The following figures refer to the speed-current and tractive effort-current characteristic of a D.C. series traction motor :

Current (amps.)	40	80	120	160	200
Speed (kmph)	51.5	33.6	28.8	26.0	24.6
Tractive effort (N)	1,200	4,200	7,400	10,700	14,000

The diameter of a car wheel is 96 cm, the gear ratio is 4 : 1. The car wheel is changed to 92 cm, the gear ratio to 3.5 : 1. Give the new characteristics.

11. Two D.C. traction motors run at speeds of 700 and 750 r.p.m. respectively when taking a current of 70 A from 500 volts main. Each motor has an effective resistance of 0.32 Ω. Calculate the speed and the potential difference across each machine when mechanically coupled and connected in series and taking a current of 50 A from 500 V mains. The resistance of each motor being unchanged.

[Ans. 350 rpm. ; 258 V ; 242 V]

12. Two D.C. series would traction motors run at 800 and 850 r.p.m. when each is connected to 700 volt D.C. mains and takes 100 A. The effective resistance of each motor is 0.5 Ω. If the two motors are connected in series to a 1,500 volt supply and draw 100 A and are also mechanically coupled, at what speed will they run and what will be the voltage across each motor.

[A.M.I.E. Sec. B. Elec. Traction, May 1970] **[Ans.** 887 r.p.m. ; 771 V ; 729 V]

13. **A locomotive exerts a tractive effort of 28,200 N in hauling a train at 48 kmph on the level track. If the motor is to haul the same train on a gradient and the tractive effort required is 44,000 N determine the power delivered by the locomotive in case it is driven by (*i*) D.C. series motors (*ii*) Induction motors.**

[Ans. **375 kW ; 470 kW]**

14. **A motor coach is required with two D.C. series motors, the characteristics of each at 675 V with 107 cm wheels are as follows :**

Speed (kmph)	**55.5**	**46.3**	**41**	**37.4**
Tractive effort (kg)	**610**	**1,120**	**1,680**	**2,240**

If the wheels driven by one motor (*A*) are 107 cm in diameter and those driven by the other motor (*B*) are 101.5 cm in diameter, determine the total tractive effort at a train speed of 43 kmph when the motors are operating in parallel at normal voltage. **[Ans. 2,625 kg]**

15. **A coach is driven by two similar series motors *A* and *B*. *A* drives 105 cm wheels and *B* drives 100 cm wheels. Characteristics of motor *A* are as follows :**

Current (amperes)	**200**	**300**	**400**	**500**
Speed (kmph)	**50.5**	**42.5**	**38.3**	**35.3**
Tractive effort (kg)	**730**	**1340**	**1970**	**2580**

Calculate the input to and tractive effort of each motor when (*a*) motors are in parallel and speed is 40 km per hour (*b*) motors are in series taking 400 A. For each motor, total resistance in the armature circuit is 0.8 Ω. Assume the motors are connected to 600 V D.C. supply system.

[Ans. (*a*) 213 kW ; 183 kW ; 1,675 kg, 1450 kg. (*b*) 117.4 kW, 122.6 kW ; 1,970 kg ; 2,068 kg]

9

Economics of Power Generation, Electric Power Supply and Utilisation

9.1. INTRODUCTION

In all fields of industry *economics plays an important role.*

In *power plant engineering* economics of power system use certain well established techniques for choosing the most suitable system. The power plant design must be made on the basis of most economical condition and not on the most efficient condition as the profit is the main basis in the design of the plant and its effectiveness is measured financially. *The main purpose of design and operation of the plant is to bring the cost of energy produced to minimum.* Among several factors, the efficiency of the plant is one of the factors that determine the energy cost. In majority of cases, unfortunately, the most thermally efficient plant is not economic one.

- In the *fields of generation, transmission, distribution and utilisation of electrical power, economic problems occur.* The electrical engineer, after studying the relative costs of possible schemes, adopts the cheapest and most convenient scheme.

9.2. TERMS AND DEFINITIONS

1. Connected load. The connected load on any system, or part of a system, *is the combined continuous rating of all the receiving apparatus on consumers' premises, which is connected to the system, or part of the system, under consideration.*

2. Demand. The demand of an installation or system *is the load that is drawn from the source of supply at the receiving terminals averaged over a suitable and specified interval of time.* Demand is expressed in kilowatts (kW), kilovolt-amperes (kVA), amperes (A), or other suitable units.

3. Maximum demand or Peak load. The maximum demand of an installation or system is the greatest of all the demands that have occurred during a given period. It is determined by measurement, according to specifications, over a prescribed interval of time.

4. Demand factor. The demand factor of any system, or part of a system, is *the ratio of maximum demand of the system, a part of the system, to the total connected load of the system, or of the part of the system, under consideration.* Expressing the definition mathematically,

$$\text{Demand factor} = \frac{\text{Maximum demand}}{\text{Connected load}} \quad ...(9.1)$$

5. Load factor. The load factor is *the ratio of the average power to the maximum demand.* In each case, the interval of maximum load and the period over which the average is taken should be definitely specified, such as a "half-hour monthly" load factor. The proper interval and period are usually *dependent upon local conditions and upon the purpose for which the load factor is to be used.* Expressing the definition mathematically,

$$\text{Load factor} = \frac{\text{Average load}}{\text{Maximum demand}} \quad ...(9.2)$$

6. Diversity factor. The diversity factor of any system, or part of a system, *is the ratio of the maximum power demands of the subdivisions of the system, or part of a system, to the maximum demand of the whole system, or part of the system, under consideration, measured at the point of supply*. Expressing the definition mathematically,

$$\text{Diversity factor} = \frac{\text{Sum of individual maximum demands}}{\text{Maximum demand of entire group}} \quad ...(9.3)$$

It is always *greater than unity.*

- The *reciprocal of the diversity factor is called the* coincidence factor. It is always *less than one.*

7. Utilization factor. The utilization factor is defined as *the ratio of the maximum generator demand to the generator capacity.*

8. Plant capacity factor. It is defined as the *ratio of actual energy produced in kilowatt hours (kWh) to the maximum possible energy that could have been produced during the same period.* Expressing the definition mathematically,

$$\text{Plant capacity factor} = \frac{E}{C \times t} \quad ...(9.4)$$

where, E = Energy produced (kWh) in a given period,

C = Capacity of the plant in kW, and

t = Total number of hours in the given period.

9. Plant use factor. It is defined as *the ratio of energy produced in a given time to the maximum possible energy that could have been produced during the actual number of hours the plant was in operation.* Expressing the definition mathematically,

$$\text{Plant use factor} = \frac{E}{C \times t'} \quad ...(9.5)$$

where, t' = Actual number of hours the plant has been in operation.

10. Installed capacity. *The total of station capacities available to supply the system-load* is called the "*installed capacity*".

11. Dump power. This term is used in hydroplants and it shows the *power in excess of the load requirements* and it is made available by *surplus water.*

12. Firm power. It is the power which *should always be available even under emergency conditions.*

13. Prime power. It is the power which may be mechanical, hydraulic or thermal that is *always available for conversion into electric power*.

14. Cold reserve. It is that *reserve generating capacity which is not in operation but can be made available for service.*

15. Hot reserve. It is that *reserve generating capacity which is in operation but not in service.*

16. Spinning reserve. It is that *reserve generating capacity which is connected to the bus and is ready to take the load.*

17. Types of loads :

(*i*) ***Residential load.*** This type of load includes domestic lights, power needed for domestic appliances such as radios, television, water heaters, refrigerators, electric cookers and small motors for pumping water.

(*ii*) ***Commercial load.*** It includes lighting for shops, advertisements and electrical appliances used in shops and restaurants etc.

(*iii*) ***Industrial load.*** It consists of load demand of various industries.

(*iv*) ***Municipal load.*** It consists of street lighting, power required for water supply and drainage purposes.

(*v*) ***Irrigation load.*** This type of load includes electrical power needed for pumps driven by electric motors to supply water to fields.

(*vi*) ***Traction load.*** It includes trams, cars, trolley, buses and railways.

18. Load curve. Refer Fig. 9.1. A load curve (or load graph) *is a graphic record showing the power demands for every instant during a certain time interval.* Such a record may cover 1 hour, in which case it would be an *hourly load graph* ; 24 hours, in which case it would be a *daily load graph* ; a month in which case it would be a *monthly load graph* ; or a year (8760 hours), in which case it would be a *yearly load graph*.

The following points are worth noting :

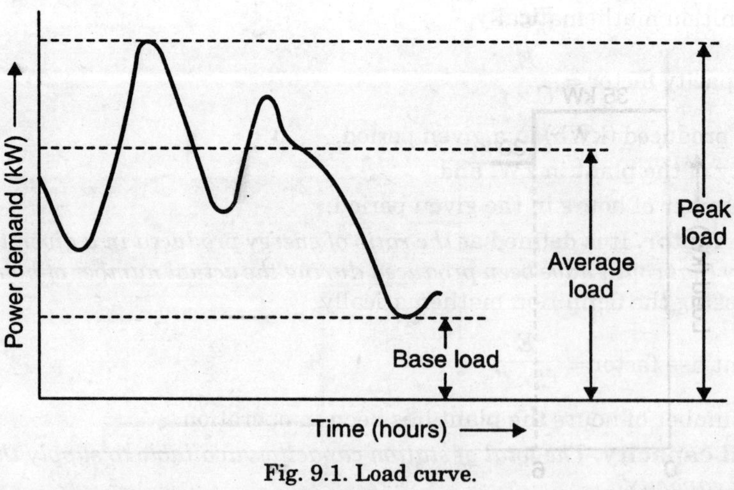

Fig. 9.1. Load curve.

(*i*) *The area under the load curve represents the energy generated in the period considered.*

(*ii*) *The area under the curve divided by the total number of hours gives the average load on the power station.*

(iii) The peak load indicated by the load curve/graph represents the maximum demand of the power station.

(iv) The ratio of the area under the load curve to the total area of rectangle in which it is contained gives the "load factor".

Significance of load curves :

- Load curves give full information about the incoming load and *help to decide the installed capacity of the power station and to decide the economical sizes of various generating units.*
- These curves also *help to estimate the generating cost and to decide the operating schedule of the power station i.e., the sequence in which different units should be run.*

19. Load duration curve. A load duration curve represents re-arrangements of all the load elements of chronological load curve in order of descending magnitude. This curve is derived from the chronological load curve.

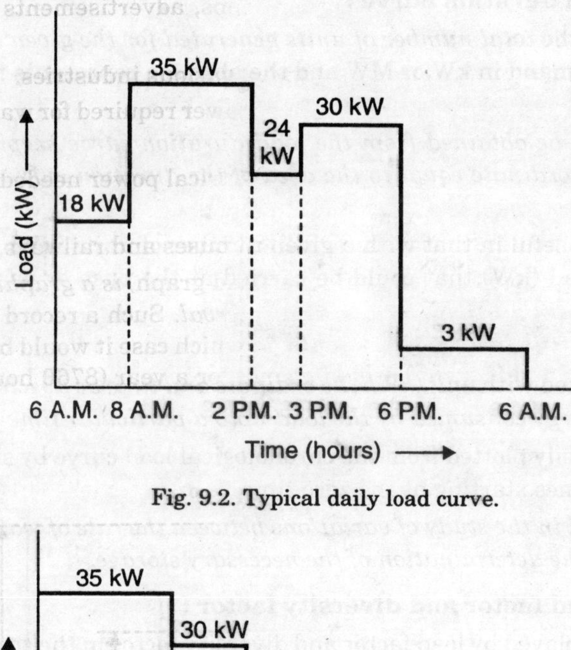

Fig. 9.2. Typical daily load curve.

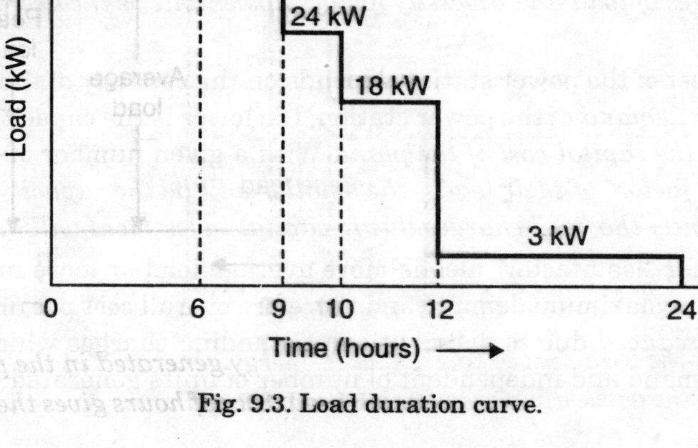

Fig. 9.3. Load duration curve.

Fig. 9.2 shows a typical daily load curve for a power station. It may be observed that the maximum load on power station is 35 kW from 8 A.M. to 2 P.M. This is plotted in Fig. 9.3. Similarly other loads of the load curve are plotted in *descending order* in the same figure. This is called *load duration curve* (Fig. 9.3).

The following points are worth noting :

(*i*) *The area under the load duration curve and the corresponding chronological load curve is equal and represents total energy delivered by the generating station.*

(*ii*) *Load duration curve gives a clear analysis of generating power economically. Proper selection of base load power plants and peak load power plants becomes easier.*

- From load duration curve the load factor of the station can also be determined.
- From these curves the distribution of load between various generating units can also be predicted.

20. Integrated load duration curve :

- This curve gives the *total number of units generated for the given demand.* The ordinate represents the demand in kW or MW and the abscissa represents the units (kWh) generated.
- *Such a curve can be obtained from the load duration curve keeping the abscissa corresponding to each ordinate equal to the area of the duration curve upto the value of that ordinate.*
- Such curves are useful in that with a given number of units (kWh) available, say from a river flow, the load (kW) that could be carried at the base or peak may be determined easily.

21. Mass curve :

- This curve is plotted with *unit* (*kWh*) *as ordinate and time as abscissa.* Thus, a *mass curve gives the total energy consumed by the load upto a particular time in a day.*

This curve can be easily plotted from the chronological load curve by summing up the energy consumed upto different times starting at the zero time.

- This curve is *used in the study of variations between the rate of water flow and the electrical load and for the determination of the necessary storage.*

Significance of load factor and diversity factor :

A significant role is played by load factor and diversity factor in the cost of supply of electrical energy. *Higher the values of load and diversity factors, lower will be the overall cost per unit generated.*

— The capital cost of the power station depends on the capacity of the power station. *Lower the maximum demand* of the power station, the lower is the capacity required and, therefore, *lower is the capital cost of the plant.* With a given number of consumers *the higher the "diversity factor" of their loads, the smaller will be the capacity of the plant required and consequently the fixed charges due to capital investment will be much reduced.*

— Further, higher "load factor" means more average load or more number of units generated for a given maximum demand and therefore overall cost per unit of electrical energy generated is *reduced* due to distribution of standing charges which are proportional to maximum demand and independent of number of units generated.

9.3. BASE LOAD AND PEAK LOAD

The load on the power station is seldom constant and varies as shown in Fig. 9.4. The load may be considered into two parts : (*i*) *Base load* ; (*ii*) *Peak load*.

Base load. It is the unvarying load which occurs *almost the whole day* on the power station.

Peak load. The various peak demands of the load *over and above the base load* of the power station is called *peak load*.

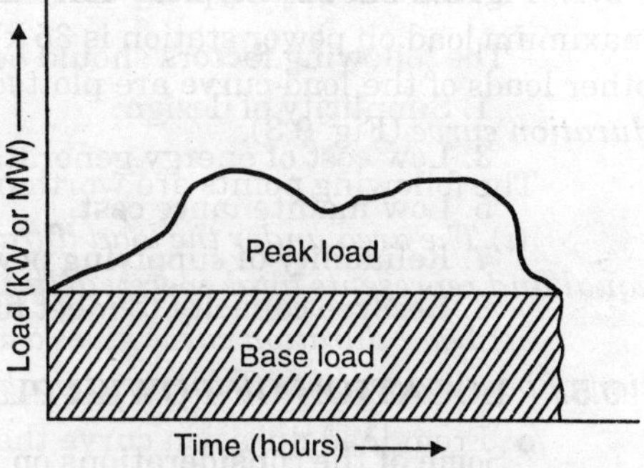

Fig. 9.4. Base load and peak load.

Load sharing between base load and peak load plants :

The *base load plants* (Hydro and nuclear power plants are usually classified as base load plants) run throughout the year and have *high load factors.* The economic characteristics of base load plants should be such that they *supply power at high capital costs but low cost of operation.*

The *peak load plants* run for a few hours in a year and work at *low load factors.* The economic characteristics of peak load plants should be such that they *supply power at low capital costs, although at high cost of operation.* The peak load plants should be *capable of starting quickly and should be inexpensive in starting and shutting down operations.*

Let the operating costs of *base load* and *peak load* plants be $Rs(a_1 \text{ kW} + b_1 \text{ kWh})$ and $Rs(a_2 \text{ kW} + b_2 \text{ kWh})$ where $a_1 > a_2$ and $b_2 > b_1$

Let, P = Maximum demand on the power plant,

x = Total number of units generated,

P_1 = Maximum demand on the base load plant, and

x_1 = Number of units generated by the base load plant.

Then, peak load on the peak load plant, $P_2 = P - P_1$

Units generated by the peak load plant, $x_2 = x - x_1$

Total annual operational cost of the system,

$$C = a_1P_1 + a_2(P - P_1) + b_1x_1 + b_2(x - x_1)$$
$$= P_1(a_1 - a_2) + a_2P + x_1(b_1 - b_2) + b_2x$$

or,
$$\frac{dC}{dP_1} = (a_1 - a_2) + \frac{dx_1}{dP_1}(b_1 - b_2)$$

For minimum cost, $\frac{dC}{dP_1} = 0$

$$\therefore \quad (a_1 - a_2) + \frac{dx_1}{dP_1}(b_1 - b_2) = 0$$

or,
$$\frac{dx_1}{dP_1} = \frac{a_1 - a_2}{b_2 - b_1} \text{ hours} \qquad ...(9.6)$$

i.e., For *economic load sharing*, the peak load plant will operate for $\frac{a_1 - a_2}{b_2 - b_1}$ hours per year.

9.4. PRINCIPLE OF POWER PLANT DESIGN

The following factors should be considered while designing a power plant :

1. Simplicity of design.
2. Low capital cost.
3. Low cost of energy generated.
4. High efficiency.
5. Low maintenance cost.
6. Low operating cost.
7. Reliability of supplying power.
8. Reserve capacity to meet future power demand.

9.5. LOCATION OF POWER PLANT

Some of the considerations on which the location of a power plant depends are :

1. Centre of electrical load. The plant should be located where there are industries and other important consumption places of electricity. There will be considerable advantage in placing the power station nearer to the centre of the load.

— There will be *saving in the cost of copper* used for transmitting electricity as the distance of transmission line is reduced.

— The cross-section of the transmission line directly depends upon the maximum current to be carried. In case of alternating current the voltage to be transmitted can be increased thus reducing the current and hence the *cross-section of the transmission line can be reduced.* This will *save the amount of copper.*

— It is desirable now to have a national grid connecting all power stations. This provides for selecting a site which has other advantages such as nearer to fuel supply, condensing water available.

2. Nearness to the fuel source. The cost of transportation of fuel may be quite high if the distance of location of the power plant is considerable. It may be advisable to locate big thermal power plants at the mouth of the coal mines. Lignite coal mines should have centralised thermal power station located in the mines itself as this type of coal cannot be transported. Such type of power stations could be located near oil fields if oil is to be used as a fuel and near gas wells where natural gas is available in abundance. In any case it has been seen that *it is cheaper to transmit electricity than to transport fuel.* Hence the power plant should be located nearer the fuel supply source.

3. Availability of water. The availability of water is of greater importance than all other factors governing station location. Water is required for a thermal power station using turbines for the following two purposes :

(*i*) To supply the make-up water which should be reasonably pure water.

(*ii*) To cool the exhaust steam. This cooling purpose is done in case of diesel engines too. For *bigger power stations* the quantity of this cooling water is tremendous and requires *some natural source of water* such as *lake, river or even sea.* Cooling towers could be used economically as the same cooling water could be used again and again. Only a part of make up water for cooling will then be required. For *small plants spray ponds* could sometimes be used. *It is economical to limit the rise in cooling-water temperature to a small value (between 6°C and 12°C), and to gain in cycle efficiency at the expense of increased cooling water pumping requirement.*

4. Type of soil available and land cost. While selecting a site for a power plant it is important to know about the character of the soil. If the soil is loose having low bearing power the pile foundations have to be used. Boring should be made at most of the projected site to have an idea of the character of the various strata as well as of the bearing power of the soil. *The best location is that for which costly and special foundation is not required.*

In case of power plants being situated near metropolitan load centres, the land there will be very costly as compared to the land at a distance from the city.

9.6. LAYOUT OF POWER PLANT BUILDING

The following points should be taken care of while deciding about power plant building and its layout :

1. The power plant structure should be simple and rugged with pleasing appearance.
2. Costly materials and ornamental work should be avoided.
3. The power plant interior should be clean, airy and attractive.
4. The exterior of the building should be impressive and attractive.
5. Generally the building should be *single storeyed.*
6. The layout of the power plant should first be made on paper, the necessary equipment well arranged and then design the covering structure. In all layout, allowances must be made for sufficient clearances and for walkways. Good clearance should be allowed around generators, boilers, heaters, condensers etc. Walkway clearances around hot objects and rapidly moving machinery should be wider than those just necessary to allow passage. Also the galleries in the neighbourhood of high tension bus bars should be sufficient as the space will permit.
7. Provision for future extension of the building should be made.
8. The height of the building should be sufficient so that overhead cranes could operate well and the overhauling of the turbines etc. is no problem. Sufficient room should be provided to lift the massive parts of the machines.
9. Each wall should receive a symmetrical treatment in window openings etc.
10. The principal materials used for building the power plant building are brick, stone, hollow tiles, concrete and steel.
11. In case of a *steam power plant*, there are distinct parts of the building *viz., boiler room, turbine room* and *electrical bays.* Head room required in the boiler room should be greater than in the others. Ventilation in boiler room presents greater difficulty because of heat liberated from the boiler surfaces. The turbine room is actually the show room of the plant. Mezzanine flooring should be used in the power plant. The chimney height should be sufficient so as to release the flue gases sufficiently high so that the atmosphere is not polluted and the nearby buildings are not affected.
12. The foundation of a power plant is one of the most important considerations. For this the bearing capacity of the sub-soil, selection of a working factor of safety and proportioning the wall footings to economical construction should be well thought of and tested. The pile foundations may have to be used where the soils have low bearing values.
13. In any power plant *machine foundation* plays an important part. The machine foundation should be able to distribute the weight of the machine, bed plate and its own weight over a safe subsoil area. It must also provide sufficient mass *to absorb machine vibrations.*
14. Sufficient room for storage of fuel should be provided indoor as well as outdoor so as to ensure against any prolonged breakdown.

9.7. COST ANALYSIS

The cost of a power system depends upon whether :

(*i*) an entirely new power system has to be set up, or

(*ii*) an existing system has to be replaced, or

(*iii*) an extension has to be provided to the existing system. The cost interalia includes :

1. Capital cost or Fixed cost. It includes the following :

(*i*) Initial cost (*ii*) Interest
(*iii*) Depreciation cost (*iv*) Taxes
(*v*) Insurance.

2. Operational cost. It includes the following :

(*i*) Fuel cost (*ii*) Operating labour cost
(*iii*) Maintenance cost (*iv*) Supplies
(*v*) Supervision (*vi*) Operating taxes.

The above mentioned costs are discussed as follows :

(*a*) **Initial cost :**

Some of the several factors on which cost of a generating station or a power plant depends are :

(*i*) Location of the plant. (*ii*) Time of construction.
(*iii*) Size of units. (*iv*) Number of main generating units.
(*v*) The type of structure to be used.

The *initial cost* of a power station includes the following :

1. Land cost 2. Building cost
3. Equipment cost 4. Installation cost
5. Overhead charges which will include the transportation cost, stores and storekeeping charges, interest during construction etc.

— To *reduce the cost of building*, it is desirable to eliminate the superstructure over the boiler house and as far as possible on turbine house also.

— The *cost on equipment can be reduced* by adopting unit system where one boiler is used for one turbogenerator. Also by simplifying the piping system and elimination of duplicate system such as steam headers and boiler feed headers. The cost can be further reduced by eliminating duplicate or stand-by auxiliaries.

— When the power plant is not situated in the proximity to the load served, the cost of a primary distribution system will be a part of the initial investment.

(*b*) **Interest**

All enterprises need investment of money and this money may be obtained as loan, through bonds and shares or from owners of personal funds. *Interest is the difference between money borrowed and money returned*. It may be charged at a simple rate expressed as % per annum or may be compounded, in which case the interest is reinvested and adds to the principal, thereby earning more interest in subsequent years. Even if the owner invests his own capital the charge of interest is necessary to cover the income that he would have derived from it through an alternative investment or fixed deposit with a bank. *Amortization* in the periodic repayment of the principal as a uniform annual expense.

(*c*) **Depreciation**

Depreciation accounts for the deterioration of the equipment and decrease in its value due to corrosion, weathering and wear and tear with use. It also covers the decrease in value of equipment due to obsolescence. With rapid improvements in design and construction of plants, obsolescence factor is of enormous importance. Availability of better models with lesser overall cost of generation makes it imperative to replace the old equipment earlier than its useful life is spent. The actual life span of the plant has, therefore, to be taken as shorter than what would be normally expected out of it.

The following methods are used to calculate the depreciation cost :

(*i*) Straight line method (*ii*) Percentage method

(*iii*) Sinking fund method (*iv*) Unit method.

(*i*) ***Straight line method.*** It is the *simplest* and *commonly used method*. The life of the equipment or the enterprise is first assessed as also the residual or salvage value of the same after the estimated life span. This salvage value is *deducted* from the initial capital cost and the balance is *divided by the life as assessed in years*. Thus, the annual value of decrease in cost of equipment is found and is set aside as depreciation annually from the income. *Thus, the rate of depreciation is uniform throughout the life of the equipment*. By the time the equipment has lived out its useful life, an amount equivalent to its net cost is accumulated which can be utilised for replacement of the plant.

(*ii*) ***Percentage method.*** In this method the deterioration in value of equipment from year to year is taken into account and the amount of depreciation calculated upon actual residual value for each year. It thus, reduces for successive years.

(*iii*) ***Sinking fund method.*** This method is based on the *conception that the annual uniform deduction from income for depreciation will accumulate to the capital value of the plant at the end of life of the plant or equipment*. In this method, the amount set aside per year consists of annual instalments and the interest earned on all the instalments.

Let, A = Amount set aside at the end of each year for n years,

n = Life of plant in years,

S = Salvage value at the end of plant life,

i = Annual rate of compound interest on the invested capital, and

P = Initial investment to install the plant.

Then, amount set aside at the end of first year = A

Amount at the end of second year

$$= A + \text{interest on } A = A + Ai = A(1 + i)$$

Amount at the end of third year

$$= A(1 + i) + \text{interest on } A(1 + i)$$

$$= A(1 + i) + A(1 + i)i$$

$$= A(1 + i)^2$$

$\therefore$ Amount at the end of nth year = $A(1 + i)^{n-1}$

Total amount accumulated in n years (say x)

= Sum of the amounts accumulated in n years

i.e., $$x = A + A(1 + i) + A(1 + i)^2 + \ldots\ldots + A(1 + i)^{n-1}$$

$$= A\,[1 + (1 + i) + (1 + i)^2 + \ldots\ldots + (1 + i)^{n-1}] \quad \ldots(i)$$

Multiplying the above equation by $(1 + i)$, we get

$$x(1 + i) = A\,[(1 + i) + (1 + i)^2 + (1 + i)^3 + \ldots\ldots + (1 + i)^n] \quad \ldots(ii)$$

Subtracting equation (*i*) from (*ii*), we get

$$x.i = [(1 + i)^n - 1]\,A$$

$$\therefore \quad x = \left[\frac{(1+i)^n - 1}{i}\right] A$$

where $x = (P - S)$

$$\therefore \qquad P - S = \left[\frac{(1+i)^n - 1}{i}\right] A \qquad \text{...(9.7)}$$

or,
$$A = \left[\frac{i}{(1+i)^n - 1}\right] (P - S) \qquad \text{...(9.8)}$$

(*iv*) ***Unit method.*** In this method *some factor is taken as a standard one and depreciation is measured by that standard.* In place of years an equipment will last, the number of hours that an equipment will last is calculated. This total number of hours is then divided by the capital value of the equipment. This constant is then multiplied by the number of actual working hours each year to get the value of depreciation for that year. In place of number of hours, the number of units of production is taken as the measuring standard.

(*d*) **Operational cost :**

The elements that make up the operating expenditure of a power plant include the following costs :

(*i*) Cost of fuels. (*ii*) Labour cost.
(*iii*) Cost of maintenance and repairs. (*iv*) Cost of stores (other than fuel).
(*v*) Supervision. (*vi*) Taxes.

Cost of fuels. In a thermal station fuel is the heaviest item of operating cost. The selection of the fuel and the maximum economy in its use are, therefore, very important considerations in thermal plant design. It is desirable to achieve the highest thermal efficiency for the plant so that fuel charges are reduced. *The cost of fuel includes not only its price at the site of purchase but its transportation and handling costs also.* In the hydroplants the absence of fuel factor in cost is responsible for lowering the operating cost. *Plant heat rate can be improved by the use of better quality of fuel or by employing better thermodynamic conditions in the plant design.*

The cost of fuel varies with the following :

(*i*) Unit price of the fuel.

(*ii*) Amount of energy produced.

(*iii*) Efficiency of the plant.

Labour cost. For plant operation labour cost is another item of operating cost. Maximum labour is needed in a thermal power plant using coal as a fuel. A hydraulic power plant or a diesel power plant of equal capacity require a lesser number of persons. In case of automatic power station the cost of labour is reduced to a great extent. However labour cost cannot be completely eliminated even with fully automatic station as they will still require some manpower for periodic inspection etc.

Cost of maintenance and repairs. In order to avoid plant breakdowns *maintenance* is necessary. *Maintenance* includes *periodic cleaning, greasing, adjustments and overhauling of equipment.* The material used for maintenance is also charged under this head. Sometimes an arbitrary percentage is assumed as maintenance cost. A good plan of maintenance would keep the sets in dependable condition and avoid the necessity of too many stand-by plants.

Repairs are necessitated when the plant breaks down or stops due to faults developing in the mechanism. The repairs may be minor, major or periodic overhauls and are charged to the depreciation fund of the equipment. This item of cost is higher for thermal plants than for hydro-plants due to complex nature of principal equipment and auxiliaries in the former.

Cost of stores (other than fuel). The items of consumable stores other than fuel include such articles as lubricating oil and greases, cotton waste, small tools, chemicals, paints and such other things. The incidence of this cost is also higher in thermal stations than in hydro-electric power stations.

Supervision. In this head the salary of supervising staff is included. A good supervision is reflected in lesser breakdowns and extended plant life. The supervising staff includes the station superintendent, chief engineer, chemist, engineers, supervisors, stores incharges, purchase officer and other establishment. Again, thermal stations, particularly coal fed, have a greater incidence of this cost than the hydro-electric power stations.

Taxes. The taxes under operating head includes the following :

(*i*) Income tax

(*ii*) Sales tax

(*iii*) Social security and employee's security etc.

9.8. SELECTION OF TYPE OF GENERATION

While choosing the type of generation the following points should be taken into consideration :

1. The type of fuel available or availability of suitable sites for water power generation.
2. Fuel transportation cost.
3. Land required.
4. Foundation cost.
5. The availability of cooling water.
6. The type of load to be taken by the power plant.
7. Reliability in operation.
8. Plant life.
9. Cost of transmitting the energy.

9.9. SELECTION OF POWER PLANT EQUIPMENT

Selection of some important power plant equipment is discussed below :

9.9.1. Selection of Boilers

It is now well known fact that only water tube boilers (fire tube boilers not suitable) should be used for all central power stations. While selecting a boiler the following points should be taken care of :

1. Type of fuel to be burnt.
2. Type of load.
3. Cost of fuel.
4. Desirability of heat-reclaiming equipment.
5. Availability of space for boiler installation.

— The design and efficiency of the boiler is considerably influenced by the *type of fuel* used in a boiler. *A high efficiency can be obtained with coal firing as compared to oil or gas firing*. This is due to increased hydrogen loss in gaseous fuels.

— *The location of the plant* will also decide the type of fuel to be used. If the plant is nearer the coal fields, coal will be cheaper. Power plants near to the oil fields and gas wells will naturally use these fuels.

— *Coal firing* will also *influence furnace design and hence the cost of boiler*. In case of low ranking fuel such as lignites etc., pulverised firing is used. Very low fusing temperatures of coal require water cooled walls and in some cases the slag tap furnaces. The yearly minimum operating cost has to be considered which may include production cost and fixed charges. In case of anthracite coal or metallurgical coke etc. the wear on pulverising machinery is relatively much higher than that of bituminous coal.

The cost of boilers vary with the following :

(*i*) Type of boiler used. (*ii*) Operating pressure.

(*iii*) Operating temperature. (*iv*) Type of firing.

(*v*) Efficiency desired.

— '*Heat-reclaiming equipment*' such as *economisers* and *air preheaters* should be provided with boilers. With the addition of economisers and air preheaters the efficiency of the boiler increases from 75% to 90% and above.

— The '*increased pressure*' *affects the cost of boiler drum, boiler tubes, headers, economisers and other accessories*. Similarly high temperatures increase the cost of superheaters as higher pressure and higher temperatures require special alloy steels. *High pressures require forced circulation also*. *This also increases the cost but this forced circulation also increases the efficiency of the boiler.*

— The *method of firing* has also an influence on the percentage of total time for which the boiler will be available and should be considered when planning boiler capacity. Stoker firing is in general *slightly less costly than pulverised fuel firing. The pulverised fuel firing increases the efficiency*.

— *Economisers* improve the boiler efficiency by 4 to 10%. The *air preheater* further improves the boiler efficiency from 6 to 8%.

— The exhaust gases should not be cooled *below 150°C*. Below this temperature the condensation of moisture may take place and when mixed with SO_2 this moisture produces a dilute solution of H_2SO_4 which is finitely detrimental to the equipment.

— While selecting the proper economiser size as well as the size of the air preheaters *fixed* as well as *operating* charges should be considered. The fixed charges include the cost of heat recovery equipment, flue work, ducts and also increased fan cost and building cost.

9.9.2. Selection of Prime-movers

For proper selection of prime-mover it is of paramount importance to construct the following curves :

(*i*) A typical daily load curve.

(*ii*) A peak-load curve.

(*iii*) A probable future-load curve.

The prime-movers to be used for generating electricity could be *diesel engines, steam engines, steam turbines, gas turbines, water turbines* etc.

— While selecting a prime-mover the *initial cost of a unit erected* has to be taken into consideration. The *efficiency of this unit at various loads* is also to be taken into consideration. *As the capacity of the unit increases there is a corresponding reduction in floor space per kW.*

— The selection of the prime-mover depends also on the type of use whether it is used for *industrial purpose* or *for central power stations*.

Prime-movers used for industrial purpose should be *non-condensing so that steam after exhausting could be used for processing*.

In case of *central power stations condensing steam turbines should be used. Diesel engines* have an advantage of higher efficiency and low cost. It also requires less labour and the initial investment is also less. But the cost of coal is less as compared to diesel oil. Also the capacity of diesel engines is limited and hence for *bigger power stations they are unsuitable.* The diesel engines are used as *standby plant* in all the central power stations whether thermal or hydro.

— In places where water is in abundance and a certain head is available hydro power plants/stations are installed. In rivers where there is a natural fall, the same could be used for driving a *water turbine* in a hydro power plant. The maintenance of hydro power plant is the cheapest.

9.9.3. Selection of Size and Number of Generating Units

There can be no hard and fast rules, but however *looking at the load curve of the station one can guess for the total generating capacity, size and number of the units. Minimum generating capacity of a plant must be more than the predicted maximum demand.* Obviously, the minimum number of generators can be one but this will not be a wise suggestion. As the load on a power station is never constant, owing to variable demands at the different times of the day, the generator will have to run continuously at variable loads, which will be much less than the rated capacity of the generator for most of the times, *without any provision for the maintenance.* So a power station which is expected to be reliable in service, must have at *least two generators, irrespective of the total capacity of the plant.*

The following points are worth noting :

(*i*) The most appropriate way of deciding the size, and number of generating sets in a station is *to select the number of sets in such a way so as to fit in the load curve as closely as possible, so that the plant capacity may be used efficiently*.

(*ii*) Extra spare capacity is not desired as it increases the capital expenditure.

(*iii*) The main aim should be to have units of different capacities which will suitably fit in the load curve so that most of the generators when in use can be operated at nearly full load.

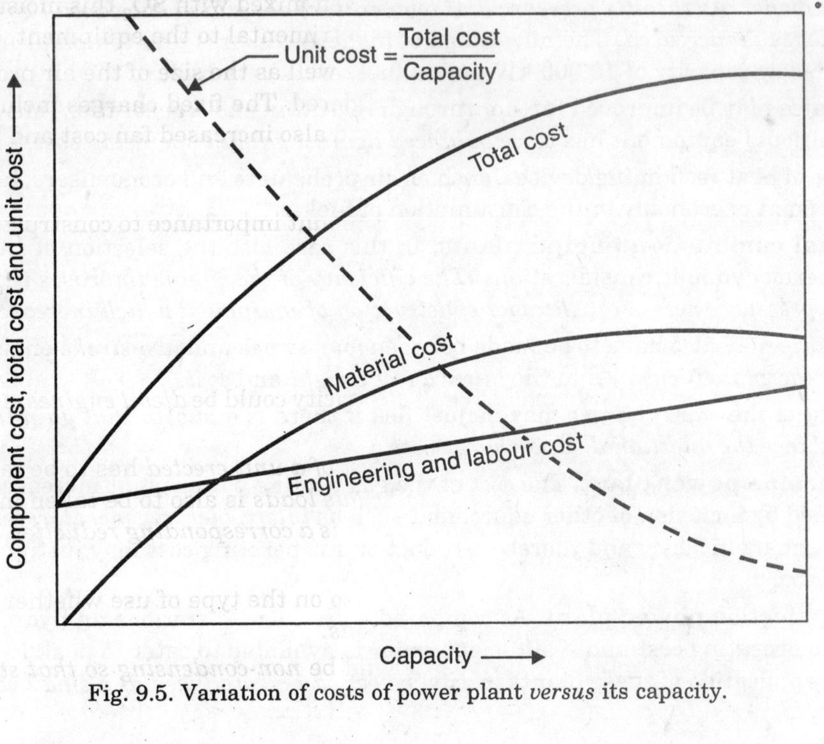

Fig. 9.5. Variation of costs of power plant *versus* its capacity.

The equipment prices are usually compared on the basis of price per unit of capacity, usually termed 'unit price'. *The unit price decreases as the capacity of the machine increases*. This is the main reason for adopting a large size generating unit in power plants. Fig. 9.5 shows the general trend and trend of the major cost components in building a given type of machine. The following points are worth noting :

— The labour and engineering cost curve *increases slightly* with the capacity of the unit.

— The material cost curve *decreases in slope* with an increase in capacity of plant.

— The total cost curve follows the pattern of material cost curve. The total cost curve shows the *positive intercept at zero capacity which represents the cost of just maintaining an organisation of men and plant ready to produce.*

— The dotted curve shows the *reduction in unit price with an increase in capacity* and this is the major argument for installing large units. *The large units are always preferred for the loads with higher load factor (0.8 to 1).*

9.10. ECONOMICS IN PLANT SELECTION

After selection of type of drive (such as steam, gas diesel or water power) which depends on availability of cheap fuels or water resources, further selection of the design and size of the equipment is primarily based upon economic consideration and a *plant that gives the lowest unit cost of production is usually chosen.* In case of all types of equipment the working *efficiency is generally higher with larger sizes of plants and with high load factor operation. Also, the capital cost per unit installation reduces as the plant is increased in size.* However, *a bigger size of plant would require greater investment, and possibilities of lower than optimum load factor usually increase with larger size of the plant.*

Steam power plants. In case of steam power plants the choice of *steam conditions* such as throttle pressure and temperature, is an important factor affecting operating costs and is, therefore, very carefully made. *As throttle pressure and temperature are raised the capital cost increases but the cycle efficiency is increased.* The advantage of higher pressures and temperatures is generally not apparent below capacity of 10,000 kW unless fuel cost is very high.

Heat rates may be improved further through *reheating* and *regeneration*, but again the capital cost of additional equipment has to be *balanced against gain in operating cost.*

The use of heat reclaiming devices, such as air preheaters and economisers, has to be considered from the point of economy in the consumption of fuel.

Internal combustion engine plants. In this case also the selection of I.C. engines also depends on thermodynamic considerations. *The efficiency of the engine improves with compression ratio but high pressures necessitate heavier construction of equipment which increases cost.*

The choice may also have to be made between *four-stroke* and *two-stroke* engines, the *former having higher thermal efficiency* and the *latter lower weight* and *cost*.

The cost of the *supercharger* may be justified if there *is a substantial gain in engine power which may balance the additional supercharge cost.*

Gas turbine power plant. The cost of the gas turbine power plant increases as the simple plant is modified by inclusion of other equipment such as *intercooler*, *regenerator, reheater*, etc. but the gain in thermal efficiency and thereby a reduction in operating cost may justify this additional expense in first cost.

Hydro-electric power plant. As compared with thermal stations an hydro-electric power plant has little operating cost and if sufficient water is available to cater to peak loads and special conditions for application of these plants justify, *power can be produced at a small cost.*

The *capital cost* per unit installed is *higher if the quantity of water is small*. Also, the unit cost of conveying water to the power house is greater if the quantity of water is small. The cost of storage per unit is also lower if the quantity of water stored is large.

An existing plant capacity may be *increased by storing additional water* through increasing the height of dam or by diverting water from other streams into the head reservoir. However, again it would be an economic study whether this additional cost of civil works would guarantee sufficient returns.

Some hydro-power plants may be made *automatic* or *remote controlled* to reduce the operating cost further, but the cost of automation has to be *balanced against the saving effected in the unit cost of generation*.

Interconnected hydro-steam system. In such a system where *peak loads are taken up by steam units*, the capacity of water turbine may be kept somewhat higher than the water flow capacity at peak loads, and lesser than or equal to maximum flow of river. This would make it possible for the water turbine to generate adequate energy at low cost during sufficient water flow.

Some of the principal characteristics of hydro-electric, steam and diesel power plants are listed below :

S. No.	*Characteristics*	*Hydro-plant*	*Steam plant*	*Diesel plant*
1.	*Planning and construction*	Difficult and takes long time	Easier than hydro-plant	Easiest
2.	*Civil works cost*	Highest	Lower than hydro-plant	Lowest
3.	*Running and maintenance cost* (as a fraction of total generation cost)	$\frac{1}{10}$	$\frac{1}{7}$	$\frac{1}{6}$
4.	*Overall generation cost*	Lowest	Lower than for diesel plants	Highest
5.	*Reliability*	Good	Good	Excellent

Advantages of interconnection :

Major advantages of interconnecting various power stations are :

1. Increased reliability of supply.
2. Reduction in total installed capacity.
3. Economic operation.
4. Operating savings.
5. Low capital and maintenance costs.
6. Peak loads of combined system can be carried at a *much lower cost* than what is possible with small individual system.

9.11. FACTORS AFFECTING ECONOMICS OF GENERATION AND DISTRIBUTION OF POWER

The economics of power plant operation is greatly influenced by :

(*i*) Load factor (*ii*) Demand factor

(*iii*) Utilisation factor.

Load factor. In a *hydro-electric power station* with water available and a fixed staff for maximum output, the cost per unit generated at 100% *load factor* would be *half* the cost per unit at 50% load factor. In a *steam power station* the difference would not be so pronounced since fuel cost constitutes the major item in operating costs and does not vary in the same proportion as load factor. The cost at 100% load factor in case of this station may, therefore, be about 2/3rd of the cost 50% load factor. For a *diesel station* the cost per unit generated at 100% load factor may be about 3/4th of the same cost at 50% load factor. From the above discussion it follows that :

(*i*) *Hydro-electric power station should be run at its maximum load continuously on all units.*

(*ii*) *Steam power station* should be run in such a way that *all its running units are economically loaded.*

(*iii*) *Diesel power station* should be worked for *fluctuating loads* or *as a stand by.*

Demand factor and utilisation factor. *A highly efficient station, if worked at low utilisation factor, may produce power at high unit cost.*

The time of maximum demand occurring in a system is also important. In an interconnected system, *a study of the curves of all stations is necessary to plan most economical operations.*

The endeavour should be to load the most efficient and cheapest power producing stations to the greatest extent possible. Such stations, called "*base load stations*" carry full load over 24 hours *i.e.,* for three shifts of 8 hours.

— The stations in the *medium range of efficiency* are operated only during the two shifts of 8 hours during 16 hours of average load.

— The older or *less efficient stations* are used *as peak or standby stations* only, and are operated *rarely* or for *short periods of time*.

Presently there is a tendency to use units of large capacities to reduce space costs and to handle larger loads. However, the *maximum economical benefit of large sets occurs only when these are run continuously at near full load. Running of large sets for long periods at lower than maximum continuous rating increases cost of unit generated.*

9.12. HOW TO REDUCE POWER GENERATION COST ?

The cost of power generation can be *reduced by* :

1. Using a plant of simple design that *does not need highly skilled personnel.*
2. Selecting equipment of *longer life and proper capacities.*
3. Carrying out *proper maintenance* of power plant equipment to avoid plant breakdowns.
4. Running the power stations at *high load factors*.
5. Increasing the efficiency of the power plant.
6. Keeping proper supervision, which ensures *less breakdowns* and extended plant life.

9.13. POWER PLANT—USEFUL LIFE

The useful life of a power plant is that *after which repairs become so frequent and extensive that it is found economical to replace the power plant by a new one*. Useful life of some of the power plants is given below :

Plant	*Useful life*
1. Conventional thermal power plant	20–25 years
2. Nuclear power plant	15–20 years
3. Diesel power plant	About 15 years.

The useful life of some of the equipment of a *steam power plant* is given below :

Equipment	*Useful life (years)*
1. Boilers	
(*i*) Fire tube	10–20
(*ii*) Water tube	20
2. Steam turbine	5–20
3. Steam turbo-generators	10–20
4. Condensers	20
5. Pumps	15–20
6. Coal and ash machinery	10–20
7. Feed water heaters	20–30
8. Stacks	10–30
9. Stokers	10–20
10. Transformers	15–20
11. Motors	20
12. Electric meters and instruments	10–15
13. Transmission lines	10–20

9.14. ECONOMICS OF HYDRO-ELECTRIC POWER PLANTS

The cost analysis of an hydro-electric power plant is different from those of the other plants in the respect that the fixed *coal is the major item of the total cost and the operating cost is relatively much smaller* whereas in the steam and other plants, the *operating cost is a large part of the total cost.*

In a hydro-plant the total annual cost can be divided into two following categories :

1. Fixed costs :

(*i*) Interest on capital (*ii*) Amortization of the capital

2. Running costs :

(*i*) Maintenance and repairs. (*ii*) Operating costs including salaries and wages.

(*iii*) Rates and taxes. (*iv*) Stores, oil and other supplies.

(*v*) Management expenses including insurance.

— The *fixed charges* of a hydro-plant are about 60 to 70% of the total cost of power and *these do not depend upon the station output.*

— The *running charges depend upon the station output, but not so much as in the thermal power plants.*

The following items go to form the *total capital outlay* or the *investment on a hydro-plant* :

(*i*) Preliminary surveys and investigations of the topography and geology of the proposed site of the plant.

(*ii*) Purchase of land (needed for adequate storage or pondage) and water rights.

(*iii*) Compensation to oustees.

(*iv*) Cost of preparation of detailed designs and specifications.

(*v*) Cost of testing the materials of construction.

(*vi*) Cost of carrying out experimental work and model tests or designs for hydraulic structures.

(*vii*) The actual cost of construction.

(*viii*) Cost relating the purchase and installation of the equipment.

(*ix*) Interest on capital during construction.

(*x*) Working capital during the period of load development.

(*xi*) Cost in respect of new roads, railway lines, residential houses and even new towns which may have to be constructed.

A typical cost analysis of a hydro-plant is as follows :

S. No.	*Components*	*Cost*
1.	Reservoir, dam and water ways	55%
2.	Land	15%
3.	Structures	10%
4.	Power plant and equipment	20%

Besides this there is an another important item called *transmission liability* which refers to the *transmission charges for conveying the electricity from the plant site to the load centre.*

— The total cost of construction of hydro-plant is *invariably higher than that of a thermal plant of equal capacity*. Therefore the annual charges for interest and depreciation are comparatively higher.

— In case of a *hydro-plant*, the *smaller the quantity of water stored higher is the cost per kW.*

9.15. ECONOMICS OF COMBINED HYDRO AND STEAM POWER PLANTS

It has been established that if a region/country is neither rich in fuel reserves nor in hydro resources then a combined operation of hydro and steam power plants give the best results in regard to generation of electricity at *the economical cost*. The following advantages accrue from combined plants :

1. Flexibility of operation.
2. Security of supply.
3. Improved utilisation of hydro-power.
4. Spare plant.

Practically all large power systems in the world have hydro-steam interconnected. *Hydro-plant may function as a capacity plant i.e., to supply system peak with minimum flow conditions*, or it may work as an *energy plant* to replace the costly steam generated electricity by *low cost hydro power.*

In such a system (interconnected) *there should be a daily or seasonal load allocation plan set prehand so that the use of the two power systems is made to the best advantage. A certain amount of forecasting of load and capacity of the system as well as flexibility are necessary for optimum results and experience is a big factor in good co-ordinated action.*

A knowledge of system plant loading schedules for minimum production cost is as important as that of reservoir levels, pond storage and stages of low flow and heavy loads.

9.16. PERFORMANCE AND OPERATING CHARACTERISTICS OF POWER PLANTS

The *performance* of generating power plants is compared by their *average efficiency* over a period of time. The average efficiency of a power plant is the *ratio of useful energy output to the total energy input during the period considered*. This *measure of performance varies with uncontrolled conditions viz.* (*i*) cooling water temperature (*ii*) quality of fuel and (*iii*) shape of load curve. Thus, unless all plant performances are corrected to the same controlled conditions it is not a satisfactory standard of comparison.

The performance of a plant can be precisely represented by the *input-output curve* from the tests conducted on individual power plant. The input-output curve is *graphical representation* between the net energy output (L) and input (I). The input is generally expressed in millions of kcal/h or kJ/h and load output is expressed as megawatts (MW). The input to hydro-plant is measured in cusecs or m^3/s of water.

In general input-output may be represented as follows :

$$I = a + bL + cL^2 + dL^3$$

where, I = Input,

L = Output, and

a, b, c, d = constants.

Input-output curve. Fig. 9.6 (*a*) shows an *input-output curve*. In order to keep the apparatus functioning at zero load, a certain input (I_0) is required *to meet frictional and heat losses.*

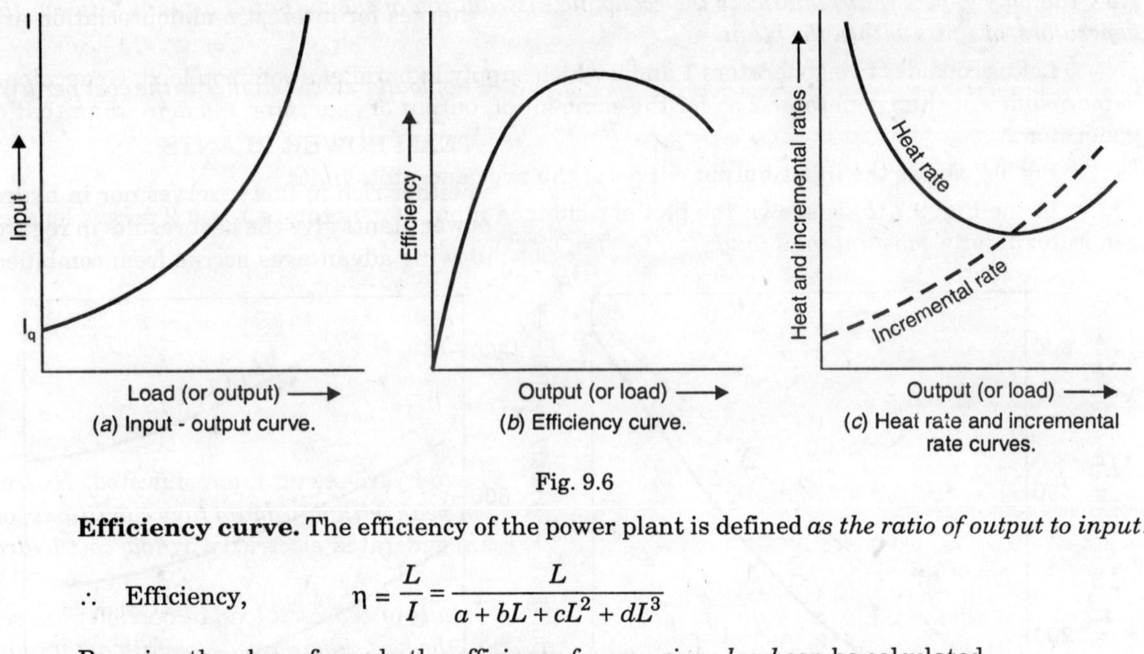

(*a*) Input - output curve. (*b*) Efficiency curve. (*c*) Heat rate and incremental rate curves.

Fig. 9.6

Efficiency curve. The efficiency of the power plant is defined *as the ratio of output to input.*

$$\therefore \quad \text{Efficiency,} \quad \eta = \frac{L}{I} = \frac{L}{a + bL + cL^2 + dL^3}$$

By using the above formula the efficiency *for any given load* can be calculated.

An *efficiency curve* is shown in Fig. 9.6 (*b*).

Heat rate and incremental rate curves. These curves can be derived from basic input-output curve.

Heat rate (H.R.) is defined as the *ratio of input to output*.

$$i.e., \quad \textit{Heat rate (H.R.)} = \frac{I}{L} = \frac{a + bL + cL^2 + dL^3}{L} = \frac{a}{L} + b + cL + dL^2$$

Heat rate curve is obtained by plotting values of heat rate against corresponding values of output. Fig. 9.6 (*c*) shows a heat rate curve.

Incremental rate (IR) is defined as the *ratio of additional input (dI) required to increase additional output (dL).*

i.e., $$\text{Incremental rate } (IR) = \frac{dI}{dL}$$

Incremental rate curve is obtained by plotting values of *IR* against corresponding values of output. Such a curve is shown in Fig. 9.6 (*c*). *This curve expresses additional energy required to produce an added unit of output at the given load.*

9.17. ECONOMIC LOAD SHARING

The primary objective of the design of all generating stations is the *economy*. For a power system to return a profit on the capital invested, proper operation of the plant is essential. As far as the efficiency of boilers, turbines, alternators etc. is concerned, engineers have been successful in increasing the efficiency continuously so that each unit added results in comparatively more efficient operation. Methods have also been devised for economic operation of plants at part loads and under variable load conditions. Attempts have been made to minimise the transmission losses too. Now the *only aspect that remains is the economic distribution of the output of a plant between the generators, or units within the plant.*

Let us consider two generators 1 and 2 which supply in parallel a common load. Generator 1 is more efficient than generator 2 as for the same input, output of generator 1 is more than that of generator 2.

Fig. 9.7 shows the input-output curves of the two generators/units.

In the Fig. 9.7 (*b*) is shown the plot of combined input of generators 1 and 2 *versus* load on generator 1, for a *constant total load.*

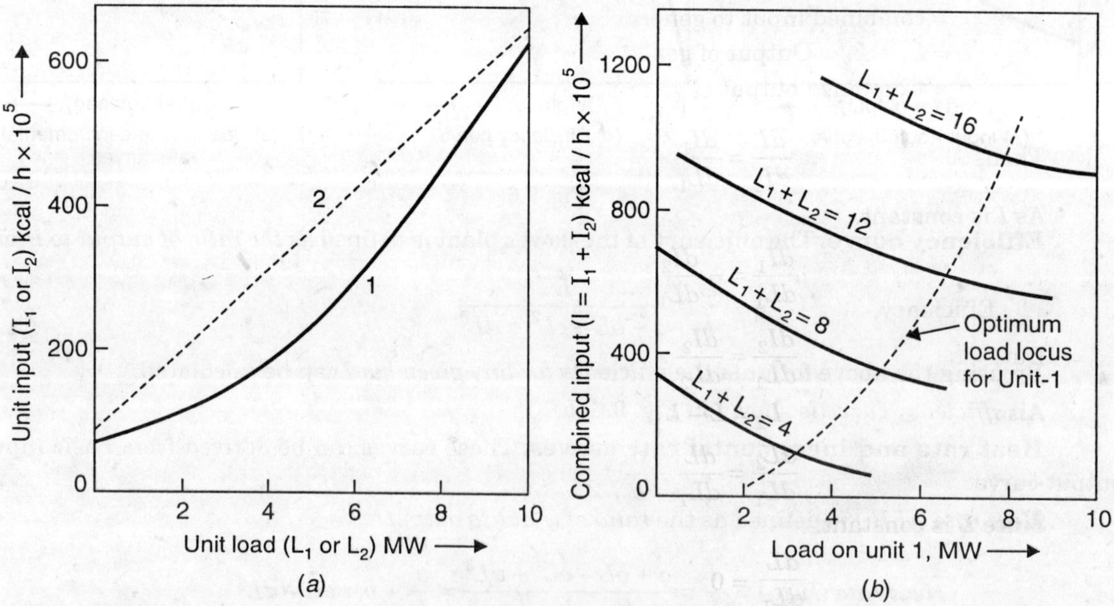

Fig. 9.7

Although generator 1 requires less input for a given output it is not essential that unit 1 should be loaded first and then generator 2. For economical loading the combined input of units

1 and 2 should be plotted against load on unit 1 for a constant total load. Let a total load of 4 MW be supplied by generators 1 and 2.

$$L = L_1 + L_2$$

where, L = Combined output,

L_1 = Output of generator 1, and

L_2 = Output of generator 2.

Let the generator 2 supply total load of 4 MW and generator 1 supply zero load. Now corresponding to zero load on generator 1 and 4 MW on generator 2 the values of input to generator 1 (I_1) and input to generator 2 (I_2) can be determined respectively from Fig. 9.7 (*a*) and thus value of ($I_1 + I_2$) can be plotted against zero load on generator 1. Again let 2 MW be supplied by generator 1 and 2 MW be supplied by generator 2 (so that total load remains 4 MW) then values of I_1 and I_2 can be determined corresponding to 2 MW load on each generator and value of ($I_1 + I_2$) can be plotted against 2 MW load on generator 1 as shown in Fig. 9.7 (*b*). In this way curve for a total load of 4 MW can be plotted corresponding to different output of generator 1.

Similarly curve for total load of 8 MW etc. can be plotted. In these curves *there is at least one point where combined input is minimum for a given total load. Corresponding to this point of minimum, the load generator 1 can be found*. Then the load on generator 2 will be difference of total load and load on generator 1. *This load saving will be the most economical.*

This method is difficult to apply in practice as such because generally stations have got more than two generators / units and in that case the application of above principles becomes a cumbersome process.

Considering any combined constant input in the Fig. 9.7 (*b*), at the point of minimum input

$$\frac{dI}{dI_1} = 0 \qquad ...(i)$$

where, $I = I_1 + I_2$ = Input of generator 1 + input of generator 2

= combined input to generators 1 and 2

$L = L_1 + L_2$ = Output of generator 1 + output of generator 2

= Combined output of generators 1 and 2

Then, $$\frac{dI}{dI_1} = \frac{dI_1}{dL_1} + \frac{dI_2}{dL_1} = 0 \qquad \left(\because \frac{dI}{dI_1} = 0\right)$$

As I is constant.

$$\therefore \quad \frac{dI_1}{dL_1} = -\frac{dI_2}{dL_1} \qquad ...(ii)$$

But, $$\frac{dI_2}{dL_1} = \frac{dI_2}{dL_2} \times \frac{dL_2}{dL_1} \qquad ...(iii)$$

Also, $$L_2 = L - L_1$$

$$\therefore \quad \frac{dL_2}{dL_1} = \frac{dL}{dL_1} - \frac{dL_1}{dL_1}$$

Since L is constant.

$$\therefore \quad \frac{dL}{dL_1} = 0$$

Hence, $$\frac{dL_2}{dL_1} = -\frac{dL_1}{dL_1} = -1 \qquad ...(iv)$$

Substituting in (*iii*), we get

$$\frac{dI_2}{dL_1} = -\frac{dI_2}{dL_2} \quad ...(v)$$

$\therefore$ From (*ii*) and (*v*)

$$\frac{dI_1}{dL_1} = \frac{dI_2}{dL_2} \quad ...(vi)$$

Thus, *for minimum combined input to carry a given combined output, the slopes of the input-output curve for each unit must be equal.*

If there are *n* units, supplying a constant load, then the required condition for the minimum input or maximum system efficiency is

$$\frac{dI_1}{dL_1} = \frac{dI_2}{dL_2} = \frac{dI_3}{dL_3} = ... = \frac{dI_n}{dL_n} \quad ...(vii)$$

Condition for maximum efficiency :

Refer Fig. 9.8. The load at which efficiency will be maximum, the heat rate will be minimum at that load as efficiency is inverse of heat.

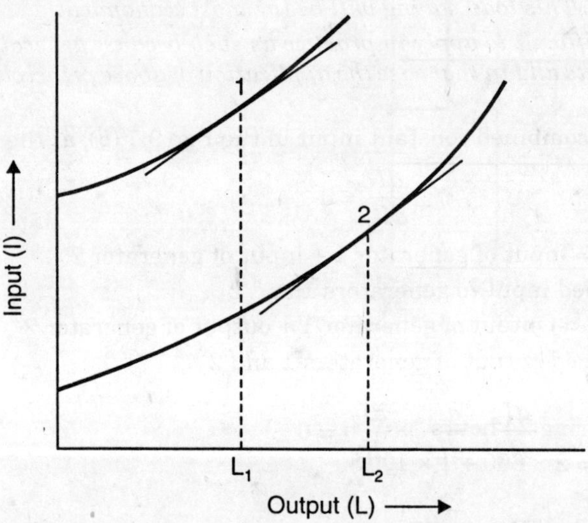

Fig. 9.8

$$\text{Efficiency} = \frac{L}{I}$$

$$\text{Heat rate } (HR) = \frac{I}{L}$$

For minimum value of heat rate,

$$\frac{d}{dL}(HR) = 0, \qquad \text{or,} \qquad \frac{d}{dL}\left(\frac{I}{L}\right)$$

$$\therefore \qquad \frac{LdI - IdL}{L^2} = 0, \qquad \text{or,} \qquad LdI = IdL \qquad \text{or} \qquad \frac{I}{L} = \frac{dI}{dL}$$

This shows that efficiency will be maximum at a load where heat rate is equal to incremental heat rate.

WORKED EXAMPLES

Example 9.1. *The following loads are supplied by a power station :*

Time hours	*6 A.M. to 8 A.M.*	*8 A.M. to 9 A.M.*	*9 A.M. to 12 Noon*	*12 Noon to 2.00 P.M.*	*2.00 P.M. to 6.00 P.M.*	*6.00 P.M. to 8.00 P.M.*	*8.00 P.M. to 9.00 P.M.*	*9.00 P.M. to 11.00 P.M*	*11.00 P.M. to 5.00 A.M.*	*5 A.M. to 6 A.M.*
Load (MW)	*600*	*1000*	*1500*	*750*	*1250*	*900*	*1000*	*500*	*250*	*400*

(*i*) *Draw the load curve and find out the load factor on the basis of 24 hours.*

(*ii*) *Choose the proper number and size of generator units to supply this load.*

(*iii*) *Find the reserve capacity of the plant and plant capacity factor.*

(*iv*) *Find the operating schedule of the units selected.*

Solution. (*i*) **Load curve and load factor :**

Load curve drawn is shown in Fig. 9.9.

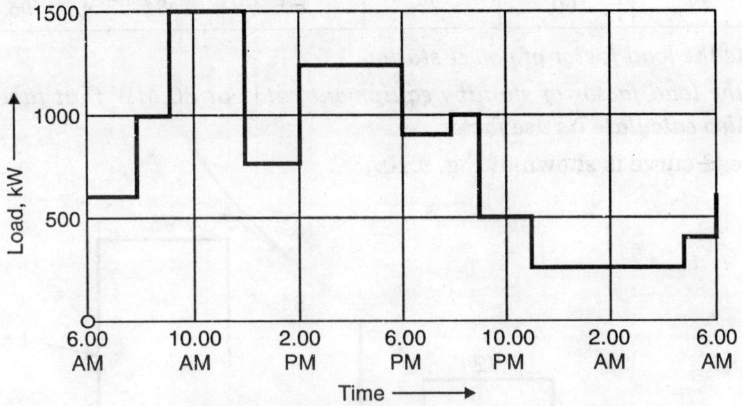

Fig. 9.9. Load curve.

Units generated during 24 hours

$$= 2 \times 600 + 1 \times 1000 + 3 \times 1500 + 2 \times 750 + 4 \times 1250 + 2 \times 900 + 1 \times 1000 + 2 \times 500 + 6 \times 250 + 1 \times 400$$

$$= 18900 \text{ kWh}$$

$$\text{Average load} = \frac{\text{Units generated}}{\text{Time in hours}} = \frac{18900}{24} = 787.5 \text{ kW}$$

$$\text{Load factor} = \frac{\text{Average load}}{\text{Maximum demand}}$$

$$= \frac{787.5}{1500} = 0.525 \text{ or } \mathbf{52.5.} \text{ (Ans.)}$$

(*ii*) **Number and size of generator units :**

For the power station 4 units of 0.5 MW each including one unit of 0.5 MW as stand-by unit may be chosen.

(*iii*) **Reserve capacity and plant capacity factor :**

Capacity of the plant $= 4 \times 0.5 = 2$ MW

Reserve capacity $= 2 - 1.5 =$ **0.5 MW. (Ans.)**

Plant capacity factor $= \dfrac{\text{Average demand}}{\text{Rated capacity of the power plant}} \times 100$

$= \dfrac{787.5}{2000} = 0.3937$ or **39.37. (Ans.)**

(*iv*) **Operating schedule of the units selected :**

The operating schedule of the units selected will be as follows :

- One set of 0.5 MW ... 24 hours
- Second set of 0.5 MW ... From 6 A.M. to 9 P.M. (15 hours)
- Third set of 0.5 MW ... From 9 A.M. to 12.00 Noon and 2 P.M. to 6 P.M. (7 hours)

(Ans.)

Example 9.2. *The maximum demand of a power station is 96000 kW and daily load curve is described as follows :*

Time hours	*0–6*	*6–8*	*8–12*	*12–14*	*14–18*	*18–22*	*22–24*
Load (MW)	*48*	*60*	*72*	*60*	*84*	*96*	*48*

(*i*) *Determine the load factor of power station.*

(*ii*) *What is the load factor of standby equipment rated at 30 MW that takes up all load in excess of 72 MW ? Also calculate its use factor.*

Solution. Load curve is shown in Fig. 9.10.

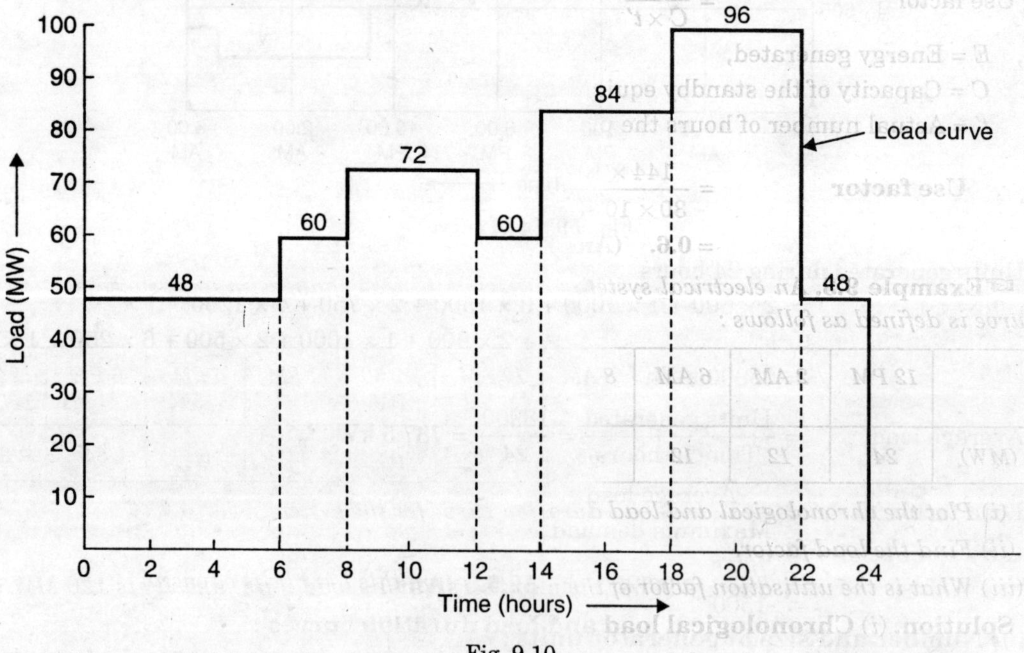

Fig. 9.10

Energy generated = Area under the load curve

$= 48 \times 6 + 60 \times 2 + 72 \times 4 + 60 \times 2 + 84 \times 4 + 96 \times 4 + 48 \times 2$

$= 1632 \text{ MWh} = 1632 \times 10^3 \text{ kWh}.$

(*i*) **Load factor :**

Average load $= \dfrac{1632 \times 10^3}{24} = 68000$ kW

Maximum demand $= 96000$ kW (given)

$\therefore$ Load factor $= \dfrac{\text{Average load}}{\text{Maximum demand}} = \dfrac{68000}{96000} = \mathbf{0.71.}$ **(Ans.)**

(*ii*) **Land factor of standby equipment :**

The standby equipment supplies

$84 - 72 = 12$ MW for 4 hours (14 – 18)

$96 - 72 = 24$ MW for 4 hours (18 – 22)

$\therefore$ Energy generated by standby equipment

$= (12 \times 4 + 24 \times 4) \times 10^3 = 144 \times 10^3$ kWh

Time for which standby equipment *remains in operation* (from the load curve)

$= 4 + 4 = 8$ hours

Average $= \dfrac{144 \times 10^3}{8} = 18 \times 10^3$ kW

Load factor $= \dfrac{18 \times 10^3}{24 \times 10^3} = \mathbf{0.75.}$ **(Ans.)**

Use factor $= \dfrac{E}{C \times t'}$

where, E = Energy generated,

C = Capacity of the standby equipment, and

t' = Actual number of hours the plant has been in operation.

$\therefore$ **Use factor** $= \dfrac{144 \times 10^3}{30 \times 10^3 \times 8}$

$= \mathbf{0.6.}$ **(Ans.)**

☞**Example 9.3.** *An electrical system experiences linear changes in load such that its daily load curve is defined as follows :*

Time	*12 PM*	*2 AM*	*6 AM*	*8 AM*	*12 AM*	*12.30 PM*	*1 PM*	*5 PM*	*6 PM*	*12 PM*
Load (MW)	*24*	*12*	*12*	*60*	*60*	*48*	*60*	*60*	*84*	*24*

(*i*) *Plot the chronological and load duration curve for the system.*

(*ii*) *Find the load factor.*

(*iii*) *What is the utilisation factor of the plant serving this load if its capacity is 120 MW ?*

Solution. (*i*) **Chronological load and load duration curves :**

Chronological load and load duration curves are drawn as shown in the Fig. 9.11 (*a*), (*b*). The procedure for constructing the load duration curve from chronological load curve is as follows :

— The abscissa of the load duration curve is laid off *equal to the number of hours in the chronological curve*, in this case 24 hours.

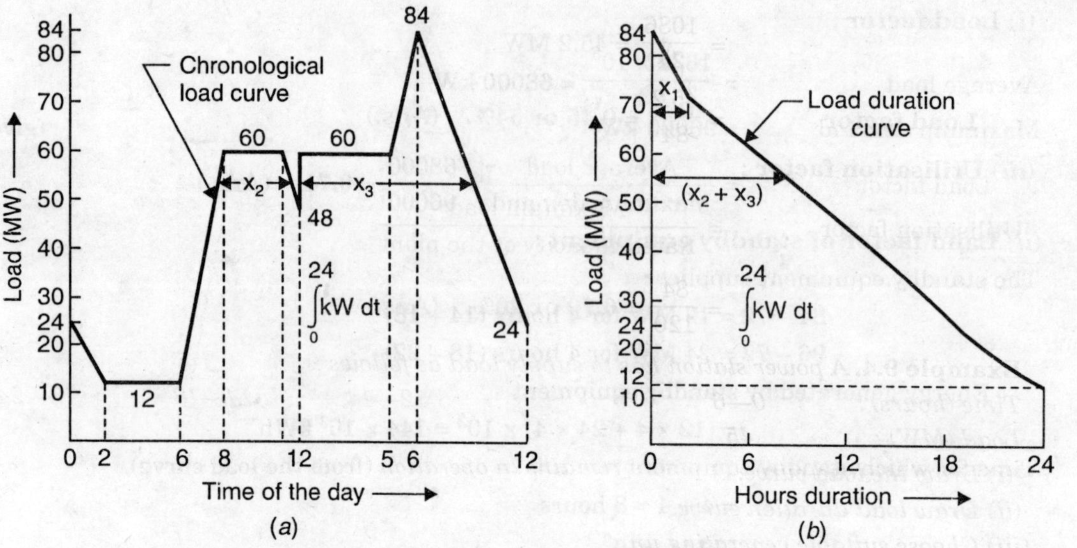

Fig. 9.11. (*a*) Chronological load curve (*b*) Derived load duration curve.

— The criterion of plotting the load duration curve *makes the abscissa at any load ordinate equal to the length of the abscissa intercepted by that load ordinate on the chronological curve.* Thus :

(*i*) At the maximum demand or peak load, the intercept is one point which will be plotted at 0 hour.

(*ii*) At 70 MW load the intercept is x_1 hours and is plotted as x_1 hour on the load duration curve.

(*iii*) At 55 MW load the intercept is a total of $(x_2 + x_3)$ and is plotted accordingly.

(*iv*) At minimum load of 12 MW the intercept covers the entire period of 24 hours.

Following points may be noted :

1. *Any point on the load duration curve is a measure of number of hours in a given period during which the given load and higher loads have prevailed.*
2. *If the chronological curve indicated a constant demand during the entire day, it would be of rectangular shape and load duration curve would be an exact duplicate.*

(*ii*) **Load factor :**

From the load duration curve, the average load can be estimated.

Average load for the period

$$= \frac{\text{Total energy in load curve for period}}{\text{Total number of hours in period}}$$

$$= \frac{\left(\frac{24+12}{2}\right)\times 2 + 12\times 4 + \left(\frac{12+60}{2}\right)\times 2 + 60\times 4 + \left(\frac{60+48}{2}\right)\times\frac{1}{2} + \left(\frac{48+60}{2}\right)\times\frac{1}{2} + 60\times 4 + \left(\frac{60+84}{2}\right)\times 1 + \left(\frac{84+24}{2}\right)\times 6}{24}$$

$$= \frac{36+48+72+240+27+27+240+72+324}{24}$$

$$= \frac{1086}{24} = 45.2 \text{ MW}$$

$$\therefore \quad \textbf{Load factor} = \frac{45.2}{84} = \mathbf{0.45 \text{ or } 54\%.} \quad \textbf{(Ans.)}$$

(*iii*) **Utilisation factor :**

$$\text{Utilisation factor} = \frac{\text{Maximum load}}{\text{Rated capacity of the plant}}$$

$$= \frac{84}{120} = \mathbf{0.70 \text{ or } 70\%.} \quad \textbf{(Ans.)}$$

Example 9.4. *A power station has to supply load as follows :*

Time (hours) :	*0—6*	*6—12*	*12—14*	*14—18*	*18—24*
Load (MW) :	*45*	*135*	*90*	*150*	*75*

(*i*) *Draw the load curve.*

(*ii*) *Draw load duration curve.*

(*iii*) *Choose suitable generating units to supply the load.*

(*iv*) *Calculate the load factor.*

(*v*) *Calculate the plant capacity factor.*

Solution. (*i*) **Load curve :**

The load curve is shown in Fig. 9.12 (*a*).

(*ii*) **Load duration curve :**

The load duration curve is shown in Fig. 9.12 (*b*).

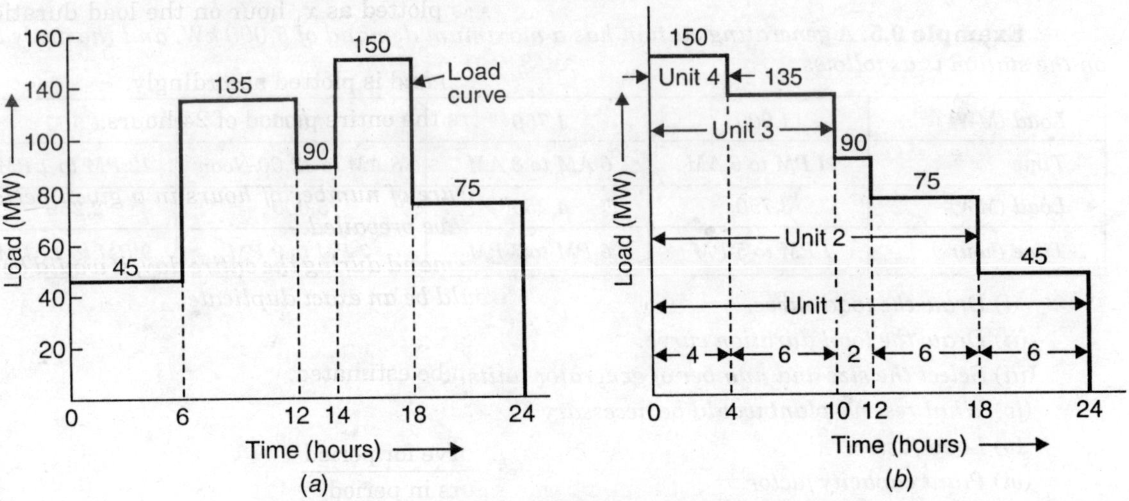

Fig. 9.12. (*a*) Load curve (*b*) Load duration curve.

(*iii*) **Selection of generating units :**

Load duration curve will indicate the operation schedule of different generating units.

1. One generating unit (unit 1) of 45 MW .... will run for **24 hours**
2. Second generating unit (unit 2) of 45 MW .... will run for **18 hours**
3. Third generating unit (unit 3) of 45 MW .... will run for **10 hours**
4. Fourth generating unit (unit 4) of 15 MW .... will run for **4 hours**

One additional unit (unit 5) should be kept as *standby*. Its capacity should be equal to the capacity of biggest set, *i.e.*, 45 MW.

(*iv*) **Load factor :**

Energy generated $= 45 \times 6 + 135 \times 6 + 90 \times 2 + 150 \times 4 + 75 \times 6$

$= 270 + 810 + 180 + 600 + 450 = 2310$ MWh

Average load $= \dfrac{2310 \times 10^3}{24}$ kW $= 96250$ kW

Maximum demand $= 150 \times 10^3 = 150000$ kW

$\therefore$ **Load factor** $= \dfrac{\text{Average load}}{\text{Maximum demand}}$

$= \dfrac{96250}{150000} =$ **0.64. (Ans.)**

(*v*) **Plant capacity factor :**

Plant capacity factor $= \dfrac{E}{C \times t}$

where, E = Energy generated (kWh),

C = Capacity of the plant (kW)

$= 45 \times 4 + 1 \times 15 = 195$ MW $= 195 \times 10^3$ kW, and

t = Number of hours in the given period = 24 hours.

$\therefore$ *Plant capacity factor* $= \dfrac{2310 \times 10^3}{195 \times 10^3 \times 24} =$ **0.49. (Ans.)**

Example 9.5. *A generating station has a maximum demand of 5,000 kW, and the daily load on the station is as follows :*

Load (MW)	*1,000*	*1,750*	*4,000*	*1,500*
Time	*11 PM to 6 AM*	*6 AM to 8 AM*	*8 AM to 12.00 Noon*	*12 PM to 1 PM*
Load (MW)	*3,750*	*4,250*	*5,000*	*2,250*
Time (hours)	*1 PM to 5 PM*	*5 PM to 7 PM*	*7 PM to 9 PM*	*9 PM to 11 PM*

(*i*) *Draw the load curve.*

(*ii*) *Draw the load duration curve.*

(*iii*) *Select the size and number of generator units.*

(*iv*) *What reserve plant would be necessary ?*

(*v*) *Load factor.*

(*vi*) *Plant capacity factor.*

Solution. (*i*) Load curve is shown in the Fig. 9.13 (*a*).

(*ii*) Load duration curve is shown in Fig. 9.13 (*b*).

(*iii*) **Size and number of generator units :**

From the load duration curve it is evident that generating sets of capacity 1000 kW, 1500 kW and 2500 kW will fulfill the requirement.

(*iv*) **Reserve capacity :**

Also, reserve capacity = Largest size of the unit in the station = **2500 kW. (Ans.)**

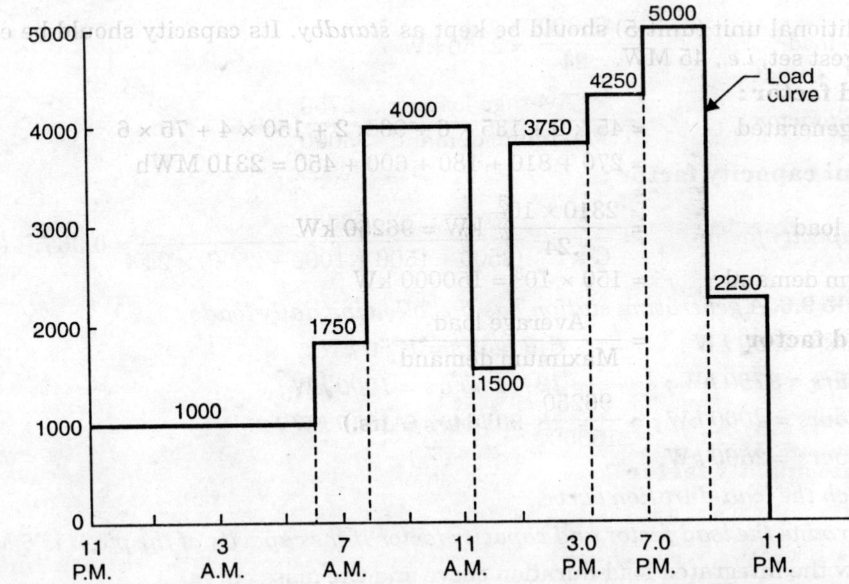

(a)

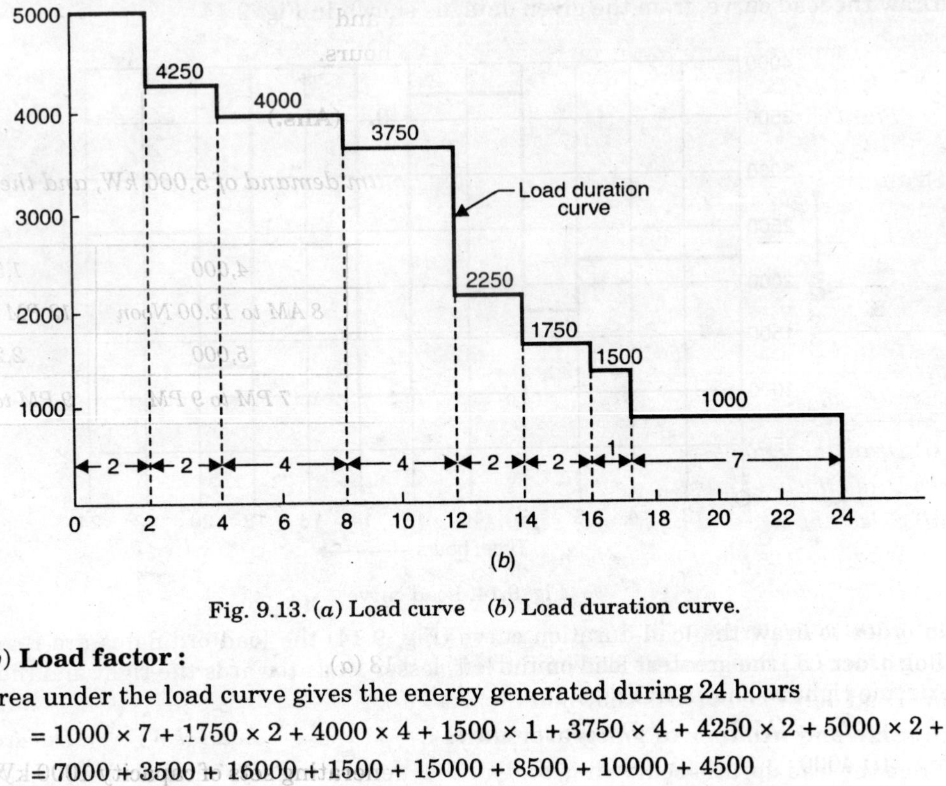

(b)

Fig. 9.13. (*a*) Load curve (*b*) Load duration curve.

(*v*) **Load factor :**

Area under the load curve gives the energy generated during 24 hours

$= 1000 \times 7 + 1750 \times 2 + 4000 \times 4 + 1500 \times 1 + 3750 \times 4 + 4250 \times 2 + 5000 \times 2 + 2250 \times 2$

$= 7000 + 3500 + 16000 + 1500 + 15000 + 8500 + 10000 + 4500$

$= 66000$ kWh

or, Average load $= \dfrac{66000}{24} = 2750$ kW

$\therefore$ *Load factor* $= \dfrac{\text{Average load}}{\text{Maximum demand}} = \dfrac{2750}{5000} = \mathbf{0.5.}$ **(Ans.)**

(*vi*) **Plant capacity factor :**

Plant capacity factor $= \dfrac{E}{C \times t} = \dfrac{66000}{(2500 + 1500 + 1000 + 2500) \times 24} = \mathbf{0.367.}$ **(Ans.)**

Example 9.6. *A generating station has the following daily loads :*

0–6 hours = 2250 kW ; *6–8 hours = 1750 kW ;*
8–12 hours = 3750 kW ; *12–14 hours = 1000 kW ;*
14–18 hours = 4000 kW ; *18–20 hours = 1250 kW ;*
20–24 hours = 2500 kW

(*i*) *Sketch the load-duration curve.*

(*ii*) *Determine the load factor and capacity factor if the capacity of the plant is 6 MW.*

(*iii*) Draw the integrated load-duration curve and the mass curve.

Solution. (*i*) **Load-duration curve :**

Draw the load curve, from the given data, as shown in Fig. 9.14.

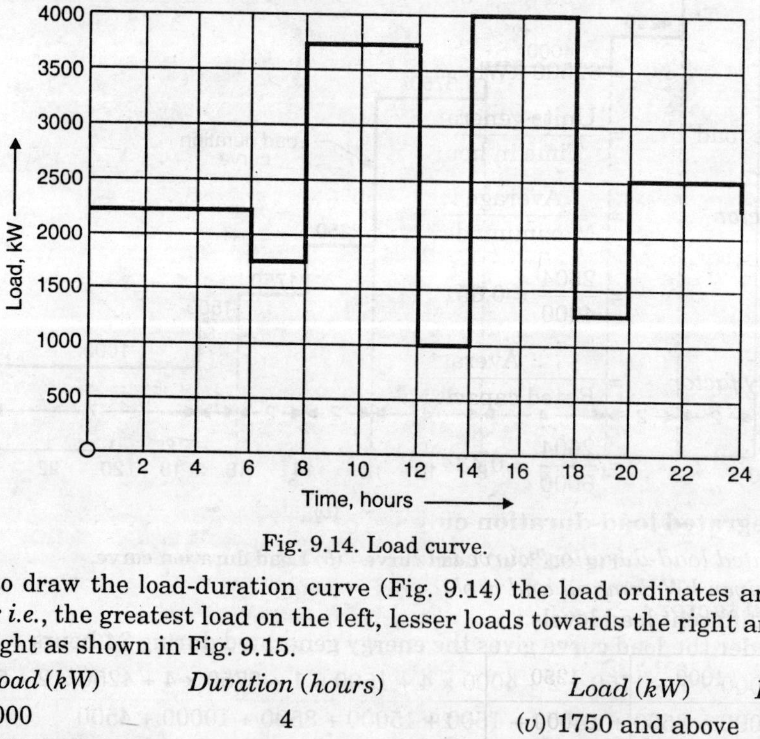

Fig. 9.14. Load curve.

In order to draw the load-duration curve (Fig. 9.14) the load ordinates are rearranged in descending order *i.e.*, the greatest load on the left, lesser loads towards the right and the least load at the extreme right as shown in Fig. 9.15.

Load (kW)	Duration (hours)
(*i*) 4000	4
(*ii*) 3750 and above	8
(*iii*) 2500 and above	12
(*iv*) 2250 and above	18
(*v*) 1750 and above	20
(*vi*) 1250 and above	22
(*vii*) 1000 and above	24

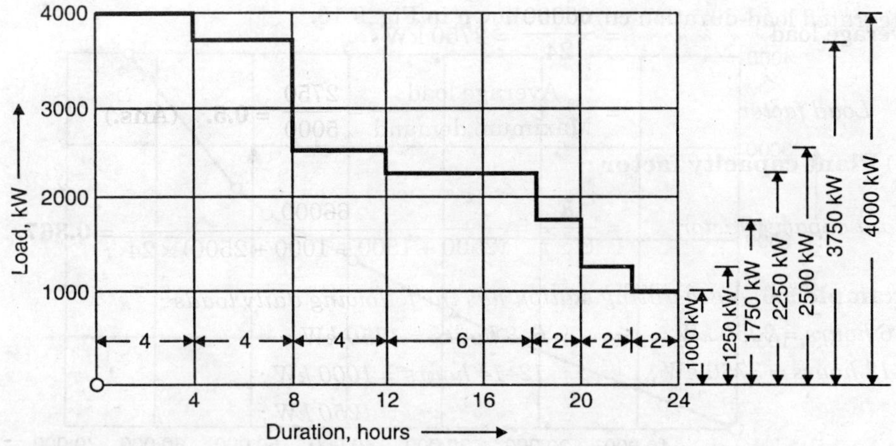

Fig. 9.15. Load-duration curve.

(*ii*) **Load factor, capacity factor :**

Capacity of the plant = 6 MW = 6000 kW

Maximum demand of the generating station = 4000 kW

Units generated in 24 hours

$$= 4000 \times 4 + 3750 \times 4 + 2500 \times 4 + 2250 \times 6 + 1750 \times 2 + 1250 \times 2 + 1000 \times 2$$

$$= 62500 \text{ kWh}$$

Average load $= \dfrac{\text{Units generated}}{\text{Time in hours}} = \dfrac{62500}{24} = 2604 \text{ kW}$

Load factor $= \dfrac{\text{Average load}}{\text{Maximum demand}}$

$$= \frac{2604}{4000} = 0.651 \text{ or } \mathbf{65.1\%.} \quad \textbf{(Ans.)}$$

Capacity factor $= \dfrac{\text{Average demand}}{\text{Rated capacity of power plant}}$

$$= \frac{2604}{6000} = 0.434 = \mathbf{43.4\%.} \quad \textbf{(Ans.)}$$

(*iii*) **Integrated load-duration cure :**

"Integrated load-duration" curve is drawn between the kW demand and the total units generated upto the given kW demand and is obtained from load-duration curve. The following table gives the energy at different load levels :

Load (kW)	1000	1250	1750	2250	2500	3750	4000
Unit generated (kWh)	24 × 1000 = 24000	24000 + 22 × 250 = 29500	29500 + 20 × 500 = 39500	39500 + 18 × 500 = 48500	48500 + 12 × 250 = 51500	51500 + 8 × 1250 = 61500	61500 + 4 × 250 = 62500

Integrated load-duration curve is shown in Fig. 9.16.

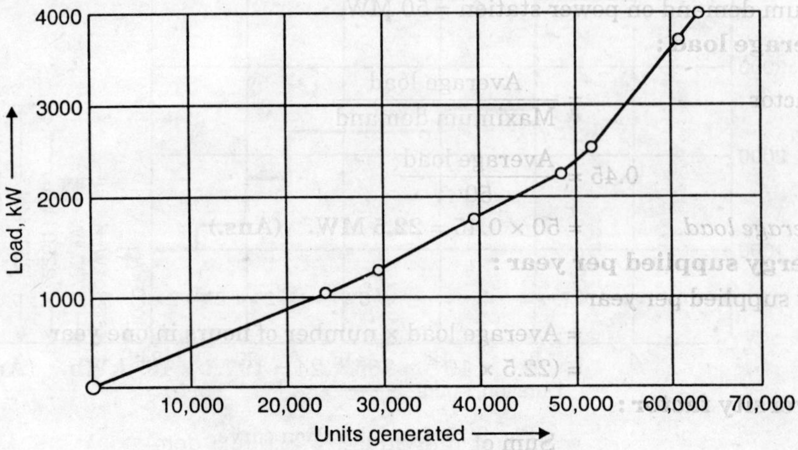

Fig. 9.16. Integrated load duration curve.

(*iv*) **Mass curve.** *"Mass curve" is plotted between the units generated upto a particular time in a day and the time.* The units generated upto different times of the day are obtained from the chronological load curve at given below :

Time	0	6 A.M.	8 A.M.	12.00 Noon	2.00 P.M.	6.00 P.M.	8.00 P.M.	12.00 Midnight
Units generated in kWh upto that time	0	6 × 2250 = 13500	13500 + 2 × 1750 = 17000	17000 + 4 × 3750 = 32000	32000 + 2 × 1000 = 34000	34000 + 4 × 4000 = 50000	50000 + 2 × 1250 52500	52500 + 4 × 2500 = 62500

The mass curve is shown in Fig. 9.17.

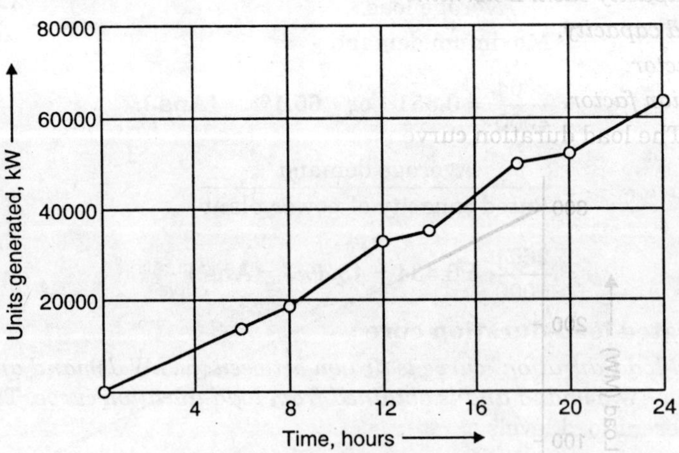

Fig. 9.17. Mass curve.

Example 9.7. *A 60 MW power station has an annual peak load of 50 MW. The power station supplies loads having maximum demands of 20 MW, 17 MW, 10 MW and 9 MW. The annual load factor is 0.45. Find :*

(i) Average load. *(ii) Energy supplied per year.*

(iii) Diversity factor. *(iv) Demand factor.*

Solution. Capacity of power station = 60 MW

Maximum demand on power station = 50 MW

(*i*) **Average load :**

$$\text{Load factor} = \frac{\text{Average load}}{\text{Maximum demand}}$$

i.e.,

$$0.45 = \frac{\text{Average load}}{50}$$

$\therefore$ *Average load* $= 50 \times 0.45 =$ **22.5 MW. (Ans.)**

(*ii*) **Energy supplied per year :**

Energy supplied per year

= Average load × number of hours in one year

$= (22.5 \times 10^3) \times 365 \times 24 =$ **197.1×10^6 kWh. (Ans.)**

(*iii*) **Diversity factor :**

$$\textit{Diversity factor} = \frac{\text{Sum of individuals maximum demands}}{\text{Simultaneous maximum demand}}$$

$$= \frac{20 + 17 + 10 + 9}{50} = \frac{56}{50} = 1.12$$

Hence, *diversity factor* = **1.12. (Ans.)**

(*iv*) **Demand factor :**

$$\textit{Demand factor} = \frac{\text{Maximum demand}}{\text{Connected load}} = \frac{50}{20 + 17 + 10 + 9} = \frac{50}{56} = 0.89$$

Hence, *demand factor* = **0.89. (Ans.)**

Example 9.8. *The yearly duration curve of a certain plant can be considered as a straight line from 300 MW to 80 MW. Power is supplied with one generating unit of 200 MW capacity and two units of 100 MW capacity each. Determine :*

(*i*) *Installed capacity.* (*ii*) *Load factor.*

(*iii*) *Plant factor.* (*iv*) *Maximum demand.*

(*v*) *Utilization factor.*

Solution. The load duration curve is shown in the Fig. 9.18.

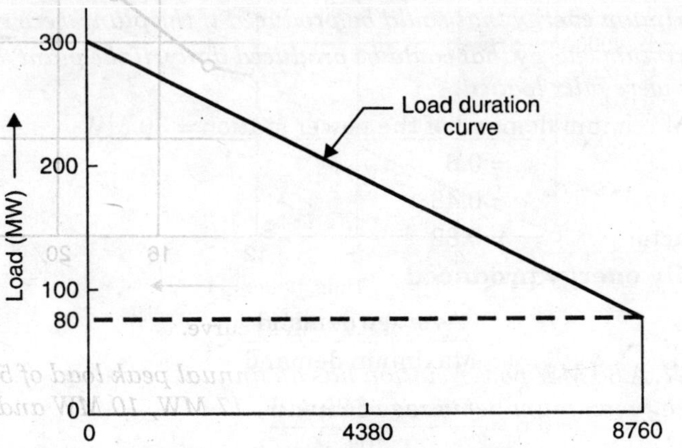

Fig. 9.18. Load duration curve.

(*i*) **Installed capacity :**

Installed capacity = 200 + 2 × 100 = **400 MW. (Ans.)**

(*ii*) **Load factor :**

$$\text{Load factor} = \frac{\text{Average load}}{\text{Maximum demand}}$$

$$\text{Average load} = \frac{\text{Total energy in load curve for the period}}{\text{Total number of hours in the period}}$$

$$= \frac{80 \times 8760 + \frac{1}{2} \times (300 - 80) \times 8760}{8760}$$

$$= \frac{8760}{8760}\left[80 + \frac{1}{2}(300 - 80)\right] = 190 \text{ MW}$$

$$\therefore \quad \textit{Load factor} = \frac{190}{300} = \mathbf{0.633.} \quad \textbf{(Ans.)}$$

(*iii*) **Plant factor :**

$$\textit{Plant factor} = \frac{\text{Average load}}{\text{Capacity of the plant}} = \frac{190}{400}$$

$$= \mathbf{0.475.} \quad \textbf{(Ans.)}$$

(*iv*) **Maximum demand = 300 MW. (Ans.)**

(*v*) **Utilization factor :**

$$\textit{Utilization factor} = \frac{\text{Maximum load}}{\text{Rated capacity of the plant}}$$

$$= \frac{300}{400} = \mathbf{0.75.} \quad \textbf{(Ans.)}$$

Example 9.9. *A generating station has a maximum demand of 30 MW, a load factor of 0.6, a plant capacity of 0.48, and a plant use factor of 0.8. Find :*

(i) The daily energy produced.

(ii) The reserve capacity of the plant.

(iii) The maximum energy that could be produced if the plant were running all the time.

(iv) The maximum energy that could be produced daily, if the plant when running according to operating schedule were fully loaded.

Solution. Maximum demand of the power station = 30 MW

Load factor = 0.6

Plant capacity = 0.48

Plant use factor = 0.82

(*i*) **The daily energy produced :**

$$\text{Load factor} = \frac{\text{Average demand}}{\text{Maximum demand}}$$

$$0.6 = \frac{\text{Average demand}}{30}$$

$\therefore$ Average demand = 30 × 0.6 = 18 MW or 18000 kW

Daily energy producted = Average demand × number of hours

$= 18000 \times 24 = \mathbf{4.32 \times 10^5}$ **kWh. (Ans.)**

(*ii*) **Reserve capacity of the plant :**

$$\text{Plant capacity factor} = \frac{\text{Average demand}}{\text{Installed capacity}}$$

$$0.48 = \frac{18000}{\text{installed capacity}}$$

$$\therefore \quad \text{Installed capacity} = \frac{18000}{0.48} = 37500 \text{ kW}$$

∴ *Reserve capacity of the plant*

= Installed capacity – maximum demand

= 37,500 – 30,000 = **7500 kW. (Ans.)**

(*iii*) **Maximum daily energy produced when running all the time**

$= \mathbf{4.32 \times 10^5}$ **kWh. (Ans.)**

(*iv*) **Maximum energy that could be produced daily :**

Maximum energy that could be produced daily, operating as per operating schedule

$$= \frac{\text{Actual energy produced}}{\text{Plant use factor}}$$

$$= \frac{4.32 \times 10^5}{0.82} = \mathbf{5.268 \times 10^5 \text{ kWh.}} \quad \textbf{(Ans.)}$$

Example 9.10. *A power station has the following loads :*

1. Residential lighting load :

Maximum demand	*= 1200 kW*
Load factor	*= 0.21*
Diversity between consumers	*= 1.32*

2. Commercial load :

Maximum demand	*= 2400 kW*
Load factor	*= 0.32*
Diversity between consumers	*= 1.2*

3. Industrial load :

Maximum demand	*= 6000 kW*
Load factor	*= 0.82*
Diversity between consumers	*= 1.22*

Overall diversity factor may be taken as 1.42.

Determine the following :

(*i*) *Maximum demand on system.*

(*ii*) *Daily energy consumption (total).*

(*iii*) *Overall load factor.*

(*iv*) *Connected load (total) assuming that demand factor for each load is unity.*

Solution. (*i*) **Maximum demand on system :**

$$\text{Group diversity factor} = \frac{\text{Sum of individual maximum demands}}{\text{Actual maximum demand of the group}}$$

$$\therefore \quad 1.42 = \frac{1200 + 2400 + 6000}{\text{Maximum demand on system}}$$

i.e., Maximum demand on system $= \dfrac{1200 + 2400 + 6000}{1.42}$

$= \mathbf{6760.5\ kW.}$ **(Ans.)**

(*ii*) **Daily energy consumption :**

Load factor $= \dfrac{\text{Average demand}}{\text{Maximum demand}}$

or, Average demand = Maximum demand × load factor

$= 1200 \times 0.21 + 2400 \times 0.32 + 6000 \times 0.82$

$= 5940$ kW

$\therefore$ *Daily energy consumption* $= 5940 \times 24 = \mathbf{142560\ kWh.}$ **(Ans.)**

(*iii*) **Overall load factor :**

Overall load factor $= \dfrac{\text{Average demand}}{\text{Maximum demand}}$

$= \dfrac{5940}{6760.5} = \mathbf{0.878.}$ **(Ans.)**

(*iv*) **Connected load :**

Maximum demand $= 1200 \times 1.32 + 2400 \times 1.2 + 6000 \times 1.22 = 11784$ kW

Connected load $= \dfrac{\text{Maximum demand}}{\text{Demand factor}} = \dfrac{11784}{1}$

$= \mathbf{11784\ kW.}$ **(Ans.)**

Example 9.11. *The following data relate to a steam power plant :*

Maximum demand	*= 30000 kW*
Load factor	*= 0.42*
Coal consumption	*= 1.1 kg/kWh*
Boiler efficiency	*= 84%*
Turbine efficiency	*= 88%*
Price of coal	*= Rs. 70 per tonne.*

Determine the following :

(i) Thermal efficiency of the plant.

(ii) Coal bill of the plant for one year.

Solution. (*i*) **Thermal efficiency of the plant :**

Thermal efficiency of the plant = Boiler efficiency × turbine efficiency

$= 0.84 \times 0.88 = \mathbf{0.7392\ or\ 73.92\%.}$ **(Ans.)**

(*ii*) **Coal bill :**

Average demand on station = Maximum demand × load factor

$= 30{,}000 \times 0.42 = 12600$ kW

$\therefore$ Energy generated per year $= 12600 \times (365 \times 24)$ kWh

$\therefore$ Coal consumption $= 12600 \times (365 \times 24) \times 1.1$ kg per year

$\therefore$ *Coal bill* $= \dfrac{12600 \times (365 \times 24) \times 1.1 \times 70}{1000} = \mathbf{Rs.\ 8498952.}$ **(Ans.)**

☞**Example 9.12.** *A power station is to supply for regions of load whose peak loads are 10 MW, 5 MW, 8 MW and 7 MW. The diversity factor of the load at the station is 1.5 and the average annual load factor is 0.6. Calculate :*

(i) Maximum demand on the station.

(ii) Annual energy supplied from the station.

Suggest the installed capacity and the number of units taking all aspects into account.

Solution. (*i*) **Maximum demand :**

$$\text{Maximum demand on power station} = \frac{\text{Sum of individual maximum demands}}{\text{Diversity factor}}$$

$$= \frac{10+5+8+7}{1.5}$$

$= \mathbf{20\ MW}$ **or** $\mathbf{20{,}000\ kW}$. **(Ans.)**

(*ii*) **Annual energy supplied :**

Average load = Maximum demand × load factor

= 20,000 × 0.6 = 12,000 kW

Annual energy supplied from the station

= Average load × (365 × 24)

= 12,000 × (365 × 24) = $\mathbf{105.12 \times 10^6}$ **kWh. (Ans.)**

Installed capacity and number of units :

Considering 50% increase in maximum demand on the power station in next five years, the *installed capacity* should be **30,000 kW or 30 MW. (Ans.)**

Select, four similar units each of 7.5 MW capacity because minimum number of spare parts will be required to be stored, at the same time *three units can supply present maximum demand and fourth unit can be taken out for routine maintenance or during breakdown without any disruption in supply*.

Example 9.13. *The peak load on a 50 MW power station is 39 MW. It supplies power through four transformers whose connected loads are 17, 12, 9 and 10 MW. The maximum demands on these transformers are 15, 10, 8 and 9 MW respectively. If the annual load factor is 50% and the plant is operating for 65% of the period in a year, find out the following :*

(i) Average load on the station *(ii) Energy supplied per year*

(iii) Demand factor *(iv) Diversity factor*

(v) Power station use factor.

Solution. Power station rated capacity = 50 MW or 50,000 kW

Maximum demand on the power station = 39 MW or 39,000 kW

Sum of connected load = 17 + 12 + 9 + 10 = 48 MW or 48,000 kW

Sum of maximum demands on the transformers = 15 + 10 + 8 + 9 = 42 MW or 42,000 kW

Annual load factor = 50% or 0.5

Plant operating period = 0.65 × (365 × 24) = 5,964 hours.

(*i*) **Average load on the station :**

Average load on the station = Maximum demand × load factor

= 39,000 × 0.5 = **19,500 kW. (Ans.)**

(*ii*) **Energy supplied per year :**

Energy supplied per year = Average load × (365 × 24)

= 19,500 × 8760 = $\mathbf{170.82 \times 10^6}$ **kWh. (Ans.)**

(*iii*) **Demand factor :**

$$\textit{Demand factor} = \frac{\text{Maximum demand}}{\text{Sum of connected load}}$$

$$= \frac{39{,}000}{48{,}000} = \mathbf{0.8125.} \quad \textbf{(Ans.)}$$

(*iv*) **Diversity factor :**

$$\textit{Diversity factor} = \frac{\text{Sum of maximum demands}}{\text{Maximum demand}}$$

$$= \frac{42{,}000}{39{,}000} = \mathbf{1.077.} \quad \textbf{(Ans.)}$$

(*v*) **Power station use factor :**

$$\textit{Use factor} = \frac{\text{Energy generated per year}}{\text{Rated capacity} \times \text{number of operating hours}}$$

$$= \frac{170.82 \times 10^6}{50{,}000 \times 5694} = \mathbf{0.6 \text{ or } 60\%.} \quad \textbf{(Ans.)}$$

Example 9.14. *A base load power station and standby power station share a common load as follows :*

Base load station annual output	*= 180 × 10^6 kWh*
Base load station capacity	*= 42 MW*
Maximum demand on base load station	*= 36 MW*
Standby station capacity	*= 22 MW*
Standby station annual output	*= 17 × 10^6 kWh*
Maximum demand (peak load) on stand by station	*= 18 MW*

Determine the following for both power stations :

(i) Load factor. *(ii) Capacity (or plant) factor.*

Solution. Base load station :

$$\text{Average load} = \frac{180 \times 10^6}{365 \times 24} = 20548 \text{ kW}$$

$$(i) \textbf{ Load factor} = \frac{\text{Average load}}{\text{Maximum demand}} = \frac{20548}{36 \times 10^3} = \mathbf{0.57.} \quad \textbf{(Ans.)}$$

$$(ii) \textbf{ Capacity factor} = \frac{\text{Energy generated}}{\text{Capacity of plant} \times (24 \times 365)}$$

$$= \frac{180 \times 10^6}{42 \times 1000 \times 24 \times 365} = \mathbf{0.489.} \quad \textbf{(Ans.)}$$

Standby power station :

$$\text{Annual average load} = \frac{17 \times 10^6}{365 \times 24} = 1940.6 \text{ kW}$$

$$(i) \textbf{ Load factor} = \frac{\text{Average load}}{\text{Maximum demand}} = \frac{1940.6}{18 \times 1000} = \mathbf{0.1078.} \quad \textbf{(Ans.)}$$

(*ii*) **Capacity factor** $= \dfrac{\text{Energy generated}}{\text{Capacity} \times (24 \times 365)}$

$$= \frac{17 \times 10^6}{22 \times 1000 \times 24 \times 365} = \mathbf{0.088.} \quad \textbf{(Ans.)}$$

Example 9.15. *A base load station having a capacity of 18 MW and a standby station having a capacity of 20 MW share a common load. Find (i) annual load factor, (ii) use factor and (iii) capacity factor of the two power stations from the following data :*

Annual standby station output $= 7.35 \times 10^6$ *kWh*
Annual base load station output $= 101.35 \times 10^6$ *kWh*
Peak load on the standby station = *12 MW*
Hours of use of standby station during the year = *2190 hours.*

Solution. Standby station :

Capacity of standby station = 20 MW or 20,000 kW
Maximum demand on standby station = 12 MW or 12,000 kW
Annual standby station output $= 7.35 \times 10^6$ kWh
Hours of use of standby station during the year = 2190 hours

Annual average load of standby station $= \dfrac{\text{Output in kWh}}{365 \times 24} = \dfrac{7.35 \times 10^6}{365 \times 24} = 839$ kW

(*i*) **Annual load factor :**

$$\textit{Annual load factor} = \frac{\text{Annual average load}}{\text{Maximum demand}}$$

$$= \frac{839}{12000} = \mathbf{0.07} \quad \textbf{or} \quad \mathbf{7\%}. \quad \textbf{(Ans.)}$$

(*ii*) **Use factor :**

$$\textit{Use factor} = \frac{\text{Total kWh generated}}{\text{Rated capacity of station} \times \text{number of operating hours}}$$

$$= \frac{7.35 \times 10^6}{20{,}000 \times 2190} = \mathbf{0.1678} \quad \textbf{or} \quad \mathbf{16.78\%}. \quad \textbf{(Ans.)}$$

(*iii*) **Capacity factor :**

$$\textit{Capacity factor} = \frac{\text{Average load}}{\text{Rated capacity}} = \frac{839}{20{,}000}$$

$$= \mathbf{0.0419} \quad \textbf{or} \quad \mathbf{4.19\%}. \quad \textbf{(Ans.)}$$

Base load station :

Capacity of base load station = 18 MW or 18000 kW
Assume maximum demand on base load station equal to its rated capacity *i.e.,* 18 MW.
Annual base load station output $= 101.35 \times 10^6$ kWh

Annual average load of base load station $= \dfrac{\text{Output in kWh}}{365 \times 24}$

$$= \frac{101.35 \times 10^6}{365 \times 24} = 11570 \text{ kW}$$

(*i*) **Annual load factor :**

$$\textit{Annual load factor} = \frac{\text{Annual average load}}{\text{Maximum demand}}$$

$$= \frac{11{,}570}{18{,}000} = \mathbf{0.643} \quad \textbf{or} \quad \mathbf{64.3\%.} \quad \textbf{(Ans.)}$$

(*ii*) **Use factor :**

$$\textit{Use factor} = \frac{\text{Total kWh generated}}{\text{Rated capacity} \times \text{number of operating hours}}$$

$$= \frac{101.35 \times 10^6}{18{,}000 \times (365 \times 24)} = \mathbf{0.643} \quad \textbf{or} \quad \mathbf{64.3\%.} \quad \textbf{(Ans.)}$$

(*iii*) **Capacity factor :**

$$\textit{Capacity factor} = \frac{\text{Average load}}{\text{Rated capacity}} = \frac{11{,}570}{18{,}000}$$

$$= \mathbf{0.643} \quad \textbf{or} \quad \mathbf{64.3\%.} \quad \textbf{(Ans.)}$$

COST ANALYSIS

Example 9.16. *Determine the annual cost of a feed water softner from the following data :*

Cost	= *Rs. 96,000*
Salvage value	= *5%*
Life	= *10 years*
Annual repair and maintenance cost	= *Rs. 3000*
Annual cost of chemicals	= *Rs. 6000*
Labour cost per month	= *Rs. 360*
Interest on sinking fund	= *5%.*

Solution. Capital cost, P = Rs. 96000.

Salvage value, $S = \frac{5}{100} \times 96000$ = Rs. 4800

Rate of interest on sinking fund, i = 5% or 0.05

Life, n = 10 years

∴ Annual sinking fund payment

$$= (P - S)\left[\frac{i}{(1+i)^n - 1}\right]$$

$$= (9600 - 4800)\left[\frac{0.05}{(1+0.05)^{10} - 1}\right] = \text{Rs. } 7250.8$$

Total cost per year :

Annual sinking fund	= Rs. 7250.8
Annual repair and maintenance cost	= Rs. 3000
Annual cost of chemicals	= Rs. 6000
Annual labour cost	= (360 × 12) = Rs. 4320
∴ *Total cost per year*	= 7250.8 + 3000 + 6000 + 4320
	= **Rs. 20,570.8. (Ans.)**

Example 9.17. *The output of a generating station is 500 × 10⁶ kWh per year and average load factor is 0.7. If the annual fixed charges are Rs. 50 per kW of installed plant and annual running charges are 5 per kWh, what is the cost per kWh of energy at the bus bar.*

Solution. Output energy per annum = 500×10^6 kWh

Average load $= \dfrac{\text{Annual average load}}{365 \times 24} = \dfrac{500 \times 10^6}{365 \times 24} = 57077$ kW

Maximum demand $= \dfrac{\text{Average load}}{\text{Load factor}} = \dfrac{57077}{0.7} = 81538$ kW

Assuming installed capacity equal to maximum demand,

Fixed charges = 50 × 81538 = Rs. 40,76,900

Running charges $= \text{Rs. } \dfrac{5}{100} \times 500 \times 10^6 = \text{Rs. } 2,50,00,000$

Total annual charges = Rs. 2,50,00,000 + Rs. 40,76,900 = Rs. 2,90,76,900

Cost of energy at bus-bar $= \dfrac{\text{Total annual charges}}{\text{Output energy per annum}}$

$= \dfrac{2,90,76,900}{500 \times 10^6}$

= **Rs. 0.058 or 5.8 p/kWh. (Ans.)**

Example 9.18. *From the following data calculate the cost of generation per unit delivered from the power plant :*

Installed capacity of the power plant	*= 200 MW*
Annual load factor	*= 0.4*
Capital cost of power plant	*= Rs. 280 lacs*
Annual cost of fuel, oil, salaries, taxation	*= Rs. 60 lacs.*
Interest and depreciation	*= 13%.*

Solution. Installed capacity of the power plant = 200 MW or 200×10^3 kW

Assuming maximum demand equal to installed capacity,

Maximum demand = 200×10^3 kW

Annual load factor = 0.4

Total units generated per annum = Maximum demand × load factor × (365 × 24)

$= 200 \times 10^3 \times 0.4 \times (365 \times 24) = 700.8 \times 10^6$ kWh

Capital cost of the power plant = Rs. 280×10^5

Annual interest and depreciation = Rs. $280 \times 10^6 \times \dfrac{13}{100}$ = Rs. 3.64×10^6

Annual cost of fuel, oil, salaries, taxation etc. = Rs. 60×10^5 or 6×10^6

Total annual cost = Rs. 3.64×10^6 + Rs. 6×10^6 = Rs. 9.64×10^6

Generating cost $= \dfrac{\text{Total annual cost}}{\text{Total units generated per annum}}$

$= \dfrac{9.64 \times 10^6}{700.8 \times 10^6}$

= **Rs. 0.0137 or 1.37 p/kWh. (Ans.)**

☞**Example 9.19.** *The following data relate to a 10 MW power station :*

Cost of plant	*= Rs. 1200 per kW*
Interest, insurances and taxes	*= 5% per annum*
Depreciation	*= 5%*
Cost of primary distribution	*= Rs. 5,00,000*
Interest, insurances, taxes and depreciation	*= 5%*
Cost of coal including transportation	*= Rs. 4.4 per kN*
Operating cost	*= Rs. 5,00,000*
Plant maintenance cost :	
(i) Fixed	*= Rs. 20,000 per annum*
(ii) Variable	*= Rs. 30,000 per annum*
Installed plant capacity	*= 10,000 kW*
Maximum demand	*= 9,000 kW*
Annual load factor	*= 0.6*
Consumption of coal	*= 255000 kN*

Determine the following :

(i) Cost of power generation per kW per year.

(ii) Cost per kWh generated.

(iii) Total cost of generation per kWh.

Transmission or primary distribution chargeable to generation.

Solution. Installed capacity of plant = 10 MW or 10,000 kW

Total cost of plant = Rs. 10,000 × 1200 = Rs. 12×10^6

Annual interest, insurances and taxes = Rs. $0.05 \times 12 \times 10^6$ = Rs. 0.06×10^6 = Rs. 600000

Annual depreciation = Rs. $0.05 \times 12 \times 10^6$ = Rs. 0.6×10^6 = Rs. 600000

Annual interest, insurance, taxes and depreciation on primary distribution

= Rs. 0.05 × 500000 = Rs. 25000

Annual plant maintenance cost (fixed) = Rs. 20000

Total fixed cost = Rs. (600000 + 600000 + 25000 + 20000) = Rs. 1245000

Annual operating cost = Rs. 500000

Annual plant maintenance cost (variable) = Rs. 30000

Annual cost of coal = Rs. 4.4 × 255000 = Rs. 1122000

Total annual running cost = Rs. (500000 + 300000 + 1122000) = Rs. 1652000

Maximum demand = 9000 kW

Annual load factor = 0.6

Average load = 9000 × 0.6 = 5400 kW

Annual energy generated = 5400 × 365 × 24 = 47.3×10^6 kWh

(*i*) **Cost of power generation per kW per year :**

$$\text{Annual cost per kW of maximum demand} = \frac{\text{Fixed cost per annum}}{\text{Maximum demand}}$$

$$= \frac{1245000}{9000} = \textbf{Rs. 138.33.} \quad \textbf{(Ans.)}$$

(*ii*) **Cost per kWh generated :**

$$\text{Annual cost/kWh} = \frac{\text{Annual running cost}}{\text{Annual energy generated}}$$

$$= \frac{1652000}{47.3 \times 10^6} = \textbf{Rs. 0.035 or 3.5 p.} \quad \textbf{(Ans.)}$$

(*iii*) **Total cost per kWh :**

$$\text{Total cost per kWh} = \frac{\text{Total annual cost}}{\text{Annual energy generated}}$$

$$= \frac{1245000 + 1652000}{47.3 \times 10^6}$$

$$= \textbf{Rs. 0.06} \quad \textbf{or} \quad \textbf{6 p.} \quad \textbf{(Ans.)}$$

Example 9.20. *Annual 200 MW steam power station is estimated to cost Rs. 350 millions. Other costs are as follows :*

Insurance and taxes = *Rs. 5 lacs per annum*
Fuel and lubricants = *Rs. 70 lacs per annum*
Transport and storage = *Rs. 10 lacs per annum*
Salaries and wages = *Rs. 12 lacs per annum*
Miscellaneous = *Rs. 2 lacs per annum*

Reckoning interest and depreciation at 15% per annum of capital cost, determine :

The cost of energy generated per unit if the power station works at an average load factor of 0.6. What would it be, if the load factor be increased to 75% with a consequent increase in fuel costs by 10%, other costs remaining the same ?

Solution. Capacity of steam power station = 200 MW or 200000 kW

Cost of power station = Rs. 350 millions = Rs. 350 × 10^6

Annual cost of insurance and taxes = Rs. 5 lacs = Rs. 0.5 × 10^6

Annual cost of fuel and lubricants = Rs 70 lacs = Rs. 7 × 10^6

Annual cost of transport and storage = Rs. 10 lacs = Rs. 1 × 10^6

Annual cost of salaries and wages = Rs. 12 lacs = Rs 1.2 × 10^6

Annual miscellaneous cost = Rs. 2 lacs = Rs. 0.2 × 10^6

$$\text{Annual interest and depreciation} = \text{Rs. } 350 \times 10^6 \times \frac{15}{100} = \text{Rs. } 52.5 \times 10^6$$

$$\text{Total annual cost} = \text{Rs. } (0.5 + 7 + 1 + 1.2 + 0.2 + 52.5) \times 10^6 = \text{Rs. } 62.4 \times 10^6$$

Total energy generated per annum

$$= \text{Maximum demand} \times \text{average load factor} \times (365 \times 24)$$

$$= 200000 \times 0.6 \times (365 \times 24) = 1051.2 \times 10^6 \text{ kWh}$$

$$\textit{Cost of generation} = \frac{\text{Total annual cost}}{\text{Total units generated}}$$

$$= \frac{62.4 \times 10^6}{1051.2 \times 10^6} = \textbf{Rs. 0.0594 per kWh or 5.94 p. per kWh.} \quad \textbf{(Ans.)}$$

If the load factor is improved to 75% :

Total energy generated per annum

$= 200000 \times 0.75 \times (365 \times 24) = 1314 \times 10^6$ kWh

Annual operating cost $=$ Rs. $(0.5 + 7 \times 1.1 \times 10^6 + 1 + 1.2 + 0.2 + 52.5) \times 10^6$

$=$ Rs. 63.1×10^6

Cost of generation $= \dfrac{\text{Total annual cost}}{\text{Total units generated}}$

$= \dfrac{63.1 \times 10^6}{1314 \times 10^6}$ = **Rs. 0.048 per kWh or 4.8 p. per kWh. (Ans.)**

Example 9.21. *A steam station has two 110 MW units. Following cost data are given :*

Particulars	*Units A*	*Units B*
Capital cost	*Rs. 2400 per kW*	*Rs. 3000 per kW*
Fixed charge rate	*10%*	*10%*
Capital factor	*0.55*	*0.60*
Fuel consumption	*1 kg / kWh*	*0.9 / kWh*
Fuel cost	*Rs. 96 per 1000 kg*	*Rs. 96 per 1000 kg*
Annual cost of operation, labour, maintenance and supplies	*20% of annual fuel cost*	*15% of annual fuel cost*
Utilisation factor	*1*	*1*

Calculate the following :

(*i*) *Annual plant cost and generation cost of unit A.*

(*ii*) *Annual plant cost and generation cost of unit B.*

(*iii*) *Overall generation cost of the station.*

Solution. (*i*) **Annual plant cost and generation cost of unit A :**

Annual fixed cost of unit A

$= \dfrac{10}{100} \times 2400 \times (100 \times 1000)$

$=$ Rs. 26.4×10^6

Annual energy output $=$ Maximum demand $\times$ capacity factor $\times$ no. of hours

$= (100 \times 1000) \times 0.55 \times (356 \times 24)$

$= 52.998 \times 10^7$ kWh

Annual fuel consumption $= 1 \times 52.998 \times 10^7$

$= 52.998 \times 10^7$ kg

Fuel cost $= \dfrac{96}{1000} \times 52.998 \times 10^7 =$ Rs. 50.87×10^7

Annual cost of operating labour, maintenance and supplies

$= 20\%$ of annual cost

$= \dfrac{20}{100} \times 50.87 \times 10^6 =$ Rs. 10.174×10^6

The annual operating cost of unit A

$=$ Annual fuel cost + annual cost of operation, labour and maintenance

$=$ Rs. $(50.87 \times 10^6 + 10.174 \times 10^6) =$ Rs. 61.044×10^6

Annual plant cost of unit A

= Annual fixed cost + annual operating cost

= Rs. $(26.4 \times 10^6 + 61.044 \times 10^6)$ = **Rs. 87.444×10^6. (Ans.)**

Generation cost of unit A

$$= \frac{\text{Annual plant cost}}{\text{Annual energy output}}$$

$$= \frac{87.444 \times 10^6}{52.998 \times 10^7} = \textbf{Rs. 0.165 or 16.5 p/kWh. (Ans.)}$$

(*ii*) **Annual plant cost and generation cost of unit B :**

Annual fixed cost of unit B

$$= \text{Rs. } \frac{10}{100} \times 3000 \times 110 \times 1000 = \text{Rs. } 33 \times 10^6$$

Expected annual energy output

$$= (110 \times 1000) \times (365 \times 24) \times 0.6 = 57.816 \times 10^7 \text{ kWh}$$

Annual fuel consumption

$$= 0.9 \times 57.816 \times 10^7 = 52.0344 \times 10^7 \text{ kg}$$

Fuel cost $= \frac{96}{100} \times 52.0344 \times 10^7 = \text{Rs. } 49.95 \times 10^6$

Annual cost of maintenance, repair etc.

$$= \text{Rs. } \frac{15}{100} \times 49.95 \times 10^6 = \text{Rs. } 7.4925 \times 10^6$$

Annual operating cost

= Fuel cost + maintenance cost

= Rs. $(49.95 \times 10^6 + 7.4925 \times 10^6)$ = Rs. 57.4425×10^6

Annual plant cost of unit B

= Fixed cost + operating cost

= Rs. $33 \times 10^6 + 57.4425 \times 10^6$

= **Rs. 90.4425×10^6. (Ans.)**

Generation cost of unit B

$$= \frac{\text{Annual plant cost}}{\text{Annual energy output}}$$

$$= \frac{90.4425 \times 10^6}{57.816 \times 10^7}$$

= **Rs. 0.1564 or 15.64 p/kWh. (Ans.)**

(*iii*) **Overall generation cost of the station**

$$= \frac{\text{Sum of annual plant cost of both units}}{\text{Sum of energy supplied}}$$

$$= \frac{87.444 \times 10^6 + 90.4425 \times 10^6}{52.998 \times 10^7 + 57.816 \times 10^7}$$

= **Rs. 0.16 or 16 p/kWh. (Ans.)**

☞**Example 9.22.** *The annual costs of operating a 15000 kW thermal power station are as follows :*

Cost of plant	*= Rs. 1080 per kW*
Interest, insurance, taxes on plant	*= 5 per cent*
Depreciation	*= 5 per cent*
Cost of primary distribution system	*= Rs. 600000*
Interest, insurance, taxes and depreciation on primary distribution system	*= 5 per cent*
Cost of secondary distribution system	*= Rs. 1080000*
Interest, taxes, insurance and depreciation on secondary distribution system	*= 5 per cent*
Maintenance of secondary distribution system	*= Rs. 216000*
Plant maintenance cost	
(i) Fixed cost	*= Rs. 36000*
(ii) Variable cost	*= Rs. 48000*
Operating costs	*= Rs. 720000*
Cost of coal	*= Rs. 7.2 per kN*
Consumption of coal	*= 300000 kN*
Dividend to stock holders	*= Rs. 1200000*
Energy loss in transmission	*= 10 per cent*
Maximum demand	*= 14000 kW*
Diversity factor	*= 1.5*
Load factor	*= 0.7*

Determine : (i) Charge per kW per year
(ii) Rate per kWh.

Solution. Maximum demand = 14000 kW

Load factor $= 0.7 = \dfrac{\text{Average load}}{\text{Maximum demand}}$

$\therefore$ Average load $= 0.7 \times 14000 = 9800$ kW

$\therefore$ Energy generated per year $= 9800 \times (365 \times 24)$

$= 85.8 \times 10^6$ kWh

Cost of plant = Capacity of plant × cost per kW

$= 15000 \times 1080 = \text{Rs. } 16.2 \times 10^6$

Interest, insurances, taxes on plant

$$= \frac{5}{100} \times 16.2 \times 10^6 = \text{Rs. } 810000$$

Plant depreciation $= \dfrac{5}{100} \times 16.2 \times 10^6 = \text{Rs. } 810000$

Cost of primary distribution system = Rs. 600000

Interest, insurance, taxes, depreciation on primary distribution system

$$= \frac{5}{100} \times 600000 = \text{Rs. } 30000$$

Cost of secondary distribution system = Rs. 1080000

$$= \frac{5}{100} \times 1080000 = \text{Rs. } 54000$$

Cost of coal = 7.2 × 300000 = Rs. 2160000.

(*i*) **Charge per kW per year :**

Fixed costs :

Interest, taxes and insurance on plant = Rs. 810000

Plant depreciation = Rs. 810000

Interest, taxes, insurance and depreciation on :

Primary distribution system = Rs. 30000

Secondary distribution system = Rs. 54000

Fixed part of plant maintenance = Rs. 36000

Dividend of stock-holder = Rs. 1200000

∴ Total fixed cost = Rs. 2940000

Sum of maximum demand of consumers

= Maximum demand × diversity factor

= 14000 × 1.5 = 21000 kW

$$\textit{\textbf{Charge per kW per year}} = \frac{2940000}{21000}$$

= **Rs. 140 per kW. (Ans.)**

(*ii*) **Rate per kWh :**

Variable costs

Cost of coal = Rs. 2160000

Plant maintenance = Rs. 48000

Operating costs = Rs. 720000

Maintenance of secondary distribution system = Rs. 216000

Total variable cost = Rs. 3144000

Energy loss in transmission = 10 percent

∴ Net energy transmitted = $0.9 \times 85.8 \times 10^6 = 77.22 \times 10^6$ kWh

∴ **Rate per kWh** $= \frac{3144000}{77.22 \times 10^6}$ = **Rs. 0.0407 or 4.07 p/kWh. (Ans.)**

Example 9.23. *It is proposed to supply a load with a maximum demand of 100 MW and a load factor of 0.4. Choice is to be made from nuclear, hydro and steam power plants. Calculate the overall cost per kWh in each scheme.*

Cost	*Nuclear power plant*	*Hydro-power plant*	*Steam power plant*
Capital per kW installed	*Rs. 6000*	*Rs. 4320*	*Rs. 2160*
Interest	*10%*	*10%*	*12%*
Depriciation	*10%*	*8%*	*12%*
Operating cost per kWh	*12 paise*	*6 paise*	*18 paise*
Transmission and distribution cost/kWh	*0.24 paise*	*0.96 paise*	*0.24 paise*

Solution. (*i*) ***Nuclear power plant :***

Capital cost $= 6000 \times (100 \times 10^3) = \text{Rs. } 60 \times 10^7$

Interest $= \frac{10}{100} \times 60 \times 10^7 = \text{Rs. } 6 \times 10^7$

Depreciation $= \text{Rs. } 6 \times 10^7$

Annual fixed cost (interest + depreciation) $= \text{Rs. } 12 \times 10^7$

Energy generated per year = Average load × (365 × 24)

= Load factor × maximum demand × (365 × 24)

$= 0.4 \times (100 \times 10^3) \times (365 \times 24)$

$= 350.4 \times 10^6$ kWh

Running cost per kWh = Operating cost per kWh + transmission and distribution cost per kWh

= 12 + 0.24 = 12.24 p

∴ **Overall cost per kWh** = Running cost/kWh + fixed cost/kWh

$$= 12.24 + \frac{12 \times 10^7}{350.4 \times 10^6} \times 100 = \mathbf{46.48\ p.} \quad \textbf{(Ans.)}$$

(*ii*) ***Hydro-electric power plant :***

Capital cost $= 4320 \times 100 \times 10^3 = \text{Rs. } 43.2 \times 10^7$

Interest $= \frac{10}{100} \times 43.2 \times 10^7 = \text{Rs. } 43.2 \times 10^6$

Depreciation $= \frac{8}{100} \times 43.2 \times 10^7 = 34.56 \times 10^6$

Annual fixed cost = Interest + depreciation

$= \text{Rs. } (43.2 + 34.56) \times 10^6 = \text{Rs. } 77.76 \times 10^6$

Running cost per kWh = (Operation cost + transmission cost)

= (6 + 0.96) p = 6.96 p

Overall cost per kWh = Running cost/kWh + annual fixed cost/kWh

$$= 6.96 + \frac{77.76 \times 10^6}{350.4 \times 10^6} \times 100$$

$$= 6.96 + 22.2 = \mathbf{29.16\ p.} \quad \textbf{(Ans.)}$$

(*iii*) ***Steam power plant :***

Capital cost $= \text{Rs. } 2160 \times 100 \times 10^3 = \text{Rs. } 21.6 \times 10^7$

Interest $= \text{Rs. } \frac{12}{100} \times 21.6 \times 10^7 = \text{Rs. } 25.92 \times 10^6$

Depreciation $= \text{Rs. } \frac{12}{100} \times 21.6 \times 10^7 = \text{Rs. } 25.92 \times 10^6$

Annual fixed cost = Interest + depreciation

$= \text{Rs. } 2 \times 25.92 \times 10^6 = \text{Rs. } 51.84 \times 10^6$

Running cost/kWh = 18 + 0.24 = 18.24 p

Overall cost/kWh = Running cost/kWh + fixed cost/kWh

$$= 18.24 + \frac{51.84 \times 10^6}{350.4 \times 10^6} \times 100$$

$= 18.24 + 14.79 =$ **33.03 p.** **(Ans.)**

From the above calculations it is concluded that overall cost/kWh is *minimum in case of hydropower plant*.

☞**Example 9.24.** *A power plant of 180 MW installed capacity has the following data :*

Capital cost = *Rs. 2160 / kW installed*
Interest and depreciation = *12 per cent*
Annual load factor = *0.6*
Annual capacity factor = *0.5*
Annual running charges = *Rs. 36 × 10⁶*
Energy consumed by power auxiliaries = *6 per cent*
Calculate : (i) Reserve capacity
(ii) Generation capacity.

Solution. Load factor $= \frac{\text{Average load}}{\text{Maximum demand}}$

and, Capacity factor $= \frac{\text{Average load}}{\text{Rated capacity}}$

$$\therefore \quad \frac{\text{Load factor}}{\text{Capacity factor}} = \frac{\text{Rated capacity}}{\text{Maximum demand}}$$

$$\frac{0.6}{0.5} = \frac{180}{\text{Maximum demand}}$$

$$\therefore \quad \text{Maximum demand} = \frac{0.5 \times 180}{0.6} = 150 \text{ MW.}$$

(*i*) **Reserved capacity :**

Reserved capacity = Installed/rated capacity – Maximum demand
$= 180 - 150 =$ **30 MW.** **(Ans.)**

(*ii*) **Generation cost :**

Average load = Load factor × maximum demand
$= 0.6 \times 150 = 90$ MW

Energy generated per annum
$= 90 \times 10^3 \times (365 \times 24) = 788.4 \times 10^6$ kWh

Energy consumed by auxiliaries

$$= \frac{6}{100} \times 788.4 \times 10^6 = 47.3 \times 10^6 \text{ kWh}$$

Net energy available $= 788.4 \times 10^6 - 47.3 \times 10^6 = 741.1 \times 10^6$ kWh

Fixed cost of generation = Interest + depreciation

$$= \text{Rs. } \frac{12}{100} \times 2160 \times 180 \times 10^3 = \text{Rs. } 46.65 \times 10^6$$

Total annual cost = Running cost + fixed cost

$= 36 \times 10^6 + 46.65 \times 10^6 = \text{Rs. } 82.65 \times 10^6$

$\therefore$ *Generation cost* $= \dfrac{82.65 \times 10^6}{741.1 \times 10^6} \times 100 =$ **11.1 p. (Ans.)**

Example 9.25. *A power station has the installed capacity of 180 MW. Calculate the cost of generation, other data pertaining to power station are given below :*

Capital cost = *Rs.* 300×10^6

Rate of interest and depreciation = *18 per cent*

Annual cost of fuel oil, salaries and taxation = *Rs.* 36×10^6

Load factor = *0.4*

Also calculate the saving in cost per kWh if the annual load factor is raised to 0.5.

Solution. Assuming maximum demand equal to the capacity of the power plant,

Load factor $= \dfrac{\text{Average load}}{\text{Maximum demand}}$

$$0.4 = \frac{\text{Average load}}{180}$$

$\therefore$ Average load $= 0.4 \times 180 = 72$ MW

Energy generated per annum $= 72 \times 10^3 \times (365 \times 24) = 630.72 \times 10^6$ kWh

Fixed cost = Interest and depreciation on capital cost

$= \dfrac{18}{100} \times 300 \times 10^6 = \text{Rs. } 54 \times 10^6$

Running (operating cost) = Cost of fuel oil, salaries and taxation

$= \text{Rs. } 36 \times 10^6$

Total annual cost = Fixed cost + operating cost

$= \text{Rs. } (54 + 36) \times 10^6$

$= \text{Rs. } 90 \times 10^6$

$\therefore$ *Cost per kWh* $= \dfrac{90 \times 10^6}{630.72 \times 10^6} \times 100 =$ **14.27 p. per kWh. (Ans.)**

When the load factor is raised to 0.5 :

Average load = Load factor × maximum demand

$= 0.5 \times 180 = 90$ MW

Energy produced per annum $= 90 \times 10^3 \times 365 \times 24$

$= 788.4 \times 10^6$ kWh

Total annual cost will not change.

$\therefore$ Cost per kWh $= \dfrac{90 \times 10^6}{788.4 \times 10^6} \times 100$

$= 11.41$ p. per kWh

$\therefore$ *Saving in cost per kWh* $= 14.27 - 11.41$

= **2.86 p. (Ans.)**

Example 9.26. *A 60 MW generating station has the following data :*

Capital cost = *Rs.* 18×10^6

Annual taxation = *Rs.* 0.48×10^6

Annual salaries and wages	$= Rs.\ 1.44 \times 10^6$
Cost of coal	$= Rs.\ 72$ *per tonne*
Calorific value (C.V.) of coal	$= 23000\ kJ/kg$
Rate of interest and depreciation	$= 12$ *per cent*
Plant heat rate	$= 138000\ kJ/kWh$ *at 100% capacity*

Calculate the generating cost / kWh at 100% capacity factor.

Solution. Maximum demand = 60 MW (= peak load)

Fixed cost = Interest and depreciation on capital cost

$$= \text{Rs. } \frac{12}{100} \times 18 \times 10^6 = \text{Rs. } 2.16 \times 10^6$$

Running cost = Annual salaries, wages and taxation

$$= \text{Rs. } (1.44 + 0.48)\ 10^6 = \text{Rs. } 1.92 \times 10^6$$

At 100% capacity factor :

Rated/installed capacity = 60 MW

$$\text{Capacity factor} = \frac{\text{Average load}}{\text{Rated capacity}} = 1$$

or, Average load = Rated capacity = 60 MW

Average energy produced per annum

$$= 60 \times 10^3 \times (365 \times 24) \text{ kWh}$$
$$= 525.6 \times 10^6 \text{ kWh}$$

Total plant heat rate $= 138000 \times 525.6 \times 10^6$ kJ

$$\text{Weight of coal required per annum} = \frac{138000 \times 525.6 \times 10^6}{23000 \times 1000} \text{ tonnes}$$
$$= 3.15 \times 10^6 \text{ tonnes}$$

Cost of fuel $= \text{Rs. } 72 \times 3.15 \times 10^6$
$= \text{Rs. } 226.8 \times 10^6$

Total annual cost $= \text{Rs. } (2.16 + 1.92 + 226.8) \times 10^6$
$= \text{Rs. } 230.88 \times 10^6$

$$\textit{Generation cost} = \frac{230.88 \times 10^6}{525.6 \times 10^6} \times 100 = \textbf{43.93 p.} \quad \textbf{(Ans.)}$$

ECONOMIC LOAD SHARING

Example 9.27. *An input-output curve of a 10 MW station is expressed as follows :*

$$I = 4 \times 10^6\ (10 + 8L + 0.4L^2)$$

where I is in kJ / hour and L is in megawatts.

(i) Without plotting any curve find the load at which the maximum efficiency occurs.

(ii) Find the increase in input required to increase station output from 3 to 5 MW by means of the input-output curve and also by incremental rate curve.

Solution. (*i*) **Load at which maximum efficiency occurs :**

$$I = 4 \times 10^6\ (10 + 8L + 0.4L^2)$$

or,
$$\frac{I}{L} = 4 \times 10^6 \left(\frac{10}{L} + 8 + 0.4L\right)$$

$$\text{Efficiency} = \frac{\text{Output}}{\text{Input}} = \frac{L}{I}$$

$\therefore$ $$\text{Efficiency, } \eta = \frac{1}{4 \times 10^6 \left(\dfrac{10}{L} + 8 + 0.4L \right)} \quad ...(i)$$

Now the *efficiency will be maximum* when $\left(\dfrac{10}{L} + 8 + 0.4L \right)$ *is minimum*

i.e., $$\frac{d}{dL}\left(\frac{10}{L} + 8 + 0.4L \right) = 0$$

$\therefore$ $$-\frac{10}{L^2} + 0.4 = 0 \quad \text{or} \quad L^2 = \frac{10}{0.4} = 25 \quad \text{or} \quad L = 5 \text{ MW}$$

Hence, *the load at which the maximum efficiency occurs* = **5 MW. (Ans.)**

(*ii*) **Increase in input :**

(*a*) ***By input output curve :***

When load, L = 3 MW

Input, $I_3 = 4 \times 10^6 (10 + 8 \times 3 + 0.4 \times 3^2)$
$= 150.4 \times 10^6$ kJ/h

When load, L = 5 MW

Input, $I_5 = 4 \times 10^6 (10 + 8 \times 5 + 0.4 \times 5^2)$
$= 240 \times 10^6$ kJ/h

Increase in input required $= I_5 - I_3$
$= (240 - 150.4) \times 10^6$
$=$ **89.6×10^6 kJ/h. (Ans.)**

(*b*) ***By incremental rate curve :***

When load varies from 3 to 5 MW, the incremental rate may be considered to be straight line and the average height of area under the curve between 3 MW and 5 MW would be

$$= \frac{3 + 5}{2} = 4 \text{ MW}$$

$$I = 4 \times 10^6 (10 + 8L + 0.4L^2)$$

Increment rate, $$IR = \frac{dI}{dL} = 4 \times 10^6 (8 + 0.8L)$$

$\therefore$ $IR = 4 \times 10^6 (8 + 0.8 \times 4)$, when load = 4 MW
$= 4 \times 10^6 (8 + 3.2) = 4 \times 10^6 \times 11.2$

Hence total increase in input $= 4 \times 10^6 \times 11.2 (5 - 3) =$ **89.6×10^6 kJ/h. (Ans.)**

This shows that increase in input required to increase the required output in both cases (*a*) and (*b*) is same. This *indicates that the incremental rate curve can be taken as straight line for small increase in output*.

Example 9.28. *The input-output curve of a 50 MW power station is given by :*

$$I = 4 \times 10^6 (8 + 8L + 0.4L^2) \text{ kJ/hour}$$

where, I is the input in kJ/hour and L is load in MW.

(*i*) *Determine the heat input per day to the power station if it works for 20 hours at full load and remaining period at no load.*

(ii) Also find the saving per kWh of energy produced if the plant works at full load for all 24 hours generating the same amount of energy.

Solution. (*i*) **Heat input per day :**

Total energy generated by the plant during 24 hours

$$= 20 \times 50 + 4 \times 0 = 1000 \text{ MWh}$$

Input to the plant when the plant is running at full load

$$I_{50} = 4 \times 10^6 (8 + 8 \times 50 + 0.4 \times 50^2) \times 20$$

$$= 4 \times 10^6 \times 1408 \times 20 \text{ kJ during 20 hours when the plant was running at } \textit{full load}.$$

Input at no load, $I_0 = 4 \times 10^6 \times 8 \times 4$

$$= 128 \times 10^6 \text{ kJ during 4 hours when the plant was running at } \textit{no load}.$$

Total input to the plant during 24 hours

$$= I_{50} + I_0 = 4 \times 10^6 \times 1408 \times 20 + 128 \times 10^6$$

$$= 10^6 (5632 \times 20 + 128) = \mathbf{112768 \times 10^6 \text{ kJ/day.}} \quad \textbf{(Ans.)}$$

(*ii*) **Saving per kWh :**

Average heat supplied per kWh generated

$$= \frac{112768 \times 10^6}{1000 \times 10^3} = 112768 \text{ kJ/kWh}$$

If the same energy is generated within 24 hours, the average load is given by :

$$\text{Average load} = \frac{1000}{24} = 41.67 \text{ MW}$$

Heat supplied during 24 hours in this case

$$I_{50} = 4 \times 10^6 (8 + 8 \times 50 + 0.4 \times 41.67^2) \times 24$$

$$= 4 \times 10^6 (8 + 400 + 694.5) \times 24$$

$$= 4 \times 10^6 \times 1102.5 \times 24 \text{ kJ/day} = 105840 \times 10^6 \text{ kJ/day}$$

Net saving per day

$$= 112768 \times 10^6 - 105840 \times 10^6 = 6928 \times 10^6 \text{ kJ/day}$$

$$\therefore \textit{ Saving per kWh} = \frac{6928 \times 10^6}{1000 \times 10^3} = \mathbf{6928 \text{ kJ/kWh.}} \quad \textbf{(Ans.)}$$

Example 9.29. *The incremental fuel costs for two generating units 1 and 2 of a power plant are given by the following equations :*

$$\frac{dF_1}{dP_1} = 0.07\, P_1 + 24$$

$$\frac{dF_2}{dP_2} = 0.075\, P_2 + 22$$

where, F is fuel cost in rupees per hour and P is power output in MW. Determine :

(i) The economic loading of the two units when the total load supplied by the power plants is 180 MW.

(ii) The loss in fuel cost per hour if the load is equally shared by both units.

Solution. (*i*) **Economic loading of two units :**

$$P_1 + P_2 = 180 \qquad \textit{(Given)} \qquad ...(i)$$

The condition required for economic loading is given by :

$$\frac{dF_1}{dP_1} = \frac{dF_2}{dP_2}$$

$\therefore \quad 0.07\,P_1 + 24 = 0.075\,P_2 + 22$ ...(*ii*)

Substituting the value of P_2 (= $180 - P_1$) from (*i*) in (*ii*), we get

$$0.07\,P_1 + 24 = 0.075\,(180 - P_1) + 22$$

or, $\quad 0.07\,P_1 + 24 = 13.5 - 0.075\,P_1 + 22$

or, $\quad 0.145\,P_1 = 11.5$

or, $\quad P_1 = \dfrac{11.5}{0.145} = \mathbf{79.3\ MW.}$ **(Ans.)**

and, $\quad P_2 = 180 - 79.3 = \mathbf{100.7\ MW.}$ **(Ans.)**

(*ii*) **Loss in fuel cost :**

If the load is equally shared by both the units $\left(\text{supplying } \dfrac{180}{2} = 90 \text{ MW each}\right)$, then the increase in cost of fuel for *unit 1* is

$$= \int_{79.3}^{90} (0.07\,P_1 + 24)\,dP_1 = \left[\frac{0.07\,P_1^2}{2} + 24P_1\right]_{79.3}^{90}$$

$$= 0.035\,(90^2 - 79.3^2) + 24\,(90 - 79.3)$$

$$= 63.4 + 256.8 = \text{Rs. } 320.2\text{/hour}$$

Increase in cost of fuel for *unit 2*

$$= \int_{100.7}^{90} (0.075\,P_2 + 22)\,dP_2 = \left[\frac{0.075 P_2^2}{2} + 22P_2\right]_{100.7}^{90}$$

$$= \frac{0.075}{2}\,(90^2 - 100.7^2) + 22\,(90 - 100.7)$$

$$= -\,76.5 - 235.4 = \text{Rs. } -\,311.9\text{/hour}$$

This indicates that the cost of fuel for unit 2 ***decreases.***

Net increase in cost (or loss in fuel cost) due to departure from economic distribution of load

$$= 320.2 - 311.9 = \textbf{Rs. 8.3/hour.} \quad \textbf{(Ans.)}$$

Example 9.30. *Two steam turbines each of 30 MW capacity take a load 45 MW. The steam consumption rates in kg per hour for both turbines are given by the following equations :*

$$S_1 = 2400 + 12L_1 - 0.00012\,L_1^2$$

$$S_2 = 1200 + 8.4L_2 - 0.00006\,L_2^2$$

L represents the load in kW and S represents the steam consumption per hour. Find the most economical loading when the load taken by both units is 45 MW.

Solution. $\quad L_1 + L_2 = 45 \text{ MW} = 45000 \text{ kW}$ ...(*i*)

For the most economical loading, the required condition is

$$\frac{dS_1}{dL_1} = \frac{dS_2}{dL_2}$$

$\therefore \quad 12 - 2 \times 0.00012\,L_1 = 8.4 - 2 \times 0.00006\,L_2$

$\quad 12 - 0.00024\,L_1 = 8.4 - 0.00012\,L_2$ ...(*ii*)

Substituting the value of L_2 (= $45000 - L_1$) from (*i*) in (*ii*), we get

$$12 - 0.00024\,L_1 = 8.4 - 0.00012\,(45000 - L_1)$$

$$12 - 0.00024\,L_1 = 8.4 - 5.4 + 0.00012\,L_1$$

$$0.00036\,L_1 = 9$$

$\therefore \quad L_1 = \dfrac{9}{0.00036} = 25000$ **kW** or **25 MW. (Ans.)**

and, $\quad L_2 = 45000 - 25000 = 20000$ **kW** or **20 MW. (Ans.)**

9.18. TARIFF FOR ELECTRICAL ENERGY

9.18.1. Introduction

The cost of generation of electrical energy consists of *fixed cost and running cost.* Since the electricity generated is to be supplied to the consumers, the total cost of generation has to be recovered from the consumers. *Tariffs or energy rates are the different methods of charging the consumers for the consumption of electricity*. It is desirable to charge the consumer according to the maximum demand (kW) and the energy consumed (kWh). *The tariff chosen should recover the fixed cost, operating cost and profit etc. incurred in generating the electrical energy*.

9.18.2. Objectives and Requirements of Tariff

Objectives of tariff :

1. Recovery of cost of capital investment in generating equipment, transmission and distribution system.
2. Recovery of the cost of operation, supplies and maintenance of the equipment.
3. Recovery of the cost of material, equipment, billing and collection cost as well as for miscellaneous services.
4. A net return on the total capital investment must be ensured.

Requirements of tariff :

1. It should be easier to understand.
2. It should provide low rates for high consumption.
3. It should be uniform over large population.
4. It should encourage the consumers having high load factors.
5. It should take into account maximum demand charges and energy charges.
6. It should provide incentive for using power during off-peak hours.
7. It should provide less charges for power connection than lighting.
8. It should have a provision of penalty for low power factors.
9. It should have a provision for higher demand charges for high loads demanded at system peaks.
10. It should apportion equitably the cost of service to the different categories of consumers.

9.18.3. General Tariff Form

A large number of tariffs have been proposed from time to time and are in use. They are all derived from the following general equation :

$$z = a.x + b.y + c \qquad ...(9.9)$$

where, z = Total amount of bill for the period considered,

x = Maximum demand in kW,

y = Energy consumed in kWh during the period considered,

a = Rate per kW of maximum demand,

b = Energy rate per kWh, and

c = Constant amount charged to the consumer during each billing period. This charge is independent of demand or total energy because a consumer that remains connected to the line incurs expenses even if he does not use energy.

Various types of tariffs :

The various types of tariffs are :

1. Flat demand rate.
2. Straight meter rate.
3. Block meter rate.
4. Hopkinson demand rate (Two-part tariff).
5. Doherty rate (Three-part tariff).
6. Wright demand rate.

1. Flat demand rate :

The flat demand rate is expressed as follows :

$$z = ax \qquad ...(9.10)$$

i.e., the bill *depends only on the maximum demand irrespective of the amount of energy consumed.* It is based on the customer's installation of energy consuming devices which is generally denoted by so many kW per month or per year. It is probably one of the early systems of charging energy rates. It was based upon the total number of lamps installed and a fixed number of hours of use per year. Hence the rate could be expressed as a price per lamp or unit of installed capacity.

Now-a-days the use of this tariff is restricted to signal system, street lighting etc., where the number of hours are fixed and energy consumption can be easily predicted. *Its use is very common to supplies to irrigation tubewells*, since the number of hours for which the tubewell feeders are switched on are fixed. The charge is made according to horse power of the motor installed.

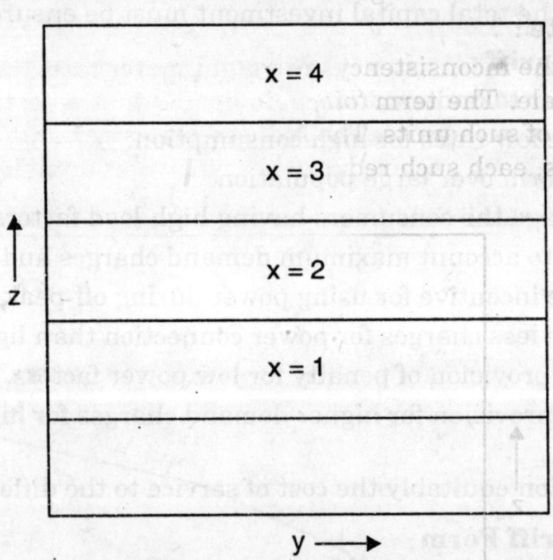

Fig. 9.19. Flat demand rate.

In this form of tariff the unit energy cost decreases progressively with an increased energy usage since the total cost remains constant. The variation in total cost and unit cost are given in Fig. 9.19.

By the use of this form of tariff the cost of metering equipment and meter reading is eliminated.

2. Straight meter rate :

The straight meter rate can be expressed in the form :

$$z = b \cdot y \qquad ...(9.11)$$

This is the simplest form of tariff. Here the charge per unit is constant. The charges depend on the energy used. This tariff is *sometimes used for residential and commercial consumer*. The variation of bill according to the variation of energy consumed is shown in Fig. 9.20.

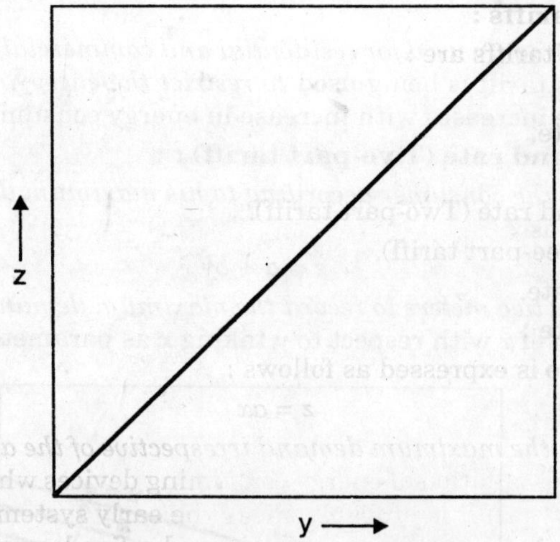

Fig. 9.20. Straight meter rate.

Advantage. *Simplicity.*

Disadvantages. 1. The consumer using no energy will not pay any amount although he has incurred some expenses to the power station.

2. This method does not encourage the use of electricity unless the tariff is very low.

3. Block meter rate :

In order to remove the inconsistency of straight meter rate, the block meter rate charges the consumers on a sliding scale. The term '*block*' indicates that a certain specified price per unit is charged for all or any part of such units. The reduced prices per unit are charged for all or any part of succeeding block of units, each such reduced price per unit applying only to a particular block or portion thereof.

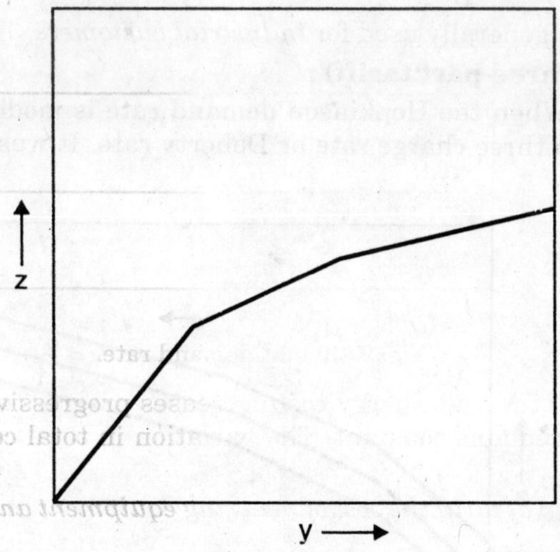

Fig. 9.21. Block meter rate.

The variation of bill according to this method is shown in Fig. 9.21.

The block meter rate accomplishes the same purpose of decreasing unit energy charges with increasing consumption as the step meter rate without its defect. Its *main defect is that it lacks a measure of the customer's demand.*

This tariff is very commonly used for residential and commercial customers. In many states of India, a *reverse form* of this tariff is being used *to restrict the energy consumption*. In this reverse form the unit energy charge increases with increase in energy consumption.

4. Hopkinson demand rate (Two-part tariff) :

This method *charges the consumer according to his maximum demand and energy consumption*. This can be expressed as

$$z = a + by \qquad \text{...(9.12)}$$

This method requires *two meters to record the maximum demand and energy consumption of the consumer*. The variation of z with respect to y taking x as parameter is shown in Fig. 9.22.

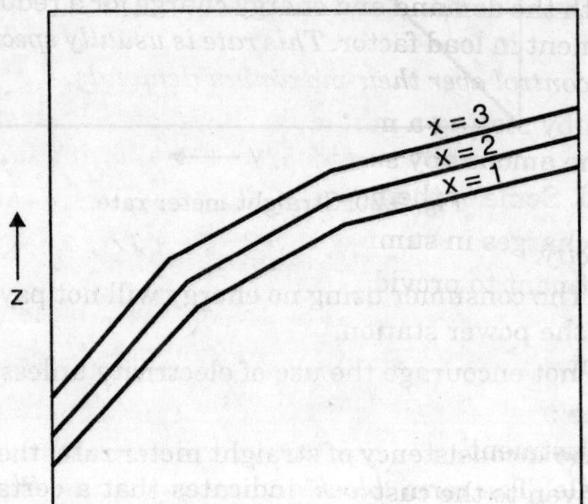

Fig. 9.22. Hopkinson demand rate (Two-part tariff).

This form of tariff is generally used for *industrial customers*.

5. Doherty rate (Three-part tariff) :

Refer to Fig. 9.23. When the Hopkinson demand rate is modified by the addition of a customer charge, it becomes a three charge rate or Doherty rate. It was first introduced by Henry L. Doherty

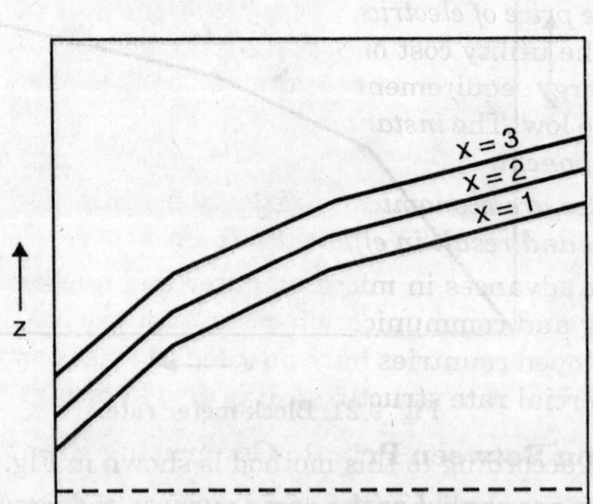

Fig. 9.23. Doherty rate (Three-part tariff).

at the beginning of twentieth century. It consists of a *customer* or *meter charge,* plus *a demand charge* plus *any energy charge*. This is expressed as follows :

$$y = ax + by + c$$

Many people consider that theoretically this is an ideal type of rate. As it requires *two meters, it is better suited for industrial than for residential customers.*

The Doherty rate is sometimes modified by specifying the minimum demand and the minimum energy consumption that must be paid for, if they are less than the minimum values specified. In this manner the customer charge is incorporated with the demand and energy component.

6. Wright demand rate :

This tariff was introduced by Arthur Wright (of England) in 1896. This rate intensifies the inducement by lowering both the demand and energy charge for a reduction in maximum demand or in other words an improvement in load factor. *This rate is usually specified for industrial consumers who have some measure of control over their maximum demands.*

The rate is modified by stating a minimum charge which must be paid if the energy for the billing period falls below the amount by such charge. For allowing fair returns some adjustment in the rate forms are provided. Some of them are :

(*i*) Higher demand charges in summer.

(*ii*) Fuel price adjustment to provide a rate change when fuel prices deviate from the standard.

(*iii*) Wage adjustment.

(*iv*) Tax adjustment.

(*v*) Power factor adjustment.

(*vi*) Discount to be given to the customers for prompt payment of bills.

9.18.4. Spot Pricing

- Spot pricing is also known as "*Load adaptive pricing*", "*Real time pricing*", "*Homeostatic control pricing*", "*Response pricing*" and "*Flexible pricing*".
- All over the world, the primary concern of electric utilities is *very large demand during peak load conditions* and *surplus energy during off peak conditions.* Thus the utilities face the problem of capacity shortages during peak load conditions and large surplus during off peak loads.

In "*spot pricing*" *the price of electricity depends on the time at which consumption occurs.* The price is varied to reflect the utility cost of providing the energy at a given time. It encourages the customer to reduce his energy requirements at times of peak load and use more energy when supply is abundant and prices are low. The *instantaneous sprt rate is the sum of incremental operating cost and quality of supply component.*

- *Spot pricing rates are economically rational because they embody the principles of marginal cost theory and result in efficient allocation of resources.*
- Owing to recent advances in microcomputer and telecommunications and ample reduction in metering and communication costs 'spot pricing' has become feasible. A large of utilities in developed countries have adopted at least partial sport pricing in their industrial and commercial rate structures. It is also known as "*Time of Day Tariff*".

9.18.5. Comparison Between Private Generating Plant and Public Supply

The industrial concerns should preferably purchase their power from public supply company rather than generating their own due to the following **reasons :**

1. Power supply from public supply company will be found *cheaper.*

2. The most distinguishing factor in favour of public supply is the *reliability of supply* since total shut down may be forced for several days due to serious breakdown in private generating plant (In case of public supply the failure occurs rarely, that too far an hour or so).

3. The management is relieved off the responsibility of generating power and as such it can concentrate its attention on the manufacturing processes.

4. Overload capacity is readily available and extensions can be made at short notices.

5. For private generation, space required is considerably large in comparison to the space required by the substation when power is purchased from public supply.

Even though above advantages accrue from public supply, yet sometimes choice for private generation is made on the following grounds :

(*i*) Use of exhaust steam.

(*ii*) Waste heat stations

(*iii*) Pit head generation

(*iv*) Diesel plant

(*v*) Emergency service.

TARIFF

Example 9.31. *Two electrical units used for same purpose are compared for their economical working :*

(i) Cost of Unit-1 is Rs. 6000 and it takes 120 kW.

(ii) Cost of Unit-2 is Rs. 16800 and it takes 72 kW.

Each of them has a useful life of 40000 hours.

Which unit will prove economical if the energy is charged at Rs. 96 per kW of maximum demand per year and 6 p. per kWh ?

Assume both units run at full load.

Solution. (*i*) **Unit-1 :**

Capital cost per hour $= \dfrac{6000}{40000} = \text{Re. } 0.15$

Maximum demand $= 120$ kW

Charge for maximum demand per hour

$$= \frac{120 \times 96}{(365 \times 24)} = \text{Rs. } 1.315$$

Energy charge per hour = Maximum demand × one hour × charge per kWh

$$= 120 \times 1 \times \frac{6}{100} = \text{Rs. } 7.2$$

$\therefore$ Total charges per hour for operation of Unit-1

$$= 0.15 + 1.135 + 7.2 = \text{Rs. } 8.485$$

(*ii*) **Unit-2 :**

Capital cost per hour $= \dfrac{16800}{40000} = \text{Re. } 0.42.$

Charge for maximum demand per hour

$$= \frac{72 \times 96}{365 \times 24} = \text{Re. } 0.789$$

Energy charge per hour $= 72 \times 1 \times \frac{6}{100} =$ Rs. 4.32

Total charges per hour for the operation of Unit-2

$= 0.42 + 0.789 + 4.32 =$ Rs. 5.529

The charges of operation for the Unit-2 per hour are less than the charges of operation for the Unit-1, therefore *Unit-2 is more economical* in this case. **(Ans.)**

Example 9.32. *The monthly electricity consumption of a residence can be approximated as under :*

Light load	*: 6 tube lights 40 watts each working for 4 hours daily*
Fan load	*: 6 fans 100 watts each working for 6 hours daily*
Refrigerator load	*: 2 kWh daily*
Miscellaneous load	*: 2 kW for 2 hours daily*

Find the monthly bill at the following tariff :

First 20 units	*Rs. 0.50/kWh*
Next 30 units	*Rs. 0.40/kWh*
Remaining units	*Rs. 0.30/kWh*
Constant charge	*Rs. 2.50 per month*

Discount for prompt payment = 5 per cent.

Solution. Total energy consumption in 30 days

$$= (6 \times 40 \times 4 \times 30 + 6 \times 100 \times 6 \times 30) \times \frac{1}{1000} + 2 \times 30 + 2 \times 2 \times 30$$

$$= (28800 + 108000) \times \frac{1}{1000} + 60 + 120 = 316.8 \text{ kWh per month}$$

The monthly bill $=$ Rs. $[(20 \times 0.5 + 30 \times 0.4 + 266.8 \times 0.3) + 2.5]$

$=$ Rs. $[(10 + 12 + 80.04) + 2.5] =$ Rs. 104.54

$\left[\because \text{ Remaining units per month } = 316.8 - 20 - 30 = 266.8\right]$

Net monthly bill if the payment is made promptly

$= 104.54 \times 0.9 =$ **Rs. 94.08. (Ans.)**

Example 9.33. *An industrial undertaking has a connected load of 220 kW. The maximum demand is 180 kW. On an average each machine works for 60% time. Find the yearly expenditure on electricity if the tariff is :*

Rs. 1200 + Rs. 120 per kW of maximum demand per year + Re. 0.15 per kWh.

Solution. Energy consumption in one year

$= 180 \times 0.6 \times (365 \times 24) = 946080$ kWh

Total electricity bill $=$ Rs. $(1200 + 120 \times 180 + 0.15 \times 946080) =$ **Rs. 164712. (Ans.)**

☞**Example 9.34.** *A Hopkinson demand rate is quoted as follows :*

Demand rates :

First 1 kW of maximum demand	*= Rs. 6/kW/month*
Next 4 kW of maximum demand	*= Rs. 5/kW/month*
Excess 5 kW of maximum demand	*= Rs. 4/kW/month*

Energy rates :

First 50 kWh	*= 7 paise / kWh*
Next 50 kWh	*= 5 paise / kWh*
Next 200 kWh	*= 4 paise / kWh*
Next 400 kWh	*= 3 paise / kWh*
Excess over 700 kWh	*= 2 paise / kWh.*

Determine : (i) The monthly bill for a total consumption of 2000 kWh and a maximum demand of 15 kW. Also find out the unit energy cost.

(ii) Lowest possible bill for a month and a corresponding unit energy cost.

Solution. (*i*) **Monthly bill and energy cost :**

Demand charges per month = Rs. (1 × 6 + 4 × 5 + 10 × 4) = Rs. 66

Energy charge $= \text{Rs. } [50 \times 7 + 50 \times 5 + 200 \times 4 + 400 \times 3 + 1300 \times 2] \times \frac{1}{100}$

$= \text{Rs. } (350 + 250 + 800 + 1200 + 2600) \times \frac{1}{100} = \text{Rs. } 52$

∴ *Monthly bill* = 66 + 52 = **Rs. 118. (Ans.)**

Average unit energy cost $= \frac{118}{2000} \times 100 =$ **5.9 paise/kWh. (Ans.)**

(*ii*) **Lowest possible bill :**

The lowest possible bill will occur when average load

= Maximum load or at 100% load factor

∴ Maximum load $= \text{Average load} = \frac{2000}{30 \times 24} = 2.77 \text{ kW}$

∴ Demand charges = Rs. (6 + 1.77 × 5) = Rs. 14.85

Energy charges will be same = Rs. 52

∴ *Minimum monthly bill* = 14.85 + 52 = **Rs. 66.85. (Ans.)**

Unity energy cost for this condition

$= \frac{66.85}{2000} \times 100 =$ **3.34 paise/kWh. (Ans.)**

TARIFF AND COST ANALYSIS

☞**Example 9.35.** *A new factory requires a maximum demand of 700 kW and load factor of 25%. The following two suppliers are available :*

(i) Public supply tariff is Rs. 48 per kW of maximum demand plus 2.4 p. per kWh.

Capital cost	*= Rs. 84000*
Interest and depreciation	*= 10 per cent*

(ii) Private oil engine generating station :

Capital cost	*= Rs. 300000*
Fuel consumption	*= 3 N / kWh*
Cost of fuel	*= Rs. 8.4 per kN*
Wages	*= 0.48 p / kWh*
Maintenance cost	*= 0.36 p / kWh*
Interest and depreciation	*= 15 per cent.*

Find which supply will be more economical ?

Solution. Load factor $= \dfrac{\text{Average load}}{\text{Maximum demand}}$

$\therefore$ Average load = Load factor × maximum demand = 0.25 × 700 = 175 kW

Energy consumed per year = 175 × (365 × 24) = 1.533 × 10^6 kWh.

(*i*) **Public supply :**

Maximum demand charges per year

$$= 48 \times 700 = \text{Rs. } 33600$$

Energy charge per year $= \dfrac{2.4}{100} \times 1.533 \times 10^6 = \text{Rs. } 36792$

Interest and depreciation $= \dfrac{10}{100} \times 84000 = \text{Rs. } 8400$

Total cost = Rs. (33600 + 36792 + 8400) = Rs. 78792

$\therefore$ *Energy cost per kWh* $= \dfrac{78792}{1.533 \times 10^6} \times 100 =$ **5.14 p.**

(*ii*) **Private oil engine generating station :**

Fuel consumption $= \dfrac{3 \times 1533 \times 10^6}{1000} = 4599$ kN

Cost of fuel = 4599 × 8.4 = Rs. 38631

Cost of wages and maintenance $= \left(\dfrac{0.48 + 0.36}{100}\right) \times 1.533 \times 10^6 = \text{Rs. } 12877$

Interest and depreciation $= \dfrac{15}{100} \times 300000 = \text{Rs. } 45000$

Total cost = Rs. (38631 + 12877 + 45000) = Rs. 96508

Energy cost per kWh $= \dfrac{96508}{1533 \times 10^6} \times 100 =$ **6.29 p.**

As the energy cost per kWh for oil engine is *less* than the public supply, the oil engine generation is more preferable. **(Ans.)**

Example 9.36. *A load having a maximum demand of 100 MW and a load factor of 0.4 may be supplied by one of the following schemes :*

(i) A steam station capable of supplying the whole load.

(ii) A steam station in conjunction with pump storage plant which is capable of supplying 130 × 10^6 kWh energy per year with a maximum output of 40 MW.

Find out the cost of energy per unit in each of the two cases mentioned above.

Use the following data :

Capital cost of steam station = *Rs. 2000/kW of installed capacity*

Capital cost of pump storage plant = *Rs. 1300/kW of installed capacity*

Operating cost of steam plant = *6 p./kWh*

Operating cost of pump storage plant = *0.5 p./kWh*

Interest and depreciation together on the capital invested should be taken as 12 per cent. Assume that no space capacity is required.

Solution. (*i*) **Steam station :**

Capital cost $= 100 \times 10^3 \times 2000 = \text{Rs. } 200 \times 10^6$

Interest and depreciation $= \frac{12}{100} \times 200 \times 10^6 = \text{Rs. } 24 \times 10^6$

Average load = Load factor × maximum demand

$= 0.4 \times 100 \times 10^3 = 40000$ kW

Energy supplied per year = Average load × (365 × 24)

$= 40000 \times 365 \times 24 = 350.4 \times 10^6$ kWh

∴ Interest and depreciation charges *per unit of energy*

$$= \frac{24 \times 10^6}{350.4 \times 10^6} \times 100 = 6.85 \text{ p/kWh}$$

∴ *Total cost per unit* $= 6 + 6.85 =$ **12.85 p/kWh. (Ans.)**

(*ii*) **Steam station in conjunction with pump-storage plant :**

The load supplied by the steam plant $= 100 - 40 = 60$ MW

∴ Capital cost of steam plant $= 60 \times 1000 \times 2000 = \text{Rs. } 120 \times 10^6$

Capital cost of pump storage plant $= 40 \times 1000 \times 1300 = \text{Rs. } 52 \times 10^6$

∴ Total capital cost of combined station $= 120 \times 10^6 + 52 \times 10^6 = \text{Rs. } 172 \times 10^6$

Interest and depreciation charges on capital investment

$$= \frac{12}{100} \times 172 \times 10^6 = \text{Rs. } 20.64 \times 10^6$$

∴ Operating cost of pump storage plant $= \frac{0.5}{100} \times 130 \times 10^6 = \text{Rs. } 0.65 \times 10^6$

The energy units supplied by steam station

= Total units required – energy units supplied by pump storage plant

$= 350.4 \times 10^6 - 130 \times 10^6 = 220.4 \times 10^6$ kWh

Operating cost of the steam station

$$= \frac{6}{100} \times 220.4 \times 10^6 = \text{Rs. } 13.22 \times 10^6$$

Total cost per year $= \text{Rs. } (20.64 \times 10^6 + 0.65 \times 10^6 + 13.22 \times 10^6) = \text{Rs. } 34.51 \times 10^6$

Total cost per unit $= \frac{34.51 \times 10^6}{350.4 \times 10^6} \times 100 =$ **9.85 p/kWh. (Ans.)**

Note. If the above example is repeated with a load factor of 0.7 it will be observed from the results that the *cost of generation becomes less with higher load factor irrespective of the type of the plant.*

Example 9.37. *The following data relate to a 2000 kW diesel power station :*

The peak load on the plant = *1500* kW

Load factor = *0.4*

Capital cost per kW installed = *Rs. 1200*

Annual costs = *15 per cent of capital*

Annual operating costs = *Rs. 50000*

Annual maintenance costs :

(*i*) *Fixed* = *Rs. 9000*

(*ii*) *Variable* = *Rs. 18000*

Cost of fuel = *Rs. 0.45 per kg*
Cost of lubricating oil = *Rs. 1.3 per kg*
Consumption of fuel = *0.45 kg/kWh*
Consumption of lubricating oil = *0.002 kg/kWh*

Determine the following :

(i) The annual energy generated.

(ii) The cost of generation per kWh.

Solution. Capital cost of the plant $= 2000 \times 1200 = \text{Rs. } 2.4 \times 10^6$ per year

Interest on capital $= \frac{15}{100} \times 2.4 \times 10^6 = \text{Re. } 0.36 \times 10^6$ per year.

(i) **Annual energy generated** = Load factor × maximum demand × (365 × 24)

$= 0.4 \times 1500 \times 365 \times 24 =$ **5.256×10^6 kWh. (Ans.)**

(ii) **Cost of generation :**

Fuel consumption $= 0.45 \times 5.256 \times 10^6 = 2.365 \times 10^6$ kg per year

Cost of fuel $= \text{Rs. } 0.45 \times 2.365 \times 10^6 = \text{Rs. } 1.064 \times 10^6$ per year

Lubricant consumption $= 0.002 \times 5.256 \times 10^6 = 10512$ kg per year

Cost of lubricating oil $= 1.3 \times 10512 = \text{Rs. } 13665$ per year

Total fixed cost = Interest + maintenance (fixed)

$= 0.36 \times 10^6 + 9000 = \text{Rs. } 369000$ per year

Total running or variable costs

= Fuel cost + lubricant cost + maintenance (running) + annual operating costs

$= 1.064 \times 10^6 + 13665 + 18000 + 50000 = \text{Rs. } 1145665$ per year

Total cost = Fixed cost + running cost = 369000 + 1145665 = Rs. 1514665

Cost of generation $= \frac{1514665}{5.256 \times 10^6} \times 100 =$ **28.8 paise/kWh. (Ans.)**

☞**Example 9.38.** *The annual costs of operating a 15 MW thermal plant are given below :*

Capital cost of plant = *Rs. 1500/kW*
Interest, insurance and depreciation = *10 per cent of plant cost*
Capital cost of primary and secondary distribution = *Rs.* 20×10^6
Interest, insurance and depreciation on the capital cost of primary and secondary distribution = *5% the capital cost*
Plant maintenance cost = *Rs.* 100×10^3 *per year*
Maintenance cost of primary and secondary equipment = *Rs.* 2.2×10^5 *per year*
Salaries and wages = *Rs.* 6.5×10^5 *per year*
Consumption of coal = 40×10^4 *kN per year*
Cost of coal = *Rs. 9 per kN*
Dividend to stockholders = *Rs.* 1.5×10^6 *per year*
Energy loss in transmission = *10 per cent*
Diversity factor = *1.5*
Load factor = *0.75*
Maximum demand = *12 MW*

(*i*) *Devise a two-part tariff.*

(*ii*) *Find the average cost per kWh.*

Solution. (*i*) Two-part tariff :

Load factor $= \dfrac{\text{Average load}}{\text{Maximum demand}}$

$\therefore$ Average load = Load factor × maximum demand

$= 0.75 \times 12 \times 10^3 = 9000$ kW

Energy generated per year $= 9000 \times (365 \times 24) = 78.84 \times 10^6$ kWh

Cost of the plant $= 15 \times 10^3 \times 1500 = \text{Rs. } 22.5 \times 10^6$

Interest, insurance and depreciation charges of the plant

$$= \frac{10}{100} \times 22.5 \times 10^6 = \text{Rs. } 2.25 \times 10^6$$

Interest, insurance and depreciation charges of primary and secondary equipments

$$= \frac{5}{100} \times 20 \times 10^6 = \text{Rs. } 1.0 \times 10^6$$

Total fixed cost = Insurance, interest and depreciation costs + dividend to stock-holders

$= \text{Rs. } (2.25 \times 10^6 + 1.5 \times 10^6) = \text{Rs. } 3.75 \times 10^6$

Sum of individual maximum demand

= Maximum demand × diversity factor

$= 12 \times 10^3 \times 1.5 = 18000$ kW

$\therefore$ *Fixed charges per kW* $= \dfrac{3.75 \times 10^6}{18000} =$ **Rs. 208.3.**

Total variable charges = All maintenance costs + salaries and wages + fuel cost

$= (100 \times 10^3 + 2.2 \times 10^5) + 6.5 \times 10^5 + 40 \times 10^4 \times 9$

$= (1 \times 10^5 + 2.2 \times 10^5) + 6.5 \times 10^5 + 36 \times 10^5$

$= \text{Rs. } 45.7 \times 10^5$ or $\text{Rs. } 4.57 \times 10^6$

Energy transmitted = Energy generated × transmission efficiency

$$= 78.84 \times 10^6 \times \left(\frac{100 - \text{energy loss in transmission}}{100}\right)$$

$$= 78.84 \times 10^6 \times \frac{90}{100} = 70.956 \times 10^6 \text{ kWh}$$

$\therefore$ *Charges for energy consumption*

$$= \frac{4.57 \times 10^6}{70.956 \times 10^6} \times 100 = \textbf{6.44 paise/kWh.}$$

$\therefore$ *Two-part tariff* = **Rs. 208.3/kW + 6.44 paise/kWh. (Ans.)**

(*ii*) **Average cost per kWh :**

Total charges = Fixed charges + variable charges

$= 3.75 \times 10^6 + 4.57 \times 10^6 = \text{Rs. } 8.32 \times 10^6$

Average cost of supply $= \dfrac{8.32 \times 10^6}{70.956 \times 10^6} \times 100 =$ **11.72 paise/kWh. (Ans.)**

Example 9.39. *A 10 MW thermal power plant has the following data :*

Peak load	*= 8 MW*
Plant annual load factor	*= 0.72*
Cost of the plant	*= Rs. 800/kW installed capacity*
Interest, insurance and depreciation	*= 10 per cent of the capital cost*
Cost of transmission and distribution system	*= Rs. 350 × 10^3*
Interest, depreciation on distribution system	*= 5 per cent*
Operating cost	*= Rs. 350 × 10^3 per year*
Cost of coal	*= Rs. 6 per kN*
Plant maintenance cost	*= Rs. 30000/year (fixed)*
	= Rs. 40000/year (running)
Coal used	*= 250000 kN/year*

Assume transmission and distribution costs are to be charged to generation

(i) Devise a two-part tariff.

(ii) Average cost of generation in paise/kWh.

Solution. (*i*) **Two-part tariff :**

S. No.	Items	Fixed cost per year (in Rs.)	Running cost per year (in Rs.)
1.	Interest, depreciation etc. of the plant	$\frac{10}{100} \times 10000 \times 800$ = Rs. 800×10^3	—
2.	Interest, depreciation etc. of the transmission and distribution	$\frac{5}{100} \times 350 \times 10^3$ $= 17.5 \times 10^3$	—
3.	Annual cost of coal	—	250000×6 $= 1500 \times 10^3$
4.	Operating cost	—	$= 350 \times 10^3$
5.	Plant maintenance cost	$= 30 \times 10^3$	$= 40 \times 10^3$
	Total cost	847.5×10^3	1890×10^3

∴ Grand total cost = Fixed cost + running cost

$= 847.5 \times 10^3 + 1890 \times 10^3 =$ Rs. 2737.5×10^3

Energy generated/year = Average load × (365 × 24)

= (Peak load × load factor) × (365 × 24)

$= (8 \times 10^3 \times 0.72) \times (365 \times 24) = 50.46 \times 10^6$ kWh

$$\therefore \text{ Two-part tariff } = \frac{\text{Fixed cost}}{\text{Maximum load}} + \frac{\text{Running cost}}{\text{Energy generated}}$$

$$= \frac{847.5 \times 10^3}{8 \times 10^3} + \frac{1890 \times 10^3}{50.46 \times 10^6} \times 100$$

= **Rs. 105.9/kW + paise 3.74/kWh. (Ans.)**

(*ii*) **Average cost generation in paise/kWh :**

$$\text{Average generation cost } = \frac{\text{Grand total cost}}{\text{Energy generated}}$$

$$= \frac{2737.5 \times 10^3}{50.46 \times 10^6} \times 100 = \textbf{5.42 paise/kWh.} \quad \textbf{(Ans.)}$$

Example 9.40. *Determine the load factor at which the cost of supplying a unit of electricity is same in Diesel station as in a steam station if the respective annual fixed and running charges are given below :*

Diesel : *Rs. (40/kW + 0.06/kWh)*

Steam : *Rs. (160/kW + 0.015/kWh).*

Solution. Let, P = Maximum load in kW

x = Load factor (same for both the stations).

∴ Average load $= P \times x$

Cost of diesel station,

$$C_{diesel} = 40\,P + 0.06 \times P \times x \times (365 \times 24)$$

Cost of steam station,

$$C_{steam} = 160\,P + 0.015 \times P \times x \times (365 \times 24)$$

As given in the problem,

Unit energy cost (diesel station) = Unit energy cost (steam station)

$$\therefore \quad \frac{40\,P + 0.06\,Px \times (365 \times 24)}{Px \times (365 \times 24)} = \frac{160\,P + 0.015\,Px \times (365 \times 24)}{Px \times (365 \times 24)}$$

$$\therefore \quad 40\,P + 0.06\,Px \times 8760 = 160\,P + 0.015\,Px \times 8760$$

or, $$40\,P + 525.6\,Px = 160\,P + 131.4\,Px$$

or, $$120\,P = 394.2\,Px \quad \text{or} \quad x = \frac{120}{394.2} = 0.3$$

i.e., *Load factor* **= 0.3. (Ans.)**

Example 9.41. *A motor of 25 H.P. connected to a condensate pump has been burnt beyond economical repairs. Two alternatives have been proposed to replace it by :*

	Cost	*η at full load*	*η at half load*
Motor A :	*Rs. 5000*	*90%*	*85%*
Motor B :	*Rs. 3500*	*86%*	*80%*

The life of each motor is 20 years and its salvage value is 12 per cent of the initial cost. The rate of interest is 5 percent annually. The motor operates at full load for 30% of time and at half load for the remaining period. The annual maintenance cost of motor A is Rs. 400 and that of motor B is Rs. 200. The energy rate is 12 paise/kWh.

Which motor will be economical ?

Solution. Motor A :

Salvage value $= \frac{12}{100} \times 5000$ = Rs. 600

Depreciation $= \frac{5000 - 600}{20}$ = Rs. 220/year

Interest $= \frac{5}{100} \times 5000$ = Rs. 250/year

Maintenance = Rs. 400

$$\text{Energy given to motor} = \frac{\text{Load on motor} \times \text{time in hours}}{\text{Efficiency of the motor}}$$

$$\therefore \text{ Energy cost} = \left[\left(25 \times 0.7355 \times (365 \times 24) \times \frac{30}{100} \times \frac{1}{0.9}\right) + \left((25 \times 0.7355) \times \frac{1}{2} \times (365 \times 24) \times \frac{70}{100} \times \frac{1}{0.85}\right)\right] \times \frac{12}{100}$$

$$= (53691.5 + 66324.8) \times \frac{12}{100} = \text{Rs. } 14402/\text{year}$$

$\therefore$ *Total cost of motor A* = 220 + 250 + 400 + 14402 = **Rs. 15272/year.**

Motor B :

$$\text{Salvage value} = \frac{12}{100} \times 3500 = \text{Rs. } 420$$

$$\text{Depreciation} = \frac{3500 - 420}{20} = \text{Rs. } 154$$

$$\text{Interest} = \frac{5}{100} \times 3500 = \text{Rs. } 175$$

Maintenance = Rs. 200

$$\text{Energy cost} = \left[\left(25 \times 0.7355 \times (365 \times 24) \times \frac{30}{100} \times \frac{1}{0.86}\right) + \left(25 \times 0.7355 \times \frac{1}{2} \times (365 \times 24) \times \frac{70}{100} \times \frac{1}{0.8}\right)\right] \times \frac{12}{100}$$

$$= (56188.8 + 70470) \times \frac{12}{100} = \text{Rs. } 15199$$

Total cost of motor B = 154 + 175 + 200 + 15199 = **Rs. 15728/year.**

Hence Motor A is economical since its annual cost is less than motor B.

Example 9.42. *The following proposals are under consideration for an industry which has a maximum demand of 45 MW and a load factor of 0.45 :*

(i) A steam plant having an initial cost of Rs. 1200/kW and maintenance cost is 2.4 paise/kWh. The coal of C.V. of 2550 kJ/N is used. The overall efficiency of the plant is 24 per cent.

(ii) An hydro-plant having a capital cost of Rs. 3600/kW and a running cost of 0.6 paise/kWh.

Assuming interest and depreciation rate of 10 per cent for steam plant and 8 per cent for hydro-plant, determine the price of coal above which steam station is uneconomical.

Solution. Energy required per year

= Peak load × load factor × (365 × 24)

$= 45 \times 10^3 \times 0.45 \times (365 \times 24) = 177.39 \times 10^6$ kWh/year

(*i*) **Steam plant :**

$$\text{Interest and depreciation} = \frac{10}{100} \times (45 \times 10^3) \times 1200 = \text{Rs. } 5.4 \times 10^6$$

$$\text{Maintenance cost} = \frac{2.4}{100} \times 177.39 \times 10^6 = \text{Rs. } 4.257 \times 10^6$$

Let, W_{coal} = Weight of coal in kN used/year, and

x = Cost of coal in Rs. per kN

$\therefore \quad W_{coal} \times 10^3 \times \text{C.V.} \times \eta_{overall} = 177.39 \times 10^6 \times 3.6 \times 10^3 \qquad (\because \quad 1 \text{ kWh} = 3.6 \times 10^3 \text{ kJ})$

or, $$W_{coal} = \frac{177.39 \times 10^6 \times (3.6 \times 10^3)}{10^3 \times 2550 \times 0.24} = 1.043 \times 10^6 \text{ kN/year}$$

Now, total cost of steam plant = Interest + maintenance cost + fuel cost

$= 5.4 \times 10^6 + 4.257 \times 10^6 + 1.043 \times 10^6 \times x \qquad ...(1)$

(*ii*) **Hydel plant :**

Interest and depreciation $= 45 \times 10^3 \times 3600 \times \frac{8}{100} = \text{Rs. } 12.96 \times 10^6$

Running cost $= \frac{0.6}{100} \times 177.39 \times 10^6 = \text{Rs. } 1.064 \times 10^6$

Total cost of hydel plant $= 12.96 \times 10^6 + 1.064 \times 10^6 = \text{Rs. } 14.024 \times 10^6 \qquad ...(2)$

The steam and hydel station will be equally economical if the total cost/year remains same.

$\therefore$ Equating the values of (1) and (2), we get

$$5.4 \times 10^6 + 4.257 \times 10^6 + 1.043 \times 10^6 \times x = 14.024 \times 10^6$$

or, $$5.4 + 4.257 + 1.043\, x = 14.024 \qquad \text{...(Dividing both sides by } 10^6)$$

$\therefore$ $$x = \frac{14.024 - 5.4 - 4.257}{1.043} = \text{Rs. } 4.19 \text{ per kN}$$

Hence, *price coal above which steam station is uneconomical* = **Rs. 4.19 per kN. (Ans.)**

Example 9.43. *An industrial consumer has a choice between low and high voltage supply available at the following rates :*

High voltage : *Rs. 50 / kW per year + paise 4 / kWh*

Low voltage : *Rs. 55 / kW per year + paise 5 / kWh*

In order to have high voltage supply, consumer has to install his own transformer which costs Rs. 110 / kW. The losses in the transformer are 4 per cent of full load. Determine the number of working hours per week above which the high voltage supply will be economical.

Assume : Interest and depreciation 12 per cent of capital, working weeks per year 50 and load of consumer as 1.5 MW.

Solution. Consumer load = 1.5 MW = 1500 kW

Required rating of transformer $= \frac{1500}{(1 - 0.4)} = 1562 \text{ kW}$

Cost of the transformer to the consumer

$= 1562 \times 110 = \text{Rs. } 171820$

Annual interest and depreciation $= \frac{12}{100} \times 171820 = \text{Rs. } 20618$

Let the number of hours for which power is required by the consumer = x hours/week

$\therefore$ Number of hours for which power is used during the year = $50x$ hours

(*i*) Number of units consumed from low voltage side if the load is connected to low voltage

$= 1500 \times 50x = 75000x \text{ kWh/year}$

(*ii*) Number of units consumed from high voltage side if the load is connected to high voltage

$= 1562 \times 50x = 78100x \text{ kWh/year}$

Total cost from low voltage supply in rupees

$$= 1500 \times 55 + 75000x \times \frac{5}{100} = 82500 + 3750x \quad ...(1)$$

Total cost from high voltage supply in rupees

$$= 1562 \times 50 + 78100x \times \frac{4}{100} + 20618$$

$$= 98718 + 3124x \quad ...(2)$$

If both the systems cost the same to the consumer, then

$$82500 + 3750x = 98718 + 3124x$$

i.e.,

$$x = \frac{98718 - 82500}{(3750 - 3124)} = \frac{16218}{626} = 25.9 \text{ hours.}$$

Hence the number of working hours above which the high voltage supply will be economical

= 25.9 hours. (Ans.)

9.19. POWER FACTOR IMPROVEMENT

9.19.1. Apparent, Active (True or Real) and Reactive Power and Power Factor

Every circuit current has two components : (*i*) Active component and (*ii*) Reactive component. *"Active component" consumes power* in the circuit while *"reactive component"* is responsible for the field which lags or leads the main current from the voltage.

In Fig. 9.24 active component is $I_{active} = I \cos \phi$, and reactive component is $I_{reactive} = I \sin \phi$.

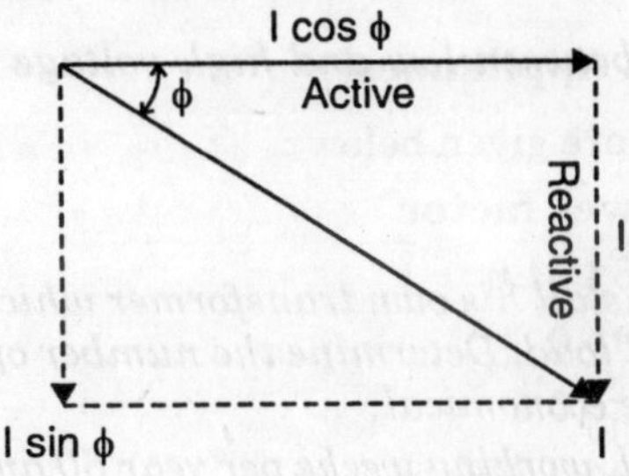

Fig. 9.24. Active and reactive components of circuit current *I*.

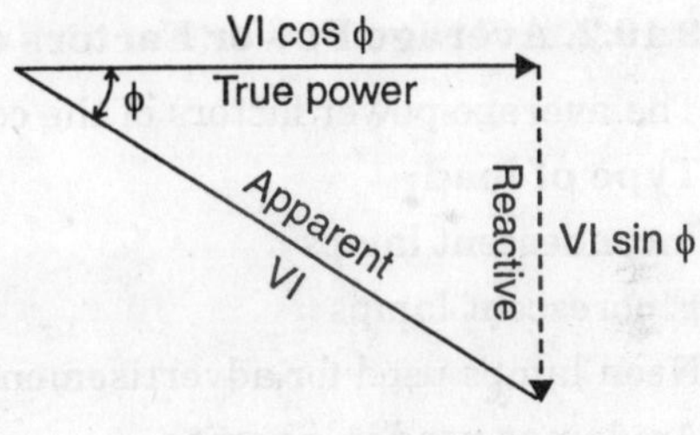

Fig. 9.25. Apparent, true and reactive power.

So,

$$I = \sqrt{(I_{active})^2 + (I_{reactive})^2} \quad ...(9.13)$$

Refer to Fig. 9.25.

(*i*) **Apparent power (S).** It *is given by the product of r.m.s. values of applied voltage and circuit current.*

$$\therefore \quad S = VI = (I \times Z) \,.\, I = I^2Z \text{ volt-amperes } (VA)$$

(*ii*) **Active or true or real power (P or W).** *It is the power which is actually dissipated in the circuit resistance.*

$$P = I^2R = VI \cos \phi \text{ watts}$$

(*iii*) **Reactive power (Q).** A pure inductor and a pure capacitor do not consume any power, since in a half cycle what so ever power is received from the source by these components the same is returned to the source. This power which flows back and forth (*i.e.,* in both directions in the circuit) or *reacts upon itself* is called *"reactive power"*.

It may be noted that the *current in phase with the voltage produces active or true or real power while the current 90° out of phase with the voltage contributes to reactive power.*

In a *R-L* circuit, reactive power which is the power developed in the inductive reactance of the circuit, is given as :

$$Q = I^2X_L = I^2Z \sin \phi = I \,.\, (IZ) \sin \phi$$
$$= VI \sin \phi \text{ volt-amperes-reactive } (VAR)$$

These three powers are shown in Fig. 9.25

Relation between *VA*, *W* and *VR*

$$W = VA \cos \phi \qquad ...(i)$$
$$VAR = VA \sin \phi \qquad ...(ii)$$

$$\therefore \quad VA = \frac{W}{\cos \phi} \qquad ...[\text{From } (i)]$$

and, $$VA = \frac{VAR}{\sin \phi} \qquad ...[\text{From } (ii)]$$

Power factor (p.f.) $$= \frac{W}{VA} = \frac{\text{True power}}{\text{Apparent power}} \qquad ...(9.14)$$

The larger bigger units of apparent, true and reactive power are *kVA* (or *MVA*), *kW* (or *MW*) and *kVAR* (or *MVAR*) respectively.

- *The power factor depends on the reactive power component. If it is made equal to the active power component, the power factor becomes unity.*

9.19.2. Average Power Factors of Some Common Appliances

The average power factors of the common appliances are given below :

Type of load	**Power factor**
Incandescent lamps	0.98 to 1.0
Fluorescent lamps	0.6 to 0.8
Neon lamps used for advertisements	0.4 to 0.5
Arc lamps used in cinemas	0.3 to 0.7
Fans	0.5 to 0.8
Induction motors	0.5 to 0.85
Fractional H.P. motors	0.4 to 0.75
Induction heaters	0.85
Resistance furnaces	0.6 to 0.9
Arc furnaces	0.85
Induction furnaces	0.6
Resistance welders	0.4 to 0.75
Arc welders	0.3 to 0.4

9.19.3. Causes of Low Power Factor

(*i*) All A.C. motors (except overexcited synchronous motors and certain types of commutators motors) and transformers operate at lagging power factor.

(*ii*) Due to typical characteristic of the arc, are lamps operate at low power factor.

(*iii*) When there is increase in supply voltage, which usually occurs during low load periods (such as lunch hours, night hours etc.) the magnetizing current of inductive reactances increases and power factor of the electrical plant as a whole decreases.

(*iv*) Arc and induction furnaces etc. operate at a very low lagging power factor.

(*v*) Due to improper maintenance and repairs of motors the power factor at which motors operate fall.

9.19.4. Effects and Disadvantages of Low Power Factor

The power factor is given by :

$$\cos\phi = \frac{\text{kW}}{\text{kVA}} \quad \text{or} \quad \text{kVA} = \frac{\text{kW}}{\cos\phi}$$

In case of *single-phase supply,*

$$\text{kVA} = \frac{VI}{1000} \quad \text{or} \quad I = \frac{1000\ \text{kVA}}{V}$$

$$\therefore \qquad I \propto \text{kVA}$$

In case of *three-phase supply,*

$$\text{kVA} = \frac{\sqrt{3}\, E_L I_L}{1000} \quad \text{or} \quad I_L = \frac{1000\ \text{kVA}}{\sqrt{3}\, E_L}$$

$$\therefore \qquad I_L \propto \text{kVA}$$

In each case, the *kVA is directly proportional to current.*

The higher current due to poor power factor affects the system and lead to the following undesirable results :

1. Rating of alternators and transformers are proportional to their output current hence inversely proportional to power factor, therefore large *generators and transformers are required to deliver same load but at a low power factor.*

2. When a load having a low lagging power is switched on, there is a large voltage drop in the supply voltage because of the increased voltage drop in the supply lines and transformers. This drop in voltage adversely affects the starting torque of motors and necessitates *expensive voltage stabilizing equipment* for keeping the consumer's voltage fluctuations within the statutory limits.

3. The cross-sectional area of the bus-bar, and the contact surface of switch gear is required to be *enlarged* for the same power to be delivered but at a low power factor.

4. To transmit same power at low power factor, more current will have to be carried by the transmission line or the distributor or cable. If the current density in the line is to be kept constant *the size of the conductor will have to be increased.* Thus *more copper is required to deliver the same load but at a low power factor.*

5. Copper losses are proportional to the square of the current hence inversely proportional to the square of the power factor *i.e.*, more copper losses at low power factor and hence *poor efficiency.*

Thus, we find that the capital cost of transformers, alternators, distributors, transmission lines etc. *increases with the low power factor.*

9.19.5. Advantages of Power Factor Improvement

The installation of power factor improvement device, to raise the power factor results in one or more of the following *effects* and *advantages* :

1. Reduction in investment in the system facilities per kW of the load supplied.

2. Reduction in a circuit current.
3. Reduction in copper losses in the system due to reduction in current.
4. Increase in voltage level at load.
5. Improvement in power factor of the generators.
6. Reduction in kVA loading of the generators and circuits.
7. Reduction in kVA demand charges for large consumers.

9.19.6. Methods of Power Factor Improvement

The various methods employed for power factor correction are :

1. Use of static capacitors.
2. Use of synchronous condensers.
3. Use of phase advancers.
4. Use of phase compensated motors.

The above methods of power factor improvement are discussed below :

1. Use of static capacitors. It is known that static capacitor/condenser takes current which leads the voltage by nearly 90°. Thus if condenser is connected across an inductive load resultant quadrature component of the whole combination will be difference of leading component of condenser current (I_c) and lagging component of lead current ($I \sin \phi_1$) as shown in Fig. 9.26. In view of reduced magnitude of quadrature component of current, p.f. of the whole combination is improved from $\cos \phi_1$ to $\cos \phi_2$.

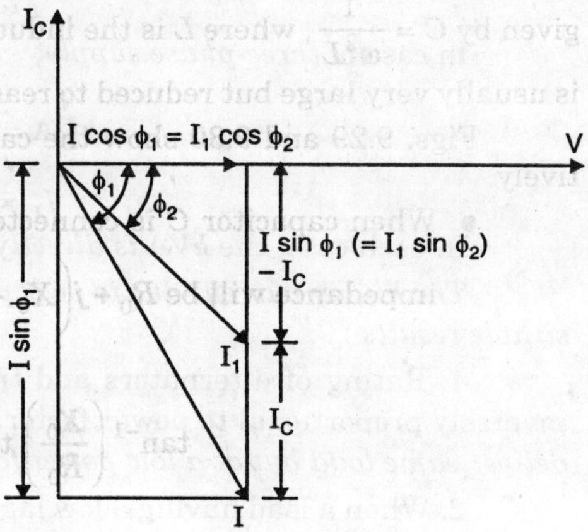

Fig. 9.26

Power factor of the system can be improved by placing static capacitors in series with the liner as shown in Fig. 9.27. Capacitors connected in series with the line neutralise the line reactance. The capacitors, when connected in series with the line, are called *"series capacitor"*, and when connected in parallel with the equipment, are called *"shunt capacitors"*.

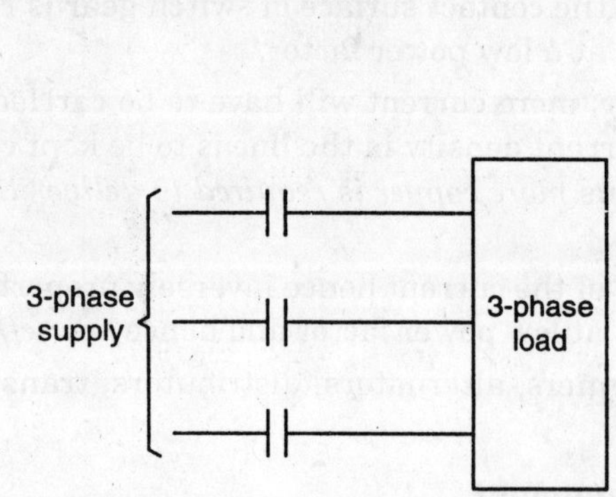

Fig. 9.27

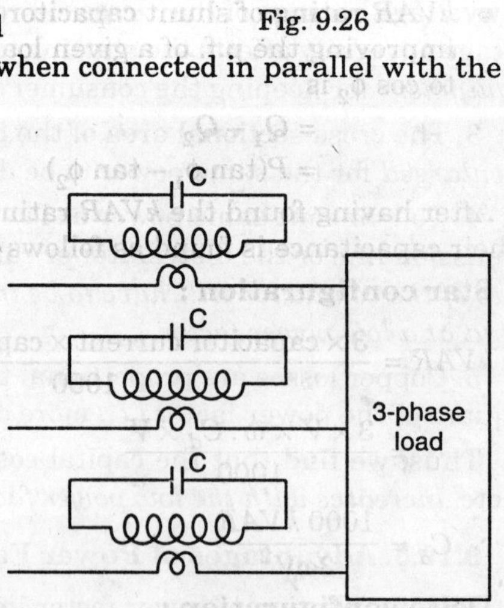

Fig. 9.28

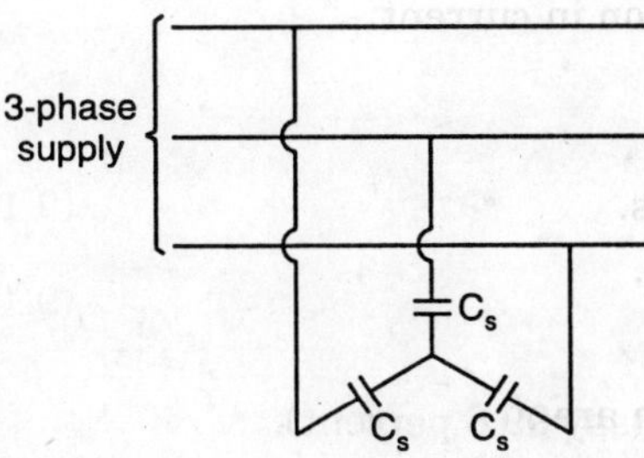

Fig. 9.29

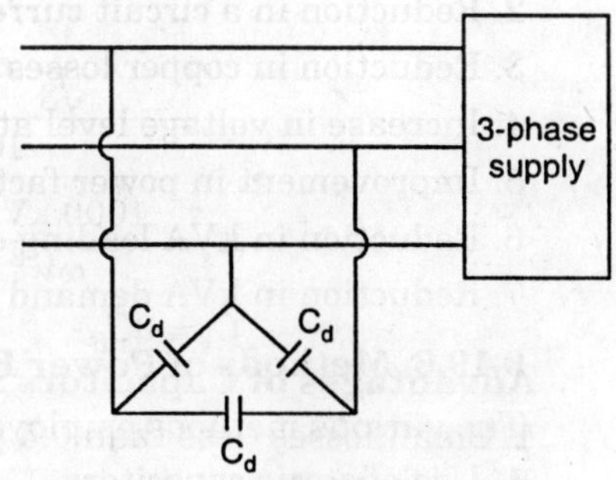

Fig. 9.30

Shunt capacitors are used in *factories, plants and also on transmission lines.*

Series capacitors *are used on long transmission lines* as they provide automatic compensation with the variations in load. The capacity of the capacitors to neutralise the line reactance is given by $C = \frac{1}{\omega^2 L}$, where L is the inductance per phase of the line. The value of reactance required is usually very large but reduced to reasonable value by use of a transformer, as shown in Fig. 9.28.

Figs. 9.29 and 9.30 show the capacitors connected in star and delta configurations respectively.

- When capacitor C is connected in series with the load impedance of $R_0 + jX_0$, resultant impedance will be $R_0 + j\left(X_0 - \frac{1}{\omega C}\right)$ and the angle of lag will reduce from angle

$$\tan^{-1}\left(\frac{X_0}{R_0}\right) \text{ to } \tan^{-1}\left(\frac{X_0 - \frac{1}{\omega C}}{R_0}\right)$$

- *kVAR* rating of shunt capacitors (Fig. 9.31) for improving the p.f. of a given load P from $\cos \phi_1$ to $\cos \phi_2$ is

$$= Q_1 - Q_2$$

$$= P(\tan \phi_1 - \tan \phi_2) \quad \text{...(9.15)}$$

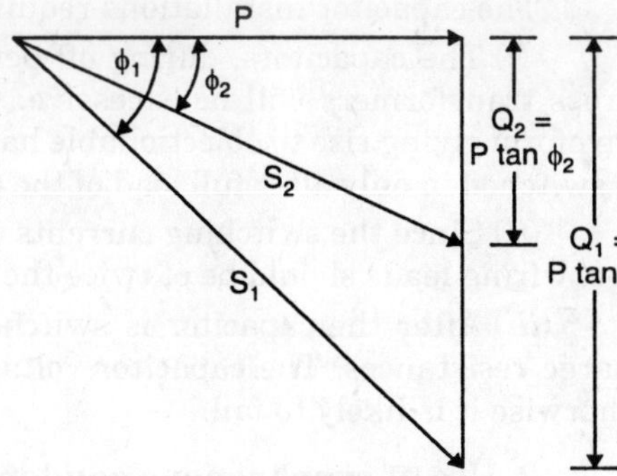

Fig. 9.31

After having found the *kVAR* rating of the capacitors, their capacitance is found as follows :

Star configuration :

$$kVAR = \frac{3 \times \text{capacitor current} \times \text{capacitor voltage}}{1000}$$

$$= \frac{3 \times V \times \omega . C_d \times V}{1000}$$

or $$C_d = \frac{1000\, kVAR}{3\omega V^2} \quad \text{...(9.16)}$$

Delta configuration :

$$kVAR = \frac{3 \times \text{capacitor current} \times \text{capacitor voltage}}{1000}$$

$$= \frac{3 \times \frac{V}{\sqrt{3}} \times \omega C_s \times \frac{V}{\sqrt{3}}}{1000}$$

or, $$C_s = \frac{1000 \text{ kVAR}}{\omega V^2} \qquad \text{...(9.17)}$$

i.e., $$C_s = 3C_d \qquad \text{...(9.18)}$$

Advantages of Capacitors :

1. Small losses (less than 0.5 percent) or higher efficiency (say 99.6 percent).
2. Low initial cost.
3. Easy installation.
4. Little maintenance.
5. Long life.
6. Greater reliability in service.
7. Flexible in operation (*kVAR* rating of capacitor can be adjusted according to the load conditions).
8. No restriction on the choice of site for capacitor and can be installed in relatively small banks located near the load (This is not possible with synchronous compensators).

Besides p.f. improvement, capacitors are employed to *perform the following functions also* :

1. To reduce losses.
2. To reduce voltage regulation of the line.
3. To meet a demand for reactive power.
4. To utilise fully the capacities of generators, transformers and transmission and distribution network.

Precautions for capacitor installations :

The capacitor installations require the following *precautions* :

(*i*) The capacitors, during off periods, should be switched off the line as otherwise voltage across transformers will be excessive. Overexcitation of transformers will result in distortion of waveform giving rise to objectionable harmonic currents. For the same reason capacitor unit should be switched in only after full load of the circuit whose power factor is to be improved, is switched on.

(*ii*) Since the switching currents of capacitors are several times the full rated current, therefore, wiring leads should be of twice the size chosen for normal current carrying capacity.

(*iii*) After the capacitor is switched off, its residual charge falls gradually through its discharge resistances. The capacitor voltage must fall to 50 V or less before it is again switched on otherwise it is likely to fail.

2. **Use of synchronous condensers :**

- *An over-excited synchronous motor running on no load is called the* ***synchronous condenser or synchronous phase advancer.*** It behaves like a capacitor, the capacitance reactance of which depends upon the motor excitation. Power factor can be improved by using synchronous condensers like shunt capacitors connected across the supply.

Refer to Fig. 9.32.

I_L = Current taken by the industrial load,

ϕ_L = Angle of lag (w.r.t. V),

I_M = Current drawn by the synchronous motor,

ϕ_M = Angle of lead,

I = Resultant current (vector sum of I_L and I_M), and

ϕ = Angle of lag (w.r.t. V).

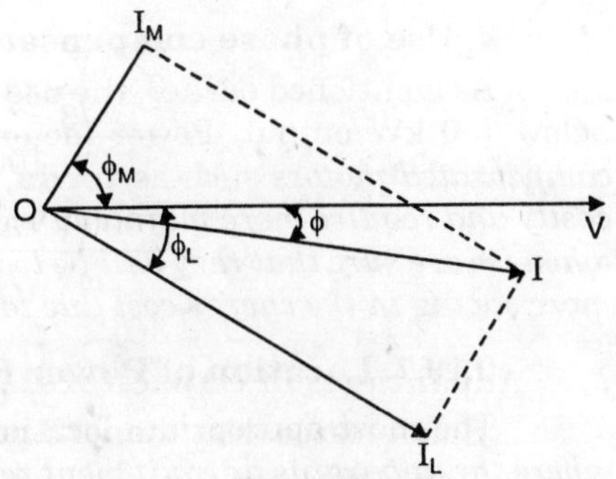

Fig. 9.32

From the Fig. 9.32 we observe that angle of lag ϕ is much smaller than ϕ_L ; thus overall factor is improved from $\cos \phi_L$ to $\cos \phi_m$ by the use of synchronous condenser.

- Synchronous condensers are usually built in large units and are employed where a large quantity of corrective *kVAR* (say, 5000 *kVAR* or more) is required.

Advantages and disadvantages of synchronous condensers :

Advantages. Following are the *advantages of synchronous condensers over static capacitors :*

1. A *Finer control* can be obtained by varying the field excitation.
2. Inherent characteristic of synchronous condensers of stabilizing variations in the line voltage and thereby automatically *aid in regulation.*
3. *Possibility of overloading a synchronous condenser for short periods.*
4. *Improvement in the system stability and reduction of the effect of sudden changes in load* owing to inertia of synchronous condenser.

Disadvantages :

1. The cost is higher than that of static capacitors of the same rating, except in size above about 5000 kVAR.
2. Higher maintenance and operating costs comparatively.
3. Comparatively lower efficiency, due to losses in rotating parts and heat losses.
4. Increase of short-circuit currents when the fault occurs near the synchronous condenser.
5. For starting synchronous condensers an auxiliary equipment is required.
6. Possibility of synchronous condensers falling out of synchronism causing interruption of supply.
7. During operation noise is produced.

Note. Synchronous motor when run without any mechanical load and solely for the purpose of power factor correction is called **synchronous compensator** or **synchronous phase modifier.** It was formerly called as **"synchronous condenser"** or **"synchronous capacitor"** which lays much stress on its function as a device taking leading current just like condenser or capacitor.

3. **Use of phase advancers :**

- The p.f. of an induction motor falls mainly due to its exciting current drawn from the A.C. supply mains, because exciting current lags behind the voltage by 90°. It may be improved by equipping the set with an "*A.C. exciter*" or "*phase advancer*" which supplies this exciting current to the rotor at slip frequency. Such an excitor may be mounted on the same shaft as the main motor or may be suitably driven from it.
- Use of phase advancer is not generally economical in connection with motors below 150 kW output but above this size, phase advancers are frequently employed.
- Shunt and series type of phase advancers are available according to whether the exciting winding of the advancer is connected in parallel or series with the rotor winding of the induction motor.

4. **Use of phase compensated motors :**

As mentioned earlier, the use of phase advancers may not be economical for induction motors below 150 kW output. Power factor improvement of the system is achieved by the *use of phase compensated motors* such as **torda, osnos** and **schrage motors.** These motors are however *very costly and require more maintenance than plain induction motors.* As such these motors *are chosen when we are sure that they will be loaded to rated output for most of the time and that they will effect more saving in the energy cost due to higher p.f. than the additional expenses incurred on them.*

9.19.7. Location of Power Factor Correction Equipment

The most appropriate location for the power factor correction equipment to be installed is *where the apparatus or equipment responsible for low p.f. is operating.*

- ***"Synchronous condensers"*** are employed at *load centres* where considerable corrective kVAR is required.

"Static capacitors" are justifiably used in smaller units and may be placed *closer to the point where the load of inductive nature is installed* and thereby relieving the distributors and feeders from carrying excessive currents due to low p.f.

- When ***"synchronous condensers"*** are to be used for p.f. improvement in the *transmission system*, then these should be *installed at the receiving end* so that not only the generators but also the transmission lines are relieved of carrying excessive current due to poor p.f. However, if synchronous condensers are installed near the generators then only generators will be relieved from the excessive lagging current component and the transmission line will have to carry it.

9.19.8. Economics of Power Factor Equipment

The improvement in p.f. (power factor) involves an expenditure on account of the p.f. correcting equipment. *Whereas p.f. improvement results in reduction in maximum demand, affecting an annual saving over the maximum demand charge, on the other hand an expenditure has to be incurred annually in the form of interest and depreciation on account of the investment made over the p.f. correcting equipment.* The limit of the p.f. at which the net saving (saving in annual maximum demand charges less annual expenditure incurred on p.f. correcting equipment) is maximum is known as *economical limit of p.f. correcting.*

Let, P = Peak load of a consumer in kW at a p.f. of $\cos \phi_1$,

x = Rate (Rs.) per kVA of maximum demand per annum, and

y = Expenditure (Rs.) per kVAR per annum of the p.f. correction equipment.

Corresponding to maximum demand of P kW at p.f. of $\cos \phi_1$

Maximum demand in kVA,

$$\text{kVA}_1 = \frac{P}{\cos \phi_1} = P \sec \phi_1$$

Reactive power, $\text{kVAR}_1 = P \tan \phi_1$

Let by installing the p.f. correction equipment (*e.g.* phase advancer) p.f. becomes $\cos \phi_2$.

New maximum demand in kVA, $\text{kVA}_2 = \dfrac{P}{\cos \phi_2} = P \sec \phi_2$

New reactive power, $\text{kVAR}_2 = P \tan \phi_2$

Annual saving in maximum demand charges

$$= \text{Rs. } x(\text{kVA}_1 - \text{kVA}_2)$$

$$= x(P \sec \phi_1 - P \sec \phi_2)$$

$$= xP(\sec \phi_1 - \sec \phi_2)$$

Leading kVAR supplied by p.f. correction equipment

$$= P(\tan \phi_1 - \tan \phi_2)$$

Annual cost of p.f. correction equipment

$$= \text{Rs. } y \times P(\tan \phi_1 - \tan \phi_2)$$

Net annual saving, $\quad S = xP(\sec \phi_1 - \sec \phi_2) - yP(\tan \phi_1 - \tan \phi_2) \quad$ ...(9.19)

This saving will be maximum when,

$$\frac{dS}{d\phi_2} = 0$$

or, $\quad \frac{d}{d\phi_2}\,[xP(\sec \phi_1 - \sec \phi_2) - yP(\tan \phi_1 - \tan \phi_2)]$

or, $\quad -xP \sec \phi_2 \tan \phi_2 + yP \sec^2 \phi_2 = 0$

or, $\quad \tan \phi_2 = \frac{y}{x} \sec \phi_2$

or, $\quad \sin \phi_2 = \frac{y}{x}$

or, $\quad \cos \phi_2 = \sqrt{1 - \sin^2 \phi_2} = \sqrt{1 - \left(\frac{y}{x}\right)^2} \quad$ ...(9.20)

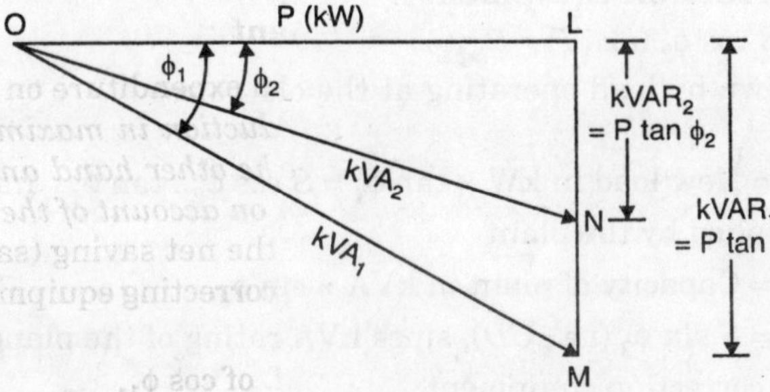

Fig. 9.33

From eqn. (9.20), *value of most economical p.f.* $\cos \phi_2$ *can be determined which is independent of original p.f.* $\cos \phi_1$ *and is governed by the relative costs of supply and p.f. correction equipment.*

Economical comparison of the methods of increasing the power supplied :

In a generating station, the increase in power demand can be met by the following *two methods* :

(*i*) *By increasing the capacity of the generating plant working at the same p.f.* ;

(*ii*) *By raising the p.f. of the system by installation of phase advancers.*

Owing to the improvement of p.f. in the beginning the saving in the generating and distributing plant outweighs the extra cost of the p.f. correction equipment in most of the cases but as the p.f. is raised further its cost begins to approximate the saving and finally any saving over the plant is obtained by incurring a greater expenditure on the p.f. correcting equipment. Thus *there is a limit beyond which it is not economical still further to improve the p.f.*

The maximum value to which the p.f. can be economically raised entirely depends upon the relative costs of the generating plant and phase advancing plant.

Let, S = kVA rating of generating plant,

P = kW load supplied by this plant at p.f. $\cos \phi_1$,

$P_{add.}$ = Additional load to be supplied,

x = Annual cost (Rs.) per kVA of generating plant, and

y = Annual cost (Rs.) per kVAR rating of p.f. correction equipment.

***(i)* Cost by increasing the capacity of the generating station :**

Refer to Fig. 9.34.

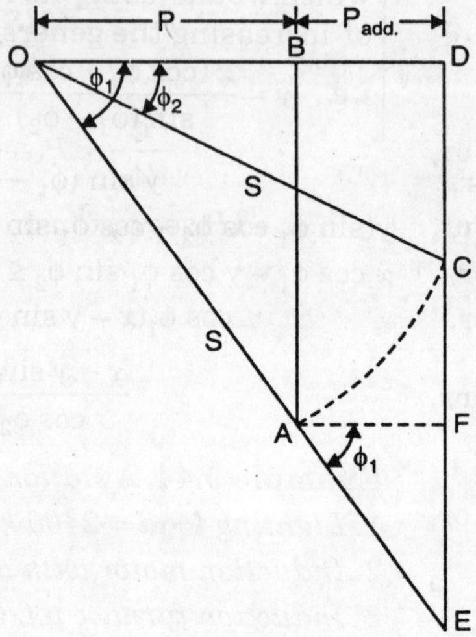

Fig. 9.34

Increase in load ($P_{add.}$) = $S(\cos \phi_2 - \cos \phi_1)$

Increase in capacity of the generating plant operating at p.f. $\cos \phi_1$ (*i.e.*, *AE*)

$$= \frac{P_{add.}\ (AF)}{\cos \phi_1} \text{ kVA}$$

$$= \frac{S(\cos \phi_2 - \cos \phi_1)}{\cos \phi_1} \text{ kVA}$$

Increase in annual cost due to increase in capacity of the plant

$$= \text{Rs. } xS\left(\frac{\cos \phi_2 - \cos \phi_1}{\cos \phi_1}\right) \quad ...(i)$$

***(ii)* Cost on p.f. correction equipment :**

New load in kW = $S \cos \phi_2$ (*i.e.*, $P + P_{add.}$)

Reactive power drawn by load operating at the old power factor of $\cos \phi_1$

$$= \text{New load in kW} \times \tan \phi_1 = S \cos \phi_2 \, . \tan \phi_1 \quad (i.e., DE)$$

Reactive power supplied by the plant

$$= \text{Capacity of plant in kVA} \times \sin \phi_2$$

$= S \sin \phi_2$ (*i.e.*, *CD*), since kVA rating of the plant remains the same

kVAR rating of p.f. correction equipment

= Reactive power drawn by the load

− Reactive power supplied by the plant (*i.e.*, *C.E*)

$$= S \cos \phi_2 \, . \tan \phi_1 - S \sin \phi_2 = S(\cos \phi_2 \, . \tan \phi_1 - \sin \phi_2)$$

Annual cost on p.f. correction equipment

$$= yS(\cos \phi_2 \, . \tan \phi_1 - \sin \phi_2) \quad ...(ii)$$

P.f. correction equipment will be *cheaper* if annual cost on p.f. correction equipment is *less* than annual cost account of increasing the generating plant capacity.

or, $$yS(\cos \phi_2 \, . \tan \phi_1 - \sin \phi_2) < xS\left(\frac{\cos \phi_2 - \cos \phi_1}{\cos \phi_1}\right)$$

or, $$y\left(\cos \phi_2 \times \frac{\sin \phi_1}{\cos \phi_1} - \sin \phi_2\right) < x\left(\frac{\cos \phi_2 - \cos \phi_1}{\cos \phi_1}\right)$$

or, $$y\left(\frac{\sin \phi_1 \cos \phi_2}{\cos \phi_1} - \sin \phi_2\right) < x\left(\frac{\cos \phi_2 - \cos \phi_1}{\cos \phi_1}\right)$$

or, $y(\sin \phi_1 \cos \phi_2 - \cos \phi_1 \sin \phi_2) < x(\cos \phi_2 - \cos \phi_1)$

or, $y [\sin (\phi_1 - \phi_2)] < x(\cos \phi_2 - \cos \phi_1)$

or, $$y < \frac{x (\cos \phi_2 - \cos \phi_1)}{\sin (\phi_1 - \phi_2)} \quad ...(9.21\ a)$$

[In the *limiting case*, the maximum cost of p.f. correction equipment, which would justify its installation, should be equal to cost on account of increasing the generating capacity,

i.e., $y = \dfrac{x (\cos \phi_2 - \cos \phi_1)}{\sin (\phi_1 - \phi_2)}$]

or, $y \sin (\phi_1 - \phi_2) \le x(\cos \phi_2 - \cos \phi_1)$

or, $y(\sin \phi_1 \cos \phi_2 - \cos \phi_1 \sin \phi_2) \le x \cos \phi_2 - x \cos \phi_1$

or, $x \cos \phi_1 - y \cos \phi_1 \sin \phi_2 \le x \cos \phi_2 - y \sin \phi_1 \cos \phi_2$

or, $\cos \phi_1(x - y \sin \phi_2) \le \cos \phi_2(x - y \sin \phi_1)$

or, $$\frac{x - y \sin \phi_2}{\cos \phi_2} \le \frac{x - y \sin \phi_1}{\cos \phi_1} \quad ...(9.21)$$

Example 9.44. *A station supplies the following loads :*

1. *Lighting load = 2400 kW*
2. *Induction motor with average p.f. 0.9 lagging = 2420 H.P. and efficiency 88.5 percent.*
3. *Induction furnace p.f. 0.8 lagging = 2112 kW.*
4. *Synchronous motor with average = 3650 H.P., p.f. 0.707 leading and efficiency 93 percent.*

(i) *Find p.f. of the station.*

(ii) *If whole of the total load is carried by single feeder cable with same heating, how much maximum load the cable will carry ?*

(iii) *If one alternator is supplying 6000 kW at unity p.f., find the load and p.f., of second alternator.*

Solution. Power factor of the station, cos ϕ :

1. *Lighting load (unity p.f.)* : $P_1 = 2400$ kW ; $Q_1 = 0$

2. *Induction motor* : $P_2 = \dfrac{2420 \times 0.7355}{0.885} = 2011$ kW ;

$\cos \phi_2 = 0.9$ or $\phi_2 = \cos^{-1}(0.9) = 25.84°$

$\therefore$ $Q_2 = P_2 \tan \phi_2 = 2011 \tan (25.84°) = - 973.9$ kVAR

3. *Induction furnace* : $P_3 = 2112$ kW ; $\cos \phi_3 = 0.8$ or $\phi_3 = \cos^{-1}(0.8) = 36.87°$;

$\therefore$ $Q_3 = P_3 \tan \phi_3 = 2112 \tan (36.87°) = 1584$ kVAR

4. *Synchronous motor* : $P_4 = \dfrac{3650 \times 0.7355}{0.93} = 2887$ kW ;

$\cos \phi_4 = 0.707$ or $\phi_4 = \cos^{-1}(0.707) = 45°$

$\therefore$ $Q_4 = P_4 \tan \phi_4 = 2887 \times \tan 45° = 2887$ kVAR

Total kW, $P_t = P_1 + P_2 + P_3 + P_4$

$= 2400 + 2011 + 2112 + 2887 = 9410$ kW

Total kVAR, $Q_t = Q_1 + Q_2 + Q_3 + Q_4$

$= 0 + (- 973.9) + (- 1584) + 2887 = 329.1$ kVAR

Now, $\tan\phi = \frac{Q_t}{P_t} = \frac{329.1}{9410} = 0.03497$ or $\phi = \tan^{-1}(0.03497) = 2°$

and p.f., $\cos\phi = \cos 2° =$ **0.9994 leading. (Ans.)**

(*i*) **Maximum load carried by single feeder cable :**

$$kVA = \sqrt{(P_t)^2 + (Q_t)^2}$$

$$= \sqrt{(9410)^2 + (329.1)^2} = \mathbf{9415.7\ kVA.}\quad \textbf{(Ans.)}$$

(*ii*) **Load and p.f. of the second alternator :**

kW supplied by the second alternator

$$= 9410 - 6000 = \mathbf{3410\ kW.}\quad \textbf{(Ans.)}$$

kVAR supplied by the second alternator

$$= 329.1 - 0 = \mathbf{329.1\ kVAR\quad (leading).}\quad \textbf{(Ans.)}$$

If p.f. angle of second alternator is ϕ', then

$$\tan\phi' = \frac{329.1}{3410} = 0.0965 \quad \text{or} \quad \phi' = 5.51°$$

$\therefore$ $\cos\phi'(\text{p.f.}) = \cos 5.51° =$ **0.9954 leading. (Ans.)**

Example 9.45. *A 40-MVA, 11 kV 3-phase alternator supplies full-load at a lagging power factor of 0.6. Find the percentage increase in earning capacity if the power factor is increased to 0.9.*

Solution. Rating of the alternator = 40 MVA

Supply voltage = 11 kV

Initial power factor = 0.6 (lagging)

Final power factor after increase = 0.9 (lagging)

Percentage increase in earning capacity :

The earning capacity is proportional to the power (in MW or kW) supplied by the alternator.

MW supplied at 0.6 lagging = 40 × 0.6 = 24

MW supplied at 0.9 lagging = 40 × 0.9 = 36

Increase in MW = 36 – 24 = 12

The increase in earning capacity is proportional to 12.

$\therefore$ *Percentage increase in earning capacity* $= \frac{12}{24} \times 100 =$ **50%. (Ans.)**

Example 9.46. *A 3-phase, 40 kW, 440 V, 50 Hz induction motor operates on full-load with an efficiency of 90 percent and at a power factor of 0.8 lagging. Calculate the total kVA rating of capacitors required to raise the full-load power factor to 0.9 lagging. What will be capacitance per phase if the capacitances are :*

(*i*) *Delta connected* (*ii*) *Star-connected.*

Solution. Motor output = 40 kW

Supply voltage = 440 V

Efficiency of motor = 90%

Motor power input, $P = \frac{\text{Output}}{\text{Efficiency}} = \frac{40}{0.9} = 44.44$ kW

Power factor 0.8 (lagging)

$$\cos \phi_1 = 0.8$$

$$\therefore \quad \phi_1 = \cos^{-1} 0.8 = 36.9°$$

$$\tan \phi_1 = 0.75$$

$$\text{Motor kVAR}_1 = P \tan \phi_1 = 44.44 \times 0.75 = 33.33$$

Power factor 0.9 (lagging)

Motor power input = 44.44 kW [As before]

[Power is same as before since the capacitors are *loss free i.e.*, they *do not absorb any power*]

$$\cos \phi_2 = 0.9$$

$$\therefore \quad \phi_2 = 25.8°$$

$$\tan \phi_2 = 0.484$$

$$\therefore \quad \text{Motor kVAR}_2 = P \tan \phi_2 = 44.44 \times 0.484 = 21.5.$$

The difference in values of kVAR is *due to the capacitors which supply leading kVAR to partially neutralize the lagging kVAR of the machine.*

∴ ***Leading kVAR supplied by capacitors is***

$$= \text{kVAR}_1 - \text{kVAR}_2 = 33.33 - 21.5$$

$$= 11.83 \text{ ...... MQ or NT. (Fig. 9.35)}$$

Since capacitors are loss-free, their kVAR is same as kVA

∴ **Total kVA rating of capacitors = 11.83. (Ans.)**

$$\text{kVAR/capacitor} = \frac{11.83}{3} = 3.943$$

∴ VAR/capacitor = 3943.

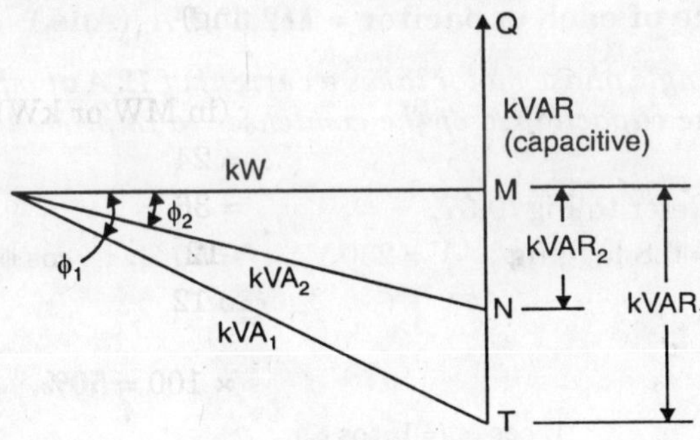

Fig. 9.35

(i) **Delta connection.** Refer to Fig. 9.36. In delta-connection, voltage across each capacitor is 440 V.

$$\text{Current drawn by each capacitor, } I_C = \frac{3943}{440} = 8.96 \text{ A}$$

Now,

$$I_C = \frac{V}{X_C} = \frac{V}{\frac{1}{2\pi fC}} = 2\pi fCV$$

$$\therefore \quad C = \frac{I_C}{2\pi fC} = \frac{6.96}{2\pi \times 50 \times 440} = 50.35 \times 10^{-6} \text{ F} = 50.35 \text{ μF}$$

Hence, **capacitance of each capacitor = 50.35 μF. (Ans.)**

(*ii*) **Star-connection.** In star-connection, voltage across each capacitor

$$= \frac{440}{\sqrt{3}} = 254 \text{ V}$$

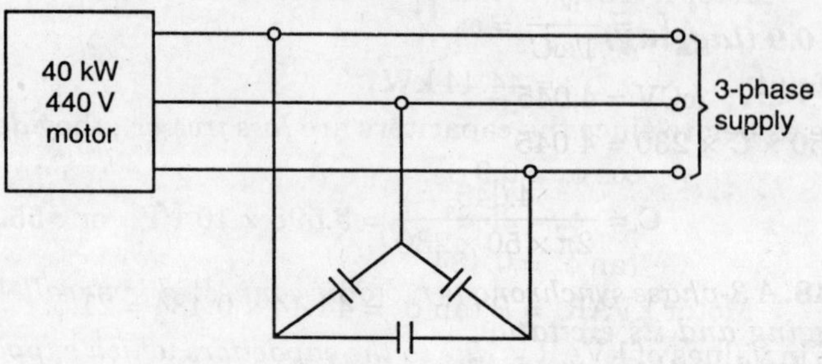

Fig. 9.36

Current drawn by each capacitor,

$$I_C = \frac{3943}{254} = 15.52 \text{ A}$$

Now,

$$I_C = \frac{V}{X_C} = \frac{V}{\frac{1}{2\pi fC}} = 2\pi fCV$$

∴

$$C = \frac{I_C}{2\pi fV} = \frac{15.52}{2\pi \times 50 \times 440} = 112.3 \times 10^{-6} \text{ F} = 112.3\ \mu\text{F}$$

Hence, **capacitance of each capacitor = 112.3 μF. (Ans.)**

Example 9.47. *A single phase motor takes a current of 12 A at p.f. of 0.8 lagging from a 230 V, 50 Hz supply. Find out the capacitance of the condenser to improve the p.f. to 0.95 lagging when connected across the load.*

Solution. *Given :* Refer to Fig. 9.37.

$I_1 = 12$ A ; $\cos \phi_1 = 0.8$ lagging ; $V = 230$ V ; $f = 50$ Hz ; $\cos \phi_2 = 0.95$ lagging.

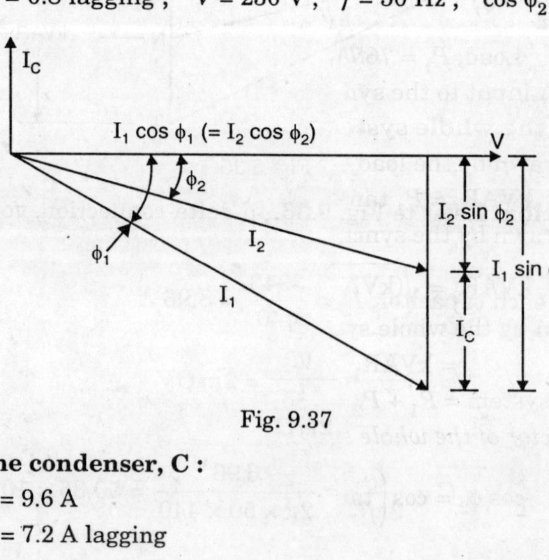

Fig. 9.37

Capacitance of the condenser, C :

$I_1 \cos \phi_1 = 12 \times 0.8 = 9.6$ A

$I_1 \sin \phi_1 = 12 \times 0.6 = 7.2$ A lagging

$$I_2 \sin \phi_2 = (I_1 \cos \phi_1) \times \tan \phi_2$$
$$= 9.6 \times \tan (\cos^{-1} 0.95) = 3.155 \text{ A}$$

In order that condenser should improve the p.f. to 0.95 lagging, it should take :

$$I_C = I_1 \sin \phi_1 - I_2 \sin \phi_2 = 7.2 - 3.155 = 4.045 \text{ A}$$

But,
$$I_C = \frac{V}{1/\omega C} = \omega CV$$

$\therefore$
$$\omega CV = 4.045$$

or,
$$2\pi \times 50 \times C \times 230 = 4.045$$

or,
$$C = \frac{4.045}{2\pi \times 50 \times 230} = 5.598 \times 10^{-5} \text{ F} \quad \text{or} \quad \mathbf{55.98\ \mu F.} \quad \textbf{(Ans.)}$$

Example 9.48. *A 3-phase synchronous motor is connected in parallel with a load to 770 kW at 0.7 power factor lagging and its excitation is adjusted until it raises the total power factor to 0.9 lagging. Mechanical load on the motor including losses is 165 kW. Find out the power factor of the synchronous motor.*

Solution. *Given* : Load, P_1 = 770 kW ; $\cos \phi_1 = 0.7$;
Motor load, P_2 = 165 kW ; $\cos \phi = 0.9$ lagging.

P.f. of the synchronous motor, $\cos \phi_2$:

Total load $= P_1 + P_2 = 770 + 165 = 935$ kW

Reactive kVA supplied by the synchronous motor

$$= P_1 \tan \phi_1 - P \tan \phi_2$$
$$= 770 \tan (\cos^{-1} 0.7) - 935 \tan (\cos^{-1} 0.9)$$
$$= 785.56 - 452.84 = 332.72 \text{ kVAR}$$

Power factor of synchronous motor,

$$\cos \phi_2 = \cos \left(\tan^{-1} \frac{332.72}{165} \right) = \cos 63.62° = \mathbf{0.444\ (leading).} \quad \textbf{(Ans.)}$$

Example 9.49. *A 3-phase synchronous motor having a mechanical load (including losses) of 183 kW is connected in parallel with a load of 765 kW at 0.8 power factor lagging. The excitation of the motor is adjusted so that the kVA input to the motor becomes 210 kVA. Determine the new power factor of the whole system.*

Solution. *Given* : Load, P_1 = 765 kW ; $\cos \phi_1 = 0.8$; kW input to the synchronous motor, P_2 = 183 kW ; kVA input to the synchronous motor = 210 kVA.

Power factor of the whole system, $\cos \phi$:

kVAR (lagging) drawn by the load,

$$\text{kVAR}_1 = P_1 \tan \phi_1 = 765 \tan (\cos^{-1} 0.8) = 573.75 \text{ kVAR}$$

kVAR (leading) drawn by the synchronous motor

$$\text{kVAR}_2 = \sqrt{(\text{kVA})^2 - (P_2)^2} = \sqrt{(210)^2 - (183)^2} = 103 \text{ kVAR}$$

Total kVAR drawn by the whole system

$$= \text{kVAR}_1 - \text{kVAR}_2 = 573.75 - 103 = 470.75 \text{ kVAR}$$

Total load on the system $= P_1 + P_2 = 765 + 183 = 948$ kW

$\therefore$ *New power factor of the whole system,*

$$\cos \phi_2 = \cos \left(\tan^{-1} \frac{470.75}{948} \right) = \mathbf{0.8956\ (lagging).} \quad \textbf{(Ans.)}$$

Example 9.50. *A 400 V, 3-phase installation draws a current of 50 A at 0.8 lagging p.f. It is desired to install a synchronous motor to improve the overall power factor to 0.95 lagging. The synchronous motor will drive a 25 H.P. (metric) load at an efficiency of 0.9 calculate :*

(i) The kVA of the synchronous motor.

(ii) The power factor of the motor. **(Bombay University)**

Solution. *Given* : Supply voltage, $V_L = 400$ V ; $I = 50$ A ; $\cos \phi_2 = 0.8$; $\cos \phi = 0.95$; $\eta = 0.9$.

(*i*) **The kVA of synchronous motor :**

kW rating of installation $= \sqrt{3} \times 400 \times 50 \times 0.8 \times 10^{-3}$ kW $= 27.7$ kW

kVA rating of installation $= \sqrt{3} \times 400 \times 50 \times 10^{-3}$ kVA $= 34.64$ kVA

kVAR rating of installation $= \sqrt{(34.64)^2 - (27.7)^2} = 20.8$ kVAR

kW input to synchronous motor $= \dfrac{25 \times 0.7355}{0.9}$ kW $= 20.43$ kW

Total kW = 27.7 + 20.43 = 48.13 kW

Overall power factor angle, $\phi = \cos^{-1}(0.95) = 18.2°$

Total kVAR = 48.13 tan (18.2°) = 15.82 kVAR (inductive)

kVAR of synchronous motor = 15.82 – 20.8 = – 4.98 or 4.98 kVAR capacitive

$\therefore$ kVA of synchronous motor $= \sqrt{(20.43)^2 + (4.98)^2} =$ **21.03 kVA. (Ans.)**

(*ii*) **The power factor of the motor, $\cos \phi_m$:**

The p.f. of the synchronous motor,

$$\cos \phi_m = \frac{20.43}{21.03} = \mathbf{0.97\ (leading).\ \ (Ans.)}$$

Example 9.51. *Power factor 0.8 lagging of a load of 600 kW is to be improved by the phase advancing plant which works at a power factor of 0.1 leading so that power factor of the whole system is now 0.9 lagging. Find the kVA rating of the phase advancing plant.*

Solution. *Given* : Refer to Fig. 9.38. Load $P = 600$ kW ; $\cos \phi_1 = 0.8$ or $\phi_1 = \cos^{-1}(0.8) = 36.86°$; $\cos \phi_2 = 0.1$ or $\phi_2 = \cos^{-1}(0.1) = 84.26°$; $\cos \phi = 0.9$ or $\phi = \cos^{-1}(0.9) = 25.84°$.

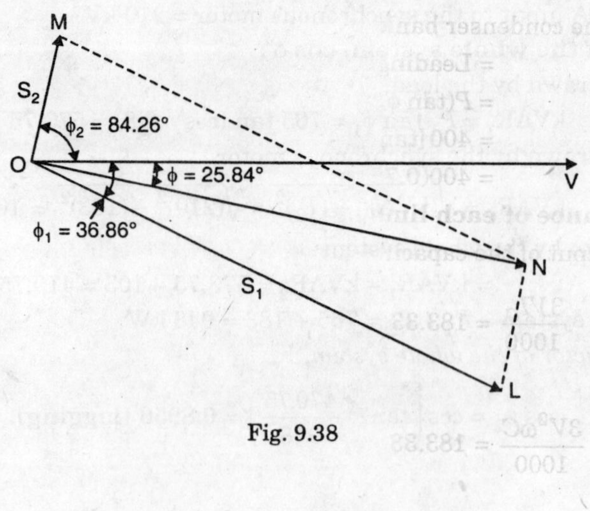

Fig. 9.38

kVA rating of the phase advancing plant, S_2 :

kVA of load, $S_1 = \frac{P}{\cos \phi_1} = \frac{600}{0.8} = 750$ kW

In the vector diagram (Fig. 9.38) *OL* represents kVA of the load at an angle of 36.86° lagging voltage vector. A line *ON* is drawn at an angle of 25.84° to the voltage vector. Line *OM* is drawn at angle 84.26° with the voltage vector. *LN* is drawn parallel to *OM* cutting *ON* at *N*. Then magnitude of kVA of phase advancing plant is given by *LN* = *OM*.

In ΔLON, $\angle LON = (36.86° - 25.84°) = 11.02°$

$\angle ONL = \angle NOM = (25.84° + 84.26°) = 110.1°$

$\angle OLN = 180° - [110.1° + (36.86° - 25.84°)] = 58.88°$

Now,
$$\frac{LN}{\sin \angle LON} = \frac{OL}{\sin \angle ONL}$$

$$\frac{S_2}{\sin 11.02°} = \frac{S_1}{\sin 110.1°}$$

or,
$$S_2 = \sin 11.02° \times \frac{750}{\sin 110.1°} = \mathbf{152.66\ kVA.} \quad \textbf{(Ans.)}$$

Example 9.52. *A 340 kW, 50 Hz, 3-phase star-connected induction motor has full load efficiency of 85 percent and p.f. of 0.8 lagging. It is desired to improve the power factor to 0.96 lagging by using a bank of three capacitors. Calculate :*

(i) The kVAR rating of the condenser bank.

(ii) The capacitance of each limbs of the condenser bank connected in delta.

(iii) The capacitance of each capacitor, if each one of the limbs of the delta-connected condenser bank is formed by using 6 similar 3300 V capacitors. **(A.M.I.E.)**

Solution. *Given* : Motor output = 340 kW ; Full efficiency of the motor = 85% ;

$\cos \phi = 0.8$ lagging ; $\cos \phi_2 = 0.96$ lagging

***(i)* The kVA rating of the condenser bank :**

Motor input, $P = \frac{340}{0.85} = 400$ kW

kVAR rating of the condenser bank

= Leading kVAR supplied by the static capacitors

$= P(\tan \phi_1 - \tan \phi_2)$

$= 400\,[\tan (\cos^{-1} 0.8) - \tan (\cos^{-1} 0.96)$

$= 400(0.75 - 0.2917) =$ **183.33 kVA. (Ans.)**

***(ii)* The capacitance of each limb of the condenser bank connected in delta :**

Reactive kVA output of the capacitor, when connected in delta is given by :

$$\frac{3VI_C}{1000} = 183.33 \quad \text{or} \quad \frac{3V \times V}{\left(\frac{1}{\omega C}\right) \times 1000} = 183.33$$

or,
$$\frac{3V^2\omega C}{1000} = 183.33 \quad \text{or} \quad \frac{3V^2 \times 2\pi f C}{1000} = 183.33$$

or, $$C = \frac{183.33 \times 1000}{3 \times (3300)^2 \times (2\pi \times 50)} \text{ F} = 1.786 \times 10^{-5} \text{ F} \quad \text{or} \quad \mathbf{17.86\ \mu F.} \quad \textbf{(Ans.)}$$

(*iii*) **The capacitance of each capacitor :**

Since 6 similar capacitors are connected in parallel (voltage rating of each capacitor being same as that of supply voltage) to form each limb of delta-connected condenser bank, therefore, capacitance of each capacitor

$$= \frac{17.86}{6} = \mathbf{2.977\ \mu F.} \quad \textbf{(Ans.)}$$

Example 9.53. *A 666.66 kW, 50 Hz, 400 V, 3-phase load has a power factor of 0.8 lagging. It is desired to improve the power factor to 0.95 lagging by using a delta-connected bank of 2200 V capacitors. Calculate :*

(*i*) *Capacitance of each capacitor.*

(*ii*) *kVA rating of capacitor bank and transformer.*

(*iii*) *Percentage reduction in line losses.*

(*iv*) *Ratio of capacitance of 2200 V capacitors to the capacitance of 400 V capacitors for the same power factor improvement.* **(Kurukashetra University)**

Solution. *Given* : 3-phase load = 666.6 kW, 50 Z, 400 V ; $\cos \phi_1 = 0.8$;

Improved power factor, $\cos \phi_2 = 0.95$

(*i*) **Capacitance of each capacitor, C :**

Load current $$= \frac{666.66 \times 1000}{\sqrt{3} \times 400 \times 0.8} = 1202.8 \text{ A}$$

Let initial current $= I_1$

$$I_1 = 1202.8(0.8 - j\,0.6) = (962.24 - j\,721.68) \text{ A}$$

Let final current at 0.95 power factor be I_2.

$$\cos \phi_2 = 0.95, \quad \tan \phi_2 = \tan (\cos^{-1} 0.95) = 0.329$$

$$I_2 = 962.24 - j\,962.24 \times 0.329$$

$$= 962.24 - j\,316.58 = 1013 \angle -18.21° \text{ A}$$

Line current of 400 V capacitor bank $= 721.68 - 316.58 = 405.1$ A

Line current of 2200 V capacitor bank $$= \frac{405.1 \times 400}{2200} = 73.65 \text{ A}$$

Phase current of capacitor bank $$= \frac{73.65}{\sqrt{3}} = 42.52 \text{ A}$$

$$X_C = \frac{2200}{42.52} = 51.74\ \Omega$$

$$\frac{1}{\omega C} = 51.74$$

or, $$\mathbf{C} = \frac{1}{2\pi \times 50 \times 51.74} = \mathbf{61.52 \times 10^{-6}\ F.} \quad \textbf{(Ans.)}$$

(*ii*) **kVA rating of capacitor bank and transformer :**

$$kVA \text{ rating of capacitor bank} = \frac{3 \times 2200 \times 42.52}{1000} = \mathbf{280.6\ kVA.} \quad \textbf{(Ans.)}$$

kVA rating of transformer to convert 400 V to 2200 V will be the **same** *as the kVA rating of capacitor bank.* **(Ans.)**

(*iii*) **Percentage reduction in line losses :**

$$\text{Percentage reduction in line losses} = \frac{I_1^2 - I_2^2}{I_1^2} \times 100$$

$$= \frac{(1202.8)^2 - (1013)^2}{(1202.8)^2} \times 100 = \mathbf{29.07\%}. \quad \textbf{(Ans.)}$$

(*iv*) **Ratio of capacitances :**

Line current of 400 V capacitor bank = 405.1 A

$$\text{Phase current} = \frac{405.1}{\sqrt{3}} = 233.9 \text{ A}$$

$$X_C = \frac{400}{233.9} = 1.71\ \Omega$$

$$\frac{1}{\omega C} = 1.71 \quad \text{or} \quad C = \frac{1}{2\pi \times 50 \times 1.71} = 1861.46 \times 10^{-6} \text{ F}$$

$$\therefore \quad \textit{Ratio of capacitances} = \frac{61.52 \times 10^{-6}}{1861.46 \times 10^{-6}} = \mathbf{0.033}. \quad \textbf{(Ans.)}$$

Example 9.54. *A 3-phase, 11 kV, 50 Hz alternator is supplying 400 kW at 0.8 p.f. lagging. A synchronous motor is put in parallel with the load to raise the power factor at alternator terminals to unity. The armature current of alternator is not to be increased.*

(*i*) *How much power can alternator supply to synchronous motor ?*

(*ii*) *How much brake horse power can the motor develop if its efficiency is 88 percent ?*

(*iii*) *What is the power factor of the synchronous motor ?*

Solution. *Given* : Supply voltage = 11 kV, 50 Hz ; Load = 400 kW ; cos ϕ = 0.8 lagging ; Final power factor = unity ; η_{motor} = 88%.

(*i*) **The power supplied by alternator to the synchronous motor :**

$$\text{Load} = 400 \text{ kW at } 0.8 \text{ p.f.}$$
$$= 500(0.8 + j\,0.6) \text{ kVA}$$
$$= 400 + j\,300 \text{ kVA}$$

When p.f. is raised to unity, the alternator can supply 500 W for the same value of armature current. *Hence it can supply* **100 W to synchronous motor. (Ans.)**

(*ii*) **Brake horse power (B.H.P.) developed :**

$$\text{B.H.P.} = \frac{100 \times 0.88}{0.7355} = \textbf{119.6 H.P.} \quad \textbf{(Ans.)}$$

(*iii*) **Power factor of the synchronous motor, cos ϕ_m :**

kVA of the motor = 100 – j 300

$$\tan \phi_m = \frac{300}{100} = 3$$

$$\therefore \quad \cos \phi_m = \cos(\tan^{-1} 3) = \textbf{0.316 leading.} \quad \textbf{(Ans.)}$$

Example 9.55. *A 50 H.P. induction motor has power factor 0.9 and efficiency 90 percent at full-load, power factor 0.6 and efficiency 70% at half load. At no-load the current is 25 percent of the full-load current and power factor 0.1. Capacitors are supplied to make the line power factor 0.8 at half-load. With these capacitors in circuit, find the line power factor at (i) full-load, and (i) no-load.* **(A.M.I.E.)**

Solution. Motor input at full load, $P_F = \dfrac{50 \times 0.7355}{0.9} = 40.86$ kW $\quad (\because \quad 1 \text{ H.P.} = 0.7355 \text{ kW})$

Lagging kVAR drawn by the motor at full-load,

$$\text{kVAR}_{F1} = P_F \tan \phi_{F1}$$
$$= 40.86 \tan (\cos^{-1} 0.9) = 19.79 \text{ kVAR}$$

Motor input at half-full load,

$$P_{HF} = \frac{(1/2) \times 50 \times 0.7355}{0.7} = 26.27 \text{ kW}$$

Lagging kVAR drawn by motor at half-full load,

$$\text{kVAR}_{HF1} = P_{HF} \tan \phi_{HF1} = 26.27 \tan (\cos^{-1} 0.6) = 35.03 \text{ kVAR}$$

Full-load current, $\quad I_F = \dfrac{50 \times 735.5}{\sqrt{3}\, V_L \times 0.9 \times 0.9} = \dfrac{26212}{V_L}$ A

Current at no-load $\quad = 0.25\, I_F = \dfrac{6553}{V_L}$ A

Input to motor at no-load,

$$P_0 = \frac{\sqrt{3}\, V_L I_0 \cos \phi_{01}}{1000} \text{ kW}$$
$$= \sqrt{3}\, V_L \times \frac{6553}{V_L} \times \frac{0.1}{1000} = 1.135 \text{ kW}$$

Lagging kVAR drawn by motor at no-load,

$$\text{kVAR}_{01} = 1.135 \tan (\cos^{-1} 0.1) = 11.293 \text{ kVAR}$$

Lagging kVAR drawn from mains at *half-load with capacitors*,

$$kVAR_{HF2} = P_{HF} \tan \phi_{HFL} = 26.27 \tan (\cos^{-1} 0.8) = 19.7 \text{ kVAR}$$

kVAR supplied by the capacitors,

$$\text{kVAR}_C = \text{kVAR}_{HF1} - \text{kVAR}_{HF2}$$
$$= 35.03 - 19.7 = 15.33 \text{ kVAR}$$

kVAR drawn from mains at *full-load with capacitors*

$$\text{kVAR}_{F2} = \text{kVAR}_{F1} - \text{kVAR}_C$$
$$= 19.79 - 15.33 = 4.46 \text{ kVAR}$$

(*i*) **Line p.f. at full-load :**

Line p.f. at full-load $\quad = \cos\left(\tan^{-1} \dfrac{\text{kVAR}_{F2}}{P_F}\right)$

$$= \cos\left(\tan^{-1} \frac{4.46}{40.86}\right) = \mathbf{0.994 \text{ lagging.}} \quad \textbf{(Ans.)}$$

(ii) Line p.f. at no-load :

kVAR drawn from mains at no load with capacitors,

$$kVAR_{02} = kVAR_{01} - kVAR_C$$
$$= 11.293 - 15.33 = -4.037$$

$\therefore$ *Line p.f. at no-load* $= \cos\left(\tan^{-1}\dfrac{-4.037}{1.135}\right)$

$= \cos(-74.3°) = $ **0.2706 (leading). (Ans.)**

Example 9.56. *Determine the value of the new power factor when the tariff is Rs. 180 per kVA of maximum demand plus a flat rate per kWh. Assume additional cost of condensers etc. at Rs. 140.00 per kVAR of such plant. Rate of interest and depreciation is together to be taken as 10 percent.*

Solution. Maximum demand charges, x = Rs. 180/kVA/anuum

Cost of phase advancing plant = Rs. 140/kVAR/annum

Expenditure on phase advancing plant, $y = 140 \times \dfrac{10}{100}$ = Rs. 14/kVAR/annum

$\therefore$ *Most economical power factor,*

$$\cos\phi_2 = \sqrt{1-\left(\frac{y}{x}\right)^2} \quad \text{...[Eqn. (9.20)]}$$

$$= \sqrt{1-\left(\frac{14}{180}\right)^2} = \mathbf{0.997\ lagging.} \quad \textbf{(Ans.)}$$

Example 9.57. *A consumer takes a steady load of 600 kW at a lagging power factor of 0.7 for 3000 hours a year. The tariff is Rs. 260 per kVA of maximum demand annually and 8 paise per kWh. The annual cost of phase advancing plant is Rs. 26.00 per kVAR. Determine the annual saving if the power factor of the load is improved.* **(Kurukshetra University)**

Solution. *Given* : Load, P = 600 kW ; $\cos\phi_1 = 0.7$, $\phi_1 = \cos^{-1} 0.7 = 45.57°$;

x = Rs. 260/kVA/annum ; y = Rs. 26/kVAR/annum.

Annual saving :

Let the p.f. be improved to most economical limit.

Most economical p.f., $\cos\phi_2 = \sqrt{1-\left(\dfrac{y}{x}\right)^2} = \sqrt{1-\left(\dfrac{26}{260}\right)^2} = 0.995$

$$\phi_2 = \cos^{-1}(0.995) = 5.73°$$

Capacity of phase advancing plant

$$= P(\tan\phi_1 - \tan\phi_2)$$
$$= 600\,(\tan 45.57° - \tan 5.73°)$$
$$= 600(1.02 - 0.1) = 552 \text{ kVAR}$$

Annual saving in demand charges

$$= \text{Rs. } 260\left(\frac{600}{0.7} - \frac{600}{0.995}\right)$$
$$= \text{Rs. } 260(857.14 - 603.01) = \text{Rs. } 66073.8$$

Annual expenditure on phase advancing plant

= Rs. 26 × 552 = Rs. 14352

Net saving in annual cost = Rs. 66073.8 – Rs. 14352 = **Rs. 51721.8. (Ans.)**

Example 9.58. *A system is working at its maximum kVA capacity with a lagging p.f. of 0.707. An anticipated increase in load can be met by (i) raising the p.f. of the system to 0.87 lagging by installation of phase advancers and (ii) installing extra generating plant, cables etc. to meet the increased power demand. The total cost of the latter method is Rs. 110 per kVA. Estimate the limiting cost per kVAR of phase advancing plant which would justify its installation.*

(Panjab University)

Solution. Let the existing rating of power plant be S kVA.

Load supplied at a p.f. of 0.707 = $0.707S$

Load supplied at a p.f. of 0.87 = $0.87S$

Increase in load = $0.87S - 0.707S = 0.163S$

Increase in capacity of generating plant

$$= \frac{\text{Increase in load}}{\text{Plant p.f.}} = \frac{0.1635}{0.707} = 0.23S$$

Cost on additional plant = Rs. $110 \times 0.23S = 25.3S$

Let the rate of interest and depreciation be z percent per annum. Then, increase in annual cost due to increase in capacity of power plant

$$= \text{Rs. } \frac{z}{100} \times 25.3S = 0.253\, zS \qquad \ldots(i)$$

Reactive power drawn by the load

$$= 0.87S \tan (\cos^{-1} 0.707) = 0.87S$$

Reactive power supplied by the plant

$$= S \sin \phi_2 = S \sin (\cos^{-1} 0.87) = 0.493S \qquad (\because \cos \phi_2 = 0.87)$$

kVAR rating of p.f. correction equipment

= *kVAR* drawn by load – *kVAR* supplied by the plant

$= 0.87S - 0.493S = 0.377S$

If y is the cost per kVAR of p.f. correction equipment then total cost on p.f. correction equipment = $0.377\, yS$.

Assume rate of interest and depreciation the same as in case of plant *i.e.*, z percent per annum.

Then, annual cost on p.f. correction equipment

$$= 0.00377\, zyS \qquad \ldots(ii)$$

In limiting case both must be equal

i.e., $0.00377\, zyS = 0.253\, zS$

or $$y = \frac{0.253}{0.00377} = \textbf{Rs. 67.11 per kVAR.} \quad \textbf{(Ans.)}$$

Alternative method :

Annual cost per kVA of generating plant,

$$x = \text{Rs. } 110 \times \frac{z}{100} = \text{Rs. } 1.1\, z$$

where, z is the percent rate of interest and depreciation per annum.

$$\cos\phi_2 = 0.87\ ; \quad \phi_2 = \cos^{-1}(0.87) = 29.54°$$
$$\cos\phi_1 = 0.707\ ; \quad \phi_1 = \cos^{-1}(0.707) = 45°$$

Annual cost per kVAR of p.f. correction equipment,

$$y = \frac{x(\cos\phi_2 - \cos\phi_1)}{\sin(\phi_1 - \phi_2)} \qquad \text{...[Eqn. (9.21 } a)]$$

$$= \frac{1.1z(0.87 - 0.707)}{\sin(45° - 29.54°)} = 0.672\ z$$

Cost per kVAR of p.f. correction equipment

$$= \text{Rs. } 0.672\ z \times \frac{100}{z} = \textbf{Rs. 67.2.} \quad \textbf{(Ans.)}$$

Choice of Equipment

Example 9.59. *The motor of a 15 H.P. condensate pump has been burnt beyond economical repairs. Two alternatives have been proposed to replace it by :*

Motor-1 : Cost = Rs. 12000 ; η at full-load = 90% ; η at half-load = 86%.

Motor-2 : Cost = Rs. 8000 ; η at full load = 85%, η at half load = 82%.

The life of each motor is 20 years and its salvage value is 10% of the initial cost. The rate of interest is 5% annually. The motor operates at full load for 25% of the time and at half load for the remaining period. The annual maintenance cost of motor-1 is Rs. 840 and that of motor-2 is Rs. 480. The energy rate is 20 P/kWh.

Which motor will you recommend ? **(Panjab University)**

Solution. ***Motor-1 :***

Capital cost = Rs. 12000

Annual interest on capital cost $= \text{Rs. } 12000 \times \frac{5}{100} = \text{Rs. } 600.00$

Annual depreciation charges $= \frac{\text{Original cost} - \text{salvage value}}{\text{Life of motor in years}}$

$$= \text{Rs. } \frac{12000 - \frac{10}{100} \times 12000}{20} = \text{Rs. } 540.00$$

Annual maintenance cost = Rs. 840

Energy input per annum $= \frac{15 \times 0.7355}{0.9} \times 0.25 \times 24 \times 365$

$$+ \frac{\frac{1}{2} \times 15 \times 0.7355}{0.86} \times 0.75 \times 24 \times 365$$

$$= 26846 + 42142 = 68988 \text{ kWh}$$

Annual energy cost $= 68988 \times \frac{20}{100} = \text{Rs. } 13798$

Total annual cost = Rs. (600 + 540 + 840 + 13798) = Rs. 15778

Motor-2 :

Capital cost = Rs. 8000

Annual interest on capital cost $= \text{Rs. } 8000 \times \frac{5}{100} = \text{Rs. } 400$

Annual depreciation charges $= \text{Rs.}\ \dfrac{8000 - \dfrac{10}{100} \times 8000}{20} = \text{Rs. } 360$

Annual maintenance charges $= \text{Rs. } 480$

Energy input per annum $= \dfrac{15 \times 0.7355}{0.85} \times 0.25 \times 24 \times 365$

$+ \dfrac{\dfrac{1}{2} \times 15 \times 0.7355}{0.82} \times 0.75 \times 24 \times 365$

$= 28425 + 44197 = 72622 \text{ kWh}$

Annual energy cost $= \text{Rs. } 72622 \times \dfrac{20}{100} = \text{Rs. } 14524$

Total annual cost $= \text{Rs. } (400 + 360 + 480 + 14524) = \text{Rs. } 15764$

Since total annual cost of motor-1 is Rs. 14 (*i.e.*, Rs. 15778 – Rs. 15764) more than that of motor-2, so **motor-2 is recommended. (Ans.)**

Example 9.60. *A consumer requires an induction motor of 50 H.P. (metric). He is offered two motors of the following specifications :*

Motor A : Efficiency = 88% ; Power factor = 0.9

Motor B : Efficiency = 90% ; Power factor = 0.81

The consumer is being charged on a two-part tariff of Rs. 70 per kVA of the maximum demand plus 5 P per unit. The power factor of the motor B is to be raised to 0.89 by installing condensers. The motor costs Rs. 150 less than motor A.

The cost of condenser is Rs. 60 per kVAR. Determine which motor is more economical and by how much. Assume rate of interest and depreciation as 10% and working hours of motors as 2400 hours in a year. **(A.M.I.E.)**

Solution. Motor A :

Output of motor *A* $= 50 \times 0.7355 = 36.775 \text{ kW}$

Input of the motor *A* $= \dfrac{36.775}{0.88} = 41.8 \text{ kW}$

Motor input in kVA $= \dfrac{41.8}{0.9} = 46.44 \text{ kVA}$

Energy consumed per annum = Motor input (kW) × working hours in a year

$= 41.8 \times 2400 = 100320 \text{ kWh}$

Annual energy charges $= \text{Rs. } 70 \times 46.44 + \dfrac{5}{100} \times 100320 = \text{Rs. } 8267$

Fixed annual charges of motor *A* in excess over that of motor *B*

$= \text{Rs. } 150 \times \dfrac{10}{100} = \text{Rs. } 15.00$

Total annual charges $= \text{Rs. } 8267 + \text{Rs. } 15.00 = \text{Rs. } 8282$

Motor B :

Motor input $= \dfrac{36.775}{0.9} = 40.86 \text{ kW}$

Motor input with condensers $= \dfrac{40.86}{0.89} = 45.91 \text{ kVA}$

Energy consumed per annum $= 40.86 \times 2400 = 98064$ kWh

Annual energy charges $= \text{Rs. } 70 \times 45.91 + \text{Rs. } 98064 \times \frac{5}{100} = \text{Rs. } 8117$

Leading *kVAR* supplied by the condensers

$$= 40.86 \; [\tan (\cos^{-1} 0.81) - \tan (\cos^{-1} 0.89)]$$
$$= 40.86(0.724 - 0.512) = 8.66 \text{ kVAR}$$

Capital cost of condensers $= \text{Rs. } 60 \times 8.66 = \text{Rs. } 520$

Annual interest and depreciation charges

$$= \text{Rs. } 520 \times \frac{10}{100} = \text{Rs. } 52$$

Total annual charges $= \text{Rs. } 8117 + \text{Rs. } 52 = \text{Rs. } 8169$

Motor B will be economical and the annual saving

$$= \text{Rs. } 8282 - \text{Rs. } 8169 = \textbf{Rs. 113.} \quad \textbf{(Ans.)}$$

Example 9.61. *It is required to choose a transformer to supply a load which varies over 24 hours period in the following manner :*

500 kVA for 4 hours ; 1000 kVA for 6 hours 1500 kVA for 12 hours and 2000 kVA for the rest of the period.

Two transformers each rated at 1500 kVA have been quoted. Transformer-1 has iron loss of 2.7 kW and full-load copper loss of 8.1 kW, while transformer-2 has an iron loss and full-load copper loss of 5.4 kW each.

(i) Calculate the annual cost of supplying losses for each transformer, if electrical energy costs 10 P/kWh.

(ii) Determine which transformer should be chosen if the capital cost of the transformer-1 is Rs. 1000 more than that of transformer-2 and annual charges of interest and depreciation are 10 per cent.

(iii) What difference in capital cost will reverse the decision made in (ii) above ? **(A.M.I.E.)**

Solution. (*i*) **Annual cost of supplying losses for each transformer :**

Transformer-1 :

Energy losses per day due to iron loss $= 2.7 \times 24 = 64.8$ kWh

Energy losses per day due to copper loss

$$= \left(\frac{500}{1500}\right)^2 \times 8.1 \times 4 + \left(\frac{1000}{1500}\right)^2 \times 8.1 \times 6 + \left(\frac{1500}{1500}\right)^2 \times 8.1 \times 12 + \left(\frac{2000}{1500}\right)^2 \times 8.1 \times (24 - 4 - 6 - 12)$$

$$= 3.6 + 21.6 + 97.2 + 28.8 = 151.2 \text{ kWh}$$

Total energy losses per annum $= (64.8 + 151.2) \times 365 = 78840$ kWh

Annual cost of supplying losses $= \text{Rs. } 78840 \times \frac{10}{100} = \textbf{Rs. 7884.}$ **(Ans.)**

Transformer-2 :

Energy loss per day due to iron loss $= 24 \times 5.4 = 129.6$ kWh

Energy loss per day due to copper loss

$$= \left(\frac{500}{1500}\right)^2 \times 5.4 \times 4 + \left(\frac{1000}{1500}\right)^2 \times 5.4 \times 6 + \left(\frac{1500}{1500}\right)^2 \times 5.4 \times 12 + \left(\frac{2000}{1500}\right)^2 \times 5.4 \times 2$$

$= 2.4 + 14.4 + 64.8 + 19.2 = 100.8$ kWh

Total energy loss per annum $= (129.6 + 100.8) \times 365 = 84096$ kWh

Annual cost of supplying losses $=$ Rs. $84096 \times \dfrac{10}{100} =$ **Rs. 8409.6. (Ans.)**

(*ii*) **Choice of the transformer :**

The capital cost of transformer-1 = Rs. 1000 more than that of transformer-2

Extra annual charges in case if transformer-1 is chosen $=$ Rs. $1000 \times \dfrac{10}{100} =$ Rs. 100.00

Annual cost in energy cost due to losses = Rs. (8409.6 – 7884) = Rs. 525.6

Since annual saving in energy cost of supplying losses is much more than extra annual charges due to extra cost of transformer-1, therefore transformer-1 will be chosen.

(*iii*) **Difference in capital cost :**

Decision taken in (*ii*) above will be reversed *i.e.*, transformer-2 will be chosen in case annual charges due to extra capital cost of transformer-1 exceeds annual saving in energy cost due to losses,

i.e., $x \times \dfrac{10}{100} >$ Rs. 525.6 where x is the capital cost of transformer-1 in excess

or, $x >$ Rs. $525.6 \times \dfrac{100}{10}$ over that of transformer-2

or, x *exceeds* **Rs. 5256. (Ans.)**

ADDITIONAL/TYPICAL EXAMPLES

Example 9.62. *A region has a maximum demand of 500 MW at a load factor of 50 percent. The load duration curve can be assumed to be a triangle. The utility has to meet this load by setting up a generating system which is partly hydro and partly thermal. The costs are as under :*

Hydroplant : Rs. 600 per kW per annum and operating cost of 3 P per kWh

Thermal plant : Rs. 300 per kW per annum and operating cost of 13 P per kWh.

Determine :

(*i*) *The capacity of hydroplant.*

(*ii*) *Energy generated annually by each station.*

(*iii*) *Overall generation cost per kWh.* **(A.M.I.E.)**

Solution. Maximum demand of the region = 500 MW = 500000 kW

Load factor = 50% or 0.5

Total energy generated per annum

$= \text{Maximum demand} \times \text{load factor} \times (365 \times 24)$

$= 500000 \times 0.5 \times (365 \times 24) = 219 \times 10^7$ kWh

The load duration curve is shown in Fig. 9.39.

Hydroplant, being cheaper in operating cost, would be used as base load power station.

(*i*) **The capacity of hydro-plant P :**

Let the capacity of hydroplant be P kW and the energy generated by it be E kWh/annum.

Then, capacity of thermal plant $= (500000 - P)$ kW

Energy generated by thermal plant $= (219 \times 10^7 - E)$ kWh

Annual generation cost of hydro-plant

= Rs. $(600\,P + 0.03\,E)$

Annual generation cost of thermal plant

= Rs. $[300(500000 - P) + 0.13(219 \times 10^7 - E)]$

$\therefore$ Total cost, C

$= 600P + 0.03E + 300\,(500000 - P) + 0.13(219 \times 10^7 - E)$

$= 600P + 0.03E + 15 \times 10^7 - 300P + 28.47 \times 10^7 - 0.13E$

$= 300P + 43.47 \times 10^7 - 0.1E$ ...(i)

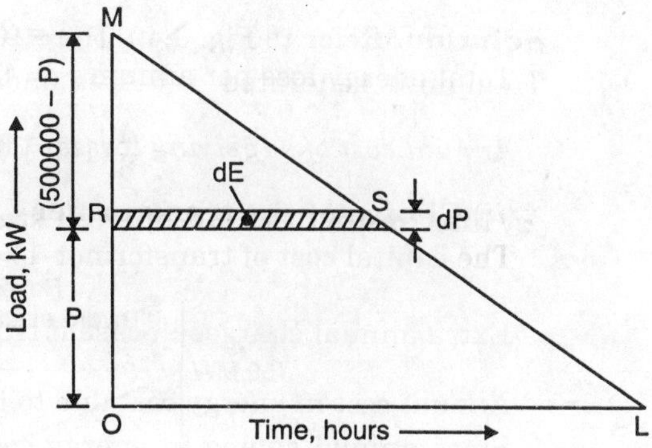

Fig. 9.39

For minimum cost, $\dfrac{dC}{dP} = 0$

$$300 - 0.1\,\frac{dE}{dP} = 0, \quad \text{or,} \quad dE = 3000\,dP$$

From Fig. 9.39, $dE = dP \times t$

$\therefore$ $t = 3000$ hours

From Δs MRS and MOL, we get

$$\frac{500000 - P}{500000} = \frac{3000}{8760}$$

or, **P = 328767 kW say 330000 kW or 330 MW. (Ans.)**

(ii) **Energy generated annually by each station :**

Capacity of thermal plant = 500 – 330 = 170 MW

Energy generated by thermal plant $= \dfrac{170 \times 10^3 \times 3000}{2} = \mathbf{255 \times 10^6}$ **kWh. (Ans.)**

Energy generated by hydroplant $= 219 \times 10^7 - 255 \times 10^6 = \mathbf{1935 \times 10^6}$ **kWh. (Ans.)**

(iii) **Overall generation cost per kWh :**

Total annual generation cost,

C = Rs. $(300 \times 330000 + 43.47 \times 10^7 - 0.1 \times 1935 \times 10^6)$ ...[From eqn. (i)]

$= 340.2 \times 10^6$

$\therefore$ Overall generation cost per kWh = Rs. $\dfrac{340.2 \times 10^6}{219 \times 10^7}$

= Rs. 0.1553 = **15.53 P/kWh. (Ans.)**

Example 9.63. *The load duration curve of a system is as shown in Fig. 9.40. The system is supplied by three stations ; steam station, a run-of-river station and a reservoir hydro-electric station. The ratios of number of units supplied by three stations are as below :*

Steam	:	*Run-of-river*	:	*Reservoir*
7	:	*4*	:	*1*

The run-of-river station is capable of generating power continuously and works as a peak load station.

Calculate :

(i) Maximum demand on each station.

(ii) Load factor of each station. **(Ranchi University)**

Solution. Refer to Fig. 9.40. Here 100 percent time will be taken as 8760 hours.

Total units generated = Area under the curve

$$= \frac{1}{2}(160 + 80) \times 10^3 \times 8760 = 1051.2 \times 10^6 \text{ kWh}$$

(*i*) **Maximum demand of each station :**

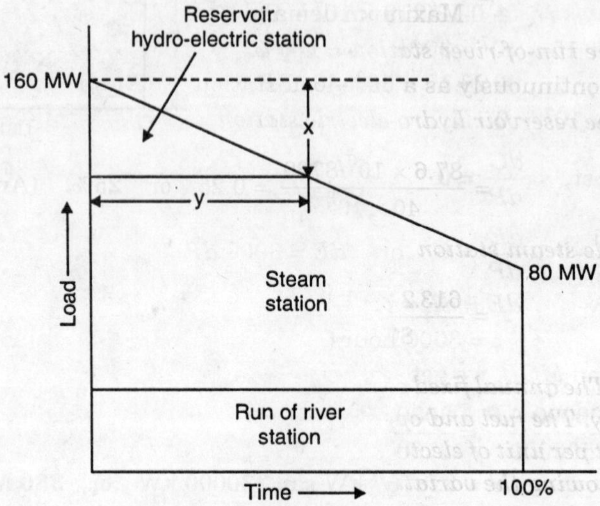

Fig. 9.40

From the given ratio, the number of units supplied by each station can be calculated.

Units generated :

Steam station $= \frac{7}{12} \times (1051.2 \times 10^6) = 613.2 \times 10^6$ kWh

Run-of-river station $= \frac{4}{12} \times (1051.2 \times 10^6) = 350.4 \times 10^6$ kWh

Reservoir hydro-electric station$= \frac{1}{12} \times (1051.2 \times 10^6) = 87.6 \times 10^6$ kWh

Maximum demand of run-of-river station $= \frac{350.4 \times 10^6}{8760} = 40000$ kW $=$ **40 MW. (Ans.)**

Let x MW be the *maximum demand of the reservoir plant*. As shown in Fig. 9.40 let it operate for y hours.

Then, $y = \frac{x}{80} \times 8760$

Area under the curve of reservoir hydro-electric station

$$= \frac{1}{2} \times y \times x = \frac{1}{2} \times \left(\frac{x}{80} \times 8760\right) \times x \times 10^3 = 54750x^2 \text{ kWh}$$

$\therefore \quad 54750x^2 = 87.6 \times 10^6$

or, $\quad$ $\boldsymbol{x = 40}$ **MW. (Ans.)**

∴ *Maximum demand of steam power station*

$= 160 - (40 + 40) =$ **80 MW.** **(Ans.)**

(*ii*) **Load factor of each station :**

$$\text{Load factor} = \frac{\text{Average load}}{\text{Maximum demand}}$$

Load factor of the run-of-river station = **100%.**

(since it works continuously as a base load station). **(Ans.)**

Load factor of the reservoir hydro-electric station

$$= \frac{87.6 \times 10^6/8760}{40 \times 10^3} = 0.25 \quad \text{or} \quad \mathbf{25\%.} \quad \textbf{(Ans.)}$$

Load factor of the steam station

$$= \frac{613.2 \times 10^6/8760}{80 \times 10^3} = 0.875 \quad \text{or} \quad \mathbf{87.5\%.} \quad \textbf{(Ans.)}$$

Example 9.64. *The annual fixed costs of a certain thermal power station amount to Rs. 400 per kW of installed capacity. The fuel and operating costs are estimated at 5.0 paise per kWh generated. Determine the total cost per unit of electrical energy generated for station load factors of 1.0, 0.8, 0.6, 0.4. Draw the graph showing the variation of cost / kWh versus the load factor.* **(A.M.I.E.)**

Solution. Let us *consider 1 kW installed capacity.*

Maximum demand on the power station = Installed capacity of 1 kW

(assuming utilisation factor to be unity)

Units generated per annum :

For load factor of **unity** $= 1 \times 8760 \times 1.0 = 8760$ kWh

For load factor of **0.8** $= 1 \times 8760 \times 0.8 = 7008$ kWh

For load factor of **0.6** $= 1 \times 8760 \times 0.6 = 5256$ kWh

For load factor of **0.4** $= 1 \times 8760 \times 0.4 = 3504$ kWh

Total cost per unit of electrical energy generated :

For load factor of **unity** = Fixed cost per kWh + fuel and operating cost per kWh

$$= \frac{400 \times 100}{8760} + 5 = \mathbf{9.566 \text{ paise per unit.}} \quad \textbf{(Ans.)}$$

For load factor of **0.8** $= \dfrac{400 \times 100}{7008} + 5$

= **10.7 paise per unit.** **(Ans.)**

For load factor of **0.6** $= \dfrac{400 \times 100}{5265} + 5$

= **12.61 paise per unit.** **(Ans.)**

For load factor of **0.4** $= \dfrac{400 \times 100}{3504} + 5$

= **16.41 paise per unit.** **(Ans.)**

Graph showing the variation of cost per kWh versus load factor is shown in Fig. 9.41.

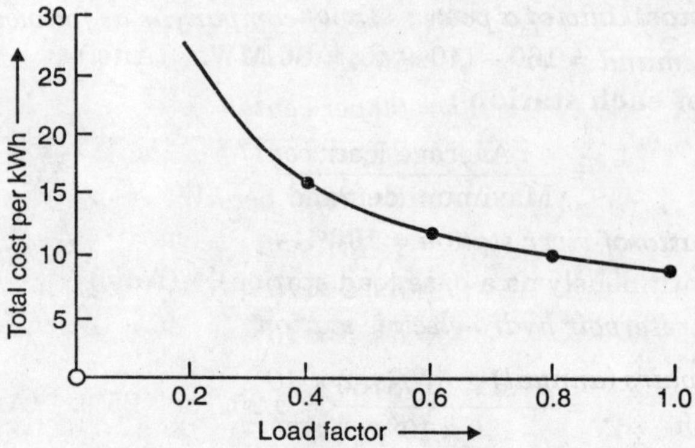

Fig. 9.41

Example 9.65. *The following data refer to a public undertaking which supplies electric energy to its consumers at a fixed tariff of 11.37 P per unit.*

Total installed capacity	*= 344 MVA*
Total capital investment	*= Rs. 22.4 crores*
Annual recurring expenses	*= Rs. 9.4 crores*
Interest charge	*= 6 per cent*
Depreciation charge	*= 5 per cent*

Calculate the annual load factor at which the system should operate so that there is neither profit nor loss to the undertaking. Assume distribution losses at 7.84 percent and the average system power factor at 0.86. **(Bombay University)**

Solution. Annual load factor $(L) = \dfrac{\text{No. of kWh supplied in a year}}{\text{Maximum no. of kWh which can be supplied}}$

$$= \frac{\text{kWh supplied per year}}{\text{Maximum output in kW} \times 8760}$$

$\therefore$ kWh supplied per year $= \text{Maximum output in kW} \times 8760 \times L$

$$= (344 \times 10^3 \times 0.86) \times 8760 \times L = 25.92 \times 10^8\ L$$

(In this case maximum demand has been taken as equal to the installed capacity)

Considering distribution losses of 7.84%, the units actually supplied

$$= (100 - 7.84) \text{ percent of } (25.92 \times 10^8 L) = 23.89 \times 10^8 L$$

Amount collected at the rate of 11.37 P/kWh

$$= \text{Rs. } 23.89 \times 10^8 L \times \frac{11.37}{100} = \text{Rs. } 271.63 \times 10^6 L$$

If there is to be no profit or gain, then this amount must just equal the fixed and running charges.

Annual interest and depreciation on capital investment

$$= 11\%(6\% + 5\%) \text{ of Rs. } 22.4 \times 10^7 = \text{Rs. } 2.46 \times 10^7$$

Total annual expenses $= \text{Rs. } 2.46 \times 10^7 + \text{Rs. } 9.4 \times 10^7 = \text{Rs. } 11.86 \times 10^7$

$\therefore$ $271.63 \times 10^6 L = 11.86 \times 10^7$

or $$\mathbf{L} = \frac{11.86 \times 10^7}{271.63 \times 10^6} = 0.437 \text{ or } \mathbf{43.7\%}. \quad \textbf{(Ans.)}$$

Example 9.66. *The cost data of a power supply company is as follows :*

Station maximum demand	= *50 MW*
Station load factor	= *60 per cent*
Reserve capacity	= *20 per cent*
Capital cost	= *Rs. 2000 per kW*
Interest and depreciation	= *12 per cent*
Salaries (annual)	= *Rs.* 5×10^5
Full cost (annual)	= *Rs.* 5×10^6
Maintenance and repairs (annual)	= *Rs.* 2×10^5
Losses in distribution	= *8 per cent*
Load diversity factor	= *1.7.*

Calculate the average cost per unit and two-part tariff, assuming 80 percent of salaries and repair and maintenance cost to be fixed. **(Bombay University)**

Solution. Station capacity $= 50 + 20\%$ of $50 = 60$ MW

Average power $= 60 \times 0.6 = 36$ MW $= 36 \times 10^3$ kW

Capital investment $=$ Rs. $(60 \times 10^3) \times 2000 =$ Rs. 12×10^7

Interest + depreciation $=$ Rs. $\dfrac{12}{100} \times 12 \times 10^7 =$ Rs. 14.4×10^6

Total cost both fixed and running

$=$ Rs. $14.4 \times 10^6 +$ Rs. $5 \times 10^5 +$ Rs. 5×10^6
$+$ Rs. $2 \times 10^5 =$ Rs. 201×10^5

No. of units generated annually $= 8760 \times 36 \times 10^3 = 315.36 \times 10^6$ kWh

$\therefore$ **Overall cost per unit** $=$ Rs. $\dfrac{201 \times 10^5}{315.6 \times 10^6} =$ Rs. 0.0637 or **6.37 paise. (Ans.)**

Fixed charges :

Annual interest and depreciation $=$ Rs. 14.4×10^6

80 percent of salaries $=$ Rs. $\dfrac{80}{100} \times 5 \times 10^5 =$ Rs. 4×10^5

80 percent of repair and maintenance $=$ Rs. $\dfrac{80}{100} \times 2 \times 10^5 =$ Rs. 1.6×10^5

Total fixed charges $=$ Rs. $(14.4 \times 10^6 + 4 \times 10^5 + 1.6 \times 10^5)$
$=$ Rs. 149.6×10^5

Aggregate maximum demand of all consumers

$=$ Maximum demand on generating station $\times$ diversity factor
$= 60 \times 10^3 \times 1.7 = 102 \times 10^3$ kW

$\therefore$ Annual cost/kW of maximum demand $= \dfrac{149.6 \times 10^5}{102 \times 10^3} =$ Rs. 146.7

Running charges :

Fuel cost $=$ Rs. 50×10^5

20% of salaries $= \dfrac{20}{100} \times 5 \times 10^5 =$ Rs. 1×10^5

20% of maintenance $= \frac{20}{100} \times 2 \times 10^5 = \text{Rs. } 0.4 \times 10^5$

Total running charges $= \text{Rs. } (50 \times 10^5 + 1 \times 10^5 + 0.4 \times 10^5) = \text{Rs. } 51.4 \times 10^5$

Considering distribution loss of 8%, cost per unit delivered to the consumer

$$= \text{Rs. } \frac{51.4 \times 10^5}{0.92 \times (8760 \times 36 \times 10^3)} = \text{Rs. } 0.0177 \quad \text{or} \quad 1.77 \text{ P}$$

Hence **two-part tariff** = **Rs. 146.7 per kW maximum demand**

+ 1.77 paise per kWh consumed. (Ans.)

Example 9.67. *The following data relate to a hydrostation : Maximum demand on the hydrostation = 160 MW at a load factor of 70 percent ; Diversity factor between consumers' maximum demand = 1.7 ; Installed capacity = 200 MW ; Capital cost of plant = Rs. 30000 per kW ; Capital cost of transmission and distribution system = Rs. 1800 × 10⁶ ; Interest, depreciation, insurance and taxes on capital investment = 11 percent ; Fixed managerial and general maintenance costs = Rs. 30 × 10⁶ per year ; Operating labour, maintenance and supplies = Rs. 236 × 10⁶ per year ; Cost of metering, billing and collection. = Rs. 90 × 10⁶ per year ; Energy consumed by auxiliaries = 5 per cent ; Energy losses in transmission and distribution = 15 per cent.*

The utility must make a profit of 25 per cent which should be proportionately charged to the fixed and operating costs.

(i) Devise a two-part tariff for the consumers fed from this utility.

(ii) The utility persuades some other consumers to use electricity during off-peak hours. This raises the load factor to 80 per cent. The operating costs for the additional energy increase in the same proportion as the increase in energy. The utility must make a profit of 50 per cent on the sale of this additional energy. Find the electricity tariff in paise per kWh for the consumers for this additional energy.

Solution. (*i*) **Two-part tariff :**

Capital cost of the plant $= \text{Rs. } 30000 \times (200 \times 10^3) = \text{Rs. } 6000 \times 10^6$

Capital cost of transmission and distribution system $= \text{Rs. } 1800 \times 10^6$

Total capital cost $= \text{Rs. } 7800 \times 10^6$

Interest, depreciation, insurances, taxes etc.

$$= \text{Rs. } 7800 \times 10^6 \times \frac{11}{100} = \text{Rs. } 858 \times 10^6$$

Sum of maximum demands of consumers

$$= 160 \times 1.7 = 272 \text{ MW}$$

Energy produced $= 160 \times 10^3 \times 8760 \times 0.7 = 98112 \times 10^4 \text{ kWh}$

Energy consumed by auxiliaries $= \frac{5}{100} \times 98112 \times 10^4 = 49056 \times 10^3 \text{ kWh}$

Energy output of plant $= 98112 \times 10^4 - 49056 \times 10^3 = 93206 \times 10^4 \text{ kWh}$

Energy sold to consumers $= 93206 \times 10^4 \times 0.85 = 792.25 \times 10^6 \text{ kWh}$

The costs can be separated into *fixed and operating costs* as under :

Fixed costs :

Interest, depreciation etc. $= \text{Rs. } 858 \times 10^6$ per year

Managerial and maintenance $= \text{Rs. } 30 \times 10^6$ per year

Total $= \text{Rs. } 888 \times 10^6$ per year

Profit @ 25% $= \text{Rs. } 222 \times 10^6$ per year

Grand total $= \text{Rs. } 1110 \times 10^6$ per year

Operating cost :

Operating labour, supplies, maintenance etc. $= \text{Rs. } 236 \times 10^6$ per year

Metering, billing etc. $= \text{Rs. } 90 \times 10^6$ per year

Total $= \text{Rs. } 326 \times 10^6$ per year

Profit @ 25% $= \text{Rs. } 81.50 \times 10^6$ per year

Grand total $= \text{Rs. } 407.50 \times 10^6$ per year

Tariff :

Cost per kW $= \text{Rs. } \dfrac{1110 \times 10^6}{272 \times 10^3} = \text{Rs. } 4080.88$

Cost per kWh $= \dfrac{407.50 \times 10^6}{792.25 \times 10^6} \times 100 = 51.4$ P/kWh

Hence *two-part tariff* **= Rs. 408.88 per kW per year + 51.4 paise/kWh. (Ans.)**

(*ii*) **Tariff in P/kWh :**

Energy produced $= (160 \times 10^3) \times 8760 \times 0.8 = 112128 \times 10^4$ kWh/year

Energy consumed by auxiliaries $= \dfrac{5}{100} \times 112128 \times 10^4$

$= 56064 \times 10^3$ kWh

Energy output of the plant $= 112128 \times 10^4 - 56064 \times 10^3 = 106521.6 \times 10^4$ kWh

Energy sold to customers $= 106521.6 \times 10^4 \times 0.85 = 90543.36 \times 10^4$ kWh

Additional energy produced $= 112128 \times 10^4 - 98112 \times 10^4 = 14016 \times 10^4$ kWh

Additional energy sold to consumers $= 90543.36 \times 10^4 - 79225 \times 10^4 = 11318.36 \times 10^4$ kWh

Additional operating cost $= 326 \times 10^3 \times \dfrac{14016 \times 10^4}{98112 \times 10^4} = \text{Rs. } 46.57 \times 10^6$ per year

Profit @ 50% $= \text{Rs. } 23.285 \times 10^6$ per year

Total cost $= \text{Rs. } (46.57 + 23.285) \times 10^6 = \text{Rs. } 69.855 \times 10^6$

Cost per kWh $= \text{Rs. } \dfrac{69.855 \times 10^6}{11318.36 \times 10^4}$

$= \text{Rs. } 0.617$ or **61.7 paise/kWh. (Ans.)**

HIGHLIGHTS

1. Types of loads : (*i*) Residential load (*ii*) Commercial load (*iii*) Industrial load (*iv*) Municipal load (*v*) Irrigation load and (*vi*) Traction load.
2. A *load curve* is a graphic record showing the power demands for every instant during a certain time interval. The area under the load curve represents the energy generated in the period considered.
3. The cost of a power system includes the following :

   A. *Capital cost or fixed cost :*

   (*i*) Initial cost (*ii*) Interest

   (*iii*) Depreciation cost (*iv*) Taxes

   (*v*) Insurance

B. *Operational cost :*

(*i*) Fuel cost (*ii*) Operating labour cost

(*iii*) Maintenance cost (*iv*) Supplies

(*v*) Supervision (*vi*) Operating taxes.

4. The following methods are used to calculate the depreciation cost :

(*i*) Straight line method. (*ii*) Percentage method.

(*iii*) Sinking fund method. (*iv*) Unit method.

5. The economics of power plant is greatly influenced by :

(*i*) Load factor (*ii*) Demand factor

(*iii*) Utilisation factor.

6. The performance of a plant can be precisely represented by the *input-output* curve from the tests conducted on individual power plant. In general input-output may be represented as follows :

$$I = a + bL + cL^2 + dL^3$$

where, I = input (in millions of kcal/h or kJ/h in case of thermal plants and m^3/s of water in case of hydro-plants)

L = output (in MW or kW)

a, b, c, d = constants.

7. For minimum combined input to carry a given combined output, the slopes of the input-output curves for each unit must be equal. If there are n units supplying a constant load, then the required condition for the minimum input or maximum system efficiency is

$$\frac{dI_1}{dL_1} = \frac{dI_2}{dL_2} = \frac{dI_3}{dL_3} = \ldots\ldots \frac{dI_n}{dL_n}.$$

8. The general tariff form is given by :

$$z = a.x + b.y + c$$

where, z = Total amount of bill for the period considered,

x = Maximum demand in kW,

y = Energy consumed in kWh during the period considered,

a = Rate per kW of maximum demand,

b = Energy rate per kWh, and

c = Constant amount charged to the consumer during each billing period. This charge is independent of demand or total energy because a consumer that remains connected to the line incurs expenses even if he does not use energy.

9. Various types of *tariff* are :

(*i*) Flat demand rate (*ii*) Straight meter rate

(*iii*) Block meter rate (*iv*) Hopkinson demand rate (Two-part tariff)

(*v*) Doherty rate (Three-part tariff) (*vi*) Wright demand rate.

10. Methods of power factor improvement :

(*i*) Use of static capacitors

(*ii*) Use of synchronous condensers

(*iii*) Use of phase advancers

(*iv*) Use of phase compensated motors.

OBJECTIVE TYPE QUESTIONS

Fill in the Blanks or Say "Yes" or "No" :

1. The main purpose of design and fabrication of the plant is to bring the cost of energy to ....

2. Diversity factor is always ...... than unity.

3. The utilisation factor is defined as the ratio of maximum generator demand to the generator capacity.
4. The total of station capacities available to supply the system-load is called the ...... capacity.
5. A load curve is a graphic record showing the power demands for every instant during a certain time interval.
6. In ...... method some factor is taken as standard one and depreciation is measured by that standard.
7. In case of a hydro-plant, the smaller the quantity of water stored lower is the cost per kW.
8. ...... are the different methods of charging the consumers for the consumption of electricity.
9. Tariff should be easier to understand.
10. Spot pricing is also known as "Response pricing".
11. The industrial concerns should preferably purchase their power from public supply company rather than generating their own.
12. The power factor depends on reactive power component.
13. An over-excited synchronous motor running on no-load is called ....
14. Static capacitors are used in smaller units.
15. The maximum value to which the power factor can be economically raised entirely depends upon the relative costs of the generating plant and phase advancing plant.

Answers

1. Minimum	2. greater	3. Yes	4. installed	5. Yes
6. unit	7. No	8. Tariffs	9. Yes	10. Yes
11. Yes	12. Yes	13. Synchronous condenser		14. Yes
15. Yes.				

THEORETICAL QUESTIONS

1. Define the following terms :
   (*i*) Connected load (*ii*) Demand
   (*iii*) Demand factor (*iv*) Load factor
   (*v*) Diversity factor (*vi*) Utilisation factor.
2. Explain briefly the following :
   (*i*) Load curve (*ii*) Load duration curve.
3. What is the significance of load curves ?
4. Enumerate various types of loads.
5. List the factors which should be considered while designing a power plant.
6. What are the considerations on which the location of a power plant depends ?
7. List the points which should be taken care of while deciding about power plant building and its layout.
8. List the various costs which go to form the total cost of a power system.
9. Explain briefly the following :
   (*i*) Capital or fixed cost (*ii*) Operational cost.
10. What do you mean by depreciation ?
11. Enumerate and explain briefly various methods used to calculate the depreciation cost.
12. Name the elements that make up the operating expenditure of a power plant.
13. What points should be considered while choosing the type of generation ?
14. Discuss the economic loading of combined steam and hydro-plants.
15. How can the power generation cost be reduced ?
16. What do you understand by the term tariff ?

17. What are the objectives and requirements of tariff ?

18. Enumerate various types of tariff and explain any two of them.

19. Explain briefly the following tariff :

(*i*) Straight meter rate (*ii*) Block meter rate
(*iii*) Doherty rate (three-part tariff).

20. Describe the straight line and sinking fund methods of determining the depreciation of an electric power plant.

21. Explain with typical examples the following types of tariffs used in electrical industry :

(*i*) Flat demand tariff (*ii*) Straight line rate tariff
(*iii*) Step rate tariff (*iv*) Block rate tariff
(*v*) Two-part tariff (*vi*) Three-part tariff.

22. What are the causes of poor power factor ? Give disadvantages of poor power factor.

23. What are the advantages of improved power factor ?

24. What do you understand by power factor ? Explain the necessity of improving power factor.

25. What are the causes of low power factor and its effect on the system operation ? Obtain an expression for the economical power factor to which a system can by raised by using power improving equipment.

26. Deduce an expression for the most economical angle of lag at which a consumer should take power when the charge is based on a fixed sum per kVA of maximum demand plus a flat rate per kWh.

27. What factors determine the economical limit of power factor correction ? Show that the economical limit to which the power factor of a lagging power factor load can be raised is independent of the original value of the power factor if the tariff consists of a fixed charge per kVA of maximum demand plus a flat rate per kWh.

28. Explain how the installation of a power factor improvement device can augment the kW capacity of a generating station of a given kVA capacity.

Derive the conditions under which the installation of a p.f. improvement plant for increasing the kW capacity of a generating station is preferable to installing additional generating equipment.

29. Write short notes on the following :

(*i*) Power factor improvement. (*ii*) Two-part tariff.
(*iii*) Causes of poor power factor. (*iv*) Location of p.f. correction equipment.
(*v*) Economic limits of p.f. improvement.

UNSOLVED EXAMPLES

1. The maximum demand of a power station is 80000 kW and the load curve is defined as follows :

Time (hours)	:	0–6	6–8	8–12	12–14	14–18	18–22	22–24
Load (MW)	:	40	50	60	50	70	80	40

(*i*) Determine the load factor of power station.

(*ii*) What is the load factor of standby equipment rated at 25 MW that takes up all load in excess of 60 MW ? Also calculate its use factor. **[Ans.** (*i*) 0.71 (*ii*) 0.75, 0.6]

2. The following load is to be supplied by a power station :

Load (MW)	:	30	90	60	100	50
Time (hours)	:	0–6	6–12	12–14	14–18	18–24

(*i*) Draw the load curve.

(*ii*) Draw the load duration curve.

(*iii*) Choose suitable generating units to supply the load.

(*iv*) Calculate the load factor.

(*v*) Calculate plant capacity factor.

[Ans. (*iii*) 30 MW (4 units including standby unit), 10 MW (one unit) (*iv*) 0.64 (*v*) 0.49]

3. The yearly duration curve of a certain plant can be considered as a straight line from 150 MW to 40 MW. Power is supplied with one generating unit of 100 MW capacity and two units of 50 MW capacity each. Determine :
(*i*) Installed capacity (*ii*) Load factor
(*iii*) Plant factor (*iv*) Maximum demand
(*v*) Utilization factor. [**Ans.** (*i*) 200 MW (*ii*) 0.633 (*iii*) 0.475 (*iv*) 150 MW (*v*) 0.75]

4. A generating station has a maximum demand of 20 MW, a load factor of 0.6, a plant capacity of 0.48 and a plant use factor of 0.80. Find :
(*i*) The daily energy produced.
(*ii*) The reserve capacity of the plant.
(*iii*) The maximum energy that could be produced if the plant were running all the time.
(*iv*) The maximum energy that could be produced daily, if the plant when running, according to operating schedule, were fully loaded.
[**Ans.** (*i*) 2.88×10^5 kWh (*ii*) 5000 kW (*iii*) 2.88×10^5 kWh (*iv*) 3.60×10^5 kWh]

5. A proposed power station has to supply load as follows :

Time (hours) :	01–08	08–12	12–17	17–20	20–23	23–01
Load (MW) :	10	20	25	18	35	20

After drawing the load curve, find out the load factor. Also choose suitable generating units to supply this load, maintaining reliability of supply. Prepare operation schedule for the machine and calculate plant use factor. [**Ans.** 0.56, 0.92]

6. A generating station supplies the following loads :
15 MW ; 12 MW ; 8 MW and 0.5 MW. The station has a maximum demand of 20 MW and the annual load factor is 0.5. Find :
(*i*) Number of units supplied annually. (*ii*) Diversity factor.
[**Ans.** (*i*) 876×10^5 kWh (*ii*) 1.775]

7. A base load power station and standby power station share a common load as follows :
Base load station annual output = 150×10^6 kWh ; Base load station capacity = 35 MW ; Maximum demand on base load station = 30 MW ; Standby station capacity = 18 MW ; Standby station annual output = 140×10^6 kWh ; Maximum demand (peak load) on standby station = 15 MW. Determine the following for both power stations :
(*i*) Load factor (*ii*) Capacity factor (plant factor).
[**Ans.** Base load station : (*i*) 0.57 (*ii*) 0.49
Standby power station : (*i*) 0.107 (*ii*) 0.09]

8. A power system has the following load particulars :

	Maximum demand	*Load factor*	*Diversity between consumers*
1. Residential load :	1000 kW	0.2	1.3
2. Commercial load :	2000 kW	0.3	1.1
3. Industrial load :	5000 kW	0.8	1.2

Overall diversity factor may be taken as 1.4.
Determine the following :
(*i*) Maximum demand on system.
(*ii*) Daily energy consumption (total).
(*iii*) Overall load factor.
(*iv*) Connected load (total) assuming that demand factor for each load is unity.

9. The following data is available for a steam power station :
Maximum demand = 25000 kW ; Load factor = 0.4 ; Coal consumption = 0.86 kg/kWh ; Boiler efficiency = 85% ; Turbine efficiency = 90% ; Price of coal = Rs. 55 per tonne.
Determine the following :
(*i*) Thermal efficiency of the station.
(*ii*) Coal bill of the plant for one year. [**Ans.** (*i*) 76.5% (*ii*) Rs. 41,43,480]

10. The daily load curve of a power plant is given by the table below :

Time :	12	2	4	6	8	10	12	2	4	6	8	10	12
Load (MW) :	2	2.5	3	4	6	6.5	6.5	5	6	8	9	5	2

(*i*) Find the daily load factor.

(*ii*) All loads in excess of 400 kW are carried out by unit No. 2 rated at 600 kW. Find its use factor.

[**Ans.** (*i*) 0.814 (*ii*) 0.417]

11. The annual peak load on a 30 MW power station is 25 MW. The power station supplies load having maximum demands of 10 MW, 8.5 MW, 5 MW and 4.5 MW. The annual load factor is 0.45. Find :

(*i*) Average load (*ii*) Energy supplied per year

(*iii* Diversity factor (*iv*) Demand factor.

[**Ans.** (*i*) 11.25 MW (*ii*) 98.55 × 10^6 kWh (*iii*) 1.12 (*iv*) 0.9]

12. A generating station supplies the following loads :

15 MW, 12 MW, 8.5 MW, 6 MW and 0.45 MW. The station has a maximum demand of 22 MW. The annual load factor of the station is 0.48. Calculate :

(*i*) The number of units supplied annually

(*ii*) The diversity factor. (*iii*) The demand factor.

[**Ans.** (*i*) 92.5 × 10^6 kWh (*ii*) 1.907 (*iii*) 0.525]

13. A power station has a maximum demand of 15 MW, a load factor of 0.7, a plant capacity factor of 0.525 and a plant use factor of 0.85. Find :

(*i*) The daily energy produced.

(*ii*) The reserve capacity of the plant.

(*iii*) The maximum energy that could be produced daily if the plant operating schedule is fully loaded when in operation. [**Ans.** (*i*) 252,000 kWh (*ii*) 5,000 kW (*iii*) 296,470 kWh]

14. Determine the annual cost of a feed water softner from the following data :

Cost = Rs. 80,000 ; Salvage value = 5%, Life = 10 years ; Annual repair and maintenance cost = Rs. 2500 ; Annual cost of chemicals = Rs. 5000 ; Labour cost per month = Rs. 300 ; Interest on sinking fund = 5%. [**Ans.** Rs. 17,140]

15. Estimate the generating cost per kWh delivered from a generating station from the following data :

Plant capacity	= 50 MW
Annual load factor	= 0.4
Capital cost	= Rs. 1.2 crores
Annual cost of wages, taxation etc:	= Rs. 4 lacs
Cost of fuel, lubrication, maintenance etc.	= 1.0 paise per kWh generated.

Interest 5% per annum, depreciation 5% per annum of initial value. [**Ans.** 1.91 paise/kWh delivered]

16. A 100 MW, steam power station is estimated to cost Rs. 20 crores. The operating expenses are estimated as follows :

Cost of fuel and oil	= Rs. 140 lacs per annum
Transportation and storage	= Rs. 20 lacs per annum
Salaries and wages	= Rs. 20 lacs per annum
Miscellaneous	= Rs. 20 lacs per annum

Reckoning interest and depreciation at 10% of the capital cost, calculate the cost of generation per unit, if the average load factor of the power station is 0.6.

What economics could be affected if the load factor was improved to 0.8, the operating expenses increasing by only 10% thereby. [**Ans.** 6 p/kWh, 21% reduction in cost of generation]

17. A steam station has two 110 MW units. Following cost data are given :

Particulars	*Unit A*	*Unit B*
Capital cost	Rs. 2000 per kW	Rs. 2500 per kW
Fixed charge rate	10 per cent	10 per cent
Capacity factor	0.55	0.6
Fuel consumption	1 kg/kWh	0.9 kg/kWh
Fuel cost	Rs. 80 per 1000 kg	Rs. 80 per 1000 kg
Annual cost of operating, labour, maintenance and supplies	20 per cent of annual cost	15 per cent of annual cost
Utilisation factor	1	1

Calculate the following :

(*i*) Annual plant cost and generation cost of unit A.

(*ii*) Annual plant cost and generation cost of unit B.

(*iii*) Overall generation cost of the station.

[Ans. (*i*) Rs. 7,28,78,080 ; 13.75 p/kWh (*ii*) Rs. 75.371648 × 10^6 ; 13.036 p/kWh (*iii*) 13.378 p/kWh]

18. The annual costs of operating a 15,000 kW thermal power station are as follows :

Cost of plant = Rs. 900 per kW
Interest, insurance, taxes on plant = 5 per cent
Depreciation = 5 per cent
Cost of primary distribution system = Rs. 5,00,000
Interest, insurance, taxes and depreciation on primary distribution system = 5 per cent
Cost of secondary distribution system = Rs. 9,00,000
Interest, taxes, insurance and depreciation on secondary distribution system = 5 per cent
Maintenance of secondary distribution system = Rs. 1,80,000
Plant maintenance cost
(*i*) Fixed cost = Rs. 30,000
(*ii*) Variable cost = Rs. 40,000
Operating costs = Rs. 600,000
Cost of coal = Rs. 60 per tonnes
Consumption of coal = 30,000 tonnes
Dividend to stock-holders = Rs. 10,00,000
Energy loss in transmission = 10 per cent
Maximum demand = 14,000 kW
Diversity factor = 1.5
Load factor = 0.7

Determine : (*i*) Charge per kW per year (*ii*) Rate per kWh. **[Ans.** (*i*) Rs. 116.6 (*ii*) 3.4 p/kWh]

19. It is proposed to supply a load with a maximum demand of 100 MW and a load factor of 0.4. Choice is to be made from steam, hydro and nuclear power plants. Calculate the overall cost per kWh in case of each machine :

Cost	*Steam power plant*	*Hydro power plant*	*Nuclear power plant*
Capital cost per kW installed	Rs. 1800	Rs. 3600	Rs. 5000
Interest	12%	10%	10%
Depreciation	12%	8%	10%
Operating cost per kWh	15 paise	5 paise	10 paise
Transmission and distribution cost/kWh	0.2 paise	0.8 paise	0.2 paise

[Ans. 27.52 p ; 23.57 p ; 38.73 p]

20. A power plant of 150 MW installed capacity has the following data :
Capital cost = Rs. 1800/kW installed ; Interest and depreciation = 12 per cent ;
Annual load factor = 0.6 ; Annual capacity factor = 0.5 ; Annual running charges = Rs. 30×10^6 ; Energy consumed by the power plant auxiliaries = 6 per cent. Calculate :
(*i*) Reserve capacity (*ii*) Generating cost. [**Ans.** (*i*) 25 MW (*ii*) 10.10 paise]

21. Compare the annual cost of supplying a factory load having a maximum demand of 1 MW at a load factor of 50% by energy obtained from
(*a*) Nuclear power plant (*b*) Public supply

Nuclear power plant

Capital cost	= Rs. 50,000
Cost of fuel	= Rs. 600 per 1000 kg
Fuel consumption	= 30 g per kWh generated
Cost of maintenance etc.	= Re. 0.005 per kWh generated
Wages	= Rs. 20,000 per annum
Interest and depreciation	= 10 per cent.

Public supply : Rs. 50 per kW + Re. 0.03 per kWh generated. [**Ans.** Rs. 1,70,740 ; Rs. 1,81,400]

22. A system with a maximum demand of 1,00,000 kW and a load factor of 30% is to be supplied by either (*a*) steam station alone or (*b*) a steam station in conjunction with a water storage scheme, the latter supplying 100 million units with a maximum output of 40,000 kW. The capital cost of steam and storage stations are Rs. 600 per kW and Rs. 1,200 per kW respectively. The corresponding operating costs are 15 paise and 3 paise per kWh respectively. The interest on capital cost is 15% per annum. Calculate the overall generating cost per kWh and state which of the two projects will be economical.
[**Ans.** 18.425 p/kWh, 15.23 p/kWh]

23. A power station has the installed capacity of 120 MW. Calculate the cost of generation, other data pertaining to power station are given below :

Capital cost	= Rs. 200×10^6
Rate of interest and depreciation	= 18 per cent
Annual cost of fuel oil, salaries and taxation	= Rs. 24×10^6
Load factor	= 0.4

Also calculate the saving in cost per kWh if the annual load factor is raised to 0.5.
[**Ans.** 14.25 paise ; 2.84 paise]

24. A 50 MW generating station has the following data :
Capital cost = Rs. 15×10^6 ; Annual taxation = Rs. 0.4×10^6 ; Annual salaries and wages = Rs. 1.2×10^6 ; Cost of coal = Rs. 65 per tonne ; Calorific value of coal = 5500 kcal/kg ; Rate of interest and depreciation = 12 per cent ; plant heat rate = 33,000 kcal/kWh at 100% capacity factor. Calculate the generating cost/kWh at 100% capacity factor. [**Ans.** 39.77 p/kWh]

25. An input output curve of a 10 MW thermal station is given by an equation

$$I = 10^6(18 + 12L + 0.5L^2) \text{ kcal/hour}$$

where I is in kcal/hour and L is the load on power plant in MW.
Find : (*i*) The load at which the efficiency of the plant will be maximum.
(*ii*) The increase in output required to increase the station output from 5 MW to 7 MW by using the input-output equation and by incremental rate curve. [**Ans.** (*i*) 6 MW (*ii*) 36×10^6 kcal/hour]

26. The input-output curve of a 60 MW power station is given by :

$$I = 10^6(8 + 8L + 0.4L^2) \text{ kcal/hour}$$

where I is the input in kcal/hour and L is load in MW.
(*i*) Determine the heat input per day to the power station if it works for 20 hours at full load and remaining period at no load.
(*ii*) Also find the saving per kWh of energy produced if the plant works at full load for all 24 hours generating the same amount of energy. [**Ans.** (*i*) 38592×10^6 kcal/day (*ii*) 4000 kcal/kWh]

27. The incremental fuel costs for two generating units 1 and 2 of a power plant are given by the following equations :

$$\frac{dF_1}{dP_1} = 0.06P_1 + 11.4$$

$$\frac{dF_2}{dP_2} = 0.07P_2 + 10$$

where P is in megawatts and F is in rupees per hour.

(*i*) Find the economic loading of the two units when the total load to be supplied by the power station is 150 MW.

(*ii*) Find the loss in fuel costs per hour if the load is equally shared by the two units.

[**Ans.** (*i*) $P_1 = 70$ MW, $P_2 = 80$ MW (*ii*) Rs. 1.63 per hour]

28. The incremental fuel costs for two generating units 1 and 2 of a power plant are given by the following equations :

$$\frac{dF_1}{dP_1} = 0.065P_1 + 25$$

$$\frac{dF_2}{dP_2} = 0.08P_2 + 20$$

where F is fuel cost in rupees per hour and P is power output in MW. Find :

(*i*) the economic loading of the two units when the total load supplied by the power plants is 160 MW.

(*ii*) the loss in fuel cost per hour if the load is equally shared by both units.

[**Ans.** $P_1 = 53.5$ MW, $P_2 = 106.5$ MW (*ii*) Rs. 35/hour]

29. Two steam turbines each of 20 MW capacity take a load of 30 MW. The steam consumption rates in kg per hour for both turbines are given by the following equations :

$$S_1 = 2000 + 10L_1 - 0.0001L_1^2$$

$$S_2 = 1000 + 7L_2 - 0.00005L_2^2$$

L represents the load in kW and S represents the steam consumption per hour.

Find the most economical loading when the load taken by both units is 30 MW.

[**Ans.** $L_1 = 20$ MW, $L_2 = 10$ MW]

30. Two electrical units used for the same purpose are compared for their economical working :

(*i*) Cost of Unit-1 is Rs. 5000 and it takes 100 kW.

(*ii*) Cost of Unit-2 is Rs. 14000 and it takes 60 kW.

Each of them has a useful life of 40,000 hours. Which unit will prove economical if the energy is charged at Rs. 80 per kW of maximum demand per year and 5 p. per kWh ? Assume both units run at full load.

[**Ans.** Unit-1 : Rs. 6.039 ; Unit-2 : Rs. 3.898, Unit-2 is more economical]

31. A new industry requires maximum demand of 800 kW at 30% load factor. The following two power supplies are available :

(*i*) *Public supply* charges Rs. 50/kW of maximum demand and 4 p. per kWh.

Capital cost	= Rs. 80,000
Interest and depreciation	= 10 per cent.

(*ii*) *Private oil engine generating station.*

Capital cost	= Rs. 30,000
Interest and depreciation	= 12 per cent
Maintenance and labour charges	= 1 p. per kWh energy generated
Fuel consumption	= 0.35 kg/kWh
Cost of fuel	= 8 paise/kg.

Find which supply will be more economical ?

[**Ans.** (*i*) 6.3 p/kWh (*ii*) 5.1 p/kWh ; oil engine generation is more preferable]

32. A load having a maximum demand of 80 MW and a load factor of 40% may be supplied by one of the following schemes :

(*i*) A steam station capable of supplying the whole load.

(*ii*) A steam station in conjunction with pump-storage plant which is capable of supplying 120×10^6 kWh energy per year with a maximum output of 30 MW.

Find out the cost of energy per unit in each of the two cases mentioned above. Use the following data :

Capital cost of steam station = Rs. 1800/kW of installed capacity ; Capital cost of pump storage plant = Rs. 1200/kW of installed capacity ; Operating cost of steam plant = 0.5 p/kWh ; Operating cost of pump storage plant = 0.4 p./kWh.

Interest and depreciation together on capital invested should be taken as 12 per cent.

Assume that no spare capacity is required. **[Ans.** (*i*) 11.16 p./kWh (*ii*) 8.42 p./kWh]

33. The monthly electricity consumption of a residence can be approximated as under :

Light load : 4 tube lights 40 watts each working for 3 hours daily ; *Fan load* : 4 fans 100 watts each working for 5 hours daily ; *Refrigerator load* : 1 kWh daily ; *Miscellaneous load* : 1 kW for one hour daily.

Find the monthly bill at the following tariff :

First 15 units : Re. 0.50/kWh, Next 25 units : Re. 0.40 per kWh ; Remaining units : Re. 0.30 per kWh ; Constant charge : Rs. 2.50 per month. Discount for prompt payment = 5% . **[Ans.** Rs. 45.20]

34. An industrial undertaking has a connected load of 110 kW. The maximum demand is 90 kW. On an average each machine works for 60 per cent time. Find the yearly expenditure on electricity if the tariff is :

Rs. 1000 + Rs. 100 per kW of maximum demand per year + Re. 0.10 per kWh. **[Ans.** Rs. 67186]

35. A Hopkinson demand rate is quoted as follows :

Demand rates

First 1 kW of maximum demand	= Rs. 5/kW/month
Next 4 kW of maximum demand	= Rs. 4/kW/month
Excess over 5 kW of maximum demand	= Rs. 3/kW/month

Energy rates

First 50 kWh	= 6 paise/kWh
Next 50 kWh	= 4 paise/kWh
Next 200 kWh	= 3 paise/kWh
Next 400 kWh	= 2.5 paise/kWh
Excess over 700 kWh	= 2 paise/kWh

Determine : (*i*) The monthly bill for a total consumption of 1500 kWh and a maximum demand of 12 kW. Also find the unit energy cost.

(*ii*) Lowest possible bill for a month and corresponding unit energy cost.

[Ans. (*i*) Rs. 79, 5.26 paise/kWh (*ii*) Rs. 46.33, 3.09 paise/kWh]

36. Find the cost of generation per kWh from the following data :

Capacity of the plant	= 120 MW
Capital cost	= Rs. 1200 per kW installed
Interest and depreciation	= 10 per cent on capital
Fuel consumption	= 1.2 kg/kWh
Fuel cost	= Rs. 40 per tonne
Salaries, wages, repairs and maintenance	= Rs. 6,00,000 per year
The maximum demand	= 80 MW
Load factor	= 40%.

[Ans. 10.18 paise/kWh]

37. The following data relate to a 2200 kW diesel power station :

The peak load on the plant	= 1600 kW
Load factor	= 45%

Capital cost per kW installed = Rs. 1000
Annual costs = 15 per cent of capital
Annual operating costs = Rs. 60,000
Annual maintenance cost :
(*i*) Fixed = Rs. 10,000
(*ii*) Variable = Rs. 20,000
Cost of fuel = Re. 0.4 per kg
Cost of lubricating oil = Rs. 1.25 per kg
C.V. of fuel = 10,000 kcal/kg
Consumption of fuel = 0.5 kg/kWh
Consumption of lubricating oil = 0.0025 kg/kWh

Determine the following :
(*i*) The annual energy generated.
(*ii*) The cost of generation (per kWh). [**Ans.** (*i*) 6.3 × 10^6 kWh/year (*ii*) 27 paise/kWh]

38. The following data relate to a steam power station of 120 MW capacity which takes 100 MW peak demand at 80% load factor :
Annual cost towards interest and depreciation = Rs. 100/kW installed ; Operating costs = Rs. 1200 × 10^3/year ; Maintenance costs = Rs. 200 × 10^3/year (fixed) and = Rs. 400 × 10^3/year (variable) ; Miscellaneous costs = Rs. 100 × 10^3/year ; Cost of coal used = Rs. 32/ton ; C.V. of fuel used = 6400 kcal/kg ; Overall efficiency of the plant = 20 per cent ; Steam consumption in kg/kWh = (0.8 + 3.5 × load factor).
Determine the following :
(*i*) Coal cost per year.
(*ii*) Overall cost of generation (paise/kWh). [**Ans.** (*i*) Rs. 15 × 10^6/year (*ii*) 4.12 paise/kWh]

39. A power system requires a maximum load of 80 MW at 35% load factor. It can be supplied by any of the following schemes :
(*i*) A steam plant capable to supply whole load.
(*ii*) A steam plant with hydel plant where energy supplied by steam plant is 120 × 10^6 kWh/year with a maximum load of 50 MW.

Plant	***Capital cost***	***Operating cost***	***Transmission cost***
Steam plant	Rs. 600/kW installed	4.8 paise/kWh	Negligible
Hydro plant	Rs. 1400/kW installed	1.2 paise/kWh	0.3 paise/kWh

Assume interest and depreciation at 12 per cent of capital for steam plant and 10 per cent of capital for hydro plant. Calculate the overall cost per kWh.
(*iii*) If the whole load is supplied by a nuclear plant, determine annual cost. Take capital cost of Rs. 2500/kW and running cost of 2.5 paise/kWh. Assume interest and depreciation as 10 per cent per annum. [**Ans.** (*i*) 7.05 paise/kWh (*ii*) 5.98 paise/kWh (*iii*) 10.67 paise/kWh]

40. The following data relate to a power plant of 120 MW capacity :
Capital cost = Rs. 1500/kW ; Interest and depreciation = 10 per cent on capital ; Annual running charges = Rs. 20 × 10^6 ; Profit to be gained = 10 per cent of the capital ; The energy consumed by the power plant auxiliaries = 5 per cent of generated ; The annual load factor = 0.6 ; Annual capacity factor = 0.5.
Calculate the following :
(*i*) The reserve capacity.
(*ii*) The cost of generation per kWh. [**Ans.** (*i*) 20 MW (*ii*) 9.3 paise/kWh]

41. A small generating unit of 5,000 kW capacity supplies the following loads :
(*i*) Street-light load with maximum demand of 200 kW at 0.3 load factor.
(*ii*) Small industrial load with maximum demand of 1800 kW at a load factor of 0.5.
(*iii*) Domestic consumers with a maximum demand of 3000 kW at a load factor of 0.2.

Find the overall energy rate for each type of consumer using the following data :

Capital cost of the plant = Rs. 1,800/kW of installed capacity

Total running cost = Rs. 6,20,000 per year

Annual rate of depreciation and interest on capital cost is 10%.

[**Ans.** (*i*) 11.4 paise/kWh (*ii*) 8.65 paise/kWh (*iii*) 14.8 p/kWh]

42. The following data relate to a 15 MW thermal plant :

Capital cost of plant = Rs. 1200/kW ; Interest, insurance and depreciation = 10% of the plant ; Capital cost of primary and secondary distribution = Rs. 15 × 10^5 ; Interest, insurance and depreciation on the capital cost of primary and secondary distribution = 5% of capital cost ; Plant maintenance cost = Rs. 80 × 10^3 per year ; Maintenance cost of primary and secondary equipments = Rs. 2 × 10^5 per year ; Salaries and wages = Rs. 6 × 10^5 per year ; Consumption of coal = 40 × 10^3 tonnes per year ; Cost of coal = Rs. 80 per tonne ; Dividend to stockholders = Rs. 12 × 10^5 per year ; Energy loss in transmission = 10% ; Diversity factor = 1.6 ; Load factor = 0.8 ; Maximum demand = 14 MW.

Devise a two-part tariff and find the average cost per kWh.

[**Ans.** Fixed charge per kW = 146, charge for energy consumption = 4.63 paise/kWh ;
Average cost of supply = 8.1 paise/kWh]

43. The following data relate to a 12 MW capacity thermal plant :

Peak load	= 10 MW
Annual load factor	= 70%
Cost of the plant	= Rs. 700/kW installed capacity
Interest, insurance and depreciation	= 10 per cent of the capital cost
Cost of transmission and distribution system	= Rs. 300 × 10^3
Interest, depreciation on distribution system	= 5 per cent
Operating cost	= Rs. 300 × 10^3 per year
Cost of coal	= Rs. 50/ton
Plant maintenance cost	= Rs. 25,000 per year (fixed)
	= Rs. 35,000 per year (running)
Coal used	= 30,000 tons/year.

Assume transmission and distribution costs are to be charged to generation.

(*i*) Design a two-part tariff.

(*ii*) Determine overall cost of generation in paise/kWh.

[**Ans.** (*i*) Rs. 88/kW + 3 paise/kWh (*ii*) 4.43 paise/kWh]

44. Determine the load factor at which the cost of supplying a unit of electricity is same in Diesel station as in a Steam station if the respective annual fixed and running charges are as given below :

Diesel : Rs. (30/kW + 0.05/kWh)

Steam : Rs. (120/kW + 0.0125/kWh).

[**Ans.** 0.275]

45. A 30 H.P. motor connected to a condensate pump has been burnt beyond economical repairs. Two alternatives have been proposed to replace it by :

	Cost	*η at full load*	*η at half load*
Motor A	Rs. 6,000	90%	86%
Motor B	Rs. 4,000	85%	82%

The life of each motor is 20 years and its salvage value is 10 per cent of the initial value. The rate of interest is 5 per cent annually. The motor operates at full load for 25 per cent of the time and at half load for the remaining period. The annual maintenance cost of motor *A* is Rs. 420 and that of motor *B* is Rs. 240. The energy rate is 10 paise/kWh.

Which motor will be economical ?

[**Ans.** Motor *A* ; Total cost : Motor *A* = Rs. 14,777/year
Motor *B* = Rs. 15135/year]

46. The following proposals are under consideration for an industry which has a maximum demand of 50 MW and a load factor of 0.4.

(*i*) A steam plant having an initial cost of Rs. 1000/kW and maintenance cost is 2 paise/kWh. The coal of C.V. of 6150 kcal/kg is used. The overall efficiency of the plant is 25 per cent.

(*ii*) A hydro-plant having a capital cost of Rs. 3000/kW and running cost of 0.5 paise/kWh.

Assuming interest and depreciation rate of 12 per cent for steam plant and 9 per cent for hydro-plant, determine the price of coal above which steam station is uneconomical. **[Ans.** Rs. 49.7/ton]

47. An industrial consumer has a choice between low and high voltage supply available at the following rates :

High voltage : Rs. 45/kW per year + paise 3.5/kWh

Low voltage : Rs. 47/kW per year + paise 4/kWh

In order to have high voltage supply, consumer has to install his own transformer which costs Rs. 100/kW. The losses in the transformer are 3 per cent of full load. Determine the number of working hours per week above which the high voltage supply will be economical. Assume interest and depreciation 10 per cent of capital and working weeks per year 50. Assume the load of the consumer as 1 MW. **[Ans.** 49 hours per week]

48. The expected annual cost of power system supplying the energy to 40,000 consumers is tabulated below :

Fixed charges	= Rs. 2400×10^3
Energy charges	= Rs. 1716×10^3
Consumer charges	= Rs. 210×10^3
Profit	= Rs. 168×10^3
Maximum demand	= 5000 kW
Diversity factor	= 4
Energy supplied	= 17×10^6 kWh.

Devise a three-part tariff allowing 25% of the profit in fixed charges, 50% in energy charges and remaining 25% in customer charges.

[Ans. Rs. 122/kW per year + 11 paise/kWh + Rs. 6.3 per consumer per year]

Power Factor Improvement

49. A 400 V, 50 Hz, 3-phase line delivers 200 kW at 0.8 power factor lagging. It is desired to raise the line power factor to unity by installing shunt capacitors. Calculate the capacitance of each unit if they are connected in (*i*) star and (*ii*) delta. **[Ans.** (*i*) 2984 μF, (*ii*) 995 μF]

50. A 3-phase synchronous motor having a mechanical load (including losses) of 122 kW is connected in parallel with a load of 510 kW at 0.8 p.f. lagging. The excitation of the motor is adjusted so that kVA input to the motor becomes 140 kVA. Determine the new p.f. of the whole system. **[Ans.** 0.8956]

51. Power factor of 0.8 lagging of a load of 400 kW is to be improved by phase advancing plant which works at power factor of 0.1 leading so that power factor of the whole system is now 0.9 lagging. Find the kVA rating of the phase advancing plant. **[Ans.** 101.77 kVA]

52. A generating station is connected with the following loads :

(*i*) A lighting load of 200 kW.

(*ii*) An induction motor of 200 h.p. (metric) power factor 0.7 lagging and efficiency 80%.

(*iii*) A synchronous motor of 100 h.p. (metric) efficiency 90%.

Calculate the power factor of the synchronous motor so that the combined p.f. becomes unity.

[Ans. 0.4 (leading)]

53. It is desired to correct the power factor to 0.95 by means of static condensers connected across each phase of a 3-phase, 400 V, 50 Hz motor installation having a maximum load of 50 kVA at a p.f. of 0.75. What must be the capacity of each delta connected condenser ? **[Ans.** 238 μF]

54. A synchronous motor having a power consumption of 50 kW is connected in parallel with a load of 200 kW having p.f. of 0.8 lagging. The excitation of the motor is adjust until the combined p.f. becomes 0.9 lagging. Calculate the kVA input to the motor. **[Ans.** 57.76 kVA]

55. Two systems of tariff are available for a factory, working 8 hours a day for 300 working days in a year.

(*i*) H.V. supply at 5 paise per unit plus Rs. 4.50 per month per kVA of maximum demand.

(*ii*) L.V. supply at Rs. 5.0 per month per kVA plus 5.5 paise per unit.

The factory has an average load of 200 kW at 0.8 p.f. and a maximum demand of 250 kW at the same power factor. The h.v. equipment costs Rs. 50.00 per kVA and losses can be taken as 4%. Interest and depreciation charges are 12%. Calculate the difference in annual cost between the two systems.

[Ans. H.T. supply is cheaper for Rs. 619.00]

56. An industrial load requiring 500 kW at 0.8 p.f. can be supplied on the following tariff rates :

(*i*) H.T. supply at Rs. 100 per kVA per annum + 4 paise per unit.

(*ii*) L.T. supply at Rs. 110 per kVA per annum + 5 paise per unit.

Transformer and switchgear for h.t. supply cost Rs. 75 per kVA, the full load losses being 2.5%. The fixed charges are 25 percent per annum of the capital cost of h.t. plant. If the installation works on full load, find the number of working hours per week above which the h.t. supply is cheaper. There are 48 weeks on a year. **[Ans.** 34.2 hours per week]

57. Calculate the minimum two part tariff to be charged from the following data :

Generating cost per unit = 4 paise

Annual generating cost per kW of maximum demand = Rs. 50/-

Units supplied annually = 4,380 × 10^4

Annual load factor of the station = 50%

Annual charges for distribution = Rs. 1.25 × 10^5

Diversity factor for distribution network = 1.25

The distribution losses are 10%.

[Ans. Rs. 50 per kW of maximum demand plus 4.44 paise per unit]

10

"Universities' Questions" (Latest)–with "Solutions"

SECTION A: SHORT ANSWER QUESTIONS

ILLUMINATION

Q. 1. *What is 'light'?*

Ans. ***Light*** is a form of *electromagnetic energy* radiated from a body which is capable of being *perceived* by the human eye.

Q. 2. *Define the term 'illumination'.*

Ans. ***Illumination*** is the luminous flux received by a surface per unit area. The unit is 'lux' and 'metre-candle'.

Q. 3. *How does 'illumination' differ from 'light'? Explain briefly.*

Ans. *Illumination* differs from *light* very much, though generally these terms are used more or less synonymously. Strictly speaking light is the cause and illumination is the result of the light on the surfaces on which it falls. Thus the illumination makes surfaces more or less bright with a certain colour and it is this brightness and colour which the eye sees and interprets as something useful or pleasant or otherwise.

Q. 4. *What is 'solid angle'?*

Ans. ***Solid angle*** is the angle generated by the surface passing through the point in space and the periphery of the area. It is denoted by ω, expressed in *steradians* and is given by the ratio of the area of the surface to the square of the distance between the area and the point.

Q. 5. *State the laws of illumination.*

Ans. The two laws of illumination are:

1. ***Law of Inversely squares.*** The illumination of a surface is *inversely proportional to the square of the distance* between the surface and the light source provided that the distance between the surface and the source is sufficiently large so that source can be regarded as a point source.
2. ***Lambert's Cosine Law.*** According to this law the illumination at any point on a surface is *proportional to the cosine of the angle between the normal at that point and the direction of luminous flux.*

Q. 6. *What is Angstrom unit (AU)*

Ans. Angstrom unit is unit of wavelength of light and is equal to 10^{-10} m.

Q. 7. *What is the relation between lux and lumens/m²?*

Ans. Lux and lumens/m^2, both are units of illumination.

1 lux = 1 lumen/m^2.

Q. 8. *What is 'space-height ratio' ?*

Ans. It is defined as the ratio of horizontal distance between adjacent lamps and height of their mountings.

Q. 9. *What is 'glare' ?*

Ans. ***Glare*** may be defined as the brightness within the field of vision of such a character as to cause annoyance, discomfort, interfere with vision or eye-fatigue.

Q. 10. *What are 'polar curves' ?*

Ans. The luminous intensity in most lamps on sources of light is not the same in all directions, because of their unsymmetrical shape. Often it is necessary to know the distribution of light in various directions to ascertain how the candle power of a light source varies in different directions. The luminous intensity in all the directions can be represented by ***polar curves.***

Q. 11. *Name the instruments which are used to measure illumination.*

Ans. Instruments employed for the measurement of illumination are:

(*i*) Trottar illumination photometer; (*ii*) Macbeth illuminometer;
(*iii*) Holophane lumeter etc.

Q. 12. *What is the difference between 'lamp efficiency' and 'specific consumption'?*

Ans. ***Lamp efficiency*** is the ratio of the luminous flux in lumens to the power input in watts while the ***specific consumption*** is the ratio of the power input in watts to the average candle power in candela.

Q. 13. *Differentiate between maintenance factor and depreciation factor.*

Ans. ***Maintenance factor*** is defined as the ratio of illumination under normal working conditions to the illumination when everything is perfectly clean. ***Depreciation factor*** is merely inverse of the maintenance factor.

Q. 14. *What are the advantages of artificial lighting produced electrically?*

Ans. The advantages of artificial lighting are:

(*i*) Cleanliness; (*ii*) Ease of control;
(*iii*) Reliability; (*iv*) Steady output; and
(*v*) Low output.

Q. 15. *What are the various types of electric lamps?*

Ans. The various types of electric lamps in common use are:

1. Arc lamps
   (*i*) Carbon arc lamp; (*ii*) Flame arc lamp;
   (*iii*) Magnetic arc lamp.
2. Incandescent lamps.
3. Gas filled lamp
4. Gaseous discharge lamps
   (*i*) Sodium vapour lamp; (*ii*) High pressure mercury vapour lamp;
   (*iii*) Neon lamp.

Q. 16. *What are the various types of lighting systems?*

Ans. ***Lighting systems:*** (*i*) Direct lighting; (*ii*) Semi-direct lighting; (*iii*) Semi-indirect lighting; (*iv*) Indirect lighting; and (*v*) General reflection.

Q. 17. *What are the characteristics which a lighting scheme should possess?*

Ans. The lighting scheme should possess the following characteristics:

1. It should provide adequate illumination.
2. It should provide light distribution all over the working plane as uniform as possible.
3. It should avoid glare and hard shadows as far as possible.
4. It should provide light of suitable colour.

Q. 18. *What are discharge lamps?*

Ans. • In these lamps, gas or vapour is made luminous by electric discharge through them. The colour and intensity of light *i.e.,* candle power power emitted depends on the nature of the gas or vapour only.

• These lamps are luminiscent-light lamps and *do not depend on temperature for higher efficiencies.* In this respect, they differ radically from incandescent lamps whose efficiency is dependent on temperature.

Q. 19. *Name the materials which are used for making filaments of incandescent lamps.*

Ans. The materials used are:

(*i*) Carbon; (*ii*) Tantalum; (*iii*) Tungsten.

Q. 20. *Why is 'tungsten' selected as filament material?*

Ans. Tungsten is the most commonly used metal for filament due to its high melting point (3400°C), high resistivity, low temperature coefficient (0.0051), low vapour pressure, being ductile and mechanically strong to withstand vibration during use.

Q. 21. *Why the glass globe of a filament lamp is completely evacuated?*

Ans. To prevent the oxidation and convection currents of the filament and also to prevent the temperature being lowered by radiation.

Q. 22. *Sodium vapour lamps are not used for general lighting, why?*

Ans. Because they have the drawback of colour discrimination.

Q. 23. *What is an 'arc lamp'?*

Ans. In an ***arc lamp*** electric current is made to flow through two electrodes in contact with each other which are drawn apart. The result is an arc being struck. The arc maintains the current, and is very efficient source of light.

• These lamps are used in search lights projection lamps and other special purpose lamps like those in flash cameras.

Q. 24. *What is a 'Neon lamp'?*

Ans. ***Neon lamps*** belong to *cold-cathode category*. The electrodes are in the form of iron shells and are coated on the inside. The colour of light emitted is red. If helium gas is used in place of neon, pinkish white light is obtained. Helium and neon through coloured glass tubing produce a variety of effects.

Q. 25. *Which factors should be considered while designing the lighting system?*

Ans. While designing the lighting system the following factors should be considered:

(*i*) Illumination level; (*ii*) Uniformity of illumination;
(*iii*) Colour of light; (*iv*) Shadows;
(*v*) Glare; (*vi*) Mounting height;
(*vii*) Spacing of luminaires; (*viii*) Colour of surrounding walls.

ELECTRIC HEATING

Q. 26. *What are various modes of heat transfer?*

Ans. ***Heat transfer*** which is defined as the transmission of energy from one region to another as a result of temperature gradient takes place by the following three methods:

(*i*) Conduction; (*ii*) Convection; (*iii*) Radiation.

Q. 27. *What is Fourier's law of heat conduction?*

Ans. Fourier's law of heat conduction is an empirical law based on observation and states as follows:

"The rate of flow of heat through a simple homogeneous solid is directly proportional to the area of the section at right angles to the direction of heat flow, and to change of temperature with respect to the length of the path of the heat flow".

Q. 28. *What is 'dielectric heating' ?*

Ans. It is also called ***high-frequency heating*** and is used for heating '*insulators*' like wood, plastics and ceramics etc. which cannot be heated easily and uniformly by other methods.

- The supply frequency required for dielectric heating is between 10–50 MHz and the applied voltage is 20 kV.
- The overall efficiency of dielectric heating is about 50%.

Q. 29. *Name the materials commonly used for heating elements.*

Ans. (*i*) For low and medium temperature services:

Alloy of *nickel* and *chromium* ... Ni = 80%, Cr = 20%

Or

Alloy of *nickel, chromium* and *iron* ... Ni = 65%, Cr = 15%, Fe = 20%.

(*ii*) For operating temperatures above 1150°C resistors are made of:

Silicon carbide, Molybdenum, Tungsten and Graphite.

Q. 30. *What is 'indirect resistance heating' ?*

Ans. In this method the current is passed through a high resistance wire known as *heating element.* The heat produced due to $I^2 R$ loss in the element is transmitted to the body to be heated by one or more mades of heat transfer viz conduction, convection and radiation

- This method of heating is used in room heaters, immersion water heaters, salt both furnaces etc.

Q. 31. *What are the properties of a good heating element?*

Ans. A good heating element should possess the following properties:

(*i*) High specific resistance; (*ii*) High melting temperatures;
(*iii*) Low temperature coefficient of resistance;
(*iv*) High oxidising temperature; (*v*) Positive temperature coefficient of resistance;
(*vi*) High ductility and flexibility; (*vii*) High mechanical strength of its own.

Q. 32. *What is an oven?*

Ans. ***Oven*** is a low temperature heating chamber with provision for ventilation.

Q. 33. *What are the causes of failure of heating elements?*

Ans. The various causes of failure of heating elements are:

(*i*) Formation of hot spots which shine brighter during operation;
(*ii*) Oxidation; (*iii*) Corrosion;
(*iv*) Mechanical failure.

Q. 34. *Give some applications of infrared heating.*

Ans. Some applications of infrared heating are:

(*i*) Preheating of plastic prior to moulding;
(*ii*) Paint stoving; (*iii*) Drying of radio-cabinets and wood furniture;
(*iv*) Softening of thermoplastic sheets etc.

Q. 35. *What are the advantages of conical shaped arc furnaces ?*

Ans. (*i*) Large surface area per unit bath volume; (*ii*) Consume less power; (*iii*) Reduced melting point; (*iv*) Reduced radiation losses.

Q. 36. *Name the types of electrodes used in arc furnaces.*

Ans. (*i*) Carbon electrodes; (*ii*) Graphite electrodes; (*iii*) Self-baking electrodes.

Q. 37. *Name the various types of arc furnaces.*

Ans. (*i*) Direct arc furnace; (*ii*) Indirect arc furnace; (*iii*) Submerged arc furnace.

Q. 38. *What is the main advantage of a 'direct arc furnace' ?*

Ans. The main advantage of a ***direct arc furnace*** is that *purer product* is obtained and composition can be exactly controlled during refining process. This is the reason that these furnaces even being costlier in initial as well as operating cost is preferred.

Q. 39. *How is power input controlled in a 'submerged arc furnace' ?*

Ans. Power is controlled by *varying distance* between electrodes or by *varying the voltage* applied to the electrodes.

Q. 40. *How is control of power affected in electric arc furnaces?*

Ans. By raising or lowering the electrodes or by varying the voltage applied to the arc furnace (by changing the tapping on supply transformer); preferably by both voltage and electrode controls.

Q. 41. *How can dielectric heating rate be varied?*

Ans. By varying either supply voltage or supply frequency.

Q. 42. *Dielectric heating is used only where other methods are impracticable or too slow, why?*

Ans. Due to very high cost of equipment required for dielectric heating.

Q. 43. *Which type of supply is used for direct core type induction furnaces?*

Ans. Single-phase, 500 Hz, 600 V supply.

Q. 44. *What are the applications of 'eddy current heating' ?*

Ans. (*i*) Surface hardening; (*ii*) Annealing; (*iii*) Soldering.

Q. 45. *What are the advantages of 'infrared or radiant heating' ?*

Ans. (*i*) Compactness of heating units; (*ii*) Rapid heating; (*iii*) Flexibility; (*iv*) Safety.

ELECTRIC WELDING

Q. 46. *What is welding?*

Ans. ***Welding*** is the process of joining two pieces of metal or non-metal at faces rendered plastic or liquid by the application of heat or pressure or both. Filler material may be used to effect the union.

All metals are weldable provided proper process and technique(s) are used. If either of these two aspects or both are overlooked, no good weld can be obtained.

Q. 47. *On which factors does the proper selection of welding process depend?*

Ans. (*i*) Types of metals to be joined; (*ii*) Cost involved; (*iii*) Nature of products to be fabricated; and (*iv*) Techniques of production adopted.

Q. 48. *What is the field of application of 'butt welding' ?*

Ans. ***Butt welding*** is being used for welding such things as steel rails whose cross-sectional area is as much as 6.25 cm^2.

Q. 49. *What are the applications of 'flash welding' ?*

Ans. ***Flash welding*** is widely used in automobile construction on the body, axles, wheels, frames and other parts. It is also employed in welding motor frames, transformer tanks and many types of sheet steel containers such as barrels and floats.

Q. 50. *Differentiate between 'plastic welding' and 'fusion welding'.*

Ans. In the ***plastic welding***, also called the pressure welding, the pieces of metals to be joined are heated to a plastic stage and then forced together by external pressure whereas in

fusion welding, also called the non-pressure welding, the material at the joint is heated to a molten state and allowed to solidify. Former one includes forge welding, resistance welding, thermit welding and gas welding, while the latter includes gas welding, arc welding and thermit welding without pressure.

Q. 51. *How does 'resistance welding' differ from 'arc welding' ?*

Ans. The resistance welding process differs from arc welding in that pressure is used but filler metal or flexes are not.

Q. 52. *Why alternate current is found most suitable for resistance welding?*

Ans. Because alternate current can provide any desired combination of current and voltage or by means of a transformer.

Q. 53. *What is 'arc blow' in electric arc welding ?*

Ans. The deflection of the arc from the weld point is called ***arc blow***. This condition is encountered only with D.C. welding sets and is especially noticeable when welding with base electrodes. It is experienced most when using currents above 200 A or below 40 A.

Q. 54. *What is an 'electrode' ?*

Ans. An ***electrode*** is a filler metal in the form of a wire or rod which is either bare or coated uniformly with the flux.

Q. 55. *On what factors does the type of electrode depends?*

Ans. (*i*) The type of metal to be welded; (*ii*) The welding position; (*iii*) The type of electric supply (A.C. or D.C.); and (*iv*) The type of the welding machine.

Q. 56. *What are the applications of 'atomic hydrogen welding' ?*

Ans. ***Atomic hydrogen welding***, being *expensive* is used mainly for high grade work on stainless steel and most non-ferrous metals.

Q. 57. *Name the type of generator which is used in electric-arc welding.*

Ans. Differential compound wound D.C. generator.

ELECTROLYTIC PROCESSES

Q. 58. *What are 'electrolytic processes' ?*

Ans. The processes based on the fact that electrical energy produce chemical changes are called ***electrolytic processes.***

Q. 59. *State the 'Faraday's laws of electrolysis'.*

Ans. *First law* states: "The mass of a substance liberated from an electrolyte in a given time is proportional to the quantity of electricity passing through the electrolyte".

Second law states: When the same quantity of electricity is passed through several electrolytes, the masses of the substances deposited are proportional to their respective chemical equivalents or equivalent weights.

Q. 60. *What are the applications of electrolytic processes?*

Ans. The electrolytic processes are widely used for:

(*i*) Extraction of pure metals; (*ii*) Refining of metals; (*iii*) Manufacturing of various chemicals; (*iv*) Electro-deposition of metals; and (*v*) Building up of worn parts in metallurgical, chemical and other industries.

Q. 61. *What is electro-deposition?*

Ans. The process of depositing a coating of one metal over another metal or non-metal electrically is called the ***electro-deposition.***

- It is used for protective, decorative and functional purposes and includes such processes as electro-plating, electro-forming, electro-typing, electro-facing, electro-metallisation etc.

Q. 62. *What is 'throwing power'?*

Ans. ***Throwing power*** is defined as the ability of the electrolyte to produce even irregular surfaces.

Q. 63. *What is 'electroplating'?*

Ans. ***Electroplating*** is an art of depositing a superior or a more noble metal on an inferior or a base metal by means of electrolysis of an aqueous solution of a suitable electrolyte.

Q. 64. *What are the factors on which the quality of electrodeposition depends?*

Ans. The quality of electrodeposition depends on the following factors:

(*i*) Nature of electrolyte; (*ii*) Current density;
(*iii*) Temperature; (*iv*) Conductivity and concentration of electrolyte;
(*v*) Addition agents; (*vi*) Throwing power;
(*vii*) Polarisation.

Q. 65. *What is 'polarisation'? How can it be reduced?*

Ans. In electrodeposition processes, when current density is used beyond a certain limit, it causes electrolysis of water and hydrogen liberation on the cathode. This hydrogen evolved on the cathode blankets the base metal which reduces the rate of metal deposition. This phenomenon is called ***polarisation***.

Q. 66. *What is 'anodizing'?*

Ans. An *anodic coating* means an oxide film deposited/created on a metal surface, with the help of an *anode* and *oxidation*.

The process of providing an *oxide film* is known as ***anodizing.***

Q. 67. *What is 'electro-forming'?*

Ans. ***Electro-forming*** is the reproduction of objects by electro-deposition on some sort of a mould or form.

Q. 68. *What is 'electro-refining'?*

Ans. ***Refining*** is the process whereby a highly concentrated mixture of metals is subjected to electro-chemical treatment for recovering not only the principal metal in pure form, but also the precious metals like gold, silver, bismith etc., which may be present in the form of minute traces.

- By electro-refining, it is possible to get metal of *about 100% purity*. This is one of most important pre-requisite expected out of copper and aluminium in order to have high electrical conductivity.

REFRIGERATION AND AIR CONDITIONING

Q. 69. *What is refrigeration?*

Ans. ***Refrigeration*** is science of producing and maintaining temperatures below that of the surrounding atmosphere.

In simple, refrigeration means the cooling or removal of heat from a system. The equipment employed to maintain the system at a low temperature is termed as *"refrigerating system"* and the system which is kept at a lower temperature is called *"refrigerated system"*

Q. 70. *What are three ways by which refrigeration is generally produced?*

Ans. The three ways are:

(*i*) By melting of a solid; (*ii*) By sublimation of a solid; and
(*iii*) By evaporation of a liquid.

- Most of the commercial refrigeration is produced by evaporation of a liquid called *"refrigerant"*. ***Mechanical refrigeration*** depends upon the evaporation of liquid refrigerant and its circuit includes the equipments, naming *'evaporator, compression'*, *'condenser'*, and *'expansion valve'*.

– "Mechanical refrigeration" is used for preservation of food, manufacture of ice, solid carbon dioxide and control of air temperature and humidity in the air-conditioning system.

Q. 71. *What are the elements of refrigeration systems?*

Ans. All refrigeration systems must include at least *Four basic units* as given below:

(*i*) A low temperature thermal "sink" to which heat will flow from the space to be cooled.

(*ii*) *Means of extracting energy* from the sink, raising the temperature level of this energy, and delivering it to a heat receiver.

(*iii*) A *receiver* to which heat will be transferred from the high temperature high pressure refrigerant.

(*iv*) *Means of reducing pressure and temperature* of the refrigerant as it returns from the receiver to the "*sink*".

Q. 72. *Enumerate the various refrigeration systems.*

Ans. The various refrigeration systems are:

(*i*) Ice refrigeration; (*ii*) Air-refrigeration system;

(*iii*) Vapour compression refrigeration system;

(*iv*) Vapour absorption refrigeration system;

(*v*) Special refrigeration systems.

Q. 73. *What is 'standard and commercial tonne of refrigeration'?*

Ans. ***Standard commercial tonne of refrigeration*** is defined as the refrigerating effect (RE) produced by melting 1 tonne of ice from and at 0°C in 24 hours. Since the latent heat of fusion of ice is 336 kJ/kg, the refrigerating effect of 336 × 1000 kJ/kg is 24 hours is rated as ***1 tonne***, *i.e.*,

$$\text{A tonne of refrigeration} = \frac{336 \times 1000}{24} = 14000 \text{ kJ/h}$$

Note: ***Ton of refrigeration (TR):*** A ton of refrigeration is basically an American unit of *RE*. It originated from the rate at which heat is required to be removed to freeze one ton of water from and at 0°C. Using American unit this is equal to removal of 200 BTU of heat per minute and in MKS units it is adapted at 50 kcal/min or 3000 kcal/hour. In S.I. units its conversion is rounded off to 3.5 kJ/(kW) or 210 kJ/min.

(1 ton of refrigeration may be taken equal to 0.9 *tonne* of refrigeration)

Q. 74. *In a simple vapour compression system which fundamental processes are completed in one cycle?*

Ans. The processes are:

(*i*) Compression; (*ii*) Condensation;

(*iii*) Expansion; (*iv*) Vaporisation.

Q. 75. *Name the various parts of simple vapour compression system.*

Ans. The various parts of a simple vapour compression system are:

(*i*) '*Compressor*'; (*ii*) Discharge line (or hot gas line),

(*iii*) '*Condenser*'; (*iv*) Receiver tank;

(*v*) Liquid line; (*vi*) '*Expansion valve*';

(*vii*) '*Evaporator*'; (*viii*) Section line.

Q. 76. *On what principle does thermoelectric refrigeration system operate?*

Ans. Principle of '*Peltier effect*'.

Q. 77. *What is 'refrigerant' ?*

Ans. A ***refrigerant*** is defined as any substance that absorbs heat through expansion or evaporation and loses it through condensation in a refrigeration system.

Q. 78. *Differentiate between 'primary' and 'secondary' refrigerants.*

Ans. ***Primary refrigerants*** are those working mediums or heat carriers which *directly take part* in the refrigeration system and cool the substance by the absorption of latent heat *e.g.*, Ammonia, Carbon dioxide, Sulphur dioxide, Methyl chloride, Ethyl chloride and Freon group etc.

Secondary refrigerants are those circulating substances which are *first cooled with the help of the primary refrigerants* and are then employed for cooling purposes, *e.g.*, ice, solid carbon dioxide, etc. These refrigerants cool substance by absorption of their sensible heat.

Q. 79. *What is 'Psychrometry' ?*

Ans. The art of measuring the moisture content is termed ***psychrometry.***

Q. 80. *Name the factor which affect the human comforts.*

Ans. (*i*) Temperature; (*ii*) Humidity; (*iii*) Air motion; and (*iv*) Air purity.

Q. 81. *What is 'effective temperature' ?*

Ans. ***Effective temperature*** is defined as a sensory index that combines into a single factor the effects of temperature, humidity and air movement on human comfort in a noise free pure air environment.

Q. 82. *What is an 'air conditioning system' ?*

Ans. ***Air conditioning system*** is an assembly of different parts of the system used to produce a specified condition of air within a required space of building.

ELECTRICAL DRIVES

Q. 83. *What is an 'electric drive' ?*

Ans. An ***electrical drive*** is a form of machine equipment designed to convert electrical energy into mechanical energy and to provide electrical control of this process.

Q. 84. *What are the major parts of an electrical drives ?*

Ans. (*i*) Load; (*ii*) Motor; (*iii*) Power modulator; (*iv*) Control unit; and (*v*) Source.

Q. 85. *How are electrical drives classified ?*

Ans. Electrical drives may be grouped into the following three categories:

(*i*) Group drive; (*ii*) Individual drive; and (*iii*) Multimotor drive.

Q. 86. *What is a group drive?*

Ans. A drive in which a single electric motor drives a line shaft by means of which an entire of group of working machines may be operated is called ***group drive.***

Q. 87. *What is 'multimotor drive' ?*

Ans. A ***multimotor drive*** consists of several individual drives each of which serves to operate one of many working members or mechanisms in some production unit.

Q. 88. *How are loads/torques classified ?*

Ans. Load torques are classified in the following two categories:

(*i*) Active or potential loads; and (*ii*) Passive loads.

Active loads: These are the loads which are due to the forces of gravity, tension or compression in a spring or any elastic body.

The active torques *act in one direction only.*

Passive torques: These torques are due to friction, cutting and deformation of inelastic bodies. They always *oppose* the motion of drive and change their signs when the direction of rotation of drive is changed.

Q. 89. *What do you mean by two-quadrant operation of a D.C. motor?*

Ans. The operation of an electric motor in forward motoring mode and in regenerating (or braking mode) is called the ***two-quadrant operation*** of the motor.

Q. 90. *In utilisation studies, why speed-torque (not torque-speed) characteristic curves are commonly used?*

Ans. In application engineering, it is usually the load torque which is known and the speed at which the motor can drive this load is unknown and is to be determined. Thus the torque is independent variable and speed is dependent variable. Consequently speed-torque characteristics curves with torque as abscissa are commonly used in utilisation studies.

Q. 91. *What are torque motors ?*

Ans. ***Torque motors*** are the D.C. motors designed to run for long periods in a stalled or a low speed condition. Some torque motors are designed to operate at low speed intermittently.

Q. 92. *Series motors are preferred for traction purposes, why?*

Ans. Because D.C. *series motor develops large starting* torque and *slows down* with the increase in load and so gets automatically relieved from heavy excessive load.

Q. 93. *D.C. motors are mostly gear motors, why?*

Ans. Because the low-speed motors are much more expensive than high-speed motors of the same kW output rating.

Q. 94. *What is a 'stepper motor' ?*

Ans. A ***stepper motor*** is an incremental motion machine (*i.e.,* the motor which turns in discrete movement called the *steps*). It does not rotate continuously as a conventional motor does.

Q. 95. *What is 'load equilisation' ?*

Ans. In several industrial drives (*e.g.,* electric hammers, presses, rolling mills, reciprocating pumps, planing machines etc.) the load fluctuates between wide limits over a span of few seconds. It is very essential to smooth out the fluctuating load, otherwise during intervals of peak load it will draw a heavy current from the supply *either producing large voltage drop in the system or requiring cables and wires of heavy section. The process of smoothing out the fluctuating load is known as* ***load equalisation.***

Q. 96. *What are the factors on which choice of drive depends?*

Ans. Choice of drive is governed by the following factors:

(*i*) Speed of driving and driven machines,
(*ii*) Convenience; (*iii*) Space available;
(*iv*) Clutching arrangement required; and
(*v*) Cost.

ELECTRIC TRACTION

Q. 97. *What is 'electric traction' ?*

Ans. The locomotive in which the driving or tractive force is obtained from electric motors is called ***electric traction.***

Q. 98. *What are the applications of electric traction?*

Ans. Electric traction is used in:

(*i*) Electric trains; (*ii*) Train cars;
(*iii*) Trolley bus; (*iv*) Diesel electric vehicles etc.

Q. 99. *Name the different systems of traction.*

Ans. The commonly used traction systems are:

(*i*) Steam engine drive; (*ii*) I.C. engine drive;
(*iii*) I.C. engine electric drive; (*iv*) Petrol-electric traction;
(*v*) Battery electric drive; (*vi*) Electric drive.

Q. 100. *What are the systems of railway electrification presently available?*

Ans. 1. D.C. system ...... 600 V, 750 V, 1500 V, 3000 V.

2. Single-phase A.C. system ...... 15 to 25 kV, $16\frac{2}{3}$, 25 and 50 Hz.

3. Three-phase A.C. system ...... 3.3 to 3.6 kV at $16\frac{2}{3}$ Hz.

4. Composite system ...... involving conversion of single-phase A.C. into three-phase A.C. or D.C.

Q. 101. *Why A.C. series motors used for traction services are designed to operate at low frequency i.e., $16\frac{2}{3}$ or 25 Hz.*

Ans. Single-phase A.C. series motors have better performance (improved p.f., higher frequency improved commutation, better starting etc.) if operated on low frequency ($16\frac{2}{3}$ Hz or 25 Hz).

Q. 102. *How is speed controlled in direct I.C. engine drive ?*

Ans. By using a gear box.

Q. 103. *What is main disadvantage of electric traction ?*

Ans. Electric traction involves a *heavy initial expenditure* for the power supply system and in the case of tramways, for the track rails and electrical return system, in addition to the cost of vehicles, depots etc.

Q. 104. *What is the value of voltage used in overhead supply for trains in India ?*

Ans. 25 kV

Q. 105. *What are major disadvantages of the trainway ?*

Ans. 1. The trainway needs laying of overhead supply system and track for its use, which is costly to maintain and constitutes a source of danger to other road users.

2. Lack of manoeuvrability in congested areas and noise.

TRAIN MOVEMENT AND ENERGY CONSUMPTION

Q. 106. *What are the various types of passenger services?*

Ans. (*i*) City or urban service; (*ii*) Suburban service;
(*iii*) Main line service.

Q. 107. *Which of the system (D.C. or A.C.) is preferred for suburban services and road transport where stops are frequent and distances of run are small?*

Ans. D.C. system.

Q. 108. *What are the most important requirements of the main line railway service?*

Ans. (*i*) High maximum (crest) speed; (*ii*) Minimum cost of overhead structure.

Q. 109. *Which system is preferred for main line railway service?*

Ans. Single-phase system.

Q. 110. *What is 'schedule speed'?*

Ans. ***Schedule speed*** is defined as the ratio of the distance curved between two stops and total time of run including time of stop.

Q. 111. *Differentiate between 'crest speed' and 'average speed'.*

Ans. ***Crest speed*** is the maximum speed attained by the vehicle during the run.

Average speed is the distance covered between two stops divided by the actual time of run.

Q. 112. *By which factors the schedule speed of a train is affected?*

Ans. (*i*) Acceleration and retardation; (*ii*) Maximum or crest speed; and (*iii*) Duration of stop.

Q. 113. *What is the adhesive weight of a train?*

Ans. The ***adhesive weight*** of a train is equal to the total weight to be carried on the driving wheels.

Q. 114. *What is 'coefficient of adhesion'?*

Ans. ***Coefficient of adhesion*** is defined as the ratio of tractive effort to slip the wheels and adhesive weight.

Q. 115. *What is the effect of increase in speed on coefficient of adhesion?*

Ans. The coefficient of adhesion decreases with increase in speed.

Q. 116. *The accelerating weight of an electric train is greater than its dead weight, why?*

Ans. With the linear acceleration of the train the rotating parts of the train such as wheels and motors also accelerate in an angular direction, and therefore, the tractive effort required is equal to the *arithmetic sum of tractive effort required to have the angular acceleration of rotating parts and tractive effort required to have the linear acceleration.* Hence the equivalent or accelerating weight of the train is taken *higher* than the dead weight.

Q. 117. *What is 'train resistance'?*

Ans. When a train runs on level track the following types of frictional forces oppose its motion:

(*i*) Internal friction (friction at bearings, guides etc.);

(*ii*) External friction (friction between wheel-flanges and rails); and

(*iii*) Air friction

All these frictional losses collectively are termed as ***train resistance.***

Q. 118. *What is 'specific energy output'?*

Ans. ***Specific energy output*** is the energy output of the driving wheel expressed in watt-hour (Wh) per tonne-km (t-km) of the train.

Q. 119. *What is 'specific energy consumption'?*

Ans. ***Specific energy consumption*** is defined as the energy consumed (Wh) per tonne mass of the train per km length of the run.

Q. 120. *What are the factors which affect the specific energy consumption of a train running at a given schedule speed?*

Ans. (*i*) Acceleration; (*ii*) Retardation; (*iii*) Maximum speed; (*iv*) Distance between stops; (*v*) Track configuration; and (*vi*) Type of train and equipment.

TRACTION DRIVES

Q. 121. *Name the motors which find application in traction.*

Ans. (*i*) D.C. series motors; (*ii*) A.C. series motors-single-phase; (*iii*) Repulsion motor; (*iv*) Three-phase induction motors; (*v*) Linear induction motors.

Q. 122. *What is the major disadvantage of D.C. motor?*

Ans. The major disadvantage of D.C. motor is the *presence of commutator and brushes*, which require *frequent maintenance* particularly when the flashovers at the commutator occur due to sharp voltage fluctuations.

Q. 123. *What are metadynes?*

Ans. The metadyne system of control estimates the energy loss and achieves a very smooth control during acceleration period. Machines with more than two brush sets per pair of poles are called ***metadynes.*** It is a device which *converts* power at constant voltage and variable current into one with constant current and variable voltage.

Q. 124. *Under what circumstances series-parallel method of speed control of D.C. series traction motors adopted?*

Ans. When motors employed are two, four, six or more than this even.

Q. 125. *What are the advantages of thyristor control of traction motors?*

Ans. (*i*) Accurate control and fast response; (*ii*) Elimination of bulky on-load tap changer and electromagnetic devices; (*iii*) Saving of energy; (*iv*) Notches control; (*v*) Increase in pulling ability of the motive power; and (*vi*) Minimum wear and tear.

Q. 126. *Why ordinary mechanical braking cannot be totally replaced by electrical braking?*

Ans. Because the electrical braking *cannot hold vehical stationary*.

Q. 127. *What is meant by 'mechanical regenerative braking'?*

Ans. During coasting a part of the stored energy (kinetic energy) is utilised in propelling the train against frictional and other resistances to motion and, therefore, speed falls. Thus coasting may be considered as a form of ***mechanical regenerative braking.***

SECTION B: CONVENTIONAL QUESTIONS WITH SOLUTIONS

Q. 10.1. ***Induction motor.*** *A rotor of a 4-pole, 50 Hz slipring induction motor has a resistance of 0.25 ohm per phase and runs at 1440 r.p.m. at full load. Calculate the external resistance per phase which must be added to lower the speed to 1200 r.p.m., the torque remaining the same.*

Solution. *Given:* $p = 4$; $f = 50$ Hz; $R_2 = 0.25\ \Omega$/phase; $N_1 = 1440$ r.p.m., $N_2 = 1200$ r.p.m.

External resistance to be added per phase, r:

Synchronous speed, $$N_s = \frac{120f}{p} = \frac{120 \times 50}{4} = 1500 \text{ r.p.m.}$$

Full load slip, $$s_1 = \frac{N_s - N_1}{N_s} = \frac{1500 - 1440}{1500} = 0.04$$

New slip, $$s_2 = \frac{N_s - N_2}{N_s} = \frac{1500 - 1200}{1500} = 0.2$$

For constant imput power to rotor and rotor current, slip $s_2 \propto R_2$

If a resistance of r ohms be connected in series per phase, then

$$\frac{R_2 + r}{R_2} = \frac{s_2}{s_1}$$

or, $$\frac{0.25 + r}{0.25} = \frac{0.2}{0.04}$$

or, $$0.04\,(0.25 + r) = 0.25 \times 0.2$$

or, $$r = 1\ \Omega$$

Hence, *external resistance/phase to be added* = **1 Ω. (Ans.)**

Q. 10.2. (*a*) *Why is it not advisable to put very hot things in the refrigerator?*

(*b*) ***Electrolysis.*** *A copper refining plant using 500 electrolytic cells carries a current of 6000 A, voltage per cell being 0.25 V. If the plant were to work for 40 hours/week. Calculate energy consumption per tonne, assuming E.C.E. of copper 0.3281 mg/coulomb of electricity.*

Solution. (*a*) It is *not* advisable to put very hot things in the refrigerator, because the system has to do *more work* to cool down the things. As a result the *system gets overloaded,* this affects life of the system badly.

(*b*) *Given:* No. of electrical cells used = 500

Voltage per cell = 0.25 V

Current carried = 6000 A

No. of working hours of the plant per week = 40 hours

E.C.E. of copper, $z = 0.3281$ mg/coulomb

$= 32.81 \times 10^{-8}$ kg/C

$= 32.81 \times 10^{-8} \times 3600$

$= 0.00118116$ kg/A-h

Total number of amperes-hours per annum

$= 500 \times 6000 \times 40 \times 52 = 624 \times 10^7$ A-h

Annual output of plant $= 0.00118116 \times 624 \times 10^7$ kg

$= 7370.4$ tonnes

Energy consumed per annum = A-h per annum × voltage per cell

$= 624 \times 10^7 \times 0.25 = 156 \times 10^7$ W-h

$= 156 \times 10^4$ kWh

$$\textit{Energy consumption per tonne} = \frac{156 \times 10^4}{7370.4} = \mathbf{211.66\ kWh/tonne.}\quad \textbf{(Ans.)}$$

Q. 10.3. ***Illumination.*** *A lamp having a uniform C.P. of 200 in all directions is provided with a reflector which directs 60% of the total light uniformly on to a circular area of 10 m diameter. The lamp is hung 6 m above the area. Calculate the illumination:*

(*i*) *At the centre and edge of the surface with and without reflector.*

(*ii*) *Over the area without the reflector.*

Solution. *Given:* C.P. = 200; $h = 6$ m; $\eta_{ref.} = 60\%$

(*i*) **The illumination at the centre without reflector:**

The illumination at the centre (without reflector)

$$= \frac{\text{C.P.}}{h^2} = \frac{200}{6^2} = \mathbf{5.56\ lux.}\quad \textbf{(Ans.)}$$

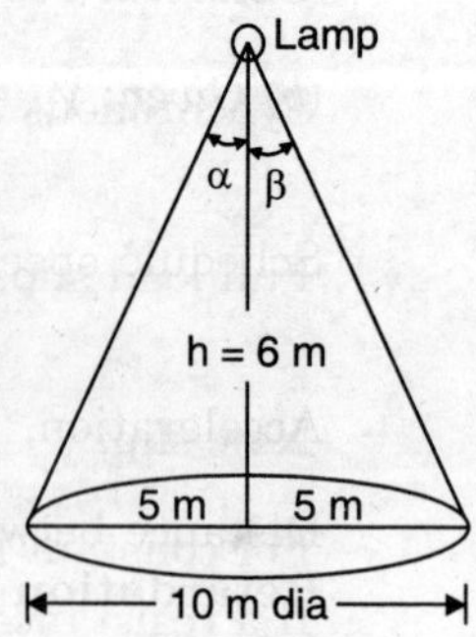

Fig. 10.1

The illumination at the edge of the surface with and without the reflector:

The illumination at the edge of the surface *without reflector*

$$= \frac{\text{C.P.}}{h^2}\cos^3\theta = \frac{200}{6^2} \times \left(\frac{6}{\sqrt{6^2 + 5^2}}\right)^3 = \mathbf{2.52\ lux.}\quad \textbf{(Ans.)}$$

With reflector the *illumination at the edge and at the centre will be same since the reflector directs the light uniformly on the surface.*

Total lumens given out, $F = 4\pi \times \text{C.P.} = 4\pi \times 200 = 800\,\pi$

$$\text{Total lumens reaching the surface} = F \times \frac{60}{100} = 800\pi \times \frac{60}{100} = 480\,\pi$$

$$\text{Total surface area} = \pi r^2 = \pi \times 5^2 = 25\pi \text{ m}^2$$

$$\therefore \quad \text{Average illumination with reflector} = \frac{480\,\pi}{25\pi} = \mathbf{19.2\ lux.} \quad \textbf{(Ans.)}$$

(*ii*) **Average illumination over the area without the reflector:** Average illumination of the surface without reflector will be found by first determining the solid angle subtended by the surface at the lamp and then finding out the luminous flux emitted in that solid angle.

Solid angle subtended by the area at the lamp,

$$\omega = 2\pi\left(1 \cos\frac{\alpha}{2}\right) = 2\pi\,(1 - \cos\theta) = 2\pi\left(1 - \frac{6}{\sqrt{6^2 + 5^2}}\right)$$

$$= 0.464\pi \text{ steradians}$$

Total flux reaching the surface $= I\omega = 200 \times 0.464\pi = 92.8\pi$ lux

$$\therefore \quad \text{Average illumination} = \frac{92.8\pi}{\pi \times 5^2} = \mathbf{3.7\ lux.} \quad \textbf{(Ans.)}$$

Q. 10.4. (*a*) *What are the applications of induction drives?*

(*b*) ***Train movement.*** *A suburban electric train has a maximum speed of 70 km/h. The schedule speed including a station stop of 30 seconds is 45 km/h. If the acceleration is 1.5 km/h/s, find the value of retardation when the average distance between stops is 4 km.*

Solution. (*a*) The applications of induction drives are:

(*i*) *Slip ring induction motor:* Generators, line shafts, lifts, pumps etc.

(*ii*) *Squirred cage induction motor:* Industrial drives, water pumps, tube wells etc.

(*iii*) *Single phase induction motor:* Fans, record players, refrigerator, washing machines, air-conditioner etc.

(*b*) *Given:* $V_m = 70 \text{ km/h} = \dfrac{70 \times 1000}{3600} = 19.44 \text{ m/s}$; Duration of stop = 30 s;

Schedule speed, $\quad V_{sch} = 45 \text{ km/h} = \dfrac{45 \times 1000}{3600} = 12.5 \text{ m/s}$

Acceleration, $\quad \alpha = 1.5 \text{ km/h/s} = \dfrac{1.5 \times 1000}{3600} = 0.417 \text{ m/s}^2$;

Distance between stops, $\quad D = 4 \text{ km} = 4000 \text{ m}$

Retardation β:

Schedule time of run, $\quad t_{sch} = \dfrac{D}{V_{sch}} = \dfrac{4000}{12.5} = 320 \text{ s}$

Actual time of run, $\quad t = t_{sch} - 30 = 320 - 30 = 290 \text{ s}$

We know that, $kV_m^2 - V_m t + D = 0$ ...(Chapter 7)

where, $\quad k = \dfrac{1}{2}\left(\dfrac{1}{\alpha} + \dfrac{1}{\beta}\right)$ or, $\quad k \times (19.44)^2 - 19.44 \times 290 + 4000 = 0$

or, $\quad k = 4.333$

or, $$k = \frac{1}{2}\left(\frac{1}{\alpha}+\frac{1}{\beta}\right) = 4.333 \quad \text{or} \quad \frac{1}{2}\left(\frac{1}{0.417}+\frac{1}{\beta}\right) = 4.333$$

or, $$\frac{1}{0.417}+\frac{1}{\beta} = 2 \times 4.333 = 8.666$$

or, $$2.398 + \frac{1}{\beta} = 8.666$$

$\therefore$ $$\beta = 0.1595 \text{ m/s}^2 = 0.1595 \times \frac{3600}{1000} = \mathbf{0.574\ km/h/s.} \quad \textbf{(Ans.)}$$

Q. 10.5. *Derive an expression for the motor torque driving a rolling mill when equipped with flywheel.*

Solution. Refer to Art. 5.16.7

Q. 10.6. *(a) Calculate sag for a trolley wire.*

(b) A trolley wire suspended from two supports 40 m apart. The diameter of the wire is 10 mm and has a weight of 0.5 kg/m. If tension is 500 kg, find:

(i) Sag, *(ii) Length of wire required.*

Solution. *(a)* Refer to Fig. 10.2. A flexible wire suspended horizontally comes down giving **"sag"** and assumes the shape of a catenary. In trolley wires, however sag is limited to 1% (max.) of the span and therefore can be considered as parabola.

Let, $2l$ = Span in metres,

w = Weight of wire in kg/m,

T = Tension of wire in kg, and

S = Sag.

Taking moments about the point P, we have

$$T \times y = w \,.\, x \times \frac{y}{2}$$

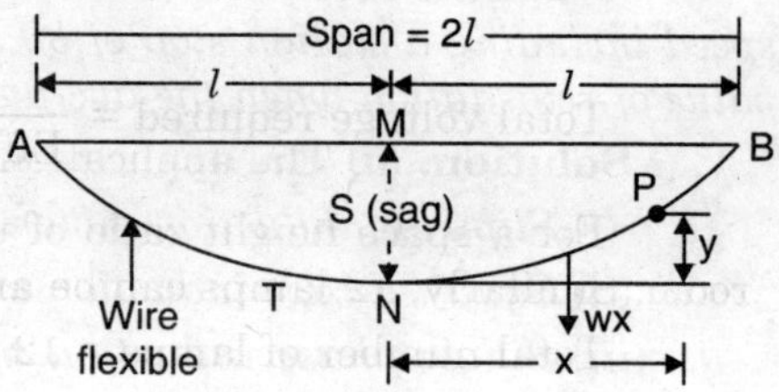

Fig. 10.2

or $$y = \frac{wx^2}{2T}$$

At support A (or B), $x = l$ and $y = S$

$\therefore$ Sag, $$S = \frac{wl^2}{2T} \quad \text{...(i)}$$

The equation of parabola is given by

$$y = \frac{Sx^2}{l^2} = \frac{w.x^2}{2T} \qquad \left(\because\ S = \frac{wl^2}{2T}\right)$$

If the length of the conductor in *half span* = l_{wire} then,

$$\frac{dl_{\text{wire}}}{dx} = \sqrt{1+\left(\frac{dy}{dx}\right)^2} = \sqrt{1+\left(\frac{wx}{T}\right)^2}$$

$$\int dl_{\text{wire}} = \int\left[1+\frac{w^2x^2}{T^2}\right]^{\frac{1}{2}} dx$$

Integrating and solving above, and of $x = l$, we get:

Length of conductor in half span,

$$l_{\text{wire}} = l + \frac{w^2l^3}{6T^2} = l + \left(\frac{2T.S}{l^2}\right)^2 \times \frac{l^3}{6T^2} = l + \frac{2S^2}{3l} \quad \text{...(ii)}$$

$$\left[\text{Substituting, } w = \frac{2TS}{l^2} \text{ from eqn. } (i)\right]$$

(*b*) *Given:* Span = $2l$ = 40 m (*i.e.*, l = 20 m); d_{wire} = 10 mm = 0.01 m; w = 0.5 kg/m; T = 500 kg.

(*i*) **Sag, S:**

$$S = \frac{wl^2}{2T} = \frac{0.5 \times 20^2}{2 \times 500} = \mathbf{0.2\ m.}\quad \textbf{(Ans.)}$$

(*ii*) **Length of wire required:**

Length of wire required for total span,

$$2l_{wire} = 2\left(l + \frac{2S^2}{3l}\right) = 2\left(20 + \frac{2 \times 0.2^2}{3 \times 20}\right) = \mathbf{40.003\ m.}\quad \textbf{(Ans.)}$$

Q. 10.7. ***Illumination.*** *An illumination on the working plane of 95 lux is required in a room 60 m × 15 m in size. The lamps are required to be hung 5 m above the work bench. Assuming a space height ratio as unity, a utilisation factor of 0.43, a lamp efficiency of 15 lumens per watt and candle power depreciation of 23% estimate the number, rating and disposition of lamps.*

Solution. *Given:* Area, A = 60 m × 15 m = 900 m^2; Illumination required; E = 75 lux (or lumens/m^2); utilisation factor, UF = 0.43; lamp efficiency = 16 lumens/watt; Maintenance factor, MF = 1 – C.P. depreciation = 1 – 0.23 = 0.77.

Number, rating and disposition of lamps:

$$\text{Gross lumens required} = \frac{E \times A}{UF \times MF} = \frac{95 \times 900}{0.43 \times 0.77} = 258230$$

$$\text{Total voltage required} = \frac{\text{Total lumens}}{\text{Efficiency of each lamp}} = \frac{258230}{15} = 17215$$

For a space height ratio of unity, only three lamps can be mounted along the width of the room. Similarly, 12 lamps can be arranged along the length of the room.

Total number of lamps = 12 × 3 = **36. (Ans.)**

$$\text{Voltage of each lamp} = \frac{17215}{36} \simeq 478 \text{ W say 500 W (nearest standard lamp).}$$

Hence, *rating of each lamp* = **500 W. (Ans.)**

The arrangement of the lamps will be as shown in Fig. 10.3.

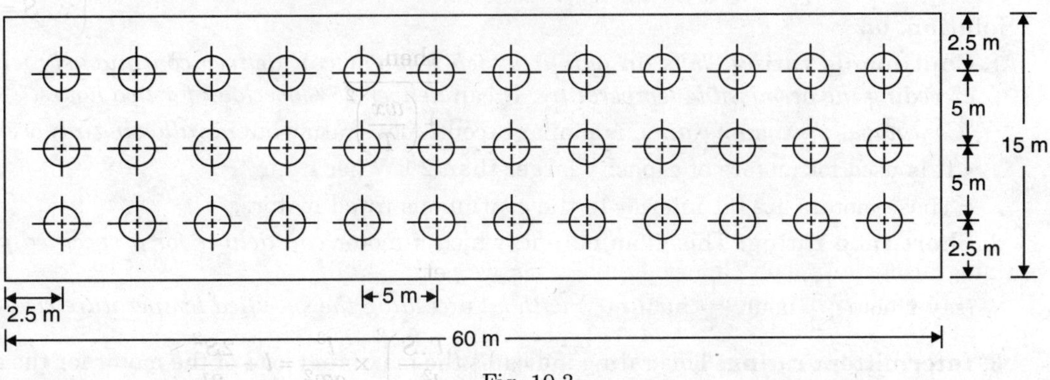

Fig. 10.3

Q. 10.8. (*a*) *Explain the various factors to be considered while designing street lighting.*

(*b*) ***Electrolysis:*** *Find the thickness of the copper deposited on a plate area of 0.00025 m*^{2} *during electrolysis if a current of one ampere is passed for 100 minutes. Density of copper is 8900 kg/m*^{3} *and E.C.E. of copper is 32.95 × 10*$^{-8}$ *kg/C.*

Solution. (*a*) Refer to Art. 1.11.

(*b*) *Given:* Plate area, A_s (surface area) = 0.00025 m^2

Current strength, $I = 1$ A

Duration of passage of 1 A current, t = 100 minutes = 100 × 60 = 6000 s

E.C.E. of copper, $z = 32.95 \times 10^{-8}$ kg/C

Density of copper = 8900 kg/m^3.

Thickness of copper deposited:

Mass of copper deposited, $m = zIt$

$$= 32.95 \times 10^{-8} \times 1 \times 6000 = 0.001977 \text{ kg}$$

$$\text{Volume of copper deposited, } V = \frac{\text{Mass}}{\text{Density}} = \frac{0.001977}{8900} = 0.222 \times 10^{-6} \text{ m}^3$$

$$\text{Thickness of copper deposited} = \frac{V}{A_s} = \frac{0.222 \times 10^{-6}}{0.00025} = \mathbf{0.888 \text{ mm.}} \quad \textbf{(Ans.)}$$

Q. 10.9. (*a*) *Explain what do you understand by 'continuous', 'short time', intermittent rating of an electric motor.*

(*b*) ***Electric heating and welding.*** *A 3-phase arc furnace is used to melt 4.3 tonnes of steel in one hour. Determine the average kW and kVA input to the furnace, arc voltage and p.f. of the current drawn from the supply, using the following data:*

Specific heat of steel	*= 444 J-kg/°C*
Latent heat of steel	*= 37.25 kJ/kg*
Melting point of steel	*= 1370°C*
Initial temperature of steel	*= 19.1°C*
Overall efficiency of furnace	*= 50% = 0.5*
Input current/phase	*= 5700 A*

Resistance of transformer referred to secondary and reactance of the transformer referred to secondary are 0.008 Ω and 0.014 Ω respectively.

Solution. (*a*)

1. **Continuous rating.** *It is an output which a motor can deliver continuously without exceeding the permissible temperature.* It can *deliver 25% overload for two hours.*

   "*Continuous maximum rating*" is similar to continuous rating but *not allowing any overload.*

   • It is used for motors of capacity larger than 2 kW per r.p.m.

   • These motors are bit inferior to the continuous-rated motors.

2. **Short time rating:** This is an output, which a motor *can deliver for a specified period* (say 1 hour, $\frac{1}{2}$ hour, $\frac{1}{4}$ hour etc.) *without exceeding the specified temperature rise.*

3. **Intermittent rating:** This rating indicates the maximum load of the motor for the specified time *followed by a no-load period during which the machine cools down to its original temperature.*

- For intermittent loads, short-time rated motors can be used if the *loading conditions corresponds to the short time rating of the motor* or if it can be assumed that the heating of the motor is proportional to the square of the current, the equivalent continuous rating can be found by taking the r.m.s. value of the load curve and choosing a motor having a continuous rating equal to this. This method is *very frequently used.*

(*b*) *Given:* Quantity of steel to be melt = 4.3 tonnes; C (specific heat) = 444 J-kg/°C = 0.444 kJ/kg°C; L = 37.25 kJ/kg; t_i = 19.1°C; $t_{melt.}$ = 1370°C; $\eta_{furnace}$ = 50% = 0.5; time required to melt the steel = 1 hour, I = 5700 A; R_{02} = 0.008 Ω; X_{02} = 0.014 Ω.

Average kW and kVA input to the furnace, are voltage and power factor:

Energy required to melt 4.3 tonnes of steel

$$= m\,[c \times (t_{melt} - t_i) + L]$$

$$= 4.3 \times 10^3\,[0.444\,(1370 - 19.1) + 37.25] = 2.739 \times 10^6 \text{ kJ}$$

$$= \frac{2.739 \times 10^6}{3600} = 760.83 \text{ kWh} \qquad (\because \quad 1 \text{ kWh} = 3600 \text{ kJ})$$

Average kW input:

$$\text{Average output} = \frac{760.83}{1} = 760.83 \text{ kW}$$

$$\therefore \qquad \text{Average kW input} = \frac{760.83}{\eta_{furnace}} = \frac{760.83}{0.5} = \mathbf{1521.66\ kW.} \quad \textbf{(Ans.)}$$

Arc voltage V_{arc}:

Voltage drop due to resistance

$$= 5700 \times 0.008 = 45.6 \text{ V}$$

Voltage drop due to reactance

$$= 5700 \times 0.014 = 79.8 \text{ V}$$

Open circuit voltage of transformer secondary (Fig. 10.4),

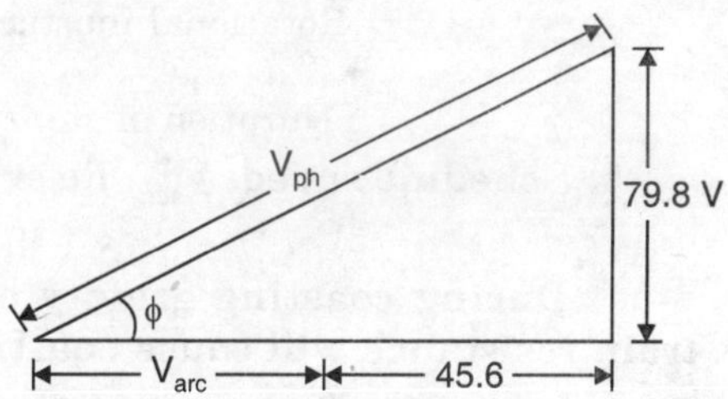

Fig. 10.4

$$V_{ph} = \sqrt{(V_{arc} + 45.6)^2 + (79.8)^2} \qquad \ldots(i)$$

$$\text{p.f. } (\cos\phi) = \frac{(V_{arc} + 45.6)}{\sqrt{(V_{arc} + 45.6)^2 + (79.8)^2}} \qquad \ldots(ii)$$

Total power input $= 3\,V_{ph}\,I_{ph}\cos\phi$

$$= 1521.66 \times 10^3$$

$$= 3 \times \sqrt{(V_{arc} + 45.6)^2 + (79.8)^2} \times 5700 \times \left[\frac{V_{arc} + 45.6}{\sqrt{(V_{arc} + 45.6)^2 + (79.8)^2}}\right]$$

From which, V_{arc} **= 43.4 V. (Ans.)**

Open circuit phase voltage (secondary)

$$V_{ph} = \sqrt{(V_{arc} + 45.6)^2 + (79.8)^2}$$

$$= \sqrt{(43.4 + 45.6)^2 + (79.8)^2} = 119.5 \text{ V}$$

Input kVA:

Input kVA taken from the supply

$$= 3 \times V_{ph} \times I_{ph} \times 10^{-3} = 3 \times 119.5 \times 5700 \times 10^{-3}$$

$$= \mathbf{2043.5\ kVA.} \quad \textbf{(Ans.)}$$

Power factor, p.f.:

$$\text{p.f.} = \frac{(V_{arc} + 45.6)}{\sqrt{(V_{arc} + 45.6)^2 + (79.8)^2}} = \frac{(43.4 + 45.6)}{\sqrt{(43.4 + 45.6)^2 + (79.8)^2}}$$

$$= \mathbf{0.744.} \quad \textbf{(Ans.)}$$

Q. 10.10. (*a*) *Explain the working of central air-conditioning system.*

(*b*) ***Train movement and energy consumption.*** *The following data relate to a 200 tonne electric train running according to the following quadrilateral speed-time curve:*

(i) Uniform acceleration from rest at 2 km/h/s for 30 seconds,

(ii) Coasting for 50 seconds, and

(iii) Duration of braking: 15 seconds.

Up-gradient = 1%; Train resistance = 40 N/t;

Rotational inertia effect = 10%; Duration of stops = 15 s.

Find the schedule speed.

Solution: *Given:* $M = 200$ tonnes; $\alpha = 2$ km/h/s;

$t_1 = 30$ s; $t_2 = 50$ s; $t_3 = 15$ s;

Up-gradient = 1%; $r = 40$ N/t;

Rotational inertia = 10% *i.e.*, $\frac{M_e}{M} = 1.1$;

Duration of stops = 15 s,

Schedule speed, V_{sch}: Refer to Fig. 10.5.

$$V_1 = \alpha \,.\, t_1 = 2 \times 30 = 60 \text{ km/h}$$

During coasting gravity component and train resistance will cause coasting retardation β_c.

$$\text{Retarding force} = (98\, MG + Mr) \text{ newton}$$
$$= 98 \times 200 \times 1 + 200 \times 40$$
$$= 27600 \text{ N}$$

Also, $277.8\, M_e \,.\, \beta_c = 27600$

or, $277.8 \times (1.1 \times 200) \times \beta_c = 27600$

i.e., $$\beta_c = \frac{27600}{277.8 \times (1.1 \times 200)}$$
$$= 0.45 \text{ km/h/s.}$$

Now, $$V_2 = V_1 + \beta_c t_2$$
$$= 60 + (-0.45) \times 50$$
$$= 37.5 \text{ km/h}$$

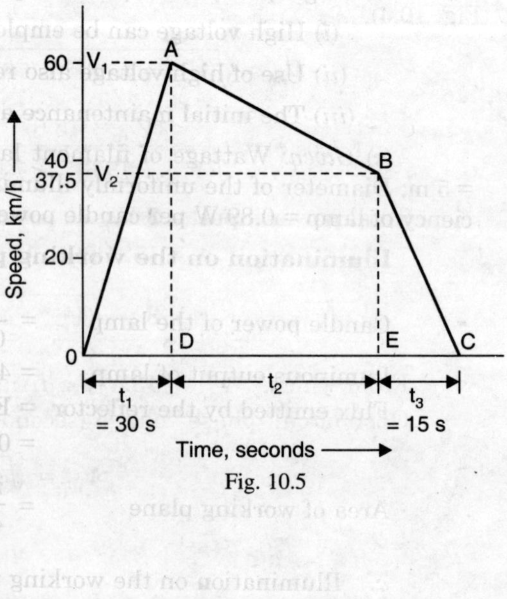

Fig. 10.5

Braking retardation, $\beta = \frac{V_2}{t_3} = \frac{37.5}{15} = 2.5$ km/h/s

Distance travelled during acceleration (area *OAD*)

$$= \frac{1}{2} V_1 . t_1 = \frac{1}{2} \times 60 \times \left(\frac{30}{3600}\right) = 0.25 \text{ km}$$

Distance travelled during coasting (area *ABED*)

$$= \left(\frac{V_1 + V_2}{2}\right) \times t_2 = \left(\frac{60 + 37.5}{2}\right) \times \frac{50}{3600} = 0.677 \text{ km}$$

Distance travelled during braking (area *BCE*)

$$= \frac{1}{2} V_2 . t_3 = \frac{1}{2} \times 37.5 \times \left(\frac{15}{3600}\right) = 0.078 \text{ km}$$

All distance travelled, $D = 0.25 + 0.677 + 0.078 = 1.005$ km

All schedule time, $t_{sch} = 30 + 50 + 15 + 15 = 110$ s

Schedule speed, $V_{sch} = \frac{D}{t_{sch}} = \frac{1.005}{(110/3600)} = $ **32.9 km/h. (Ans.)**

Q. 10.11. (*a*) *Suggest suitable type of motors with reasons for spinning mills and cement mills.*

(b) Why single phase system is preferred for main line railway service?

(c) ***Illumination.*** *A filament lamp of 20 W is suspended at a height of 5 metres above working plane and gives uniform illumination over an area of 4 m diameter. Assume efficiency of reflector as 50%. Determine the illumination on the working plane. Efficiency of the lamp is 0.89 watt per candle power.*

Solution.

(*a*) For **spinning work,** moderate starting torque and smooth acceleration is required, so **"squirrel cage motor,"** may be used.

- Starting torque in **cement mills** is 120% of the full-load torque. So, **"slip ring induction motors"** may be used.

(*b*) Single phase A.C. system is preferred for main line for the following **reasons**:

(*i*) High voltage can be employed, due to which, cost of distribution is reduced.

(*ii*) Use of high voltage also reduces the number of substations.

(*iii*) The initial maintenance and operating costs of single phase system is less.

(*c*) *Given:* Wattage of filament lamp = 20 W; Height of the lamp above the working plane = 5 m; Diameter of the uniformly illuminated area = 4 m; Efficiency of reflector = 50% = 0.5; Efficiency of lamp = 0.89 W per candle power.

Illumination on the working plane:

Candle power of the lamp $= \frac{20}{0.89} = 22.472$ C.P.

Luminous output of lamp $= 4\pi \times 22.472 = 89.89\pi$

Flux emitted by the reflector = Efficiency of reflector × total lumens output of the lamp

$= 0.5 \times 89.89\,\pi = 44.945\,\pi$ lumens

Area of working plane $= \frac{\pi}{4} \times (4)^2 = 4\pi$

$\therefore$ Illumination on the working plane $= \frac{44.945\pi}{4\pi} =$ **11.24 lumens/m². (Ans.)**

Q. 10.12. (*a*) *What will happen if automatic starter relay stops to operate in refrigerator?*

(*b*) *What system of lighting should be recommended for drawing office?*

(*c*) ***Train movement and energy consumption.*** *An electric train has an average speed of 42 km/h on a level track between stops 400 m apart. It is accelerated at 1.7 km/h/s and is braked at 3.3 km/h/s. Draw the speed time curve for the run.*

Solution. (*a*) The relay is provided to connect auxiliary winding at the start of the motor and disconnect the same, when the motor picks up the speed.

(*b*) For drawing office, "**direct lighting system**" may be adopted on the working table. Light should be concentrated by reflector and shadow and glare should be avoided.

(*c*) Refer to Example 7.6.

Q. 10.13. (*a*) *Explain briefly the trolley bus system of electric traction.*

(*b*) ***Electric traction.*** *A train runs between two stations 1.8 km apart at an average speed of 40 km/h. The run is to be made according to simplified quadrilateral speed-time curve. The maximum speed is limited to 60 km/h, acceleration to 2 km/h/s, coasting retardation to 0.15 km/h/s and braking retardation 3 km/h/s.*

Determine the duration of acceleration, coasting and braking periods.

Solution. *Given:* Distance between two stations, $D = 1.8$ km = 1800 m;

$$V_a = 40 \text{ km/h} = \frac{40 \times 1000}{60 \times 60} = 11.11 \text{ m/s};$$

$$V_1 = 60 \text{ km/h} = \frac{60 \times 1000}{60 \times 60} = 16.67 \text{ m/s};$$

$$\alpha = 2 \text{ km/h/s} = \frac{2 \times 1000}{60 \times 60} = 0.555 \text{ m/s}^2;$$

$$\beta_c = 0.15 \text{ km/h/s} = \frac{0.15 \times 1000}{3600} = 0.042 \text{ m/s}^2;$$

$$\beta = 3 \text{ km/h/s} = \frac{3 \times 1000}{60 \times 60} = 0.833 \text{ m/s}^2.$$

Duration of acceleration, coasting and braking periods, t_1, t_2, t_3:

Refer to Fig. 10.6.

Duration of acceleration

$$t_1 = \frac{V_1}{\alpha} = \frac{16.67}{0.555} = \mathbf{30\ s.} \quad \textbf{(Ans.)}$$

Actual time of run,

$$t = \frac{D}{V_a} = \frac{1800}{11.11} = 162 \text{ s}$$

Let the speed at the end of coasting period be V_2.

Then duration of coasting,

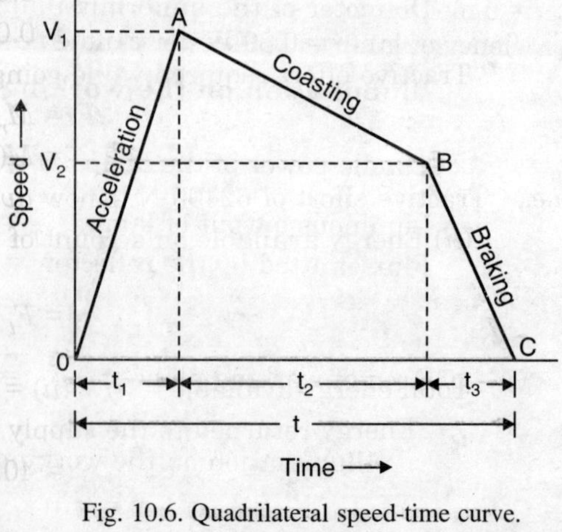

Fig. 10.6. Quadrilateral speed-time curve.

$$t_2 = \frac{V_1 - V_2}{\beta_c} = \frac{16.67 - V_2}{0.042} \text{ sec.}$$

Braking period, $t_3 = \dfrac{V_2}{\beta} = \dfrac{V_2}{0.833}$ sec.

Now, $t = t_1 + t_2 + t_3$

or, $162 = 30 + \dfrac{16.67 - V_2}{0.042} + \dfrac{V_2}{0.833}$

or, $162 = 30 + 396.9 - 23.8\,V_2 + 1.2\,V_2$

or, $V_2 = 11.72$ m/s

$\therefore$ *Coasting period,* $t_2 = \dfrac{16.67 - 11.72}{0.042} \simeq$ **117.9 s. (Ans.)**

and, *Braking period,* $t_3 = \dfrac{V_2}{0.833} = \dfrac{11.72}{0.833} \simeq$ **14.1 s. (Ans.)**

Q. 10.14. *(a) Why are A.C. drives preferred?*

(b) ***Traction drives.*** *A 400 tonnes train proceeds down a gradient of 2%. Its speed is reduced by applying regenerative braking from 40 km/h to 20 km/h while going a distance of 2 km. If the tractive resistance is 40 N/tonne, rotational inertia allowed is 10% and conversion efficiency is 75%, calculate energy returned to supply.*

Solution. *(a)* The A.C. drives are preferred because A.C. motors:

(i) are compact and have higher efficiency;

(ii) have a uniform torque;

(iii) have more torque as compared to D.C. motors of the same rating;

(iv) do not need a commutator which is most troubled spot in D.C. motors.

(b) Given: $M = 400$ tonnes; $G = 2\%$; Distance, $s = 2$ km $= 2000$ m;
$V_1 = 40$ km/h; $V_2 = 20$ km/h; tractive resistance,
$r = 40$ N/tonne; $\dfrac{M_e}{M} = 1.1$, $\eta = 75\%$

Energy returned to the supply:

(i) Energy available due to reduction in speed

$= 0.01702\,M_e\,(V_1^2 - V_2^2)$ ...[Refer to Art. 8.10.6]

$= 0.01702 \times 400 \times 1.1\,(40^2 - 20^2) = 5660$ Wh or 5.66 kWh

Tractive effort required while going down the gradient,

$F_t = M_r - 98$ MG

$= M(r - 98G) = 400\,(40 - 98 \times 2) = -62400$ N

i.e., Tractive effort of 62400 N is now *available.*

(ii) Energy available on account of moving down the gradient over a distance of 2000 m

$$= F_t \times D = \frac{62400 \times 2000}{1000 \times 3600} = 34.67 \text{ kWh}$$

Total energy available = *(i)* + *(ii)* = 5.66 + 34.67 = 40.33 kWh

$\therefore$ Energy returned to the supply = 40.33 × η

= 40.33 × 0.75 = **30.25 kWh. (Ans.)**

Q. 10.15. *(a) Enlist the advantages of electric heating.*

(b) A resistance oven employing nichrome wire is to be operated from 220 V single-phase supply and is to be rated at 16 W. Find the length and diameter of the element wire, using the data given below:

The temperature of the element to be limited = 1170°C; average temperature of the charge = 500°C; Specific resistance of nichrome = 102 × 10^{-8} Ωm; Emissivity = 0.9; Radiating efficiency = 0.57.

Solution. *(a)* **Advantages of electric heating:**

The advantages of electric heating are:

(*i*) Cleanliness;
(*ii*) Ease of control
(*iii*) Low alternation and maintenance cost;
(*iv*) Automatic control;
(*v*) Quickness;
(*vi*) Uniform heating;
(*vii*) Absence of flue gases;
(*viii*) Automatic protection;
(*ix*) High efficiency;
(*x*) No upper limit of temperature;
(*xi*) Heating insulating materials.

(*b*) *Given:*

$V = 220$ volts; $P = 16$ kW;

$T_1 = 1170 + 273 = 1443$ K; $T_2 = 500 + 273 = 773$ K;

$\rho = 109 \times 10^{-8}$ Ωm; $e = 0.9$; $\eta_{rad} = 0.57$

Length and diameter of element wire: l, d:

We know that,

$$\frac{l}{d^2} = \frac{\pi V^2}{4\rho P} \quad \text{...[Eqn. (2.9)]}$$

$$= \frac{\pi \times 220^2}{4 \times 109 \times 10^{-8} \times (16 \times 10^3)} = 2179660 \quad ...(i)$$

Now, heat radiated,

$$H = 5.67\, \eta_{rad}\, e \left[\left(\frac{T_1}{100}\right)^4 - \left(\frac{T_2}{100}\right)^4\right] \text{ W/m}^2 \quad \text{...[Eqn. (2.7)]}$$

$$= 5.67 \times 0.57 \times 0.9 \left[\left(\frac{1443}{100}\right)^4 - \left(\frac{773}{100}\right)^4\right] = 115729 \text{ W/m}^2$$

Also, total heat dissipated/sec. = Electrical power input

$\therefore \quad (\pi d) \times l \times 115729 = 16 \times 10^3 \quad \therefore \quad dl = 0.044$

or, $\quad d^2 l^2 = 0.001936 \quad ...(ii)$

Multiplying (*i*) and (*ii*), we have:

$$\frac{l}{d^2} \times d^2 l^2 = 2179660 \times 0.001936$$

or, $\quad l =$ **16.16 m. (Ans.)**

and $\quad d = \dfrac{0.044}{16.16} = 2.723 \times 10^{-3}$ m = **2.723 mm. (Ans.)**

Q. 10.16. *(a) What is negative resistance characteristic of arc?*

(b) In the case hardening of a steel pulley, the depth of penetration required is 1.25 mm. If the relative permeability is unity and the specific resistivity of steel is 5 × 10^{-7} Ω m, determine the frequency required.

Solution. (*a*) **Negative resistance characteristic of arc:**

The arc has a *negative* temperature of resistance, *i.e.,* resistance of the arc decreases with increase of temperature (or current). The decreased resistance further increases the current (Fig. 10.7). Therefore, it is not possible to obtain a steady arc. To obtain a steady arc, supply source should have drooping characteristics (Fig. 10.8).

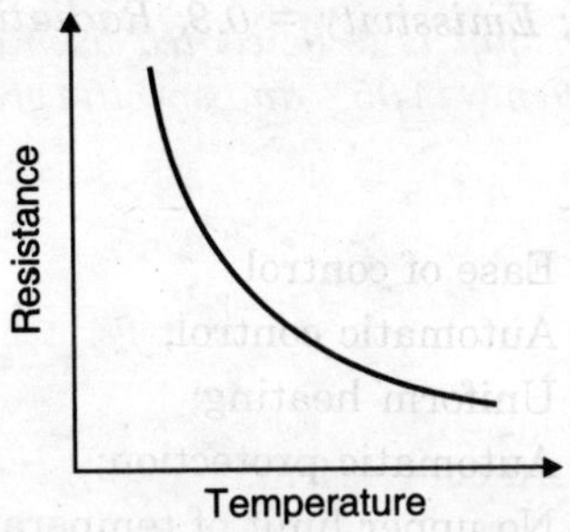

Fig. 10.7

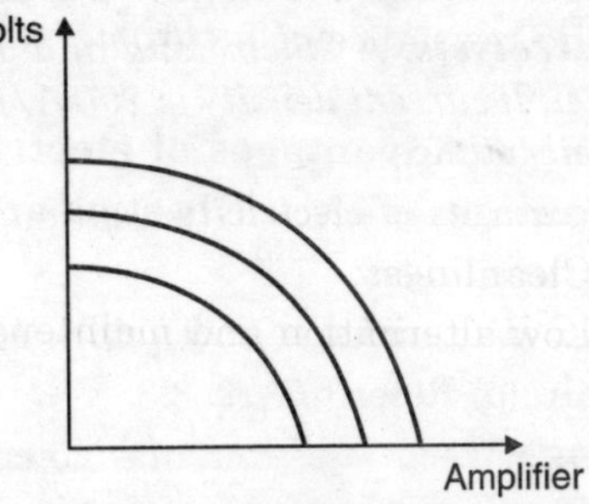

Fig. 10.8

In order to obtain a steady arc for a D.C. welding set, D.C. compound generator is used, the drooping characteristics is obtained by using reactors.

(*b*) *Given:* Depth of penetration, $d = 1.25$ mm $= 0.125$ cm; $\mu_r = 1$; $\rho = 5 \times 10^{-7}$ Ωm.

Frequency, *f*:

Using the relation: $$d = \frac{1}{2\pi}\sqrt{\frac{\rho \times 10^9}{\mu_r . f}} \text{ cm} \quad \text{...[Eqn. (2.18)]}$$

Squaring both sides and simplifying, we have:

$$f = \frac{\rho \times 10^9}{d^2 \times 4\pi^2 \times \mu_r} = \frac{(5 \times 10^{-7} \times 100) \times 10^9}{(0.125)^2 \times 4\pi^2 \times 1} = \mathbf{81056 \text{ Hz.}} \quad \textbf{(Ans.)}$$

Q. 10.17. (*a*) *What is vapour compression refrigeration system? Explain its cycle.*

(*b*) *What are the requirements of an air-conditioning system?*

Solution. (*a*) Refer to Art. 4.8.

(*b*) **Air-conditioning** is the technique to "*condition*" the air or a room or space so that it is comfortable to human being and helps in the satisfactory operation of the equipment placed there in.

An **air-conditioning system** is defined as an assembly of different parts of the system used to produce a specified condition of air within a required space of building.

"Air-conditioning" involves the control of the following factors:

(*i*) **Temperature control:** The major function of an air-condition is to maintain comfortable temperature for human beings in the room. The comfortable temperature for human being is taken as 22°C to 26°C. To maintain this temperature the "summer air-conditioner" removes heat from the room and the "winter air-conditioner" adds heat in the room.

(*ii*) **Humidity control:** The air-conditioner maintains humidity (content of water vapours in the air) for comfort of the human being and for quality production of certain commodities. The comfortable percentage humidity for human beings in the air is taken as 50%. To maintain this, the "summer air-conditioner" uses *humidifier* (which adds humidity) and winter air-conditioner" uses *dehumidifier* (which removes humidity) from the air. Humidification involves spray of cold water and dehumidification involves use of water absorbing materials such as silica gel, calcium chloride etc.

(*iii*) **Air movement control:** To maintain comfortable temperature control of air movement in the room is necessary.

(*iv*) **Air quality control:** The air entering in the room should be free from dust, dist, pollen grains, bacteria etc. In other words, the air should be healthy. So the air entering the room is filtered, cleaned and purified, otherwise it can cause various diseases.

Q. 10.18. (*a*) *State and explain Faraday's laws of electrolysis.*

(*b*) ***Electrolysis:*** *A 25 cm long and 10 cm diameter shaft is to be coated with a thickness of 1.5 mm of nickel. If current density is 200 A/m², current efficiency is 95% and specific gravity of nickel is 8.9 g/cm, determine,*

(i) The quantity of electricity required, and

(ii) The time taken

Take E.C.E. of nickel = 1.09 kg per 1000 amp. hr.

Solution. (*a*) Refer to Art. 3.3

(*b*) *Given:*

$$l = 25 \text{ cm} = 0.25 \text{ m}; d = 10 \text{ cm} = 0.1 \text{ m};$$
$$t = 1.5 \text{ mm} = 0.0015 \text{ m};$$
$$\text{current density}, \delta = 200 \text{ A/m}^2; \text{Density of nickel} = 8.9 \times 10^3 \text{ kg/m}^3$$
$$\text{E.C.E}, Z = 30.4 \times 10^{-8} \text{ kg/C}.$$

(*i*) **Quantity of electricity required Q:**

Mass of nickel to be deposited,

$$m = \text{Surface area} \times \text{thickness of coating} \times \text{density of nickel}$$
$$= \pi dl \times 0.0015 \times 8.9 \times 10^3$$
$$= \pi \times 0.1 \times 0.25 \times 0.0015 \times 8.9 \times 10^3 = 1.0485 \text{ kg}$$

Theoretical value of quantity of electricity required,

$$Q = \frac{m}{z} = \frac{1.0485}{30.4 \times 10^{-8}} \text{ A.s} = \frac{1.0485}{30.4 \times 10^{-8} \times 3600} = \mathbf{958 \text{ A-h.}} \quad \textbf{(Ans.)}$$

(*ii*) **Time taken, t:**

Current density, $$\delta = \frac{\text{Current } (I)}{\text{Surface area } (A_s)}$$

or, $$200 = \frac{I}{\pi \times 0.1 \times 0.25} = 12.732I$$

or, $$I = 15.71 \text{ A}$$

Also, $$Q = I \times t$$

$\therefore$ $$t = \frac{Q}{I} = \frac{958}{15.71} = \mathbf{61 \text{ h.}} \quad \textbf{(Ans.)}$$

Q. 10.19. (*a*) *Define the following:*

(i) Space height ratio; *(ii) Utilisation factor;*

(iii) Depreciation factor.

(*b*) ***Electric heating and welding.*** *A slab of insulating material 125 cm² in area and 1 cm thick is to be heated by dielectric heating. The power required is 360 W at 30 MHz. Material has a relative permittivity of 5 and p.f. of 0.05. Absolute permittivity = 8.854 × 10⁻¹² F/m. Determine the voltage required.*

Solution. (*a*) Refer to Art. 1.2.

(*b*) *Given:* $A = 125\text{ cm}^2 = 125 \times 10^{-4}\text{ m}^2$;

$t = 1\text{ cm} = 0.01\text{ m}$; $P = 360\text{ W}$;

$f = 30\text{ MHz}$; $\varepsilon_r = 5$; p.f. = 0.05;

$\varepsilon_0 = 8.854 \times 10^{-12}\text{ F/m}$.

Voltage required, V:

Capacitance, $$C = \frac{\varepsilon_0 \varepsilon_r A}{t} = \frac{8.854 \times 10^{-12} \times 5 \times 125 \times 10^{-4}}{0.01} = 55.34 \times 10^{-12}\text{ F}$$

Also, $$P = 2\pi f\, CV^2 \cos\phi \qquad \text{...[Eqn. (2.19)]}$$

$$360 = 2\pi \times 30 \times 10^6 \times 55.34 \times 10^{-12} \times V^2 \times 0.05$$

or, $$V^2 = \frac{360}{2\pi \times 30 \times 10^6 \times 55.34 \times 10^{-12} \times 0.05} = 6902.27$$

or, $V =$ **830 volts. (Ans.)**

Q. 10.20. (*a*) *List the types of motors generally used for electrical drives.*

(*b*) ***Electrical drives.*** *A 6-pole, 50 Hz synchronous motor coupled to a load has a moment of inertia of 594 kg-m². If the load torque is independent of speed and the frictional torque is 310 Nm, calculate the time taken by the motor to come to stop and number of revolutions made during Rheostatic braking, the field current and braking resistance being kept constant at values which give initial electrical braking torque of 5800 Nm.*

Solution. (*a*) **Motors generally used for electrical drives:**

Following types of motors are generally used for electrical drive:

(*i*) D.C. shunt, series and compound motors.
(*ii*) Three-phase induction motor.
(*iii*) Compensated induction motor.
(*iv*) Schrage motor.
(*v*) Stator-fed commutator motor.
(*vi*) Three-phase series motor.
(*vii*) Synchronous and synchronous induction motor.
(*viii*) Single-phase series motor.
(*ix*) Repulsion motor.
(*x*) Single-phase induction motor.

(*b*) *Given:* $p = 6$, $f = 50$ Hz; Inertia,

$J = 594\text{ kg-m}^2$; frictional torque,

$T_f = 310$ Nm; Initial electrical braking torque,

$T = K\omega_1 = 5800$ Nm, where K is constant.

Time taken by the motor to come to stop, t_r:

Initial braking torque, $T = K\omega_1 = 5800$ Nm

Now, $$\omega_1 = \frac{2\pi N}{60} = \frac{2\pi\left(\frac{120f}{p}\right)}{60} = \frac{2\pi \times \left(\frac{120 \times 50}{6}\right)}{60} = 104.72\text{ rad/s}$$

$\therefore$ $$K = \frac{T}{\omega_1} = \frac{5800}{104.72} = 55.39$$

Time taken by the motor to stop,

$$t_r = \frac{J}{K} \ln \left(\frac{K\omega_1 + T_f}{T_f} \right) \qquad \text{...[Eqn. (5.92)]}$$

or,
$$t_r = \frac{594}{55.39} \ln \left(\frac{5800 + 310}{310} \right) = \mathbf{31.97 \ s.} \quad \textbf{(Ans.)}$$

Number of revolutions made before coming to rest, N_r:

$$N_r = \frac{1}{2\pi K} \left[\frac{J}{K} (K\omega_1 + T_f)(1 - e^{-Kt_r/J}) - T_f . t_r \right] \qquad \text{...[Eqn. (5.94)]}$$

$$= \frac{1}{2\pi \times 55.39} \left[\frac{594}{55.39} (55.39 \times 104.72 + 310)(1 - e^{(-55.39 \times 31.97/594)}) - 310 \times 31.97 \right]$$

$$= 0.002873 \, [65523 \, (1 - e^{-2.981}) - 9910.7] = 150.2$$

i.e., N_r = **150.2. (Ans.)**

Q. 10.21. (*a*) *Define the following:*

(i) Connected load; (ii) Load factor; (iii) Diversity factor.

(b) The peak load on a 50 MW power station is 40 MW. It supplies power through four transformers whose connected loads are 19, 11, 9, 10 MW. The maximum demands on these transformers are 16, 9, 8, 9 MW respectively. If the annual load factor is 55% and the plant is operating for 70% of the period in a year, find out the following:

(i) Average load on the station; *(ii) Energy supplied per year;*
(iii) Demand factor; *(iv) Diversity factor;*
(v) Power station use factor.

Solution. (*a*) (*i*) **Connected load.** The connected load on any system, or part of a system, is the *combined continuous rating of all the receiving apparatus on consumers' premises, which is connected to the system, or part of the system, under consideration.*

(*ii*) **Load factor.** *It is the ratio of the average power to the maximum demand.*

i.e.,
$$\text{Load factor} = \frac{\text{Average load}}{\text{Maximum load}}$$

(*iii*) **Diversity factor.** *It is the ratio of the maximum power demands of the subdivisions of the system, or part of the system, to the maximum demand of the whole system.*

i.e.,
$$\text{Diversity factor} = \frac{\text{Sum of individual maximum demands}}{\text{Maximum demand of entire group}}$$

It is always *greater than unity.*

(*b*) Rated capacity of the power station = 50 MW or 50,000 kW
Maximum demand on the power station = 40 MW or 40,000 kW
Sum of connected load = 19 + 11 + 9 + 10 = 49 MW or 49,000 kW
Sum of maximum demands on the transformers
= 16 + 9 + 8 + 9 = 42 MW or 42,000 kW
Annual load factor = 55% or 0.55
Plant operating period = 0.7 × (365 × 24) = 6132 h

(*i*) **Average load on the station:**

Average load on the station = Maximum load × load factor

= 40,000 × 0.55 = **22,000 kW. (Ans.)**

(*ii*) **Energy supplied per year:**

Energy supplied per year = Average load × (365 × 24)

= 22,000 × 365 × 24 = **192.72 × 10⁶ kWh. (Ans.)**

(*iii*) **Demand factor:**

$$\textit{Demand factor} = \frac{\text{Maximum demand}}{\text{Sum of connected load}}$$

$$= \frac{40,000}{49,000} = \mathbf{0.8163.} \quad \textbf{(Ans.)}$$

(*iv*) **Diversity factor:**

$$\textit{Diversity factor} = \frac{\text{Sum of maximum demands}}{\text{Maximum demand}} = \frac{42,000}{40,000} = \mathbf{1.05.} \quad \textbf{(Ans.)}$$

(*v*) **Power station use factor:**

$$\textit{Power station use factor} = \frac{\text{Energy generated per year}}{\text{Rated capacity} \times \text{Number of operating hours}}$$

$$= \frac{192.72 \times 10^6}{50,000 \times 6132} = \mathbf{0.629 \text{ or } 62.9\%.} \quad \textbf{(Ans.)}$$

Q. 10.22. *Illumination.* *(a) What are polar curves? Explain briefly.*

(b) A lamp of 200 candela is placed 1.2 m below a plane mirror which reflects 82% of light falling on it. The lamp is hung 5 m above the ground. Find the illumination at a point on the ground 3.5 m away from the point vertically below the lamp.

Solution. (*a*) Refer to Art. 1.4.

(*b*) The lamp *L* and the mirror arrangement is shown in Fig. 10.9. The lamp *L* produces an image *L′* as far behind the mirror as it is in front.

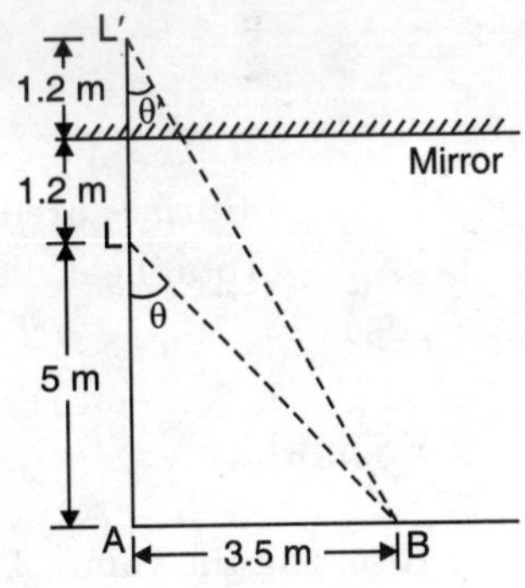

Fig. 10.9

Height of the image from the ground = 5 + 1.2 + 1.2 = 7.4 m

L′ acts as the secondary source of light and its candle power = 0.82 × 200 = 164 candela. Illumination at point *B* (3.5 m away from the point vertically below the lamp)

= Illumination due to *L* + illumination due to *L′*

i.e.,

$$E_B = \frac{200}{(5)^2}\cos^3\theta + \frac{164}{(7.4)^2}\cos^3\theta$$

$$= \frac{200}{5^2}\left(\frac{5}{\sqrt{5^2+3.5^2}}\right)^3 + \frac{164}{(7.4)^2}\left(\frac{7.4}{\sqrt{7.4^2+3.5^2}}\right)^3$$

= 4.399 + 2.212 = **6.61 lux. (Ans.)**

Q. 10.23. *Electric heating and welding.* *(a) Explain briefly Tungsten Inert Gas (TIG) welding, stating its advantages and applications.*

(b) A 24 kW, 3-phase, 400 V resistance oven is to employ nickel-chrome strip 0.25 mm thick for the three start-connected heating elements. If the temperature of the strip is to be 950°C and that of the charge be 600°C estimate a suitable width for the strip. Assume: Emissivity = 0.9; radiating efficiency = 0.5; Resistivity of the strip material = 101.6 × 10^{-8} Ωm.

Solution: *(a)* Refer to Art. 2.2.6.

(b) Given:

$$P_{ph} = \frac{24}{3} = 8 \text{ kW};\ V_{ph} = \frac{400}{\sqrt{3}} = 231 \text{ V};\ t = 0.25 \text{ mm} = 0.25 \times 10^{-3} \text{ m};$$

$$T_1 = 950° + 273 = 1223 \text{ K};\ T_2 = 600 + 273 = 873 \text{ K};\ e = 0.9,$$

$$\eta_{rad.} = 0.5;\ \rho = 101.6 \times 10^{-8}\ \Omega\text{m}.$$

Width of strip, w:

Let, the resistance of the strip be R, then

$$R = \frac{V_{ph}^2}{P} = \frac{(231)^2}{8 \times 10^3} = 6.67\ \Omega$$

Also, Resistance of the strip, $R = 6.67 = \frac{\rho l}{a} = \frac{\rho l}{w.t}$

or,
$$\frac{l}{w} = \frac{6.67 \times t}{\rho} = \frac{6.67 \times (0.25 \times 10^{-3})}{101.6 \times 10^{-8}} = 1641.24 \quad ...(i)$$

Heat dissipated from surface of strip (H) is given as:

$$H = 5.67\ \eta_{rad.} \times e\left[\left(\frac{T_1}{100}\right)^4 - \left(\frac{T_2}{100}\right)^4\right]$$

$$= 5.67 \times 0.5 \times 0.9\left[\left(\frac{1223}{100}\right)^4 - \left(\frac{873}{100}\right)^4\right] = 42262 \text{ W/m}^2$$

Surface area of strip = $2wl$

∴ Total heat dissipated = $2wl$ × 42262

Also, $2wl \times 42262 = 8 \times 10^3$

From which,
$$wl = \frac{8 \times 10^3}{2 \times 42262} = 0.09465 \quad ...(ii)$$

Inserting the value of l (= 1641.24 w) from *(i)*, in *(ii)*, we get:

$$w \times 1641.2420 = 0.09465$$

or,
$$w = \left(\frac{0.09465}{1641.24}\right)^{1/2} = \mathbf{7.59 \times 10^{-3}\ m\ or\ 7.59\ mm.} \quad \textbf{(Ans.)}$$

Q. 10.24. *Electrical drives* *(a) Explain briefly speed control of D.C. motor with thyristor.*

(b) A 240 V series motor takes 45 A when going its rated output at 1500 r.p.m. Its resistance is 0.28 Ω. Find what resistance must be added to obtain rated torque: (i) at starting, and (ii) at 1000 r.p.m.

Solution: (*a*) Refer to Art. 5.12.5.3.

(*b*) *Given*: $V = 240$ volts; $I\ (=I_a) = 45$ A; $N = 1500$ r.p.m.; $R = 0.28\ \Omega$.

(*i*) **Resistance to be added to obtain rated torque at starting, R_{add}:**

Since the torque remains the same in both the cases, it is obvious that the current drawn by the motor remains constant at 45 A.

$$\therefore \qquad 45 = \frac{240}{R_{add} + 0.28}$$

$$\text{or,} \qquad R_{add} = \frac{240}{45} - 0.28 = \mathbf{5.05\ \Omega.} \quad \textbf{(Ans.)}$$

(*ii*) **Resistance to be added to obtain rated torque at 1000 r.p.m., R_{add}:**

$$\text{We know that,} \qquad \frac{N_2}{N_1} = \frac{E_{b_2}}{E_{b_1}}$$

$$E_{b_1} = 240 - 45 \times 0.28 = 227.4 \text{ V}$$

$$\therefore \qquad \frac{1000}{1500} = \frac{E_{b_2}}{227.4}$$

$$\text{or,} \qquad E_{b_2} = \frac{1000 \times 227.4}{1500} = 151.6 \text{ V}$$

$$\text{Also,} \qquad E_{b_2} = V - 45\,(R_{add} + 0.28)$$

$$\text{or,} \qquad 151.6 = 240 - 45\,(R_{add} + 0.28)$$

$$\therefore \qquad R_{add} = \frac{240 - 151.6}{45} - 0.28 = \mathbf{1.68\ \Omega.} \quad \textbf{(Ans.)}$$

Q. 10.25. *Electrical drives.* (*a*) *What is load equilisation? Explain.*

(*b*) *A 25 H.P., 3-phase, 10-pole, 50 Hz induction motor provided with a flywheel has to supply a load torque of 750 Nm for 10 seconds followed by a no-load period during which the flywheel regains the full speed. The full-load slip of the motor is 4% and the torque-speed curve may be assumed linear over the working range. Find the moment of inertia of the flywheel if the motor torque is not to exceed the 1.8 times the full-load torque.*

Solution: (*a*) Refer to Art. 5.16.7.

(*b*) *Given:* Power of the motor = 25 H.P.; $p = 10$;

$f = 50$ Hz; $T_{load} = 750$ Nm for 10 seconds;

slip, $s_f = 4\%$; $T_{motor} = 1.8\ T_f$.

Moment of inertia of the flywheel, J:

$$\text{Synchronous speed,} \qquad N_s = \frac{120f}{p} = \frac{120 \times 50}{10} = 600 \text{ r.p.m.}$$

$$\text{Full load speed,} \qquad N_f = N_s\,(1 - s_f) = 600\,(1 - 0.04) = 576 \text{ r.p.m.}$$

$$\text{Full load torque,} \qquad T_f = \frac{25 \times 735.5}{(2\pi \times 576/60)} = 304.8 \text{ Nm}$$

$$T_{motor} = 1.8\ T_f = 1.8 \times 304.8 = 548.64 \text{ Nm}$$

Slip speed, $s = 0.04 \times 600 = 24$ r.p.m. $= \dfrac{24 \times 2\pi}{60} = 2.5133$ rad/s

As the slip at full load is proportional to full-load torque,

$$K = \frac{s}{T_f} = \frac{2.5133}{304.8} = 0.008246$$

Also, $T_{motor} = T_{load} - (T_{load} - \tau_0)\, e^{-t/KJ}$ ...[Eqn. (5.130)]

$= T_{load}\,(1 - e^{-t/KJ})$ $(\because\ T_0 = 0)$

or, $548.64 = 750\,[1 - e^{-10/(0.008246 \times J)}]$

or, $0.732 = 1 - e^{(-1212.7/J)}$

or, $e^{-1212.7/J} = 1 - 0.732 = 0.268$

or, $e^{1212.7/J} = \dfrac{1}{0.268} = 3.73$

$\therefore$ **$J = 921.2$ kg-m². (Ans.)**

Q. 10.26. (*a*) *Explain briefly, with a neat diagram, the working of central system of air-conditioning.*

(*b*) ***Train movement and energy consumption:*** *A 200-tonne motor coach driven by four motors takes 18 seconds to attain a speed of 40 km/h, starting from rest on an ascending gradient of 1 in 80. The gear ratio is 3.2, gear efficiency 92%, wheel diameter 92 cm, train resistance 45 N/tonne and rotational inertia 10% of the dead weight. Find the torque developed by each motor.*

Solution. (*a*) Refer to Art. 4.15.4.

(*b*) *Given:* $M = 200$ tonnes; $t_1 = 18$ s;

$V_m = 40$ km/h; $\%\ G = \dfrac{1}{80} \times 100 = 1.25$;

Gear ratio, $\gamma = 3.2$; $\eta = 92\%$;

Wheel dia., $D = 92$ cm $= 0.92$ m; Train resistance, $r = 45$ N/tonne; $M_e = 1.1$ M.

Torque developed by each motor, T:

Total tractive effort,

$F_t = (277.8 \times M_e + 98\ MG + Mr)$ newton ...[Eqn. (7.15)]

In this equation M is in tonnes, $\alpha \left(= \dfrac{V_m}{t_1} \right)$ in km/h/s, G in percent and r in N/tonne of train mass.

Substituting the values in the above equation, we get:

$$F_t = \left[277.8 \times \frac{40}{18} \times (1.1 \times 200) + 98 \times 200 \times 1.25 + 200 \times 45 \right]$$

$$= 135813 + 24500 + 9000 = 169313 \text{ N}$$

Also, $F_t = 2\gamma\eta \dfrac{T}{D}$ ...[Eqn. (7.7)]

$$169313 = 2 \times 3.2 \times 0.92 \times \frac{T}{0.92}$$

i.e., $$T = \frac{169313 \times 0.92}{2 \times 3.2 \times 0.92} = 26455 \text{ Nm}$$

$\therefore$ Torque developed by each motor $= \frac{26455}{4} =$ **6613.75 Nm. (Ans.)**

Q. 10.27. (*a*) *State the requirements of an ideal traction system.*

(*b*) ***Traction drive.*** *A 400-tonnes train has its speed reduced by regenerative braking from 60 to 40 km/h over a distance of 1.8 km along down gradient of 1.4%. Assuming specific train resistance as 45 N/tonne, rotational inertia effect 10% and conversion efficiency of 72 percent, calculate:*

(*i*) *Electrical energy returned to the line.*

(*ii*) *Average power returned to the line.*

Solution. (*a*) Refer to Art. 6.3

(*b*) *Given:* $M = 400$ tonnes; $V_1 = 60$ km/h; $V_2 = 40$ km/h;

Distance, $S = 1.8$ km; $G = 1.4\%$; $r = 45$ N/tonne; $\frac{M_e}{M} = 1.1$; $\eta = 72\%$

(*i*) **Electrical energy returned to the line:**

Electrical energy returned to the line

$$= \eta \left[0.01072 \frac{M_e}{M} (V_1^2 - V_2^2) + S(27.25G - 0.2778r) \right] \text{ Wh/tonne} \quad \text{...[Eqn. (8.7)]}$$

$$= 0.72 \, [0.01072 \times 1.1 \times (60^2 - 40^2) + 1.8 \, (27.25 \times 1.4 - 0.2778 \times 45)]$$
$$= 0.72 \, (23.58 + 46.17) = 50.22 \text{ Wh/tonne} = 50.22 \times 400 \times 10^{-3}$$

= **20.09 kWh. (Ans.)**

(*ii*) **Average power returned to the line:**

$$\text{Average speed} = \frac{60 + 40}{2} = 50 \text{ km/h}$$

$$\therefore \quad \text{Time taken} = \frac{\text{Distance } (s)}{\text{Average speed}} = \frac{1.8}{50} = 0.036 \text{ h}$$

Hence, power returned $= \frac{20.09}{0.036} =$ **558 kW. (Ans.)**

Index